# man
# microbes
# and matter

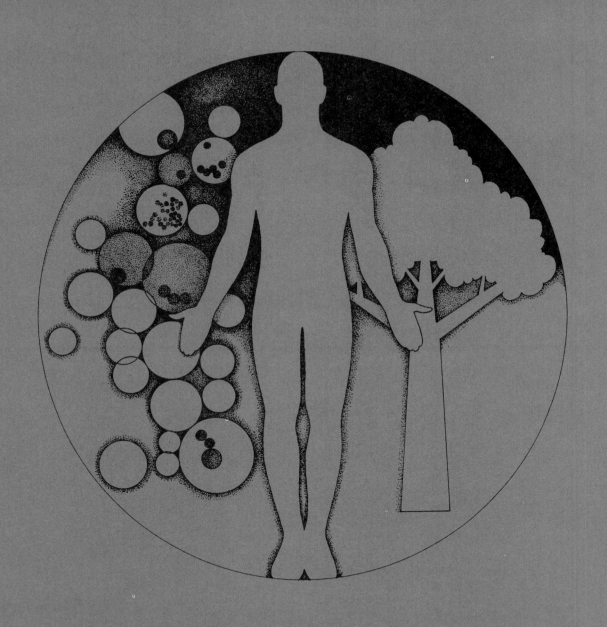

# man microbes and matter

**BarTLEY C. BLOCK**
Assistant Professor of Biology
University of Bridgeport

**JACQUES DUCAS**
Art for Education
Illustrator

**McGRAW-HILL BOOK COMPANY**

**A BLAKISTON PUBLICATION**

New York   St. Louis   San Francisco   Auckland   Düsseldorf   Johannesburg
Kuala Lumpur   London   Mexico   Montreal   New Delhi   Panama
Paris   São Paulo   Singapore   Sydney   Tokyo   Toronto

# man
# microbes
# and matter

1 2 3 4 5 6 7 8 9 0 VHVH 7 9 8 7 6 5

Library of Congress Cataloging in Publication Data

Block, Bartley C
    Man, microbes, and matter.

    1.  Human biology.  I.  Title.  [DNLM: 1.  Environ-
mental health—Popular works.   2.   Medicine—Popular
works.  3.  Microbiology—Popular works.  WB120  B651m]
I.  Title.
QP34.5.B57        612        74-17225
ISBN 0-07-005913-6

This book was set in Helvetica by Textbook Services, Inc.
The editors were Cathy Dilworth, Stuart D. Boynton,
and Ida Abrams Wolfson;
the designer was J. E. O'Connor;
the production supervisor was Leroy A. Young.
Von Hoffmann Press, Inc., was printer and binder.

# ILLUSTRATION CREDITS

**CHAPTER 5**

*Fig. 5-2* From L. Langley, I. Telford, and J. Christensen, *Dynamic Anatomy and Physiology,* 4th ed., McGraw-Hill, Inc., 1974.

**CHAPTER 11**

*Fig. 11-13* From L. Langley, I. Telford, and J. Christensen, *Dynamic Anatomy and Physiology,* 4th ed., McGraw-Hill, Inc., 1974.

*Figs. 11-1, 11-5, 11-6, 11-7, 11-8, 11-9, 11-10, 11-11, 11-12* From B. Melloni, P. Stone, and J. Hurd, *Anatomy and Physiology* (overhead transparency series), Book No. III, McGraw-Hill, Inc., 1971.

**CHAPTER 13**

*Fig. 13-2* From B. Melloni, P. Stone, and J. Hurd, *Anatomy and Physiology* (overhead transparency series), Book No. II, McGraw-Hill, Inc., 1971.

*Fig. 13-7* From L. Langley, I. Telford, and J. Christensen, *Dynamic Anatomy and Physiology,* 4th ed., McGraw-Hill, Inc., 1974.

**CHAPTER 14**

*Figs. 14-1, 14-2, 14-3, 14-4, 14-5, 14-6, 14-7, 14-9, 14-12, 14-13, 14-14, 14-15, 14-16* From B. Melloni, P. Stone, and J. Hurd, *Anatomy and Physiology* (overhead transparency series), Book No. II, McGraw-Hill, Inc., 1971.

**CHAPTER 19**

*Figs. 19-1, 19-2, 19-3, 19-4, 19-5, 19-6, 19-7* From B. Melloni, P. Stone, and J. Hurd, *Anatomy and Physiology* (overhead transparency series), Book No. II, McGraw-Hill, Inc., 1971.

*Figs. 19-8, 19-9, 19-10, 19-12* From L. Langley, I. Telford, and J. Christensen, *Dynamic Anatomy and Physiology,* 4th ed., McGraw-Hill, Inc., 1974.

*Fig. 19-11* From R. M. DeCoursey, *The Human Organism,* 4th ed., McGraw-Hill, Inc., 1974.

**CHAPTER 22**

*Fig. 22-1* From B. Melloni, P. Stone, and J. Hurd, *Anatomy and Physiology* (overhead transparency series), Book No. II, McGraw-Hill, Inc., 1971.

**CHAPTER 23**

*Figs. 23-1, 23-5, 23-6, 23-7, 23-8, 23-9, 23-10, 23-11, 23-12, 23-13, 23-14, 23-15, 23-16, 23-17, 23-18, 23-19, 23-22, 23-23* From B. Melloni, P. Stone, and J. Hurd, *Anatomy and Physiology* (overhead transparency series), Book No. I, McGraw-Hill, Inc., 1971.

*Fig. 23-2* From L. Langley, I. Telford, and J. Christensen, *Dynamic Anatomy and Physiology,* 4th ed., McGraw-Hill, Inc., 1974.

*Figs. 23-20, 23-21* From R. M. DeCoursey, *The Human Organism,* 4th ed., McGraw-Hill, Inc., 1974.

**CHAPTER 24**

*Figs. 24-1, 24-2, 24-5, 24-10* From L. Langley, I. Telford, and J. Christensen, *Dynamic Anatomy and Physiology,* 4th ed., McGraw-Hill, Inc., 1974.

*Figs. 24-11, 24-12, 24-13, 24-14* From B. Melloni, P. Stone, and J. Hurd, *Anatomy and Physiology* (overhead transparency series), Book No. I, McGraw-Hill, Inc., 1971.

*Figs. 24-15, 24-16, 24-17, 24-18, 24-19, 24-20, 24-21, 24-22* From B. Melloni, P. Stone, and J. Hurd, *Anatomy and Physiology* (overhead transparency series), Book No. II, McGraw-Hill, Inc., 1971.

**CHAPTER 25**

*Fig. 25-1* From B. Melloni, P. Stone, and J. Hurd, *Anatomy and Physiology* (overhead transparency series), Book No. I, McGraw-Hill, Inc., 1971.

**CHAPTER 26**

*Fig. 26-1* From B. Melloni, P. Stone, and J. Hurd, *Anatomy and Physiology* (overhead transparency series), Book No. II, McGraw-Hill, Inc., 1971.

*Figs. 26-2, 26-3* From B. Melloni, P. Stone, and J. Hurd, *Anatomy and Physiology* (overhead transparency series), Book No. I, McGraw-Hill, Inc., 1971.

*Figs. 26-4, 26-6, 26-7, 26-8, 26-9, 26-10, 26-11, 26-12* From B. Melloni, P. Stone, and J. Hurd, *Anatomy and Physiology* (overhead transparency series), Book No. IV, McGraw-Hill, Inc., 1971.

**CHAPTER 28**

*Figs. 28-1, 28-2, 28-4, 28-5, 28-6* From B. Melloni, P. Stone, and J. Hurd, *Anatomy and Physiology* (overhead transparency series), Book No. IV, McGraw-Hill, Inc., 1971.

**CHAPTER 30**

*Fig. 30-2* From R. M. DeCoursey, *The Human Organism,* 4th ed., McGraw-Hill, Inc., 1974.

*Figs. 30-3, 30-4, 30-5, 30-6* From B. Melloni, P. Stone, and J. Hurd, *Anatomy and Physiology* (overhead transparency series), Book No. IV, McGraw-Hill, Inc., 1971.

*Figs. 30-8, 30-9* From *Blakiston's Gould Medical Dictionary,* 3rd ed., McGraw-Hill, Inc., 1972.

*Fig. 30-12* From B. Melloni, P. Stone, and J. Hurd, *Anatomy and Physiology* (overhead transparency series), Book No. III, McGraw-Hill, Inc., 1971.

TO
JAN,
KENNY, DEBBIE,
AND STEVE

# contents

# Preface

*Man, Microbes, and Matter* fills a long-felt need for a textbook which truly integrates the basic sciences for students in allied health—related curricula. The textbook is intended to satisfy all basic science requirements of college-level freshmen in programs such as nursing; courses for medical secretaries and medical assistants; physical, occupational, and inhalation therapy; medical, dental, and radiation technology; dental hygiene; and dietetics.

The sciences integrated include environmental science, microbiology, physics, chemistry, basic anatomy and physiology, and pathophysiology. Integration of the subject matter is fashioned from a biological perspective.

Objectives include a comprehension of and an appreciation for

1   Man's interactions with his environment and their implications for public health
2   The microbiologic principles essential for the effective practice of hygiene and sterilization
3   The structural organization of the human body
4   The physiochemical principles behind the physiologic processes which occur in cells, tissues, organs, and organ systems

5   The principles relevant to selected diagnostic and therapeutic techniques in medicine
6   The anatomic and physiologic disruptions which give rise to medically or economically important diseases

*Man, Microbes, and Matter* is organized around three basic biologic concepts—homeostasis, perpetuation, and metabolism. Homeostasis refers to the dynamic equilibria or steady states maintained within the body, despite large changes in either the internal or external environment. Adaptive mechanisms include integration, excretion, support, and movement. These perpetuative mechanisms maintain the existence of the individual. Reproduction promotes species perpetuation.

Metabolism is defined as the sum total of all the chemical and physical reactions occurring in the body. Liberating energy for both synthetic and degradative reactions, metabolism is sustained by nutritive and respiratory functions and is promoted through transport and exchange mechanisms.

Both perpetuation and metabolism are controlled by homeostatic processes. Homeostatic disruptions result in disease (lack of ease).

These ideas are developed fully in the six parts into which the book is divided. Part One examines some of the general homeostatic mechanisms which tend to maintain constant external and internal environments. Part Two deals with the physiochemical principles required for an understanding of the physiologic bases of life. The last four parts examine various aspects of perpetuation and metabolism and the specific homeostatic control mechanisms which maintain these functions.

The following are among the distinctive features of *Man, Microbes, and Matter:*

1   The breadth of subject matter permits its use as an instructional tool in a wide variety of health-related curricula.
2   The sciences are integrated by building methodically on the concepts and principles developed in each chapter.
3   Difficult concepts, especially in physics and chemistry, are explained by analogy to readily grasped examples.
4   The abundant illustrations, most of which are original, are central to an understanding of the textual explanations, and many of them are rendered in two colors used functionally.
5   Many of the important spinal tracts are illustrated as they would appear in the intact central nervous system, eliminating the need for traditional "wiring" diagrams.
6   Each chapter on the normal anatomy and physiology of an organ system is followed by a chapter on its pathophysiology, thereby reinforcing a knowledge, and uncluttering explanations, of the "normal."
7   Throughout the textbook, diagrams of vicious circles help to visualize cause-and-effect relationships during disease causation and its course.
8   The annotated collateral reading lists at the end of each chapter, many of which cite pertinent *Scientific American* articles, permit the student to pursue supplementary reading on a more rational basis than would otherwise be the case.
9   Science is presented as a grouping of interrelated disciplines where knowledge is finite and controversy abounds.
10  Anthropomorphism (attributing human characteristics to nonhumans) and teleology (attributing purpose to nature) are avoided during physiologic explanations.

Several other instructional tools are designed for use with this textbook. A coordinated laboratory manual is available which permits a "hands-on" approach to the subject matter. In addition, a four-volume set of overhead transparencies, entitled *Anatomy and Physiology*, by Biagio J. Melloni, Peter Stone, and Jane A. Hurd, will be found a useful adjunct in both lectures and laboratories. The series contains most of the anatomic illustrations used in this textbook. The series was created by Biagio Melloni, Chairman, and Peter Stone and Jane Hurd, Medical Illustrators, of the Department of Medical-Dental Communication, Georgetown University School of Medicine and Dentistry, Washington, D.C., and was published by McGraw-Hill Book Company in 1971.

Many people and institutions have been involved in giving birth to this textbook. I am pleased to acknowledge their assistance.

I am indebted, both literally and figuratively, to the following people for their generous financial support for the development of original illustrations: my parents, David and Anne; and my wife's parents, Paula Jacobs and her late husband Barrett; and Mr. and Mrs. Bernard Weissman. The University of Bridgeport has also supported these efforts.

Mr. Jacques Ducas is my major illustrator. His fine hand, meticulous attention to detail, creative imagination, and searching intellect have made it a delight and privilege to collaborate with him.

The following people have provided critical reviews of various chapters in this textbook: Dr. Charles Spiltoir, Jr., Dr. John Poluhowich, and Professor Cynthia Kaufman of the University of Bridgeport; Professor Peter Tenerowicz of Southern Connecticut State College; Professor Charles Naden of Quinnipiac College; and Dr. Theodore Ducas of the Massachusetts Institute of Technology. I am grateful to my reviewers and, where possible, have implemented their thoughtful suggestions. Nevertheless, I accept full responsibility for any errors of fact or interpretation which may have been committed.

The chairman of the biology department at the University of Bridgeport, Dr. Michael E. Somers, managed to pave my way as an author by means of courtesies too numerous to mention. His catalytic contributions are deeply appreciated.

The manuscript was typed by my mother-in-law Paula and my wife Janet. The Joseph C. Day Co., Milford, Connecticut, provided technical assistance.

Finally, the support of my family is gratefully acknowledged. Their encouragement and enthusiasm have made it easier for me to negotiate the many hurdles routinely encountered in a project of this magnitude.

*Bartley C. Block*

# Part one
# The maintenance
# of health

# CHAPTER 1
# man's external environment

Human beings have always been beset with environmental and behavioral problems since emerging as a distinct species more than 4 million years ago. Some basic problems, e.g., those stemming from a lack of food, shelter, health, and security, have been reduced in severity with the development of civilization. During the same time span, man's aggressive behavior, finding expression in violence and other forms of antisocial activity, has either shifted its emphasis or increased in intensity. Still other influences with the potential to influence profoundly the destinies of all living things have been created as the result of significant advances in civilization.

This chapter examines the causes of the world's current environmental problems, their effects on natural processes, as well as their possible impact on the future of the human species. The examination is made from an ecological view, bearing in mind that Homo sapiens is just one species among several million which must live in natural balance, or equilibrium.

Implicit in our exploration is the question of whether, indeed, human beings face quantitatively greater and qualitatively different problems now than at any previous time during recorded history. Specifically, the review covers the steady-state controls for, and disruptions in, population growth, available living space and natural resources, environmental quality, and social balances. The effectiveness of past and present attempts to reduce either the consequences or causes of environmental disruptions is also assessed. Finally, the range of human adaptability to these disruptions and their consequences are explored, in an attempt to shed more light on their long-term implications.

## THE EVOLUTION OF ENVIRONMENTAL PROBLEMS

### THE DEVELOPMENT OF CIVILIZATION

Civilization arose from influences which caused hominoids to aggregate in groups. A hostile environment and the security and protection of numbers and improvised shelters led to various forms of cooperation between people. Primitive human beings were endowed with the biologic and intellectual capacities to take advantage of acculturation. Thus, the resultant cultural revolution which occurred between 1 million and 100,000 years ago led to the first sustained proliferation of the human population.

The next major transition in the development of civilization came when human beings gradually shifted from hunting societies of low efficiency to agricultural societies of higher efficiency. Agricultural communities replaced family hunting units, and population centers began to spring up around the cultivated land.

The agricultural revolution, a product of these shifting cultural tides, provided the impetus for the second sustained increase in the world's population. This revolution occurred between 10,000 and 5,000 years ago.

With the emergence of the Renaissance from the Dark Ages and the subsequent scientific-industrial revolution of the nineteenth and twentieth centuries, much of the human population was emancipated from the burden of being self-sustaining. This newfound freedom further increased the operational efficiency of social organization but decreased the functional capability of the individual. As a result, human beings became the recipient of a third and easily the most forceful surge in population growth. The economic, political, and social changes necessitated by the scientific-industrial revolution aided and, indeed, encouraged human propagation.

The world's population is currently undergoing still another sharp increase in numbers, a phenomenon which we shall refer to as the "population revolution." The current revolution is the combined result of increasing numbers of people and a decreasing death rate.

The effects of these influences on the future development of civilization are by no means clear but give cause for concern. The "doubling time" for the approximately 3.9 billion people inhabiting this planet is now estimated at a little more than 30 years. This figure is expected to decrease as the world population continues to spiral. At the very least, future population trends are not encouraging.

The population surges which have characterized the significant advances in the development of civilization, along with future projections, are shown graphically in Fig. 1-1. The curves above and below the solid line are upper and lower limits of the population estimates. The four pronounced population increases are correlated with the *cultural revolution* (*A*), the *agricultural revolution* (*B*), the Renaissance and ensuing *scientific-industrial revolution* (*C*), and the projected *population revolution* (*D*).

As with all pollution problems, proposed solutions to the current population explosion lie in social, economic, and political spheres. Such problems will not be eliminated or significantly reduced in the foreseeable future by advances in scientific knowledge or developing technology alone.

## SOCIETY AND HEALTH

The revolutions of civilization are largely matched by the type of society characteristic of each period. Although intergradations and overlappings occur among contemporary societies in various parts of the world, five distinct states of society may be recognized. Theoretically, all societies pass through these states.

The first, or *traditional,* state consists largely of a rural environment and is characterized by high rates of birth and death. Contamination, infestations, infections, and nutritional deficiencies are common health problems. Medical practices are based mainly on tradition.

The second state in the development of a society, *early transitional,* is also characterized by a largely rural environment and possesses a high birth rate but an intermediate death rate. The health problems are similar to those of a traditional state of a society. Nevertheless, medical and health practices are extended to the routine control of endemic diseases, the application of environmental sanitation, the treatment of minority groups, and the nutritional improvement of diet.

The *late transitional* state of a society, third in chro-

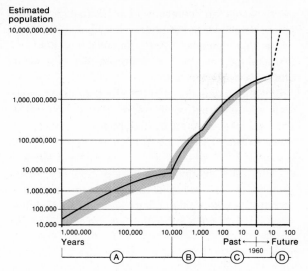

*Figure 1-1* *Actual and projected increases in the world's population growth during the past million years and the next fifty years, respectively. Note that the scales are not linear or arithmetic. Year 0 is 1960. The colored curves above and below the solid line are upper and lower limits of the estimate.*

nologic order, differs from the early transitional state in two respects: the environment contains both rural and urban components, and the birth rate is intermediate while the death rate is low. Comprehensive and integrated health care and medical services are applied to all diseases and illnesses. Further advances in nutrition and action-oriented health programs are hallmarks of this societal state.

*Modern affluent* societal states are characterized by a largely urban environment with its attendant increases in physical and mental illnesses, general biospheric pollution, and other societal disruptions. Rates of birth and death are low, in contrast to those of other societal states. Medical and health services are provided to all geographic regions and societal segments but are increasingly impersonal and inefficient. While there is some overnutrition because of general affluence and personal indulgence, apathy and overpopulation contribute to widespread malnutrition and undernutrition.

If these trends continue, the state of society evolv-

ing may consist almost entirely of supercities, or megalopolises, possessing low birth rates but high death rates. Because of general biospheric pollution and overpopulation, food and water may be frequently contaminated, with corresponding periodic scarcities of these natural resources. Further exacerbations of other societal disruptions such as riots and civil disturbances may also be evident. To accommodate to spiraling health and medical costs and increasing numbers of patients, socialized medicine may become widely accepted. A growing emphasis may be placed on fertility control, to achieve zero population growth. Critical reevaluations of medical goals will undoubtedly be undertaken, to permit changes in the legal definitions of birth and death. Malnutrition and undernutrition are likely to increase, causing periodic famines of alarming proportions. To offset these effects, novel foods will be increasingly used by the world's human population.

To return from projections of the evolving state of society to the actualities of modern society, the most forceful statement we can make is that the human population is drastically increasing, a condition which has no precedent in previous societal states. Thus, nature's steering wheel is probably not now entirely under human control.

The currently burgeoning population is continually interacting with its environment in ways that are obvious, subtle, or unknown. The reciprocity of these interactions and some important natural laws that govern them will now be considered, so that we can appreciate more fully the operation of nature's accelerator and brake.

## FACTORS REGULATING POPULATION GROWTH

### CONTROLS AND DISRUPTIONS OF STEADY STATES

A close correspondence exists between functional efficiency, structural complexity, and total energy expenditure. These relationships hold true for individuals as well as populations. Disturbances in one or more of these factors may lead to organismic diseases or societal disruptions.

To emphasize this important concept, let us follow

the interactions of the living world and the inanimate environment over a time span sufficient to observe demonstrable changes in one or both of these components. The time span must be long enough to cause physical and biologic changes in the same geographic region.

The interactions between the biologic and physical components of the environment are diagrammed in Fig. 1-2. The components are intermeshed with each other; changes in one cause changes in the other. These reciprocal changes may become increasingly perceptible with the passage of time. For example, consider the human population as the living (biotic) component enlarging on the left-hand side of the figure. Lacking an ability to apply effective counterforces, the environment will become increasingly polluted by the combined influences of human endeavor, metabolic demands, and waste products. In turn, the changing nature of the environment will more profoundly affect human health. The result of these in-

teractions is an increasing environmental deterioration, as shown at the right in the figure. The sharp rise in environmental effects as the population continues to increase can be stemmed only by a frontal attack on overpopulation.

Biospheric pollution and its attendant health problems are a direct result of overpopulation. Overpopulation is reflected by disruptions in natural balances between the living and nonliving worlds. The key to all environmental and social crises resides in the judicious manipulation of the resulting imbalances so as to reestablish or restore an equilibrium compatible with continued existence. The practical resolutions to such problems are obviously not so simple.

## THE LAWS OF THERMODYNAMICS

Because of the direct relation between energy consumption and population size, an understanding of

**Figure 1-2**  *The interdependences between the nonliving (abiotic) and living (biotic) components of the environment. Increase in one component occurs only at the expense of the other and may result in organismic diseases or societal disruptions.*

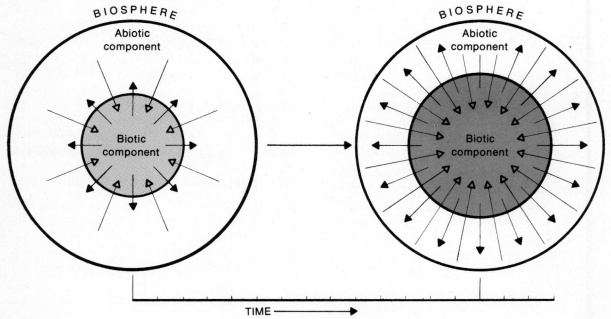

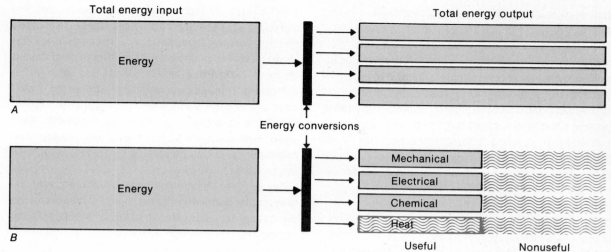

**Figure 1-3** *The biologic constraints imposed by the laws of thermodynamics.* A. *First law.* B. *Second law.*

the natural balance between the living and nonliving worlds depends upon an appreciation of the origin, conversion, and utilization of energy.

The ultimate origin of all earthbound energy is, of course, the sun. This energy must be converted into forms which we can use. As the population continues to increase, its rate of energy use will ultimately exceed its rate of energy production. Thus, we must grasp the mechanics of energy conversion to appreciate why it is a growth-limiting factor.

The natural laws which describe the mechanics of energy conversion are known as the *first* and *second laws of thermodynamics.* The first law states that energy can be neither created nor destroyed but can be changed from one form to another. The statement simply means that the quantities of energy before and after processing are exactly the same, if both usable and nonusable forms are accounted for. The first law also informs us that interconversions between types of energy are possible. Embodied in the first law of thermodynamics, these concepts are summarized in Fig. 1-3*A*.

The second law describes what happens to a given amount of energy during conversion from one form to another. The law states that there is always a loss of energy during the conversion process. The energy is

lost through heat evolution. The second law of thermodynamics is illustrated in Fig. 1-3*B*.

Because heat losses may be reduced but never completely eliminated, perpetual motion is impossible. For a similar reason, energy is continually required to maintain the structural organization of any type of matter. More highly organized structures require a continual supply of a greater amount of energy to maintain their integrity than those less well ordered physically. For living things, energy expenditures and resultant heat losses increase with structural complexity.

Stated in still another way, the second law explains that matter must go from a state of order to a state of disorder if the necessary energy inputs are not maintained. The material organization of a room, the mechanical functioning of an automobile, and, indeed, human biologic and social structures all obey the second law of thermodynamics. This fundamental principle imposes restrictions on operational efficiency and ultimately defines its limits.

The ramifications of the first and second laws are profound. Many proposed solutions to our environmental and social problems must ultimately meet an impasse because of the inexorable consequences of these laws.

## TRENDS IN CURRENT ENVIRONMENTAL PROBLEMS

If present population trends continue, their major effects will be felt, and in fact are already being felt, in the areas of available living space, per capita food production, mineral consumption and corresponding energy production, biospheric pollution, and other societal disruptions. The impact of continuing population growth on these factors will have profound deleterious effects on the overall quality of life for everyone.

### LIVING SPACE

As the number of people populating the earth increases, the amount of land per individual is reduced.

The increase in population numbers and resulting decrease in livable land area have been perpetuated through the four revolutions which characterize the development of civilization. This inverse relationship between people and land has held true despite the fact that the human population has spread to all portions of the earth from its African origins during this time. Indeed, increasing population pressures have been correlated throughout recorded history with migrations and emigrations to, or invasions of, new land masses.

Figure 1-4 summarizes these relationships by graphically portraying the estimated increases in the density per square kilometer of the world's population during the past 1 million years and the next 50 years. The curve reflects the four revolutions undergone by civilization.

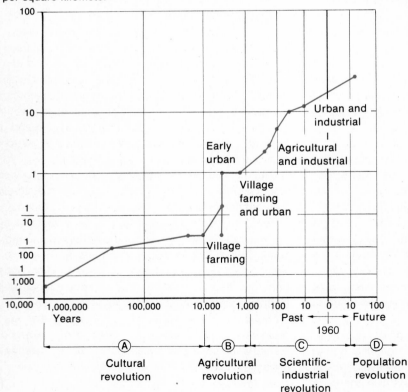

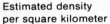

**Figure 1-4** *Estimated increases in the density per square kilometer of the world's population during the past 1 million years and the next 50 years, relative to 1960.*

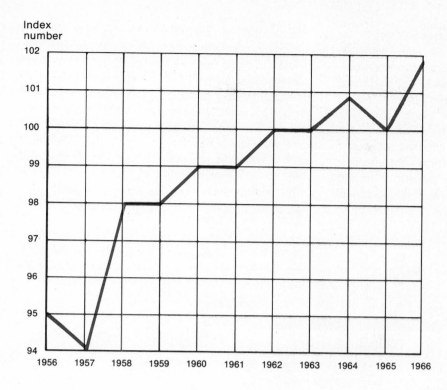

**Figure 1-5** *Index numbers of world per capita food production during the 10-year period 1956–1966. The year 1963 is set equal to 100. The indexes are derived ratios of food production to population size.*

## PER CAPITA FOOD SUPPLY

The historical relation between human population growth and world food supply is not as easily traceable as that between population density and available living space. Figure 1-4 suggests that agricultural techniques were relatively unknown prior to 6000 B.C. Before that time, food was secured mainly by hunting. The growth in population that resulted during and after the transition of hunting societies to agricultural societies was in part due to an increase in food production. Similarly, the next sustained increase in population, resulting in the transition from an agricultural-industrial society to an urban-industrial society, may also be explained by improvements in agricultural technology and efficiency. However, the historical picture is blurred by periodic population eruptions, famines, epidemics and pandemics, wars, and other events that have had disruptive effects on food production, distribution, and consumption.

Available figures for the current population revolution suggest a gradual, continuing decline in world per capita food production. In Fig. 1-5, the ratios of food production to population size are displayed for the period 1956–1966. Although there are fluctuations in the curve which complicate interpretations, an increase of five index numbers occurred during the first five years, followed by an increase of only three index numbers in the succeeding five years. Because both agricultural technology and cultivation of arable land increased during the above 10-year period, the decline is mainly attributable to increasing population pressures. Undoubtedly, the variability in food distribution, caused by socioeconomic factors and political considerations, has contributed to the differences in nutritional adequacy between nations. The long-term consequence of these trends is a decreasing average nutrition for the world's population.

Surprisingly, world per capita food production for

the years from 1967 to 1970 indicates a dramatic rise of about 7 to 14 percent, even for many underdeveloped countries. The relative increase in food abundance is attributed to the cultivation of the newer varieties of high-yield wheat and grains. The increased yields in the last few years have been labeled the "green revolution." How green remains to be seen, however, since many authorities believe the revolution to be a technological artifact, the effects of which will be swamped by further population growth.

The decrease in the available per capita food supply on a worldwide basis has been sharpened by an accumulation of persistent pesticides such as DDT and heavy metals such as mercury in animals and animal products used for human consumption. Such contaminated animals include shellfish, game fish, game birds, and cattle. The persistence of certain toxic substances in the environment is related to the inability of microbes to degrade these materials into nontoxic, simpler chemicals. It is for this reason that such substances are referred to as *nonbiodegradable*.

The phenomenon which has caused an accumulation of toxic, nonbiodegradable substances in ani-

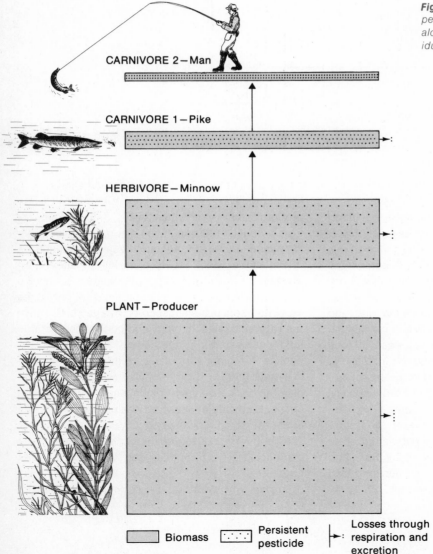

**Figure 1-6** *Trophic magnification of persistent pesticide residues passed along a simple food chain. The toxic residues are indicated by dots.*

CARNIVORE 2 — Man

CARNIVORE 1 — Pike

HERBIVORE — Minnow

PLANT — Producer

Biomass  Persistent pesticide  Losses through respiration and excretion

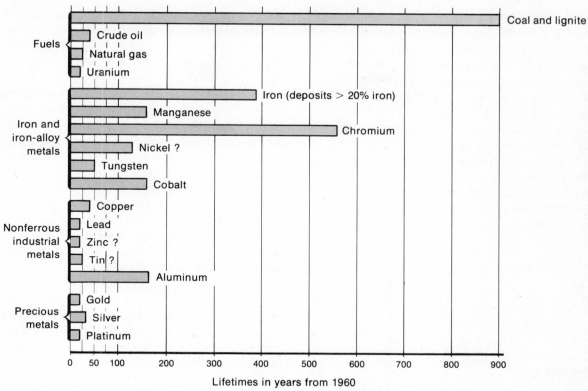

**Figure 1-7**  *Remaining lifetimes, estimated from 1960, of mineral resources mined with present techniques.*

mals ultimately destined for human consumption is called *trophic magnification.* Trophic magnification is diagramed in Fig. 1-6, using DDT as an example. Trophic magnification refers to a relative increase in the concentration of a toxic, nonbiodegradable substance released into the environment and passed from plant to animal when contaminated plants are eaten and from animal to animal when contaminated animals or their products serve as food. As a consequence of the second law of thermodynamics discussed earlier, it takes a larger mass of plant life to provide sufficient energy to support a given mass of animal life. The same natural principle holds when dealing with smaller animals eaten by larger animals. Thus, trophic magnification concentrates toxic, nonbiodegradable substances as they pass successively from contaminated plant to small animal to large animal. In other words, although the amount of persistent toxicant does not change in a food chain consist-

ing of plants, small animals, and large animals, the amount of toxicant undergoes a relative increase in concentration because of an actual decrease in the respective masses of each higher species in the food chain. Persistent toxicants are concentrated most in the animals at the top of the food chain. These animals obviously include human beings.

## MINERAL CONSUMPTION

The population surge during and following the scientific-industrial revolution and the present population revolution have increased world demands for mineral resources. These resources include fossil and nuclear fuels, iron and iron alloys, nonferrous industrial metals, and precious metals. Figure 1-7 indicates the estimated remaining lifetimes of these resources, starting from 1960.

To estimate remaining reserves, technology, population, and consumption were held constant. In addition, future discoveries of unknown deposits and possible uses of presently submarginal ores were disregarded. With these qualifications and an awareness of the inherent uncertainties about projections of this nature, 11 of the 18 minerals listed will almost certainly be completely depleted within 100 years. Coal and lignite have the longest remaining lifetimes, estimated at more than 900 years; uranium, lead, zinc, gold, and platinum have projected reserves of less than 25 years. Iron and iron alloys possess the longest remaining lifetime average of the four major groups of minerals. However, the figure shows clearly that the most important minerals are by no means inexhaust-

**Figure 1-8** *The estimated time required for, and the effects of, measurable increases in the average world temperature due to the energy consumption of man.*

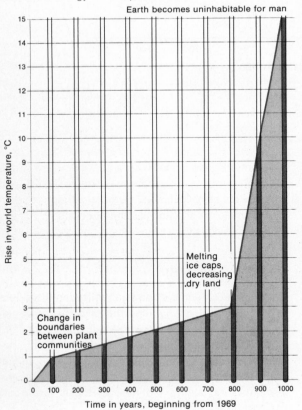

ible. Their reserves are being depleted in accordance with the laws of supply and demand.

## ENERGY PRODUCTION

The increasing demands for fossil and nuclear fuels have had environmental repercussions other than the imminent depletion of these materials. Fuel consumption produces both useful and nonuseful energy, a concept described by the second law of thermodynamics discussed earlier. The nonuseful energy is that of heat transmitted to the environment. The larger the amount of energy produced, the greater the amount of heat evolved. At some point, as energy production increases in response to demand, there should be a perceptible rise in the atmospheric or aquatic temperatures at the surface of the earth.

Assuming that energy production by man is increasing at a constant rate of 7 percent per year and assuming a 1969 average temperature for the surface of the earth of 15°C (degrees Celsius, formerly known as degrees centigrade), the times required for measurable rises in the average world temperature are given in Fig. 1-8. According to this graph, an average rise of 1°C may take less than 100 years. This temperature rise will cause definite changes in the boundaries between various plant communities. An average rise of 3°C will likely occur in under 800 years, causing a decrease in dry land mass through the melting of ice caps. An average increase of 15°C, raising the temperature above the thermal death point for man, is probable before nine centuries have elapsed.

Thus, significant ecological effects should become apparent within the next century unless steps are taken to decrease energy consumption or to increase the efficiency with which energy is utilized. Although important, temperature elevation is just one aspect of thermal pollution, a problem to which we shall return later in this chapter.

## BIOSPHERIC POLLUTION

We have reviewed four of the major effects of overpopulation: average decreases in living space, food

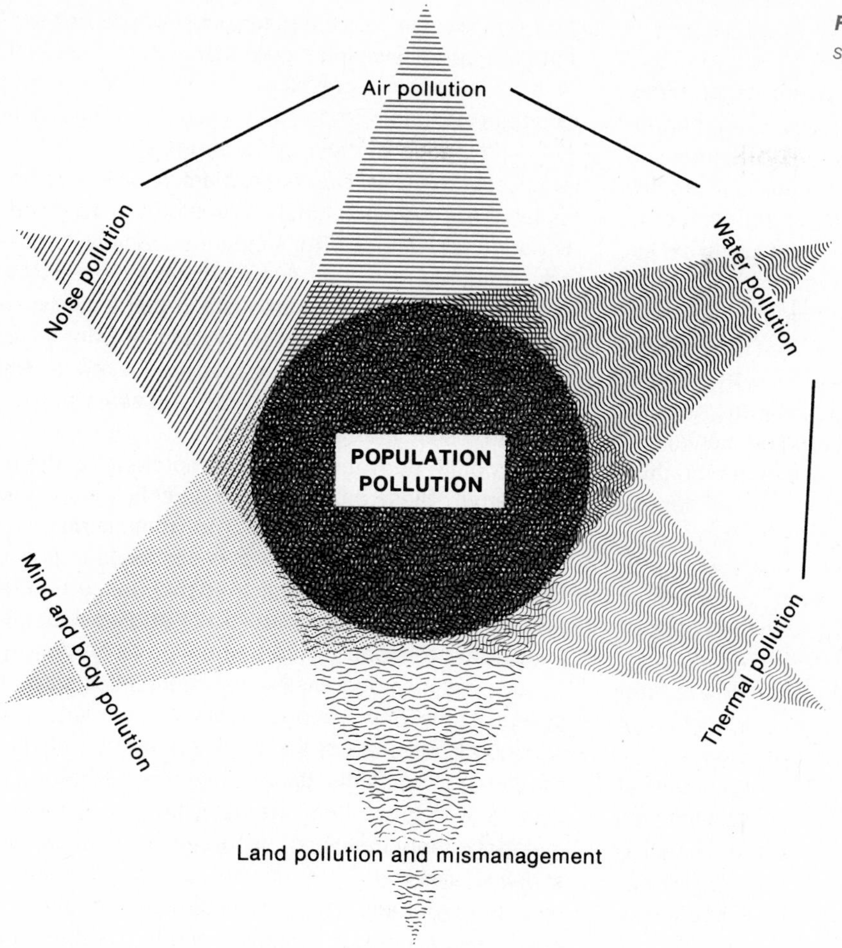

**Figure 1-9** *The major types of biospheric pollution.*

Air pollution

Noise pollution

Water pollution

**POPULATION POLLUTION**

Mind and body pollution

Thermal pollution

Land pollution and mismanagement

production, and available mineral reserves, and an average increase in energy consumption and heat evolution. We will now examine general biospheric pollution, still another major result of overpopulation. We will look at the causes and consequences of environmental pollution and their possible but complex solutions.

## TYPES OF BIOSPHERIC POLLUTION

Six major types of pollution have resulted from population pollution. Three of these, land pollution, air pollution, and water pollution, occur in different types of en-

vironments. These environments have definite physical characteristics which separate one from the other. Two other types of pollution, noise pollution and thermal pollution, are not restricted to specific portions of the environment. These forms of pollution actually represent types of energy, namely, sound and heat. Noise pollution is usually transmitted through the atmosphere but may also originate under water or in the ground, while thermal pollution is atmospheric or aquatic.

The relationships existing among the various types of pollution are summarized diagrammatically in Fig. 1-9.

## "MIND AND BODY" POLLUTION

The use of drugs for hedonistic purposes probably precedes recorded history. Therefore, drug dependence is not a new development. Nevertheless, the current drug dependence problem differs in quality from previous drug usage because of the number of drug users, the changing social contexts in which drugs are used, and the long-term effects of currently used psychogenic, or "mind-affecting," drugs on human beings.

Fundamentally, drug dependence is an indirect result of overpopulation. Coupled to socioeconomic motivations, overpopulation has created centers of high population densities called cities. Although drug usage is not limited to cities, high population density seems always to be a factor among the many causes. However, whether drug dependence, in common with many other complex societal phenomena, is simply supported by, or an effect of, overcrowding is difficult to state unequivocally.

Other common factors in the epidemiology of drug dependence are psychological stress, alienation of family or society, and, possibly, the development of anticultures or countercultures. The spectrum of socially adaptive and maladaptive adjustments which may be made in response to the sociopathic stress of overpopulation is postulated in Fig. 1-10.

We shall recognize four categories of hedonistic (pleasure-giving) drug: (1) those which are legal only on prescription and may be addictive, e.g., the opiate narcotics morphine, codeine, and heroin, and the barbiturates; (2) those which are legal only on prescription but not strictly addictive, such as cocaine and amphetamine; (3) those which are legal without prescription and may be addictive, such as alcohol, caffeine, and nicotine; and (4) those which are legal only for investigational use and hallucinogenic, such as mescaline, psilocybin, lysergic acid diethylamide (LSD), and marijuana. The chemical structures, plant sources where applicable, and some general effects of the above drugs are tabulated in Fig. 1-11.

Drugs by definition are chemical agents that have physiologic or psychological effects useful in the treatment of disease. It is therefore not surprising to find a large number of natural and synthetic chemicals included with the hedonistic drugs. However, what is surprising chemically is the multiple ring structure containing an indole grouping (shaded in Fig. 1-11) found in most of the drugs obtained from plant sources. These diverse plant substances are called *alkaloids.* Their functions in plants are generally poorly understood. The similarities in their chemical structures, however, suggest similarities in biochemical pathways of metabolism in plants and man as well as related pharmacologic actions. The structural similarities are especially pronounced in the opium derivatives as well as in marijuana, caffeine, nicotine, psilocybin, and LSD.

The opium derivatives include morphine, codeine, and heroin. These drugs are true narcotics, acting as nervous system depressants. The administration of opium narcotics and barbiturates can result in physiologic tolerance and physical dependence. These two properties are characteristic of *addictive* drugs. *Physiologic tolerance* occurs when the same dose of a drug produces decreasing behavioral effects over a period of time. *Physical dependence* results when the continued presence of the drug in the body causes physiologic adjustments in metabolism. When the drug is withdrawn from the user, the drug-induced metabolic equilibrium is disturbed. Called an *abstinence syndrome,* the physiologic disturbance is reflected by characteristic symptoms and signs. The abstinence syndrome can be relieved either by readministering the same drug or by administering another drug which can metabolically substitute for it.

The addictive nature of morphine is complicated by observations that addiction usually results only from the active, voluntary use of the drug. Passive, involuntary administration, e.g., during the normal clinical practice of medicine, does not ordinarily lead to addiction. The mode of administration may therefore be an important determinant of addiction.

Cocaine stimulates the nervous system and may become habit-forming. However, the symptoms of withdrawal characteristic of the opium derivatives following cessation of their use do not occur when cocaine is removed. Therefore, cocaine is not strictly addictive.

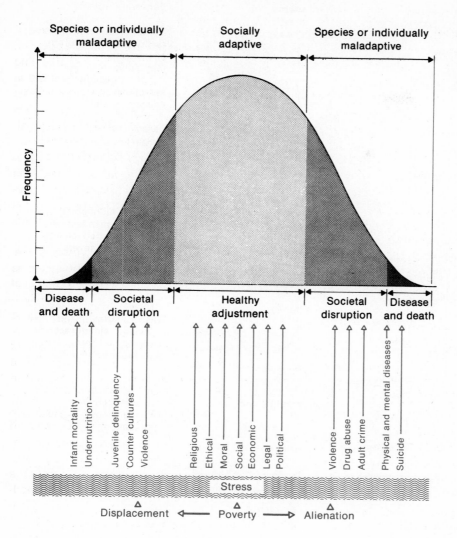

*Figure 1-10* *The spectrum of socially adaptive and maladaptive adjustments which may be provoked as a consequence of overpopulation.*

All the potentially addictive nonprescription drugs except alcohol possess at least one ring in their chemical structures. The barbiturates are related chemically, as are caffeine and nicotine. Three of the compounds—alcohol, caffeine, and nicotine—come from plant sources; the others are synthetic. Amphetamine and caffeine are stimulants. Alcohol is a depressant of the central nervous system, although its behavioral effects are initially stimulating and secondarily depressing.

The hallucinogens include mescaline, psilocybin,

LSD, and marijuana. Psilocybin and LSD possess an indole group and in this respect are similar to psychogenic drugs derived from other plant sources. Most hallucinogens have at least one ring in their chemical makeup. In addition, all are natural products obtained from plant sources.

Use of hallucinogens gives rise to sensory distorting effects, especially with regard to space, form, time, color, and sound. Such chemicals are also habit-forming, similar in that respect to cocaine. Furthermore, LSD has been implicated as a causative

| Compound | Chemical structure | Plant source | Physiologic and behavioral effects |
|---|---|---|---|
| Opium derivatives: Morphine | Morphine 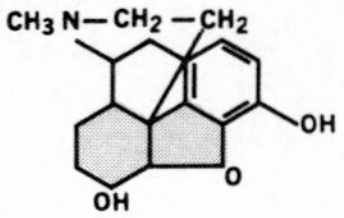 $CH_3N-CH_2-CH_2$ OH OH | Opium poppy *Papayer somniferum*  | Narcotic, reducing aversive responses to painful stimuli; causes pupillary constriction; may cause respiratory arrest; impairs digestion by inhibiting secretions and reflexes; may cause true addiction |
| Codeine | Methylmorphine 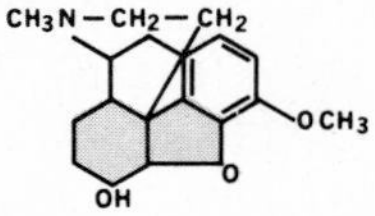 $CH_3N-CH_2-CH_2$ $OCH_3$ OH | | Less potent narcotic than morphine; pharmacologic effects similar to morphine |
| Heroin | Diacetylmorphine 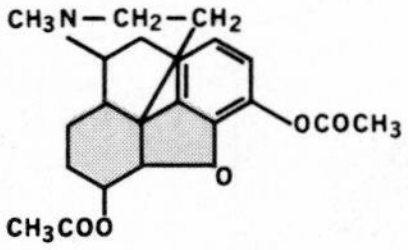 $CH_3N-CH_2-CH_2$ $OCOCH_3$ $CH_3COO$ | | Narcotic; more potent than morphine but effects are of shorter duration; causes euphoria and severe addiction |
| Cocaine | Cocaine 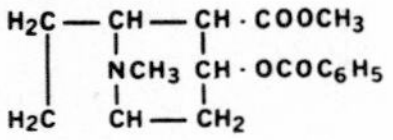 $H_2C-CH-CH \cdot COOCH_3$ $NCH_3 \cdot OCOC_6H_5$ $H_2C-CH-CH_2$ | Coca leaves  *Erythroxylon* spp. | CNS stimulant; causes vasoconstriction and other sympathomimetic actions; causes short-term exhilaration; habit-forming |
| Marijuana and derivatives: Hashish Dagga Charas Bhang | 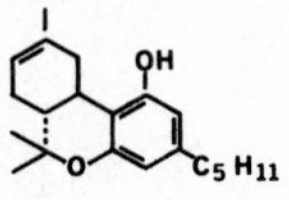 Tetrahydrocannabinol 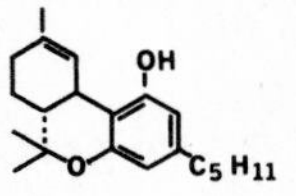 OH $C_5H_{11}$ Major component OH $C_5H_{11}$ Minor component | Hemp and marijuana leaves and flower bracts  *Cannabis sativa* | Unreliable euphoriant and hallucinogen; impairment of recent memory; may lead to habituation |
| Barbiturates: Amobarbital (Amytal) | 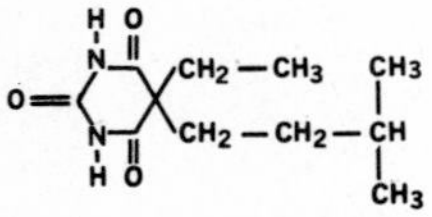 $CH_2-CH_3$ $CH_3$ $CH_2-CH_2-CH$ $CH_3$ | | Sedative and hypnotic; may cause intoxication and hangover; may be addictive; moderately long-lasting |
| Phenobarbital (Luminal) | 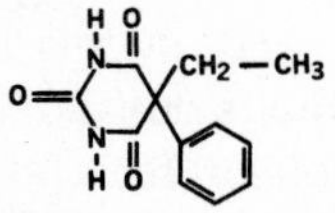 $CH_2-CH_3$ | | Sedative, hypnotic, and anticonvulsant; action similar to other barbiturates; long-acting |
| Amphetamine (Benzedrine) | Methylphenethylamine 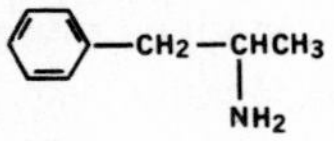 $-CH_2-CHCH_3$ $NH_2$ | | Stimulant, causing restlessness and insomnia; causes initial pleasurable sensations; may be addictive |

**Figure 1-11** *Some of the most common psychogenic drugs used by man, together with their chemical structures, plant sources, and some physiologic and behavioral effects.*

| Compound | Chemical structure | Plant source | Physiologic and behavioral effects |
|---|---|---|---|
| Alcohol (anhydrous) | Ethanol<br><br>CH₃<br>CH₂OH | Cereal grasses | Initial repression of inhibitions; CNS depressant; unreliable euphoriant; causes diuresis; may damage CNS and liver; may be addictive |
| Caffeine | Trimethylxanthine | Coffee plant seed — *Coffea arabica*  Tea plant leaf — *Camellia sinensis* | Mild CNS stimulant; stimulates heart and skeletal muscle at high doses; relaxes vascular musculature, except in brain; causes diuresis; restores prefatigue levels of physical and mental competence |
| Nicotine | Pyridyl methylpyrrolidine | Tobacco plant stem and leaf  *Nicotiana* spp. | CNS and peripheral nervous system stimulant, mimicking acetylcholine and releasing epinephrine; secondary depressant; inhibits hunger pangs; reduces $O_2$-carrying ability of blood |
| Mescaline | Trimethoxyphenethylamine | Peyote cactus flesh  *Lophophora williamsii* | Produces visual and color hallucinations and euphoria; habit-forming (?) |
| Psilocybin | 3-[2-(Dimethylamino)ethyl]-indol-4-ol | Mushroom  1. *Psilocybe mexicana*  2. *Stropharia cubensis* | Causes pupillary dilatation and elevates blood pressure; produces visual and auditory hallucinations; causes euphoria and relaxation; habit-forming (?) |
| LSD | Lysergic acid diethylamide | Rye ergot fungus  *Claviceps purpurea* | Causes pupillary dilatation and elevates blood pressure; produces visual, auditory, and temporal hallucinations; may be teratogenic; habit-forming (?) |

factor in chromosomal damage and developmental defects in the progeny of its users, but the evidence is controversial (see Chap. 21).

Marijuana has a number of derivatives: hashish, dagga, charas, and bhang. The names of the derivatives are related to the part of the plant used for extraction of the natural product. Chemically, the derivatives belong to a family called the cannabinoids. The family is named for the plant *Cannabis sativa,* from which marijuana is derived.

The chemistry of marijuana is complex, with at least two pharmacologically active components, one major and the other minor (see Fig. 1-11). To what degree these components interact biochemically and physiologically is not yet known. Indeed, whether the active components may be metabolized into a more pharmacologically potent product after entry into the body is an open question.

Collectively, these drugs pollute the drug-dependent mind and body just as surely as other substances pollute land, air, and water.

## LAND POLLUTION AND MISMANAGEMENT

Land is becoming increasingly polluted. The pollution is caused primarily by accumulations of municipal solid wastes, discarded consumer products such as motor vehicles, bottles, cans, and plastics, and atmospheric pollutants such as persistent pesticides, radioactive materials, and other specific contaminants which settle on the ground. Added to these sources of pollution are practices in the areas of agriculture, timber, wildlife, and recreational and natural land usage that are inimical to resource conservation. Other man-made and natural disturbances also contribute to the per capita decrease in land utilization.

The effects of mistreating land, whether unavoidably, accidentally, or otherwise, are wide-ranging and incompletely understood. Their impact is in some cases subtle and may result in either reduced recreational and esthetic opportunities or increased psychological tensions. In other cases, the impact is physical and may cause erosion, leaching, and siltation of soil or contamination and spoilage of crops.

Expanded research, educational efforts, technol-ogy, and legal constraints are needed to deal with the increasingly serious land pollution and mismanagement problems we now face. However, these combined efforts will represent only a partial solution. Basic changes in individual, institutional, and societal attitudes must occur before even currently available remedies can be successfully implemented on a large scale.

## AIR POLLUTION

Perhaps more is known about the subject of air pollution than about any other form of environmental contamination. Its causes and effects are in many cases easily identifiable and have pronounced consequences on large populations over wide geographic areas. However, notwithstanding the rapidly accumulating literature on air pollution, there are still many gaps in our basic knowledge of its multiple and complex causes, effects, and solutions.

Atmospheric pollutants are either primary or secondary emissions. *Primary emissions* are produced by identifiable sources such as chimneys, exhausts, and cigarettes. Such sources are discrete, easy to locate, and relatively stable. *Secondary emissions,* on the other hand, are produced by the interactions of primary pollutants with either normal atmospheric constituents or other primary pollutants. The interactions may occur with or without photoactivation of the chainlike chemical reactions which characterize them. Secondary emissions are not produced directly from discrete sources, are difficult to isolate, and are short-lived.

In addition, primary and secondary emissions may interact with each other and with other natural constituents, producing a spectrum of products and effects. These interactions are only now beginning to be critically studied. The results of such investigations should shed more light on cause-and-effect relationships and, hopefully, potential remedies.

If the primary emissions are less than one ten-thousandth of a meter (1/10,000 m), or less than 100 micrometers (100 $\mu$m), they are classified as fine aerosols. When greater than this size, they are considered coarse particles. Primary emissions generally belong to one of four families: sulfur, organic compounds, nitrogen, and oxygen. Sulfur, nitrogen, and oxygen compounds contain one or more atoms of these three elements. Organic compounds are those containing one or more carbon atoms but exclude carbon dioxide and carbon monoxide.

Secondary emissions are products formed during the following reactions: (1) *thermal reactions,* requiring heat energy; (2) *catalytic reactions,* requiring chemical or physical agents which bring the reactants into close physical proximity; (3) *energy degradation processes,* such as oxidation, which result in a decrease in the total energy of the chemical system; or (4) *photochemical activation processes*, initiating atomic excitation by light energy.

Figure 1-12 summarizes one type of light-activated chemical chain reaction. The reaction forms a variety of secondary emissions which interact with other primary emissions. The chain reaction is initiated by the activation of nitrogen dioxide with sunlight. The products may produce haze or smog and other undesirable or harmful effects.

Urban smog depends upon a relatively concentrated, stagnant mass of atmospheric pollutants above a city. One way in which these conditions are commonly met is through the formation of a thermal inversion which traps cool air and its particulate contaminants under a mass of warm air. This situation is the reverse of the normal. The stagnant cold air mass is finally removed when the warm air rises, usually

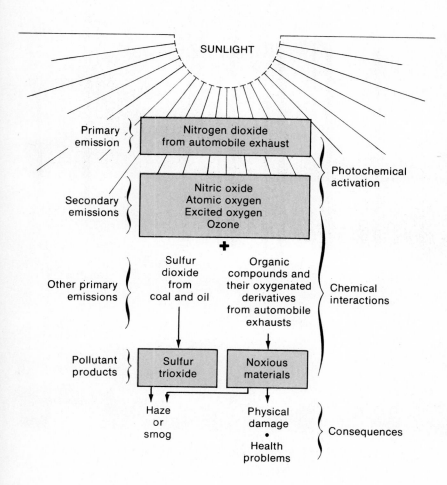

*Figure 1-12* One type of photochemically activated reaction based on the production of primary emissions from fuel combustion. Nitrogen dioxide, a primary emission from automobiles, is photochemically activated with sunlight. The products of this reaction, nitric oxide, atomic oxygen, excited oxygen, and ozone, are secondary emissions. They, in turn, may interact with the primary emissions sulfur dioxide or organic compounds and produce sulfur trioxide and noxious materials, respectively. Haze or smog, physical damage, and health problems may result from these interactions.

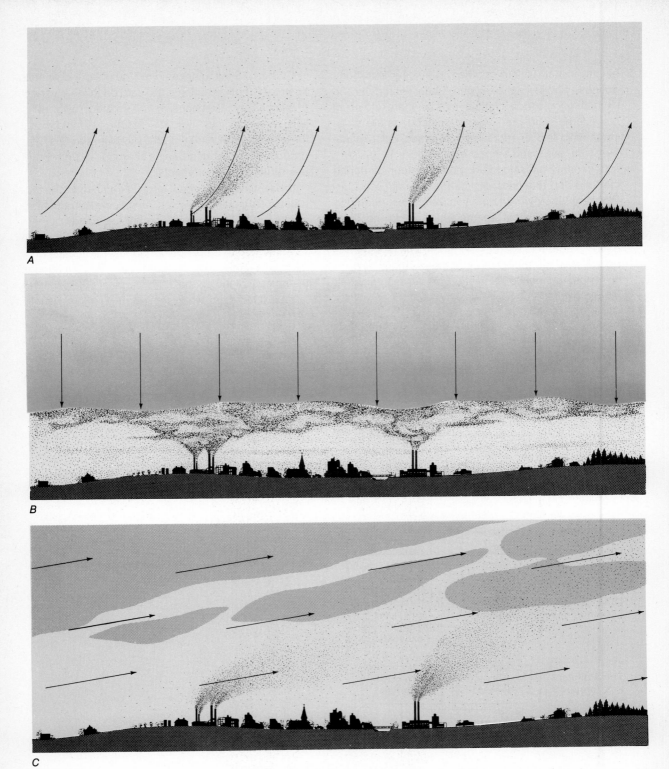

**Figure 1-13** *The formation of smog by thermal inversion. Darker tones indicate cooler air. A. Normal condition. B. Formation of thermal inversion. C. Dissipation of thermal inversion.*

because of wind action. The steps in the formation and dissipation of a thermal inversion are traced in Fig. 1-13.

The detrimental effects of air pollution, although in many cases subtle, are nevertheless detectable. For example, the rhythmic beating of hairlike projections called *cilia*, characteristic of cells which line the upper respiratory tract, are reduced or inhibited by inhaled tobacco smoke. The ciliary cells ordinarily transport inspired atmospheric contaminants on a mucous layer to the back, or posterior, part of the mouth. This mechanism is termed a *respiratory escalator*. The inspired particles are then swallowed and rendered harmless. When the respiratory escalator breaks down, the results are irritation of the lungs and an episode of coughing.

Some particulates in smoke are also believed to inhibit subcellular enzyme systems. Enzymes are biocatalysts which affect the rate of intracellular reactions. When one or more enzymes become inoperative, various chemical products within the cells cannot be further degraded or synthesized. Disturbances of metabolism may result from this type of disruption.

The increase in the number of respiratory cells with atypical nuclei in smokers, compared with nonsmokers, is consistent with the theory of enzymatic inhibition. The validity of the theory, however, remains to be established.

Smoke, especially that from tobacco with its phenols, tars, and nicotine, may also cause deleterious effects to the tissues of the lungs. The effects may or may not be detectable. Changes in localized rates and volumes of blood flow in various parts of the body are some of the subclinical effects. The clinical effects include complaints such as coughing, shortness of breath, fatigue, and insomnia, as well as diseases such as heart disease, lung cancer, and assorted respiratory disorders. It is not yet clear whether the primary causes of the above-listed complaints and diseases are due to the nicotine, tars, and phenols found in tobacco smoke, other atmospheric pollutants drawn in large quantities through the cigarette, cigar, or pipe and ignited during inhalation, other subcellular influences only indirectly related to smoking, or still other factors.

That tobacco smoke is not the only atmospheric

agent with wide-ranging public health ramifications is also well documented. Figure 1-14 shows the effects of bronchitis-emphysema on the lower respiratory tract. This syndrome consists primarily of an irritation of the lungs, a constriction in the respiratory passageways which causes a resistance to airflow, an increase in heart strain as a result of an attempt to aerate the blood adequately, and a corresponding buildup of fluid congestion within the lungs. The severity of these symptoms is statistically increased by living in geographic pockets of chronic air pollution. Sublethal

**Figure 1-14** *Bronchitis-emphysema, a chronic lung disease apparently aggravated by air pollution. A. Normal lung. B. Constriction in the diameters of the airways in diseased lung. C. Reduction in amount of gas exchange membrane in diseased lung.*

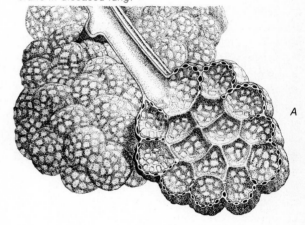

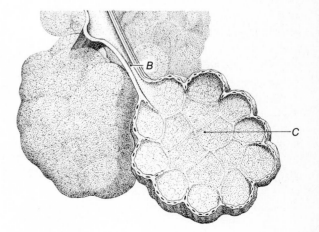

poisoning due to localized increases in the atmospheric concentrations of certain volatile gases and other noxious products may have similar effects.

Worldwide fatalities from lethal fogs have been increasing in frequency during this century, especially among people with a medical history of pulmonary and cardiovascular involvement. In addition, psychosomatic stress reactions are also increasing in frequency in response to the threat of, or repeated experiences in, polluted air.

The present warming trend in climate is also partially a result of air pollution. While causing an actual decrease in the length of time the sun shines, atmospheric pollutants scatter radiation from the sun, causing the temperature to rise.

Because most primary and secondary emissions depend ultimately on identifiable sources for their production, solutions to air pollution require methods of reducing or replacing objectionable sources or removing or recycling their by-product contaminants. The internal-combustion engine is one prominent example of such a discrete source. It produces carbon monoxide, carbon dioxide, nitrogen dioxide, and lead compounds, some of which are potentially toxic primary emissions. The automobile engine also creates a substrate for photochemically activated secondary emissions which may result in haze or smog, physical damage, and other complications.

Solutions to air pollution are easy to suggest. However, their implementation is intertwined with politicoeconomic decisions and prevailing social attitudes. For example, to replace the internal-combustion engine is a solution formidable to an industry the very existence of which is based on it and to the consumer who also has an appreciable investment in it. Furthermore, the technology of engines using other sources of power, such as electricity and solar energy, does not yet yield the performance characteristics obtained with gasoline engines.

In attempts to offset the undesirable features of internal-combustion engines, automotive manufacturers have been forced by law to install devices which reduce the concentrations of particulates emitted from exhaust manifolds. The petroleum industry is under pressure to explore the feasibility of reducing or completely eliminating the lead content of high-grade gasoline. However, federal legislation was necessary in both cases before industry would act while the legislation has so far been either ineffective or nonenforceable. Furthermore, consumer preferences still seem to be related to performance, convenience, and comfort rather than to physical and environmental safety and efficiency. Collectively, the tides of business, political power, and consumer gratification demonstrate the forces resisting effective action against consumer products which are not in the long-term interest of the public.

## NOISE POLLUTION

Increases in the ambient, or background, noise level to which people are subjected are inversely related to living space and directly related to population density and developing technology. Industries, transportation, crowding, and the cost of suppressing noise have contributed to the din of megalopolises, cities, and other urban centers.

One by-product of transportation technology, the sonic boom, while not a significant noise factor now, seems destined to increase in frequency if not severity as commercial airlines begin to use supersonic jet transport aircraft. The electronically amplified sounds of rock bands in discotheques provide still another example of a significant but localized increase in noise level.

The examples can be compounded, but their effects are similar. First, an increased chronic stress is experienced by most people exposed to unusually high noise levels. Chronic exposure to stress may lead to sociopathic disruptions and disease. Second, structural damage to either material possessions or hearing is a real threat arising from exposure to localized but high air pressures. Finally, in all cases of noise pollution, the invasion of privacy which is caused by unwanted sounds further erodes individual freedom and decreases the quality of life for everyone.

## WATER POLLUTION

The causes of water pollution are primarily man-made and, in most cases, easy to identify. Water can be polluted by industrial and agricultural wastes, natural products, municipal sewage, atmospheric pollutants, and runoff water from natural and artificial sources.

Examples of adverse effects on the quality of water are (1) the growing contamination of domestic water supplies; (2) the spread of, and increase in, diseases such as infectious hepatitis, diarrhea, dysentery, and poisoning by lead, mercury, and other heavy metals; and (3) economic and esthetic losses to commercial fisheries, recreational industries, and other private interests based on the use of water.

Water pollution abatement is hampered by a lack of effective moral and legal controls. In addition, basic knowledge and its corresponding technology are currently inadequate. For example, our knowledge does not yet permit the effective recycling or utilization of waste effluents or the development of alternative, inexpensive sources of industrial and agricultural waters through desalinization. These difficulties are tempered by a growing concern for the cleanliness of water, the effects of which should be increasingly felt in the future.

## THERMAL POLLUTION

Urbanization, industrialization, and the accompanying technology of modern society have created rising demands for electricity and the nuclear power plants to partially satisfy those needs. We have already noted the relation between an increasing power consumption and a corresponding increase in the dissipation of heat energy. The resulting thermal pollution is a cumulative function of industrial, municipal, and individual demands for electric power.

The rates of heat conduction, convection, and radiation and the influences of eddy currents, winds, density, viscosity, and other physical factors must be taken into account when considering the environmental effects of thermal pollution. These factors determine horizontal stratifications and vertical gradients of heat and are usually more demonstrable in water than in air. There are not only easily measurable stratifications and gradients in water, but also cyclic oscillations of temperature with time. These variations may be caused by seasonal changes in the output of

a nuclear power plant or by any large-scale industrial process of a periodic nature. Thus, the effects of thermal pollution are the result of a complex interaction between the organism exposed to the increased quantity of heat energy, the medium or environment in which the organism is located, and the distribution of heat within the environment.

Solutions to thermal pollution are generally similar to those discussed for water pollution. Such solutions include an improved technology and enforceable legislation to cope with the increasing thermal burden of the environment and raised social consciousness to provide a base for remedial action.

## THE FUTURE

### SOLUTIONS WITHOUT FERTILITY CONTROL

If overpopulation is at the root of environmental crises, then reasonably rapid achievement of zero population growth should at least reduce the severity of these crises. On what evidence is this premise based? The answer may be approached by weighing the effects of proposed solutions to overpopulation that do not involve fertility control.

Because the population explosion and available living space are inversely related to each other, these problems will be considered together. Excluding fertility control, proposed solutions require natural or man-made disasters and developing technology. However, these solutions seem self-defeating. Unchecked human propagation will most likely continue to overwhelm the reductions in population caused by disaster or by extensions of geographic horizons accomplished through extraordinary technological feats. In addition, not only is there the question of the immorality of relying on solutions based on population diminution through disasters or wars, but such an approach simply leaves these matters to chance.

The solutions to food production emphasize improvement in yields and distribution, increase in the amount of arable land, and harvesting food from the sea. All these solutions are contingent upon a continually developing technology. Expanded and more effective health education, while obviously important, suffers from most of the inadequacies attached to solutions of the population problem, which are fertility control-free. Population pressures will inevitably swamp the ability of developing technology to keep pace.

To increase energy production and the reserves of mineral resources and decrease the amount of energy utilization, expanded technology and exploration have sometimes been offered as panaceas. Continued exploration is indeed expected to uncover new sources of nonrenewable resources, but the global reserves of minerals are limited. Likewise, technology will surely develop alternative sources of power such as solar, tidal, geothermal, and nuclear energy to competitively substitute for fossil fuels and water power. However, vested politicoeconomic interests and adverse environmental side effects will severely limit their usefulness.

The direct causes of biospheric pollution and mismanagement may possibly be reduced by raising our ecological consciousness, restructuring political priorities to provide for sufficient funding in environmental research and regulation, and increasing scientific knowledge, technology, and education in these areas. That these solutions have so far been largely ineffective is attested to by the continuing deterioration of health and environment caused by pollutants. In any event, the above mechanisms cannot completely eliminate these problems, because population pollution will continue in the absence of fertility control measures beyond those already in use.

The severity of other societal disruptions may also be reduced by implementing these remedial measures, although the ultimate fate of societal disruptions depends upon social forces in the politicoeconomic spheres. For example, serious attempts to increase living space and possibly reduce urban tensions by locating new urban centers outside existing megalopolises may create more problems than they solve. Vested interests, institutional inertia, measures perceived as threats to traditional ways of doing things, and the effects of displacement or relocation combine to militate against these types of solutions. Attempts

to encourage the evacuation of people from geographic regions having high incidences of natural disasters or to improve the systems of legal and social justice meet with many similar forms of resistance, justified or not.

## THE RANGE OF HUMAN ADAPTABILITY

That overpopulation has significantly contributed to biospheric pollution and to environmental and societal disruptions is abundantly clear. If organisms including human beings cannot adapt to these growing imbalances, the alternatives are as follows: displacement or migration of a population or a portion of a population to a potentially more congenial environment, death of the maladapted individuals, or extinction of the species. Of these alternatives, displacement and migration may themselves be adaptive, depending upon the probability of survival of the emigrants.

*Adaptation,* upon which successful existence hinges, takes two related and parallel pathways. One pathway is *short-term* and transient, lasting from several minutes to no longer than a few days. Short-term adaptation results in acclimatization to the environmental influence or disturbance. Altered cardiovascular, respiratory, and other metabolic functions, usually resulting from hormonal changes, are characteristic acclimatization responses.

The other pathway is *long-term* and may be ontogenic or genetic. Ontogenic responses are the result of an individual's chronic exposure to a changed environment and are not inherited. Adaptive responses are genetic when the potential for a particular adaptation is inherited. Ontogenic and genetic adaptations are in most cases difficult to separate functionally. Nevertheless, decreases in the rate of growth, skeletal maturation, and metabolism, and an increase in skin pigmentation are some examples of long-term biologically adaptive responses. These characteristics are exhibited by certain racial and ethnic groups chronically exposed to hot temperatures.

Adaptation for man has a cultural component as well as a biologic base. The cultural responses

**Table 1-1**   The range of human adaptability to selected physical and cultural stresses

| Chronic stress | Biologic responses | | Cultural responses |
| | Short-term (transient and acclimatization) | Long-term (ontogenic and genetic) | |
|---|---|---|---|
| Altitude (atmosphere and temperature) | More rapid breathing and pulse rate; increased hemoglobin concentrations | Increase in ability to propagate in high altitude habitats; decreased rate of heartbeat; retardation of growth and skeletal maturation; rise in basal metabolic rate; increase in thoracic volume and lung capacity; increase in erythrocyte and hemoglobin concentrations | Use of alcohol and cocaine, supplemental oxygen supplies, clothing |
| Cold temperature | Shivering and general rise in basal metabolic rate; reduction in circulation to arms and legs; decreased heat loss from body | Cyclic reduction in circulation to arms and legs; increased cooling of arms and legs, relative to trunk, and decreased average body temperature during sleep; seasonal rise in basal metabolic rate during colder part of year; increase in body height or volume; decrease in width of nose (probably not adaptive); increase in subcutaneous fat | Use of natural and artificial shelters, clothing, fires, tobacco, and alkaloids |
| Hot temperature | Increased sweating and peripheral circulation; decreased cardiovascular strain; increased fluid and salt intake | Decrease in body size and girth; increase in skin pigmentation; increase in or retention of head hair (probably not adaptive) | Use of devices preventing, counteracting, or reducing heat absorption |
| Biospheric pollution | Increase in avoidance responses to polluted food, air, and water; increases in tearing of eyes, coughing, bronchitis-emphysema, and other "urban pathologic conditions"; detoxification, elimination, or neutralization of ingested pollutants | Too early to assess effects accurately; difficult to find current control populations in which environments are unpolluted | Use of moral, legal, and economic mechanisms to reduce population or kinds and numbers of biospheric pollutants |
| High population density | General adaptation syndrome, consisting of hormonally induced increases in rate of heartbeat, respiration, and metabolism and temporary inhibition of digestion, excretion, and reproduction | Too early to assess effects accurately | Screening out of extraneous sensory inputs; possible formation of dominance-subordination hierarchies; violence and other forms of aggressive behavior (probably not adaptive) |

*(Continued)*

**Table 1-1** (Continued)

| Chronic stress | Biologic responses | | Cultural responses |
| --- | --- | --- | --- |
| | Short-term (transient and acclimatization) | Long-term (ontogenic and genetic) | |
| Undernutrition | Increase in acceptability of any food source containing nutrients deficient in diet | Lowered basal metabolic rate; decrease in body size | Increased imports and altered agricultural and dietary practices to secure lacking nutrients |
| Other societal disruptions | General adaptation syndrome | Too early to assess effects accurately | Violence and other forms of aggressive behavior (probably not adaptive); development of countercultures and increase in drug usage; attempts to solve disruptions through existing mechanisms |

usually increase or facilitate the biologic responses. For instance, clothing, shelter, and drugs are used to neutralize the stressful effects of particular environments. Occasionally, the cultural response may be diametrically opposed to normal social adjustments. Countercultural or suicidal mechanisms employed for removal from stressful situations are some specific examples of this type of response.

Cultural adaptations may be individual, familial, or institutional. An example of an individual cultural adaptation is the unique ability on the part of urbanites to screen out extraneous sensory inputs and to focus almost entirely on information necessary for the execution of daily tasks. At the familial level, sibling rivalries may lead to the formation of dominance-subordination hierarchies, or human "peck orders." The adaptive significance of such hierarchies resides in the possibility that some subordinate individuals, forced to live and work beyond the sphere of dominant influence, may actually thrive in the new environment. Institutional and social adaptations consist primarily of moral and religious, legal and ethical, and economic and political mechanisms designed to reestablish balances compatible with a viable society. Church or social sanctions and pressures, and legislation, enforcement, and taxation at the local, state, and federal levels are among the culturally

adaptive mechanisms. Nevertheless, these mechanisms may not always achieve their objectives. In fact, institutional objectives are sometimes diametrically opposed to the good of society. A society which condones the use of alcohol and cigarettes, largely ignores the safety hazards of certain food additives, current coal-mining practices, and automobiles, and is mostly apathetic to the environmental threats posed by biospheric pollution and still unchecked population growth belies its faith in the values of institutional protections.

Violence is both a biologic and cultural response. While the aggressive behavior associated with violence certainly had an adaptive significance in the Stone Age, many authorities believe the response is maladaptive in modern society. That individual and mass violence still persists is not only mute testimony to man's ancestral origins but a symptom of serious and chronic societal disruptions.

The short-term and long-term biologic responses to selected chronic stresses, together with some accompanying cultural responses, are summarized in Table 1-1. These responses suggest that man is capable of a wide plasticity of adjustment to environmental and social stresses. If this is so, plasticity may provide an adaptive mechanism for man's survival as the external environment continues to deteriorate.

## COLLATERAL READING

Adams, Elijah, "Barbiturates," *Scientific American,* January 1958. The different effects of increasing concentrations of barbiturates are reviewed.

Baker, Jeffrey J. W., "Science, Birth Control and the Roman Catholic Church," *BioScience,* 20(3):143–151, 1970. Origin and analysis of the positions taken by the Roman Catholic Church on birth control.

Barron, Frank, Murray E. Jarvek, and Sterling Bunnell, Jr., "The Hallucinogenic Drugs," *Scientific American,* April 1964. A discussion of various physiologic and psychologic effects of the major hallucinogens.

Casarett, Alison P., *Radiation Biology,* Englewood Cliffs, N.J.: Prentice-Hall, Inc., 1968. An excellent description of the environmental and biologic effects of radiation.

Chamber, Leslie A., "Classification and Extent of Air Pollution Problems," in C. E. Johnson (ed.), *Human Biology: Contemporary Readings,* New York: Van Nostrand Reinhold Company, 1970, pp. 205–223. Systematizes the types of atmospheric emissions and explores air pollution from a historical perspective.

Christian, John J., "Social Subordination, Population Density, and Mammalian Evolution," *Science,* 168:84–90, 1970. The dynamics of social behavior are discussed as an adaptive force in mammalian evolution.

Cole, LaMont C., "Thermal Pollution," *BioScience,* 19(11):989–992, 1969. An analysis of the bioenergetics of an increasing energy consumption by the human population.

Dasmann, Raymond F., *Environmental Conservation,* 2d ed., New York: John Wiley & Sons, Inc., 1968. An introduction to ecology from a conservationist's point of view.

Ehrlich, Paul R., and John P. Holdren, "Population and Panaceas: A Technological Perspective," *BioScience,* 19(912):1065–1071, 1969. Emphasizes that technology alone will not solve our environmental crises and proposes zero population growth as a solution.

Gilula, Marshall F., and David N. Daniels, "Violence and Man's Struggle to Adapt," *Science,* 164:396–405, 1969. Gilula and Daniels conclude that the use of violence in modern society is no longer adaptive.

Haagen-Smit, A. J., "The Control of Air Pollution," *Scientific American,* January 1964. Provides an in-depth analysis of the origin, effects, and control of Los Angeles smog.

Hammond, E. Cuyler, "The Effects of Smoking," *Scientific American,* July 1962. Statistical and biologic evidence reveals a deleterious relationship between cigarette smoking and health.

Hennigan, Robert D., "Water Pollution," *BioScience,* 19(11):976–978, 1969. A general assessment of the entire water pollution problem.

Hulett, H. R., "Optimum World Population," *BioScience,* 20(3):160–161, 1970. Hulett concludes that about 1 billion is the optimum number of people that the world can support adequately.

Lasker, Gabriel W., "Human Biological Adaptability," *Science,* 166:1480–1486, 1969. A review of various adaptations to stress by human beings and an attempt to assess the importance of short-term plasticity.

McDermott, Walsh, "Air Pollution and Public Health," *Scientific American,* October 1961. McDermott concludes from emission-disease correlations that the cumulative effects of air pollution have harmful effects on human health.

Milgram, Stanley, "The Experience of Living in Cities," *Science,* 167:1461–1468, 1970. Human adaptations to regions of high population density are considered from a behavioral perspective.

Nichols, John R., "How Opiates Change Behavior," *Scientific American,* February 1965. The results of experiments with rats suggest that human morphine addiction depends on whether the user is an active or passive recipient of the drug.

Paddock, William C., "How Green Is the Green Revolution?" *BioScience,* 20(16):897–902, 1970. Paddock reviews the reasons for the green revolution and weighs the effects of food production and human reproduction through the next decade.

Robinson, Trevor, "Alkaloids," *Scientific American,* July 1959. A discussion of the chemical interrelationships and varied physiologic effects of this diverse group of natural products.

Spengler, Joseph J., "Population Problem: In Search of a Solution," *Science,* 166:1233–1238, 1969. Explores the role of human motivation in population control and proposes a fertility control system incorporating financial incentives.

Wager, J. Alan, "Growth versus the Quality of Life," *Science,* 168:1179–1184, 1970. Analyzes the economics of "the good life" and relates the findings to present population trends.

Woodwell, G. M., "Toxic Substances and Ecological Cycles," *Scientific American,* March 1967. Emphasizes the dangers of trophic magnification of certain biospheric pollutants.

Woodwell, G. M. "Effects of Pollution on the Structure and Physiology of Ecosystems," *Science,* 168:429–433, 1970. A critical review of the literature identifies similarities in the effects of various pollutants on human beings and their external environment.

Wynne-Edwards, V. C., "Population Control in Animals," *Scientific American,* August 1964. Forms of social behavior that limit reproduction and avoid overexploitation of food reserves are reviewed for animals other than human.

# CHAPTER 2
# LIFE WITH MICROBES

Our environment is literally teeming with microbial life. Bacteria, fungi, and other microbes are normal constituents of the air, water, and land which surround us. Therefore, it is natural enough to find many types of microbe living on or in the body. Their relationship with human beings may be beneficial, neutral, or harmful. These relationships are referred to as *mutualistic, commensalistic,* and *parasitic,* respectively.

Any of the above associations between two organisms of different species is called *symbiotic.* Symbionts usually live in intimate association with each other, i.e., one living on or in the body of the other. The various kinds of symbiotic relationship are summarized in Fig. 2-1. Obviously, symbionts of medical interest are parasitic.

In a parasitic relationship, there is an advantage to the parasite in reducing its adverse or harmful effects on the host. A pathogen which kills its host also kills itself. Unless there is a stable mechanism to transmit viable, living pathogens from a dead or dying host to a new host, the pathogenic species is in danger of extinction. Reduction in virulence of the pathogen allows for peaceful coexistence. Adaptation usually encourages a progression from parasitism to mutualism.

We first review microbial diversity and classification, especially noting the myriad sizes, shapes, adaptations, and interrelationships of microbes. Microbial nutrition, respiration, and reproduction are examined next, to gain some appreciation of the unique requirements and life styles of microbes. Microbial effects, both beneficial and harmful to human beings, are then examined, so that the roles of microbes in nature can be placed in perspective. The activities of nonsymbiotic microbes in biospheric pollution and material damage form a substantial part of the survey. The discussion then focuses on symbionts of human beings, dealing with modes of transmission to, and portals of entry into and exit out of, the body for both commonly occurring and potentially pathogenic mi-

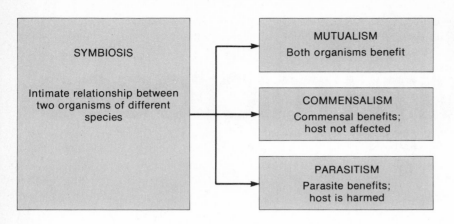

**Figure 2-1** *The types of symbiotic relationships.*

crobes. Finally, microbial control measures and the general mechanisms by which microbes become resistant to toxic chemicals are discussed. Emphasizing the continual and rapid adjustments of pathogens to specific controls, the review underscores the perpetual warfare and accommodations made between human beings and the microbial parasites which plague their bodies.

## DIVERSITY

### CLASSIFICATION

The microbes of human beings belong to eight major groups based on structural and functional differences: bacteria, rickettsias, viruses, fungi, protozoa, flatworms, roundworms, and arthropods. This is indeed a diverse assemblage of organisms!

The subdivisions which make up each of these groups are given in the upper part of Fig. 2-2. Bacteria belong to the phylum Schizophyta, divided into at least five subgroups possessing parasitic representatives. The classification of viruses is even more complicated than that of bacteria, with at least 11 groups causing human disease. Fungi belong to the phylum Mycophyta. Parasitic representatives are restricted to the yeasts and molds, members of the Ascomycetes. Parasitic protozoa belong to four subgroups: sarcodinians, or ameboid forms; mastigophorans, or flagellates; ciliophorans, or ciliates;

and sporozoa, possessing no active method of locomotion.

Three multicellular groups contain parasitic symbionts of human beings: flatworms, roundworms, and arthropods. The flatworms and roundworms belong to the Platyhelminthes and Aschelminthes, respectively. These two phyla are referred to simply as helminths, or "worms." The flatworms include parasitic flukes, or trematodes, and tapeworms, or cestodes; the parasitic roundworms belong to the nematodes. Finally, acarine (acarian) mites and insects such as lice, fleas, and piercing and biting flies are examples of parasitic arthropods.

Figure 2-2 represents one view of the relationships which exist among parasitic symbionts. Its virtues lie in interrelating a structurally and functionally diverse grouping of organisms, even if the classification may ultimately be proved incorrect in one or more particulars.

From a biologic viewpoint, we assume that life began in the form of a primordial cell, the progenitor of everything living, which successfully reproduced again and again. This event happened at some time in the distant past, perhaps 2 billion years ago or longer. The primordial cell, most biologists theorize, did not have a discrete nucleus.

Evolutionary theory suggests that the original cell gave rise to two cell lines, the *prokaryotes* (before nuclei) and *eukaryotes* (true nuclei). All prokaryotes remained without discrete nuclei; all eukaryotes acquired nuclei. From these lines, four major groups are believed to have arisen: Monera, Protista, Metaphyta (multicellular plants), and Metazoa (multicellular animals). The Monera are thought to have given rise to bacteria and blue-green algae. The rickettsias probably arose at an earlier time from the same prokaryotic cell line. The groups believed to have descended from the Protista include the algae, fungi, protozoa, and slime molds. The Metaphyta and Metazoa probably descended separately from the eukaryotic line.

## STRUCTURAL CHARACTERISTICS

Also illustrated in Fig. 2-2 are the general structural attributes of the parasitic symbionts of man. The following generalizations may be drawn among the various groups:

1 Parasitic symbionts range in structural complexity between noncellular viruses and metazoans which contain numerous organ systems.
2 Size closely parallels the degree of structural complexity, varying from the ultramicroscopic viruses to the macroscopic metazoans.
3 The shapes or symmetries of the monerans and protists are more variable than those of the metazoans, with spherical, spiral, and filamentous forms predominating in noncellular and unicellular groups and bilateral forms predominating in multicellular groups.
4 The outside covering of most parasitic symbionts is distinctive, consisting of cell walls in many unicellular groups and noncellular cuticles in helminths.
5 The cytoplasmic organization of the unicellular groups increases in complexity from rickettsias through protozoa.
6 Complex life cycles, arthropod vectors which carry the microbe to its host, and other specific structural and functional specializations are among the hallmarks of numerous pathogens.
7 The smaller the microbe, the more physiologic and biochemical its diagnostic features become. However, the structure of the smallest microbes is still important and indeed requires closer study in the viruses and bacteria.

Viral symmetry is either spherical, helical, or complex. The shape of the nucleic acid core confers the helical or spiral symmetry characteristic of plant viruses. Animal viruses are typically spherical, while complex or irregular symmetry is an attribute of bacterial viruses, i.e., viruses which differentially attack bacterial cells.

Bacteria have three basic body shapes—spherical, cylindrical, and spiral—although there are gradations between them. These shapes are illustrated in Fig. 2-3 and are referred to as *cocci, bacilli,* and *spirilla,* respectively. Bacteria which cannot be easily placed into one of these three categories are called *pleomorphic.*

Thus, a coccus is a spherical bacterium. However, bacteria tend to aggregate and cocci are no exception. Coccal aggregations may be paired or in tetrads, chains, irregular masses, and packets and are correspondingly called diplococci, tetracocci, streptococci, micrococci or staphylococci, and sarcinae. In addition, cocci usually do not move actively or form spores and are characteristically gram-positive. The Gram stain and its uses are explained in Chap. 3.

A bacillus is a rod-shaped or cylindrical bacterium. Generally speaking, spore-forming bacilli are gram-positive; non-spore-forming bacilli are gram-negative. The ability to move actively is an attribute of some but not all bacilli.

A spirillum is a spiral rod-shaped bacterium. There are two general types: spirilla with rigid cell walls and spirochetes with nonrigid, flexible cell walls. Both types are motile.

Pleomorphic bacteria are either intergrades between cocci and bacilli called coccobacilli or between bacilli and spirilla called spirillobacilli. A number of rickettsias are similarly pleomorphic.

## METABOLIC AND PERPETUATIVE MECHANISMS

### NUTRITION AND RESPIRATION

Microbes have developed multitudinous ways of obtaining nourishment and securing energy. All the methods are inextricably intertwined with their ways

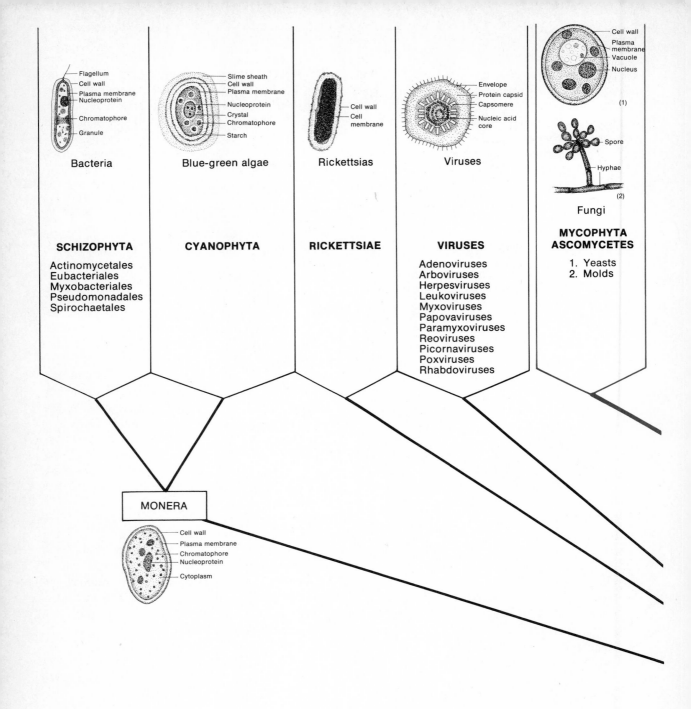

**Figure 2-2** *General structural characteristics of, and relationships among, the microbial pathogens and parasites of human beings.*

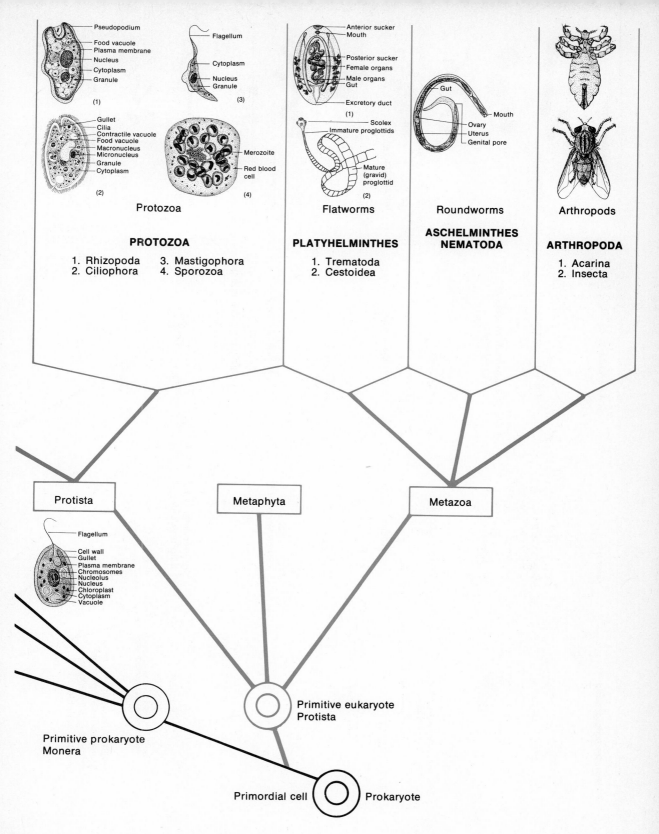

**PROTOZOA**

1. Rhizopoda   3. Mastigophora
2. Ciliophora   4. Sporozoa

**PLATYHELMINTHES**

1. Trematoda
2. Cestoidea

**ASCHELMINTHES NEMATODA**

**ARTHROPODA**

1. Acarina
2. Insecta

*Figure 2-3* The shapes and characteristics of common bacterial types.

| Morphologic groups | Genera | Characteristics | Representative forms |
|---|---|---|---|
| Coccus | | | |
| | Diplococcus | Paired; lance-shaped | Diplococcus |
| | Neisseria | Gram-negative; paired; bean-shaped | 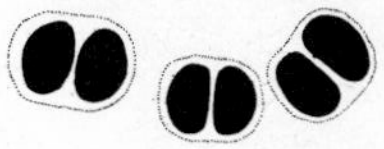 |
| Spherical; nonmotile; non-spore-forming; gram-positive | Streptococcus | In chains | Streptococcus  |
| | Tetracoccus | In groups of four | Tetracoccus 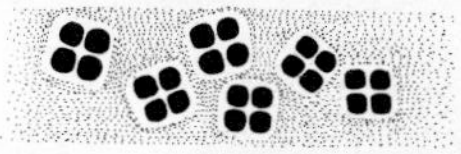 |
| | Sarcina | In packets | Sarcina  |
| | Micrococcus (Staphylococcus) | In irregular masses | Micrococcus  |
| Coccobacillus (pleomorphic) | | Intermediate morphology | |
| Bacillus 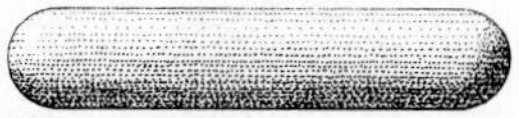 | Corynebacterium (diphtheroids) | Gram-positive | Diphtheroid bacilli |
| | | Granules or bars | Clostridium  |
| Straight rods; spore formers gram-positive; nonspore formers gram-negative; motility variable | | | Corynebacterium  |

| Bacillospirillum (pleomorphic) | | Intermediate morphology | |
|---|---|---|---|
| Spirillum  Rigid spiral rods; motile; non-spore-forming | *Spirillum* | Long |  *Spirillum* |
| | *Vibrio* | Short |  *Vibrio* |
| Spirochete  Nonrigid spiral rods; motile | *Treponema* | Tightly coiled; regularly spiraled |  *Treponema* |
| | *Borrelia* | Loosely coiled; irregularly spiraled |  *Borrelia* |
| | *Leptospira* | Finely coiled; one or both ends hooked |  *Leptospira* |
| *Moldlike*  Nonmotile; lack endospores; may form fine filaments and nonsporelike conidia | *Actinomyces* | Fine branching filaments fragmenting to short rods |  *Actinomyces* |
| | *Streptomyces* | Fine branching filaments; bears chains of conidia |  *Streptomyces* |
| | *Mycobacterium* (acid-fast bacillus) | Acid-fast rods; branching and conidia usually lacking |  *Mycobacterium* |

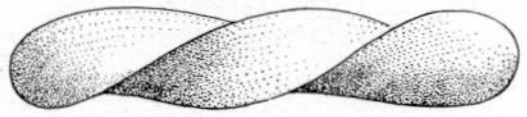

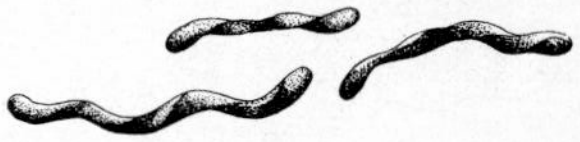

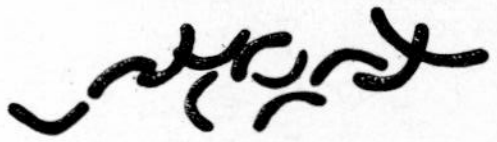

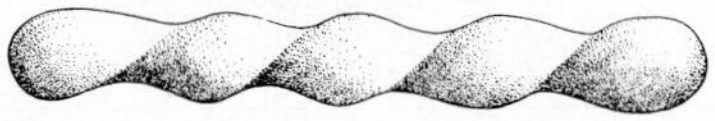

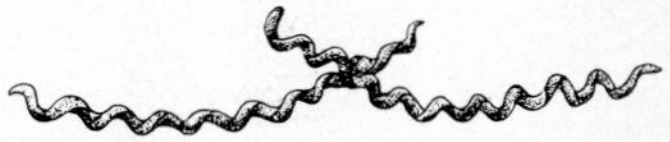

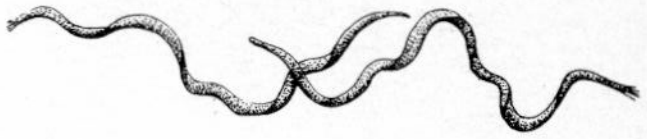

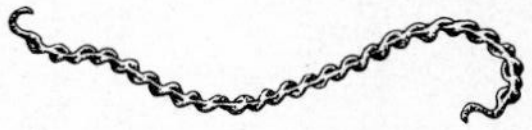

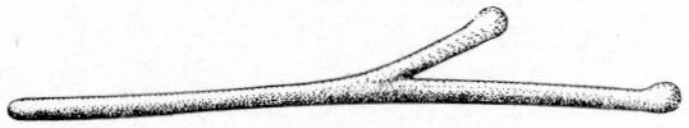

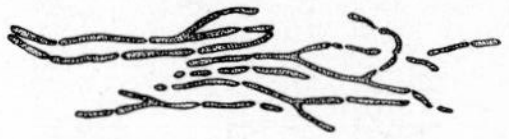

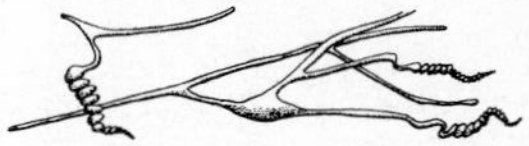

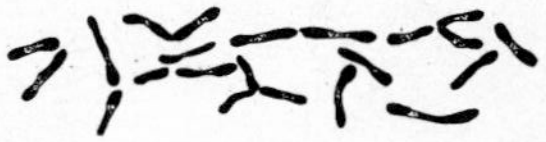

of life. Basically, there are two methods of obtaining energy for cellular synthesis and two classes of raw materials used either as fuels or as raw materials for food manufacture.

The energy source driving cellular metabolism may be sunlight or chemicals. Therefore microbes are either photosynthetic or chemosynthetic. On the other hand, the chemicals utilized as foodstuffs are manufactured from raw materials or oxidized directly. If inorganic chemicals, i.e., compounds not containing carbon atoms, are used for food manufacture or for oxidation, the organisms are autotrophic. If prefabricated organic compounds, i.e., chemicals made of two or more carbon atoms linked together, are required, the organisms are heterotrophic. By simultaneously considering the energy source for cellular metabolism and the chemical nature of the food synthesized or oxidized, four fundamental metabolic plans emerge: photosynthetic autotrophy, chemosynthetic autotrophy, photosynthetic heterotrophy, and chemosynthetic heterotrophy.

*Photosynthetic autotrophs* literally synthesize with light and are self-nourishing. Freely translated, this means that such organisms are capable of manufacturing foods using special photosynthetic pigments known as *chlorophylls*. Whatever the type of chlorophyll employed, they trap light energy to split inorganic molecules containing hydrogen into molecular hydrogen ($H_2$) and other by-products. Following carbon dioxide fixation, which provides a carbon source for the synthesis of organic foodstuffs, the liberated hydrogen and carbon dioxide interact chemically in a complex series of reactions yielding carbohydrate. If the pigment is a bacteriochlorophyll, then hydrogen sulfide is used as the inorganic raw material, and sulfur is one of the by-products of the overall reaction. The purple and green sulfur bacteria exhibit this type of metabolism. Among the microbes that use chlorophyll to split water into molecular hydrogen and derivatives of oxygen are the blue-green algae.

In *chemosynthetic autotrophy*, the primary energy source and the raw material oxidized are the same. To obtain energy, certain autotrophs oxidize inorganic chemicals such as hydrogen, hydrogen sulfide or sul-

fur, ferrous iron, ammonia or nitrite, and carbon monoxide. Chemosynthetic autotrophs which use these foodstuffs include the hydrogen, sulfur, iron, nitrifying, and carbon monoxide–oxidizing bacteria.

The sulfur and iron bacteria are anaerobes, not requiring oxygen for the reactions which liberate energy locked within chemical bonds of the fuel molecules. The other chemosynthetic autotrophic bacteria are aerobic, requiring oxygen for the oxidative reactions which yield energy.

Chemosynthetic autotrophs such as iron bacteria cause the corrosion of iron pipes. Other microbes in this group such as sulfur and nitrifying bacteria contribute to pollution. Sulfur bacteria may form sulfuric acid as a natural pollutant of water; nitrifying bacteria may contribute to the leaching of soil by depleting its supply of natural or artificial nitrogenous fertilizer.

In addition, all the photosynthetic and chemosynthetic bacteria play important roles in the natural recycling of carbon, oxygen, hydrogen, nitrogen, sulfur, and several other elements. Required by all living things, as well as in geologic processes and agricultural and industrial technology, these elements are made available in part through microbial action.

In *photosynthetic heterotrophy,* organic chemicals are used as raw materials for photosynthesis. Rather than serving as fuels to be oxidized, these chemicals provide a source of hydrogen for food manufacture. Instead of organic carbon sources, inorganic carbon dioxide is used. Given hydrogen and carbon dioxide, carbohydrate is manufactured photosynthetically. Only one group of organisms, the purple nonsulfur bacteria, is capable of this type of energy metabolism.

In *chemosynthetic heterotrophy,* an organic compound serves as both the primary energy source and the oxidizable foodstuff. The compound may be degraded in the presence or absence of oxygen. Among the organisms capable of anaerobic oxidation are saprotrophic and symbiotic bacteria and fungi. Without engaging in prior digestion, *saprotrophs* are capable of absorbing organic energy nutrients directly from the medium in which they live. Among the organisms which oxidize aerobically are saprotrophic, symbiotic, and holozoic protozoa and metazoans. Protozoa and metazoans are one-celled and multicellular animals, respectively. *Holozoans* are animals which ingest part or all of other animals. Chemosynthetic heterotrophs include human beings and their microbial parasites.

Of all the diverse organisms that inhabit the earth, only bacteria possess all the types of energy metabolism described above. Furthermore, two of the four types, photosynthetic autotrophs and chemosynthetic heterotrophs, contain all the other known microbial forms, excluding viruses. Because they are noncellular, viruses cannot be conveniently pigeonholed into any of the categories based on energy source and raw materials and will be dealt with separately.

## REPRODUCTION

If microbial metabolism consists largely of nutrition and cellular respiration, microbial perpetuation consists mainly of reproduction. Pathogenic microbes and multicellular parasites have developed a veritable arsenal of reproductive methods as adaptive mechanisms to ensure the perpetuation of their parasitic modes of existence. Among the numerous methods used by at least one group of parasites are (1) intracellular replication; (2) asexual methods such as fission, budding, sporulation, and immature or larval multiplication; and (3) sexual methods such as sporulation, conjugation, and bisexuality.

*Intracellular replication* is performed by viruses. Viral nucleic acids within a host cell may preempt its genetic or protein synthetic machinery. When this happens, new directives are provided to the cell, resulting in the manufacture of viral proteins rather than essential host proteins. Its own metabolic machinery rendered inoperative, the host cell fills with newly manufactured virus particles and ultimately bursts. Cytolysis, or the destruction of the host cell, releases the entrapped viruses, which are then free to reinfect new host cells. Therefore, viruses really provide alternative instructions to their host cells and cannot be said to have a cellular method of reproduction. However, as we all know from experience, viruses are extremely successful at making carbon copies of themselves. This function explains why viruses are among the most abundant of our microbial parasites.

*Fission,* or the splitting of one cell into two or more daughter cells, is the most common reproductive method used by cellular microbes. The prokaryotic bacteria and rickettsias belonging to the Monera as well as fungi and protozoa belonging to the Protista are examples of microbes which reproduce by fission. Fission is accomplished by mitosis, a mechanism of cell division described in Chap. 6.

*Sporulation* is a process which forms one or more reproductive cells called *spores.* Each of the spores is capable of developing directly into an adult or engaging in the exchange of genetic material with another strain of the same species. Sporulation is found in some species of bacteria, fungi, and protozoa.

*Budding,* a form of fission, is engaged in by some fungi and several flatworms. Progeny form through the development of branches from the main body of, or segments added to, the parental organism. Budding is characteristic of yeasts and tapeworms.

*Immature,* or *larval, multiplication* is a unique type of reproductive method found in some flukes. The progeny are themselves embryonic and must go through one or more developmental stages in secondary aquatic hosts before reaching sexual maturity in the primary host.

*Conjugation* is a sexual form of reproduction found in bacteria and protozoa. During conjugation, genetic material is either transferred from one individual to another or exchanged between both individuals. The genetic transfer or exchange makes the method sexual. Not all bacteria engage in conjugation, and among those which do the method is used sporadically. In addition and just as unpredictably, a number of ciliates and sporozoa reproduce by some form of conjugation.

Sexual reproduction based on *bisexuality* is employed by all eukaryotic metazoans, including helminths and arthropods. In virtually all flatworms, the reproductive systems of both sexes are in the same individual. The sexes are separate in the roundworms and arthropods.

## EFFECTS OF NONSYMBIOTIC MICROORGANISMS

Nonsymbiotic microorganisms are either beneficial or harmful to human beings, although the distinction is not always clear. Beneficial ones act in the geochemical cycles of various elements, decomposing organic compounds into inorganic substances and providing an important natural mechanism for their eventual reuse. At the other extreme, some nonpathogenic microorganisms contribute to biospheric pollution and material damage. Beneficial and harmful activities are not mutually exclusive, and some aspects of elemental recycling contribute to environmental contamination, to economic losses, or to both.

Beneficial microbes participate primarily in oxidative reactions involving carbon, oxygen, hydrogen, nitrogen, sulfur, calcium, iron, and manganese. Atoms of these elements are chemically bonded to various molecules oxidized by microorganisms.

Microorganisms also contribute to atmospheric pollution. Many form gaseous pollutants such as carbon monoxide and dioxides of carbon, nitrogen, and sulfur.

Aquatic pollutants produced by microorganisms include mercury compounds, ammonia, hydrogen sulfide, methane, manganese dioxide, and ferric hydroxide. All except the last-named compound are soluble in water and therefore dissolved. Ferric hydroxide is an insoluble precipitate. In addition, ammonia, hydrogen sulfide, and methane are volatile and readily enter the atmosphere from water.

The involvement of microorganisms in pollution and material damage is a consequence of their ways of life. Certain microorganisms, especially aquatic blue-green algae, pollute by fouling water with large amounts of organic matter. Some microorganisms cause fermentation when they oxidize carbohydrates under anaerobic conditions.

Nevertheless, nonpathogenic microorganisms are of inestimable importance in the balance, the economy, and the continuation of natural processes. They serve as food sources for innumerable organisms; decompose organic compounds, including dead plants and animals; demineralize rock, contributing to its disintegration and the subsequent development of soil; participate in the formation of fossil fuels; and are instrumental in the recycling of many bound elements. In addition, certain symbiotic but ordinarily nonpathogenic microorganisms are commonly involved in several normal physiologic processes in the human body.

## BIOSPHERIC POLLUTION

Directly or indirectly, various microorganisms contribute significantly to pollution. Carbon dioxide, methane, and sulfur compounds are among the atmospheric pollutants that are formed through microbial action. By sheer number, some aquatic microbes foul the water. Others release ammonia, methylated mercury, sulfuric acid, and other compounds that have deleterious effects on marine or freshwater life or on the various organisms that eat them. Certain microbes remove nitrogen from the soil through leaching; many cause decay. Most of these microorganisms are commonly found and widely distributed in nature.

The contribution of microorganisms to carbon dioxide pollution is indirect. Fossil fuels are formed from the products of incomplete microbial oxidations and high pressures, conditions which prevail in the depths of the earth's crust. Carbon dioxide is an end product of the combustion of coal, oil, or natural gas.

Methane ($CH_4$), the most simple organic compound and hydrocarbon, is found during the oxidative metabolism of microorganisms such as blue-green algae. Methane has a characteristic odor and is frequently detected around marshlands, giving rise to its common name, marsh gas. The gas is flammable and may create eerie images as it burns. Although formed by natural events, methane is an atmospheric pollutant.

Sulfurous pollution occurs during the microbial oxidation of elemental sulfur or sulfides to sulfates such as sulfuric acid ($H_2SO_4$). The high acidity created by $H_2SO_4$ is injurious to many forms of life. Microbial reduction of sulfuric acid to sulfides such as hydrogen sulfide ($H_2S$) creates another source of pollution. $H_2S$ is toxic to plant life. While most of these microbial reactions occur in water, some of the sulfur dioxide

(SO₂) in the atmosphere is formed through the action of sulfur-oxidizing bacteria on reduced forms of sulfur.

Certain microorganisms, especially algae, bloom when the water in which they live undergoes organic enrichment. Organic enrichment, or *eutrophication,* may result from siltation, sewage, or other forms of pollution. The marked proliferation of algae in response to eutrophication permits the support of larger organisms in the water. The ensuing invasion by higher plants and animals eventually dries up the water, and the water basin fills with soil. The consequences of this natural process can be economically and esthetically devastating.

Ammonia is another potential aquatic pollutant. Although poisonous to life, ammonia is ordinarily oxidized or reduced microbially to less toxic or nontoxic compounds. Where the water is already heavily polluted, ammonia may accumulate in concentrations that are biologically harmful. Such waters are commonly found downstream from industries and municipalities using them as dumping grounds for solid and liquid wastes.

Mercury (Hg), one of the heavy metals, is rapidly becoming a serious threat to fish, wildlife, and human beings through the action of specific microbes and the natural seasonal turnover of water. Using the slogan "The solution to pollution is dilution," many industries have been pouring effluents, or waste discharges, containing mercury into streams, rivers, and lakes. Since mercury is inert in water, the offending industries have reasoned that the element would settle to the bottom mud and pose no problems to life. However, we now know that one species of anaerobic bacterium living in mud is capable of detoxifying mercury through the addition of a methyl (CH₃) group. The detoxification mechanism was apparently developed in self-defense by these bacteria. Methylated mercury is soluble in water and is released by the bacteria following its formation.

The next step in the natural processing of mercury is regulated by seasonal changes in water temperature. In the fall the surface waters cool, become more dense, and sink to the bottom. The deeper, warmer waters rise to the surface. In the spring the process is reversed. Thus, seasonal turnover brings methylated mercury from deeper to more shallow waters, where the compound is ingested by game fish and other organisms. Because mercury is nonbiodegradable, the element undergoes trophic magnification as food contaminated with it is ingested by higher organisms.

The toxicity of mercury is due to its inhibitory action on enzymes. Dangerously high levels of mercury and symptoms of poisoning have been reported in shellfish and game fish as well as in human beings who have unwittingly consumed these contaminated seafoods over extended periods of time.

Microbial utilization of nitrogen may accelerate the removal of much of this element from soil. This type of leaching is especially true where there is a large, sustained runoff of surface waters carrying soluble ammonia. Leaching is an example of bad land management, a topic discussed in Chap. 1.

## MATERIAL DAMAGE

Spoilage of food and material destruction through mildew, erosion, and corrosion are among the more important types of physical damage caused by nonsymbiotic microorganisms and impose a significant drain on the world economy. In chronically moist areas, for example, molds cause mildew in fabrics such as clothing, upholstery, and draperies, in nonsolid-state electronic circuitry, and in works of art such as books and paintings.

Erosion has a microbial origin when sulfur from the pollution-laden air of urban centers is oxidized biologically. Whether formed biologically or spontaneously, the oxidation product, sulfuric acid, causes the slow crumbling of sculptures and the architectural facades of buildings or other man-made structures. The crumbling is evidenced by pitting and flaking. This type of deterioration is evident on numerous marble statues in Milan, Venice, and other industrialized cities of the world.

The process of corrosion is also strongly influenced by microorganisms. For example, iron bacteria oxidize iron to obtain energy. Oxidized iron is insoluble and precipitated as ferric hydroxide. The iron deposits formed in this manner may clog the insides of pipes and eventually lead to their total disintegration. The ferrobacilli which oxidize iron act only in acidic water lacking oxygen.

## MICROBIAL SYMBIONTS OF MAN

### TRANSMISSION

The human body is continually exposed to microbes. Exposure may be from the atmosphere, contaminated nutrients, arthropod and other animal bites, or contact with people, animals, soil, and objects.

Microbes may be transported directly through the air or by way of airborne droplet nuclei from a sneeze or cough. When inspired, contaminated atmospheric particulates may also serve as a vehicle of microbial transmission.

Microbes may also be ingested in contaminated nutrients. For instance, microbial contaminants may be accidentally ingested with foods. Microbes may also be swallowed with water during swimming or by drinking polluted water.

Sometimes arthropods and other animals may transmit microbes when they or their products penetrate the skin. For example, dogs, cats, bats, and other mammals may infect human beings when they bite or scratch. Insects such as mosquitoes and other flies as well as fleas, lice, and bugs may serve as vectors of microbes. *Vectors* transmit potentially pathogenic microbes when they bite through, or defecate on, the skin. Fecal contaminants infect man when they are scratched into or otherwise enter the skin.

Microbes can also be transmitted by direct contact between two people, one of whom is a carrier. The *carrier* usually does not exhibit clinical symptoms of a microbial disease but harbors a potentially pathogenic microbe and is capable of transmitting it to others.

Contact with contaminated mouths, hides, or products of domestic animals also provides a mechanism for the spread of microbes. As examples, we may cite the contact which occurs when a pet animal licks or rubs against a human being, when a rabbit is being prepared for human consumption, or when soil or objects contaminated by animal feces or urine are touched by the hands or other parts of the body. In these examples, the microbes either bore directly through the skin, the skin after contamination is placed in or rubbed against the mouth, or the microbes are transferred directly to the mouth, nose, or other natural orifices of the human body by the contaminated animal.

### INVASIVENESS AND HOST DAMAGE

Many microbes to which human beings are exposed are nonpathogenic and do not cause disease. Some microbes are commensalistic or mutualistic symbionts; others as accidental passengers on insects and particulates are simply victims of circumstance.

Of the actual pathogens that gain a foothold in human beings, only a portion cause clinical symptoms of disease in their hosts. The foothold may be on the outside or inside of the body. An external invasion is termed an *infestation*. Examples are microbial infestations of the epidermis and the lumina of the digestive tract, including those of the mouth, stomach, intestines, and rectum. An internal invasion is called an *infection*. Infections of tissues below the epidermis of the skin and of organs, glands, and body fluids are examples.

The *invasiveness* of a parasite is not necessarily synonymous with its pathogenicity, or *virulence.* For example, a pathogen may be highly infectious but weakly virulent, or vice versa. We will consider the functional interactions between a host and its parasite in Chap. 4.

In general, host damage is due to the effects of toxins, allergens, nutritional injury, and related tissue destruction or organ obstruction. Although the mere presence of microbes within the body may cause one or more of the above-named effects, severe pathologic responses by the host are ordinarily provoked by an accumulation of toxins secreted or otherwise released by the parasites. The mechanisms which cause microbial damage are shown in Table 2-1.

Most gram-positive bacteria produce *exotoxins* secreted during the lives of the pathogens. Each type of exotoxin is distinctive, producing its own characteristic symptoms by interfering with a specific aspect of cellular metabolism.

**Table 2-1** Factors in microbial pathogenicity

| Pathogenic mechanisms | Types of microbes and their characteristics | Characteristics of disease-producing factors | Host responses |
|---|---|---|---|
| **Toxins** exotoxins | Gram-positive bacteria; pathogens stay at focus of infection; pathogens stimulate host production of antitoxins (antibodies); number of species of pathogens is limited | Secretions are released to the extracellular environment; each exotoxin produces its own characteristic symptoms; exotoxins block some aspect of cellular metabolism | Paralysis; hemolysis or cytolysis; vomiting and diarrhea; tissue damage; fever or hypothermia |
| Endotoxins | Gram-negative bacteria | Secretions are bound to the cell wall and released during bacteriolysis; all endotoxins produce a similar syndrome of effects | Fever; diarrhea; hemorrhagic shock; tissue damage |
| Allergies (hypersensitivities) | Bacteria; mites associated with house dust | Cell wall antigens, waste products, or other secretions; body fragments are inhaled | Skin reactions; nervous system reactions; respiratory reactions |
| Nutritional injury | Bacteria and viruses; entamebas and helminths | Large amounts of host nutrients are required; blood cells and soluble nutrients are engulfed and absorbed, respectively | Weakness and apathy; weight loss; tissue damage; hemolysis; diarrhea |

*Endotoxin* production, on the other hand, is characteristic of gram-negative bacteria. Endotoxins are bound to the bacterial cell wall and released only during bacteriolysis, or death of the bacterium. All endotoxins produce a similar disease pattern.

*Allergies*, or hypersensitivities, may be elicited by literally thousands of foreign materials that come in contact with the body. Such materials act as weak antigens and induce the formation of incomplete antibodies. Although antigen-antibody interactions will be thoroughly discussed in Chap. 4, we note now that microbes even as small as viruses are mixtures of many different antigens and that host responses are highly variable. Parasitic invasion exposes the immunogenic machinery of the host to the antigens of the microbe. The antigens are found along the surface of the microbes or as parts of microbial secretions and excretions, setting the stage for possible allergic reactions.

*Nutritional injuries* and corresponding destruction of tissues may result directly from the demands of the parasites for host nutrients and the consequent creation of nutritional imbalances. Toxin production and general host debility may contribute indirectly to abnormal nutritional states through vomiting, diarrhea, and lack of appetite. Before such effects are recognized clinically, large numbers of microbial pathogens are usually required. Nutritional injury is much more likely to be associated with diseases caused by multicellular, ordinarily macroscopic helminths.

## PORTALS OF ENTRY AND EXIT

Whether or not potentially pathogenic, most microbes invade the body through fixed portals of entry, and those which leave do so through fixed portals of exit.

*Portals of entry* are primarily associated with the integumentary (skin), respiratory, digestive, and urogenital systems. Such gateways are primarily natural openings of the major organ systems that lead into the body.

The skin serves as an effective barrier against microbial infection. However, when penetrated because of a scratch, cut, wound, or animal bite, the barrier is vulnerable to the onslaught of omnipresent microbial contaminants. In addition, several fungi in chronic infections are capable of reaching the blood through initially intact skin.

Air which enters the nose and mouth may carry microbial contaminants to other portions of the respiratory system. Because the atmosphere is one of the most common modes of microbial transmission, the nose, mouth, paranasal sinuses, air passageways, and lungs are particularly susceptible to microbial invasions.

The digestive system contains two openings which serve as portals of entry for certain microbes. The much more commonly used and exposed of the two is the mouth. The other is the anus when contaminated by unsterile materials such as toilet paper or by human or animal contact.

The urogenital system also provides several natural entrances for microbial infection. Contamination of the urethral orifice in males may result in urogenital infections. Contamination of the same orifice in females may lead to urinary tract infections, while microbial contamination of walls surrounding the vaginal orifice or lumen may lead to infections of the reproductive system.

Sometimes conjunctival membranes, especially at the moist corners of the eyes, or less obvious natural openings such as the orifices of the endolymphatic ducts in the inner ear, are used as portals of entry. Occasionally, individual microbes may take advantage of the opportunity afforded by a novel portal in gaining entrance. Once inside the body, numerous microbes can be transported through the blood and lymph of the circulatory system, may bore through the voluntary muscles, or in other ways may infect tissues, organs, and systems not associated with their portals of entry.

Microbial *portals of exit* are similar to the portals of entry identified above, the integumentary, respiratory, digestive, and urogenital systems serving as escape routes. Some microbes are transferred to new hosts by person-to-person contact. The microbes may be rubbed off intact skin when shaking hands or contracted from the purulent discharges of a festering wound or open sore. Infection may result from the transfer of the microbial pathogens while kissing, during sexual intercourse with an infected person or from contact with their discharges, and from other forms of interpersonal contact.

Numerous microbial parasites leave the body when an arthropod vector such as a mosquito imbibes a blood meal through the perforated skin of an infected host. This portal of exit is integumentary because the microbes leave through the skin.

Other microbes are discharged in the droplet nuclei produced by a sneeze. Transmission is *direct* when airborne droplets are inspired. Transmission is *indirect* when dust and other small particulates containing the microbial contents of dried droplets are inhaled.

Numerous microbial symbionts leave in the spray from a cough or in vomitus, feces, and exudates located around the anus. These microbes use a digestive portal of exit, because they leave in materials produced or processed by the digestive system.

Several medically important microbes commonly leave the body by way of the urogenital orifices. Some take advantage of the opportunity presented by direct contact with another host. Others escape in urine and in genital exudates.

Portals of entry and exit and mechanisms of transmission are not necessarily identical or consistent among microbial pathogens. Only in the bacteria and arthropods are the portals of entrance and exit the same. In addition, specific species may use portals of exit different from those of entry. Pathogenic rickettsias require arthropod vectors, while most symbiotic fungi are widely distributed in nature and therefore almost impossible to avoid. Bacteria and viruses use the greatest number of portals and are the most versatile of man's symbionts. Excluding the rickettsias and fungi, pathogens with the least number of portals available to them include the flatworms, roundworms, and arthropods. A more detailed discussion of specific portals of entry and exit is presented in Chap. 3.

## COMMONLY FOUND MICROBES

Because of the frequency of their association with human beings, microbial symbionts are considered to be part of the normal flora and fauna of the human body. That they usually live on or in the body, however, does not necessarily mean they are completely nonpathogenic, or harmless. Indeed, about one-third are potentially pathogenic.

Virulence depends upon the numbers of microbes present, their invasiveness and ability to spread from the initial locus of infection, the defense and antimicrobial mechanisms employed by the host, and the ability of the pathogen to be transmitted through the environment from one host to another. Pathogenicity is a relative term and not the inevitable consequence of the association between a microbial parasite and the human body.

Some microbes may be transmitted from mother to child during pregnancy. If the microbe is virulent, the disease contracted by the child is said to be *congenital*. Congenital syphilis is one example.

Ordinarily, however, the flora and fauna of the human body are acquired after birth when a baby is first exposed to microbial contamination. Some commonly found bacteria and fungi of babies, together with their modes of transmission and portals of entry, are identified in Fig. 2-4.

Among the most common bacteria that gain a foothold in infants are streptococci, staphylococci, and lactobacilli. *Candida albicans,* a universally distributed fungus, is also a common contaminant. All these microorganisms are spread either through the atmosphere, by contact with contaminated objects, or through ingestion of contaminated food and water. Therefore, portals of entry in the newborn include the respiratory, integumentary, and digestive systems.

*Staphylococcus aureus* and *Candida albicans* both require a high moisture and low organic acid content. These conditions usually prevail on the outer layer of skin called the stratum corneum. *Staphylococcus aureus* is a good secondary invader but ordinarily does not gain a foothold unless a congenial niche or habitat is vacant on the skin. The newborn possess a number of available niches for the potential growth and development of such microorganisms. *Candida albicans* thrives especially well on chronically moist skin, a condition not uncommon under the chin and in the inguinal region of babies.

On the other hand, streptococci and lactobacilli generally survive a wide range of oxygen tensions and, in fact, usually oxidize anaerobically. Such bacteria proliferate in the mouth. Although the oral cavity is ordinarily exposed to air containing oxygen, anaerobic conditions may prevail in localized regions. Such conditions are maintained by the presence of aerobic microbes which continually deplete the available oxygen.

While human adults may normally harbor all the above-mentioned microorganisms, their microbial flora is at least five times as extensive as that possessed during the first few months after birth. The potential pathogenicity of *S. aureus* and *C. albicans* is diminished with age, because the moisture content of the adult stratum corneum is normally reduced while its organic acid content is elevated, compared with the situation during childhood. *Staphylococcus aureus* usually becomes a secondary invader in adults, gaining a foothold only in hosts whose resistance is lowered because of another disease or whose flora is temporarily changed because of the prophylactic use of antibiotics. *Candida albicans* must wait for chronically moist conditions generally resulting from some pathologic condition or occupational hazard.

In place of, or in addition to, the above-mentioned microorganisms, many other commonly found bacteria and fungi may take over as potential pathogens. Among such microbes are *Corynebacterium acnes,* the causative agent of acne; *Corynebacterium diphtheriae, Diplococcus pneumoniae,* and *Mycobacterium tuberculosis,* all of which employ the respiratory portal of entry and are the causative agents of diphtheria, pneumococcal pneumonia, and tuberculosis, respectively; and *Leptotrichia buccalis,* a causative agent of dental plaque.

Some commonly found bacteria and fungi of adults,

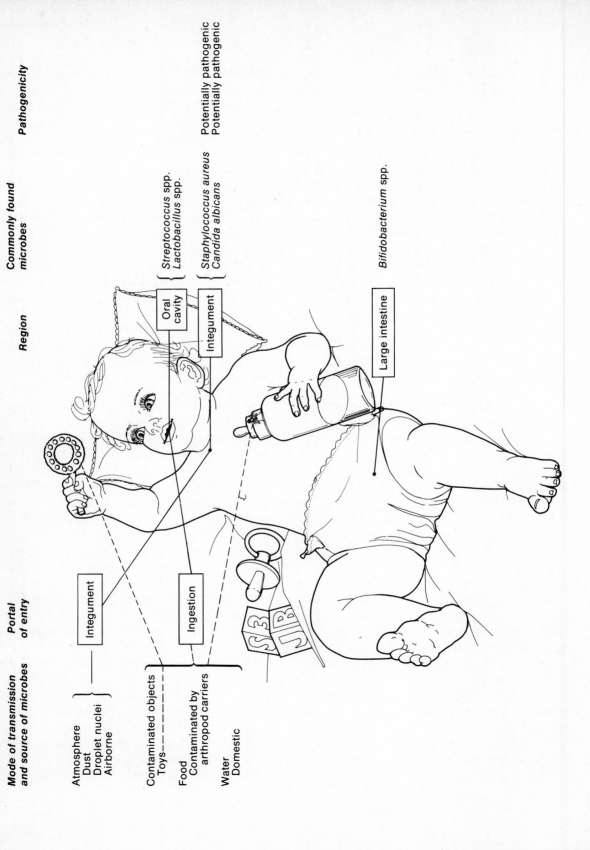

**Figure 2-4** *Some commonly found microbes of infants.*

| Mode of transmission and source of microbes | Portal of entry | Region | Commonly found microbes | Pathogenicity |
|---|---|---|---|---|
| Atmosphere<br>Dust<br>Droplet nuclei<br>Airborne | Integument | | | |
| Contaminated objects<br>Toys<br>Food<br>Contaminated by arthropod carriers<br>Water<br>Domestic | Ingestion | Oral cavity | *Streptococcus* spp.<br>*Lactobacillus* spp. | |
| | | Integument | *Staphylococcus aureus*<br>*Candida albicans* | Potentially pathogenic<br>Potentially pathogenic |
| | | Large intestine | *Bifidobacterium* spp. | |

*Figure 2-5*  Some commonly found microbes of adults.

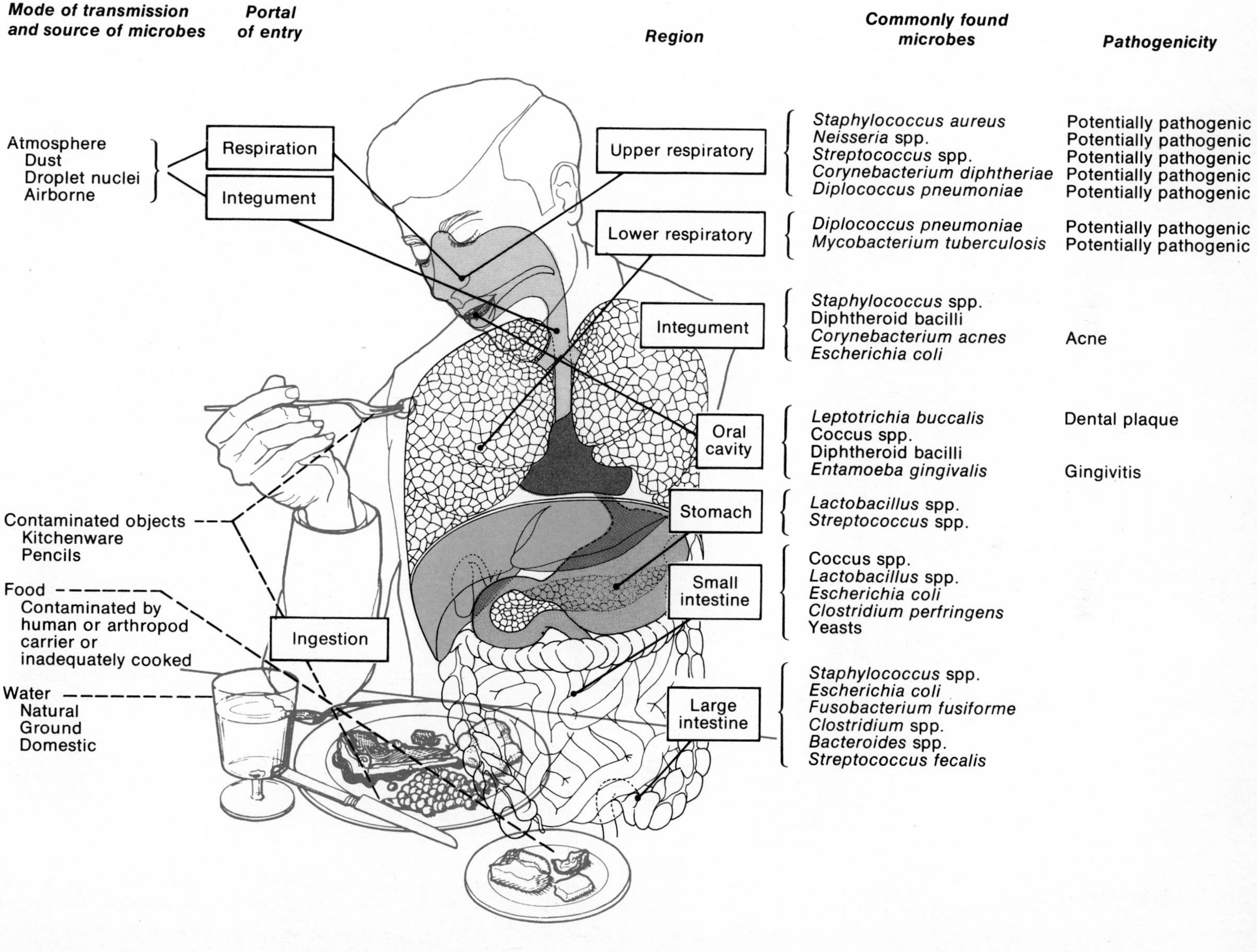

**Mode of transmission and source of microbes**

Atmosphere
Dust
Droplet nuclei
Airborne

Contaminated objects
Kitchenware
Pencils

Food
Contaminated by human or arthropod carrier or inadequately cooked

Water
Natural
Ground
Domestic

**Portal of entry**

Respiration
Integument

Ingestion

**Region**

Upper respiratory
Lower respiratory
Integument
Oral cavity
Stomach
Small intestine
Large intestine

**Commonly found microbes**

*Staphylococcus aureus*
*Neisseria* spp.
*Streptococcus* spp.
*Corynebacterium diphtheriae*
*Diplococcus pneumoniae*

*Diplococcus pneumoniae*
*Mycobacterium tuberculosis*

*Staphylococcus* spp.
Diphtheroid bacilli
*Corynebacterium acnes*
*Escherichia coli*

*Leptotrichia buccalis*
*Coccus* spp.
Diphtheroid bacilli
*Entamoeba gingivalis*

*Lactobacillus* spp.
*Streptococcus* spp.

*Coccus* spp.
*Lactobacillus* spp.
*Escherichia coli*
*Clostridium perfringens*
Yeasts

*Staphylococcus* spp.
*Escherichia coli*
*Fusobacterium fusiforme*
*Clostridium* spp.
*Bacteroides* spp.
*Streptococcus fecalis*

**Pathogenicity**

Potentially pathogenic
Potentially pathogenic
Potentially pathogenic
Potentially pathogenic
Potentially pathogenic

Potentially pathogenic
Potentially pathogenic

Acne

Dental plaque

Gingivitis

45

together with their modes of transmission and portals of entry, are given in Fig. 2-5. The potential pathogens cited above and the other commonly found microorganisms of adults, like their counterparts in infants, are transmitted through the atmosphere or via contaminated nutrients and objects. The larger diversity of commonly found microorganisms in the average adult, compared with the average infant, is explained by the greater and more extensive adult contact with these microbes.

Predictable variations in specific flora occur in response to significant changes in reproductive life as well as because of physiologic differences between the sexes. For example, *Escherichia coli* and diphtheroid bacilli are among the commonly found vaginal symbionts before puberty and after menopause. Between these periods, *E. coli,* lactobacilli, streptococci, and yeasts ordinarily predominate. The latter are generally capable of fermenting glycogen, a carbohydrate found in abundance in the vagina during this time. Differences also show up between the sexes when, for example, the microbial contents of the urethras are examined. Allied much more commonly with the male urethra than with that of the female are diphtheroid bacilli and *Corynebacterium diphtheriae.*

## DIAGNOSIS OF MICROBIAL DISEASES

Fundamental to the medical treatment of a disease caused by a microbial parasite is an accurate identification of the causative agent. That it is not always possible to pinpoint the microbial culprit is common knowledge. Obviously, between theory and practice there are large gaps in knowledge and technology, or in the ability to apply both effectively.

The steps required for the positive identification of bacterial pathogens were first articulated successfully by Robert Koch and have since come to be known as *Koch's postulates.* An analogous set of steps has subsequently been developed for use with viral pathogens. A comparison of these steps and their attendant difficulties is given in Fig. 2-6.

Careful study of Fig. 2-6 leads to the following general conclusions: (1) The positive identification of a viral pathogen is more complex than that of a bacterial pathogen and requires a larger number of steps

and more highly sophisticated equipment and techniques; (2) difficulties in positively identifying a presumptive bacterial pathogen are related to an insufficient sensitivity of the method, mistaking secondary invaders for the primary pathogen, an inability to grow the pathogen in laboratory cultures or animals, and host variability in susceptibility and symptomatology; (3) while sensitivity of the method, laboratory culturing techniques, and host variability are sometimes limiting factors in identifying presumptive viral pathogens, positive identification is also confounded by the peculiarities of viral activities; and (4) because of these difficulties, host animals are only infrequently used for the culturing of presumptive viral pathogens; instead, symptomatology and the results of chemical tests are usually employed. The case against specific viral pathogens is usually circumstantial rather than definitive.

## ANTIMICROBIAL MEASURES

If man must coexist with microbes, many of which are pathogenic, he can at least attempt to shift the balance of power in his direction and decrease the probability of contracting disease. In a practical sense, favorable shifts in power balance are useful control measures. Antimicrobial control measures may be prophylactic, physical, or chemical.

Prophylaxis is preventive and is usually instituted before potentially pathogenic microbes have an opportunity to invade. Physical and chemical antimicrobial measures, on the other hand, are generally initiated after pathogens contaminate nutrients, drugs, and other materials or cause the appearance of disease symptoms. In many cases, a considerable overlap exists between prophylactic and other antimicrobial measures as, for example, when chemotherapy prevents the recurrence of a microbial disease already contracted.

## PROPHYLAXIS

Prophylaxis against a potentially pathogenic microbe is based on its modes of transmission and portals of entry and exit. In other words, protection may be achieved by preventing its transmission through the environment or by blocking its entry into the body.

In practice, prophylaxis takes various forms. It may result simply from the observance of proper hygienic and sanitary procedures. Such measures reduce the possibility of microbial contamination of the skin and the digestive and urogenital orifices.

A number of flatworm and roundworm infections can be avoided by eating only properly prepared meat and fish. Adequate cooking permits the destruction of any heat-resistant parasites within these foods.

One of the best defenses against invasion by communicable disease microbes is the avoidance of human carriers or people with infectious diseases. These people should be identified and isolated, if at all feasible, and their direct contact with others should be prevented.

Various pathogenic microbes that require arthropod vectors may be avoided by either destroying the vectors or bypassing geographic regions in which the vectors are endemic. Some microbial pathogens may be controlled by judiciously manipulating the population size and geographic distribution of animals in which the parasite may be harbored. Other parasites are more susceptible to control measures instituted during specific stages in their life cycles. Immunity to numerous medically important microbes can be obtained through the use of vaccines or immunogenic drugs. Protection against all microbes is enhanced by maintaining a high level of body resistance through proper nutrition and adequate rest.

Prophylaxis is obviously not a cure-all but does help to reduce the incidence of microbially induced disease. However, that the foregoing steps are not always practical or easy to obey is also clear. Some measure of these difficulties may be gained by examining our efforts to acquire immunity to respiratory illness.

Basically, the difficulties impeding immunoprophylactic progress are due to the interactions between the causative agents and their hosts and the nature of available vaccines and detection methods. The complexities of the problem are suggested by the fact that over one hundred microbes can cause respiratory disease. The disease pattern is in part determined by the age of the host and environmental con-

ditions. These variables may be impossible to control. In addition, genetic changes in the nature of the microbial antigens and the prevalence of some microbial pathogens vary in an apparently unpredictable fashion and may neutralize the action of a previously effective vaccine.

The host also contributes his share to the complexities of successful immunoprophylaxis. For instance, a host may respond to certain respiratory pathogens by a transient resistance, only to be followed by reinfection at a later time. Nevertheless, the presence or absence of natural immunogenic factors in the host will determine the effectiveness of the immune response. The immune response is discussed in Chap. 4.

The limitations of viral vaccines and microbial detection methods impose further burdens on our already frail ability to immunize ourselves against respiratory pathogens. In the former case, nonpathogenic viral vaccines may suddenly become virulent through spontaneous changes in their molecular makeups. In the latter case, a microbial infection may be present in a form which cannot be detected by the most sensitive assays available. In either case, the usefulness of immunization is severely hampered.

## PHYSICAL CONTROL

Physical antimicrobial methods create environmental conditions that either kill microbes outright or arrest their growth and development. Control may be obtained through the manipulation of temperature or some other component of the microbial environment. Selected physical control methods and their effects are given in Table 2-2.

*Moist heat* is secured by boiling water or through the use of a pressure cooker or autoclave. Heated water under a pressure higher than that of the atmosphere has a greater germicidal efficiency than the same water under normal atmospheric pressures (see Chap. 8). Heat-resistant spores which are not destroyed by boiling water alone are usually lysed by using pressurized steam for a long enough period of time. The germicidal action of moist heat is a consequence of microbial protein coagulation.

*Dry heat* is produced when air is heated. The use of a hot-air stream or incinerator are some practical

## A. Bacterial identification

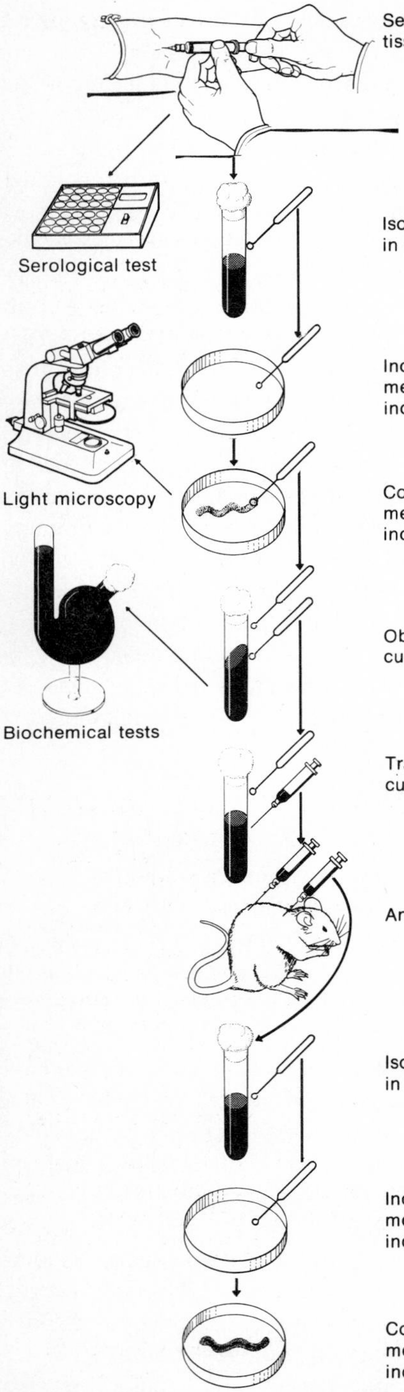

| Steps | Koch's postulates | Difficulties |
|---|---|---|
| Securing infected tissue | The organism must be found in every case of the disease | The pathogen may be in low enough concentration to escape detection; an abundant secondary invader may be misidentified as the pathogen |
| Serological test | | |
| Isolation of pathogen in culture medium | The organism must be isolated and grown in pure culture | Some pathogens cannot be grown in culture media |
| Inoculation of culture medium before incubation | | |
| Light microscopy | | |
| Colony on culture medium before incubation | | |
| Obtaining pure culture | | |
| Biochemical tests | | |
| Transfer to another culture medium | The organism from a pure culture must cause the disease when inoculated into a susceptible animal | Some pathogens cannot be grown in animals or do not produce characteristic disease symptoms; not all hosts are equally susceptible |
| Animal inoculation | | |
| Isolation of pathogen in culture medium | The same organism must be recovered from the inoculated animal | Finding the same organism does not necessarily mean that it caused the original disease symptoms |
| Inoculation of culture medium before incubation | | |
| Colony on culture medium after incubation | | |

**Figure 2-6** *The steps required for the positive identification of microbial pathogens. A. Bacterial identification. B. Viral identification.*

## B. Viral identification

| Steps | Viral infections | Difficulties |
|---|---|---|

| | | |
|---|---|---|
| **Infected host** | Infected tissues must not show microorganisms with the standard methods; chemical diagnosis must correlate with viral infection | The viral infection may be latent and therefore undetected |
| **Cell-free extract** | Isolation and identification of virus | Virus may become attenuated during laboratory culture; death of host may occur before viral replication begins; toxins may be impossible to identify, even with the most sensitive assays |

**Chick embryo cultivation**

**Tissue culture cultivation**

**Replication to ensure presence of only one type**

**Diluted extract in tissue culture**

**Extract plus agar before incubation**

**Plaques after incubation**

**Electron microscopy**

| | | |
|---|---|---|
| **Plaques plus tissue culture** | After filtration of infected tissues through bacteria-proof filters, filtrate must still be capable of transmitting the disease symptoms to susceptible animals (infrequently used) | Some viruses cannot be grown in laboratory animals; the laboratory animals harbor latent viruses not readily separated from the viruses under study; not all hosts are equally susceptible, and some not infected in nature may be easily infected in the laboratory |
| **Inoculation causes infection in susceptible host** | | |
| **Cell-free extract** | Similar filtrate from infected animals must be capable of transmitting disease symptoms to other animals (infrequently used) | Some viruses cannot be grown in laboratory animals; the laboratory animal may harbor latent viruses not readily separated from the virus under study |
| **Inoculation causes infection in another susceptible host** | | Laboratory animals not infected in nature may be easily infected in the laboratory |

**Table 2-2** Common physical antimicrobial methods

| Method | Mechanism of action | Uses and applications |
| --- | --- | --- |
| Moist heat: | | |
|   Autoclave | Protein coagulation | Sterilization—medical equipment and supplies<br>Preservation—food |
|   Boiling water | Protein coagulation | Sterilization—kitchenware and medical supplies |
|   Steam under pressure | Protein coagulation | Sterilization—food and microbiologic media |
| Dry heat: | | |
|   Hot air | Oxidation | Sterilization—glassware and powders |
|   Incineration | Complete oxidation | Destruction—contaminated materials and<br>dead organisms |
| Cold (e.g., refrigeration) | Metabolic retardation<br>Cytolysis at low enough<br>temperatures | Preservation by bacteriostasis—foods and drugs |
| Drying (e.g., desiccation) | Sorption inhibition | Preservation—foods<br>Bacteriostasis—microbiologic materials |
| Pasteurization | Protein coagulation | Sterilization—milk |
| Radiation (e.g., ultraviolet light) | Oxidation | Bactericide—atmospheric microbes |
| Hydrostatic pressure | Plasmolysis | Preservation—food |

applications of the principle. The germicidal action of these techniques is related to their oxidizing abilities.

Both moist and dry heat are used as *sterilants,* preventing the contamination of consumables and objects by living microbes or their products.

*Cold* ordinarily acts by drastically reducing the overall metabolism of microorganisms, preventing their growth and development. This action is referred to as *bacteriostasis* when the microorganism is bacterial. Bacteriostasis maintains the status quo but usually does not kill. Refrigerators and freezers are common examples of this principle.

Aside from the antimicrobial effects of temperature extremes, physical control may be achieved by drying or the application of radiation or pressure. *Drying* inhibits the uptake of water vapor by bacteria, resulting in the preservation of food through bacteriostatic action. Desiccators are based on this principle. *Radiation,* on the other hand, is bactericidal and causes oxidation. An example is the use of ultra-

violet light. This method is especially useful against microorganisms transmitted through air. The application of *pressure* is also an effective antimicrobial technique. For instance, by manipulating osmotic and hydrostatic pressures, large batches of food are preserved during industrial processing. The application of pressure causes microbial plasmolysis. Plasmolysis is a disruption of cellular protoplasm.

Most of the above-mentioned methods are also used to preserve drugs and microbiologic media.

CHEMICAL CONTROL

Excluding the use of antibiotics, common chemical antimicrobial methods are summarized in Table 2-3. Included among the chemicals are heavy metals and their derivatives, halogens and their derivatives, alcohols, phenols and their derivatives, quaternary ammonium compounds, food preservatives, and a number of other agents. Considered as a group, these chemicals adversely affect some aspect of protein function or serve as oxidizing or reducing agents.

The chemicals act as antiseptics, disinfectants, sterilants, and bacteriostatic agents. *Antiseptics* inhibit the development of microbes without killing them. *Disinfectants* kill all pathogenic microorganisms and inactivate their products. Disinfectants are stronger and more potent chemicals than antiseptics, although sometimes these terms are used interchangeably.

Heavy metals are the active ingredients of a number of antimicrobial chemicals. These chemicals

**Table 2-3** Common chemical antimicrobial methods

| Compounds | Mechanisms of action | Uses and applications |
|---|---|---|
| Heavy metals and derivatives: | | |
| Mercuric chloride (mercury bichloride) | Protein inhibition | Disinfectant—skin<br>Antiseptic |
| Organic mercurials | Protein inhibition | Antiseptic—skin |
| Silver nitrate | Protein precipitation | Antiseptic—eyes of newborn<br>(prevents gonorrhea) |
| Copper sulfate | | Algicide—water |
| Halogens and derivatives: | | |
| Iodine | Protein inhibition | Antiseptic—skin<br>Disinfectant—medical instruments |
| Chlorine gas | Protein oxidation | Disinfectant—drinking water,<br>general purposes |
| Alcohols: | | |
| Ethanol | Protein denaturation | Antiseptic—skin |
| Isopropanol | Protein denaturation | Disinfectant—medical instruments |
| Phenols and their derivatives | Surface-active<br>Protein denaturation | Disinfectant<br>Preservative |
| Hexachlorophene | | Bacteriostat—soaps and lotions |
| Quaternary ammonium compounds | Surface-active<br>Protein denaturation | Antiseptic—soaps<br>Disinfectant—medical instruments,<br>general purposes |
| Formaldehyde | Reduction | Disinfectant<br>Preservative |
| Hydrogen peroxide | Oxidation | Disinfectant |
| Ethylene oxide | Oxidation | Sterilant without heat |
| Food preservatives: | | |
| Salts of proprionic acid | | Bacteriostat—bread<br>Fungistat—bread |
| Sodium benzoate | | Bacteriostat—beverages, juices, preserves<br>Fungistat—beverages, juices, preserves |
| Sorbic acid | | Fungistat—cheese and salads |

cause the inhibition of biologically active proteins, including enzymes. The uses of heavy metals include antisepsis, disinfection, and sterilization.

Among the antimicrobial compounds containing heavy metals are mercury compounds, silver nitrate, and copper sulfate. Silver nitrate is routinely applied to the eyes of newborn infants as a prophylactic against the bacteria which cause gonorrhea. Copper sulfate is frequently used as an algicide in public water supplies.

Among the halogens used as antimicrobial compounds are chlorine and iodine. Their actions are similar to those of heavy metals, interfering with or disrupting biologically active proteins. The halogens may serve as antiseptics or disinfectants, depending upon their concentrations.

Alcohols and phenols also destroy the biologic nature and activity of proteins. This action is called *protein denaturation* and is not limited to these two groups of antimicrobial chemicals. Ethanol and isopropanol, two of the more common alcohols, are used for antisepsis and disinfection, respectively. As a group, phenols are either bacteriostatic or bactericidal and make excellent preservatives. The phenol hexachlorophene, formerly a bacteriostat commonly used in soaps and skin lotions, has been banned because of its potentially damaging effects on the central nervous system after being absorbed directly through the skin.

Formaldehyde, hydrogen peroxide, ethylene oxide, and quaternary ammonium compounds are other antimicrobial chemicals. Quaternary ammonium compounds are surface-active agents that cause protein denaturation. In function, these compounds resemble phenols. Ammonium compounds are used as antiseptics in soaps and detergents and as general purpose disinfectants. Formaldehyde is a reducing agent commonly used to preserve dead animals. Diametrically opposite in action to formaldehyde are hydrogen peroxide and ethylene oxide, both of which are oxidizing agents. Hydrogen peroxide in dilute concentrations acts as a mild skin disinfectant.

*Food preservatives* are chemical food additives that are either bacteriostatic or fungistatic. Among the major food preservatives are salts of proprionic acid, sodium benzoate, and sorbic acid. Proprionic acid derivatives are used as both bacteriostats and fungistats in bread, while sodium benzoate acts in the same dual roles in beverages, juices, and preserves. Sorbic acid is a fungistat commonly used in cheese and salads.

*Antibiotics* are antimicrobial chemicals extracted or derived from microorganisms such as bacteria and fungi. In some cases, antibiotics are synthetic derivatives of microbial natural products. Some common antibiotics, together with their sources and applications, are summarized in Table 2-4.

Most of the antibiotics listed are obtained from various species of *Streptomyces,* a bacterial genus with many germicidal natural products. One of the famous antibiotics, penicillin, is produced by the mold *Penicillium notatum.*

Antibiotics adversely affect the cell wall, membrane function, or protein synthesis of various microbes. The specificity of cell wall receptor sites to particular antibiotics is uniquely prokaryotic. Antibiotics that bind to such sites cause bacteriolysis. Protein synthesis seems to be more vulnerable to antibiotic attack in prokaryotes than in eukaryotes. In any case, antibiotics are either restricted in their effects to one or more related species or act on a broad spectrum of species.

## MICROBIAL RESISTANCE

### FACTORS IN RESISTANCE

The mechanisms of antibiotic action provide clues to why drug-sensitive microbes suddenly become resistant. The development of microbial resistance to drugs is called *infectious drug resistance.* First, infectious drug resistance can develop if there is a structural modification at, or functional loss of, the site of antibiotic action. These changes involve microbial cell wall receptor sites of the enzymes required for antibiotic action. Second, resistance may result from a partial or complete loss of microbial permeability to an antibiotic. The permeability effects may result from an increase in thickness of the bacterial cell wall or

**Table 2-4** Some common antibiotics

| Antibiotics | Antibiotic-producing group | Applications |
|---|---|---|
| Penicillins | Fungus—*Penicillium* spp. | Prokaryotes:<br>Cocci<br>Spirochetes |
| Polymyxins, circulin | Bacterium—*Bacillus* spp. | Prokaryotes and eukaryotes, especially *Pseudomonas* spp. |
| Nystatin | Bacterium—*Streptomyces* spp. | Eukaryotes, especially *Candida albicans* |
| Filipin | Bacterium—*Streptomyces* spp. | Eukaryotes |
| Cycloheximide | Bacterium—*Streptomyces* spp. | Eukaryotes |
| Chloramphenicol | Bacterium—*Streptomyces* spp. | Prokaryotes, especially *Salmonella typhosa* |
| Tetracycline | Bacterium—*Streptomyces* spp., chemical synthesis | Prokaryotes |
| Erythromycin | Bacterium—*Streptomyces* spp. | Prokaryotes, especially coccus spp., *Bacillus anthracis*, *Clostridium* spp. |
| Streptomycin | Bacterium—*Streptomyces* spp. | Prokaryotes, especially *Pasteurella* spp., *Shigella* spp., *Salmonella* spp. |
| Neomycin | Bacterium—*Streptomyces* spp. | Prokaryotes, especially *Mycobacterium tuberculosis* |

the functional loss of chemical systems within the cell wall for the transport of specific antibiotic molecules. Third, resistance can be secured if the antibiotic is functionally inactivated. Inactivation is usually achieved by enzymatic hydrolysis of biologically active portions of the antibiotic, according to the general scheme

Reactant                Products

$$A \xrightarrow[(+)H_2O]{enzyme} B + C$$

The antibiotic A is split in the presence of a specific enzyme and water into inactive products. The general mechanisms of antibiotic resistance are summarized in Table 2-5.

## GENETIC CORRELATES

The foregoing mechanisms all involve genetic changes rather than acquired characteristics. Resistance results from the transfer of genetic material containing a gene for drug resistance from one cell to another. There are three methods by which these transfers normally occur in bacteria: transformation,

**Table 2-5** Mechanisms of antibiotic resistance

| Antibiotic resistance | Mechanism |
|---|---|
| Structural modification at site of antibiotic action | Loss of specific enzyme |
| Functional loss at site of antibiotic action | Loss of cell wall receptor site |
| Decrease in or loss of antibiotic permeability | Increase in surface barrier<br>Loss of transport system |
| Functional inactivation of antibiotic | Enzymatic hydrolysis of active portion of antibiotic molecule |

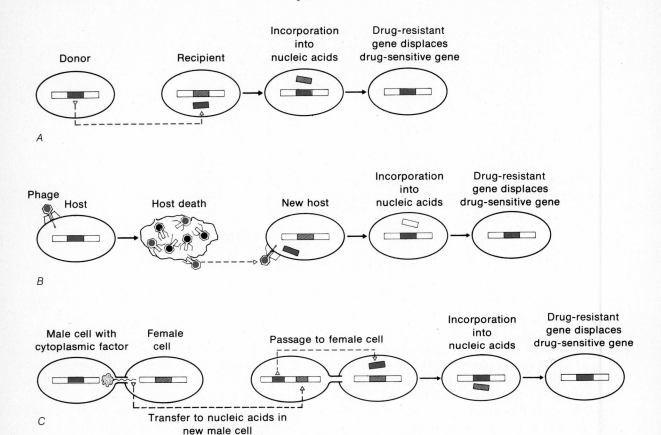

**Figure 2-7** *Methods of transferring genetic material from one bacterial cell to another.*
A. *Transformation.* B. *Transduction.* C. *Conjugation.*

transduction, and conjugation. These methods are summarized in Fig. 2-7.

In *transformation*, genetic material called *deoxyribonucleic acid* (DNA) is extracted from donor cells or excreted during bacteriolysis. The DNA must contain a mutant gene providing the potential for drug resistance. The DNA is added to, or directly enters, the medium in which drug-sensitive bacterial cells are growing. These cells must differentially absorb the foreign DNA from their surroundings and incorporate it into their nucleic acid structures. Transformation is completed when the gene for drug resistance replaces or preempts the gene for drug sensitivity.

*Transduction* requires the intervention of *bacteriophages*. By infecting only bacteria, bacteriophages serve as vehicles for the genetic transfer of

DNA located in their nucleic acid cores. Some of the bacteriophage DNA may have been obtained originally from bacteria in which the phages replicated, or multiplied, parasitically.

The attachment of a T4 bacteriophage to the cell wall of *Escherichia coli* and the subsequent injection of its nucleic acid core into the bacterial protoplasm are illustrated in Fig. 2-8. Complexly symmetric in shape, T4 phage is an assembly of protein components divided into three basic regions: head, neck, and tail. The head contains 30 facets. The DNA core is enclosed by the facets. Attached by a neck and collar to the head, the tail is a contractile sheath surrounding a hollow core. At the base of the tail is a spiked end plate and six fibers.

When a complementary binding site on the bac-

A

B

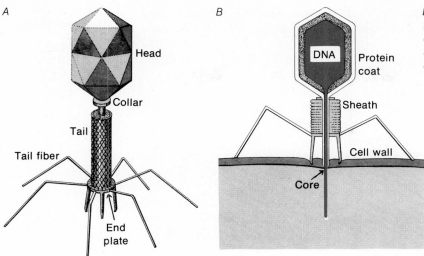

*Figure 2-8*  A. *Bacteriophage.* B. *At-tachment of a bacteriophage to the cell wall of a bacterium and the subsequent injection of its nucleic acid core.*

terium has been identified, the bacteriophage attaches to the outside of the bacterial cell wall by means of the spiked end plate. Like the landing pods on a lunar module, the six tail fibers provide stability during the process of attachment. They also assist after attachment, when the bacteriophage tail sheath contracts. This action drives the DNA core through the bacterial cell wall and into its protoplasm.

Once in the cell, the viral nucleic acid may divert the metabolic machinery in the cytoplasm from the synthesis of bacterial protein, substituting instead instructions for the manufacture of viral protein. Viral protein is then synthesized. Some of the newly manufactured virus particles may incorporate into their structure bacterial DNA with a mutated gene for drug resistance. The infected bacterial cells ultimately disintegrate, releasing the entrapped bacteriophages. The phage particles with mutant genes may

then infect drug-sensitive bacteria. Transduction is successfully completed when the drug-resistant genes replace the drug-sensitive genes.

In *conjugation,* a sexual method of reproduction in bacteria, cytoplasmic factors from male donor cells are unilaterally transferred to female recipient cells during direct contact. The cytoplasmic factors may then be incorporated into the nucleic acid structure of the recipient cells. Now acting as donors containing genes for drug resistance, such cells subsequently conjugate with certain drug-sensitive recipient cells. The gene for drug resistance must replace the drug-sensitive gene as the last step in the acquisition of infectious drug resistance.

While transformation, transduction, and conjugation all involve genetic transfer, only conjugation is a bona fide method of reproduction. However, all three methods of genetic transfer occur naturally.

## COLLATERAL READING

Braude, A. T., "Bacterial Endotoxins," *Scientific American,* March 1964. Endotoxins are characterized both physiologically and immunologically.

Brock, Thomas D., *Biology of Microorganisms,* Englewood Cliffs, N.J.: Prentice-Hall, Inc., 1970. A well-written text and reference for advanced undergraduates, this book emphasizes normal biology rather than pathogenicity.

Broecker, Wallace S., "Man's Oxygen Reserves," *Science,* 168:1537–1538, 1970. Evidence is presented suggesting that molecular oxygen is a virtually unlimited natural resource.

Chanock, R. M., "Control of Acute Mycoplasmal and Viral Respiratory Tract Disease," *Science,* 169:248–256, 1970. An optimistic view of the eventual success of vaccination in the prophylaxis of respiratory disease.

Echlin, Patrick, "The Blue-Green Algae," *Scientific American,* June 1966. A description of the beneficial and harmful effects of blue-green algae, together with their extraordinary range of habitats and close resemblance to bacteria.

Emerson, Ralph, "Molds and Men," *Scientific American,* January 1952. The profound cultural influences of molds and their wide distribution are interestingly authenticated.

Frobisher, Martin, Lucille Sommermeyer, and Ernest H. Blaustein, *Microbiology for Nurses,* Philadelphia: W.B. Saunders Company, 1967. Chapter 13 is devoted to sterilization and disinfection in nursing. Chapter 36 discusses nursing responsibilities in communicable disease cases.

Gorini, Luigi, "Antibiotics and the Genetic Code," *Scientific American,* April 1966. Streptomycin alters the structure of ribosomes and adversely affects protein synthesis.

Horne, R. W., "The Structure of Viruses," *Scientific American,* January 1963. A description of the basic shapes and kinds of building blocks found in viruses.

Hotchkiss, Rollin D., and Esther Weiss, "Transformed Bacteria," *Scientific American,* November 1956. The inheritance of mutant characteristics in pneumococci is investigated by extracting and purifying DNA from one strain and permitting its incorporation into another strain.

Hutchinson, G. Evelyn, "The Biosphere," *Scientific American,* September 1970. This article leads an entire issue devoted to the biosphere and describes in broad strokes the major cycles of energy and elements.

Sharon, Nathan, "The Bacterial Cell Wall," *Scientific American,* May 1969. A relation between bacterial virulence and the integrity of the cell wall of a bacterium is uncovered during studies with antibiotics that interfere with the synthesis of the cell wall.

Stewart, Sarah E., "The Polyoma Virus," *Scientific American,* November 1960. An excellent description of the steps leading to the discovery and subsequent culturing of any viral pathogen.

Watanabe, Tsutomu, "Infectious Drug Resistance," *Scientific American,* December 1967. A discussion of the genetics of sudden bacterial resistance to a previously effective antibiotic.

Wollman, Elie L., and François Jacob, "Sexuality in Bacteria," *Scientific American,* July 1956. The mechanism and significance of bacterial conjugation are explored.

Zinder, Norton D., "'Transduction' in Bacteria," *Scientific American,* November 1958. A study of the viral transfer of genetic material from one species of bacterium to another.

# CHAPTER 3
# man's microbial
# SYMBIONTS

In the last chapter, we gained some appreciation of microbial ways of life and the roles which microbes play in nature. The stage is now set for a more detailed analysis of each of the major pathogenic groups which devastate human beings.

The pathogens reviewed in this chapter are major in the sense that they are medically and economically important to man. Their medical interest is determined primarily by the nature of the disease elicited. Economically important microbes are costly burdens to society in terms of both work hours lost through their debilitating effects and the financial budgets allocated for their control and medical management each year. The biologic significance of microbial pathogens is based on the promise of developing practical control measures against them and the possibility that the knowledge gleaned from their study will provide deeper insights into host-parasite interactions in general. Ordinarily, because a medically

important pathogen is also economically noteworthy and biologically intriguing, these interests are not mutually exclusive.

Pathogens of the human body are found among bacteria, rickettsias, viruses, fungi, protozoa, flatworms, roundworms, and arthropods, i.e., the eight groups to which all microbial parasites belong. The parasites will be considered individually in the groups of which they are members.

The usual information on each pathogenic microbe reviewed here consists of its generic (genus and species) and common names, the medical and common names and symptoms of the disease elicited, the pathogen's portals of entry and exit, and the methods used in laboratory identification and general prophylaxis. In numerous cases, illustrations are provided of the causative agents and, where applicable, of their arthropod vectors. Chemotherapeutic measures, including the use of specific antibiotics and most an-

**Table 3-1** Major diseases of man caused by bacteria

| Causative agent and disease | Symptoms | Characteristics and detection* | Prophylaxis |
|---|---|---|---|
| *Salmonella spp:* Typhoid fever Paratyphoid fever (more mild than typhoid) Food poisoning | Infection of lymphatic tissues and spleen ulcerating hemorrhages and perforations of intestinal wall leading to peritonitis Headache, fever, anorexia, abdominal pain, weakness, diarrhea, vomiting, and dehydration | (−) bacillus; identification of causative agent from blood, feces, or urine; demonstration of antibodies in blood serum | Sanitary control of food, water, and sewage disposal; typhoid vaccine |
| *Escherichia coli:* Diarrhea and other infections | Diarrhea and dehydration; urinary tract infections and associated pain | (−) bacillus; identification of causative agent from locus of infection; serologic tests | Asepsis |
| *Corynebacterium diphtheriae:* Diphtheria | Pharyngeal infections, leading to toxemia; formation of ulcers and fibrous exudates (pseudomembranes) may block respiratory passageways; nerve and heart muscle degeneration | (−) bacillus; identification of typical diphtheroids; toxin production test for virulence | Pasteurization of milk; fluid toxoid and diphtheria antitoxin |
| *Neisseria gonorrhoeae:* Gonorrhea Acute infectious conjunctivitis Gonorrheal vulvovaginitis | Inflammation of, and purulent discharge from, genitals; functional sterility in extreme cases; gonococcal arthritis, endocarditis, and conjunctivitis | (−) diplococcus; smears and cultures | Personal hygiene and sanitation |
| *Clostridium tetani:* Tetanus (lockjaw) | Spasms of voluntary muscular contractions | (+) bacillus; clinical diagnosis; identification of causative agent and its toxin | Tetanus toxoid and antitoxin |
| *Clostridium perfringens:* Gas gangrene | Fulminating infection causing necrosis, gas production, and severe toxemia | (+) bacillus; identification of causative agent; clinical diagnosis | Asepsis; antitoxin |
| *Bacillus anthracis:* Anthrax | Malignant pustule in skin is first sign; infection of blood and lymph | (+) bacillus; identification of causative agent through smears | Hygiene, asepsis, and sterilization; animal vaccines; antiserum |

| Organism / Disease | Symptoms | Identification | Control |
|---|---|---|---|
| *Pasteurella pestis:* Plague (black death) Bubonic Septicemic | Enlarged inguinal lymph nodes or buboes are first signs Septicemia | (−) bacillus; smears and cultures; animal inoculation; serologic tests | Rodent and flea control; hygiene and sanitation; bacterial vaccine |
| *Clostridium botulinum:* Botulism | Food poisoning, resulting in headache and fatigue; difficulties in swallowing and speech; nervous system paralysis affecting respiration | (+) bacillus; identification of causative agent | Antitoxins and toxoid |
| *Hemophilus pertussis* (*Bordetella pertussis*): Pertussis (whooping cough) | Infections of mucous membranes of respiratory tract and eyes; acute bronchitis, with characteristic cough including violent inspiration; bronchopneumonia may be secondary complication | (−) coccobacillus; identification of causative agent through smears | Vaccination with killed cultures |
| *Mycobacterium tuberculosis:* Tuberculosis | Primary tubercle is first sign, accompanied by mild fever; chronic pulmonary infections containing pus and development of hypersensitivity; alveolar destruction, causing coughing and expectoration; weight loss and fatigue | (+) bacillus; identification of causative agent; skin testing | Pasteurization of milk; disease control in dairy cattle; isolation of active human cases |
| *Mycobacterium leprae:* Leprosy | Lesions of skin, mucous membranes of mouth, and peripheral nerves | (+) bacillus; differential staining of causative agent collected from locus of infection; skin testing | Isolation of contagious cases |
| *Treponema* spp.: Syphilis (lues) Yaws Bejel Pinta | Chancre at locus of infection is first sign; secondary lesions of skin and mucous membranes; tertiary involvement of brain, heart, skin, bones, and viscera, with formation of ulcers or gummas | (−) spirochete; identification of causative agent; serologic tests; complement-fixation test | Hygiene and sanitation; avoidance of direct sexual contact with human carriers |

*(−) = gram-negative; (+) = gram-positive.

timicrobial drugs, have been deliberately omitted to avoid the almost immediate obsolescence which the rapid expansion of this field engenders.

## PATHOGENIC PROKARYOTES

### BACTERIA

As a group, bacteria are probably the best known of all microbial pathogens. Bacterial cells were among the first microbes observed microscopically as well as among the first correlated with specific pathologic conditions in human beings. The origins of disinfection, sterilization, asepsis, and, indeed, all of medical bacteriology can be traced to the gradual realization that bacteria cause disease.

The cultural influences of bacterial pathogens, although traditionally underestimated in a historical sense, are now receiving increasing recognition as a significant social force. Bacterial disease epidemics have influenced the resolution of many past military conflicts and wars; led to the establishment of certain customs based on superstition, folk medicine, and science; and ultimately affected societal structure and function. For these reasons, the prevalence of particular bacterial diseases in a society is not only a reflection of its economics and politics but also a measure of its organization, development, types of morality condoned, and attitudes upheld.

Diseases   Bacteria are the causative agents of diseases as historically important as plague, cholera, diphtheria, and typhoid; as medically important as anthrax, tetanus, tuberculosis, pneumonia, gas gangrene, leprosy, meningitis, and infectious jaundice; as common as food poisoning, diarrhea, dysentery, scarlet fever, the whooping cough, and urinary infections; and as reproductively significant as gonorrhea and syphilis. The causative agents of these and other major bacterial diseases, their portals of entry, and their symptomatology, diagnosis, and control are given in Table 3-1.

Cocci, bacilli, and spirilla all have pathogenic representatives. Over half of the major bacterial diseases are caused by bacilli.

Almost three-fourths of all bacterial pathogens are *gram-negative*; the rest are *gram-positive*. The steps involved in the application of the Gram stain are outlined in Table 3-2.

The virulence of a bacterium is indirectly measured by the results of the Gram stain. Virulence is related to the ability of a bacterium to be decolorized after being stained with crystal violet. Decolorization is attempted after first adding iodine to make the stain water-insoluble and then adding a decolorization solvent in which the stain-iodine complex is soluble. The thicker the cell wall of the bacterium, the less permeable it is to the solvent, and the more blue it will remain after attempted decolorization. Virulent, gram-negative bacteria have relatively thin cell walls, as evidenced by their permeability to the decolorization solvent and the subsequent need for counterstaining.

Symptoms   Many bacterial diseases are marked by heat, edema, redness, and pain in the infected area. These are the so-called cardinal signs of inflammation. Although most parts of the body are vulnerable to bacterial infection, moist lining tissues such as the meninges, peritoneum, and mucous membranes of the mouth, large intestine, and urogenital systems are particularly susceptible.

Other common symptoms include fever, weakness, malaise or uneasiness, and digestive upsets. Gastrointestinal disturbances frequently lead to vomiting or diarrhea and subsequent dehydration. Peripheral rashes, sores, and lesions such as ulcers and abscesses are not uncommon in the advanced stages of bacterial diseases. It follows that necrosis, or tissue death, often accompanied by a purulent discharge, is a characteristic sign.

A number of pathogenic bacteria are transported in the blood, creating a condition called *bacteremia*, and are capable of causing *toxemia* or *septicemia* when the blood is actually poisoned by their products. Several pathogens differentially attack the lymphatic vessels or are transported from the blood to the lymph.

Unique symptoms are associated with specific bacterial pathogens. For example *Corynebacterium diphtheriae*, the causative agent of diphtheria, forms

**Table 3-2**  Differential staining of bacteria with the Gram stain

| Steps | Chemicals | Colors | Differentiation |
|-------|-----------|--------|-----------------|
| Staining | Crystal violet<br>Iodine | Blue | |
| Decolorization | A solvent such as alcohol<br>or acetone | Blue<br>Colorless | Gram-positive<br>Gram-negative |
| Counterstaining | A red acidic dye such as<br>acid fuchsin | Blue<br>Red | Gram-positive<br>Gram-negative |

fibrous exudates called pseudomembranes in the respiratory passageways. *Pasteurella pestis,* the plague organism, may cause the formation of diagnostic skin swellings called buboes. The first sign of tuberculosis is the formation, usually within the lymph nodes, of primary tubercles. A primary tubercle is a wall of cells enclosing the tubercle bacilli *Mycobacterium tuberculosis.* Chancroids, or soft, nonsyphilitic ulcers, are formed by *Hemophilus ducreyi*; chancres, or hard syphilitic ulcers, are induced by *Treponema pallidum.* The cough and violent inspirations characteristic of the whooping cough are evoked by *Hemophilus pertussis* (*Bordetella pertussis*). Spirochetal jaundice is the result of infection by *Leptospira* spp. and causes the deposition of bile pigments in the blood, yellowing the skin.

Several different species within the same genus may each cause a distinctive disease. *Salmonella* spp. are responsible for typhoid and paratyphoid fevers and certain types of food poisoning. *Treponema* spp. are the causative agents of syphilis, yaws, bejel, and pinta. Of course, the most famous (or infamous) is *T. pallidum,* the causative agent of syphilis. Other genera with multiple pathogenic species include *Shigella, Brucella, Micrococcus, Actinomyces,* and *Leptospira,* causing bacillary dysentery, undulant fever, skin infections, actinomycosis, and leptospirosis, respectively.

Finally, two different bacteria may cause distinctive effects in their host only after joining forces as, for example, in the case of trench mouth. This disease is due to the combined action of two genera, the paired bacillus *Fusobacterium fusiforme* and the spirochete *Borrelia vincentii.*

**Portals of entry and exit**  Bacterial portals of entry and exit are illustrated in Fig. 3-1. The organisms listed are among the most important bacterial pathogens of human beings. Notice that the digestive tract is heavily used as a route of entry and exit, while the urinary tract is infrequently used. Only a few bacterial pathogens reside exclusively on or in the skin.

**Diagnosis**  Identifying a pathogenic bacterium requires its isolation either from the initial site or the definitive locus of infection or from the blood, feces, urine, or purulent exudates of the host. Attempts are then made to establish pure cultures of the pathogen in the laboratory. The pathogen is finally characterized in both the initial isolate and subsequent cultures by differential staining reactions, culture characteristics, bioassays, and chemical tests.

The laboratory diagnosis of a bacterial pathogen depends in large part on the adequacy of the sample collected from the patient. The method of collection, in turn, determines the type of diagnostic tests which can be employed. Blood smears or *biopsy tissues,* i.e., pieces of presumably infected tissues, are commonly examined for evidence of the causative agent. Selective and *differential stains* are routinely used in laboratory diagnosis. Size, shape, color, texture, and other physical characteristics of the bacterial colonies are also diagnostic.

*Bioassays* usually involve the injection of an inoculum containing the presumed pathogen into a test population of laboratory animals. The infectivity and virulence of the pathogen are then measured. Tests based on bacteriophage-specific infectivity for a particular strain of pathogen, a technique termed *phage typing*, is routinely used where multiple strains make precise identification difficult.

*Chemical tests* frequently rely on some type of antigen-antibody interaction or the presence or absence of reactions catalyzed by specific bacterial enzymes.

Tests based on the presence of specific antibodies include complement fixation; agglutination; precipitation; capsular swelling, or Neufeld typing (Neufeld quellung test); and various skin sensitivity reactions. Many of these reactions are considered in Chap. 4. The *complement-fixation reaction*, one of the most important diagnostic tests involving antigen-antibody interactions, is detailed in Table 3-3.

**Control** General prophylactic measures against bacterial pathogens are based on asepsis, sanitation, and animal control. *Asepsis* refers to the elimination from a given area of all disease-causing microbes (disinfection) or, in a broader context, the elimination of all living microorganisms (sterilization). Medical and surgical practices strive to achieve the former and the latter, respectively. *Sanitation*, which has the same objectives but is less rigorous, revolves around personal hygiene and public health measures which arrest the growth of, or prevent invasions by, potential pathogens. In this sense, *prophylaxis* encompasses (1) the production, preparation, and consumption of food, including the pasteurization of milk; (2) the preparation and use of water for drinking and swimming; (3) the application of sanitary methods to sewage and waste disposal; and (4) the regulation of arthropod vectors, animal reservoirs, and human carriers that transmit and harbor pathogenic bacteria. In some cases, the cosmopolitan distribution of a microbe may make its avoidance almost impossible, nullifying the beneficial effects of general prophylactic practices.

Vaccines, toxoids, and antitoxins, all of which are antibacterial substances (see Chap. 4), are immunoprophylactic tools. At the present time, immunization is usually effective against more than 30 percent of the major bacterial pathogens that afflict human beings. The percentage is expected to increase as time goes on, due to the anticipated discovery of new vaccines and the introduction of more pow-

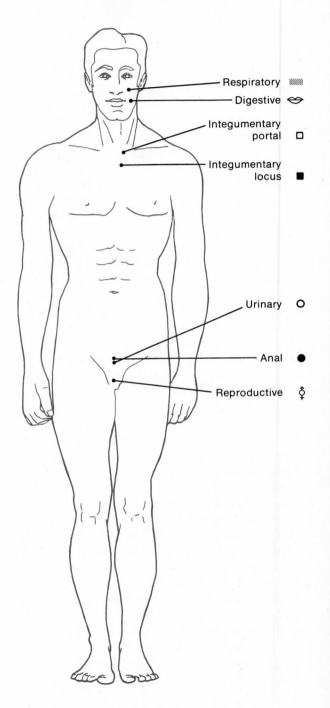

**Figure 3-1** *Modes of transmission and portals of entry and exit of the major bacterial pathogens of man.*

# Portals of entry / Major bacterial pathogens / Portals of exit

Legend of symbols used: ♂♀ = Reproductive · ○ = Urinary (open circle) · ■ = Integumentary locus (filled square) · □ = Integumentary portal (open square) · ⬭ = Digestive (mouth) · ▒ = Respiratory (shaded) · ● = filled circle

| Reproductive | Urinary | Integ. locus | Integ. portal | Digestive | Respiratory | Major bacterial pathogens | Respiratory | Digestive | Integ. portal | Urinary | Anal | Reproductive |
|---|---|---|---|---|---|---|---|---|---|---|---|---|
| | | | | | ▒ | *Mycoplasma mycoides* infections | | | | | | |
| | | | | | ▒ | *Diplococcus pneumoniae* pneumococcal pneumonia | ▒ | | | | | |
| | | | | ⬭ | ▒ | *Corynebacterium diphtheriae* diphtheria | | | | | | |
| | | | | ⬭ | ▒ | *Hemophilus pertussis* pertussus or whooping cough | | ⬭ | | | | |
| | | | | ⬭ | ▒ | *Actinomyces* spp. actinomycosis or ''lumpy jaw'' — Airborne, not person-to-person | | | | | | |
| | | ■ | □ | ⬭ | ▒ | *Pseudomonas aeruginosa* infections — Normal inhabitants of soil and water; not person-to-person | | | | | | |
| ♂♀ | ○ | ■ | □ | | ▒ | *Staphylococcus aureus* boils and carbuncles | ▒ | ⬭ | □ | | | |
| | | | □ | | ▒ | *Bacillus anthracis* anthrax — From animal contact | | | | | | |
| | | | □ | | ▒ | *Pasteurella pestis* plague or black death — Also from rat flea bite | ▒ | | | | | |
| ♂♀ | ○ | | □ | ⬭ | ▒ | *Mycobacterium tuberculosis* tuberculosis — Also from cow's milk | ▒ | | | | | |
| | | | □ | ⬭ | ▒ | *Mycobacterium leprae* leprosy — Not highly contagious | ▒ | | □ | | | |
| | | | | ⬭ | | *Salmonella* spp. typhoid, paratyphoid, food poisoning | | | | ○ | ● | |
| | | | | | | *Shigella* spp. shigellosis or bacillary dysentery | | | | | ● | |
| | | | | | | *Escherichia coli* intestinal and urinary infections | | | | | ● | |
| | | | | | | *Vibrio comma* Asiatic cholera | | ⬭ | | | ● | |
| | | | | | | *Fusobacterium fusiforme* and *Borrelia vincentii* trench mouth or Vincent's angina | | ⬭ | | | | |
| | | | | | | *Clostridium botulinum* botulism — From food; unable to grow in body | | | | | | |
| | | | | ⬭ | | *Brucella* spp. brucellosis or undulant fever — From animal contact | | | | | | |
| | | | □ | ⬭ | | *Pasteurella tularensis* tularemia, deer fly, or rabbit fever — From animal bite or contact; not person-to-person | | | | | | |
| | | | □ | ⬭ | | *Leptospira* spp. leptospirosis or infectious jaundice — From animal urine | | | | | | |
| | | | □ | | | *Clostridium tetani* tetanus or lockjaw — Also from soil or dust | | | | | ● | |
| | | | □ | | | *Clostridium perfringens* gas gangrene — Also from soil, dust, water | | | | | ● | |
| | | | | | | *Spirillum minus* and *Streptobacillus moniliformis* rat bite or Haverhill fever — From rat bite | | | | | | |
| ♂♀ | | | □ | | ▒ | *Streptococcus pyogenes* and *Micrococcus* spp. infections | ▒ | | | | ● | |
| | | | | | | *Neisseria gonorrhoeae* gonorrhea | | | | | ● | ♂♀ |
| | | | | | | *Mima polymorpha* and *Neisseria meningitidis* infections | ▒ | | | | | |
| ♂♀ | | | | | | *Hemophilus ducreyi* chancroid or soft chancre | | | | | | |
| | | | | | | *Calymmatobacterium granulomatis* granuloma inguinale | | | | | | |
| | | | | | | *Treponema pallidum* and other *Treponema* spp. syphilis or lues | | | | | | ♂♀ |

**Table 3-3** Testing for the presence of specific antibodies with the complement-fixation reaction

| Steps | Materials* | Procedures | Results |
|---|---|---|---|
| Complement destruction | (1) Patient's serum<br>(2) Animal serum | Heat separately at 56°C | Complement in the test serum is destroyed |
| Preparation of sensitizing mixture | (3) Animal erythrocytes<br>(2) Animal serum | Combine (2) and (3) | Hemolysis does not occur |
| Complement fixation | (1) Patient's serum<br>(4) Test antigen<br>(5) Animal complement | Combine (1), (4), and (5) | Complement is fixed when the three substances combine if the patient's serum contains the specific antibody; no visual indication results |
| Test for hemolysis | (2 + 3) Sensitized mixture<br>(1 + 4 + 5) Complement-fixation mixture | Combine mixture (2 + 3) with (1 + 4 + 5) and heat at 37°C | If complement fixation has already occurred, no hemolysis takes place, and the cells settle beneath a clear serum; if complement fixation has not occurred, hemolysis takes place and yields a clear red fluid but no sediment |

*Numbers in parentheses identify the different materials used in the test.

erful biochemical and biophysical techniques of diagnosis.

## RICKETTSIAS

Only a limited number of rickettsias are pathogens of man. Some, however, cause well-known diseases such as Rocky Mountain spotted fever, the various kinds of typhus, and rickettsialpox. Although all the rickettsial pathogens require an arthropod vector, the causative agent of Q fever, *Coxiella burnetii,* can also be inhaled with dust. The causative agents of the major rickettsial diseases, their portals of entry, symptomatology, diagnosis, and control are summarized in Fig. 3-2.

Portals of entry   Of the two portals of entry used by rickettsias, the skin portal is the rule and the respiratory route is an exception. Entrance is gained when the skin of the host is perforated by the bite of a vector, when the feces of the vector contaminate the site of a bite, or when a vector is crushed into the skin during an episode of scratching by the host.

Symptoms   Almost all pathogenic rickettsias cause disease with similar symptoms. The typical syndrome of effects begins with a rash, or lesion, at the site of a bite on the skin. This sign is usually followed by headaches and fever, sometimes accompanied by muscle aches and pains and a prolonged illness.

Diagnosis   Detection of specific antibody in the patient's serum is usually used to confirm the clinical diagnosis. The *Weil-Felix agglutination test* is a widely used, nonspecific diagnostic method for the detection of rickettsias. Excluding rickettsialpox, trench fever, and Q fever, if the disease is rickettsial the patient's serum will agglutinate bacteria of the genus *Proteus* when *Proteus* organisms and the serum are combined. Presumably *Proteus* species and most rickettsias have similar antigens. Complement-fixation reactions resembling those described for bacteria are also frequently used in the laboratory diagnosis of rickettsias. While having the advantage of species specificity, complement fixation is more expensive and time-consuming to perform, compared with agglutination.

Control  Prophylaxis is based on hygiene and sanitation, effective regulation of arthropod vectors and animal reservoirs, vaccination, or some combination of these practices. Because of the central role played by vectors in the transmission of rickettsias, control of ticks, mites, lice, and fleas is an important prophylactic measure. The role of rickettsial vectors can also be neutralized by controlling their rodent reservoirs such as rats and mice. Because vaccines now have been developed against almost 40 percent of the pathogenic rickettsias of human beings, immunization is also a significant antirickettsial measure.

## VIRUSES

Viruses are not only the most widespread of all microbes but also the most subtle and sophisticated in achieving their own self-serving ends within the bodies of their hosts. The classification of viruses is limited only by human ability to adequately characterize their finer architectural differences. While these molecular differences are just now being described, host specificity is itself a self-limiting mechanism which imposes constraints on both viral diversity and distribution.

The names of viruses are primarily derived from (1) the diseases which they cause (e.g., common cold viruses), (2) the body structure differentially affected (e.g., adenoviruses, which inflame the adenoids), (3) the geographic region from which the pathogens were initially obtained (e.g., the Coxsackie virus, first isolated in Coxsackie, New York), or (4) a combination of letters indicative of the state of knowledge regarding the virus at the time of assignment (e.g., echo viruses, short for enteric cytopathogenic human orphan). Aside from names and letters, current viral nomenclature uses arabic and roman numerals to differentiate strains within particular groupings. Information on the major viral diseases of human beings is summarized in Table 3-4.

Diseases  Viruses are responsible for an impressive array of diseases. Over half of the infectious human respiratory diseases have a viral origin or cause. In addition, viruses have been implicated as the cause of a few types of animal cancer and are suspected by some authorities of contributing to all forms of human cancer. A causal link between viruses and human cancer is a controversial idea that has been reconsidered several times since the relationship was first suggested.

Excluding their roles in the induction of respiratory diseases or common colds and in carcinomas, or cancers of the skin and other epithelial tissues, viruses are definitely responsible for such well-known afflictions as mumps, measles and German measles, chickenpox and smallpox, influenza, yellow fever, rabies, infectious and serum hepatitises, and poliomyelitis. Of these diseases, roughly one-third are easily recognized as typical of, but not restricted to, childhood. One of them, German measles, takes on added significance if contracted by women in the early stages of pregnancy because of the increased probability of developmental anomalies in their children (see Chap. 21). Periodic epidemics and pandemics of influenza, however, have their greatest impact on older people, especially those with prior respiratory impairment.

Portals of entry and exit  Portals of entry and exit of the major pathogenic viruses of human beings are diagramed in Fig. 3-3. Over 20 viruses are identified, along with the diseases they cause.

A number of interesting similarities and differences are revealed when viral portals are compared with bacterial portals. First, the respiratory tract replaces the digestive tract as the most heavily used portal of entry. Second, like the bacteria, viruses apparently do not or only infrequently use the urogenital orifices as either portals of entry or exit. Third, although over one-third of the viral pathogens cause skin lesions and rashes, not one is confined exclusively to the skin, in contrast to bacteria. Finally, although almost half of the pathogenic viruses are capable of leaving the body from more than one portal, only one uses multiple portals of exit.

Symptoms  Many symptoms caused by viruses are similar. They include an inflammation at the locus of infection as well as some combination of fever, headache, backache, or other muscle aches. These complaints culminate in a generalized malaise and fatigue. The fever is usually accompanied by peripheral or systemic lesions such as eruptions, rashes, and ulcers. The inflammation may remain localized or can spread to the lymph nodes, meninges, mucous membranes, liver, or respiratory and nervous systems. Inflammation of the lymph nodes is a characteristic sign in cat-scratch fever, German measles, and mumps. Aseptic meningitis may occur as a consequence of poliomyelitis, influenza (the grippe), and summer diarrhea. Conjunctivitis and inflammation of mucous membranes are frequent components in pharyngoconjunctivitis fever, influenza, and smallpox. The liver is inflamed, creating a jaundiced condition, in the hepatitises and yellow fever. Involvement of the upper or lower respiratory tract occurs during the common cold, pharyngitis, influenza, measles, primary atypical pneumonia, and psittacosis. Nervous system effects are found in poliomyelitis, aseptic meningitis, rabies, epidemic encephalitis, and the shingles. The voluntary muscles are indirectly affected through viral actions on nervous tissues. Muscle abnormalities may result after contracting poliomyelitis, influenza, yellow fever, dengue fever, or epidemic encephalitis.

Diagnosis   Presumptive identification of a viral pathogen is based primarily on symptomatology. The diagnosis is confirmed by *serologic tests*, which check for a rise in the concentration of specific antibody. Such tests include complement fixation, agglutination, and neutralization. Skin tests can be used in diagnosing cat-scratch fever. *Bioassays* (biological assays) in which laboratory animals are tested with an extract of infected tissue are performed when rabies and yellow fever are suspected. Other tests which help to differentiate between different viral pathogens include liver function tests, biopsies of presumably infected tissues, and the identification of unique structures formed by the microbe. These tests are used in the diagnosis of infectious hepatitis and rabies.

Control   At the present time, protection against

| Causative agent and disease | | Portals of entry |
|---|---|---|
| *Rickettsia rickettsii* 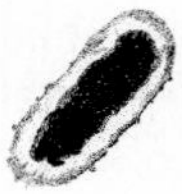 | Rocky Mountain spotted fever Boutonneuse fever | ☐ |
| *Rickettsia prowazekii*  | Epidemic typhus fever | ☐ |
| *Rickettsia typhi*  | Endemic (murine) typhus | ☐ |
| *Rickettsia tsutsugamushi* | Scrub typhus (tsutsugamushi fever) | ☐ |
| *Rickettsia akari* | Rickettsialpox | ☐ |
| *Rickettsia quintana* | Trench fever | ☐ |
| *Coxiella burnetii*  | Q fever | ▦ ☐ |

**Figure 3-2**   *Major diseases of man caused by rickettsias.*

| Symptoms | Detection | Prophylaxis | Vectors |
|---|---|---|---|
| Fever; headache; rash on hands and feet; prolonged illness | Weil-Felix agglutination test; complement-fixation test | Tick control; rickettsial vaccine | *Dermacentor andersoni*—tick |
| Fever; headache; malaise; muscular pains; rash | Weil-Felix agglutination test; complement-fixation test | Louse control; hygiene and sanitation; rickettsial vaccine | *Pediculus humanus*—human louse |
| Similar to above but less severe | Weil-Felix agglutination test; complement-fixation test | Rat and flea control | *Nosopsyllus fasciatus*—rat flea |
| A lesion, called an eschar, at locus of infection; rash; headache; fever | Weil-Felix agglutination test; complement-fixation test | Mite control | *Trombicula akamushi*—chigger mite |
| Lesion at locus of infection; rash; fever | Complement-fixation test | Mouse and mite control | *Bdellonyssus bacoti*—tropical rat mite |
| Rash; headache; fever | Complement-fixation test | Louse control; hygiene and sanitation | Same as vector illustrated for epidemic typhus fever |
| Mild pneumonia unaccompanied by rash | Isolation of causative agent | Tick control; rickettsial vaccine | Same as vector illustrated for Rocky Mountain spotted fever |

Integumentary  ☐
Respiratory  ▨

**Table 3-4** Major diseases of man caused by viruses

| Causative agent and disease | Symptoms | Detection | Prophylaxis |
|---|---|---|---|
| Poliomyelitis virus: Poliomyelitis | Short illness accompanied by fever and headache is first sign; nonparalytic, aseptic meningitis, accompanied by stiff neck and back; chronic flaccid paralysis of one or more muscle groups | Isolation of causative agent from fecal specimens; serologic test for rise in antibody concentration | Viral vaccines |
| Infectious hepatitis virus: Infectious hepatitis | Inflammation of the liver, usually accompanied by jaundice | Clinical symptoms; liver function tests | Sanitary control in sewage disposal; asepsis and sterilization |
| Serum hepatitis virus: Serum hepatitis | Symptoms similar to those of infectious hepatitis | Clinical symptoms | Sterilization of syringes and needles |
| Rabies virus: Rabies (hydrophobia) | Encephalitis, accompanied by paralysis, delirium, and convulsions | Identification of Negri bodies in nerve cells; laboratory bioassay of brains of presumptively rabid animals | Rabies control in animals; viral vaccine antiserum; rapid treatment of infected individuals |
| Yellow fever virus: Yellow fever | Headache, backache, fever, nausea, and vomiting; liver damage, accompanied by jaundice | Blood bioassay with mice | Mosquito control; viral vaccine |
| Epidemic encephalitis virus: Epidemic encephalitis | Short illness accompanied by fever and headache is first sign; inflammation of the brain, accompanied by headache, fever, and mental confusion; spastic paralysis and muscular tremors in severe cases | Serologic tests including complement fixation, hemagglutination, and neutralization | Mosquito control; viral vaccine; immune serum |

| Virus: Disease | Description | Diagnosis | Control |
|---|---|---|---|
| Common cold viruses: Common cold | Mild upper respiratory infections | Clinical diagnosis; isolation of causative agent; serologic test for rise in antibody concentration | |
| Influenza virus: Influenza | Infections of epithelial tissues lining respiratory passages, sometimes accompanied by respiratory complications | Culture; serologic test for rise in antibody concentration | Viral vaccine |
| Mumps virus: Mumps (epidemic parotitis) | Acute inflammation of the parotid salivary glands and the reproductive organs | Culture; serologic test for rise in antibody concentration | Isolation of infected individuals |
| Measles virus: Measles (rubeola) | Acute respiratory infection is first sign; oral eruptions called Koplik's spots follow; measles rash | Clinical symptoms; culture; serologic tests | Viral vaccine |
| German measles virus: German measles (rubella) | Fever and rash, both usually mild | Clinical symptoms | Isolation of infected individuals |
| Chickenpox virus: Chickenpox (varicella) | Mild, communicable disease with symptoms similar to those of smallpox | Clinical symptoms | |
| Smallpox virus: Smallpox (variola major and minor) | Fever, with ulceration of mucous membranes; skin lesions and eruptions | Clinical symptoms and laboratory diagnosis, including complement-fixation test | Viral vaccine |

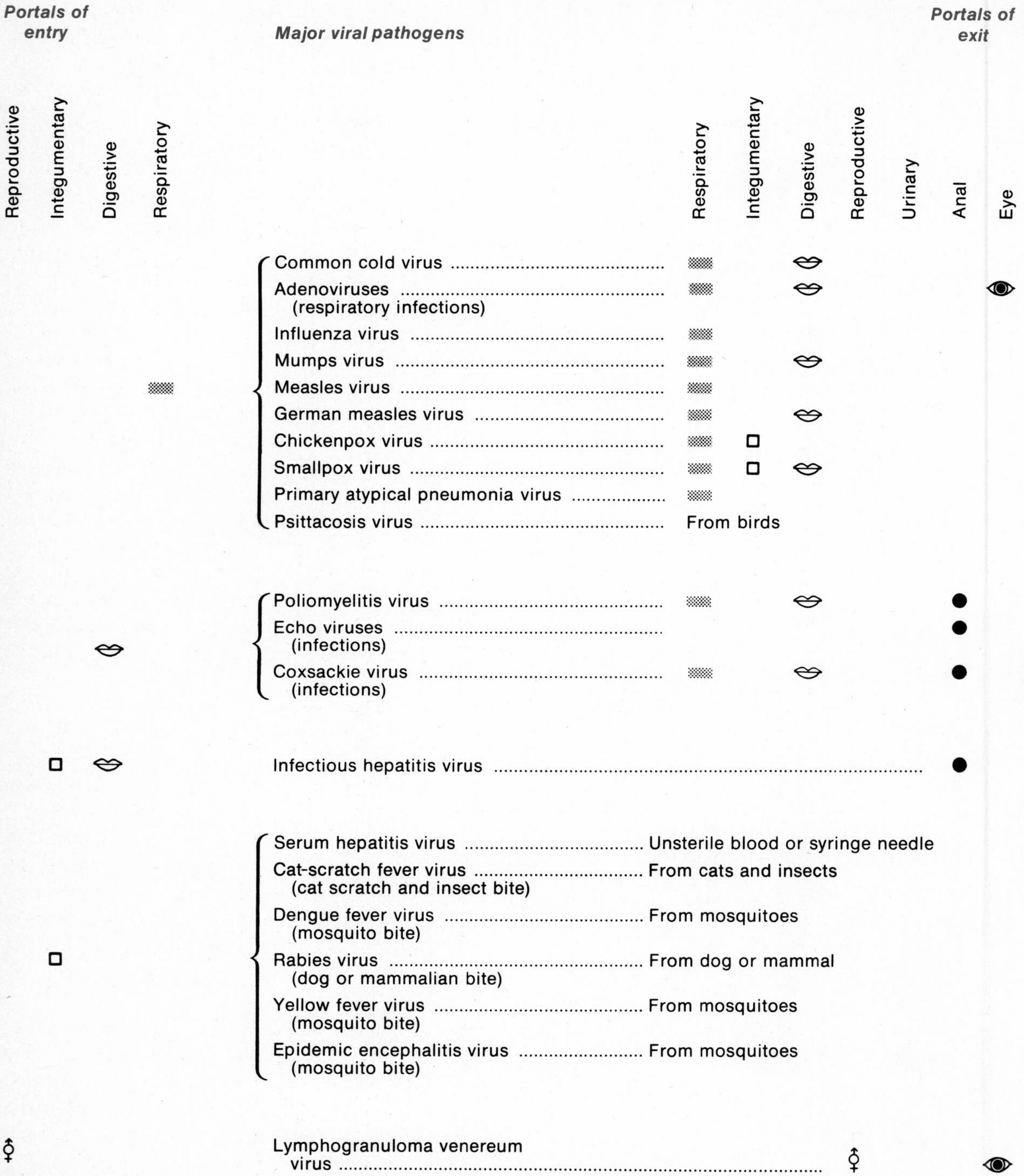

**Portals of entry**

**Major viral pathogens**

**Portals of exit**

| Reproductive | Integumentary | Digestive | Respiratory | Major viral pathogens | Respiratory | Integumentary | Digestive | Reproductive | Urinary | Anal | Eye |
|---|---|---|---|---|---|---|---|---|---|---|---|
| | | | ▨ | Common cold virus | ▨ | | ⬡ | | | | ◉ |
| | | | | Adenoviruses (respiratory infections) | ▨ | | ⬡ | | | | ◉ |
| | | | | Influenza virus | ▨ | | | | | | |
| | | | | Mumps virus | ▨ | | ⬡ | | | | |
| | | | | Measles virus | ▨ | | | | | | |
| | | | | German measles virus | ▨ | | ⬡ | | | | |
| | | | | Chickenpox virus | ▨ | □ | | | | | |
| | | | | Smallpox virus | ▨ | □ | ⬡ | | | | |
| | | | | Primary atypical pneumonia virus | ▨ | | | | | | |
| | | | | Psittacosis virus | From birds | | | | | | |
| | | | | Poliomyelitis virus | ▨ | | ⬡ | | | ● | |
| | | ⬡ | | Echo viruses (infections) | | | | | | ● | |
| | | | | Coxsackie virus (infections) | ▨ | | ⬡ | | | ● | |
| | □ ⬡ | | | Infectious hepatitis virus | | | | | | ● | |
| | | | | Serum hepatitis virus | Unsterile blood or syringe needle | | | | | | |
| | | | | Cat-scratch fever virus (cat scratch and insect bite) | From cats and insects | | | | | | |
| | | | | Dengue fever virus (mosquito bite) | From mosquitoes | | | | | | |
| □ | | | | Rabies virus (dog or mammalian bite) | From dog or mammal | | | | | | |
| | | | | Yellow fever virus (mosquito bite) | From mosquitoes | | | | | | |
| | | | | Epidemic encephalitis virus (mosquito bite) | From mosquitoes | | | | | | |
| ⚥ | | | | Lymphogranuloma venereum virus | | | | ⚥ | | | ◉ |

**Figure 3-3** *Modes of transmission and portals of entry and exit of the major viral pathogens of man.*

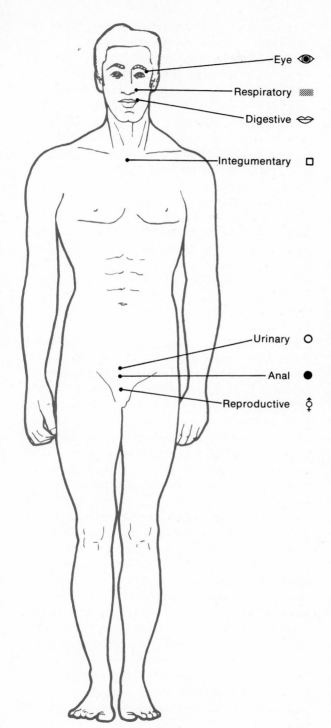

Eye

Respiratory

Digestive

Integumentary

Urinary

Anal

Reproductive

viruses relies upon asepsis and sterilization, sanitation, the control and avoidance of arthropod vectors, animal reservoirs, and infected human beings, in addition to procedures which heighten the natural immunogenic capabilities of the host.

Asepsis and sterilization are essential in avoiding the hepatitises. In fact, serum hepatitis results solely from the use of unsterile syringe needles.

Sanitary control in the disposal of sewage is an important public health measure against several viral diseases. These diseases include influenza, summer diarrhea, aseptic meningitis, and infectious hepatitis.

The mosquito *Aëdes aegypti* is primarily responsible for transmitting the viruses which cause yellow fever, dengue fever, and epidemic encephalitis; therefore the control of this arthropod vector is fundamental in the prophylaxis of these diseases.

Animal reservoirs can also transmit viruses to human beings, as emphasized in the following examples: First, the rabies virus is spread when someone is bitten by a rabid dog, a bat, or one of a variety of other mammals. Second, cat-scratch fever may be contracted following skin penetration by the claws of felines. Third, birds such as parrots and their products disseminate psittacosis viruses among human beings.

Although avoidance of infected animals and human beings is the best insurance against contracting the above-named diseases, it is usually an unrealistic recommendation when the life styles and values of people are considered. To physically avoid people who have active viral infections and are therefore capable of transmitting the microbes to others is difficult, as we all know. Nevertheless, isolation or quarantines have been and still are prescribed in cases of the more infectious communicable diseases caused by viruses, especially where children are likely to be attacked.

Vaccines and immune sera can be used against almost half of the viruses which cause human diseases. Vaccines are much more commonly used than immune sera. The usefulness of these immunoprophylactic measures is limited by various characteristics unique to viruses.

The promise of effective immunoprophylaxis

against all viruses received a substantial boost by the fairly recent discovery of *interferon*, a natural product common to most cells. Interferon prevents intracellular replication by viruses during the early stages of infection. In these initial stages, antibody concentrations are just building up. The widespread use of interferon in the future may therefore provide a significant component to the vigorously sought-after shield against viral virulence. The mechanism of action of interferon is discussed in Chap. 4.

## PATHOGENIC EUKARYOTES

### FUNGI

Pathogenic fungi are relatively few in number of species but are nevertheless extremely common. They cause a variety of diseases affecting the skin and respiratory system. Fungal diseases are more or less debilitating because of their inflammatory nature.

Among the more common human diseases caused by *yeasts* and *molds* are candidiasis, various forms of ringworm, histoplasmosis, and coccidioidomycosis. Information on these and other major fungal diseases is summarized in Fig. 3-4.

Transmission   Most fungi are widely disseminated as spores in the atmosphere and soil. In order to germinate, they usually require moisture. Moist conditions are found on the skin and within the respiratory tract under various circumstances, e.g., skin contamination by wet surfaces, saliva, and fluid exudates or spore contamination of the natural orifices and surrounding mucous membranes of the body.

Portals of entry and exit   All fungal pathogens use the skin as an invasion beachhead. With the exception of *Sporotrichum schenckii,* the causative agent of sporotrichosis, each pathogen is also capable of using one or more alternative routes such as orifices of the respiratory, digestive, and urogenital systems. Portals of exit are relatively unimportant, because of the cosmopolitan distribution of fungal spores. However, communicable transmission probably does occur to a limited extent.

Symptoms   Lesions are characteristic signs of fungal infections and may take the form of ulcerating nodules, scaly discolorations, painful cracks on the skin, or calcified cysts in the lungs. Sporotrichosis, blastomycosis, and aspergillosis may become generalized after spreading from their original sites of infection. Several pathogens which use a respiratory portal of entry, especially the causative agent of histoplasmosis, *Histoplasma capsulatum,* create symptoms closely mimicking those of tuberculosis. Such infections are frequently misdiagnosed and therefore treated incorrectly.

Diagnosis   Positive identification of the causative agent requires its isolation from the locus of infection or a lesion. In most cases, examining smears for the presence of fungi and physically characterizing the colony following culturing are sufficient for reaching a laboratory diagnosis.

Control   Several practices are useful in attempting to avoid the harmful effects of pathogenic fungi. First, the proper observance of sanitation is obviously important in preventing invasions of soil fungi. Contact of bare feet with soil or the floor of public establishments and showers should be avoided, where practical to do so. The possibility of skin contamination looms large when this practice is ignored. Second, because fecal material of aggregated and nesting birds serves as a nutrient medium for the growth of *H. capsulatum* and *Coccidioides immitis,* wide berth should be given to trees, open attics, bell towers, and other shelters serving as roosts for large numbers of birds. The feces accumulate and dry in and around such roosts, giving rise to large amounts of fungal spores. The spores may be inhaled when the air is stirred. If it is necessary to enter such areas, face masks should be worn. Third, prophylaxis can be improved by refraining from the use of antibiotics unless absolutely necessary for medical reasons. Antibiotics inevitably decrease the numbers and species diversity of normally occurring microbes in the body. The use of antibiotics paves the way for invasions by various potentially pathogenic fungi which benefit by reduced competition.

## PROTOZOA

Unlike many of their prokaryotic brethren and pathogenic fungi, protozoal parasites as a group are uniquely adapted to their human hosts. Their wide distribution and successful perpetuation are assured by (1) life cycle stages requiring human, mammalian, or avian blood, and (2) arthropod vectors serving both as hosts of certain protozoa and parasites of human beings and their domesticated animals. The untold human suffering inflicted by protozoal pathogens is difficult to assess but must rank near the top of the list of microbially caused misery.

Among the more well-known protozoal diseases are malaria, African sleeping sickness (African trypanosomiasis), and amebic dysentery. Information about these and other major protozoal diseases is presented in Fig. 3-5.

**Transmission and portals** The most important mode of transmission is via an insect vector. At least half of all parasitic protozoa are disseminated by arthropods. Second in importance is the transmission of the pathogens in food and water. Although relatively minor, aerial transmission of encysted amebas and skin contamination by pathogenic protozoa also play roles in dispersal.

Because modes of transmission determine portals of entry, the skin and mouth are among the most important entry routes of unicellular animal parasites.

**Symptoms** Disease symptoms elicited by protozoa closely resemble those caused by bacteria. Among the common complaints are inflammation coupled with fever and gastrointestinal upsets of varying severity. All the protozoa carried by insects are transported through, and release toxins in, human blood. Toxin production results in septicemia. Other frequently found clinical signs include peripheral or systemic lesions such as swellings, rashes, ulcers, and abscesses. Body involvement in general infections is variable but almost always encompasses either the blood and lymph, the moist lining tissues of the intestines or urethra, as well as specific digestive, excretory, or nervous system glands and organs.

**Diagnosis** Laboratory diagnosis is based on identifying one or more life cycle stages of the protozoa in samples taken from the host. Microscopical examinations are performed on smears or other preparations from the locus of infection and associated exudates, as well as from the blood and lymph, biopsy tissues, or the feces. The laboratory diagnosis is confirmed by inducing the disease in laboratory animals with an inoculum from the host.

**Control** Where the mouth serves as a portal of entry, effective prophylaxis is based on food, water, and waste sanitation. The control and regulation of insect vectors and animal reservoirs are indicated when the skin is used for entry. Feminine hygiene is necessary to prevent the spread of *Escherichia coli* from contaminated vaginal discharges.

## FLATWORMS

All pathogenic flatworms are multicellular animal parasites. In human beings, who serve as the primary, or definitive hosts, flatworms reproduce *sexually*. Life cycles of the parasites commonly include one or more secondary, or intermediate, hosts, either invertebrate or vertebrate, in which the parasites reproduce *asexually*. Most flukes undergo an alternation of sexual and asexual stages, a phenomenon unknown among tapeworms.

The fecundity of flatworms is almost legendary. The extraordinarily large number of eggs laid is a consequence of the intermeshing of both male and female reproductive systems in the same individual, a condition called *hermaphrodism*. The potential of this arrangement is augmented by a sexually reproductive life usually measured in years. Fecundity is promoted by asexual larval reproduction in flukes of stages called *rediae* and *cercariae* and by adult budding in tapeworms of whole hermaphroditic units called *proglottids*. The result from one individual is generally a plethora of eggs totaling 30 million or more during its life. A fivefold increase in numbers is attributable to asexual methods of reproduction.

In spite of their egg-laying competence, parasitic flatworms are in fact relatively rare finds, even when

| Causative agent and disease | | Portals of entry | Symptoms |
|---|---|---|---|
| *Candida albicans*  | Moniliasis<br>Thrush<br>Candidiasis | 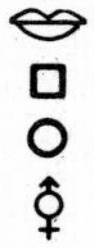 ⬭<br>◻<br>◯<br>⚲ | Ringworm-like lesions on chronically moist skin<br><br>Chronic bronchitis via respiratory portal |
| 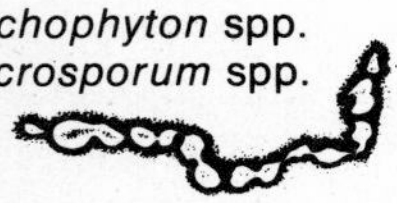 *Epidermophyton* spp. | Ringworm<br><br>*A* Tinea pedis<br>*B* Tinea corporis<br><br>*C* Tinea capitis<br>*D* Tinea unguium | ◻ | Integumentary infection of varying severity:<br><br>A Ringworm of foot<br>B Ringworm of smooth skin<br><br>C Ringworm of scalp<br>D Ringworm of nails |
| *Trichophyton* spp.<br>*Microsporum* spp. | | | |
| *Microsporum audouini*<br>*Microsporum furfur*  | Ringworm | ◻ | Infections of outer epidermis, causing brownish, scaly discolorations |
| *Sporotrichum schenckii*  | Sporotrichosis | ◻ | Ulcerating nodules on skin, paralleling lymphatics; general body infection possible |
| *Cryptococcus neoformans* 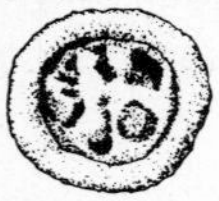 | Cryptococcosis | ▦ | Dermal ulcers or meningitis |
| *Coccidioides immitis* 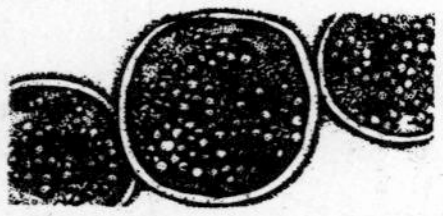 | Coccidioidomycosis | | Respiratory attack with possible skin and pulmonary lesions |
| *Histoplasma capsulatum*  | Histoplasmosis | | Similar to those of tuberculosis, leaving calcified lesions |
| *Blastomyces dermatitidis*  | Blastomycosis | | Chronic suppurating lesions on skin; also respiratory and general body infections |
| *Aspergillus fumigatus* | Aspergillosis | | Dermal, respiratory, and general body infections |

**Figure 3-4**  *Major diseases of man caused by fungi.*

| Detection | Prophylaxis |
|---|---|
| Smears and cultures | Avoid disturbing normal microbial flora |
| Isolation and identification of causative agents; cultures | Hygiene and sanitation, especially in shower |
| Isolation and identification of causative agents from scales or lesions | Hygiene and sanitation, especially in shower |
| Cultures | Difficult to avoid |
| Isolation and identification of causative agent from locus of infection | Difficult to avoid |

- ▨ Respiratory
- 👄 Digestive
- ▢ Integumentary
- ⭘ Urinary
- ☿ Reproductive

actively sought. The reasons, while not readily apparent, are not too difficult to understand. First, larval stages are extremely small and soft-bodied. Second, intermediate, or secondary, hosts such as infected snails and fish are generally inaccessible or appear uninfested even when examined. Third, infestations in human beings are similarly difficult to diagnose, because host-parasite interactions have led to the evolution of a worm difficult to detect and subtle in the damage caused to its definitive host. Fourth, high mortality of the eggs and subsequent developmental stages also takes its toll, as indicated by the relatively stable numbers of parasites which actually reach sexual maturity from one generation to another. When we consider that diverse intermediate hosts must be found by free-swimming, aquatic larvae, that intermediate and definitive host paths generally do not cross in ways that reasonably ensure the transmission of the worm from one to the other, and that intermediate and definitive host habitats are specific and vary greatly, then we must conclude that the reproductive success enjoyed by parasitic flatworms is truly an amazing feat.

*Fluke* and *tapeworm* invasions cause host damage in essentially three ways: (1) nutritional injury, including necrosis; (2) poisoning and hypersensitive reactions from toxins or other secretions and excretions; and (3) obstructing lumina, especially of the cystic and bile ducts. Penetration of the skin and mucous membranes, e.g., in schistosomiasis, exposes the body to potential secondary invasions by bacteria and other microorganisms. Further information on the major platyhelminthic diseases of human beings is presented in Fig. 3-6.

**Transmission and portals**  Only the blood fluke *Schistosoma*, the causative agent of schistosomiasis, actively invades its host by boring through the skin. Other flatworms are passive and usually transmitted when a person eats contaminated intermediate hosts which have been poorly and inadequately prepared as food or ingested raw. Although the eggs of most parasitic flatworms are voided in the feces, those of *Schistosoma* may also escape in the urine. On the other hand, the lung fluke *Paragonimus* frequently becomes encysted, lost, or otherwise imprisoned in

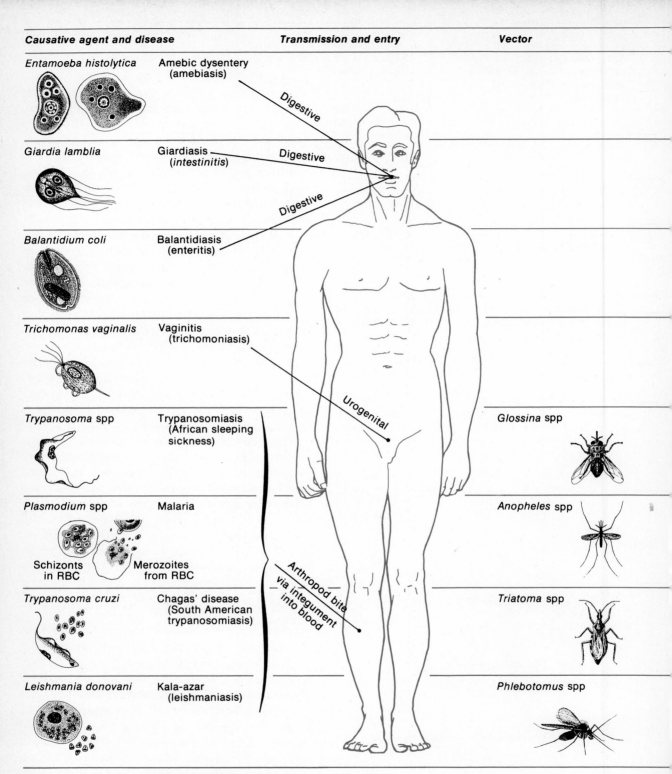

| Causative agent and disease | Transmission and entry | Vector |
|---|---|---|
| *Entamoeba histolytica* — Amebic dysentery (amebiasis) | Digestive | |
| *Giardia lamblia* — Giardiasis (*intestinitis*) | Digestive | |
| *Balantidium coli* — Balantidiasis (enteritis) | Digestive | |
| *Trichomonas vaginalis* — Vaginitis (trichomoniasis) | Urogenital | |
| *Trypanosoma* spp — Trypanosomiasis (African sleeping sickness) | | *Glossina* spp |
| *Plasmodium* spp — Malaria | | *Anopheles* spp |
| Schizonts in RBC / Merozoites from RBC | Arthropod bite via integument into blood | |
| *Trypanosoma cruzi* — Chagas' disease (South American trypanosomiasis) | | *Triatoma* spp |
| *Leishmania donovani* — Kala-azar (leishmaniasis) | | *Phlebotomus* spp |

**Figure 3-5** *Major diseases of man caused by protozoa.*

| Symptoms | Detection | Prophylaxis |
|---|---|---|
| Ulcerations in mucosa of large intestine; liver, lung, and brain abscesses in severe cases | Identification of cyst and motile stages in feces | Food, water, and waste sanitation |
| Indigestion and dietary deficiencies | Identification of motile and cyst stages in feces | Food, water, and waste sanitation |
| Mild diarrhea; severe ulcerations in mucosa of large intestine | Identification of motile or cyst stages in feces | Food, water, and waste sanitation |
| Infections of urethra and prostate gland; persistent vaginal or urethral discharge and inflammation | Identification of causative agent in discharge or at site of infection | Personal hygiene and sanitation in disposal of vaginal discharges |
| Intense headache, fever, insomnia, rash, and anemia initially; drowsiness, tremors, delusions, and lethargy subsequently | Identification of causative agent in blood smears and biopsies of infected lymphatic tissue; induction of disease through animal inoculation | Protection from, and control of, tsetse fly vector; control of animal reservoirs |
| Chills and fever at regular intervals, followed by profuse sweating | Identification of life-cycle stages in blood smears | Protection from, and control of, mosquito vector; control of animal reservoirs |
| Swollen lymph nodes, fever, and anemia, primarily in children | Identification of causative agent in blood smears; induction of disease through animal inoculation | Protection from, and control of, bug vector; control of animal reservoirs |
| Visceral and skin lesions | Identification of rounded, nonmotile intracellular forms in blood smears; induction of disease through animal inoculation | Protection from, and control of, sandflies; control of animal reservoirs |

its human host and does not ordinarily exit from the body. These parasitic "flaws" strongly suggest that human beings are not their normal definitive hosts.

**Migrations** Thumbnail sketches of the migrations undergone by pathogenic flatworms within their definitive hosts are given in the second column of Fig. 3-6. While *Schistosoma*, *Opisthorchis,* and *Paragonimus* are blood, liver, and lung flukes, respectively, their larval stages of development are commonly found elsewhere in the body, e.g., the brain, liver, or duodenum. In the case of *Schistosoma,* however, reproductively mature adults migrate from the hepatic portal vein, their normal habitat, to the bladder or rectum for sexual reproduction, paving the way for a subsequent reinfestation via the anus.

**Symptoms** The disease symptoms created by the presence of parasitic flatworms in the body reflect the three basic platyhelminthic mechanisms by which host damage is produced. Intestinal worms cause gastrointestinal upsets. Their secretions and excretions may adversely affect the nervous system in advanced cases. The liver fluke *Opisthorchis sinensis* is also capable of poisoning its host through the secretion of toxins. Associated with tapeworm infestations are induced vitamin deficiencies and loss of appetite. The vitamin deficiencies precipitate anemia. The loss of appetite causes a lowered vitality.

Characteristic of schistosomiasis are skin irritations caused by parasitic penetration, allergic responses produced by the eggs of the parasite, and obstruction of the peripheral lymphatic circulation by the adults. Irritation, allergy, and obstruction are also seen in opisthorchiasis, but here the primary effects, excluding those caused by physical blockage, are due mostly to the toxic secretions of the parasites. The deposited eggs of *Schistosoma* may spread through the circulation to distant parts of the body, such as the brain and lungs, causing lesions and functional impairment. *Paragonimus* differentially affects the lungs, eliciting symptoms not unlike those of tuberculosis.

**Diagnosis** The clinical diagnosis of most platyhelminthic disorders is confirmed by identifying one or

| Causative agent and disease | Mode of transmission, portal of entry, and foci of infection | |
| --- | --- | --- |
| *Schistosoma* spp. Fluke  Male  Female | Schistosomiasis Bilharziasis <br><br> Integumentary portal <br> Early aggregation <br> Development in portal veins <br> Sexual reproduction in bladder or rectum | 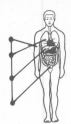 |
| *Opisthorchis sinensis* Fluke  | Opisthorchiasis <br><br> Digestive portal <br><br> Definitive infection in bile passages | 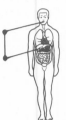 |
| *Paragonimus* spp. Fluke  | Paragonimiasis <br><br> Digestive portal <br> Excyst in duodenum <br><br> Definitive infection in lungs | 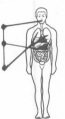 |
| *Taenia* spp. Tapeworm  | Taeniasis <br><br> Digestive portal <br><br> Brain and liver developmental sites | 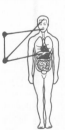 |
| *Dibothriocephalus latus* Tapeworm  | Dibothriocephaliasis <br><br> Digestive portal <br><br> Definitive infection in intestine | 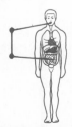 |

**Figure 3-6** *Major diseases of man caused by flatworms.*

| Symptoms | Detection | Prophylaxis |
|---|---|---|
| Dermatitis when entering skin; diarrhea, cystitis, and hematuria; tumors and allergic reactions to presence of eggs | Identify ova in feces or ova and larvae in urine; complement-fixation and skin tests for chronic cases | Snail control; avoid swimming in water where snail hosts are present; avoid water contamination with schistosome eggs |
| Irritation; secondary effects from toxic secretions | Identify ova in feces | Snail control; avoid raw or improperly cooked fish; sanitation in waste disposal |
| Tuberculosis-like infection; intermittent coughing and blood-stained sputum | Chemical tests and x-ray diagnosis | Avoid raw or improperly cooked crabs or crayfish; snail control; water sanitation |
| Gastrointestinal and nervous upsets; anemia and lowered vitality | Identify ova or proglottids in feces | Sanitation with food and water and in waste disposal; avoid contact with infected animals |
| Gastrointestinal and nervous upsets; anemia, induced vitamin deficiencies; anorexia, lowered vitality | Identify ova or proglottids in feces | Sanitation with food and water and in waste disposal; avoid contact with infected fish; avoid eating raw or improperly cooked fish |

more life cycle stages, body fragments, or proglottids in the feces. Usually the ova are diagnostic in size, shape, color, texture, and the presence or absence of both an embryo and an operculum, or cap, through which the embryo escapes. Sometimes, however, the ova are not detected during smear examinations and require more difficult separation techniques, such as flotation or sedimentation. The most distal segments of tapeworms continually slough off in the intestinal lumen and are readily found, usually still squirming, in the feces. Chronic schistosomiasis is occasionally diagnosed with antigen-antibody tests such as complement-fixation and specific skin reactions, while x-ray diagnosis usually reveals the presence of *Paragonimus* in the lungs.

Control Prophylaxis against parasitic platyhelminths is a two-pronged effort based mainly on the adequacy of food preparation and the control and regulation of intermediate hosts. Snails, crabs, crayfish, and various bony fish serve as intermediate hosts of liver and lung flukes. Sheep and swine serve similar roles for tapeworms. Because these animals may harbor the parasites, their human consumption should be either avoided or preceded by sufficient cooking to kill any encysted, immature flatworms.

Divergent cultural attitudes, however, can be potent forces resisting such well-meant recommendations. Where this situation occurs, of course, fluke and tapeworm infestations are usually endemic to a geographic region and its human population.

Several other prophylactic measures may then be instituted. For instance, attempts may be made to kill the larval stages of the parasite either within the intermediate host population or in the surrounding, usually aquatic, environment. The parasite may also be eliminated from a geographically localized human population if one or more of the intermediate hosts is completely exterminated. Which treatment, if any, is undertaken depends in part on how extensively the intermediate target host is used as food and how close to human habitation the intermediate hosts are found.

Prophylaxis by avoidance is also a useful measure. Aside from avoiding food infested with the larval stages of parasitic flatworms, avoidance may necessitate the strict observance of rules prohibiting swimming in water where intermediate hosts such as snails are normally found. Bypassing the geographic area completely may also be effective, depending on the likelihood of the human population's serving as a definitive host. When the intermediate hosts are domestic animals such as dogs, cats, and sheep, avoidance may take the form of preventing direct contact between the carrier and the mouth. This type of avoidance prevents the transfer of eggs between hosts, a common mode of transmission of tapeworms to man. Obviously, any measures which help to avoid contacting or ingesting water, sewage, and soil contaminated with the larvae of parasitic flatworms will have some prophylactic value.

## ROUNDWORMS

The wormlike appearance, complexity of life cycles, reproductive fecundity, tortuous migrations within human beings, portals of entry and of exit, general symptoms, laboratory diagnosis, and prophylaxis of all flatworms and roundworms are similar. Where group differences do occur, they are more quantitative than qualitative, i.e., in the roundworms, a shift in the relative importance of the portals of entry as reflected by the increased usage made of insect vectors, a larger variety of disease symptoms and a greater frequency of skin reactions, and corresponding changes in the emphasis placed on specific prophylactic techniques. The sexes are also separate in all parasitic roundworms, in contrast to the uniformly bisexual nature of parasitic flatworms.

Among the more common or well-known diseases of man caused by parasitic roundworms are trichinosis, hookworm disease, filariasis, ascariasis, and enterobiasis, or pinworm disease. Information about these and other major aschelminthic diseases is compiled in Fig. 3-7.

Portals of entry and exit Over half of the parasitic aschelminths gain entry into the body through the skin. Skin penetration is accomplished by active boring, e.g., by the hookworms *Ancyclostoma duodenale* and *Necator americanus,* or during the acts of biting by an insect vector, e.g., by microfilarial worms. The remainder enter through the mouth into the digestive tract. Here transmission relies upon the ingestion of contaminated food, water, dirt, or other debris on fingers placed in the mouth. Several flies serve as intermediate hosts of parasitic roundworms. In lieu of true intermediate hosts, other roundworms utilize alternative hosts or are spread from person to person either directly or indirectly.

The most common portal of exit is the anal orifice from the digestive tract. Well over half of the parasitic roundworms, whether as eggs, larvae, or adults, use an anal escape route almost exclusively. Other, more accidental departures sometimes occur. Examples include exiting from opened skin nodules in the case of *O. volvulus,* escaping through the mouth during a cough in the cases of *Ascaris lumbricoides, Trichinella spiralis,* and *Strongyloides stercoralis,* or leaving from the genitals in the case of *Wuchereria bancrofti.*

Migrations  Once infective larvae enter the body, they migrate extensively before reaching their definitive habitats. The parasites reach sexual maturity in these locations. The migrations are indicative of the high degree of physiologic and behavioral specializations which are hallmarks of successful parasitism. In diseases such as hookworm, strongyloidiasis, and ascariasis, the larvae after hatching from the eggs ultimately enter the blood. The larvae are then transported to the lungs and are either coughed out of the body or swallowed again. In the lumen of the small intestine, they reproduce.

The trichina worm, *Trichinella spiralis,* enters the digestive tract in improperly cooked pork. Thus, pigs are alternative but not true intermediate hosts in which the infective larvae exist until eaten. Once ingested by human beings, the larvae mature into adults and reproduce in the small intestine. The progeny find their way into and through the blood to all parts of the body, although they ordinarily become encysted in the cells that make up the voluntary muscles. Such encystment closes the door to future reproductive opportunities, although the larvae are reportedly capable of remaining viable for several years in this condition. The larvae rather than the

adults are the culprits that create the havoc of trichinosis through pathologic conditions resulting from nutritive injuries, physical damage, and toxic excretions.

The pinworm *Enterobius vermicularis* and the whipworm *Trichuris trichiura* experience less complicated migrations than those described above. Both enterobiasis and trichuriasis are contracted either by ingesting water containing embryonated eggs or from hands contaminated from contact with moist soil. *Enterobius vermicularis* reaches maturity and reproduces in the small intestine. Oviposition by the fertilized females requires their migration to the perianal region, in a manner reminiscent of the parasitic flatworm *Schistosoma.*

The internal development of the whipworm is even more direct, albeit slow. After being swallowed, the larvae hatch from the eggs, reach sexual maturity in the intestines, and finally mate in the cecum of the large intestine.

Following entry into their definitive host, microfilariae are usually found in peripheral blood. The larvae, however, are not directly associated with the production of disease symptoms. Clinical manifestations of disease are produced primarily by adult filarial worms. The lymphatics and connective tissue are the loci of infection for *W. bancrofti.* Skin nodules on the face and neck are commonly formed by *O. volvulus.* The subcutaneous connective tissue, including the conjunctiva of the eye, is the intended or accidental target of *Loa loa.*

Symptoms  Numerous disease symptoms are elicited in human beings by aschelminthic parasites. The symptoms involve both peripheral and deep tissues and organs. Those parasites using a portal of entry on the skin initially provoke inflammatory and allergic skin responses such as itching, edematous swellings, and localized eruptions or nodules. Some of the filarial worms may cause a creeping itch by burrowing beneath the skin, irritate or damage the conjunctiva of the eye, possibly causing blindness, or promote secondary invasions by pathogenic bacteria. The causative agents of hookworm disease, although entering through the skin, produce the host's major

Figure 3-7 *Major diseases of man caused by roundworms.*

| Causative agent and disease | Mode of transmission, portal of entry, and foci of infection | Symptoms | Detection | Prophylaxis | Vector |
|---|---|---|---|---|---|
|  **Ascaris lumbricoides** Ascariasis<br><br>Adult female | Digestive portal<br><br>Immature stages migrate from blood to lungs and back to intestine<br><br>Definitive loci of infestation | 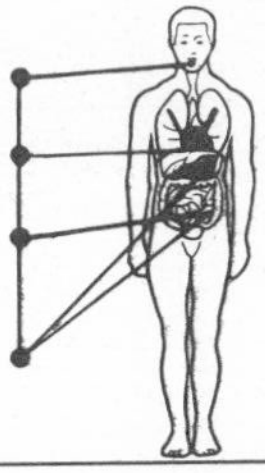 Pneumonitis; anemia and fever, if heavily infected | Identify ova in feces or find adults during treatment | Sanitation in waste disposal | |
| 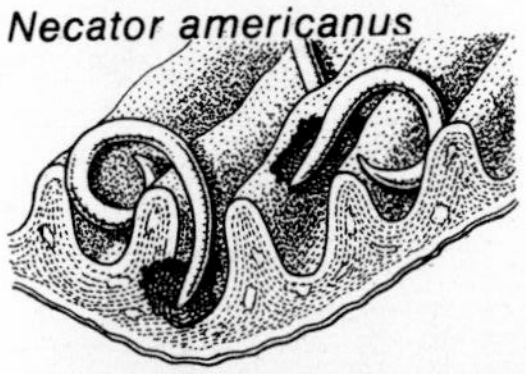 **Ancylostoma duodenale** **Necator americanus** Hookworm disease | Integumentary portal<br><br>Immature stages migrate from blood to lungs<br><br>Definitive locus of infestation | 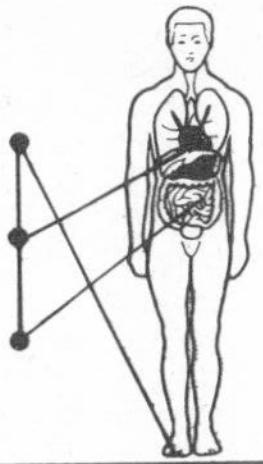 Gound itch and localized eruption at site of entry; anemia, when severely infested; abscesses at site of intestinal attachment | Identify ova in feces | Sanitation in waste disposal; avoid walking barefoot | |
| Hookworms causing intestinal lesions | | | | | |
| **Trichinella spiralis** Trichinosis<br><br>Encysted larvae in striated muscle | Digestive portal<br><br>Locus of development and reproduction<br><br>Progeny migrate from blood to muscles<br><br>Muscle encystment of progeny | 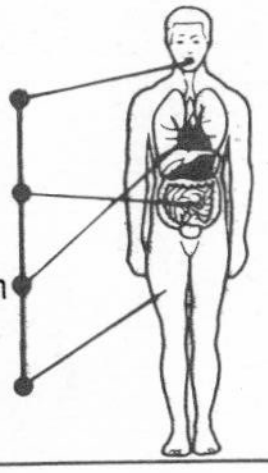 Abdominal pains, nausea, vomiting, and diarrhea; fever, edema, and eosinophilia; muscular pains and pneumonia | Identify causative agent in feces, suspected food, or in muscle biopsy tissue | Avoid eating improperly cooked pork | |
| **Strongyloides stercoralis** 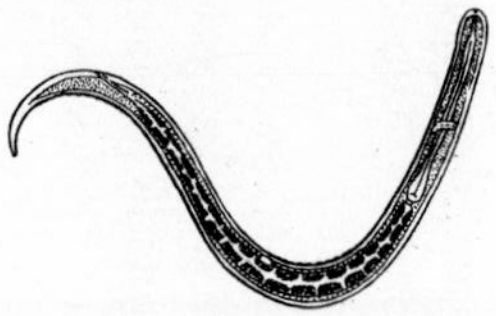<br><br>Mature larva | Integumentary portal<br><br>Immature stages migrate from blood to lungs<br><br>Immature stages leave lungs and are swallowed<br><br>Definitive locus of infestation and reproduction | 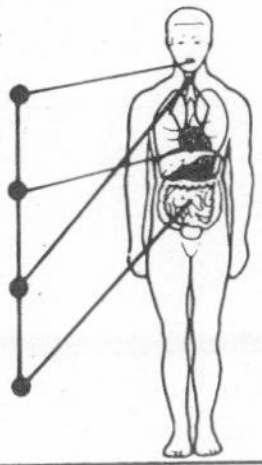 Redness and intense itching; pulmonitis and epigastric pains; eosinophilia, nausea, vomiting, diarrhea, and weight loss | Identify larvae in feces | Sanitation in waste disposal; avoid skin contact with contaminated soil | |

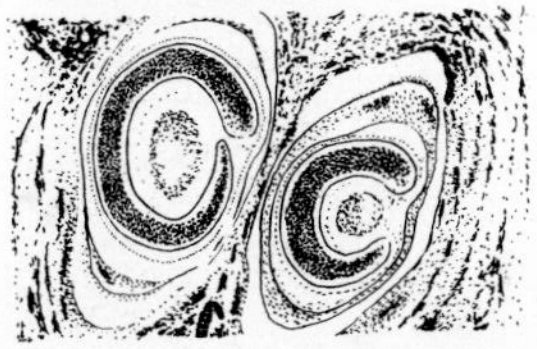

| *Enterobius vermicularis* Pinworm | *Trichuris trichiura* Whipworm | *Wuchereria* spp. | *Onchocerca volvulus* | *Loa loa* in eye |
|---|---|---|---|---|
| Enterobiasis | Trichuriasis | Filariasis | Onchocerciasis Filariasis | Loiasis Filariasis |
| Digestive portal | Digestive portal | Integumentary portal via mosquito bite | Integumentary portal via black fly bite | Integumentary portal |
| Immature development and reproduction | Immature development | Microfilariae in peripheral blood | Definitive locus of infection | Definitive locus of infection in subcutaneous connective tissue |
| Oviposition by fertilized female in perianal region | | Adults in lymphatics and connective tissue | | |
| Rectal irritation and anal itching, especially in children | Nutritional and digestive disturbances; anemia and eosinophilia; muscular aches and dizziness; abdominal pain and discomfort | Enlargement and inflammation of lymphatics; elephantiasis and fever in acute cases | Nodules on face and neck and itching skin; possible blindness from secondary infection | Itchy, edematous swellings at foci of infection; skin and eye irritations and creeping pains |
| Identify ova in feces or adults in perianal region | Identify ova in feces | Identify microfilariae in blood and lymph | Identify microfilariae in skin or nodules | Identify microfilariae in blood |
| Sanitation with food and in waste disposal | Sanitation in waste disposal; personal hygiene | Mosquito control; treatment of carriers | Black fly control | Mango fly control |
| | | *Culex* spp. | *Simulium* spp. | *Chrysops dimidiata* |

Embryonated eggs around anus

Adult male

Microfilarial worm in blood

Open nodule with worms

Microfilarial worm in eye

disease symptoms by sucking blood while attached to the mucosa of the small intestine. Physical damage to the tissue lining the lumen leads to necrosis and the formation of abscesses at the sites of attachment. The continual drainage of blood from the host results in metabolic imbalances, leading especially to anemia and lowered vitality.

Three overlapping syndromes of symptoms are associated with those aschelminthic parasites using the digestive portal of entry. Larval migration from the tissues through the blood to and within the lungs may be responsible for pneumonitis and pneumonia. Intestinal roundworms, irrespective of their portal of entry, may cause nutritional and digestive disturbances including epigastric and abdominal pains and discomfort, nausea, vomiting, diarrhea, and subsequent dehydration and loss of weight. Parasites which encyst in the muscles or aggregate in large numbers at particular loci of infection such as the cecum are capable of inducing fever, muscle aches and pains, and possible nervous system effects resulting in dizziness. Although pinworms are swallowed and enter the digestive tract, their most frequent effects are a rectal irritation and an anal itching. These conditions are caused from abrasions of the perianal mucosa by the eggs which the females deposit after mating in the intestine.

Diagnosis Laboratory diagnosis of aschelminthic parasitism is based mainly on the isolation and identification of the eggs in fecal smears or from fecal samples after flotation or sedimentation. Sometimes the larvae and adults may be found in either the feces, perianal region, skin nodules, or sputum coughed up from the throat. Microfilarial worms can occasionally be recovered from the blood, lymph, or skin. Severe cases of trichinosis are diagnosed by performing a biopsy on a portion of voluntary muscle, such as the diaphragm, but lighter infections with less pronounced symptoms may yield negative results. Therefore, the inability to recover the causative agent from the host does not necessarily support the conclusion that the disease is not the result of parasitism. For this reason, laboratory diagnosis may profit from examinations of presumably contaminated food, water, or soil with which the host has come in contact, if such samples are available.

Certain roundworm diseases are contracted as a by-product of poverty and corresponding undernutrition and inadequate clothing. The effects of these parasitic diseases contribute to carelessness in both personal habits and cleanliness as well as a general lethargy. These factors combine in a vicious circle to further increase the degradation of the afflicted.

Hookworm disease literally gains its foothold when hookworm larvae in the topsoil bore through the bare feet of children, who make up most of its victims. Ascaris and whipworm infections arise from the ingestion of moist soil containing eggs with embryos of these parasites. The soil may literally be eaten by poverty-stricken youngsters or taken in the mouth with the fingers or in contaminated water.

Control Effective prophylaxis against hookworms may be secured by wearing shoes or some protective covering for just the heels and soles. Preventing the contamination of soil by human feces with the use of minimum sanitary facilities such as outhouses will reduce the incidence of both ascaris and whipworm infections. What is not so obvious is the prophylactic effect of an adequate and well-balanced diet in warding off potential hookworm infections or in reducing the severity of infections already contracted.

Avoidance is as important for the prevention of roundworm invasions as it is against flatworms. For example, infection by *Trichinella spiralis* may be avoided by abstaining from eating pork and pork products or refusing such foods when uncooked or otherwise improperly prepared. The probability of infection increases with the frequency of pork meals eaten and with the purchase of pork from local slaughterhouses or farms.

The insect vectors of the various microfilarial worms are amenable to control. Control may be effected with (1) the use of insecticidal sprays, (2) the elimination of stagnant ponds and pools and other aquatic environments which form their breeding grounds, (3) the use in endemic areas of protective materials such as mosquito netting and repellents applied to the skin, and (4) isolation of the insect vectors from human habitation. The long-term adverse environmental effects of applications of insecticides must always be weighed against the shorter-term public health and economic benefits which result from their use.

One other prophylactic measure of some significance is the prompt treatment of human carriers. Culicine mosquitoes, for example, become infected when sucking human blood harboring filarial worms. Treatment of such carriers would prevent the infective stage of the parasite from ever developing in the mosquito. Pinworm infections are another case in point. After contaminating their hands with eggs by scratching irritated rectal regions, children infected with *Enterobius vermicularis* either reinfect themselves when biting their nails or infect others when shaking hands or by other forms of interpersonal contact. Another common mode of transmission of *Enterobius vermicularis* is by contact with clothing, bedding, and air contaminated with its minute but abundant eggs. One highly infected carrier can disseminate the parasite to an entire family. Early treatment of such carriers is obviously essential.

## ARTHROPODS

Excluding their roles in microbial disease transmission, entomophobia, or fear of insects, and poisoning with defensive secretions, various insects and other arthropods are capable of directly causing pathological conditions. Such conditions appear as (1) urticarias, or irritations and itchings of various sorts on the skin; (2) swellings or wheals at the site of an arthropod bite; and (3) accidental and opportunistic infestations of fly maggots, a condition known as *myiasis,* within the nostrils or along the urogenital orifices and other moist mucous membranes.

Pathogenic arthropods are either obligatory or facultative ectoparasites or endoparasites. *Obligatory parasites* must continually obtain their energy at the expense of the host. For *facultative parasites* the host is not the sole food source. *Ectoparasites* are found on the surface of the skin; *endoparasites* live symbiotically in the skin or within the body. Arthropod parasites induce adverse reactions through secreted or excreted toxins, introduced allergens, mechanical injury or obstruction, and nutritional injury.

Among the more well-known arthropod diseases are the itch or mange, pediculosis, and various dermatoses. Information on these and other major arthropod diseases of human beings is reviewed in Fig. 3-8.

**Transmission** The list of causative agents is notorious, as suggested by the range of human emotions elicited when mentioning names such as the human lice, *Pediculus humanus capitis* and *corporis*; the human flea, *Pulex irritans*; the bedbug, *Cimex lectularius*; and the chigger, *Eutrombicula alfreddugesi.* Bedbugs, fleas, and piercing and biting flies including mosquitoes are facultative ectoparasites. Flies which lay their eggs on or in the body are also facultative parasites. In contrast, mites and lice which cause the various arthropod-linked dermatoses are obligatory parasites tenaciously clinging to their human hosts after successful transmission.

Pathogenic arthropods, even if small and therefore not necessarily observed, are nevertheless all macroscopic. *Sarcoptes scabiei,* the itch, or mange, mite, and *Demodex folliculorum,* the follicle mite, are just noticeable to the human eye. The mange mite burrows in the skin where the female lays its eggs. The follicle mite resides in hair follicles beneath the surface of the skin. The chigger and human louse not only take advantage of their small size but capitalize on the inaccessibility or concealment provided by their habitats on the body. Fleas and bedbugs have other mechanisms to maintain anonymity. Fleas can hide in host hairs, especially on the head and in the inguinal region, moving rapidly in response to their displacement. Bedbugs invade their sleeping hosts from the mattress. Flies which occasionally lay their eggs on or in the body usually exploit the host's insensitivity. The eggs may be laid while the host, usually naked and uncovered, is asleep, unconscious, or suffering from some type of incapacitation. Occasionally the eggs are attached by the flies to the body hairs of a host in spite of their host's apparently alert and wakeful state. The larvae, or maggots, which hatch from the eggs are exceedingly small and ordinarily go undiscovered

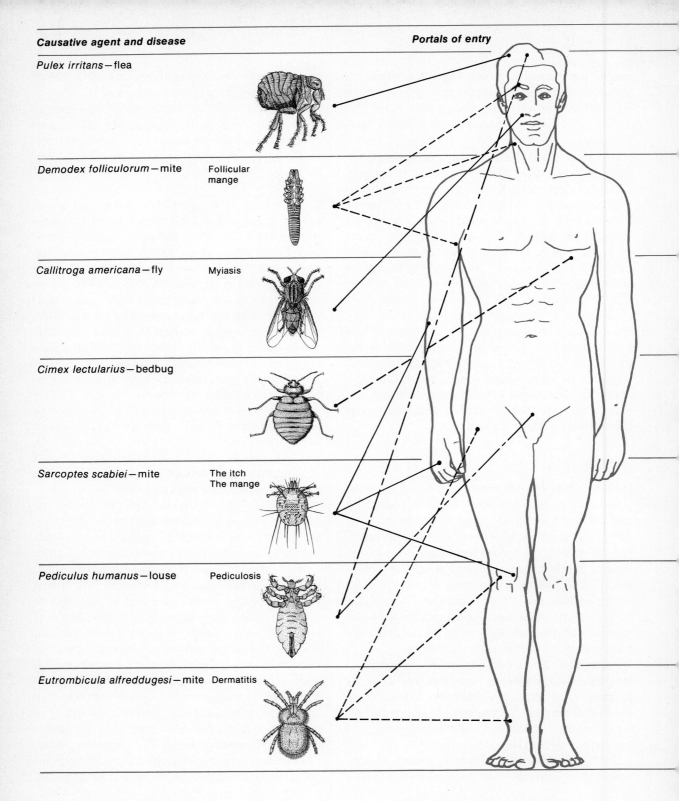

**Causative agent and disease**

*Pulex irritans* — flea

*Demodex folliculorum* — mite    Follicular mange

*Callitroga americana* — fly    Myiasis

*Cimex lectularius* — bedbug

*Sarcoptes scabiei* — mite    The itch The mange

*Pediculus humanus* — louse    Pediculosis

*Eutrombicula alfreddugesi* — mite    Dermatitis

**Portals of entry**

| Symptoms | Detection | Prophylaxis |
|---|---|---|
| Painful bites and skin irritation after lying on infested rug or mattress | Identify fleas on skin | Flea control in animals; personal hygiene and sanitation |
| Inflammation of the sebaceous glands; production of acne-like conditions | Detection is difficult because follicle mites are deep within skin | Personal hygiene and sanitation |
| Nasopharyngeal and wound infestations of larvae hatched from eggs laid at sites of putrefaction or from accidental contact with infested animal | Identify maggots at site of infestation | Fly control in animals; personal hygiene and sanitation |
| Marked swellings and considerable irritation; secondary bacterial infection at site of wheal; nervous and digestive disorders | | Bedbug control in animals and contaminated materials; personal hygiene and sanitation |
| Irritation and hypersensitivity; secondary bacterial infection at site of lesion | Identify characteristic burrows in skin | Avoid skin contact with carriers |
| Irritation and hypersensitivity; secondary bacterial infection at site of lesion | Identify ova or adults in hair, body, or clothing | Personal hygiene and sanitation |
| Intolerable itching; severe dermatitis with pustules and wheals | Identify bright red mite barely visible to naked eye; may have 3 or 4 pairs of legs | Chigger control on lawns and in gardens; chigger repellents |

until they have molted several times, each time increasing in size. The parasites are finally noticed when they obstruct or otherwise interfere with essential physiologic processes.

The injuries caused by parasitic arthropods occur mostly after they pierce or bite the host's skin with their mouth parts. In some cases of myiasis, fly maggots gain entry into the body from the nose, urogenital orifices, or exposed wounds.

**Symptoms** Arthropod diseases run a course not unlike those of other microbial pathologic conditions, except that the signs and symptoms are more relegated to the skin and less likely to become general infections. Usually the first sign is a localized inflammation coupled to an irritation which may take the form of a mild to severe itching. The skin reactions are the result of hypersensitivities, or allergies. Lesions, pustules, and wheals may appear where the parasite penetrates the skin. The bite itself can be painful, aside from effects attributable to injected toxins and introduced allergens. In addition, skin penetration opens up a Pandora's box of secondary microbial invaders, the most prominent of which are bacteria. Secondary infection is therefore a significant indirect consequence of an arthropod bite.

**Diagnosis** The laboratory diagnosis of arthropod infestations and infections is built around the recovery of the causative agent from preparations obtained from the host. Samples for microscopic analysis are usually secured from the site of the bite or from burrows, hair follicles, head or body hairs, and other suspected habitats on or in the host. When mites, lice, and fleas are suspected, the host's clothing, including hats and underwear, may also be examined. Indirect evidence of piercing and biting arthropods is usually possible to ascertain, e.g., when characteristic skin burrows are observed or when unmistakable puncture marks and diagnostic swellings are found.

**Control** The best prophylactic measure, avoidance of potentially harmful arthropods, is also the most impractical to apply. Mosquitoes, biting flies, bugs, fleas, lice, ticks, and mites are, after all, widely dispersed and generally close to human habitation. However, there are certain actions that may reduce the probability of coming in direct contact with them. First, offensive measures can be instituted such as treatment of pest insect and acarine breeding grounds as well as of lawns and gardens with appropriate pesticides and other pest control measures. Control of their population size and geographic distribution and of their animal reservoirs is ordinarily an integral part of the total program. Second, defensive measures of a more personal nature may at the same time be taken to avoid situations with a high infestation potential, such as sleeping on a bed in a deteriorating rooming house, walking through woodlands with considerable brush without adequate protective clothing, or coming in direct contact with potentially parasitized animals without observing the fundamental rules of personal hygiene. Against mosquitoes and biting flies, repellents applied to the skin and protective materials such as mosquito netting are among the defensive tools available.

Taken collectively, prophylaxis against parasitic arthropods is effective to the extent that long-standing balances are established between human hosts and their arthropod parasites. Equilibrium, however, is not static. Thus, a perennial battle continues in which human intelligence is pitted against an almost infinite number of arthropods and their high levels of genetic plasticity.

## COLLATERAL READING

Bernarde, Melvin A., *Our Precarious Habitat,* New York: W. W. Norton & Company, Inc., 1970. Several chapters are devoted to the roles of microbes in the ecology of health and disease and of bacteria in the causes of food poisoning.

Chandler, Asa C., *Introduction to Parasitology,* 8th ed., New York: John Wiley & Sons, Inc., 1950. This authoritative and witty textbook reviews parasites in general, giving detailed coverage to protozoa, helminths, and arthropods.

DeKruif, Paul, *Microbe Hunters,* New York: Harcourt, Brace & World, Inc., 1953. A description of the achievements of a selected group of microbiologists that conveys some of the excitement surrounding the discoveries.

Dulbecco, Renato, "The Induction of Cancer by Viruses," *Scientific American*, April 1967. Tissue culture techniques are used to study cancer induction by several viruses, each of which has less than 10 genes.

Greenberg, Bernard, "Flies and Disease," *Scientific American,* July 1965. Statistical and laboratory evidence is enlisted to support the view that sucking flies spread disease to human beings by passively carrying a variety of pathogens.

Herms, William B., *Medical Entomology,* 4th ed., New York: The Macmillan Company, 1953. Chapter 3 discusses parasites and parasitism. Chapter 4 explains how arthropods cause disease and carry microbial pathogens.

Kadis, Solomon, Thomas C. Montie, and Samuel J. Ajl, "Plague Toxin," *Scientific American,* March 1969. This endotoxin apparently impairs the ability of a cell to oxidize aerobically. Some of the experiments that lead to this conclusion are described in an exciting manner but at an advanced level.

Rapp, Fred, and Joseph L. Melnick, "The Footprints of Tumor Viruses," *Scientific American,* March 1966. A review of the role of viruses as carriers of foreign genetic information and its implications in the causes of malignant growth.

Rothschild, Miriam, "Fleas," *Scientific American,* December 1965. A discussion of experiments with the European rabbit flea which demonstrates its dependence on the sex hormone cycle of the host.

Wheeler, Margaret F., and Wesley A. Volk, *Basic Microbiology,* Philadelphia: J. B. Lippincott Company, 1964. A textbook that emphasizes the portals of entry of microbial pathogens.

Young, G., *Witton's Microbiology,* New York: McGraw-Hill Book Company, 1961. An elementary microbiology textbook that stresses nursing applications.

# CHaPTer 4
# BODY DeFense mechanisms

In earlier chapters, we traced the weaknesses which make man vulnerable to various kinds of biospheric pollution and, in addition, stressed the modes of transmission and portals of entry of pathogenic microbes which exploit these weaknesses. Aside from the potential threats posed by physical, chemical, and biologic agents, the body is subjected to mechanical injuries from bruises, abrasions, cuts, wounds, and other physical abuses of various dimensions. In most cases, the damage is a normal consequence of physical activity and repaired without awareness or medical intervention. Sometimes the injury is serious enough to require medical attention and may result in secondary microbial infection, in temporary or permanent loss of function of the damaged part, or in both types of impairment.

Following physical injury, body defense and homeostasis are accomplished by

1  The various chemical factors released by damaged tissue cells and blood platelets which initiate and sustain inflammation and clotting

2  The phagocytic actions of the numerous ameboid leukocytes during the inflammatory response
3  Interferon released from host cells following viral invasion
4  Antibodies synthesized by plasma cells in response to the entry of foreign particles, or antigens
5  The secretion of collagen by fibroblasts and the formation of new epidermis by epidermal cells that migrate into the wound, processes that occur during tissue regeneration
6  Hormones released from the hypothalamus, the adenohypophysis (anterior lobe of the hypophysis, discussed more fully later in this chapter), and the adrenal glands in response to neural stimuli of a stressful nature

Independent of, but supplementary to, these actions are the phagocytic activities of fixed macrophages lining the sinusoids of diverse lymphoid tissues. The entire collection of fixed macrophages is known collectively as the *reticuloendothelial system* (see

"Phagocytosis," later in this chapter).

Thus, there are basically three types of defense and homeostatic mechanisms: nonspecific, specific, and reparative. Among the *nonspecific mechanisms* are the skin in its role as the primary defense barrier of the body; the reticuloendothelial system for the filtration of blood and lymph; and the inflammatory response, including phagocytosis, for the cleansing and removal of damaged tissues and the neutralization of antigens. *Specific mechanisms* include the inhibition of viral replication by interferon prior to significant antibody induction as well as immunologic responses based on antigen-antibody reactions. Blood clotting, wound cleansing, and tissue regeneration are basic to the healing process and together make up the *reparative mechanism*.

The three processes outlined above overlap in time, space, and effect. These mechanisms form the subject matter of this chapter.

### THE INFECTIOUS DISEASE PROCESS: STEPS AND DETERMINANTS

Nowhere are the vulnerability and homeostatic mechanisms of the body more intertwined than during the invasion of a pathogenic microbe and the subsequent course of the host's response. The pathogen must be transferred to a new host through the environment. Resistance to its invasion is determined by the degree of exposure and the physiologic condition of the host, as well as by the integrity of the skin.

An intact skin is the primary defense barrier of the body, even when other portals of entry may be used as invasion routes. If necessary nutrients and physical conditions are present on the skin, beachheads are established for the subsequent growth of the pathogen.

The infectivity of the pathogen is determined by its ability to spread through the body and establish foci in specific target organs. The phagocytic actions of the reticuloendothelial macrophages and the ameboid leukocytes mobilized during the inflammatory response are among the nonspecific defense mechanisms quickly invoked by the host. These actions usually neutralize or destroy the pathogen and its potentially toxic products.

Virulence is in part reflected by the seriousness of the clinical symptoms. During this phase of the disease process, the host uses immunologic responses against the pathogen and its products, including the release of antiviral interferon from dying cells and the release from plasma cells of antigen-specific antibodies. These actions lead to the acquisition of a certain degree of immunity to further exposures of the same pathogen and to an attenuation of its immediate effects.

### THE SKIN ECOSYSTEM: CHARACTERISTICS AND INHABITANTS

The skin, or integument, is the first line of defense of the body. Actually a collection of specialized organs and tissues, the skin is a true organ system. Consisting of two major layers, an outer epidermis and an inner dermis, or corium, the skin contains a number of discrete environments, each with its own topography, climate, and inhabitants.

A generalized section of human skin is shown in Fig. 4-1. The *epidermis* is divided into two strata: an outer *stratum corneum* and an inner *stratum germinativum*. The stratum germinativum, or germinal layer, is a region of active cellular proliferation. All cells within the epidermis arise from the stratum germinativum. *Melanocytes* which give the skin its pigmentation are found in the basal layer. The cells in the variable layers of the stratum corneum above the stratum germinativum are either dead or dying and in various stages of cornification. The hardening and sloughing off of the outermost layers of the stratum corneum are due to normal wear and tear.

The *dermis* consists of connective tissue in which are found the bases of hairs and associated follicular structures, eccrine and apocrine sweat glands, sebaceous glands, blood and lymph vessels, nerve fibers,

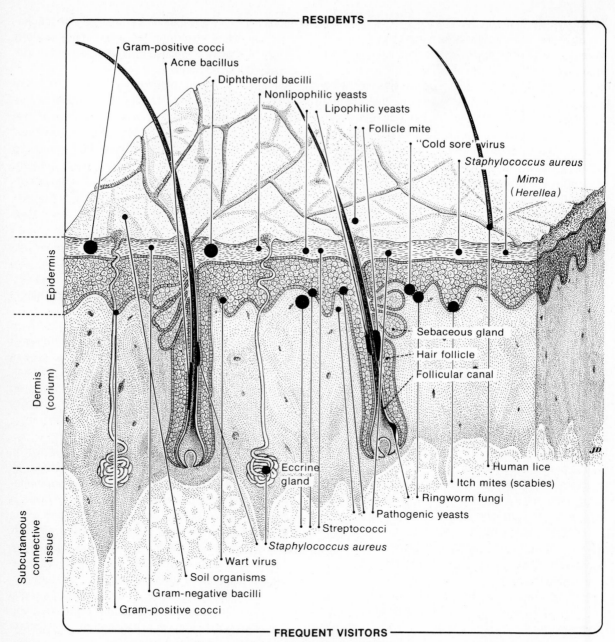

**Figure 4-1** *The major microbial symbionts found on or in the skin of human beings.*

and a smooth muscle called the arrector pili for the movement of each hair. *Subcutaneous connective tissue* separates the dermis from the underlying skeletal musculature.

Aside from defense, the skin functions in (1) the reception of stimuli such as touch, pressure, heat, cold, and pain; (2) the excretion from eccrine and apocrine glands of sweat containing water, salt, and nitrogenous materials such as urea; (3) the secretion from the sebaceous glands of sebum for the lubrication of integumentary hairs; and (4) the regulation of excess heat through the control of heat radiation from the surface of the skin.

As shown in Fig. 4-1, the normal flora and fauna of the human skin are either residents or frequent visitors. Such microbes are also divided into those which are usually harmless and those which are potentially pathogenic. Because these differences reflect probabilities and not necessarily actualities, the distinctions are quantitative rather than qualitative.

Among the usually harmless residents are bacteria such as gram-positive cocci and diphtheroid and acne bacilli, fungi such as *Candida albicans,* and arthropods such as *Demodex follicularum.* Potentially pathogenic residents include the bacteria *Mima polymorpha* and *Staphylococcus aureus* as well as the cold-sore virus. Various bacteria and soil organisms frequently visit the skin but generally do little or no damage. Equally frequent but more potentially damaging visitations are paid by staphylococci, streptococci, and other bacteria. Other potential pathogens include the fungi which cause ringworm, parasitic arthropods such as *Sarcoptes scabiei* and *Pediculus humanus* and the virus which causes warts.

Of the microbial species commonly associated with the epidermis, most are found exclusively in the stratum corneum. Within the dermis, the ducts of the eccrine glands, the follicular canals surrounding the papillae and shafts of the hairs, and the surrounding connective tissue may each contain one or more species of microorganism, although no microbe is confined entirely to this layer. In fact, over one-third of the microbes listed in Fig. 4-1 are capable of infecting more than one stratum of the skin. The most popular regions of microbial growth, the epidermis and follicular canals, are also those which contain the largest numbers of microbial species.

## WOUND HEALING

### BLEEDING AND COAGULATION

When the skin is broken, an exquisite and sequential series of chemical and physical reactions automatically takes over to heal the wound. The first steps in the repair process are local vasoconstriction, blood clotting, and clot retraction. These steps are illustrated in Fig. 4-2.

The sharp instrument shown on the left has severed a small blood vessel in the dermis. Within the first few seconds following the cut, the smooth muscle walls of the vessel contract spasmodically, reducing blood loss to a minimum. *Vasoconstriction* is under the control of the sympathetic nervous system (a division of the autonomic nervous system) and the chemical serotonin. Serotonin acts directly on the smooth muscle of blood vessels following its release from disintegrating blood platelets.

Within three to six minutes, the actions of a complex chemical mechanism result in blood clotting. The chemicals which participate are factors liberated from both perivascular tissue cells and blood platelets; specific blood proteins acting as precursors, activators, or enzymes during the clotting reactions; and certain minerals in the blood which also serve as activators.

Other chemicals prevent clotting under conditions where blood vessels are not damaged. Some of these chemicals are released from white blood corpuscles called *mast cells* in the connective tissue. Other chemicals which inhibit clotting are present in the blood. Some of these chemicals constitute a system for the enzymatic digestion of the fibrous clots as rapidly as they normally form.

Blood platelets are fragments of large white blood corpuscles called *megakaryocytes* located in the red bone marrow. In an intact blood vessel, platelets do not ordinarily adhere to the endothelial wall which

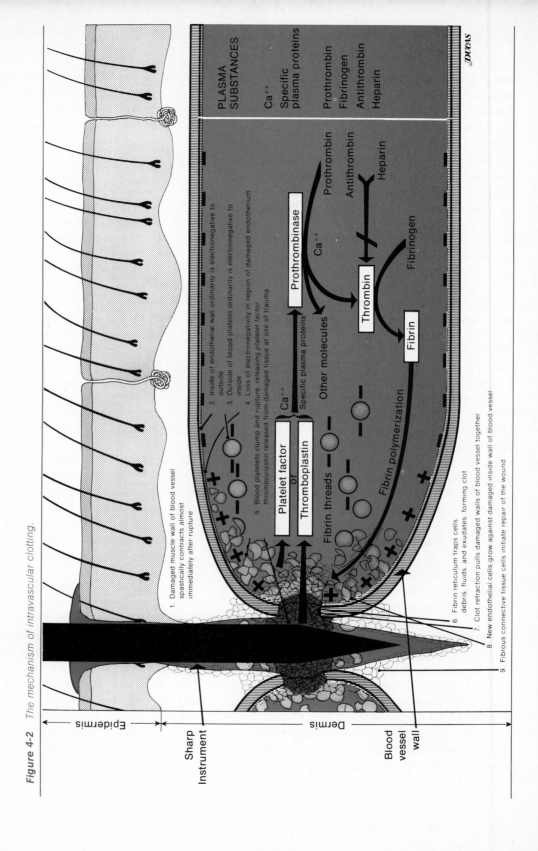

**Figure 4-2** The mechanism of intravascular clotting.

PLASMA SUBSTANCES

Ca++
Specific plasma proteins

Prothrombin
Fibrinogen
Antithrombin
Heparin

Prothrombin
Antithrombin
Heparin

Prothrombinase

Ca++

Thrombin

Fibrinogen

Fibrin

Platelet factor
or
Thromboplastin

Ca++

Specific plasma proteins

Other molecules

Fibrin threads

Fibrin polymerization

1. Damaged muscle wall of blood vessel spastically contracts almost immediately after rupture

2. Inside of endothelial wall ordinarily is electronegative to outside

3. Outside of blood platelet ordinarily is electronegative to inside

4. Loss of electronegativity in region of damaged endothelium

5. Blood platelets clump and rupture, releasing *platelet factor thromboplastin* released from damaged tissue at site of trauma

6. Fibrin reticulum traps cells, debris, fluids, and exudates, forming clot

7. Clot retraction pulls damaged walls of blood vessel together

8. New endothelial cells grow against damaged inside wall of blood vessel

9. Fibrous connective tissue cells initiate repair of the wound

Epidermis

Dermis

Sharp Instrument

Blood vessel wall

lines the lumen, because both bear negative electrostatic charges. When a vessel ruptures, however, the charge of its damaged endothelium is reversed, becoming positive and attracting the electronegative platelets. The pathologic adhesion of platelets to the endothelial wall leads to their agglutination, or clumping, and subsequent platelet rupture. Destruction results in the release of a substance called *platelet factor* into the blood plasma, in addition to serotonin, mentioned previously. Simultaneously, the damaged perivascular tissues release a factor called *thromboplastin*.

Each of these factors interacts with specific plasma proteins, in the presence of *calcium ions* ($Ca^{++}$) which activate the reaction, forming the active enzyme *prothrombinase*. When both prothrombinase and $Ca^{++}$ are available in adequate amounts, the inactive plasma protein, *prothrombin,* is converted into its active counterpart, *thrombin.* Because vitamin K is necessary for the synthesis of prothrombin, this vitamin is called the *antihemorrhagic factor.* Under the enzymatic control of thrombin, small protein molecules of *fibrin* form from their inactive plasma protein precursor, *fibrinogen.*

Intravascular clotting is normally prevented by the electrostatic repulsion between the platelets and the endothelial walls of the vessels, by the combined anticoagulant effects of heparin and antithrombin, and by the system of enzymes involved in destroying naturally formed clots. *Heparin* is a polysaccharide, or small carbohydrate, secreted in large concentrations by mast cells in the surrounding connective tissue following the release of platelet factor. *Antithrombin* is a blood protein which ordinarily prevents intravascular clotting by inhibiting the action of thrombin and thereby preventing the conversion of fibrinogen to fibrin. Heparin synergizes the action of antithrombin, in addition to inhibiting thrombin directly.

The fibrils of fibrin which result from the clotting reactions are only one molecule long. They subsequently polymerize, or link together, into *fibrin threads* consisting of gigantic chains of identical fibrin molecules. These threads contribute to a strongly cohesive, interlacing meshwork at the site of intravascular trauma, or damage. As the formed elements, fluids of the blood, and cellular debris are increasingly caught within its substance, the meshwork is progressively converted into a *clot.* When blood serum filters through and escapes from the meshwork, the clot retracts inwardly, drawing the injured walls of the blood vessel together. Excluding blood vascular effects, a similar mechanism forms a clot on the surface of the skin.

Both clots effectively wall off the site of the wound, closing the abnormal portals of entry and exit and preventing those microbes and other contaminants already introduced from spreading through the body. Oxidation and evaporation will cause a hardening and darkening of the surface clot, giving rise to a well-formed *scab* within a week following the injury. In the same time span, the ingrowth of fibrous connective tissue cells and new endothelial cells will repair the tissue damage directly surrounding the intravascular clot.

## INFLAMMATION

Blood coagulation, clotting, and the physical isolation of the wound and its contaminants are followed closely by a dramatic physiologic response called *inflammation.* The most readily detectable signs of inflammation are an elevated temperature, or *fever; redness* of the skin; fluid congestion, or *edema;* and *pain.* Internally and invisibly, a great number of inflammatory cells start a phagocytic cleansing of the wound following their mobilization from the blood and an ameboid, chemically directed attraction through the tissues to the site of injury.

The initial inflammation and later immunologic responses are augmented and adjusted by certain cells and natural products of the body. The chronological appearance of the various inflammatory cells, all of which are types of phagocytic leukocytes, and the plethora of chemicals which participate in the inflammatory response attest to the drill-like precision with which biologic search-and-destroy and cleanup operations are executed. The inflammatory response also contributes to the subsequent effectiveness of the reactions against foreign antigens which enter during or after the original injury.

The inflammatory and synergistic cells are illustrated in Fig. 4-3. The account which follows describes the inflammatory process and emphasizes its reliance on the interplay between molecules and cells.

The first generation of phagocytes to reach the scene of destruction from nearby blood capillaries are *polymorphonuclear neutrophils,* so-called because their nuclei are lobate and their small cytoplasmic granules are impervious to both acidic and basic dyes. The motile neutrophils phagocytically engulf and intracellularly digest about five antigens apiece before undergoing lysis because of their increased volume and depleted energy reserves. Neutrophils differentially respond to acute injury and infection. The resulting neutrophilia, or increased number of neutrophils, is of diagnostic significance in determining the severity of the insult sustained by the body.

On the cellular heels of the neutrophils come the second generation of ameboid inflammatory cells consisting of *lymphocytes* and *monocytes.* Lymphocytes possess spherical nuclei. Compared with them, monocytes are larger and contain more cytoplasm. Both cells are agranular. These cells swell to become tissue *macrophages* after phagocytizing cellular debris and foreign antigens. Monocytes are unique among inflammatory cells in that they have the ability to engulf and digest up to 25 foreign particles as well as the capacity to increase in number by cellular division at the site of damage. Both actions are adaptive and increase their overall operational efficiency. In contrast to the trouble-shooting role assigned to neutrophils, monocytes act in cases of chronic, or long-term, inflammation.

Microbial parasites are ordinarily attacked by inflammatory cells called *eosinophils.* Therefore, eosinophilia is diagnostic of parasitic infections. Like neutrophils, eosinophils are granulocytes. However, their granules are larger than those of neutrophils and differentially stain with the red dye eosin.

When antigens are present, *mast cells* intensify the immunologic response of the host by releasing a

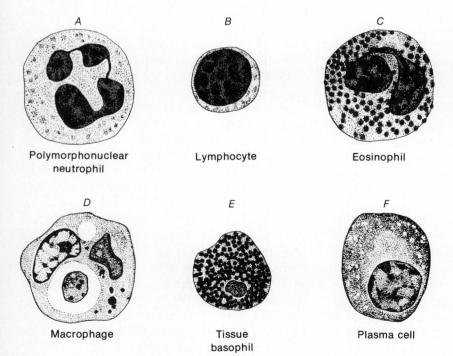

**Figure 4-3** *The cast of inflammatory white blood corpuscles. A. Neutrophil. B. Lymphocyte. C. Eosinophil. D. Macrophage. E. Mast cell, or tissue basophil. F. Plasma cell. All the cells are drawn to the same scale.*

Polymorphonuclear neutrophil

Lymphocyte

Eosinophil

Macrophage

Tissue basophil

Plasma cell

natural product called *leukocytosis-promoting factor* at the site of inflammation. The basophils lyse in response to the action of *histamine,* a chemical released by dead and dying cells. The basophilic mast cells are also granulocytic. Their cytoplasmic granules are the largest of those found in granulocytes and stain blue with basic or alkaline dyes.

*Plasma cells* differentiate from lymphocytic macrophages. Plasma cells proliferate in the tissues to form clones (colonies) of identical cells committed to the manufacture and release of specific antibodies, when the body is challenged by the introduction of antigens complementary to them.

The mobilization and subsequent migration of the inflammatory cells following injury are under chemical control. These chemical regulators include serotonin, which is released from disintegrating blood platelets as well as the chemicals necrosin, leukocytosis-promoting factor, and leukotaxine. These chemicals are liberated by damaged tissue cells.

*Serotonin* causes a decrease in the blood pressure of vessels near the site of trauma. The drop in blood pressure results in a congestion and localization of the blood in blood vessels near the injury and a corresponding retardation in the local blood flow rate. These effects aid in the mobilization of the inflammatory cells and their subsequent migration within the blood to the inside walls of the capillaries which enclose them. This peripheral migration is termed *margination.* The ameboid cells then actively crawl through the pores between contiguous endothelial cells. The active escape of the inflammatory cells through the walls of the blood vessels is termed *diapedesis.*

Under the influence of *leukotaxine,* the phagocytes, now located in the tissues, are directed chemotactically toward the source of injury. *Chemotaxis* is a general term for an attraction toward a chemical stimulus.

When tissues are damaged, *necrosin* is also released. Necrosin increases the permeability of blood capillaries. The altered permeability permits blood serum and plasma proteins to flood the extracellular tissue spaces, causing a local edema. This action is responsible for the tissue fluid congestion which ordinarily accompanies inflammation. The increase in extracellular tissue fluid provides raw materials and energy nutrients for phagocytosis, immunologic responses, and tissue regeneration.

The red bone marrow is also affected chemically during inflammation. The release of leukocytosis-promoting factor causes leukocytopoiesis, an increase in the production of white blood corpuscles by the red bone marrow. The resulting leukocytosis ensures an adequate inflammatory response.

The course of the inflammatory response is summarized in Fig. 4-4.

## PHAGOCYTOSIS

The business of the inflammatory cells is *phagocytosis,* the active ingestion of foreign materials. The ingested particles are neutralized or enzymatically destroyed within the phagocytes. Acting before immunologic responses are effective, the ameboid white blood corpuscles cleanse the wound of debris and antigens. Nevertheless, antigen-antibody reactions are still necessary to completely eliminate the antigenic residue left after the inflammatory response runs its course.

The ameboid movements of phagocytes are accomplished by pseudopodia, streaming protoplasmic extensions of the cell itself. Although the protoplasmic transformations which result in differential streaming are extremely complex and as yet not completely understood, the results of this type of locomotion are easily observed. Phagocytosis is diagrammed in Fig. 4-5. Several pseudopodia from different regions of the phagocyte project toward the antigen. The progressive displacement of the cytoplasm draws the rest of the cell in the same general direction. Cellular contact with the particle causes it to be surrounded and finally completely engulfed by several pseudopodia. A thin film of water, taken in during the ingestive process, encircles the antigen within the cytoplasm of the phagocyte. This action is repeated with other antigens until the phagocyte bursts and, in its turn, becomes cellular debris for further phagocytic actions.

Phagocytosis is enhanced by rough surfaces,

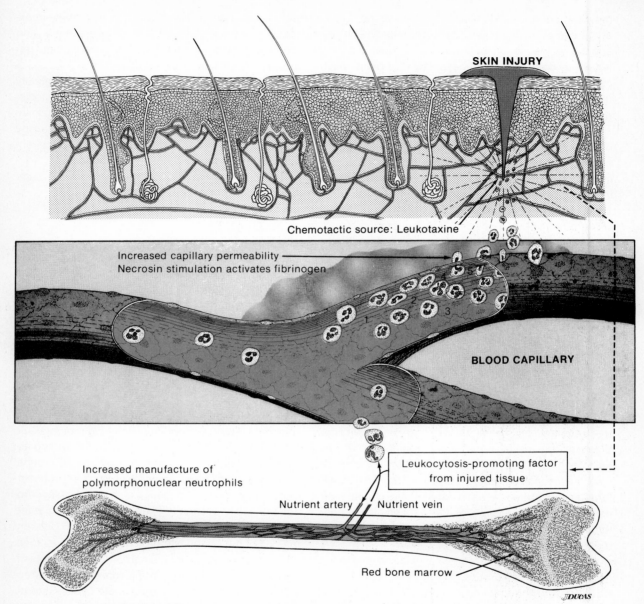

**SKIN INJURY**

Chemotactic source: Leukotaxine

Increased capillary permeability
Necrosin stimulation activates fibrinogen

**BLOOD CAPILLARY**

Increased manufacture of
polymorphonuclear neutrophils

Leukocytosis-promoting factor
from injured tissue

Nutrient artery | Nutrient vein

Red bone marrow

JDUCAS

**Figure 4-4**  *The course of the inflammatory response. A damaged blood vessel is exposed longitudinally. To emphasize the vascular and chemical events which occur during inflammation, the skin, blood vessel, and bone have not been drawn to the same scale.*

against which foreign materials can be pinned by phagocytes prior to engulfment. This type of cellular ingestion is called *surface phagocytosis*. The relative lack of rough surfaces in ulcerous or abscessed tissues explains why the phagocytic cleansing of such tissues is usually not very effective. On the other hand, surface phagocytosis within the blood is of primary importance in reducing certain types of septicemia.

Encapsulated bacteria present smooth surfaces to phagocytes, making their engulfment difficult. To offset this limitation, type-specific antibodies are manufactured and secreted. Called *opsonins*, they increase the phagocytic abilities of macrophages. Through the hydrolytic actions exerted by their lysosome enzymes, macrophages are frequently able to destroy antigens bound to opsonin antibodies. Opsonins have no apparent effect on encapsulated bacteria by themselves, however, and can exert their in-

**Figure 4-5**    *The mechanism of phagocytosis. A. A phagocyte and foreign particle. B. and C. The phagocyte moves toward the antigen, engulfing it phagocytically. D. The antigen is destroyed intracellularly.*

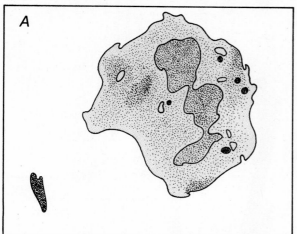

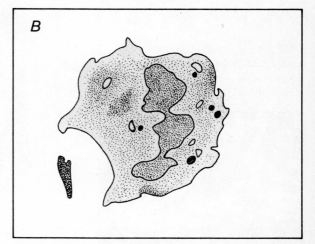

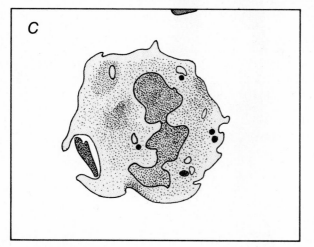

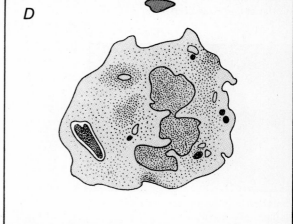

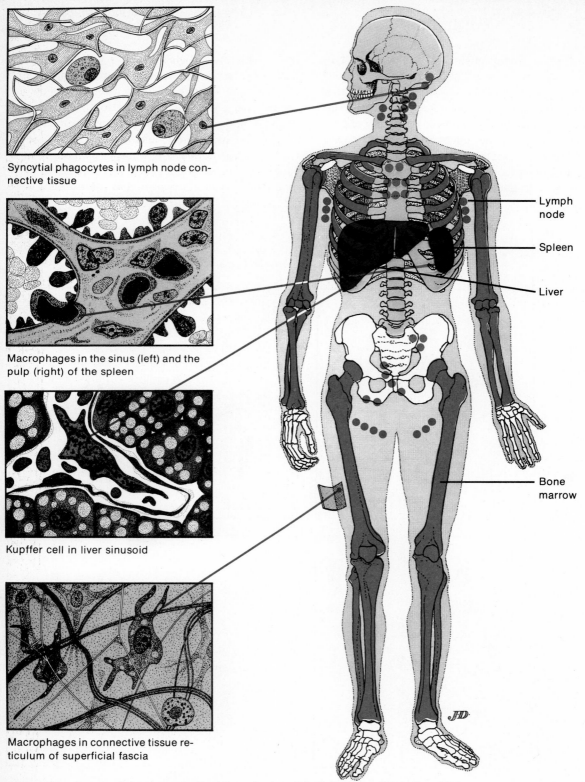

Syncytial phagocytes in lymph node connective tissue

Macrophages in the sinus (left) and the pulp (right) of the spleen

Kupffer cell in liver sinusoid

Macrophages in connective tissue reticulum of superficial fascia

Lymph node

Spleen

Liver

Bone marrow

**Figure 4-6** *Some cellular components of the reticuloendothelial system. The cells, while ameboid, remain in place.*

fluence with or without complement, a nonspecific blood factor.

In addition to the inflammatory macrophages, the body possesses a widely spread collection of fixed, or nonmotile, macrophages, the *reticuloendothelial system*. These macrophages are usually associated with structures that filter foreign materials from the body fluids before they can enter the general circulation. The components of the reticuloendothelial system are illustrated in Fig. 4-6 and include Kupffer cells lining the hepatic sinusoids, syncytial phagocytes or reticulum cells within the lymph nodes, tissue histiocytes associated with the bone marrow, and macrophages of the spleen and superficial fascia. Because the main entry routes to the systemic circulation are guarded by this diverse filtration system, the blood is ordinarily kept in a sterile condition.

## INHIBITION OF VIRAL REPLICATION

If an infection is caused by a virus, antibodies manufactured specifically for its antigenic determinants will be secreted by clones, or colonies, of committed plasma cells. However, the proliferation of plasma cells and the induction of antibodies and their subsequent increase in blood concentration require a significant period of time, relative to the onset of the critical phase of a serious infection. During this black-

out period of from three to seven days in which the body is not insured immunologically against the ravages of intracellular viral parasites, temporary protection is provided by a chemical called *interferon*. Released by most cells following their viral-induced cytolysis, interferon prevents the destruction of neighboring cells by temporarily inhibiting intracellular viral replication after their infection.

Figure 4-7 offers a possible mechanism by which interferon temporarily protects cells against the effects of viruses until antibodies can neutralize or destroy them. The entry of a virus into a cell stimulates either its nucleus or its cytoplasm to synthesize and excrete interferon. Simultaneously, the virus replicates intracellularly, ultimately lysing the cell. The phage particles escape, to reinfect other cells. The liberated interferon molecules become bound to the newly infected cells and induce the production of new *messenger ribonucleic acid* (mRNA) bearing instructions for the cytoplasmic synthesis of antiviral protein. The production of viral nucleic acids and virus-specific proteins is correspondingly reduced because of the increasing competition between the virus and interferon for available raw materials. Thus, viral replication is effectively thwarted.

In the meantime, antibody blood concentrations gradually rise to levels which are effective against the viral antigens. In response to the increasing level of

**Figure 4-7**  *One possible mechanism of action of interferon.*

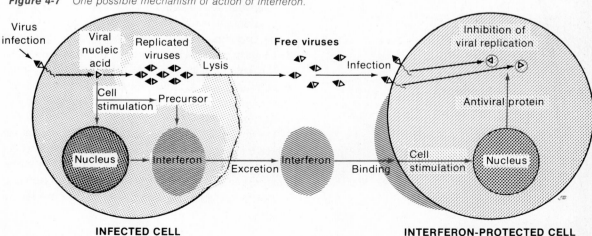

**INFECTED CELL**                                                                    **INTERFERON-PROTECTED CELL**

antigen-antibody reaction, the concentration of viral particles gradually tapers off. The concentration of interferon drops precipitously as the concentration of circulating antibody begins to rise noticeably. The temporal relationships among the relative concentrations of virus, interferon, and antibody are graphically plotted in Fig. 4-8.

## SPECIFIC IMMUNOLOGIC RESPONSES

As the phagocytic and unselective cleansing of the damaged tissues proceeds according to a genetically predetermined program, specific immunologic responses are increasingly stimulated by the continued presence of foreign materials, toxins, and microbes. Antigens may remain at the initial locus of infection or spread either actively or passively to definitive loci in other portions of the body. One of literally tens of thousands of different antigenic molecules, the invading antigen is nevertheless neutralized or destroyed by an antibody tailored to its structural specifications. This remarkable defensive ability not only reduces or eliminates the immediate threat posed by the introduction of the antigen but also

provides a mechanism for withstanding the future entry of the same antigen without serious effect. Immunologic competence and antigenic exposure confer an immunity against the disease syndrome generally elicited by the antigen.

Highly specific, and usually beneficial, immunities may be short-lived or long-lasting but also have the puzzling distinction of forming the basis for the development of hypersensitivities, or allergies, as well as tissue and organ rejection. The ability to respond with antibody induction to self-antigens under certain conditions results in what is called an *autoimmune response.*

**The nature and action of antibodies** Antibodies are protein molecules that neutralize or destroy antigens. Those antibodies manufactured in the bone marrow, lymph nodes, and spleen are *humoral* antibodies; those synthesized in the thymus gland are *carrier* antibodies. The self-nonself discriminatory roles of these two types of antibodies are complementary. Circulating antibodies have a transient existence. In the tissues, on the other hand, antibodies are fixed and probably last indefinitely when not bound to antigens. Those antibodies participating in allergic reactions are ordinarily not fully formed. Such incomplete molecules are referred to as *reagins.*

An antibody functions by capturing one or more of the same antigens. The number of entrapped antigens determines the valence of the antibody; monovalent, bivalent, and polyvalent antibodies complex (combine) with one, two, or more antigens, respectively. Bivalent antibodies are the most common type of immunoglobulin. Antibodies are highly specific, apparently combining with only one type of antigen or, at most, with several similar but not identical antigens.

The structural composition of the most common type of antibody provides some insights into the action of antibodies in general. Such an antibody is illustrated in Fig. 4-9.

The molecule consists of four protein chains, of which the paired A chains are heavy and long and the paired B chains are light and short. The entire molecule is bilaterally symmetric, the A chains opposing each other and forming the dividing line between its

**Figure 4-8** *The relative abundance of two antiviral agents, antibody and interferon, after the onset of a viral infection.*

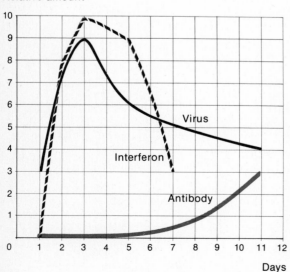

Relative amount

Days

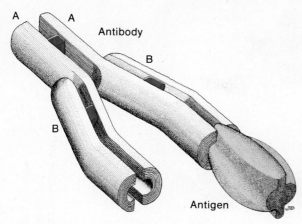

A    A
Antibody
B

B

Antigen

***Figure 4-9*** *Antigenic neutralization by a specific gamma immunoglobulin antibody molecule.*

two halves. Consisting of packets of amino acids wound together flexibly, the A and B chains are located on each side of the globulin molecule. The A chains are fastened to each other. The B chains are attached to the A chains by chemical bonds between adjacent sulfur atoms of the contiguous segments.

The foreign particles referred to as antigens in Fig. 4-9 are actually combinations of many different antigenic molecules. Each molecule of this mixture may contain one or more regions of antigenic determinants to which specific antibodies combine. Each antigenic determinant possesses a dominant portion which defines its specificity for an antibody combining site. The nature and number of chemical signals resulting from the combination between the antigenic determinants and the antibody combining sites determine whether the interaction will lead to antibody induction or its lack.

Each of the two antigens trapped by the bivalent antibody in our example interacts with the terminal portions of adjacent A and B chains. These two chains also interact with each other, conforming to the shapes of the antigens. The flexibility of the chains during the interaction enables the antibody to capture a spectrum of closely related antigens. Ultimately, the antigen-antibody complex is phagocytized by a macrophage and destroyed within its cytoplasm. However, a small fraction of the antigens becomes bound to RNA (ribonucleic acid) in the cytoplasm and some-

how acquires protection against enzymatic hydrolysis. Macrophages with RNA-bound antigens are probably "memory" cells responsible for the relatively rapid immunologic responses beginning with the second introduction of the same antigen.

**Types of antigen-antibody interaction**   As mentioned earlier, phagocytosis is enhanced by antibodies called opsonins. *Opsonization* is one type of antigen-antibody reaction. Other antigen-antibody interactions include neutralization, precipitation, agglutination, and lysis. Because all these responses to antigens (see Chap. 2) occur in blood serum, they are called *serologic reactions*. The mechanisms by which the various serologic reactions operate in the removal of antigens from the blood are depicted in Fig. 4-10.

*Neutralization* occurs when an antigen is prevented from combining with binding sites on the outside of a cell wall or membrane. The blockage is due to antibodies which preemptorily occupy the antigenic determinants, or active sites of the antigen. The neutralization of toxins by antitoxins provides one example of this type of antigen-antibody reaction.

*Precipitation* of soluble and therefore dissolved antigens takes place following their capture by bivalent antibodies. The two ends of each antigen become linked to specific antibodies, ultimately forming chains in which the antigen-antibody components alternate with each other. As the chains become longer, they also get heavier and finally settle out of solution as a precipitate.

*Agglutination,* or clumping, is caused when *agglutinin* antibodies combine with *agglutinogen* antigens. The antigenic determinants are located either on the surface of a cellular microbe or along protoplasmic extensions of its body such as flagella. The agglutinins have an opsonizing effect in that after agglutination the phagocytic engulfment of agglutinogens is enhanced.

*Cytolysis* of bacteria is called *bacteriolysis*. The lysis of gram-negative forms is accomplished through the combined efforts of specific antibodies called *cytolysins* and nonspecific complement. The mechanism is complex and involves the absorption of cytolysins to antigenic binding sites on the bacterium,

**Figure 4-10** *The mechanisms of antigen-antibody interaction in the blood.*

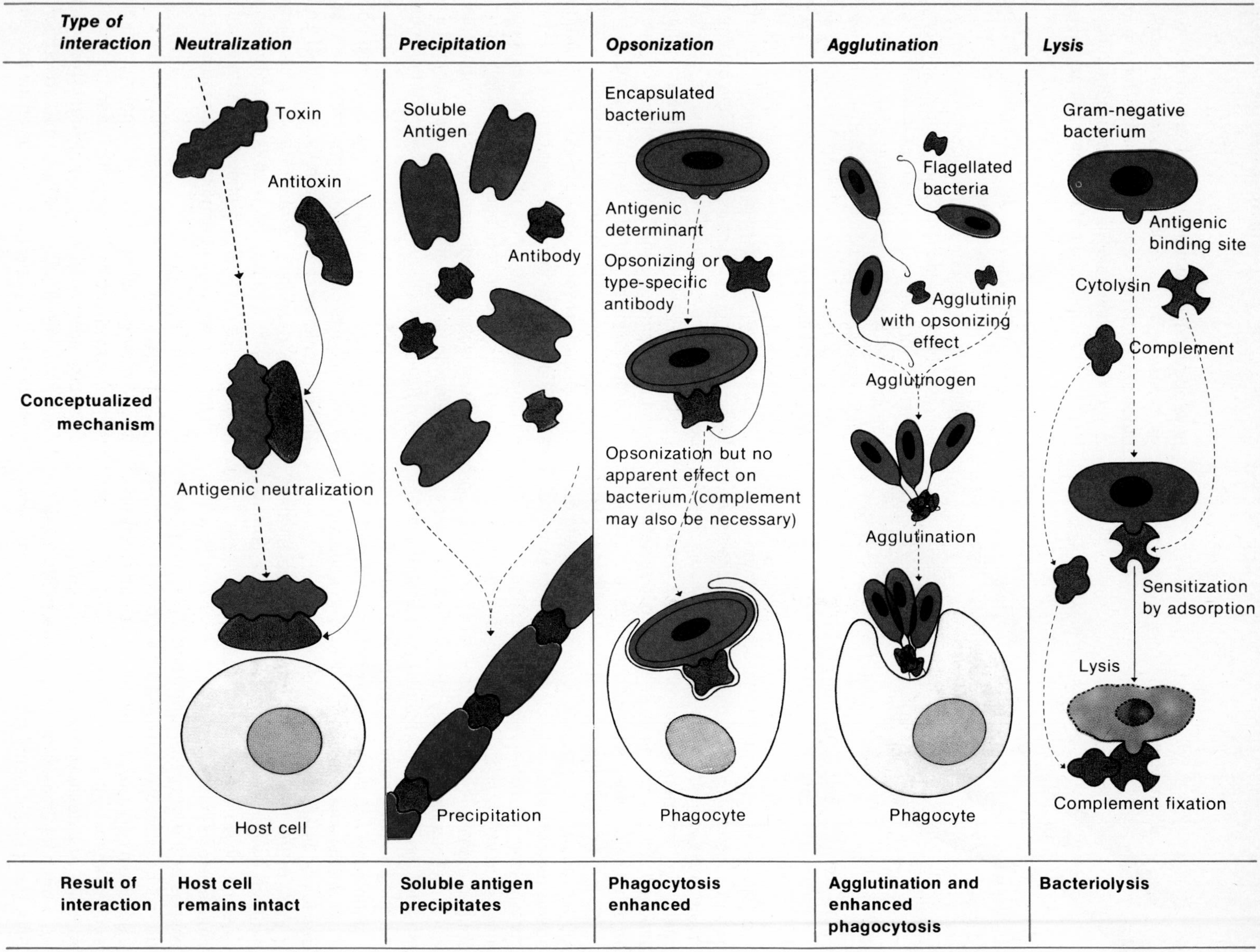

| Type of interaction | Neutralization | Precipitation | Opsonization | Agglutination | Lysis |
|---|---|---|---|---|---|
| Conceptualized mechanism | Toxin / Antitoxin / Antigenic neutralization / Host cell | Soluble Antigen / Antibody / Precipitation | Encapsulated bacterium / Antigenic determinant / Opsonizing or type-specific antibody / Opsonization but no apparent effect on bacterium (complement may also be necessary) / Phagocyte | Flagellated bacteria / Agglutinin with opsonizing effect / Agglutinogen / Agglutination / Phagocyte | Gram-negative bacterium / Antigenic binding site / Cytolysin / Complement / Sensitization by adsorption / Lysis / Complement fixation |
| Result of interaction | Host cell remains intact | Soluble antigen precipitates | Phagocytosis enhanced | Agglutination and enhanced phagocytosis | Bacteriolysis |

104

sensitizing or priming it for subsequent lysis. Following the fixation of complement to cytolysin, the bacterial cell wall ruptures and releases the protoplasm.

The fractions of blood serum in which circulating antibodies are found  Blood proteins include both antibodies and nonimmunogenic molecules. The blood proteins can be divided into two major categories: serum *albumins* and *globulins.* Albumins do not function in immunity. Globulins, on the other hand, are immunogenic and can be subdivided into alpha, beta, and gamma fractions. Most antibodies are of the gamma variety. *Gamma globulins,* in turn, can be separated into at least five different subfractions called gamma-G, A, M, D, and E globulins. Over 80 percent of the immunogenic proteins belong to the gamma-G globulins. This most common of all antibodies is called *gamma-G immunoglobulin,* abbreviated IgG. The relationships between the various blood and immunogenic proteins are summarized in Fig. 4-11.

The relation of lymphocytes to plasma cells  Humoral antibodies are manufactured by, and released into the blood from, structural variants of macro-phages called plasma cells, mentioned earlier. The macrophages are formed from swollen lymphocytic and monocytic phagocytes. These inflammatory cells originate from circulating lymphocytes. In addition to their immunogenic role, lymphocytes may enter the red bone marrow to proliferate into erythrocytes or granular leukocytes such as eosinophils and basophils or may specialize as embryonic connective tissue cells called *fibroblasts* for the deposition of collagen and other proteinaceous tissue components. The potential fates of circulating lymphocytes are traced in Fig. 4-12.

The role of the thymus gland in the development of immunologic competence  The thymus gland is implicated in the acquisition of self-nonself discrimination, i.e., the ability to form antibodies against antigens which have entered from exogenous sources while exhibiting immunity against endogenous natural antigenic chemicals.

Possessing two lobes embedded in surrounding fatty tissues, the thymus gland lies above and in front of the heart, partially obscured by the lungs. Although relatively large during prenatal life and the first 10 years after birth, the thymus ultimately atrophies by

**Figure 4-11** *The relationship of the gamma immunoglobulins to other proteins in blood serum.*

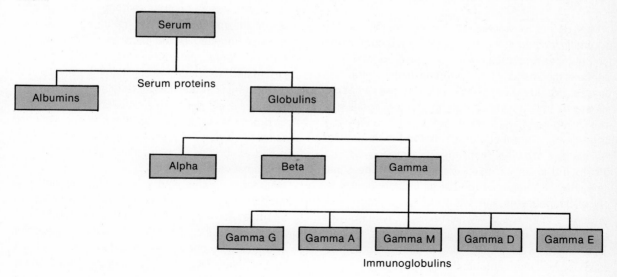

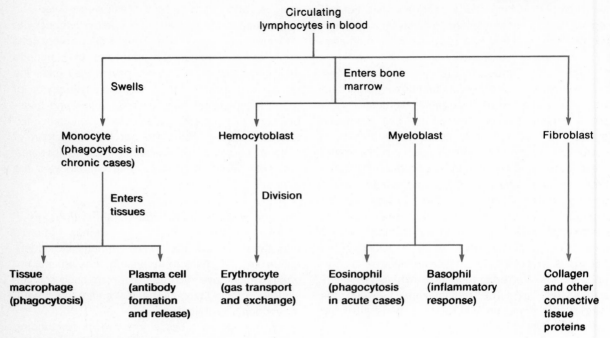

**Figure 4-12** *The potential fates of circulating lymphocytes.*

involution at puberty, or sexual maturity. After involution, it remains vestigial and inconspicuous for the rest of life. Immunologic competence of the antigen-sensitive cells of the body is gained after exposure to *thymic hormone* during the first few days following birth.

Most definitive lymphocytes come from thymocytes originally located in the thymus gland, although some lymphocytes and cells destined to take up later residence in the thymus gland originate in other lymphoid tissues. Excluding the thymus gland, these lymphoid structures include the spleen, the lymph nodes, the tonsils and adenoids in the throat, and the solitary and aggregated lymph nodules embedded in the intestinal walls. Following birth, thymic lymphocytes are distributed to all the lymphoid organs. At the same time, the lymph nodes produce an independent supply of lymphocytes and undifferentiated stem cells. Under the influence of thymic hormone, lymphocytes exported from the thymus gland as well as lymphocytes and stem cells originating in other lymphoid tissues become immunologically competent to respond

to foreign antigens. Simultaneously, immunity is acquired toward self-antigens, ordinarily eliminating the possibility of an autoimmune response.

The later entry of a foreign antigen will then cause the competent but still uncommitted immunogenic cells to increase in number, forming clones, or colonies, of plasma cells. Some of the plasma cells become committed to the production of antibodies complementary to the introduced antigen. The other uncommitted plasma cells quickly disintegrate, spewing their cellular contents into the region surrounding the antigens. This cellular debris presumably promotes the further proliferation of committed plasma cells and their secretory products, antigen-specific antibodies. These relationships are shown diagrammatically in Fig. 4-13.

Most foreign antigens are proteinaceous, although fats and carbohydrates may also possess antigenic properties. As a result of specific pathologic conditions, however, the antigens attacked by antibodies may actually be one's own proteins or self-antigens. This condition arises from diseases such as rheuma-

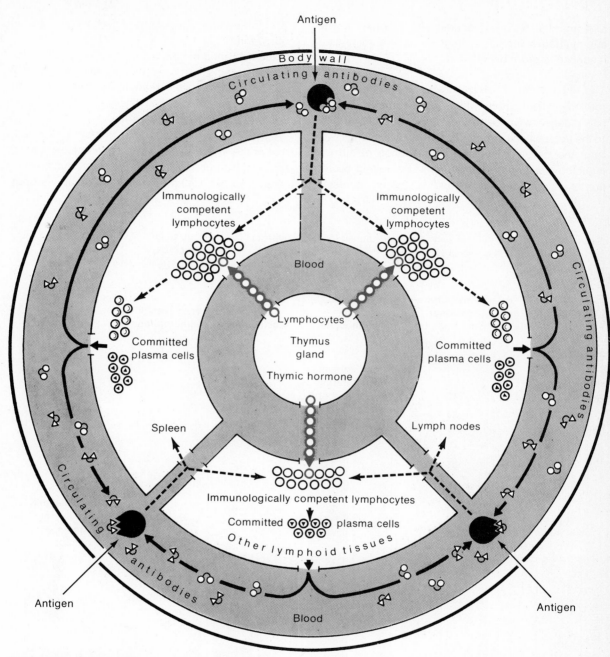

**Figure 4-13** *The role of the thymus gland in the development of the immunologic response.*

toid arthritis, histolytic anemia, myasthenia gravis, and multiple sclerosis, in all of which an autoimmune response is probably elicited.

**The development of hypersensitization and immunity** Inflammatory white blood corpuscles not only cleanse wounds, but they also contribute directly to the acquisition of immunity. The contribution is twofold: the development and maintenance of the immunologic response and the manufacture and secretion of antigen-specific antibodies. The mechanisms by which these contributions are accomplished are outlined in Fig. 4-14.

When an antigen enters the body for the first time, the body reacts with a specific defense mechanism called the *primary immunologic response*. With certain exceptions, this response is similar to the inflammatory reaction discussed earlier.

As the primary immunologic response proceeds, most of the invading antigens are ultimately neutralized by macrophages. The primary response is therefore largely cell-mediated, rather than antibody-

**Figure 4-14** *The role of the inflammatory cells during the development of hypersensitization and immunity. The primary immunologic response is shown at the left; the secondary response is shown at the right for each type of cell. The height of each bar is proportional to the intensity of the response.*

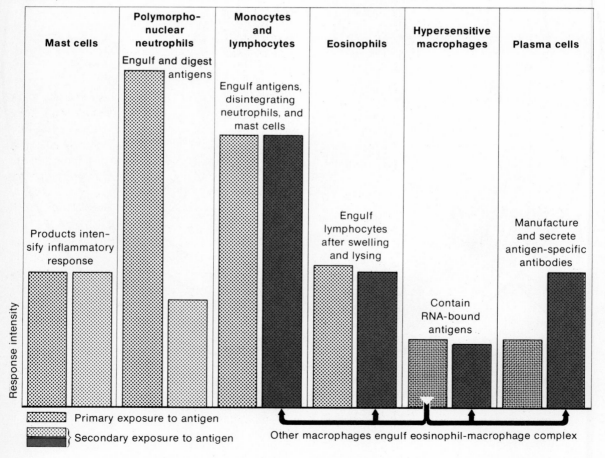

mediated. However, some of the antigens bind specifically to RNA within the cytoplasm of the macrophages. Although the invading antigens and the RNA-bound antigens are functionally inactivated, the bound antigens are not actually neutralized. By virtue of containing RNA-bound antigens, such macrophages are hypersensitized to future doses of the same antigen. Hypersensitive, "memory" macrophages remain in the body indefinitely after migrating into the deep portions of various lymphoid tissues.

A subsequent entry of the same foreign material elicits a *secondary immunologic response*. Neutrophils are once again mobilized to phagocytically neutralize the antigens, but the effect is not as great as that obtained during the primary response. The weak neutrophilic reaction is followed, however, by a vigorous mobilization of monocytes and lymphocytes. Many macrophages are formed as a consequence of their phagocytic activities. The phagocytosis of new antigens by hypersensitized macrophages somehow frees the old antigens bound to macrophage RNA, although such antigens still remain on the RNA. The liberation of old antigens and the engulfment of new antigens finally lyses the hypersensitive macrophages. Eosinophils then migrate into the wound to destroy the disintegrating cells. Other macrophages also enter the wound. Some of the latter engulf both disintegrating eosinophils and macrophages. Within the active macrophages, proteolytic enzymes hydrolyze the ingested cellular debris. Among the materials destroyed are both old and new antigens.

Following the destruction of the antigens on it, the RNA is released and becomes active once again. Macrophages possessing active RNA become plasma cells and begin secreting large quantities of antigen-specific antibody. At the same time, another fraction of the macrophages becomes newly hypersensitized, ensuring the future continuation of a rapid antibody-mediated secondary immunologic response.

**The characteristics of the basic types of immunologic response** Following the acquisition of immunologic competence, the mere presence of a foreign antigen initiates the synthesis of specific an-

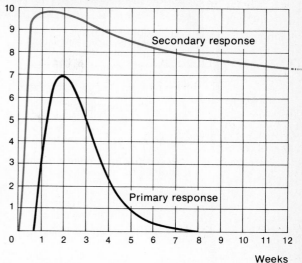

**Figure 4-15** *The fluctuations of circulating antibody concentrations during the primary and secondary immunologic responses. The graph plots relative antibody concentrations on the Y axis and time in weeks on the X axis.*

tibodies in clones of committed plasma cells. As a result of the initial exposure to an antigen after birth, memory cells containing RNA-bound antigens are formed. Such cells are rapidly mobilized during a subsequent exposure to the same antigens. The establishment of memory cells is the main reason for quantitative differences in the nature of the primary and secondary immunologic responses. These differences are graphically portrayed in Fig. 4-15.

From the above description and an examination of Fig. 4-15, we conclude that the primary immunologic response is slow, is not very strong, and disappears relatively rapidly, while the secondary immunologic response is rapid, vigorous, and long-lasting. In addition to the enhancing effect of memory cells on the secondary response, there are both absolutely more committed plasma cells and a greater cytoplasmic volume for protein synthesis in each plasma cell following the completion of the primary immunologic response. The rapidity of the secondary response is a reflection of the combined influence of the above factors.

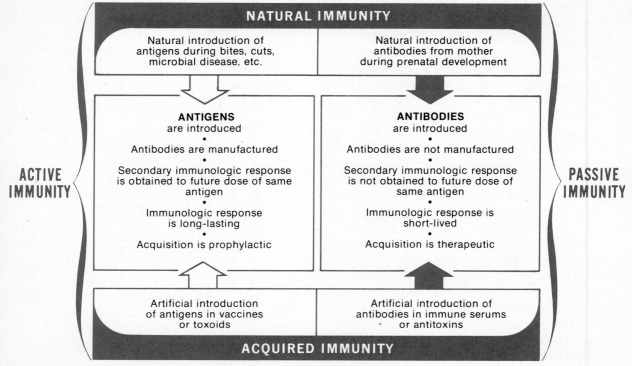

**Figure 4-16**  *The development and consequences of active and passive immunity.*

**Functional distinctions among the various types of immunity**  Immunity may be acquired naturally or artificially as well as actively or passively. There are therefore four basic types of immunity. The functional distinctions between types are conceptualized in Fig. 4-16. *Natural immunity* may result from the involuntary introduction of either antigens or antibodies. Antigens may enter when man is bitten, cut, or otherwise invaded by microbes. Maternal antibodies are introduced when they cross the placental barrier during intrauterine life. Artificially *acquired immunity* requires the voluntary exposure of the body to antigens or their derivatives in vaccines and toxoids or to antibodies in immune sera and antitoxins. In other words, active or passive immunity may follow the introduction of antigens or antibodies, the distinction being whether the exposure was voluntary or involuntary.

The introduction of antigens permits the development of relatively long-lasting immunologic responses and confers an *active immunity* with respect to the antigen. This principle is embodied medically in the technique of vaccination, a fundamental prophylactic measure against various microbial pathogens.

When antibodies are introduced into the body either voluntarily or involuntarily, the resulting immunity is passive and short-lived. *Passive immunity* does not evoke a secondary immunologic response, because antibody production is not stimulated. Nevertheless, immunotherapy is based on passive immunity and emphasizes its medical importance. This form of therapy is commonly used to carry a patient over the critical period of an infectious disease or toxic reaction.

## TISSUE REGENERATION

The operation of the various defense mechanisms and the progressive cleansing of the wound pave the way for the final phase of the healing process, regeneration of the dead and dying tissues. Tissue regeneration can be divided into several overlapping steps: the synthesis and remodeling of scar tissue, and final repairs at the site of the former wound. Significant landmarks in the healing region include the anastomosis of blood vessels; the formation, dissolution, and re-formation of collagen; the ingrowth of epidermal cells; and the sloughing off of the protective scab.

Within a short time after an injury, *fibroblasts* from the dermis are mobilized and begin to surround the wound. These embryonic connective tissue cells are destined to produce collagen and other dermal proteins which collectively form the structural components of scar tissue. While increasing numbers of fibroblasts aggregate around the injured area, the cut edge of the epidermis on each side of the surface clot starts to intrude into its interior. Some of the nearby epidermal cells which remain intact begin to undergo cellular division, eventually replacing the displaced epidermal cells. Such cells migrate as one sheet of tissue, since they are held together by local thickenings of their contiguous plasma membranes. Called *desmosomes,* these thickenings are responsible for the fusion of the cut ends of the epidermis, an event which is completed about two days after the body sustains the original injury. At this point, the newly regenerated epidermis is still below a protective scab. About one week after the initial insult to the body, the scab sloughs off.

Meanwhile, the surrounding fibroblasts migrate into the wound. There, as collagen factories, they collectively demonstrate their synthetic prowess. Raw materials for collagen manufacture are initially supplied from the copious edematous fluids formed during the inflammatory response. Subsequently, raw materials are provided directly from the blood following an extensive proliferation of a network of blood capillaries at the site of the wound.

The functional characteristics of *collagen* are determined by its molecular structure. The gross (general) chemical makeup of collagen and its orientation in connective tissue are shown in Fig. 4-17.

A collagen molecule contains four contiguous subunits, each of which is made of three spiral and intertwined chains of amino acids in identical sequences. Following its secretion from a fibroblast, the collagen molecule is known as *tropocollagen. Native collagen* fibrils take shape when strings of tropocollagen molecules become stacked into layers. Each layer of native collagen overlaps its neighbors by one-quarter of the length of a tropocollagen molecule, so that vertical rows of tropocollagen have a decided slant. In addition, the fibrils of native collagen become organized into layers of variable thicknesses, each layer at right angles to the other. Thus native collagen is composed of alternating stacks of fibrils possessing both the strength and resilience necessary in connective tissue.

As a consequence, the scar tissue which forms is actually more dense than the connective tissue which it replaces. Although a few inflammatory cells may remain at the site of the former wound, healing is now essentially complete.

## A SUMMARY OF WOUND HEALING

The major processes by which wounds are healed are illustrated in Fig. 4-18. Basically, the processes are protective, cleansing, and regenerative, in that approximate order of occurrence. Protective processes include those which prevent the loss of blood and attenuate the effects of introduced microbes and other antigens. Cleansing occurs during the inflammatory response. Regeneration is a long-term process, temporally overlapping the protective screen provided by the immunologic responses of the body.

## THE GENERAL ADAPTATION SYNDROME

Up to this point, we have reviewed the various protective and defensive mechanisms mobilized by the body in response to an injury. However, our story would not be complete without discussing the adaptive mechanisms evoked by psychic and physiologic stresses. The syndrome is entirely nonspecific and consists basically of (1) metabolic effects, increasing the efficiency with which energy is utilized, (2) anti-

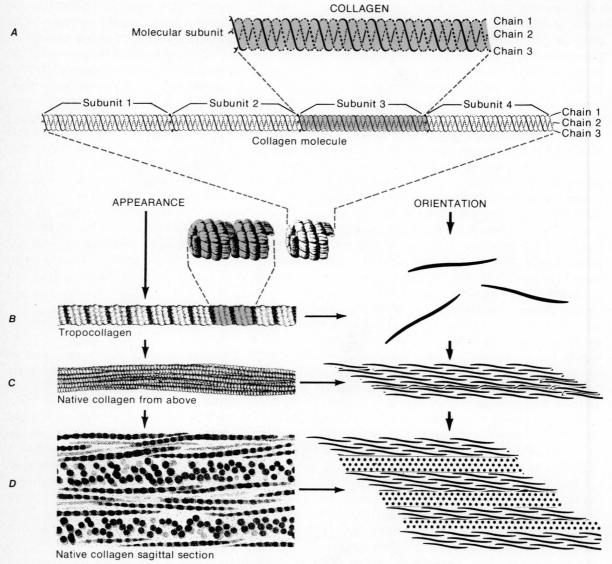

**Figure 4-17** *The development and orientation of collagen fibrils in connective tissue. A. One collagen molecule. B. Tropocollagen. C. Native collagen fibrils. D. The layering of collagen fibrils.*

inflammatory effects, reducing tissue damage and promoting healing, and (3) other widespread physiologic changes, improving the capacity of the body to respond to perceived emergencies.

There are two essential and interrelated systems overseeing adaptive responses to stress. One is referred to as the *pituitary-adrenal* (hypophyseoa-drenal) axis; the other is commonly known as the *sympatheticoadrenal* axis. The pituitary-adrenal system is mediated by interactions between adrenocorticotropic hormone (ACTH) secreted by the adenohypophysis, which is the anterior lobe of the hypophysis (pituitary gland), and glucocorticoids, including the hormone cortisol (hydrocortisone), secreted by the outer cortical portion of the adrenal gland. The sympatheticoadrenal system is regulated by epinephrine (Adrenalin) and norepinephrine, hormones collectively referred to as catecholamines. These hormones are secreted by the inner medullary portions of the adrenals.

Together, these two hormonally based mechanisms, each involving the adrenal gland, enable the body to successfully negotiate the stresses and strains encountered during life. The catecholamines exert long-term integrative effects on behavior. These profound changes affecting appetitive behavior, learning, and the degree of excitation of the central nervous system are only now beginning to be understood. The normal operation of both axes results in a complex of effects known as the *general adaptation syndrome* (GAS). Under the influence of chronic stress, GAS has been shown to contain three sequential steps: *alarm reactions* (AR), resistance, and finally exhaustion. Preparing the body for "flight or fight," the ARs are perhaps the best known of the overt effects during GAS. Exhaustion leads to death, a form of suicide in which the overactive defense mechanisms are literally consumed by the rest of the body. Between these two extremes is the broad plateau on which adaptation to stress is found.

## THE PITUITARY-ADRENAL AXIS

The pituitary-adrenal axis revolves around the combined actions of the nervous system, including the neurohumor corticotropin-releasing factor (CRF), and two or more related hormones of the endocrine system, ACTH and cortisol. Their secretion and the subsequent adaptive responses of the body are initiated by neural stimulation to the brain. The stressful stimuli may result from injury, disease, harmful environmental conditions, or psychogenic factors, but the responses of the body are similar.

Stimuli which initiate GAS are conducted as nerve impulses to the hypothalamus, a region at the hind part of the forebrain, immediately below the thalamus. Responding to the signals, neurosecretory cells of the hypothalamus release CRF. CRF is a hormone manufactured by the glandular ends of certain hypothalamic axons. Axons are fibrous extensions conducting either nerve impulses or secretory products away from the main body of the nerve cells. The neurohumor CRF is transported through the blood to its target structure, the nearby hypophysis (pituitary gland).

The hypophysis is situated at the base of the brain. Under the influence of CRF, the adenohypophysis (anterior lobe of the hypophysis) liberates ACTH. ACTH, in turn, is carried in the blood to all parts of the body but especially affects the cortices of the adrenal glands, initiating the release of cortisol and its derivatives. The increasing blood concentrations of the adrenocorticoids reduce the synthesis in the hypothalamus of CRF, thereby preventing the further release of adenohypophyseal ACTH. These events are summarized in Fig. 4-19.

The adaptive responses of GAS are in part elicited by the cooperative efforts of ACTH and cortisol. ACTH stimulates the synthesis of glucocorticoids in the adrenal cortices by promoting protein synthesis and glucose uptake. In addition, ACTH has other metabolic effects such as the inhibition of protein synthesis and the stimulation of fat breakdown in fatty tissues. Although not influencing the adrenal glands directly, these metabolic effects initiate oxidative processes which tap the energy locked chemically in the fatty acids derived from fats. ACTH also causes an increase in motivated, goal-oriented behavior requiring a high degree of attention. The elevation in appetitive or goal-directed behavior improves perceptual abilities. For these reasons, the operation of the pituitary-adrenal axis following exposure to stress is adaptive.

The glucocorticoids represent some 20 to 30 similar steroid or lipid-derived molecules, some of which are probably only precursors or intermediates in the synthetic pathways of one or more cortical hormones. Resembling in chemical structure the reproductive hormones in the ovaries and testes, the glucocorticoids primarily influence carbohydrate metabolism and secondarily affect protein and fat metabolism. These metabolic changes are influential in the mediation of inflammation and tissue regeneration.

The primary influence of excessive concentrations of glucocorticoids stems from hyperglycemia, an increase in the concentration of blood glucose, following glycogenolysis in the liver. The breakdown

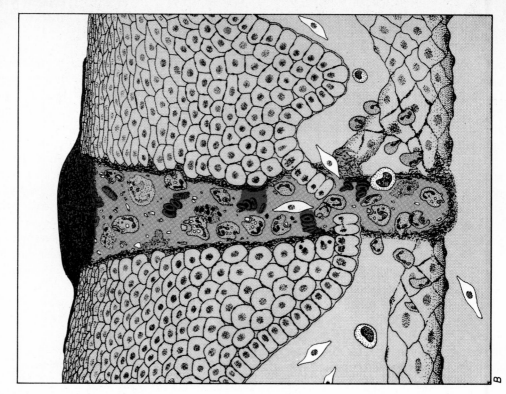

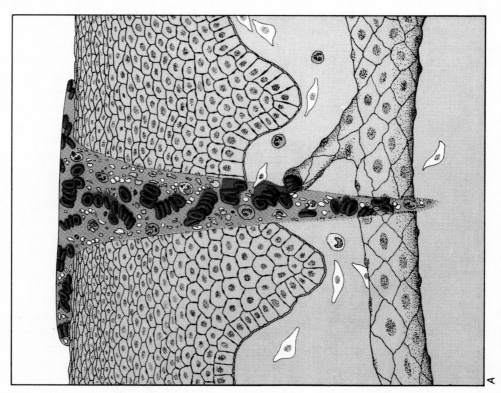

**Figure 4-18** The events that occur during wound healing. A. Initial injury. B. Early inflammation. C. Late inflammation. D. Tissue regeneration.

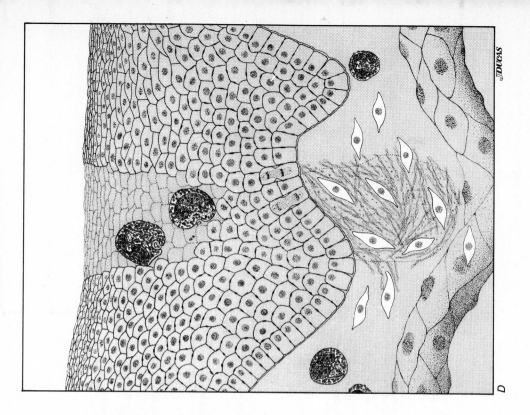

D

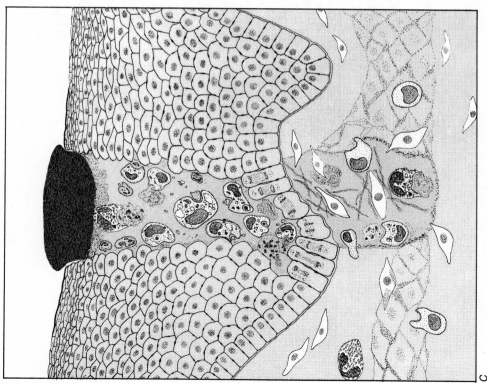

C

**Figure 4-19** *The pituitary-adrenal (hypophyseoadrenal) and sympatheticoadrenal axes of the general adaptation syndrome (GAS).*

| Structural scheme | Sequence of events | Nature of component | Functions of chemical |
|---|---|---|---|
| (see figure) | Neural stimulation of brain | | |
| | Secretion of corticotropin-releasing factor (CRF) from hypothalamus | Polypeptide ACTH | Stimulates glucocorticoid synthesis by adrenal cortices; stimulates adrenal protein synthesis and glucose uptake; inhibits protein synthesis in fatty tissues; stimulates fat breakdown; increases appetitive behavior |
| | CRF carried in blood to pituitary gland (hypophysis) | | |
| | Secretion of adrenocorticotropic hormone (ACTH) from anterior lobe of pituitary | 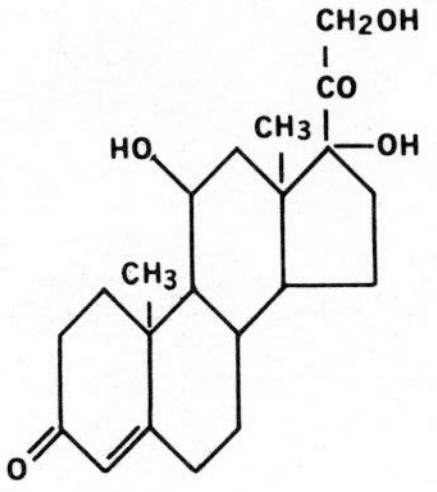 Cortisol | Anti-inflammatory agent; promotes healing; initiates glycogen synthesis in liver; causes redistribution of fat in body; alters nitrogen balance; causes changes in types and numbers of white blood cells; necessary for normal muscular functioning; alters central nervous system threshold of excitation; affects connective tissue differentiation; induces synthesis of new enzymes in liver |
| | Secretion of glucocorticoid hormones from adrenal cortices | | |
| | Glucocorticoids in blood modulate CRF production in feedback loop to hypothalamus | | |
| | Neural stimulation of adrenal medullae | Electrochemical nerve impulses of the sympathetic autonomic nervous system | Increases rate and force of heartbeat; causes arteriolar constriction in skin and visceral regions; causes dilatation of arterioles in skeletal muscles; promotes glycogen breakdown in liver; causes bronchial dilatation; decreases coagulation time |
| | Secretion of catecholamines from adrenal medullae | 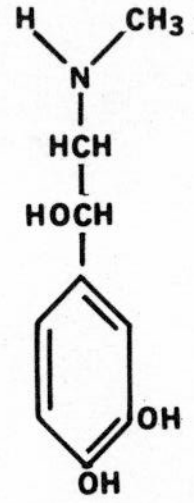 1. Epinephrine | 1. Increases rate and force of heartbeat; increases the systolic pressure of the heart; causes arterial constriction and dilatation, depending on location of blood vessels; affects smooth musculature; decreases central nervous system threshold of excitation; promotes skeletal muscle contraction; elevates sugar content of blood; promotes glycogen breakdown in liver |
| | Epinephrine and norepinephrine modulate their further production in feedback loop to adrenal medullae through blood | 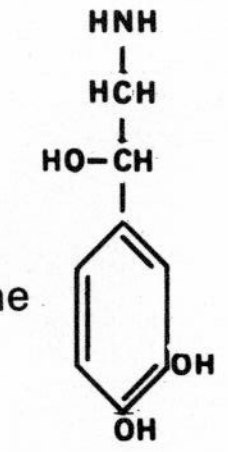 2. Norepinephrine | 2. Causes arteriolar constriction only; increases both systolic and diastolic pressures of heart; excluding metabolism, all other effects are similar to those in 1 above |
| | Neural feedback modulates further excitation of adrenal medullae | | |

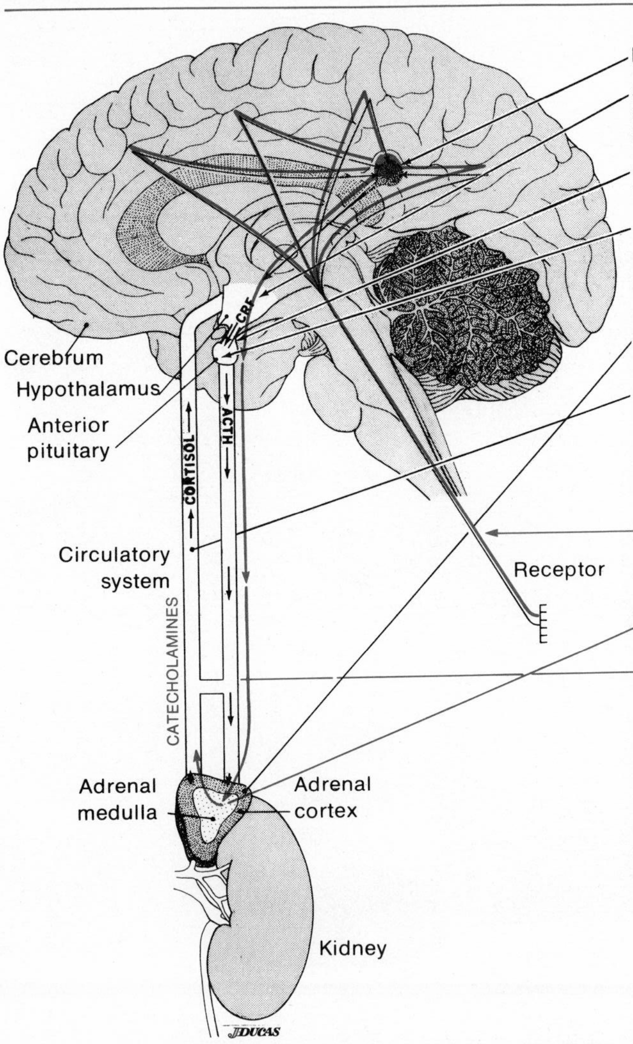

of glycogen into glucose is also promoted in the voluntary muscles and leads to an increase in the intramuscular content of lactic acid. The accumulating lactic acid is transported through the blood to the liver, where it is used in the resynthesis of glycogen or, when adequate amounts of oxygen are restored to the muscles, is completely oxidized. As a consequence of these actions, the primary energy nutrient of the body, glucose, is made available in greater amounts in the blood for transport to existing and potential cellular sites of energy utilization.

During stress, protein metabolism becomes completely degradative, with no synthetic component. This condition results in a *negative nitrogen balance,* reflecting an increase in the production of nitrogenous waste products, including gases.

An excess of glucocorticoids or their synthetic analogues affects the metabolism of fat and its distribution in the body. Prolonged, hypersecretory activity by the adrenal cortex leads to the development of a moon face, to humpback (kyphosis), and to an increase in the deposition of fat in the axillary and inguinal regions, symptoms characteristic of Cushing's disease (see Chap. 18).

The glucocorticoids are also *anti-inflammatory,* promoting healing by reducing the damaging effects of the inflammatory defense mechanism on tissues. The inflammatory cells and the lymphoid tissues, including the thymus gland, are differentially destroyed by an overabundance of adrenocorticoids. Because suppression of the inflammatory response increases the vulnerability of the body to infection, the anti-inflammatory abilities of glucocorticoids are a mixed blessing. However, cortisol therapy is prescribed medically to reduce the inflammation accompanying severe allergies, rheumatoid arthritis, major surgery, and any other inflammatory condition which does not respond to conventional treatment. The healing effects of cortisol are also evidenced by the differentiation of new connective tissue through the mobilization of fibroblasts to secrete collagen.

Administration of the adrenocorticoids, especially cortisol, is also used to counteract the effects of adrenal insufficiency, the clinical manifestations of which are collectively known as Addison's disease (see Chap. 18). The symptoms of Addison's disease all follow from a decrease in glycogenolysis and a corresponding development of hypoglycemia, or low blood-sugar levels. These effects result in muscle weakness and fatigue, poor circulation or hypotension, and a general decrease in appetitive behavior.

The glucocorticoids are synthesized from derivatives of cholesterol in the presence of vitamin C, or ascorbic acid. The cortical synthesis of cortisol and other related molecules reduces the amounts of cholesterol and vitamin C within the adrenals. The role of vitamin C in the synthesis of the glucocorticoids is not yet clear, but adrenocortical insufficiency results from either its absence or its presence in inadequate amounts. The relation between vitamin C, glucocorticoid synthesis, and GAS has led to suggestions that a substantially increased ingestion of vitamin C may improve the overall health and vigor of the body. The theory has yet to receive rigorous experimental verification, but the meager clinical data bearing on this question seem to support the claim. The role of vitamin C is discussed further in Chap. 27.

As noted earlier, the pituitary-adrenal axis is summarized in Fig. 4-19.

## THE SYMPATHETICOADRENAL AXIS

The pituitary-adrenal system is designed for the long haul, the persistence of chronic stress over an extended period of time, and does not adequately meet the exigencies of an imminent threat or actual emergency. The rapid ARs (alarm reactions) of GAS are instead mediated by the sympatheticoadrenal system, although both short- and long-term responses are usually provoked by the same stimuli. Consisting of the coordinated activities of the central nervous system, the sympathetic nervous system, and epinephrine and norepinephrine from the adrenal medullae, the system normally prepares the body to meet the physical and mental exertions resulting from immediate duress.

The system becomes operative when neural stimuli, after undergoing integration within the central nervous system, are conducted through the sympathetic nervous system (part of the autonomic nervous system) to the adrenal medulla. Here, neural stimulation induces the secretion of *catecholamines.*

The adrenal medulla is derived from neural ectoderm, whereas the surrounding cortex is derived from presumptive gonadal endoderm. In other words, the adrenal gland is formed from two of the three primary tissue layers originally laid down in the embryo, a fact which is suggestive of the functional divergence between its two parts.

The adrenal cortex is not alone in secreting catecholamines. The terminal axons of sympathetic, postganglionic neurons lying outside the central nervous system also secrete a neurohumor similar to the catecholamines. Because of the resemblance of the neurohumor to epinephrine (Adrenalin), such nerve fibers are functionally termed *adrenergic.*

The functional specializations and interrelations between the adrenal cortex and medulla, on the one hand, and the medulla and the sympathetic nervous system, on the other hand, provide the physiologic foundation for the normal operation of GAS.

The actions of *epinephrine* and *norepinephrine* are sympathomimetic; i.e., they act in concert with the sympathetic nervous system, of which they are hormonal adjuncts and synergists. In response to increasing blood concentrations of the catecholamines and decreased neural stimulation, their production by the adrenal medullae is reduced. These regulatory mechanisms between the blood, nervous system, and adrenal medullae maintain relatively constant blood concentrations of the catecholamines. The events which take place during the action of the sympatheticoadrenal axis are summarized in Fig. 4-19.

The catecholamines influence the heart and blood vessels, involuntary and voluntary muscles, respiration, the central nervous system, and, as a consequence, metabolism. Together, these effects comprise the ARs of the body to perceived threats.

The cardiovascular changes include an increase in the rate and force of the heartbeat and a constriction of the arterioles (vasoconstriction) of the body, resulting in a rise in blood pressure. However, in response to the differential action of epinephrine, vasodilatation occurs in the skeletal muscles, heart muscle, and lungs, organs which require large amounts of nu-

trients and oxygen during stressful periods. Vasodilatation permits a greater volume of blood and, therefore, a greater concentration of nutrients and oxygen, to flow within these organs.

Smooth muscle effects include both excitation and inhibition. The smooth muscle walls of the spleen, the radial muscle of the iris, the muscular sphincters guarding the passages of the digestive and urinary tracts, and the arrector pili muscles associated with the hairs of the skin are among the involuntary muscles stimulated by the catecholamines to contract. This last reaction is involved in the formation of "goose pimples" on the skin in response to cold temperatures. However, the smooth muscles associated with the walls of the digestive tract, the bronchioles, and the urinary bladder are inhibited and remain relaxed. These actions are correlated with a reduction in visceral functions and an increased awareness and defensive posture.

The performance of skeletal muscles is enhanced by the catecholamines, increasing the strength and duration of contraction while decreasing its threshold for excitation. In addition, local vasodilatation in active skeletal muscles increases the supply of nutrients and oxygen at a time when reserves of these essential metabolites have diminished. The result of these changes is best exemplified by the extraordinary strength not uncommonly exhibited by a highly agitated person. Sometimes anecdotal observations of these phenomena, with appropriate embellishments, enter into the folklore of a culture as myths or fanciful stories. Their origin, however, can usually be traced back to the successful action of the sympatheticoadrenal axis.

Oxygen utilization, rising to almost $1\frac{1}{3}$ times the basal rate in response to an excess of catecholamines, indirectly increases the rate and depth of respiration. Because the smooth muscles of the bronchioles are inhibited by epinephrine, its use in asthma ameliorates the effects of bronchoconstriction and reduces dyspnea, or shortness of breath.

The central nervous system effects of epinephrine and norepinephrine result in a psychic state of arousal. The increased awareness of environmental stimuli is mediated by the reticular formation of the brainstem and is obviously an adaptive advantage when one is faced with a threatening situation.

The metabolic effects of the catecholamines are simply ramifications of the numerous and diverse physiologic changes described above. A rise in metabolic rate shows up as an increase in heat evolution by the body of as much as 20 percent. The heat may be dissipated by the cooling action derived from sweating, a common experience of individuals under stress.

## COLLATERAL READING

Armelagos, George J., and John R. Dewey, "Evolutionary Response to Human Infectious Diseases," *BioScience* 20(5):271–275, 1970. An examination of the role of culture in reducing the threat of disease.

Burnet, Sir MacFarlane, "The Mechanism of Immunity," *Scientific American,* January 1961. Evidence is marshaled to support the view that antigens actually commit specific plasma cells to the manufacture of their antibodies.

Constantinides, P. C., and Niall Carey, "The Alarm Reactions," *Scientific American,* March 1949. A discussion of experiments with rats which led to the development of the general adaptation syndrome.

Edelman, Gerald M., "The Structure and Function of Antibodies," *Scientific American,* August 1970. An analysis of the complete amino acid sequence of gamma immunoglobulin.

Gross, Jerome, "Collagen," *Scientific American,* May 1961. The formation and reconstruction of collagen are described.

Guyton, Arthur C., *Function of the Human Body,* 3d ed., Philadelphia: W. B. Saunders Company, 1969. An excellent introduction to selected areas of human physiology. Chapter 9 discusses the reticuloendothelial system and immunity.

Harris, Maureen, "Interferon: Chemical Application of Molecular Biology," *Science,* 170:1065–1070, 1970. This topical review briefly explains the molecular biology of interferon, its mechanism of antiviral action, and its potential clinical applications in antiviral therapy.

Levey, Raphael H., "The Thymus Hormone," *Scientific American,* July 1964. A review of experiments with mice in which their thymus glands are removed at birth.

Levine, Seymour, "Stress and Behavior," *Scientific American,* January 1971. A review of the functional organization of the pituitary-adrenal axis, citing relevant stress experiments with rats.

Levine, Seymour, and F. Richard Nichol, "Interferon Inducers," *BioScience,* 20(12):696–701, 1970. The role of interferon in the natural recovery from viral infection is reviewed by examining the mechanisms by which interferon is formed.

Li, Choh Hao, "The ACTH Molecule," *Scientific American,* July 1963. The structure of the polypeptide ACTH is chemically analyzed into 39 amino acid units.

Marples, Mary J., "Life on the Human Skin," *Scientific American,* January 1969. This important and interesting article describes the human skin as a series of biomes such as the tundra and tropical rain forest.

Nossai, G. J. V., "How Cells Make Antibodies," *Scientific American,* December 1964. Experiments with tissue cultures of cells producing single antibodies indicate that their manufacture is directed genetically.

Piliero, Sam J., "Immune Suppressants," *BioScience,* 20(12):710–714, 1970. A review of the significance of current efforts concerned with the control of the immunologic responses and a description of the different classes and uses of immunosuppressive agents.

Porter, R. R., "The Structure of Antibodies," *Scientific American,* October 1967. A description of the molecular structure and functional flexibility of gamma immunoglobulin.

Ross, Russell, "Wound Healing," *Scientific American,* June 1969. A review of the elaborate process by which inflammatory cells assist in the regeneration of injured tissue.

Sela, Michari, "Antigenicity: Some Molecular Aspects," *Science,* 166:1365–1374, 1961. A difficult article that examines the current state of knowledge regarding the molecular features of antigens and their relationships to antigen-antibody interactions.

Speirs, Robert S., "How Cells Attack Antigens," *Scientific American,* February 1964. Focuses on the specialized functions of the various inflammatory cells.

Wood, W. Barry, Jr., "White Blood Cells v. Bacteria," *Scientific American,* February 1951. Experimental data lead to the conclusion that phagocytes are more important in acute inflammation than in chronic infections.

Verzar, Frederic, "The Aging of Collagen," *Scientific American,* April 1963. The aging process of collagen is traced.

# CHAPTER 5
# BODY BRICKS

Your body is composed of many trillions of cells that perform cooperatively in various aspects of metabolic, homeostatic, and reproductive processes. Cells are the bricks of the body.

Two concepts are fundamental to our understanding of cells: (1) The cell is the basic unit of biologic organization, just as an atom is the basic unit of physical organization, and (2) each cell is derived from a preexisting cell.

The first premise suggests that an understanding of health and disease must ultimately be based on a knowledge of cell structure and function. The second proposition indicates that the uniqueness of a cell is a consequence of its biologic origin.

In this chapter, we will review the fine structure and nonspecific functions of tissue cells. A more detailed consideration of cellular activities is reserved for the next chapter.

Although the 5 to 10 trillion cells in your body are of many sizes and shapes and may contain unique structural specializations, practically all cells carry on similar basic metabolic and perpetuative activities. Each tissue cell must engage in the following overlapping tasks:

1 Securing metabolic fuels from external sources
2 "Burning" these fuels in the presence of oxygen to obtain enough energy to maintain the cell's existence
3 Trapping and transporting the chemical energy to cellular sites of utilization
4 Synthesizing structural and enzymatic proteins, lipids, nucleic acids, and various other products as needed
5 Repairing and replacing portions of its membrane system, spent enzymes, nucleic acids, and phospholipids
6 Homeostatically regulating the electrolyte and fluid balance and the molecular content of the intracellular environment

7 Eliminating solid, liquid, and gaseous waste products of metabolism

8 Participating cooperatively in the activities of the tissue of which it is a member

9 Recognizing and responding appropriately to neural, hormonal, and other chemical signals to which it is exposed

While this is no small order for a single tissue cell, the molecular interactions which permit these activities to proceed smoothly in literally trillions of cells is an impressive phenomenon.

## RELATION BETWEEN CELL STRUCTURE AND FUNCTION

In any cell, structure and function are inextricably intertwined. In one sense, a cell is a bag of interacting chemicals highly integrated in both space and time. The molecular organization and, therefore, the general shape of every structure in a cell is based on physiochemical interactions.

Basically, a cell consists of a membrane system which surrounds and runs through the substance of the protoplasm. The membrane system forms the outer boundary, or plasma membrane, of the cell, gives rise to intracellular partitions of several types within the cytoplasm, and surrounds a specialized region of protoplasm called the *nucleus*. The intracellular and nuclear partitions consist of a double membrane which encloses a continuous space running from the nucleus to the outside of the cell.

Thus, there are three ways of thinking about the basic structure of a cell: (1) as a bag of chemicals, (2) as a special substance called *protoplasm* enclosed and partitioned by a membrane system within which is a space continuous with the external environment, and (3) as a primary structural unit consisting of a nucleus, cytoplasm, and plasma membrane. These concepts of cell structure are diagramed in Fig. 5-1. Although all these concepts are valid, they fall short

**Figure 5-1** *Conceptions of the structure of a tissue cell.*

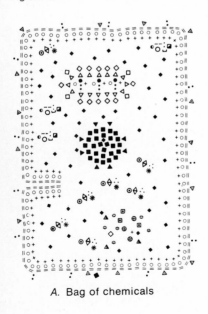

A. Bag of chemicals

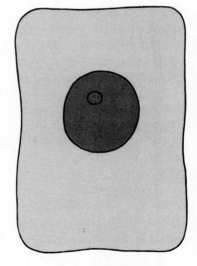

B. Protoplasm enclosed by membrane system containing a channel

C. Nucleus surrounded by cytoplasm bounded by plasma membrane

of contributing significantly to our understanding of the unique qualities which separate living cells from other types of matter. To gain this degree of understanding requires a more critical examination of cellular structure and a comprehensive attempt to correlate structure with function.

## CELLULAR ORGANELLES AND STRUCTURES

The complexities of cellular functions can be appreciated only by analyzing each of the cellular components. This is a formidable task because of the extremely small size of cell structures and the intricacies of their molecular organization. Nevertheless, some quite precise microscopic, cytologic, and biochemical techniques have been developed and applied to cells in continuing attempts to learn how they metabolize and perpetuate. Our knowledge in this area is increasing at a rapid rate. However, significant gaps in our understanding of cellular structure and function exist at every turn in the road, as we shall see.

Cells are composed of a large array of structures. Some of these components consist of nothing more than aggregations of storage or secretory products or simple accumulations of large crystals or other substances which have entered from the extracellular tissue fluid. Such cellular components are called *inclusions*. Other intracellular components are much more complex, subserving one or more functions at the cellular level which organs ordinarily perform at the organismic level. Discrete cellular structures which perform essential functions are called *organelles*. Frequently, cytoplasmic organelles are surrounded by one or more membranes. Nevertheless, numerous organelles lack membranes. Discrete regions of the membrane system of the cell also qualify as organelles.

The structure of a generalized animal cell is shown in Fig. 5-2. Surrounding the cell are enlarged perspective drawings of various organelles and portions of the membrane system. The fine structures of these organelles are based on interpretations of pho-

tomicrographs obtained with the electron microscope. The conventional structure of a cell visible with a light microscope is depicted in the upper right-hand corner of the illustration.

Among the structural components of a cell are (1) a membrane system including the plasma membrane, the endoplasmic reticulum, and the Golgi apparatus; (2) membrane-lined organelles such as the nucleus, mitochondrion, and lysosome; (3) organelles which lack membranes, e.g., the ribosome and centriole; and (4) inclusions such as granules, globules, vacuoles, and crystals.

### THE PLASMA MEMBRANE

Deceptively simple in structural appearance, the plasma membrane contains many different kinds of molecules. Various portions of the plasma membrane turn inward and are continuous with the membranous structure of the endoplasmic reticulum. Therefore, the interstitial space surrounding the cell is continuous with the space between the double membrane of the endoplasmic reticulum within the cell. The outer and inner surfaces of the plasma membrane are by no means smooth. Both membrane surfaces consist of a random array of protoplasmic projections and depressions of various sizes and shapes.

The structure of a plasma membrane is characteristic of biologic membranes in general. There are three layers to the plasma membrane: an outer and an inner layer of compound proteins permeable to water-soluble particles such as simple sugars and amino acids, and a middle layer consisting of two sublayers of phospholipids impermeable to water-soluble compounds but permeable to lipid-soluble materials.

Analogous to a "butter sandwich" in which the bread is protein, the universality of this membrane organization has led to its designation as a *unit membrane*. A portion of a unit membrane is shown in Fig. 5-3.

Each lipid in the phospholipid layer is composed of two fatty acids attached to a glycerol "backbone." The glycerol portions of the lipids are in turn linked to phosphate-containing groups lying next to the proteinaceous portions of the membrane. The phospholi-

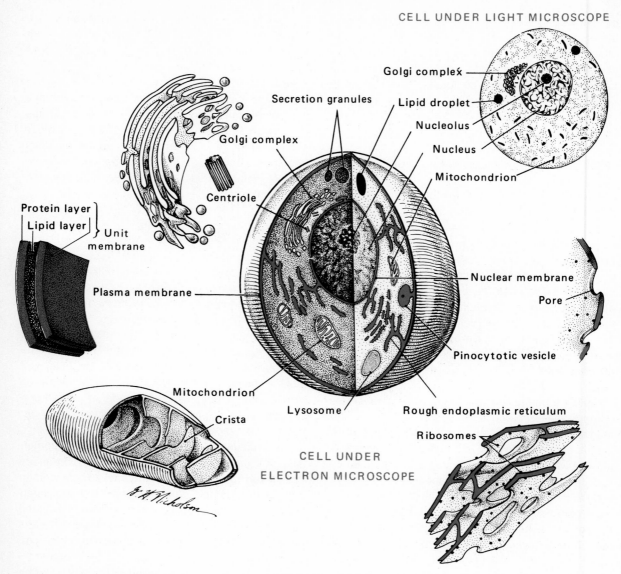

CELL UNDER LIGHT MICROSCOPE

Golgi complex

Lipid droplet

Nucleolus

Nucleus

Mitochondrion

Secretion granules

Golgi complex

Centriole

Protein layer
Lipid layer
Unit membrane

Plasma membrane

Nuclear membrane

Pore

Pinocytotic vesicle

Mitochondrion

Crista

Lysosome

Rough endoplasmic reticulum

Ribosomes

CELL UNDER
ELECTRON MICROSCOPE

**Figure 5-2** *The gross and fine structure of a cell. Cellular detail seen through a light micro-scope is shown at the upper right. The fine structure of a cell is reconstructed from electron photomicrographs in the center. The drawings which surround the central cell are enlarge-ments showing some of the structural details of various organelles.* (From Langley, Telford, and Christensen.)

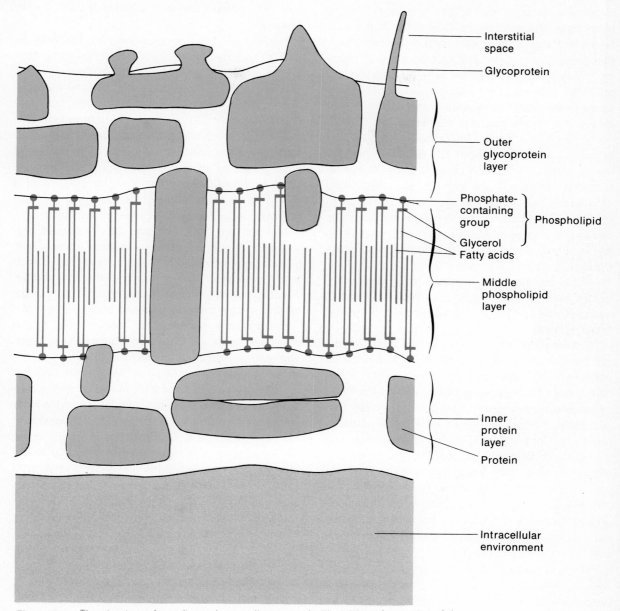

**Figure 5-3** *The structure of a unit membrane: diagrammatic illustration of a portion of the plasma membrane.*

pids are oriented at right angles to the plane of the membrane. The ends of the fatty acids in the sublayers face each other in the center of the membrane. The phosphate-containing groups are polar or asymmetrically charged, while the fatty acid chains are nonpolar. Because of the nonpolar nature of the fatty acids, water-soluble molecules cannot readily diffuse through the middle region of the membrane.

The membrane proteins are probably not restricted to the outer and inner layers but penetrate partially into, or completely through, portions of the phospholipid layer. Moreover, the proteins are most likely aggregated into functional groups by complexing (combining) with each other. The outer surface of the membrane also seems to be structurally unique, in that it contains a variety of *glycoproteins*—proteins containing side chains consisting of sugar residues. The sugar-containing side chains project varying distances into the interstitial spaces, forming a sea of carbohydrate that more or less obscures the membrane surface.

There are countless *pores* in the fabric of the plasma membrane. Communicating between the intracellular and extracellular environments, these holes have a diameter about one-twelfth the thickness of the plasma membrane. The pores are lined with positively charged particles which electrostatically repel small ions of similar charge, thereby impeding their diffusion through these openings. Pores are constantly forming and disappearing throughout the plasma membrane. The number of pores present at any one time is probably determined by the nature and concentration of the ionic environment surrounding the cell. The role of the pores in membrane transport is considered in Chap. 6.

Various specialized structures form from, or are associated with, the plasma membrane. These structures include pinocytotic vesicles, cilia, microvilli, and desmosomes.

*Pinocytotic vesicles* are spherical invaginations of the plasma membrane which form in response to contact with large particles. A portion of the plasma membrane surrounds the particle, and both disappear into the cytoplasm of the cell. The fate of a pinocytotic vesicle will be discussed shortly. This form of membrane transport is described in detail in Chap. 6.

A *cilium* is a short protoplasmic outgrowth of the cell covered by plasma membrane. Possessing contractile properties, a cilium is capable of waving back and forth. Ciliary movements create water currents. The currents carry extracellular particles from one place to another. Cilia are derived indirectly from centrioles (see later) which take up positions immediately below the plasma membrane. Cilia consist of nine groupings of radially arranged microtubules. Each microtubular group in a cilium consists of a pair, or doublet. In addition, there are two separate microtubules in the central portion of a cilium.

*Microvilli* are short, thin, and closely packed protoplasmic projections of the cell surface covered with plasma membrane. Microvilli increase the surface area for the absorption of soluble nutrients and other metabolic raw materials. These structures are characteristic of intestinal epithelial and secretory cells.

*Desmosomes* are structural modifications in the plasma membranes of two contiguous tissue cells. The cell membranes thicken in the desmosome region, causing an increase in the distance between the cells. A system of parallel microfilaments penetrates some distance into the cytoplasm from the inner surface of the thickened plasma membrane in each of the contiguous cells. Desmosomes are believed to link associating cells together more firmly than would otherwise be possible, adding strength and stability to the physical union. The structure of desmosomes is shown in Fig. 5-4.

Plasma membranes perform the following functions:

1   The passive transport of small materials through the pores by diffusion in either direction
2   The facilitated transport of water-soluble nutrients across the nonpolar portion of the double phospholipid layer
3   The active transport of materials from regions in which they are in short supply to regions in which they are in abundance or under the opposite conditions when the rate of transport must be increased to meet metabolic demands

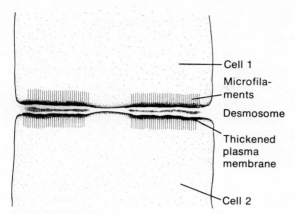

**Figure 5-4**  *The structure of a desmosome.*

4  The pinocytotic engulfment of particles too large to enter by any of the above mechanisms as well as the pinocytotic egestion of indigestible residues from food vacuoles within the cytoplasm

5  The binding of information-rich molecules such as hormones to complementary binding sites on the outer surface of the membrane

6  The recognition of similar cells during tissue differentiation by the formation of complexes between glycoproteins on the outer surfaces of adjacent membranes

7  The reinforcement of intercellular association between tissue cells by means of specialized membrane modifications such as desmosomes

8  The synthesis within the membrane of new phospholipids and the incorporation of new membrane proteins

Mechanisms of cell membrane transport are discussed in Chap. 6.

## ENDOPLASMIC RETICULUM

Located throughout the cytoplasm, the endoplasmic reticulum is continuous with the nuclear membrane, plasma membrane, and Golgi apparatus. When the endoplasmic reticulum was first named, it was thought to exist only in the inner endoplasmic portion of the cytoplasm. We now know that the endoplasmic reticulum is present in the outer ectoplasm as well as in the inner endoplasm, although the original name persists.

The organelle consists of a highly branched grouping of double membranes in the shape of *lamellae* or *vesicles.* Each of the membranes is slightly thinner than the plasma membrane. An extremely small space called a *cistern* separates the two membranes of each lamella or vesicle. Membranous partitions randomly connect the membranes of adjacent lamellae. Each membrane is pock-marked with cavernous openings called *fenestrations* which communicate with the inner cistern. The space between the double membrane opens to the outside of the cell at various points along the plasma membrane. The general organization of the endoplasmic reticulum is diagramed in Fig. 5-2.

The endoplasmic reticulum exists in two forms: (1) When ribosomes are attached to the outer surfaces of the membranes, the endoplasmic reticulum is referred to as *rough,* or granular and (2) in the absence

**Figure 5-5**  *The organization of the endoplasmic reticulum. This portion of the membrane system extends from the nuclear membrane to the plasma membrane.*

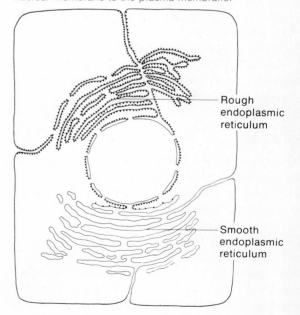

of membrane-bound ribosomes, the endoplasmic reticulum is called *smooth,* or nongranular. Both types of membrane may exist simultaneously in the same cell, depending upon the synthetic activities in which the cell is engaged. The difference between these two forms of reticulum is illustrated in Fig. 5-5.

The endoplasmic reticulum is highly labile, or unstable, some portions disappearing and other parts being built up all the time. The organelle also moves in response to protoplasmic streaming, or *cyclosis.* Participating in various ways in the metabolic activities of a cell, the endoplasmic reticulum may be restricted to one region of a cell or alternatively spread throughout its cytoplasm.

The counterpart of the endoplasmic reticulum in muscle fibers is called the *sarcoplasmic reticulum.* The sarcoplasmic reticulum is involved in various aspects of the contractile process (see Chap. 24).

The following functions have been attributed to the endoplasmic reticulum:

1  The direct transport of various materials through the space between the double membranes from one region of the cell to another and from the intracellular environment to the interstitial space
2  The storage of export proteins, which will be subsequently packaged within the Golgi apparatus before secretion
3  The provision of a membranous surface upon which chemical reactions can take place
4  The possible maintenance of an internal structural organization within the cytoplasm

## RIBOSOME

Ribosomes are the primary chemical factories on which proteins are synthesized. Chains of amino acids called *polypeptides* are built up on the surfaces of ribosomes. If the chain is long enough to qualify as a protein, then a protein has indeed been synthesized on the ribosome. More often than not, however, proteins consist of multiple chains of polypeptides. In these cases, proteins must somehow be assembled from various polypeptide chains synthesized on different ribosomes. How the chains are put together into a complex protein is not known with certainty.

Ribosomes are extremely small organelles, even by cellular standards. In some cells or cellular regions, the ribosomes are bound to the outer portions of the membranes of the endoplasmic reticulum. In other cells or cellular regions, the ribosomes exist free in the cytoplasm.

Ribosomes are themselves assembled in the cytoplasm from several types of nucleic acid known as *ribosomal ribonucleic acid* (rRNA) and from many different kinds of protein molecules. Since they are composed of rRNAs and proteins, ribosomes are also called *ribonucleoprotein particles.* The evidence to date indicates that the nucleic acids are in the core of each particle, surrounded by the various ribosomal proteins.

Ribosomes are composed of two subunits—a large particle and a small particle. Each of these subunits consists of about one-third protein and two-thirds rRNA. Thirty different kinds of protein and two types of rRNA have been identified in the large subunit, while twenty different proteins and one specific rRNA have so far been discovered in the makeup of the small subunit. The rRNAs of the ribosome are synthesized from nuclear DNAs (deoxyribonucleic acids) of chromosomes which course through the substance of the nucleolus. The rRNAs are presumably stored in the nucleolus and must enter the cytoplasm before the subunits of the ribosome can be constructed. Some of these ideas are conceptualized in Fig. 6-4.

Once the subunits are put together, the ribosome can participate in protein synthesis. The details of protein synthesis are described in Chap. 6. Here we shall point out that the ribosomes "read" the genetic messages transcribed from genes on the chromosomes to RNAs (ribonucleic acids) called *messenger RNAs* (mRNAs), mentioned in Chap. 4.

During the translation of the instructions carried by one mRNA, many ribosomes line up in single file and move across the surface of the mRNA. During this stage of protein synthesis, ribosomes are often called *polyribosomes,* or simply *polysomes.* The amino acids used in the synthesis of proteins are carried to the ribosomes by still another type of RNA called *transfer RNA* (tRNA). The small subunit of the ribosome contains a binding site for mRNA, while its large subunit contains binding sites for tRNA. The large subunit also binds to rough endoplasmic reticulum. Presumably, ribosomes also have a role in recognizing where to begin reading the genetic message transcribed onto mRNA.

Although the roles of the ribosomal proteins are not completely understood, many of them are necessary for the accurate functioning of the ribosomes during protein synthesis. Some of the ribosomal proteins are probably responsible for the ability of ribosomes to move along the length of mRNAs during the translation of genetic messages.

## THE GOLGI APPARATUS

Cells synthesize proteins for three basic purposes: (1) to form structural components of the cell, (2) to provide enzymes for the mediation of chemical reactions in the cell, and (3) to form products including hormones, enzymes, and antibodies for export to other parts of the body. The Golgi apparatus is involved in the structural modification of various proteins and their packaging in preparation for subsequent secretion or structural incorporation into the plasma membrane.

The Golgi apparatus is a prominent organelle found next to the nuclei in secretory cells. Cuplike in shape, this organelle consists of a series of concentrically arranged and discontinuous membranes. The packaging of the secretory proteins takes place inside the "cup." The Golgi apparatus and its role in protein export are diagramed in Fig. 5-6.

Structural and enzymatic proteins are synthesized

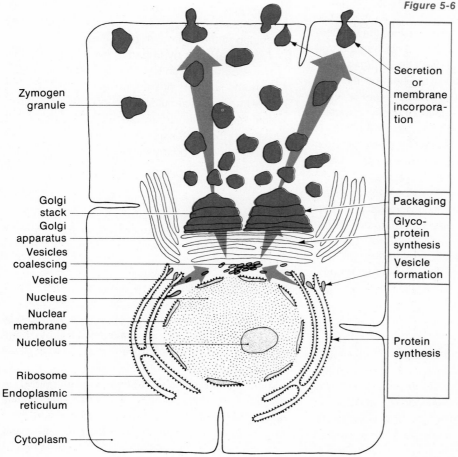

**Figure 5-6** *The structure and functions of the Golgi apparatus.*

Zymogen granule

Golgi stack
Golgi apparatus
Vesicles coalescing
Vesicle
Nucleus
Nuclear membrane
Nucleolus
Ribosome
Endoplasmic reticulum
Cytoplasm

Secretion or membrane incorporation

Packaging
Glycoprotein synthesis
Vesicle formation

Protein synthesis

on smooth endoplasmic reticulum, while membrane and secretory proteins are manufactured on membrane-bound ribosomes—the so-called rough endoplasmic reticulum. Following their synthesis, secretory proteins are secreted into the space between the double membrane of the rough endoplasmic reticulum. As proteins accumulate in these spaces, portions of the endoplasmic reticulum surround them, forming vesicles which are ultimately set free in the cytoplasm. Numerous individual vesicles coalesce, or come together, near the base of the Golgi apparatus. The vesicular aggregate enters the Golgi apparatus for physical and chemical processing. The proteins are condensed, thereby becoming concentrated into much smaller packets. Sugar residues are also added to most or all of the processed proteins, forming compound molecules called *glycoproteins*. Glycoproteins chemically resemble both carbohydrates and proteins.

Following these stages of processing, the concentrated and chemically modified proteins undergo packaging in unique structures called *Golgi stacks*. The Golgi stacks are located in the basin of the cup-shaped Golgi apparatus. The lowest layers are thin and wide. The layers of the stacks become progressively thicker and shorter as the top of the cup is approached. Packets of glycoprotein called *zymogen* or *mucus granules* split away from the upper layers of the Golgi stacks and enter the cytoplasm. As the granules accumulate, some make contact and fuse with the plasma membrane. Following fusion, a portion of the glycoproteins may be incorporated into the structure of the plasma membrane. The remainder of the protein is secreted from the cell as the granules pass pinocytotically through the plasma membrane.

## MITOCHONDRION

A mitochondrion is a sausage- or spherically shaped organelle distributed throughout the cytoplasm, especially in regions where there are high energy demands. The number of mitochondria in a cell is highly variable and is determined by its average energy requirements. For example, motile cells or gland cells continually engaged in synthetic reactions have a larger number of mitochondria than quiescent, connective tissue cells. Each tissue cell possesses about 200 mitochondria on an average, although the actual number per cell ranges between 20 and 2,000.

Each mitochondrion is bounded by a double membrane. Between the two membranes is a small space which serves as both corridor and storeroom for chemicals entering from the cytoplasm. The inner membrane contains incomplete shelflike partitions called *cristae* which project into a fluid matrix. Important and highly complex metabolic reactions take place on enzymes located on the outer membrane and along the cristae of the inner membrane.

Mitochondria have the following functions:

1 The aerobic degradation of intermediary metabolites is mediated by enzymes located on the cristae of the inner membrane.
2 The oxidation of hydrogen is mediated by oxidative enzymes which project horizontally in stacks from the surfaces of the cristae.
3 The trapping of about 95 percent of the total chemical energy utilized by a cell is accomplished through the synthesis of adenosine triphosphate (ATP) during a poorly understood synthetic reaction which presumably occurs on the surface of the inner membrane.
4 Synthetic reactions such as the assembly of lipids from appropriate raw materials in the presence of ATP are believed to take place on the surface of the outer membrane.

The primary role played by mitochondria in releasing and trapping chemical energy locked in the bonds of metabolic fuels has conferred upon these organelles the nickname "powerhouse of the cell." The biochemical activities of mitochondria are considered in detail in Chap. 6.

Mitochondria are *self-replicating* organelles. In other words, mitochondria reproduce themselves without the entire cell's undergoing division. This unique cellular power plant contains DNA, RNA, and ribosomes for the synthesis of numerous proteins incorporated into its structure. The cell nucleus is also known to provide genetic information to mitochondria. The semiautonomous nature of mitochondria has led to speculation that they were once parasitic microbes which, in the course of time, have become indispensable mutualistic symbionts within their host cells. Other authorities take the view that mitochondria evolved from infoldings of the plasma membrane of some ancestral eukaryotic cell. The infoldings presumably formed specialized vesicles which subsequently acquired other structural advances to meet functional demands. These demands necessitated an increased surface area on which to carry out aerobic respiration. According to this theory, DNA and RNA were incorporated to circumvent the relative impermeability caused by the presence of membranes. This controversy cannot be resolved on the basis of the present evidence.

## LYSOSOME

Lysosomes seem to be common to all animal cells. Variable in size and shape, lysosomes are only about half as large as mitochondria. These peculiar organelles are surrounded by a single membrane and contain a variety of enzymes for the digestion of proteins, fats, carbohydrates, and nucleic acids. Why the lysosomal membrane is not digested by these enzymes is not known. Presumably, some property of the lysosomal membrane prevents enzymatic digestion of the surrounding protoplasm.

Lysosomes have been implicated in the following important cellular functions:

1 The intracellular digestion of large foreign particles which enter the cytoplasm pinocytotically and the pinocytotic egestion of indigestible food residues
2 The intracellular digestion of portions of the cytoplasm to provide a source of metabolic fuels under conditions of cellular starvation
3 The autolysis, or self-destruction, of dead and dying tissue cells and of phagocytes following ingestion of a full complement of antigens
4 The extracellular secretion of lysosomal enzymes for the erosion of adjacent tissue cells during their rebuilding, as in the case of bone, or the provision of a pathway for cellular penetration, as in the case of fertilization of a developing ovum by a spermatozoon

The intracellular digestive role of lysosomes is depicted in Fig. 5-7. Assume that a large particle enters a tissue cell pinocytotically. After entering the cytoplasm, the pinocytotic vesicle containing the foreign particle is called a *food vacuole*. The contents of the food vacuole are digested by lysosomal enzymes.

The lysosomal enzymes are synthesized on rough endoplasmic reticulum. Following their synthesis, the enzyme molecules are packaged within a membrane, giving rise to an organelle called a *lysosome*.

Digestion of the particle within the food vacuole begins when the lysosome contacts and fuses with the vacuole. As a consequence of this union, the enzymes of the lysosome enter the food vacuole and begin to degrade the foreign particle contained within. The intact membrane of the compound structure still prevents the lysosomal enzymes from entering the cytoplasm.

As the particle within the food vacuole is broken down by the lysosomal enzymes, soluble end products of digestion are absorbed into the cytoplasm by diffusion. Finally, the digestive vacuole contacts and fuses with the plasma membrane. The indigestible waste products and spent enzymes are subsequently eliminated from the cell by a process which resembles pinocytosis but takes place in the reverse direction.

Autolysis occurs when the lysosomal membranes are disrupted. A sharp decrease in the amount of oxygen supplied to a cell and the presence of certain chemicals are known to promote lysosomal membrane rupture. Other chemicals seem to increase the structural integrity of lysosomal membranes. The significance of these diverse influences on cellular homeostasis has yet to be determined.

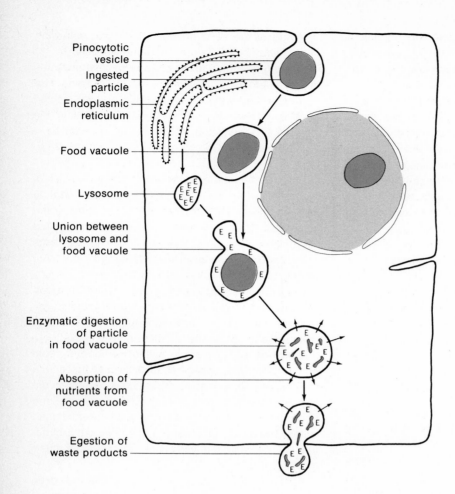

Pinocytotic vesicle

Ingested particle

Endoplasmic reticulum

Food vacuole

Lysosome

Union between lysosome and food vacuole

Enzymatic digestion of particle in food vacuole

Absorption of nutrients from food vacuole

Egestion of waste products

**Figure 5-7** *The intracellular digestion by lysosomal enzymes of large particles which enter by pinocytosis.*

## CENTRIOLE AND SPINDLE

Two centrioles and the spindle associated with them play a prominent role during cellular division in animal cells generally. Their functions in mitosis are described in detail in Chap. 6. The centrioles and the spindle are both derived from aggregates of submicroscopic structures called *microfilaments*. Microfilaments seem to be present in the cytoplasm of all tissue cells examined under the electron microscope. Microfilaments are somehow responsible for protoplasmic movements of various sorts and may also contribute to the cytoskeleton of a cell by providing internal support and rigidity.

The structure of a centriole is illustrated in Fig. 5-8.

A centriole is a self-replicating organelle located in the cytoplasm next to the nucleus. Two centrioles can be observed usually at right angles to each other. Each centriole is a small cylinder, closed at one end and open at the other, composed of nine radially arranged groupings of *microtubules*. Each radial group consists of three microtubules which run in parallel. In other words, a centriole is made of 27 microtubules, arranged in nine radial triplet groupings. There is some evidence to indicate that each microtubule is comprised of an aggregate of 10 to 15 microfilaments. A series of spoke-like microtubules or similar structures usually radiates in all directions in the region surrounding each centriole. Called *asters* because of their star-shaped appearance, these structures become visible under a light microscope during cellular division.

During mitosis, the centrioles migrate toward opposite poles of the cell and are involved in organizing other microfilaments in the cytoplasm to form a series of parallel microtubules. Coursing through the center of the cell between the two asters, the microtubules are called *spindle fibers.* The spindle fibers are referred to collectively as a *spindle.* The spindle is believed to organize the chromosomes in the center of the cell, thereby ensuring that one complete set of chromosomes will be allocated to each of the two cells resulting from division.

## NUCLEUS

This organelle contains practically all the hereditary or genetic information required to direct the metabolic and perpetuative activities of a cell. Although variable in size, shape, and intracellular location, the nucleus is frequently large, spheroid, and centrally located.

The outer boundary of a nucleus consists of a dou-

ble membrane continuous with those of the endoplasmic reticulum (see Fig. 5-5). The space between the two parts of the double membrane is called the *perinuclear space.* The double nuclear membrane is frequently referred to as a *nuclear envelope.* The nuclear envelope is discontinuous, giving rise to relatively large openings called *nuclear pores.* Presumably, the nuclear pores permit a two-way flow of materials between the nucleus and cytoplasm.

The protoplasm contained within the nuclear envelope is called *nucleoplasm.* Because of the nuclear pores and the continuity between the nuclear envelope and the endoplasmic reticulum, the nucleoplasm is continuous with the cytoplasm. Within human tissue cell nucleoplasm are located 46 *chromosomes* (23 pairs) and one or more *nucleoli* (plural of *nucleolus*).

In a cell not engaged in division, the chromosomes appear as thin elongated threads distributed throughout the nucleoplasm. During cellular division, the chromosomes become short, thick, and discrete (separate). Hereditary units called *genes* are arranged in linear order along the length of each chromosome. A chromosome consists of a type of nucleic acid called

**Figure 5-8** *The structure of a centriole. The two centrioles near the nucleus of a cell are ordinarily at right angles to each other, as shown. The left-hand centriole is viewed from lateral aspect. The dark bands are triplet groupings of microtubules. Note that one end of the centriole is closed, the other end open. The right-hand centriole is viewed on end. There are nine groupings of triplet microtubules on the periphery of a centriole.*

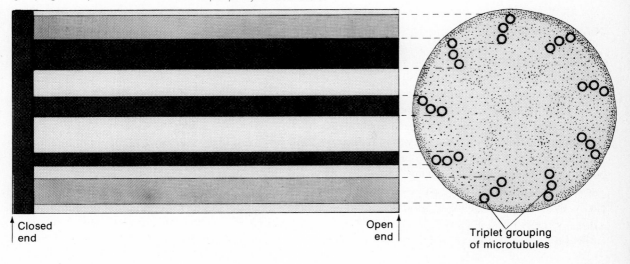

Closed end
Open end
Triplet grouping of microtubules

*deoxyribonucleic acid* (DNA) and protein. Genetic information is encoded in the sequence of nitrogenous purine and pyrimidine bases contained in the DNA molecule of each gene. The DNA portion of each chromosome synthesizes an exact replica of itself prior to cellular division, thereby giving rise to a duplicate set of chromosomes. or synthesizes slightly different molecules, the *messenger ribonucleic acids* (mRNAs) discussed earlier, each of which carries to the cytoplasm coded instructions for the synthesis of one or several specific proteins. Therefore, the chromosomes not only play a primary role during cellular division but also are responsible for the nature of the metabolic activities taking place in the cytoplasm. The manner in which these activities take place is discussed in Chap. 6.

One or more dark-staining nucleoli are found in the nuclei of tissue cells. The nucleolus is not an organelle. Instead, it gives the appearance of an aggregation of loosely organized particles. The nucleolus is a storage center for two types of nucleic acid called *transfer ribonucleic acid* (tRNA) and *ribosomal ribonucleic acid* (rRNA), both of which were mentioned earlier in this chapter. These RNAs perform important functions during protein synthesis in the cytoplasm (see "Protein Synthesis," Chap. 6).

Nucleoli disappear during the onset of cell division and reappear when division has been completed. The regeneration of a nucleolus is apparently under the control of nucleolus-forming centers located on one of the 23 pairs of chromosomes.

## OTHER CELLULAR INCLUSIONS

Various cells contain one or more types of inclusion in the cytoplasm. These inclusions often reflect the nature of the metabolic activities performed by the cell or the kinds of substances in the interstitial fluid surrounding it. The following are among the diverse groupings of cellular inclusions:

1 *Zymogen* or *mucus granules,* in which enzymes or secretory chemicals are contained, respectively (see "The Golgi Apparatus," above)
2 *Fat globules* in certain secretory and adipose connective tissue cells
3 Various types of *vacuoles,* presumably formed by a union of pinocytotic vesicles and lysosomes (see "Lysosome," above)
4 Numerous kinds of *granules,* such as glycogen granules in hepatic cells
5 *Crystals,* either free in the cytoplasm or enveloped by lysosomes

## COLLATERAL READING

Brachet, Jean, "The Living Cell," *Scientific American,* September 1961. A general review of the fine structure of a cell.

Culliton, Barbara J., "Cell Membranes: A New Look at How They Work," *Science,* 175:1348–1350, 1972. While dealing mainly with the molecular structure of cell membranes, this review emphasizes the problems encountered in attempting to formulate models of membranes which measure about 1/100,000 millimeter in thickness.

De Duve, Christian, "The Lysosome," *Scientific American,* May 1963. A general review of the structure and function of lysosomes.

Fox, C. Fred, "The Structure of Cell Membranes," *Scientific American,* February 1972. An excellent summary of current knowledge regarding cell membrane structure.

Goodenough, Ursula W., and R. P. Levine, "The Genetic Activity of Mitochondria and Chloroplasts," *Scientific American,* November 1970. Goodenough and Levine suggest that mitochondria may have originated as microbial parasites which persisted by evolving a mutualistic relationship with their host cells.

Green, David E., "The Mitochondrion," *Scientific American,* January 1964. A lucid review of the molecular structure of this subcellular organelle, emphasizing its role in electron or hydrogen transport and the synthesis of ATP.

Holter, Heinz, "How Things Get into Cells," *Scientific American,* September 1961. A review of basic passive and active membrane transport mechanisms. Holter regards pinocytosis as a mechanism to increase the efficiency of differential transport inside the cell, rather than as an alternative to standard transport mechanisms at the cell membrane.

Kurkland, C. G., "Ribosome Structure and Function Emergent," *Science,* 169: 1171–1177, 1970. An advanced review of ribosomal organization.

Raff, Rudolf A., and Henry R. Mahler, "The Non-Symbiotic Origin of Mitochondria," *Science,* 177:575–582, 1972. Although it is widely held that mitochondria arose from intracellular symbionts, Raff and Mahler offer the following alternative hypothesis: (1) that nuclear-cytoplasmic and nuclear-mitochondrial interactions arise in prokaryotic cells and are maintained for different purposes in eukaryotic cells and (2) that the different regulatory functions of these two systems require compartmentalization.

Robertson, J. David, "The Membrane of the Living Cell," *Scientific American,* April 1962. Evidence from electron microscopy suggests that cells are composed of three phases—cytoplasm; the cavity of the endoplasmic reticulum; and the unit membrane which delimits the nucleus, cytoplasmic compartments, and the outer boundary.

Satir, Peter, "Cilia," *Scientific American,* February 1961. A general review of the structure of a cilium.

Solomon, Arthur K., "Pores in the Cell Membrane," *Scientific American,* December 1960. Using some extremely clever techniques, Solomon concludes that the pore diameter in biologic membranes is approximately 8 Å (angstroms).

Wessells, Norman K., "How Living Cells Change Shape," *Scientific American,* October 1971. Microtubules and microfilaments discovered in various types of cells serve as skeleton and muscle, respectively, during cellular movement. Wessells describes the activation of this system by the release of $Ca^{++}$.

Whaley, W. Gordon, Marianne Dauwalder, and Joyce E. Kephart, "Golgi Apparatus: Influence on Cell Surfaces," *Science,* 175:596–599, 1972. The authors conclude that the Golgi apparatus probably synthesizes carbohydrates and definitely links carbohydrate groups to proteins and possibly to lipids, thereby forming informationally rich macromolecules.

# CHAPTER 6
# CELLULAR ACTIVITIES
# AND ASSOCIATIONS

To sustain life, each cell must possess mechanisms which provide for both energy and information flow. Thus, the qualities which characterize living things revolve around the processing of energy and information.

Since each cell is a highly organized "bag" of chemicals, we must look at the regulatory activities of intracellular molecules and ions for answers to how cells carry on metabolic activities, i.e., the provision of energy and the biosynthesis of organic compounds, and perpetuative activities, e.g., cellular division. The complexity of the molecular events which occur routinely in cells may be appreciated by noting that there are probably between 2,000 and 10,000 different types of enzyme and approximately 1 million to 2 million genes on the 46 chromosomes in every tissue (somatic) cell in the body. The regulation of the flow of energy and information is inextricably intertwined with the actions of enzymes and genes, respectively.

The tissue cells of the human body associate and communicate with each other. Both these abilities are promoted by specific chemicals. The coordinated activities in a mass of similar cells and between several groupings of different cells have led to the establishment of tissues and the development of organs. The functions of the human body in health and disease are determined by the nature of the reciprocal interactions between the groupings of organs which make up its organ systems.

The tissues, organs, and organ systems maintain homeostasis of the human body, just as genes, enzymes, and other regulatory chemicals initiate and sustain cellular metabolism and perpetuation. Homeostatic disruptions, whether mild or severe, at any one level of organismic organization cause reverberations at all other levels.

In this chapter, we shall explore in more detail the mechanisms by which tissue cells maintain homeostasis. An examination of the metabolic and perpetuative activities of individual cells should provide a general appreciation of how cells interact cooperatively as members of tissues.

## CELLULAR HOMEOSTASIS

### ENERGY FLOW

Cells require energy to sustain both metabolism and perpetuation. Energy is obtained at various times during the intracellular degradation of soluble foodstuffs. Simple sugars, fatty acids and glycerol, and amino acids are the raw materials used as metabolic fuels. The fuels are oxidized, liberating energy which is trapped chemically and subsequently transported to cellular sites of utilization.

The production and utilization of energy    That aspect of metabolism in which raw materials are degraded intracellularly is called *catabolism,* while the opposite process is called *anabolism.* The energy released during the breakdown of metabolic fuels is trapped chemically and utilized for various types of biologically useful work. The regulation of energy flow within each cell is highly complex and depends in part upon the availability of metabolic raw materials, oxygen, enzymes, vitamins, and other regulatory chemicals which participate in the diverse reactions by which energy is liberated.

To obtain useful energy during the combustion of fuels, tissue cells have had to overcome two basic problems: (1) to sustain combustion reactions at body temperature and (2) to trap the energy in a form which can be utilized for useful work. That cells have been admirably successful in surmounting these problems cannot be denied. The first stumbling block was circumvented by developing a chemical mechanism for the incremental release of energy from available fuels rather than the release of all the energy at one time. The second hurdle was surmounted by trapping a portion of the liberated energy in a chemical bond within a special molecule used universally as the "energy currency of cells," just as money is used universally for the purchase of goods and services. These two innovations enable cells, even while engaged in combustion reactions, to operate at room temperature, to trap and transport energy to sites of energy utilization, and to match energy supply with energy demands.

*Overview*    The fuels from which tissue cells derive energy include simple sugars, amino acids, and fatty acids and glycerol, or their derivatives. These fuels are degraded intracellularly during a complex series of chemical reactions referred to sequentially as *glycolysis* and the *citric acid cycle,* or Krebs cycle (tricarboxylic acid cycle). Glycolysis occurs in the cytoplasm; the citric acid cycle takes place in the mitochondria. Each of the individual reactions is mediated by a specific enzyme. Numerous reactions in the degradative sequence require coenzymes or vitamins and mineral activators (see Chap. 26).

During degradation, organic fuels usually containing six carbon atoms are reduced to the waste product $CO_2$, an inorganic compound containing one carbon atom. Many of the reactions result in the removal of hydrogen ($H_2$) from the fuel and the binding of hydrogen to special molecules called *hydrogen carriers.* Within the mitochondria, hydrogen is shuttled through a series of oxidative enzymes. The ensuing sequence of oxidation-reduction reactions releases large amounts of energy. Almost half of the energy is used to synthesize ATP (adenosine triphosphate). Synthesized from adenosine diphosphate (ADP), ATP is the energy currency of cells. The oxidation of hydrogen is completed when it contributes to the formation of $H_2O$ (water). The $H_2O$ enters the general metabolic pool of the cell. Since the $O_2$ of metabolically formed $H_2O$ is obtained through external respiration, the oxidation of $H_2$ is called *cellular respiration.* Most of the energy released from a fuel molecule occurs during oxidation within the mitochondria. Only a small amount of ATP is ordinarily synthesized in the cytoplasm, except in certain specialized cells.

The overall reaction during the oxidation of a fuel such as glucose is expressed as

$$C_6H_{12}O_6 + 6O_2 \xrightarrow[\text{hydrogen carriers, ADP}]{\text{enzymes, coenzymes, mineral activators}}$$

$$6CO_2 + 6H_2O + 38 \text{ ATP} + \text{heat} \quad (6\text{-}1)$$

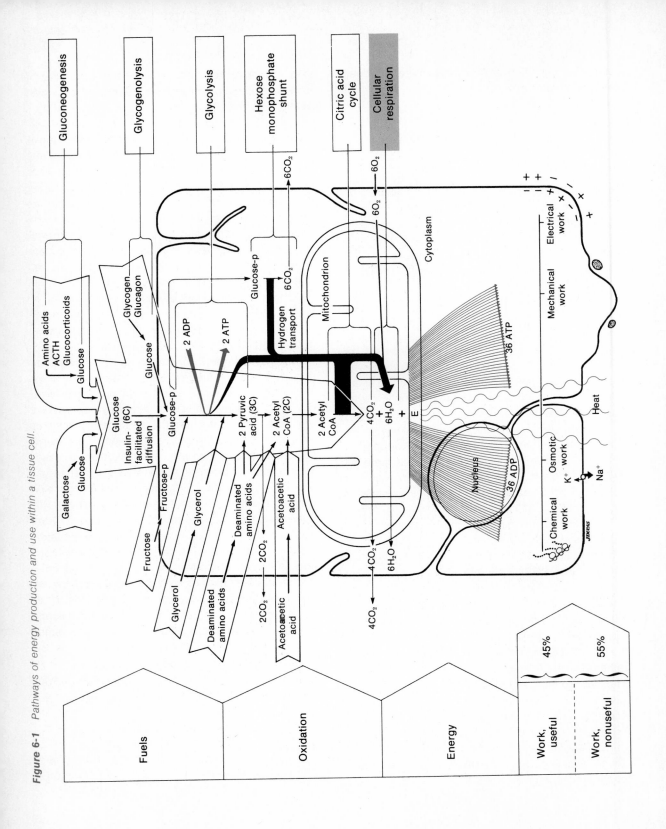

**Figure 6-1** Pathways of energy production and use within a tissue cell.

This summary of cellular respiration states that six parts of oxygen are used in the oxidation of one part of glucose, resulting in the formation of six parts of both carbon dioxide and water, the synthesis of 38 parts of adenosine triphosphate, and the liberation of heat. The sequence of some 30 chemical reactions summarized in Eq. (6-1) occurs in about one second, attesting to the high degree of molecular organization found within cells.

Although the heat energy maintains a constant body temperature, it cannot be utilized metabolically and therefore is essentially wasted. The ATP represents useful energy and is used in various types of biologic work including biosynthetic reactions, active transport, muscle contraction, and nerve impulse conduction.

The pathways of energy production and use within a tissue cell are given in Fig. 6-1. We will now examine these pathways more closely.

*The metabolic fuels*   Glucose is the primary energy nutrient of tissue cells. This 6-carbon organic compound is absorbed from the digestive tract. After processing in the liver, glucose is transported through the blood to all the tissue cells of the body. Glucose moves through plasma membranes in response to a concentration gradient by a passive transport mechanism called *facilitated diffusion.*

During transport, glucose is bound to a carrier molecule which is soluble in the middle lipoid region of the membrane. Since glucose is not lipid-soluble, it enjoys a free ride through an otherwise impermeable barrier. The secretion of insulin from the islets of Langerhans of the pancreas facilitates the uptake of glucose. Facilitated diffusion is discussed in detail later in this chapter.

Once within a tissue cell, glucose is enzymatically phosphorylated by the addition of an inorganic phosphate group (-p). *Phosphorylation* activates the glucose molecule, thereby preventing it from diffusing out of the cell and preparing it for subsequent degradation.

Two other simple sugars, *fructose* and *galactose,* are also absorbed in abundance directly from the digestive tract. Fructose, a 5-carbon sugar, can move from the blood into tissue cells by facilitated diffusion and, following phosphorylation, can enter the glycolytic pathway of metabolism to be discussed below. Galactose, however, must be converted by the liver into glucose before it can be utilized intracellularly for energy.

Under stress or during an insufficiency of blood glucose, glycogen is broken down into glucose. The process is called *glycogenolysis.* Glycogenolysis results in the progressive splitting off of the terminal glucose residues from glycogen molecules following their phosphorylation. Phosphorylation occurs in response to the activation of adenylate cyclase (see Chap. 17).

Glycogenolysis is promoted by the hormones epinephrine and glucagon. Secreted by the adrenal medullae in response to stress, epinephrine promotes glycogenolysis in all the tissue cells of the body. This adaptive response mobilizes body resources to meet the physiologic demands imposed by stress. *Glucagon,* on the other hand, promotes glycogenolysis differentially in the liver. Glucagon is secreted by the islets of Langerhans of the pancreas when the blood glucose concentration drops below the normal level of 90 mg/100 ml blood.

During a depression in the concentration of blood glucose, the liver converts other soluble end products of digestion such as amino acids and glycerol into glucose. Called *gluconeogenesis,* this process requires the secretion of three hormones—corticotropin-releasing factor (CRF) from the hypothalamus, adrenocorticotropic hormone (ACTH) from the adenohypophysis (anterior lobe of the pituitary gland), and the glucocorticosteroids, especially cortisol, from the adrenal cortex. Cortisol promotes the degradation of proteins, causing an increase in the blood concentration of amino acids. These amino acids are then processed in the liver and subsequently used in the synthesis of glucose. Part of the general adaptation syndrome (GAS), this adaptive response is reviewed in Chap. 4.

Amino acids can also be used by cells as energy metabolites. Before amino acids can be oxidized, they must be deaminated in the liver. During *deamination,* the amino groups ($NH_2$) containing nitrogen atoms are removed, forming ammonia ($NH_3$). Ammonia then combines with carbon dioxide ($CO_2$), giving rise to urea, the main nitrogenous constituent of urine. Following their release into the blood, the processed amino acids are actively transported into tissue cells and added at several points to the intracellular pathways which yield energy, as shown in Fig. 6-1.

The soluble end products of lipid digestion ordinarily provide almost half of the energy liberated by cells. After diffusing into tissue cells, glycerol is incorporated into the glycolytic pathway in the cytoplasm. However, fatty acids must be processed by the liver before they can be used in energy metabolism. In the liver, 2-carbon fragments of fatty acids containing 16 or 18 carbons in a chain are progressively split off by a process called *beta oxidation*. The 2-carbon fragments resulting from beta oxidation combine with a molecule called *coenzyme A* (CoA), forming an acetylated molecule known as *acetyl CoA*. In the presence of $H_2O$, two acetyl CoA molecules are converted into one molecule of *acetoacetic acid*. These molecules rapidly diffuse from the liver cells into the blood and thence into the tissue cells. Within the tissue cells, acetoacetic acid is reconverted into two molecules of acetyl CoA, as shown in Fig. 6-1. Acetyl CoA enters the main pathway of energy metabolism during its participation in the formation of citric acid within the mitochondria.

*Glycolysis* Phosphorylated glucose and certain other metabolic fuels, as pointed out above, enter into a series of enzymatically mediated chemical reactions called *glycolysis*. Glycolysis occurs in the cytoplasm of tissue cells. During glycolysis, each molecule of glucose-p (phosphorylated glucose), a 6-carbon sugar, is transformed into two molecules of *pyruvic acid*, a 3-carbon intermediary metabolite. As a result of glycolysis, four molecules of ATP are synthesized. The synthesis occurs through the transfer of high-energy phosphate groups from several of the intermediary metabolites to ADP. Since one molecule of ATP is consumed in the original phosphorylation of glucose and another ATP is used in one of the glycolytic reactions, a net of two molecules of ATP is generated.

During glycolysis, four hydrogen atoms are removed from one of the intermediary metabolites and bound to a cytoplasmic hydrogen carrier molecule. In the presence of $O_2$, these hydrogen atoms are released from their carriers and processed in the mitochondria. The oxidations in which hydrogen participates liberate enough energy to cause the synthesis of six additional ATPs, on an average.

Under anaerobic conditions, pyruvic acid is transformed into *lactic acid,* which, in lieu of $O_2$, serves as the final hydrogen acceptor. Thus, anaerobic glycolysis yields a net of two ATPs. This quantity of ATP is only about 5 percent of the total amount that can be obtained during the oxidation of glucose under aerobic conditions. Although a highly inefficient process, anaerobic glycolysis becomes important when vigorous physical exercise forces the skeletal muscles to incur an $O_2$ debt (see Chap. 24). Lactic acid is reconverted into pyruvic acid when $O_2$ is again made available in adequate amounts.

*The formation of acetyl CoA* Each molecule of *pyruvic acid,* a 3-carbon compound, is converted into one molecule of *acetyl CoA*, a 2-carbon intermediate. Requiring the participation of *thiamine* (vitamin $B_1$—see Chap. 26), this complex oxidation results in the removal of two hydrogen atoms and the liberation of one molecule of $CO_2$ from each pyruvic acid molecule processed, or four hydrogen atoms and two molecules of $CO_2$ per molecule of glucose. Under aerobic conditions, the hydrogen is accepted by hydrogen carrier molecules and oxidized enzymatically within the mitochondria. The oxidation of four hydrogen atoms results in the synthesis of six ATPs. Within the mitochondria, the 2-carbon acetyl group of acetyl CoA combines with *oxaloacetic acid*, a 4-carbon acid, forming a 6-carbon intermediary metabolite known as *citric acid*.

*The citric acid cycle* During this cycle of reactions

mediated by highly ordered enzymes located on the cristae in the mitochondria, the 2-carbon acetyl group from acetyl CoA is degraded enzymatically to two molecules of $CO_2$. Since two molecules of acetyl CoA are generated per molecule of glucose degraded, four molecules of $CO_2$ are produced during two complete revolutions of the citric acid cycle. CoA is set free once the acetyl group is transferred to oxaloacetic acid. Each full swing of the cycle regenerates oxaloacetic acid. Sixteen hydrogen atoms are liberated from various intermediary metabolites during two full swings of the citric acid cycle. These hydrogen atoms are oxidized in the presence of $O_2$ by a series of oxidative enzymes also located on the cristae in the mitochondria. These oxidative reactions yield enough energy for the synthesis of 24 ATPs.

*Hydrogen transport*　Hydrogen atoms are released during glycolysis, the conversion of pyruvic acid to acetyl CoA, and the citric acid cycle. These hydrogen atoms are all transported by appropriate hydrogen carriers to a system of oxidative enzymes called *cytochromes* located in the mitochondria.

The cytochromes are protein catalysts, even though they are nonspecific and act only in the presence of other specific oxidizing enzymes. Indeed, some authors call the cytochromes *coenzymes*, because they aid the functions of other enzymes. The iron-containing portions of the cytochromes are responsible for their oxidizing abilities.

During its transport by the cytochrome system, each hydrogen atom ionizes into a *proton* and *electron*. The electrons are shuttled from one oxidative enzyme to the next in the mitochondrial transport system, during which time each enzyme in succession is alternately oxidized and reduced. During this sequence, about 95 percent of the energy locked in the chemical bonds of the original fuel is released. In the presence of ADP, almost half of the liberated energy is trapped in the high-energy phosphate bonds of ATP. *Oxygen* is the final electron acceptor, forming *hydroxyl* ions ($OH^-$). Each $OH^-$ readily combines with a free proton, resulting in the formation of $H_2O$.

The energy yield during the oxidation of glucose can be summarized as follows:

| Process | Net energy yield | |
| --- | --- | --- |
| | Cytoplasmic | Mitochondrial |
| *1*　Anaerobic glycolysis | 2 ATPs | |
| *2a*　Aerobic glycolysis | 2 ATPs | 6 ATPs |
| *b*　Pyruvic acid ⟶ acetyl CoA | | 6 ATPs |
| *c*　Citric acid cycle | | 24 ATPs |

Thus, one molecule of glucose oxidized anaerobically yields 2 ATPs, while the aerobic oxidation of the same molecule yields a net of 38 ATPs.

*The hexose monophosphate shunt*　In liver and brain cells, there is an alternative pathway of energy metabolism which bypasses both glycolysis and the citric acid cycle. The pathway is called the *hexose monophosphate shunt*. Glucose is completely degraded in the cytoplasm via this pathway, forming six molecules of $CO_2$ and releasing 24 atoms of hydrogen. The hydrogen is transported by appropriate hydrogen carriers to the mitochondria and oxidized by the cytochrome system in the presence of $O_2$. Because the two ATPs generated cytoplasmically during glycolysis do not appear and one ATP is used in the phosphorylation of glucose, the energy yield by this pathway is slightly less than that resulting from glycolysis and the citric acid cycle. Nevertheless, the hexose monophosphate shunt may normally provide as much as 30 percent of the useful energy in certain cells.

*The efficiency of cellular respiration*　The aerobic oxidation of 1 molecule of glucose yields 38 molecules of ATP. Thus, 1 mole (atomic weight in grams) of glucose will yield 38 moles of ATP. One mole of glucose equals 180 Gm. When 1 mole of glucose undergoes complete combustion in a furnace, 686,000 cal of heat energy is liberated. Each gram of glucose therefore liberates 3800 cal heat energy. When 38 moles of ATP are utilized biologically, approximately 304,000 cal of energy is liberated. Thus, each mole of

ATP liberates 8000 cal of energy. The efficiency of cellular respiration is determined by dividing the biologic output, 304,000 cal, by the physical input, 686,000 cal. Slightly better than 45 percent, this efficiency rivals that of our best internal combustion engines!

*Energy utilization*   Energy released during the oxidation of foodstuffs is used in various types of physical and chemical work, including (1) the *biosynthesis* of proteins, fats, carbohydrates, and nucleic acids as well as the degradation of metabolic fuels; (2) the *active transport* of ions and molecules across cellular membranes, a phenomenon referred to as *osmotic work;* (3) the *contraction* of muscle fibers and all *movements* exhibited by other types of cells; and (4) the conduction of nerve impulses and the generation of *bioelectricity.*

The remainder of the energy liberated from glucose is evolved as *heat.* Although not available for metabolic uses, the liberated heat energy holds the body temperature at an average of 98.6°F. The maintenance of a constant body temperature is discussed in Chap. 8.

Synthetic reactions   Among the molecules synthesized by cells are proteins, lipids, carbohydrates, nucleic acids, and derivatives of these natural products. These materials are incorporated into the structural fabric of cells, become mediators of intracellular chemical reactions, or are processed and packaged for export to other parts of the body.

*Protein synthesis*   Proteins are synthesized biochemically by linking 50 or more *amino acids* together or by joining together two or more chains of *polypeptides.* As the carboxyl group (COOH) of one amino acid is joined to the amino group ($NH_2$) of the next amino acid in the growing chain, a molecule of $H_2O$ is formed, establishing a *peptide bond* (OC-NH) between two adjacent amino acids. Each successive amino acid in turn is linked to the chain in a similar manner. Most biologically important proteins are composed of 100 to 300 amino acid residues.

Almost all protein synthetic reactions are carried out in the cytoplasm based on instructions provided by the nucleus. The process is complex and requires the participation of chromosomes, three different products which they synthesize, various synthetic enzymes, and ATP to drive the reaction. The details of protein synthesis are described later in the chapter.

*Lipid synthesis*   Lipids or glycerides consist of one, two, or three *fatty acids* [$CH_3(CH_2)_nCOOH$] joined to a *glycerol* backbone. The biosynthesis of lipids occurs by a pathway different from that used for their oxidation, thereby physically separating these pathways from each other.

The pathway of intracellular lipid synthesis is diagramed in Fig. 6-2. Basically, lipid synthesis involves two steps: (1) the synthesis of fatty acids in the cytoplasm and (2) the assembly of lipids on the mitochondria, provided that the necessary raw materials and energy to drive the reactions are present.

Fatty acids are synthesized in the cytoplasm by the successive addition of 2-carbon fragments to an initial intermediate formed by the interaction between two acetyl CoA's. Each step in the synthesis is complex, requiring the formation of chemical intermediates, reduction of the intermediates through the addition of hydrogen supplied by hydrogen carriers, and the formation of various by-products including CoA, $CO_2$, and $H_2O$. In this manner, even-numbered fatty acids containing 16 to 18 carbon atoms are formed in the cytoplasm in close proximity to the mitochondrial assembly plants where they will be utilized in the formation of lipids.

How the mitochondrion participates in the assembly of lipids is by no means clear. However, the first step in lipid assembly requires the phosphorylation of glycerol in the presence of ATP. One fatty acid is added to each phosphorylated position on the glycerol molecule, splitting off inorganic phosphate. The result is a lipid or fat consisting of a monoglyceride, diglyceride, or triglyceride, depending on whether one, two, or three fatty acid residues are added to the glycerol backbone.

*Carbohydrate synthesis*   Carbohydrates are synthesized biochemically by linking together many simple

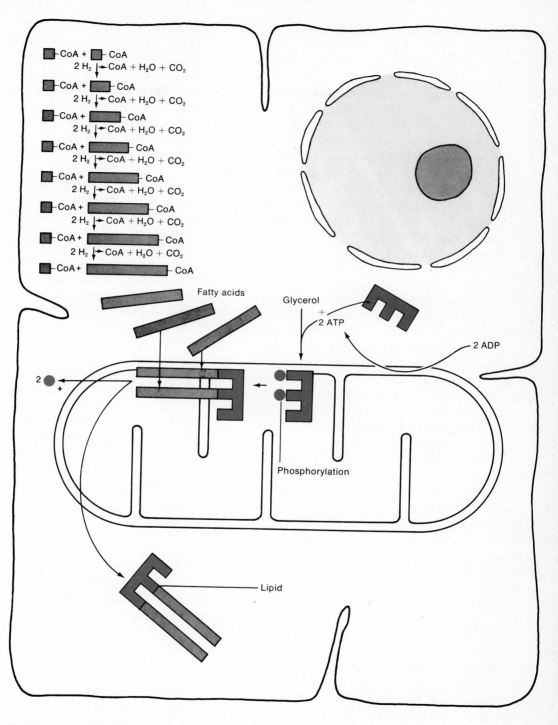

**Figure 6-2**  *The intracellular synthesis of lipids.* ▨ *= 2-carbon fragment*

sugars called *monosaccharides*. A molecule of $H_2O$ is removed when two monosaccharides are joined together by an oxygen bridge referred to as a *glycosidic bond*. The double sugar is called a *disaccharide*. Each successive simple sugar in turn is joined to the rest of the molecule in a similar manner, forming a progressively larger *polysaccharide*. Thus, the glycosidic bonds between the sugar residues of carbohydrates rank in importance with the peptide bonds between the amino acid residues of proteins.

The synthesis of carbohydrate requires the availability of simple sugars, synthetic enzymes, and high-energy phosphate. This type of reaction is of especial importance in liver cells, where glycogen is stored when there is an excess of blood glucose.

*Nucleic acid synthesis*   Nucleic acids consist of chains of nucleotides. Each *nucleotide* is composed of a *sugar-phosphate* group attached to a purine or pyrimidine base. A nucleic acid is called *deoxyribonucleic acid* (DNA) when the sugar is deoxyribose and *ribonucleic acid* (RNA) when the sugar is ribose. There are also slight differences in the base content between DNA and RNA. DNAs are double strands of helically arranged nucleotides. RNAs are single-stranded helices. The adjacent bases between each strand of a double helix of DNA are linked—one purine to one pyrimidine—by two or three *hydrogen bonds*. Hydrogen bonds are discussed in Chap. 10.

Most DNAs and RNAs are synthesized by the chromosomes within the nucleus of a cell. The DNAs of each chromosome are replicated prior to cell division. RNAs are synthesized on portions of chromosomes, using DNAs as templates. The DNAs contain the hereditary information of a cell. The RNAs participate in various aspects of protein synthesis.

## INFORMATION FLOW

The metabolic and perpetuative activities of a cell are highly controlled rather than chaotic. The controlled flow of *cellular energy* requires a controlled flow of *cellular information*. Homeostasis is disrupted when the normal flow of either cellular information or energy is altered.

The primary regulation of information flow within a cell is carried out by *genes* on the *chromosomes* in the nucleus. The nature of the genetic instructions supplied by genes is reflected by the kinds of proteins synthesized in the cytoplasm. Many of the cellular proteins are enzymatic mediators of intracellular chemical reactions. *Enzymes* regulate the flow of information in the cytoplasm through their control of degradative and synthetic pathways of metabolism.

Many other cellular proteins are structural. The presence of specific structural proteins is responsible for cellular differentiation and specialization. Still other proteins are packaged for export, carrying molecular information to far-flung parts of the body.

Through their mediation of biochemical reactions, enzymes also regulate the synthesis of other structural and information-bearing molecules besides proteins. These molecules establish the differential permeability of the plasma membrane, provide for the recognition and binding of specific molecules brought to the cell, and enable a cell to recognize and associate with similar cells during tissue formation.

Thus, the key to an understanding of information flow in a cell requires an appreciation of how the hereditary information in genes is translated into the synthesis of specific proteins. The mechanisms by which the hereditary information ordinarily contained in a gene can change with the passage of time and the factors which differentially activate and inhibit genes must also be understood, if we are to obtain a complete picture of information flow at the molecular level.

Protein synthesis   The instructions for the manufacture of proteins in a cell are provided by the chromosomes in its nucleus. The information-bearing portions of the chromosomes are contained in unique molecules, the DNAs (deoxyribonucleic acids). Presumably, each DNA molecule on one chromosome bears hereditary instructions for the synthesis of one or, at most, a few protein molecules.

The structure of a portion of chromosomal DNA is shown schematically in Fig. 6-3. Since the genes are in linear order on chromosomes, the entire DNA in one chromosome contains many individual molecules of DNA. Chromosomal DNA consists of two helical strands of DNAs linked to each other by hydrogen bonds, forming a *double helix* of DNA.

The DNA double helix has been likened to a spiral staircase in which the steps of the staircase carry the genetic information. The railings of the staircase—the backbones of the DNA macromolecule—contain alternating groupings of sugar-phosphate residues. The sugar is invariably *deoxyribose,* a 5-carbon molecule which lacks an oxygen atom found in *ribose.* The steps of the staircase consist of a pair of nitrogenous bases, each linked to the other by two or three hydrogen bonds. The opposite end of each base is chemically bonded to a sugar of an adjacent deoxyribose molecule. Thus, each strand in a DNA double helix consists of repeated units containing a base, a sugar, and a phosphate. Each base-sugar-phosphate group is called a *nucleotide.*

Only four possible base molecules are used to form the steps of the staircase. Two of the bases are called *purines;* the other two are referred to as *pyrimidines.* One of the purines is always specifically bonded by hydrogen to one of the pyrimidines during the formation of each step.

The consistency in specific purine-pyrimidine or pyrimidine-purine base pairing between adjacent nucleotides of the two single DNA strands establishes a four-letter *genetic code.* The code corresponds to the four possible combinations in which two specific purines may pair with their respective pyrimidine partners from adjacent strands of the double helix. Every three consecutive bases on a DNA molecule or gene correspond to one of about twenty amino acids.

For this reason, each consecutive triplet grouping of bases is termed a *codon.* The sequence of codons in a gene therefore prescribes a specific sequence of amino acids to be incorporated into a protein.

Proteins are synthesized in the cytoplasm. Yet the double helices of DNA within the chromosomes do not leave the nucleus. Thus, the genetic instructions contained in the codons of DNA must be transcribed into expendable messenger molecules which do enter the cytoplasm.

*Transcription*  Transcription refers to the molecular process by which DNA molecules in the chromosomes transfer their genetic messages to mRNAs (messenger ribonucleic acids) during their synthesis. Transcription is conceptualized in Fig. 6-4. During transcription, a portion of a DNA double helix corresponding to one gene somehow unwinds. After unwinding, each of the two DNA partial strands, although still helical, is separate from the other. Then one of the two strands synthesizes a molecule of mRNA from raw materials available in the nucleoplasm. The factors which initiate transcription in only one of the two strands are not understood.

The raw materials required for mRNA synthesis include many molecules of the following substances: (1) two purine and two pyrimidine bases, one of which is different from and consistently substitutes for one of the four bases in the DNA molecule; (2) the sugar ribose rather than the deoxyribose sugar found in the DNA molecule; and (3) a triphosphate group. The raw materials are assembled into nucleotides which pair with the bases that project from the sugar-phosphate backbone of the template DNA strand. Base pairing

*Figure 6-3*  *An abstract representation of the double helical structure of deoxyribonucleic acid (DNA).*

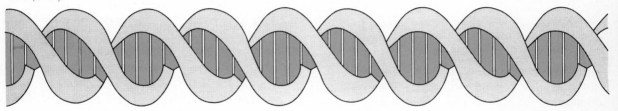

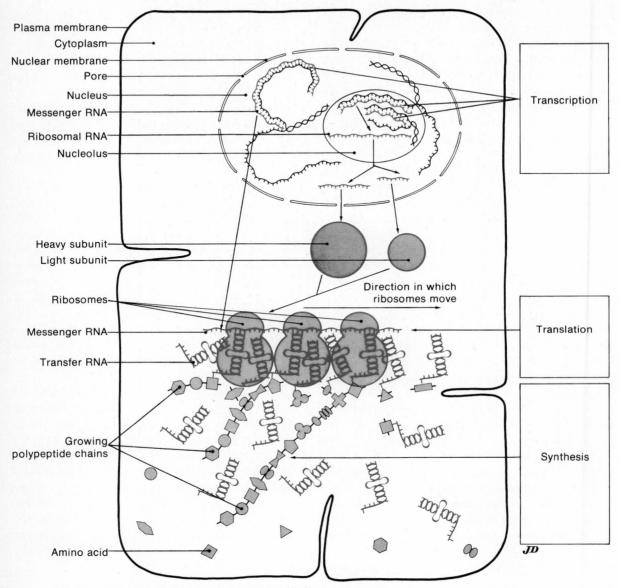

Plasma membrane
Cytoplasm
Nuclear membrane
Pore
Nucleus
Messenger RNA
Ribosomal RNA
Nucleolus

Transcription

Heavy subunit
Light subunit

Direction in which
ribosomes move

Ribosomes

Messenger RNA

Transfer RNA

Translation

Growing
polypeptide chains

Synthesis

Amino acid

**Figure 6-4** *The mechanism by which proteins are synthesized.*

ordinarily ensures that the genetic message is transcribed without error. The *polymerization* reaction which follows requires ATP and a specific synthetic enzyme. Polymerization joins the individual nucleotides together, giving rise to a molecule of mRNA.

After the synthesis of mRNA is complete, the genetic messenger is detached from the parent DNA

strand. As the two DNA strands somehow reestablish a double helical configuration, the newly synthesized mRNA moves through one of the nuclear pores into the cytoplasm.

*Translation* Translation refers to the molecular process by which proteins are synthesized in the cytoplasm based on genetic instructions provided by mRNA. The synthesis of proteins requires the assistance of two other types of RNA—rRNAs (ribosomal RNAs) and tRNAs (transfer RNAs). *Ribosomal RNAs*, in conjunction with specific proteins, make up ribosomes which read the genetic messages embodied in the triplet codons of mRNA. *Transfer RNAs* recognize, bind, and transport specific amino acids to the correct codons on an mRNA molecule, permitting the amino acids which they carry to join the growing polypeptide chain in the order transcribed onto mRNA and translated by ribosomes.

Ribosomal RNA is synthesized on portions of chromosomes located within the nucleolus. Ribosomal RNA is manufactured by a molecular mechanism similar to that which synthesizes mRNA. Following its synthesis, rRNA is stored in the nucleolus. When needed, rRNA enters the nucleoplasm, dividing into a large and a small strand which are delivered to the cytoplasm. These events are summarized in Fig. 6-4.

In the cytoplasm, the rRNAs and ribosomal proteins are assembled into two particles which subsequently fuse to form a ribosome. A more detailed account of the structure of a ribosome is given in Chap. 5.

During the synthesis of proteins, a group of ribosomes collectively called a *polysome* travels in single file across the length of an mRNA, translating into proteins the genetic message encoded in its consecutive codons of triplet bases. The small subunit of a ribosome binds to mRNA. Each ribosome is also presumably involved in the recognition of where to begin reading the genetic message borne on mRNA. Ribosomes attached to the outer surface of the endoplasmic reticulum are held by binding sites located on the large subunits. Finally, the large subunits also contain binding sites and release mechanisms for tRNAs delivering the amino acids used as raw materials.

Transfer RNAs are synthesized on chromosomal segments of DNA inside the nucleolus. Presumably, there are as many tRNAs as there are different kinds of amino acids, i.e., about 20 different types of tRNA.

Transfer RNAs are single-stranded molecules that come in a variety of helical and twisted configurations. One end of a tRNA molecule contains a triplet sequence of bases complementary to a codon on mRNA which specifies the amino acid which it carries. The other end of a tRNA molecule contains a binding site for the amino acid it recognizes. The origin and general structure of tRNAs are shown in Fig. 6-4.

During protein synthesis, each type of tRNA recognizes and binds with one of twenty-odd amino acids which have entered the cell as raw materials. The tRNAs then transport their bound amino acids to an mRNA molecule active in protein synthesis. As each ribosome translates the consecutive codons on an mRNA molecule, tRNAs complementary to the codons being read attach by base pairing to the mRNA. Thus, numerous identical proteins, each in varying stages of completion, are synthesized simultaneously along one molecule of mRNA.

The amino acid carried by a newly attached tRNA molecule links to the closest amino acid in the growing polypeptide chain. Since the most recent amino acid addition remains attached to its tRNA, the entire chain is transferred to the newly arrived tRNA from an adjacent tRNA bound to the previous codon. Following release of its peptide chain, the empty tRNA is set free in the cytoplasm to bind once again to its specific amino acid.

The addition of an amino acid is repeated at each successive codon until the entire polypeptide or protein has been assembled on a ribosome. The polymerization reaction by which the amino acids are linked together requires a synthetic enzyme and a source of ATP. The newly synthesized protein is then detached from the ribosome and set free in the cytoplasm. The entire translation process, from its inception to its completion, requires about one minute. Translation of the genetic message transcribed onto mRNA by nuclear DNA is conceptualized in Fig. 6-4.

*Mutations*   A mutation is a potential biologic change which may appear because of an alteration in a gene on a chromosome. Mutations are due either to chemical mistakes made during the assembly of bases in the nucleotide chains of formative DNA molecules or to damage of chromosome segments in which already existing DNAs are located. Thus, a mutation is actually a change in the sequence of nitrogenous bases in the double helix of DNA.

When a mutated portion of a single strand of DNA synthesizes mRNA, novel genetic instructions are transcribed and subsequently sent to the cytoplasm. Depending upon the specific alteration in the base sequence of mRNA during transcription, the ribosomes translate the altered message by synthesizing one of the following: (1) a new but inactive protein, (2) a new protein even more biologically effective than the old protein, (3) a slightly different protein possessing biologic activity similar to that of the old protein, or (4) the same protein as before. Of course, severe chromosomal damage may completely prevent translation and protein synthesis.

Thus, the mutation of a gene appears as a change in the nature or the complete absence of a specific protein synthesized in the cytoplasm. The synthesis of an inactive protein may be disastrous to the cell if the old protein was a metabolically important enzyme or may be only slightly debilitating or even unnoticeable if the old protein was ordinarily inactive or relatively unimportant in cellular metabolism. When the new protein is more biologically active or effective than the old protein, the mutation is actually beneficial. When an identical protein is synthesized in the cytoplasm, irrespective of the altered sequence of bases on mRNA, the new and old base sequence most likely code for the same amino acids. When the new protein is somewhat different in amino acid sequence but equal in activity to the old protein, the amino acid sequences in the biologically active portions of both molecules are the same. In other words, the molecular structure of the new protein is altered in a region which does not possess biologic activity. Thus, mutations are deleterious or beneficial or have no observable effect. Most visible mutations are deleterious only because nature has already incorporated into the genetic structure of the human species practically all the possible mutations which tend to promote its survival under prevailing environmental conditions.

Mutations are induced by natural causes or man-made factors. Natural background radiation due to the continual bombardment of the earth by high-speed ionizing particles from outer space establishes the base level rates of genetic mutation. Mutations are promoted by various artificial causes such as exposure to medical x-rays and other man-made sources of ionizing radiation or the injection or ingestion of mutagenic chemicals. The effects of ionizing radiation and chemical mutagens on chromosomes and the nature of the mutations which result from these influences are discussed in Chap. 9.

Mutations occur in both somatic, or body, cells and germinal, or sex, cells. Let us assume that a mutation occurs during the replication of a DNA strand prior to cell division in both types of cell. Since somatic cells are not involved in reproduction, the effect of a *somatic mutation* is restricted to the original cell in which the mutation occurs and to its mitotic descendants. Thus, any adverse effects of a somatic mutation will not extend to the progeny of the individual in which the mutation occurred.

During the formation of sex cells, however, a newly mutated chromosome is distributed to one of four possible spermatozoa or ova. Thus, there is one chance in four that a gamete carrying a mutation will participate in fertilization, thereby passing new genetic instructions to the developing child. Even then, if the single mutation is recessive, its effects may not appear in the progeny. Nevertheless, *germinal mutations* at least have the potential of being transmitted to the next generation.

*The regulation of protein synthesis* Different proteins are synthesized by different kinds of cells as well as by the same cell at different times. These variations are responsible for the specialization of undifferentiated cells into the numerous types of *epithelial, connective, muscular,* and *nervous tissue cells* present in the body. These differences are also reflected in fluctuations in the degrees and types of metabolic activities in which cells engage. Yet all human tissue cells, with the exception of reproductive cells, contain 46 chromosomes with identical genetic information in the base sequences of their DNA double helices. Apparently, a number of factors are capable of turning numerous genes on or off differentially. These factors presumably control cellular differentiation and the nature of ongoing cellular metabolism through the regulation of protein synthesis.

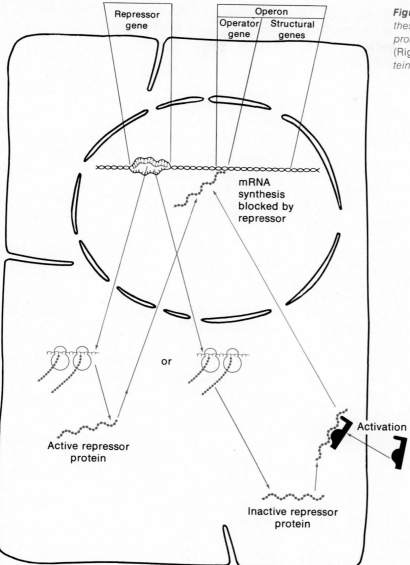

**Figure 6-5** *The regulation of protein synthesis.* (Left) *The action of active repressor proteins after their cytoplasmic synthesis.* (Right) *The action of inactive repressor proteins following their cytoplasmic synthesis.*

Repressor gene

Operon

Operator gene | Structural genes

mRNA synthesis blocked by repressor

or

Activation

Active repressor protein

Inactive repressor protein

Figure 6-5 postulates several mechanisms by which the kinds of proteins synthesized in the cytoplasm are influenced genetically. There is ample evidence to indicate that portions of some chromosomes are organized functionally. Such chromosomes have been found to possess functional units consisting of repressor, operator, and structural genes. The *structural genes*, when active, are responsible for the synthesis of an entire group of mRNAs which contain the genetic instructions for the synthesis of a corresponding group of proteins. The structural genes are active only when a specific *operator gene* is active. An operator gene is believed to control the activities of about 10 structural genes.

The operator gene and the structural genes which it supervises are collectively termed an *operon*. The regulation of the operator gene is, in turn, controlled by a *repressor gene* located on the same chromosome as the operon.

Apparently a repressor gene may exert its effects in one of two ways. Both mechanisms involve the repression or activation of the operator gene, based on the presence or absence of certain chemicals in the cell, but their details differ.

In the mechanism shown at the left in Fig. 6-5, the repressor gene synthesizes repressor mRNA. In the cytoplasm, repressor mRNA serves as a template for the synthesis of a repressor protein capable of repressing the operator gene. In the absence of a particular metabolite ordinarily carried to the cell through the blood, the operator is inactivated, and an entire metabolic pathway controlled by the structural genes remains dormant. When the metabolite is present, however, it complexes (combines) with the repressor protein. The formation of a chemical complex alters the structure of the repressor protein, thereby causing its inactivation and preventing it from combining with the operator gene. In the absence of a brake on its activities, the operator gene activates the structural genes. Thereupon, the structural genes provide the genetic instructions necessary to establish the metabolic pathway in which the available metabolite is utilized.

In the mechanism illustrated at the right in Fig. 6-5, the repressor protein synthesized on the template provided by mRNA is initially inactive. If a specific metabolite is unavailable, then the operator gene is active by default, and the metabolic pathway controlled by the structural genes is present. In the presence of the metabolite, however, the repressor protein is activated and consequently represses the activities of the operator gene. This chain of events results in the inactivation of the operon.

These genetic control mechanisms have been demonstrated in bacteria. How widely applicable the mechanisms are in higher organisms, including human beings, remains to be determined. At least these mechanisms are models of how protein may be differentially synthesized. Other influences may also act as limiting factors in determining whether an operon is activated or inhibited.

Chromosome replication   The formation of new chromosomes in preparation for cellular division requires the replication of the double helices of DNA. The initiation of DNA replication assumes the existence of some sort of cellular signal. That these mechanisms are exact and efficient is obvious. Most types of cells divide with regularity under normal conditions. Chromosome replication is ordinarily executed faithfully so that each of the daughter cells resulting from mitotic division receives an identical set of chromosomes.

*The regulation of DNA replication*   Before a cell can give rise to two identical cells, each of its chromosomes must exist in duplicate. For a human tissue cell, this means that there must be $46 \times 2$, or 92, chromosomes in the cell before division is initiated. What regulatory factors are involved in inducing chromosome replication prior to division?

Frankly, we have no knowledge of what these regulatory controls might be on the human level. The only evidence bearing on the problem so far comes from studies of bacteria and bacterial chromosomes. However, because many molecular phenomena discovered first in bacteria have been found to apply in somewhat modified form to many other organisms, including human beings, it is worthwhile to review briefly what little we do know about bacterial control of DNA replication. We will tentatively extend the results to human tissue cells.

The mechanism by which a cell regulates the replication of DNA is suggested in Fig. 6-6. Presumably, a *regulatory gene* bearing the genetic message for the initiation of DNA replication synthesizes mRNA which serves as a template for the synthesis of inactive *initiator protein*. This specialized protein is always present in the cytoplasm but must be activated before it can initiate DNA replication. Although the nature of the activator remains unknown, it is presumably a chemical signal. The activated initiator complexes with a special replicator gene. Analogous to an operator gene of an operon, the activated replica-

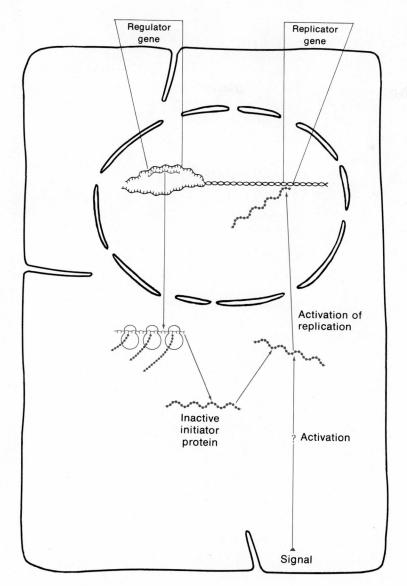

**Figure 6-6** *Control of the replication of DNA.*

tor gene initiates DNA replication. According to this scheme, the only external stimulation this self-regulating process requires is the presence of an appropriate cellular signal that will activate the initiator.

***The molecular basis***  Once DNA replication is initiated, what happens? Basically, the DNA double helix in each chromosome progressively unwinds, and each single strand of DNA, in the presence of the necessary raw materials and enzymes, gives rise to a mirror image of itself. The replication process takes approximately one minute. The result of DNA replication is the formation of two identical double helices of DNA from one DNA double helix, i.e., each chromosome duplicates. The process of DNA replication is conceptualized in Fig. 6-7.

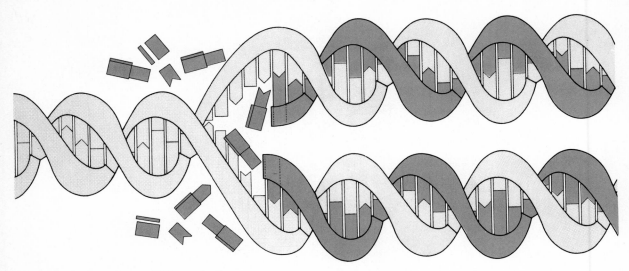

***Figure 6-7*** *The process of DNA replication.*

Following activation of the replicator gene, each of the DNA double helices in the nucleus of the cell progressively unwinds. In contrast to the synthesis of mRNA on one of two partially unwound strands of DNA, both DNA strands of the double helix unwind completely, and each strand replicates in its entirety during DNA synthesis.

DNA replication requires the following raw materials: the 5-carbon sugar deoxyribose, two kinds of nitrogenous purine and pyrimidine bases, and triphosphate groups. These raw materials are assembled in the nucleoplasm into nucleotides. The nucleotides attach through complementary base pairing to appropriate positions on both the parent strands of DNA. Each nucleotide in both formative strands is joined enzymatically to the next in line in a polymerization reaction. During this reaction, two of the three phosphate groups are split off of each attaching nucleotide in turn. The reaction gives rise to two new strands of DNA, each of which is complementary to the strand with which it is base-paired. Each new strand is identical in base sequence to the opposite strand of old DNA. Thus, the two double helices which form during DNA replication are mirror images of each other and possess identical sequences of base pairs. Since the genetic information is encoded in the sequence of base pairs, each of the two new chromosomes will ordinarily contain identical genetic information.

DNA replication provides a mechanism for the exact duplication of each chromosome in the cell. The next problem the cell faces is that of ensuring that one complete set of paired chromosomes will be distributed to each daughter cell during cell division. This problem is solved through the operation of a cellular mechanism called mitosis.

***Mitosis*** The process by which one tissue cell divides into two cells is called *mitosis*. As the result of mitosis, each of the two daughter cells acquires exactly the same number and kinds of chromosomes that are found in the original parent cell. Although the time required for cell division is highly variable, an average mitotic division lasts one hour. A mitotic division of a generalized animal cell is illustrated in Fig. 6-8.

1   *Interphase.* During interphase, which is the stage between cell divisions, the cell engages in normal metabolism. The premitotic parent cell offers no visible clues to its impending division. The chromosomes within the nucleus are maximally elongated, giving the appearance of thin, extremely long strands distributed throughout the nucleoplasm. However, chromosome replication is initiated and completed during the last part of interphase

2   *Prophase.* As the cell enters a mitotic cycle, the following sequence of events occurs: (*a*) The

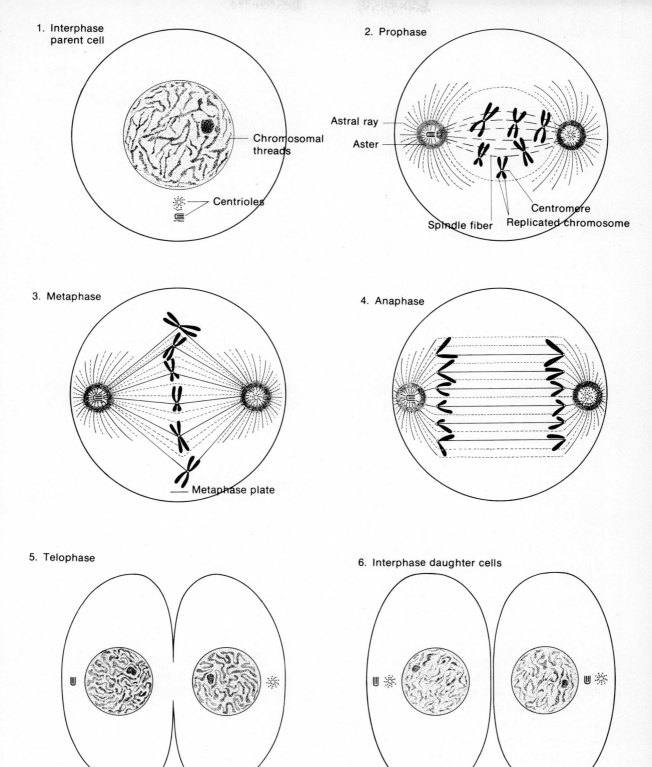

**Figure 6-8** *The events during mitosis in a hypothetical cell containing three pairs of chromosomes.*

replicated chromosomes shorten tremendously by coiling, becoming thick enough to be seen under a light microscope following the use of appropriate staining techniques; (b) the centrioles, by a poorly understood mechanism, begin a migration which will carry them to opposite poles of the cell; (c) the nuclear envelope and nucleolus gradually disappear; (d) once at the poles, each centriole initiates the formation of astral rays and spindle fibers which traverse the center of the cell; (e) a centrally located spindle fiber from each centriole makes contact with the point of attachment called a *centromere* between each pair of replicated chromosomes, presumably to serve as an orienting mechanism to align the chromosomes correctly during their migration; and (f) the chromosomes begin to line up along the metaphase plate (see below), or equatorial plane, an imaginary line perpendicular to a line drawn between the two centrioles at opposite poles of the cell. Prophase is terminated when the replicated chromosomes are aligned in the plane of the metaphase plate.

3   *Metaphase.* The chromosomes, although pointing in various directions, are on the metaphase plate. Because each chromosome is replicated, there are in fact double the species number of chromosomes in the cell. Each centromere receives two spindle fibers, one from each aster. The centrioles have completed their poleward migration. The asters which surround the centrioles are large and well defined. Metaphase is obviously a transient stage of cell division, serving as a transition between prophase and anaphase.

4   *Anaphase.* The replicated chromosomes separate from each other and migrate to their respective poles. Spindle fibers attach between the centromeres of homologous or identical chromosomes as well as between the centromeres and the astral rays of the asters nearest to them. As poleward migration proceeds, the chromosomes bend from the centromere region, forming arms which point toward the metaphase plate. The formation of chromosomal arms suggests that the chromosomes are exposed to a force which pulls them toward their respective poles. Since the spindle fibers are attached to the centromeres, it is logical to assume that they are responsible for the imposed force. However, experiments in which the spindle fibers are severed with a microscalpel during anaphase indicate that the spindle fibers are not necessary for the completion of anaphase. The cause of chromosomal bending therefore still remains unexplained. When each complete set of chromosomes is bunched together next to its aster, anaphase is completed.

5   *Telophase.* The onset of telophase is marked by a pinching in of the plasma membrane from each side in the plane of the former metaphase plate. The formative plasma membrane from each side will ultimately meet in the midline of the cell, resulting in the separation of the original parent cell into two daughter cells and terminating telophase. While cellular partitioning takes place, the chromosomes and centrioles in each of the formative daughter cells undergo changes which are essentially the reverse of those which occurred during prophase: (a) The spindle fibers which surround the aster of each centriole disappear; (b) a nuclear envelope surrounds each set of bunched chromosomes; (c) a nucleolus reappears in each of the newly formed nuclei; (d) the chromosomes in each nucleus become thin, elongated strands which are distributed throughout the entire nucleoplasm; and (e) the single centriole in each cell replicates, giving rise to two daughter centrioles. Mitosis is now complete, and the daughter cells enter interphase.

6   *Interphase.* Each of the daughter cells is about half the volume of the parent cell. Growth will occur as the daughter cells resume normal metabolic functions. Apparently, because of a relationship between cellular volume and the rate of diffusion of nutrients and oxygen, when the daughter cells reach the size of the parent cell, they in turn will undergo mitotic division.

## MEMBRANE TRANSPORT MECHANISMS

Maintenance of steady states requires that cells regulate the external supply of nutrients, oxygen, and raw materials used in synthesis, eliminate waste products of metabolism, and adjust their fluid and electrolyte content to changing external conditions. Therefore, it is not surprising to find that cells have developed a variety of transport methods to cope with various demands regulated by differential permeability. The operation of some of these mechanisms simply requires the existence of a concentration gradient of the material across the membrane. Other methods require the assistance of membrane-bound carrier molecules. One method, which demands the expenditure of cellular energy, is capable of transporting certain materials against a concentration gradient, i.e., from a region in which the material is in lesser concentration to a region in which the same material is already in higher concentration. Other, even more complex mechanisms are used by certain cells to increase the flexibility with which they can pass materials across their differentially permeable membranes.

Based on energy requirements and other considerations, there are fundamentally three qualitatively different kinds of membrane transport mechanism—passive transport, active transport, and pinocytotic transport.

Passive transport    A cell does not expend energy in the form of ATP during the passive movement of diffusible materials across its plasma membrane. Usually, all that is required for the passive transport of such a material is a difference in its concentration on both sides of the membrane. Under these conditions, the material will move from the region in which it is in higher concentration to the region in which it is in lower concentration; i.e., the material will be transported along its *concentration gradient.* The net movement of the material will cease when its average concentration on both sides of the membrane is equal. At this point, which is called *equilibrium,* there are equal numbers of diffusible particles moving in opposite directions across the membrane. Note that the concentration—the number of particles per unit volume—on both sides of the membrane is the limiting factor in determining whether or not there will be a differential movement of a diffusible material.

*Diffusion*    Diffusion occurs because molecules are in constant motion above a temperature of zero kelvin (see Chap. 8). Diffusion is the passive movement of a solute or solvent in response to a concentration gradient. Frequently, the term is applied specifically to the movement of solute. In either case, the diffusible particles will be transported through the pores of a plasma membrane from a region in which the particles are in greater concentration into a region in which they are in lesser concentration. At equilibrium, the average number of diffusible particles on both sides of the membrane will be equal, even though some particles continue to move passively through the pores in both directions.

The process of diffusion is illustrated in Fig. 6-9. The diagram at the left shows a large concentration of solute particles in the interstitial space and none in the intracellular fluid before diffusion begins. Assume that the solute particles are dissolved in the solvent water and that the solute and solvent are both diffusible, i.e., capable of moving passively through the pores of the plasma membrane in response to a concentration gradient. Because there is a greater concentration of diffusible solute in the interstitial fluid, compared with the cytoplasm of the cell, the solute will be transported through the pores of the membrane into the cell.

Since the solvent is also capable of diffusing, let us now consider what happens to it during the establishment of equilibrium. Initially, per unit volume, there is a greater concentration of solvent inside the cell, compared with the solvent concentration in the interstitial fluid. In other words, the presence of solute in the interstitial fluid decreases the number of solvent particles which can occupy a given unit of interstitial volume. Before equilibrium is established, solvent particles will diffuse from the less concentrated solution (the cytoplasm) to the more concentrated solution (the interstitial fluid). Notice that the relative concentration of a solution depends upon its solute concentration and not upon the size of the solute particles. Therefore, to say that a diffusible solvent will move from a region in which it is in greater concentration to a region in which it is in lesser concentration is equivalent to stating that the solvent will diffuse from the less concentrated, or more dilute, solution to the more concentrated, or less dilute, solution. At equilibrium,

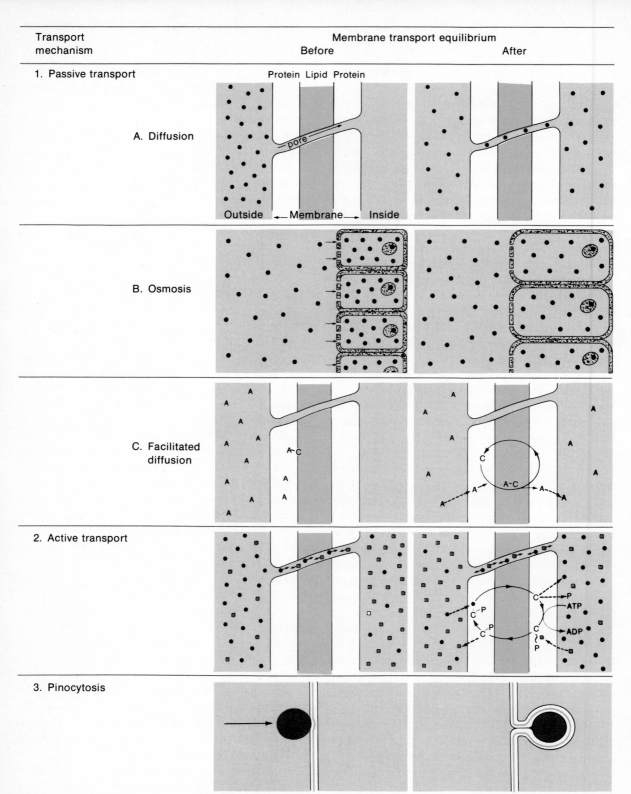

**Figure 6-9** *Mechanisms of cell transport, showing the movement of materials between the interstitial and intracellular fluids across a differentially permeable plasma membrane.*

there will be equal numbers of solute and solvent particles diffusing in both directions across the pores in the membrane. Following equilibrium, the net rate of diffusion is zero.

Assuming the existence of a concentration gradient and a temperature above absolute zero, particles are capable of diffusing through the pores of a plasma membrane under the following conditions: (1) The particles must be smaller in size than the diameter of the pores, and (2) the particles must be either electrically neutral or bear an electric charge of opposite sign to those which may line the walls of the pores and the outer and inner surfaces of the membrane.

Frequently, the walls of the pores are positively charged, causing a retardation in the rate of diffusion of small positively charged particles or completely blocking their transport. In addition, the outside surfaces of plasma membranes are positively charged, while their inside surfaces are negatively charged, giving rise to electric gradients across the membranes.

When diffusible solute particles such as small ions and nondiffusible solute particles such as proteins are both present in an aqueous solution on one side of a differentially permeable membrane, diffusible particles will be transported passively across the membrane, while the nondiffusible particles will be left behind. Because protoplasm is an aqueous solution of proteins, this condition normally prevails during the diffusion of particles across plasma membranes. Now consider, however, a situation in which the differentially permeable membrane is man-made, as in the case of the "artificial kidney," and in which the diffusible particles can be continually removed from the solution on the opposite side of the membrane to prevent the establishment of an equilibrium. It should be clear that when this method is properly regulated, all of the diffusible solute can be removed or separated from the nondiffusible solute. Called *dialysis,* the method has numerous clinical and research applications in the separation of large and small particles.

When a mechanical force is responsible for promoting the movement of diffusible solute and solvent particles across a differentially permeable membrane, the process is called *filtration.* Filtration depends on a mechanical pressure difference across a membrane.

For example, filtration is responsible for the diffusion of materials from capillary blood to the interstitial fluid.

*Osmosis*   Osmosis is the passive movement of a solvent in response to a concentration gradient. Since the solvent of cells is $H_2O$ (water), osmosis in a biologic context refers to the diffusion of $H_2O$ across a differentially permeable membrane. Notice that, in principle, osmosis does not differ from diffusion.

Membrane transport by osmosis is diagramed in Fig. 6-9. The solute particles are represented by dots. The solute is non-diffusible because it is larger than the diameter of the pores in the plasma membranes. Solvent particles are not indicated. Before osmosis begins (at the left in Fig. 6-9), the intracellular fluids are more concentrated than the interstitial fluid bathing the cells. Thus, water will begin to diffuse through the pores of the membranes to a greater degree from the interstitial fluid to the intracellular fluids than in the reverse direction.

The number of particles which strike a wall or membrane in a unit of time at a given temperature depends on the concentration of the particles. The greater the particle concentration, the greater the frequency of collision of the particles with the membrane which confines them. Simply because there are more of them, water molecules in the interstitial space will initially strike the membranes with a greater frequency than water molecules in the intracellular fluid. Since particle collisions on both sides of the membranes are randomly distributed, more water molecules, on an average, will be striking and entering pores on the outer surfaces of the membranes, compared to their inner surfaces. Therefore, although water molecules will be diffusing in both directions across the membranes, initially more molecules per unit time will diffuse into the cells than in the reverse direction. The number of water molecules which diffuse into the cells will progressively decrease with time as more water molecules accumulate intracellularly and increase the frequency of molecular strikes against the inside surfaces of the membranes.

Nevertheless, as osmosis proceeds toward equilibrium, an increasing number of water molecules enters the cells from the interstitial fluid, even though the net number entering per unit time decreases. In response to the increased volume of intracellular fluid, the cells swell. Swelling is a consequence of an increased intracellular fluid pressure which causes the plasma membranes to stretch outward in all directions. Osmotic equilibrium is achieved when the average number of strikes of the water molecules against both sides of the membranes is equal (see Fig. 6-9). Following equilibrium, the average number of water molecules diffusing in both directions across the membranes and the average number of nondiffusible solute particles per unit volume on both sides of the membranes are equal.

In the above example, we noted that a diffusible solvent moves osmotically across a membrane permeable to the solvent but not to the solute. The more concentrated solution is said to exert an *osmotic pressure* which "pulls" diffusible solvent away from the less concentrated solution. Before osmotic equilibrium is established, the cells are *hyperosmotic* or *hypertonic* to the interstitial fluid. In other words, the cells have a greater osmotic pressure or a greater number of nondiffusible particles per unit volume, compared with the interstitial fluid. Conversely, the interstitial fluid is *hyposmotic* or *hypotonic* to the intracellular fluids. At equilibrium, the cells and the interstitial fluid which bathes them are at the same osmotic pressure and therefore *isosmotic* (*isotonic*) to each other.

As stated above, osmotic equilibrium depends upon achieving the same average number of solvent strikes against both sides of the membranes, and the number and force of the molecular collisions against the membranes in a given time are a measure of the fluid pressure. Thus, the initial osmotic pressure of the cells can be expressed by measuring the intracellular hydrostatic pressure which will exactly counterbalance the inward movement of solvent from the less concentrated solution in the interstitial space.

*Facilitated diffusion* Facilitated diffusion is a form of passive transport in which a lipid-soluble carrier molecule participates in moving a water-soluble molecule across the phospholipid layer of the plasma membrane. Since the carrier molecule is soluble in the phospholipid layer, the attached water-soluble molecule is carried across an otherwise impenetrable layer of the membrane.

The mechanics of facilitated diffusion are diagramed in Fig. 6-9. The molecules of A in the interstitial space are water-soluble. While the outer and inner portions of the membrane are freely permeable to A, the middle phospholipid layer represents an impermeable barrier, blocking its diffusion into the cell. However, after diffusing into the outer layer of the plasma membrane, A binds to a membrane carrier C. Since C is lipid-soluble, it diffuses with ease through the phospholipid layer, carrying A along with it. As A penetrates the inner layer of the membrane, its solubility characteristics cause it to be released from C, which is then free to complex with another molecule of A at the interface between the outer protein and middle phospholipid layers. Once A is set free in the inner layer of the membrane, it diffuses into the intracellular fluid because of its concentration gradient.

This method of membrane transport relies on the existence of a concentration gradient across the membrane. In other words, facilitated transport will move certain water-soluble particles from a region in which they are in greater concentration to a region in which they are in lesser concentration. At equilibrium, the average concentrations of water-soluble particles on both sides of the membrane are equal.

*Active transport* Probably the most important and common method of membrane transport employed by human tissue cells, this form of carrier-mediated transport requires an energy expenditure in the form of ATP. Active transport is employed for two general purposes: (1) to move a material across a membrane against its concentration gradient, e.g., to transport sugars and amino acids from a less concentrated to a more concentrated region and (2) to increase the rate of uptake of a material by a cell beyond that which can be accomplished by diffusion alone, e.g., to maintain high concentrations of extracellular $Na^+$ and intracellular $K^+$.

Active transport requires the presence in the

plasma membrane of a chemical "pump" consisting of (1) a carrier protein to bind with and transport the required material or substrate across the membrane and (2) high-energy phosphate in the form of ATP to drive the reaction. The carrier protein may accomplish the task of active transport by shortening when bound to its substrate and, when the substrate is released at the other side of the membrane, elongating again. If so, rapid changes in size through repetitive substrate binding and release cause the carrier to swivel back and forth between the two inside surfaces of the membrane, each time delivering a specific ion or molecule across the membrane. This or some similar type of molecular behavior endows active transport with a high degree of efficiency.

Active transport across a plasma membrane is conceptualized in Fig. 6-9. Assume initially that there is a high extracellular concentration of $K^+$ and a high intracellular concentration of $Na^+$. This condition normally obtains in a nerve fiber immediately following the conduction of a nerve impulse (see Chap. 13). Ionic equilibrium requires the reestablishment of high concentrations of extracellular $Na^+$ and intracellular $K^+$. Equilibrium is restored through the activation of a chemical pump in the plasma membrane which transports $Na^+$ out of the cell. The same chemical pump passively transports $K^+$ into the cell. The operation of this chemical pump concentrates one of these ions on each side of the membrane during the establishment of an electrical equilibrium. However, active transport prevents the establishment of an equilibrium based on diffusion.

Notice that the operation of the chemical pump in the membrane requires several steps: First, the carrier interacts with ATP, giving rise to ADP and a carrier $\sim$ high-energy phosphate complex (abbreviated $C \sim P$ where $\sim$ represents the high-energy phosphate bond). ADP will be used in the resynthesis of ATP; $C \sim P$ then combines with $Na^+$ and transports it from the inner to the outer layer of the plasma membrane. This reaction utilizes the energy stored in the high-energy phosphate bond. As $Na^+$ is released from the outer layer, the C-P complex, now minus its high-energy bond, binds to $K^+$ and transports it to the inner layer. $K^+$ and inorganic phosphate separately diffuse

into the intracellular fluid, following their release in the inner layer. Thus, C is again available to actively transport another $Na^+$ to the outer layer of the membrane. The process continues and rapidly restores ionic equilibrium. Following equilibrium, active transport simply neutralizes the effects of an ionic imbalance resulting from the tendency of extracellular $Na^+$ and intracellular $K^+$ to diffuse through the pores of the membrane in response to their concentration gradients.

**Pinocytosis** Pinocytosis is a unique method of membrane transport and apparently a more complex method than any of those mechanisms discussed heretofore. Combining elements of both passive and active transport, pinocytosis is a nonselective method of passing large molecules, especially proteins and other particles adsorbed to the proteins, into a cell as well as of passing indigestible residues in food vacuoles out of a cell. Thus, pinocytosis can occur in either direction across a plasma membrane. Because this mechanism does not differentiate among the various particles being moved, pinocytosis is not a valid example of active transport. Nevertheless, energy is expended in accomplishing the movement. Indeed, the exceptional nature of pinocytosis permits it to transport large particles either along or against their concentration gradients. However, not all cells are capable of pinocytotic transport.

The transport of a large particle by pinocytosis is illustrated in Fig. 6-9. Apparently, macromolecules to be moved pinocytotically are adsorbed on, or "stick" to, the surface of the plasma membrane. Some cells may possess specialized membrane structures to which the large proteins are bound prior to pinocytosis. Any other particles adsorbed to the proteins will of course also undergo pinocytotic transport.

The presence of adsorbed proteins causes the plasma membrane at that point to wrinkle inward. The reason for the infolding is not known with certainty but seems to involve colloidal gelation, or an increased viscosity in the cytoplasm beneath the membrane. The infolding progressively increases in extent, resulting in the formation of a pinocytotic vesicle consisting of a spherical region of plasma membrane surrounding the proteins. Although still attached to the main portion of the plasma membrane, the membrane of the vesicle projects into the cytoplasm. As inpouching continues, the vesicle forms a neck consisting of opposed membranes attached to the main portion of the plasma membrane. Finally the neck constricts sufficiently to break, releasing the pinocytotic vesicle into the cytoplasm. The plasma membrane in the region of the neck repairs itself immediately. The intracellular fate of pinocytotic vesicles and their contents is discussed in Chap. 5.

## COLLATERAL READING

Berlin, Richard D., "Specificities of Transport Systems and Enzymes," *Science,* 168:1539–1545, 1970. Evidence is presented which suggests that active transport of a small molecule consists of two steps: (1) carrier-mediated transport across the membrane, during which time the transported molecule is protected against enzymatic attack, and (2) specificity of a particular enzyme to the molecule by differentially binding to reactive sites. This sequential dual selection tends to limit active transport to molecules capable of undergoing enzymatic attack.

Changeuk, Jean-Pierre, "The Control of Biochemical Reactions," *Scientific American,* April 1965. Changeuk concludes that an enzyme molecule undergoes automatic structural modifications in response to the presence of "signal" molecules. The nature of the change in molecular structure of an enzyme determines the rate of its catalysis.

Clark, Brian F. C., and Kjeld A. Marcker, "How Proteins Start," *Scientific American,* January 1968. The role of tRNA in protein synthesis is described.

Crick, F. H. S., "The Genetic Code: III," *Scientific American,* October 1966. A succinct summary of how chromosomal instructions provided by the nucleus of a cell are translated into proteins in its cytoplasm.

Green, David E., "The Synthesis of Fat," *Scientific American,* February 1960. How fatty acids are synthesized from acetic acid fragments and how lipids are synthesized from a glycerol backbone, fatty acids, and other distinctive chemical groups.

Kleinsmith, Lewis J., "Molecular Mechanisms for the Regulation of Cell Function," *BioScience,* 22(6):343–348, 1972. Kleinsmith argues that the total regulation of cell function can be explained by control at four basic levels—transcription, translation, enzymatic protein activity, and protein degradation.

Lehninger, Albert, "Energy Transformation in the Cell," *Scientific American,* May 1960. The flow of energy is followed during aerobic oxidation in mitochondria.

Loewenstein, Werner R., "Intercellular Communication," *Scientific American,* May 1970. Specialized contact junctions composed of molecules between two adjacent cells of certain tissues permit the two-way passage of chemical and electrical information. These junctions may be passageways for growth-regulating substances, according to Loewenstein.

Mazia, Daniel, "How Cells Divide," *Scientific American,* September 1961. An account of the "ballet" performed by specialized components of a cell during mitosis.

Meister, Alton, "On the Enzymology of Amino Acid Transport," *Science,* 180:33–39, 1973. Experiments are described which indicate that any amino acid may bind to a specific binding site on the outer surface of the plasma membrane; a membrane-bound transpeptidase is activated by the addition of a carrier group derived from the chemical glutathione; the activated transpeptidase removes the amino acid from its binding site and actively transports it across the membrane to the intracellular environment; and an intracellular enzyme releases the amino acid from its carrier.

Miller, O. L., Jr., "The Visualization of Genes in Action," *Scientific American,* March 1973. Electron photomicrographs of cells during transcription and translation are correlated with the postulated mechanisms of protein synthesis. The magnificent illustrations are highly instructive.

Moscona, A. A., "How Cells Associate," *Scientific American,* September 1961. A discussion of the protein-bound carbohydrates which, as surface-active materials on the cell membranes, bind individual cells together into tissues.

Phillips, David C., "The Three-Dimensional Structure of an Enzyme Molecule," *Scientific American,* November 1966. This fascinating account of how the structure of a complicated enzyme called lysozyme was deciphered suggests an atomic mechanism by which it hydrolyzes polysaccharides.

Segal, Harold L., "Enzymatic Interconversion of Active and Inactive Forms of Enzymes," *Science,* 180:25–32, 1973. How the activity of one class of enzymes which includes dehydrogenases, phosphorylases, synthetases, and polymerases can be switched on and off by another group of enzymes. The switching signals are provided by hormones. By their interconversion between active and inactive forms, the switching enzymes enable a cell to meet its own metabolic needs as well as to fulfill metabolic requests from other parts of the body.

Spiegelman, S., "Hybrid Nucleic Acids," *Scientific American,* May 1964. Spiegelman's experiments indicate that only one strand of double-stranded DNA serves as a template for RNA synthesis.

# PART TWO
# THE FOUNDATIONS
# OF NATURAL SCIENCE

# CHAPTER 7
# THE NATURE
# OF THINGS

The panoply of human life, from conception through birth to death, is as awesome as it is familiar, as mysterious as it is comprehensible, and as magnificent as it is inspiring. It is a canvas painted with light and other forms of energy, sprinkled with stars from this and other galaxies, strewn with life, rocks, water, and air, and made of particles so small that nearly all our knowledge of them is indirect.

If our senses are boggled when we attempt to comprehend the meaning of the canvas, it is because ordinary reality is that which we can perceive as well as agree upon in a social context. It stretches the imagination to consider the phenomena which occur routinely in the ordinarily invisible world of atoms and molecules. However, these particles which populate the cells of our bodies are so important biologically that a consideration of their physical behavior is precisely what we will undertake in this chapter.

Since the basic unit of matter is the atom, just as the basic unit of life is a cell, we will begin by studying the nuclear and orbital structures of atoms. The insights gained from an understanding of atomic structure will lay the foundation for a later study of the chemical behavior of elements and compounds. The physical nature of matter is the province of this chapter; chemical interactions are reviewed in Chap. 10.

Next, we will examine the three basic states of matter—solids, liquids, and gases—and the manifold combinations of matter which exist in nature. This review is capped by a consideration of the unique qualities of living matter, focusing on the behavior of water and colloids in the cell interior.

## THE PHYSICAL CHARACTERISTICS OF MATTER

*Matter* is defined as anything that has mass and takes up space. The second point is easier to clarify than the first. That two objects cannot occupy the same space at the same time hardly bears repetition. One object must be displaced by the other when competing for the same physical position. The concept of mass requires a more lengthy explanation.

We are all familiar with the expression of *weight*. How much each of us weighs is determined by our *mass* and the force of *gravity* exerted on us. Since the earth is flattened at the poles (polar diameter = 7,899.95 mi) and bulges at the equator (equatorial diameter = 7,926.68 mi), an individual weighs slightly more at either pole than at the equator. In other words, a greater gravitational force is exerted on matter at the poles, compared with the equator. The relative forces of gravity at these geographic locations are shown in Fig. 7-1.

Yet mass does not change just because we happen to be at a pole, at the equator, or even in outer space! Mass is determined by the number and kinds of atoms of which a given volume of matter is composed. The atoms may be packed close together or far apart. In addition, atoms may be heavy (large) or light (small).

The preferred unit for expressing mass is the *gram* (Gm). One gram is equal to $\frac{1}{453.6}$ lb. Stated another way, 453.6 Gm is the same as 1 lb. It should be borne in mind, however, that these units are ordinarily used

**Figure 7-1** *The differences between the force of gravity at the poles and at the equator. The relative differences in the gravitational forces of attraction are suggested by the arrows.*

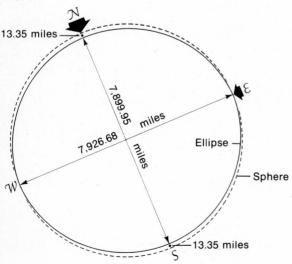

to express a relative weight rather than an absolute mass. The units of volume, on the other hand, are the *liter* and the *cubic meter* (cu m, or m³). One liter is equivalent to 1.06 quarts (qt); 1 meter (m) is the same as 39.37 inches (in.). Therefore, the volume 1 m³ is equal to a cube each side of which is 39.37 in., or 1 m, in length. In a biologic context, the cubic meter and the liter are large units. In practice, we use a unit 100 times smaller than a cubic meter, i.e., 1/100 of a cubic meter, or 1 *cubic centimeter* (cc), or a unit 1,000 times smaller than a liter, i.e., $\frac{1}{1,000}$ liter or 1 *milliliter* (ml). The volume difference between 1 cc and 1 ml is relatively small.

How may we compare different masses occupying various volumes? The answer is in terms of the densities of the objects being compared. *Density* provides us with information on how closely packed or how heavy the atoms are which occupy a unit of volume. Density relates mass to volume by means of the following equation:

$$D\text{(density)} = \frac{M\text{(mass)}}{V\text{(volume)}} \quad \text{or} \quad D = \frac{M}{V} \qquad (7\text{-}1)$$

If mass is in grams, the volume is ordinarily in milliliters, and density is then expressed in grams per milliliter (Gm/ml).

Suppose the density of an object is 1 Gm/ml. Now consider a larger mass of the same object, say 100 Gm. In order for the density to remain constant, 100 Gm of the same type of matter must of necessity occupy 100 ml. Thus, provided that the type of matter does not change, its density is the same, irrespective of its mass.

Next, assume that we have determined that a given object has a mass of 453.6 Gm but have not measured the amount of space, or volume, the object occupies. Under these circumstances, there is no way to express the density of the object. We would meet a similar obstacle if we knew the volume of the object but not its mass.

Let us return once again to the basic relationship, $D = M/V$. If $M$ increases while $V$ is held constant, then $D$ also increases. Conversely, if $V$ increases in relation to $M$, then $D$ must decrease. These conclusions follow from an understanding that the number above the line (the numerator, symbolized by $M$ in this equation) is divided by the number below the line (the denominator, here symbolized by $V$) to determine the density (the quotient, symbolized by $D$). In a practical sense, a lead brick has a high density and a correspondingly large mass, compared with a block of wood of identical dimensions. In a similar vein, bone is more dense per unit volume than connective tissue. Of course, we all knew these facts without the benefit of the formula $D = M/V$. The point is, density relates mass to volume and permits precise comparisons of these qualities among the multitudinous kinds of matter of which the earth is composed.

## THE STRUCTURE OF ATOMS

As we noted earlier, the fundamental unit of matter is the *atom*. That is, the smallest level of physical organization is atomic, just as the lowest level of biologic organization is cellular. However, this does not imply that atoms cannot be broken down into smaller parts or that cells do not contain various types of inclusions. All it means is that atoms give unique characteristics to the various kinds of matter with which we are familiar. Thus when atomic structure is disrupted, there is a change in the composition of the affected matter. In an analogous manner, when cells are disrupted, their protoplasmic qualities are altered.

To carry our analogy between atoms and cells even further, let us compare their gross (general) structural similarities and differences. A cell consists of a nucleus surrounded by cytoplasm. An atom contains a nucleus bounded by one to seven shells, or energy levels. A cell contains highly complex structures called organelles and other inclusions. An atom is constructed of three basic types of particles—protons, neutrons, and electrons. The nucleus of a cell contains chromosomes and one or more dark-staining bodies called nucleoli. An atomic nucleus bears one or more protons and neutrons. The cytoplasmic portion of a cell contains organelles such as the endoplasmic reticulum, the Golgi apparatus, mitochondria, centrosomes, and lysosomes, as well as other inclusions such as fat globules and crystals. The energy levels outside the nucleus of an atom contain nothing but electrons. The gross structure of a cell and an atom are compared in Fig. 7-2.

Although atoms are a much simpler form of matter than cells, their behavior is fantastically complex.

**Figure 7-2** *A comparison of the gross structure of a cell* (A) *and an atom* (B).

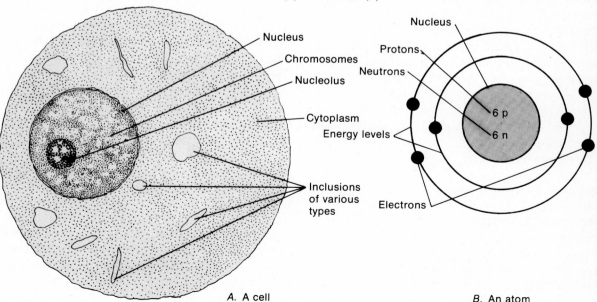

Nucleus
Chromosomes
Nucleolus
Cytoplasm
Inclusions of various types

Nucleus
Protons
Neutrons
Energy levels
Electrons
6 p
6 n

*A. A cell*

*B. An atom*

Each atom is a mass of particles in motion, its electrons orbiting around the nucleus at unbelievably great velocities as well as spinning on their own axes. Although we know the number of electrons in each energy level, or orbit, it is impossible to pinpoint their exact locations because of their orbital velocities.

The atomic nucleus is also somewhat of a mystery. Although we can specify the number of protons and neutrons in the nucleus of any atom with precision, we do not know how they are arranged. To complicate matters further, certain elementary particles called *mesons* constantly shuttle back and forth between the central portions of the protons and neutrons, providing a kind of nuclear cement which holds these particles together. A cloud of mesons therefore orbits around, and forms the outer part of, each proton and neutron. In addition, these particles spin on their axes.

With these points in mind, let us probe the structure of atoms more carefully. First, we will examine atomic nuclei. Then we will consider the electronic energy levels outside atomic nuclei.

**The nucleus**  Compared with the size of an entire atom, the diameter of an atomic nucleus is infinitesimally small. Its small size, of course, makes the nucleus extremely difficult to study. As a consequence, nuclear organization and behavior are poorly understood.

The nucleus of any atom consists of subatomic particles. Of the various kinds of particles which have been identified as nuclear in origin, only two particles, the *proton* (p) and *neutron* (n), are fundamental to our understanding of the unique characteristics of different types of atoms. The reason is that the number of protons and neutrons in an atomic nucleus determines the physical qualities and chemical behavior of the entire atom.

**Protons**  Being particulate, protons possess mass. This mass ($1.672 \times 10^{-27}$ kilograms, or 0.000, 000,000,000,000,000,000,001,672 kg) is extraordinarily small and extremely cumbersome to use in practice. Therefore what is usually done is to express the mass of subatomic particles of any atom in relation

to $\frac{1}{12}$ the absolute mass of an atom of carbon containing six protons and six neutrons. When carbon is used as the standard for comparison, the mass is expressed in *atomic mass units* (amu). Atomic mass units make comparisons in the size range of atoms less awkward and more useful than measurements of mass based on the gram. The mass of a proton is 1.007 amu. In practice, however, an atomic mass unit of 1 is ordinarily used.

Each proton bears a positive (+) electric charge. The number of positively charged protons in the nucleus of an atom is balanced by the number of negatively charged electrons outside the nucleus. Since the number of positive and negative charges is identical, *electrical neutrality* is preserved. Because an atom with an excess or deficiency of one or more electric charges is technically an ion and not an atom, electrical neutrality is an essential atomic requirement. More importantly, as will be shown later, the number of orbital electrons determines the chemical properties of atoms.

The number of protons in an atom is referred to as its *atomic number*. All atoms which bear the same atomic number (and therefore the same number of electrons) belong to the same element, irrespective of the number of neutrons present.

There are 103 generally accepted elements which, singly or in combination, make up all the matter of which the universe is composed. The atom is the smallest unit of an element which still possesses its basic physical properties and chemical behavior. Thus, which of the 103 elements an atom belongs to is determined by its proton, or atomic, number. The element hydrogen has an atomic number of 1, and the element lawrencium possesses an atomic number of 103. The atomic numbers of all the other elements range between these two extremes.

**Mesons**  The core of a proton is thought to be identical with the core of a neutron. The central part of a proton or neutron is called a *nucleon*. Each nucleon is surrounded by a cloud of whirling mesons, as we have already noted. Every so often, one or more mesons are spewed out from the cloud, to be absorbed by a meson cloud of another proton or neutron during a reciprocal exchange of mesons. The mesons contribute to the nuclear force which binds the nucleons together. In addition, a proton differs from a neutron only in the composition and mass of the *meson cloud* which envelops each.

The binding force which holds the nucleons together is determined by the relative positions of all the interacting mesons. Because the positions of mesons are constantly changing in relation to one another, the nuclear force is constantly fluctuating. If mesons from two adjacent nucleons approach each other too closely, they are repelled; when they are a little farther apart, they are mutually attracted. However, the nuclear forces are not electrostatic; i.e., they are not due to any electric charges which the particles may possess. Rather, the nuclear force is due to the angular spin and orbital momentum of each meson.

As two mesons change their positions, the force they exert on each other changes in a manner analogous to a force change resulting from a shift in position of two opposed bar magnets. The magnetic force created by the magnets is determined by the alignment of their north and south poles. The nuclear force created by adjacent mesons is determined by the directions of their respective spins and orbital paths. The similarity in force changes with shifts of position for both mesons and bar magnets is illustrated conceptually in Fig. 7-3.

*Neutrons*  The mass of a neutron, $1.674 \times 10^{-27}$ kg (0.000,000,000,000,000,000,000,000,001,674  kg), is the same as 1.0087 amu, when compared with $\frac{1}{12}$ the mass of a carbon atom. Compared with the mass of a proton, this number is larger by 0.0009 ($\frac{9}{10,000}$, or nine ten-thousandths) amu. The difference in mass between a proton and a neutron is, of course, due to a difference in the mass of the mesons which envelop the nucleons. Neutrons possess mesons of slightly greater mass than those of protons.

In addition to the mass difference between a proton and neutron, there is also a difference in charge. Whereas a proton bears a positive electric charge, a neutron does not possess an electric charge.

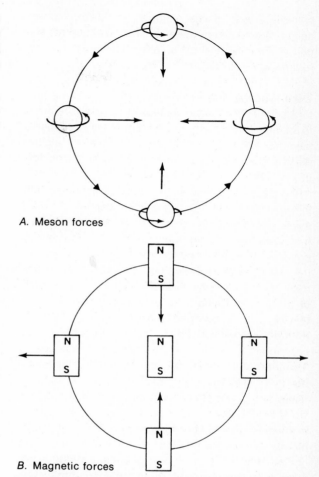

*A.* Meson forces

*B.* Magnetic forces

**Figure 7-3**  *A comparison of the forces exerted between mesons* (A) *and bar magnets* (B). *Lengths and directions of arrows indicate force magnitudes and directions during interactions at various positions.*

The significance of neutrons resides in the masses which they confer on atoms in which they are nuclear components. However, so long as the proton or atomic number remains constant, atoms may differ in the number of neutrons they possess but still belong to the same parent element. In other words, although their atomic mass numbers are different, the *chemical properties* of such atoms are basically the same. When atoms differ only in neutron number, they are called *isotopes*. Some isotopes are stable; others, called *radioisotopes*, spontaneously disintegrate because of nuclear instability. We will have more to say regarding radioisotopes in Chap. 9.

*Other nuclear characteristics*   The sum of the protons and neutrons in the nucleus of an atom is referred to as the *atomic mass number*. Since many elements have numerous isotopes, all with different atomic mass numbers, the term is useful only when considering the mass of a particular isotope.

Atomic mass numbers should not be confused with the atomic weights. The *atomic weight* refers to the average mass of any element, compared with a carbon standard. The average mass of an element is calculated in the following manner: First, the atomic mass number of each isotope in the family of the element is multiplied by its abundance, compared with all the other isotopes in the family; next, the values obtained for all the isotopes from the preceding step are added together; and, finally, the sum obtained in the second step is divided by the number of isotopes in the family. The atomic weight calculated in this manner is an average value and does not necessarily relate to a given atom of the element. Often, atomic weight indicates the number of protons and neutrons in the most abundant isotope of an element. Although atomic weights are ordinarily not whole numbers but contain fractions, they are nevertheless frequently close to whole numbers. Atomic mass numbers, on the other hand, are always whole numbers, because the figure simply represents the sum of the protons and neutrons in a specific atom.

At this point, you may be wondering why the electrons in the energy levels outside the nucleus are ignored in determinations of the atomic weights of elements. Surely electrons, being particulate, have mass! However, the mass of an electron is close to 2,000 times less than the mass of a nucleon. Therefore, even in a heavy atom containing many electrons, the electron mass is negligible, compared with the combined mass of nuclear protons and neutrons. If it were possible to eliminate all the electrons from every atom in your body, you would be only ounces lighter.

The actual physical structure of an atomic nucleus is the subject of much debate. This mass of matter is thought by some to functionally resemble a cloudy crystal ball and by others to be an irregular grouping of protons and neutrons, while still others look at the nucleus as a deep well with straight sides and a flat bottom. All these conceptual models have some element of truth in them. It remains to be seen which theory, if any, is actually correct. We do know that the density of an atomic nucleus is greatest in its central portion and diminishes progressively toward the periphery. Even here, our understanding is limited because there seem to be peaks and troughs of density between the center of the nucleus and its outer boundary, suggesting some internal organization.

*Energy levels*   The entire nucleus of an atom is also enveloped in a cloud, in a fashion analogous to nucleons. However, instead of meson clouds, the cloud is composed of between 1 and 103 electrons. Each electron spins on its own axis and also rotates around the nucleus in a defined path, or orbit.

Each electron has a mass of $9.1 \times 10^{-31}$ kg (0.000,000,000,000,000,000,000,000,000,910 kg). This mass is 1,836 times less than 1 amu or the approximate mass of a proton or neutron. Also, each electron bears a negative electric charge equal in magnitude to the positive charge of a proton. Since unlike electric charges attract each other, the electrons, because of their negligible masses, are electrostatically attracted toward the nucleus. This electrostatic force holds the electrons in discrete orbits outside the atomic nucleus.

Because it is impossible, even in principle, to simultaneously measure the position and velocity of an electron in orbit, it is best to think of an orbital electron as a shapeless, cloudlike mass. The orbits which the electrons describe have a variety of shapes, such as spherical and ellipsoid, and are altered by magnetic forces of attraction and repulsion in predictable ways.

The paths described by orbital electrons are located in one of seven separate *shells*, or *energy levels*. The energy level closest to the nucleus is called the K shell. This shell is occupied by a maximum of two electrons. The next energy level outward from the nucleus is the L shell and can hold up to eight electrons. The other energy levels outward from the nucleus and the maximum number of electrons which they hold are as follows: M (18), N (32), O (32), P (9), and Q (2).

The energy levels and their theoretical electron numbers are summarized diagrammatically in Fig. 7-4. All seven energy levels are represented, and each shell is filled to its maximum capacity. The shell designations are given at the right; the maximum electron capacity of each shell is shown at the left. All the electrons in any one shell possess similar energies, although the shapes of their orbital paths around the nucleus may differ. The shells are arranged in an orderly way, and it is possible to predict the location of each energy level possessed by an atom if the total number of electrons which it possesses is known. Thus, even though we cannot specify both the instantaneous position and velocity of an orbital electron, we can predict the approximate location of the discrete (separate) energy level it occupies.

The atomic model illustrated in Fig. 7-4A has severe limitations of which we should be aware. Although this conception of an atom permits us to visualize the number of electrons contained in each energy level and the relative distance from the nucleus to each shell, it allows us to do nothing more. For example, the drawing tells us nothing about the shape of the individual orbits described by each of the electrons, the location of the individual electrons in each energy level (an impossibility in any working model), or the exact distance of each energy level from the nucleus. In addition, the relative sizes of the nucleus and the energy levels are grossly inaccurate. The fact is the nucleus is so infinitesimally small, compared with the

**Figure 7-4** *The energy levels and electron capacities of atoms. A. A hypothetical atom showing the energy levels and the maximum electron capacities of each shell based on atomic theory. B. The relative density of the electron cloud surrounding the nucleus of an atom of lawrencium (atomic number 103).*

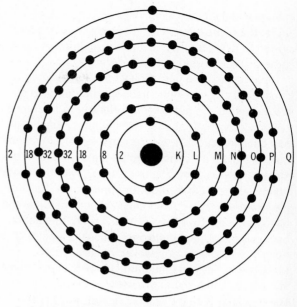

*A.* A conceptual diagram of the energy levels and their electron numbers in a hypothetical atom.

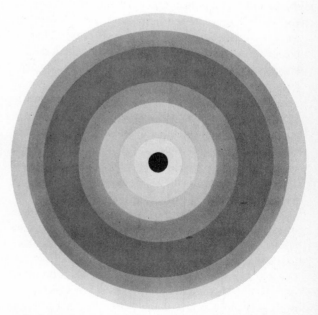

*B.* A conceptual diagram of the cloud of electrons of an atom of lawrencium, atomic no. 103.

diameter of the entire atom, it would be invisible even under the highest resolution of electron microscopes currently available, if the atom examined were as large as the drawing in Fig. 7-4A.

Nevertheless, we can achieve more realism in our concept of an atom if the electrons are depicted as a cloud around the nucleus. The density of the cloud, i.e., how dark it will appear, is then proportional to the number of electrons in each energy level. Using these guidelines, let us conceptualize the structure of the element lawrencium. Shells K and Q, with two electrons each, should be light in appearance, because the two electrons in each shell are far apart. On the other hand, energy levels N and O, with 32 electrons each, should appear quite dark, since the electrons in each of these shells are packed closely together. The other shells, with intermediate numbers of electrons, should assume a shading somewhat between energy levels K and Q, on the one hand, and shells N and O, on the other hand.

Finally, in a further attempt to interject realism into our visualization of an atom, the nucleus should be drawn as small as reasonably possible and should be very dark in appearance, compared with the electron cloud. A small size is more compatible than a larger size with the actual point size of the nucleus. A deeply shaded nucleus in this type of model is symbolic of its high density.

Figure 7-4B, showing an atom of lawrencium, is based on these considerations. The important point to appreciate when looking at this atom is that its electron density increases from the nucleus to what would be the N and O energy levels and then decreases through the P and Q shells.

Isotopes  The lightest element is hydrogen. An atom of ordinary hydrogen has only one proton in its nucleus. No neutrons are present. Since the protons and electrons of any atom are balanced, there is just one electron in hydrogen. The orbital electron is located in the lowest, or K, energy level. The positioning of this single electron follows from the fact that the electrons of any element always occupy the energy levels closest to the nucleus.

Thus, an atom of ordinary hydrogen possesses

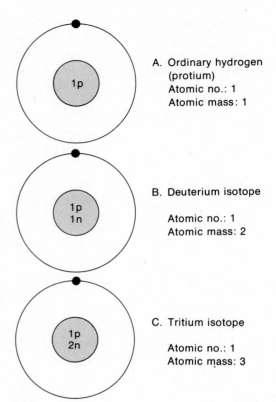

**Figure 7-5**  *A comparison of the atomic structure of ordinary hydrogen (protium) and its two isotopes, deuterium and tritium.*

one positively charged proton in its nucleus and one negatively charged electron outside the nucleus. However, the element hydrogen also has two isotopes, deuterium and tritium. Each of these isotopes has one proton and one electron, just as does ordinary hydrogen. However, the nucleus of deuterium contains one neutron while that of tritium contains two neutrons. Therefore, the difference between these isotopes and the parent element resides in the neutron number. That is, the atomic mass numbers of protium, deuterium, and tritium are 1, 2, and 3, respectively, although each of their atomic numbers is 1. Since the chemical properties of atoms are determined by their atomic numbers, protium and its two isotopes possess an identical chemical behavior. In addition, the physical characteristics of the three types of hydrogen atoms are very similar.

Such properties as size, physical state at room temperature, color, and odor do not change just because the neutron number varies. In fact, only their masses differ. The atomic structures of protium, deuterium, and tritium are compared diagrammatically in Fig. 7-5.

**An overview of atomic organization**  The next lightest element after hydrogen is helium. Atoms of helium possess two protons and two neutrons in, and two electrons outside, their nuclei. The two electrons completely fill the K energy level. The next heavier element must add its third electron to the L energy level. The electronic structure of atoms will be discussed in more detail in Chap. 10, when we consider the organization of the periodic table of the elements. For now, we note that the outermost energy level of any atom can contain no more than eight electrons. Any atom may contain less than eight electrons in its outermost shell, but no atom possesses more than this number.

The electrons in the outermost energy level of an atom are the ones which ordinarily interact chemically with other atoms. The behavior of the outermost electrons permits predictable chemical associations to occur among particular atoms.

Let us return again to helium, the second lightest of the elements known to man. Because this element possesses two protons and two neutrons in its nucleus, helium has an atomic number of 2 and an atomic mass number of 4. The atomic structure of helium is shown in the insert in Fig. 7-6.

The larger drawing of an atom of helium in Fig. 7-6 will serve to review our knowledge of atomic structure. The drawing, of course, is a conceptualization of the helium atom and is only partially correct. The nucleus is shown as a cloudy crystal ball. The "cloudiness" is suggestive of different regions of density. The protons and neutrons are illustrated as large dots surrounded by meson clouds. The dots represent nucleons. The wavelike arrows in both directions between the protons and neutrons indicate reciprocal meson exchanges between nucleons. The electrons are shown as a cloud surrounding the nucleus. In a helium atom, this electron cloud constitutes the K energy level. The negatively charged electrons exactly balance the positively charged protons. The electrons are held in orbit by electrostatic forces of attraction between them and the nuclear protons.

## BULK MATTER

One hundred and three elements make up the stuff of the universe. The elements may be pure or mixed, or combined chemically. The molecules formed through chemical combinations of atoms may themselves be of the same type or of different types mixed together. Every object we look at or touch, every odor we smell—in short, anything that has mass—is composed of atoms and molecules combined chemically and mixed physically.

If we wish to gain a better understanding of the human body, then we must delve further into the organization of protoplasm. Protoplasm is the epitome of a mixture of different compounds, although this statement is by no means its definition. Nevertheless, the characteristics of life are established and maintained by the interactions of the diverse compounds in the cell interior. To understand protoplasm better, we must have a working knowledge of its physical constitution. But first the foundations must be laid. We have to know more about the unique qualities of matter in bulk. A knowledge of the properties of bulk matter will enable us to examine the behavior of cellular molecules more critically than we otherwise could.

**Similarities and differences between solids, liquids, and gases**  Matter exists in three states: *solids, liquids,* and *gases.* This, of course, is no revelation to anyone, because we have all had intimate personal contact for many years with trees, water, air, and other types of matter in these three states. Our concern here is to compare the physical properties of solids, liquids, and gases.

*Shape*  The shape of most solids is fixed and can be deformed only by exerting a sufficiently large force. Liquids, on the other hand, conform to the shapes of their containers, seeking the lowest levels in which the force of gravity is the same at any point along their uppermost surfaces. By contrast, gases may expand or contract without limit, completely filling the

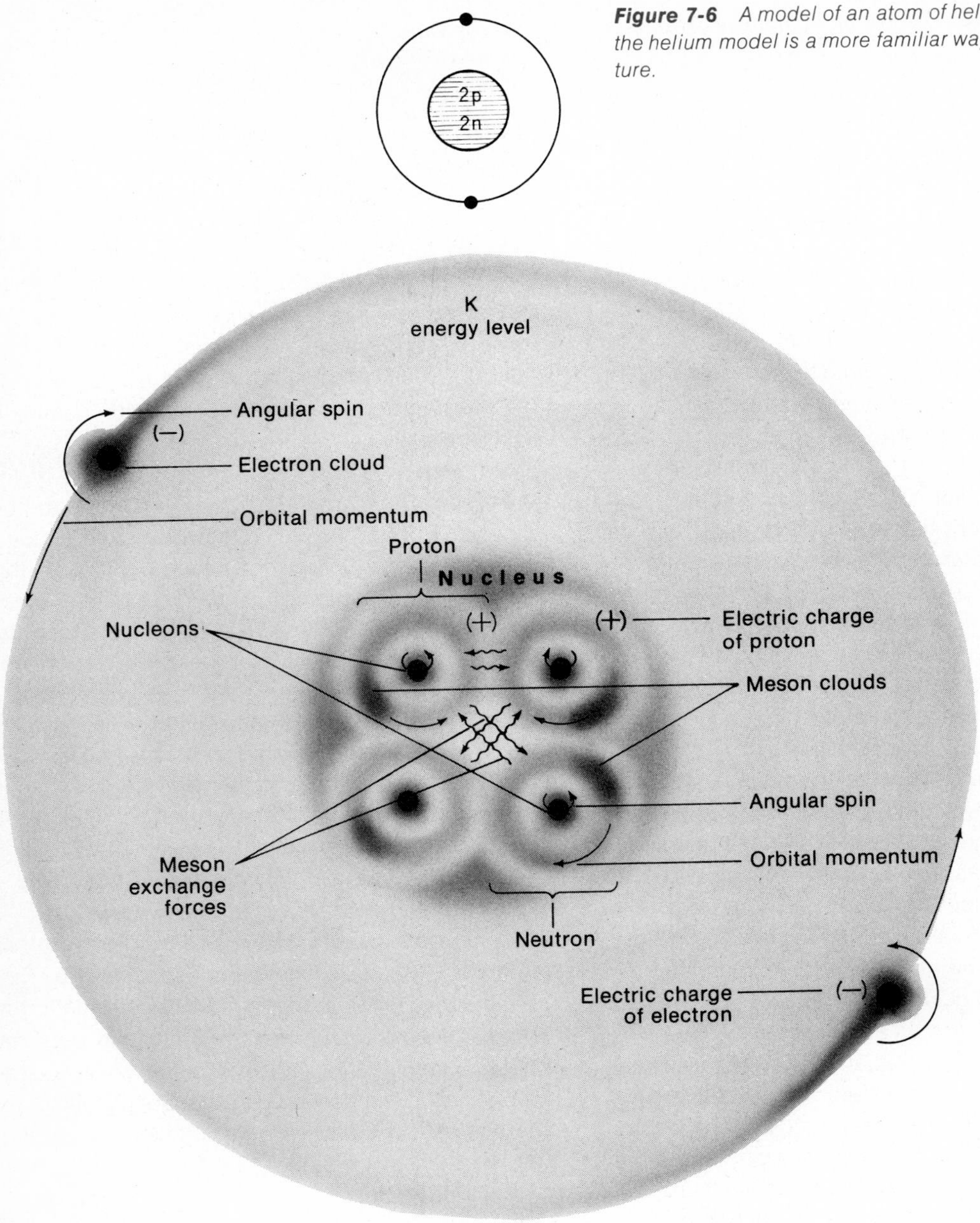

**Figure 7-6** *A model of an atom of helium. The insert above the helium model is a more familiar way of showing its structure.*

volumes of the containers which enclose them. These points are illustrated diagrammatically in Fig. 7-7.

To appreciate why each state of matter is unique, we must look more closely at their atomic structures.

*Atomic organization* The atoms or molecules in solids, liquids, and gases are arranged in different

ways. Differences in their structural organization account in large part for the qualities which are unique to each of these three states of matter.

Atoms and molecules of solids are packed closely together, forming a relatively dense mass. Spaces between atoms are few and far between.

The particles which make up liquids are farther

apart than those of solids. Although there are notable exceptions, such as mercury, the density of liquids is usually considerably less than of solids. In addition, atoms of liquids link together to form chains of distinctive multi-ring-like structures. Each of the rings is multifaceted, or polyhedral, in shape. These ring structures play an important role in determining the way in which materials move through liquids, as will be shown shortly.

By way of contrast, gaseous atoms or molecules are separated from each other by relatively large distances, and they tend to be uniformly distributed in their dispersing medium. That is, the gaps between gaseous particles are wide, and the particles are more or less equidistant from each other.

Differences in the basic arrangement of particles in solids, liquids, and gases are illustrated in Fig. 7-8.

*Particle movement* The particles of which solids, liquids, and gases are constructed also move in different ways. Atoms or molecules in solids vibrate, or oscillate, on their own axes, without net displacement. The axis of vibration of each particle may be haphazard, or all the particles may oscillate along the same general plane. The oscillatory behavior of some solids, such as crystals and magnets, is made use of for various purposes.

The movement of liquid particles, on the other hand, is much more complicated than that of solids. The branching chains of interconnected molecules as well as free molecules are in constant flux, changing their spatial position from one moment to another. However, since the molecules of each ring are fixed to a particular chain of ring structures, the position of the molecules relative to each other does not vary appreciably. Therefore, liquids have a structural integrity with a more or less defined shape. In this respect, liquids resemble solids more closely than they do gases, since gases do not have an identifiable structure.

A. Solid

Gravity

B. Liquid

*Figure 7-7* *A comparison of the shapes of solids, liquids, and gases. A. The vise exerts strong compressional forces (inwardly directed arrows) on the block of wood without breaking it. B. The liquid remains level in the tipped glass when it is being poured. C. The small container of volume a holds ten particles of gas. When the container is expanded to volume* b, *the gas particles distribute themselves uniformly throughout the new volume. The same phenomenon occurs when the container is enlarged to volume* c.

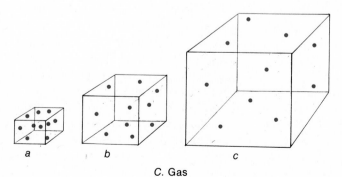

a          b          c

*C.* Gas

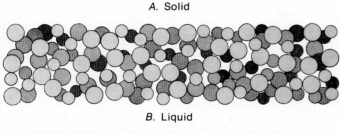

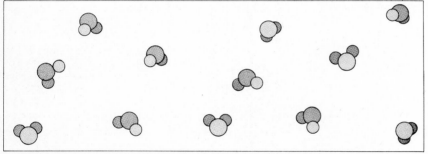

*A.* Solid

*B.* Liquid

*C.* Gas

**Figure 7-8**  *A comparison of the atomic organization of matter. The atoms are shown as spheres of various sizes. A. Atoms of solids are densely packed. B. Atoms of liquids are somewhat more spread out than those of solids. C. Molecules of gases are separated from each other by relatively large distances, and they tend to be distributed uniformly.*

The structural organization of liquids has profound biologic implications. The molecular rings entrap particles. In technical jargon, we say that the particles are dissolved in solution. Because the rings shift their positions constantly, the enclosed particles are transported from one point to another in solution. This method of transport is used widely by particles dispersed in water inside cells.

Gaseous particles move in a random, zigzag manner due to thermal agitation. In other words, heat energy is exemplified in the motion of gaseous molecules. The erratic motion characteristic of gases is called *Brownian movement*. The unpredictable shifts in the travel path of a gaseous particle are due to collisions between it and other particles. If the particles with which the gaseous particles collide are extremely small, they will move ever so slightly in response to each bombardment. Such movements of particulates may be observed under the high-power objective of a standard laboratory microscope. While we commonly say that the particles exhibit Brownian movement, what we really mean is that the particles move as the result of being hit by gaseous molecules undergoing Brownian movement.

The characteristic movements of solid, liquid, and gaseous particles are shown in Fig. 7-9.

*Compressibility*   The nature of the atomic or molecular organization of solids, liquids, and gases explains their characteristic differences in compressibility. Solids, as we are well aware, resist deformation because the particles of which they are composed are densely packed. For the same reason, solids are only slightly elastic and resist stretch. Nevertheless, there are some notable exceptions. For example, gold can be hammered into extremely thin sheets of foil. Rubber and its synthetic substitutes are capable of a high degree of stretch.

Liquids, on the other hand, are more compressible than most solids, because the spaces between liquid particles are somewhat farther apart. Under a high enough pressure, many liquids may be squeezed to about 75 percent of their original volume. Other liquids of higher density are more resistant to compression, exhibiting a behavior more like that of solids than of liquids.

By way of contrast to both solids and liquids, gases are infinitely compressible, because gaseous particles are widely scattered throughout the space available to them. The relatively large gaps between particles of a gas permit it to be squeezed with ease into a vanishingly small volume.

The differences in compressibility between solids, liquids, and gases are summarized in Fig. 7-10.

*Pressure*   Because of density differences, the internal pressures of solids, liquids, and gases differ in

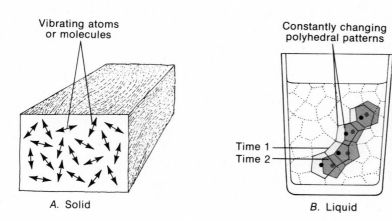

Vibrating atoms
or molecules

*A.* Solid

Constantly changing
polyhedral patterns

Time 1
Time 2

*B.* Liquid

**Figure 7-9**   *A comparison of the movement of particles in solids, liquids, and gases. A. The particles of solids vibrate, or oscillate, about their own axes. B. The characteristic ring structures which molecules of liquids assume are constantly shifting position. A segment of rings formed by linked molecules is shown in black at time 1. The same segment is shown in color at time 2. C. Gaseous molecules move erratically through space.*

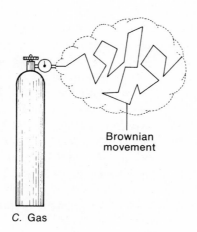

Brownian
movement

*C.* Gas

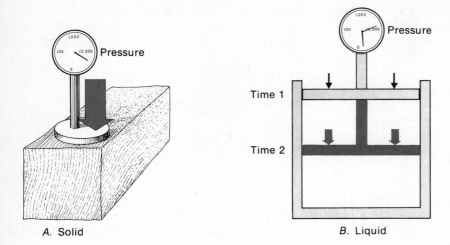

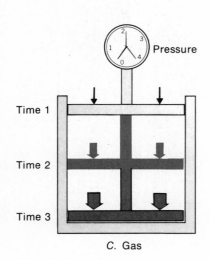

*A. Solid*

*B. Liquid*

*C. Gas*

**Figure 7-10** *A comparison of the compressibility of solids, liquids, and gases. A. The shape of a solid may be slightly distorted as a result of pressure exerted on it, but its total volume is not noticeably affected. B. The pressure of the piston at time 1 (black) is due only to its weight and the pressure of the atmosphere on top of the liquid. Its pressure at time 2 (color) is due to these factors plus a large externally applied force. C. The external pressure on a gas enclosed in the cylinder is progressively increased from time 1 to time 3. The downwardly directed arrows are proportional to the magnitudes of the externally applied forces.*

certain quantitative aspects. Before looking at the specific differences in internal pressures between these three states of matter, we must understand the concept of pressure and the nature of the forces which contribute to it.

*Pressure* is defined as a force acting on a unit of area of any mass. More precisely:

$$P(\text{pressure}) = \frac{F(\text{force})}{A(\text{area})} \quad \text{or} \quad P = \frac{F}{A} \qquad (7\text{-}2)$$

If force is expressed in pounds (lb), then area is expressed in square inches (sq in., or in.²). Using these units, the pressure is stated in pounds per square inch (lb/in.²).

The air we breathe is a mixture of gases and contaminants. Gas particles possess mass. Because the atmosphere extends many miles above sea level, air has considerable weight due to the force of gravity. We are not aware of the pressure exerted by the atmosphere, simply because the pressures on the inside and the outside of the body are balanced.

However, when the air pressure changes abruptly,

we are certainly aware of its effects. In fact, our behavior may be temporarily modified as, for example, during an impending thunderstorm due to the rapid approach of a low-pressure weather front. It is not uncommon to react to a sharp drop in barometric pressure by exhibiting moody, sluggish, and irritable behavior. These behavioral effects are readily discernible among some patients in mental institutions. We react to rapid changes in altitude by "popping" our ears. Chewing gum or otherwise moving the jaws during airplane takeoffs and landings, when in a rapidly moving express elevator, or when driving up a steep mountainous grade relieves the pressure against the eardrums.

The atmospheric pressure is measured with a barometer. In principle, a barometer consists of a calibrated glass column evacuated of air and immersed in a dish of mercury (Hg). The height to which the mercury rises in the column is determined by the weight of the overlying air mass on the mercury in the dish. At equilibrium, the downward pressure exerted by the mercury in the column is exactly balanced by the pressure of the air. Since barometers are calibrated in units of distance, barometric pressures are indirect measures of the force per unit area, i.e., the pressure of the air mass.

At sea level and 0°C, the barometric pressure is 760 millimeters of mercury (mmHg). This pressure is the same as 76 cmHg or 29.92 in.Hg, since 1 cm = 10 mm and 1 in. = 2.54 cm. In other words, if the pressure is expressed in millimeters of mercury and the pressure is desired in centimeters of mercury (cmHg), simply move the decimal point (760.) one place to the left (76.0). To express the pressure in inches of mercury (in.Hg) when the original pressure is in millimeters of mercury, convert the millimeter distance into centimeters, as before, and then divide this derived number by 2.54 centimeters per inch (in our example, 76 cm/2.54 cm/in.). In ordinary pressure units, a pressure of 760 mmHg is the same as 14.7 lb/in.². This number means that on every square inch of surface area, there is a force of 14.7 lb. A pressure of 14.7 lb/in.² is called 1 atmosphere (atm).

The barometric pressure is due to the density of the overlying air mass. Recall that density is equal to mass divided by volume ($D = M/V$). As the mass of air increases, so does its density, provided that the volume is held constant. The density of air is greatest at sea level and decreases with altitude. (What effect would the temperature of an air mass have upon its density and barometric pressure?)

The pressure decreases from 1 atm at sea level to somewhat less than $\frac{1}{3}$ atm at 50,000 ft of altitude. In the vernacular, the air is "thinner" at 50,000 ft than at sea level. What we really mean is that the distance between air particles is greater at the former than at the latter height. The change in atmospheric pressure with increasing height is relatively gradual.

The pressure at any depth below the surface of water is made up of two components: the air pressure exerted on the surface of the water and the weight due to the overlying water mass at the depth in question. At sea level, of course, the air pressure is equal to 1 atm. Starting from sea level, the fluid pressure increases by 1 atm, or 14.7 lb/in.², for every $33\frac{1}{3}$ ft of water depth. Therefore, if the water surface is at sea level, the pressure $33\frac{1}{3}$ ft below sea level is 2 atm (1 atm of air pressure + 1 atm of fluid pressure). At 100 ft below the surface of the water, the pressure is equal to 4 atm.

The change in water pressure is approximately 1,200 times that of air pressure, per unit vertical distance. The pressure of water imposes lower limits on the depths to which a swimmer in a free dive may travel before chest pains become excruciating or to which deepwater submergibles may descend before their specially reinforced steel hulls are crushed.

The weight of a mass of air, water, or any physical object is determined by the gravitational force with which the mass is attracted toward the center of the earth. The gravitational force exerted on your body is your weight. You, in turn, exert an equal but opposite force on the earth. For example, if you weigh 110 lb, you are being pulled toward the center of the earth with a gravitational force of 110 lb and are exerting 110 lb of force on the earth in the opposite direction. The earth, of course, is the winner in the tug-of-war, because of the difference in the magnitudes of the two masses.

Weight is determined by the location of the object

on, as well as its height above, the center of the earth. Mass, on the other hand, remains constant, irrespective of latitude or altitude. However, since weight is determined by the force of gravity, weight must be equal to the product of the mass of an object and the gravitational force which acts on it. These relationships are quantitatively expressed in the following manner:

$$W(\text{weight}) = M(\text{mass}) \times g(\text{gravitational force}) \text{ or } W = Mg$$
$$(7\text{-}3)$$

If we eliminate the gravitational force, $g$, then the expression becomes: Weight is proportional to mass, or $W \propto M$, where $\propto$ is the symbol for the phrase *is proportional to*.

Because solids are relatively dense, the external pressures along their surfaces can be unequal without affecting shape or volume. In other words, although the internal pressures of a solid may vary considerably, the solid can still remain unaltered physically.

An externally applied mechanical force on a liquid, on the other hand, will cause an immediate deformation. Under these conditions, the fluid is said to flow. However, even in the absence of a mechanical force, gravity affects the internal pressure of liquids. Recall that fluid pressure increases with depth, because of the additive effects of the weight of the overlying air and fluid masses. Therefore, when a container of water is perforated with holes at different depths, the rate of fluid flow is proportional to the depth. That is, water will flow with greater force from a container as the vertical distance increases between the outlet and the surface. Nevertheless, the pressure measured from any angle at any one point within a liquid remains constant. In other words, a liquid exerts an equal force in any direction, regardless of the position with respect to the vertical from which the measurement is made.

The forces which determine the pressure of gases are similar to those which influence fluid pressures. However, because of the relatively low densities of gases, differences in gas pressure at various depths in a closed container are negligible, in a practical sense. Therefore, when gas outlets are opened at dif-

ferent heights, the enclosed gas flows from each outlet with approximately the same force.

Figure 7-11 summarizes the important differences between the internal pressures exhibited by solids, liquids, and gases.

**The kinetic behavior of gases**  Gases exert pressure against the inside walls of their enclosing containers. These pressures are due to the Brownian movement of gaseous particles. Brownian movement, in turn, is produced by their kinetic energy. The average force with which the gas particles strike the walls is the pressure of the gas. Therefore, gas pressure is proportional to the average kinetic energy of the gas molecules. The relationship between the temperature of a gas and its kinetic energy and pressure is known as the *kinetic-molecular theory*.

An example of the behavior of a gas at different temperatures is illustrated in Fig. 7-12A. Assume that the volume of gas ($V_1$) in container 1 is equal to the gas volume ($V_2$) in container 2, i.e., $V_1 = V_2$. While volume is held constant, container 2 is heated; container 1 remains at room temperature. After the passage of time, the temperature $T_2$ of container 2 is higher than the temperature $T_1$ of container 1. In other words, $T_2 >$ (is greater than) $T_1$. A comparison of the gas pressure between the two containers indicates that container 2 is at a higher pressure, or $P_2 > P_1$.

We may conceptualize these findings by suggesting that the gas molecules are in motion because they possess kinetic energy. The random meandering of gas particles is the Brownian movement which we mentioned earlier. Let us assume for simplicity that six particles, on an average, are striking the walls of both containers in the same unit of time. Since $P_2 > P_1$ and $V_1 = V_2$, the particles in container 2 must be striking the walls with more force than the particles in container 1. In Fig. 7-12A, the differences in the kinetic energy of the gas particles in the two containers are indicated by the relative lengths and thicknesses of the arrows.

The pressure of a gas held at constant volume changes by $\frac{1}{273}$ (approximately 0.0037 percent) for every degree Celsius change in temperature. For example, the pressure of any gas is $\frac{1}{273}$ more at 1°C than

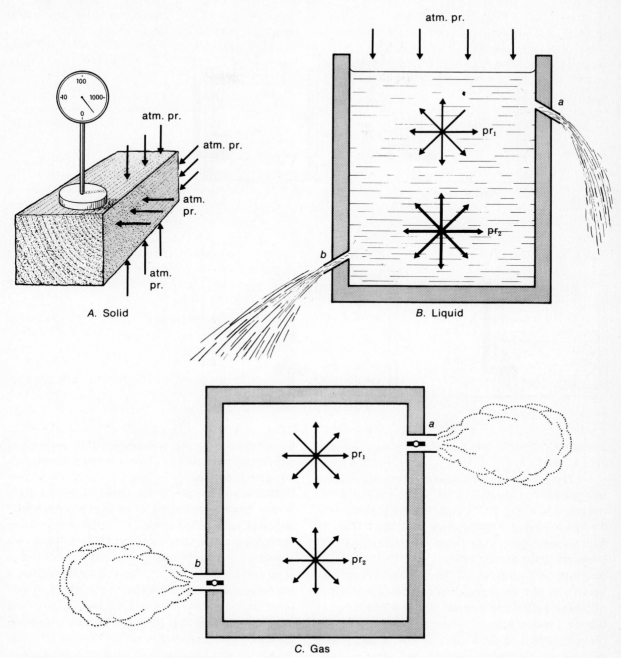

**Figure 7-11**  *comparison of the internal pressures within solids, liquids, and gases. A. An external force is applied to one portion of a solid. Other parts of the solid are exposed only to the pressure of the atmosphere. B. A liquid flows from outlet b with a greater force than from outlet a, due to the larger mass of overlying liquid. C. When outlets a and b are opened, the rates of gas flow are practically identical.*

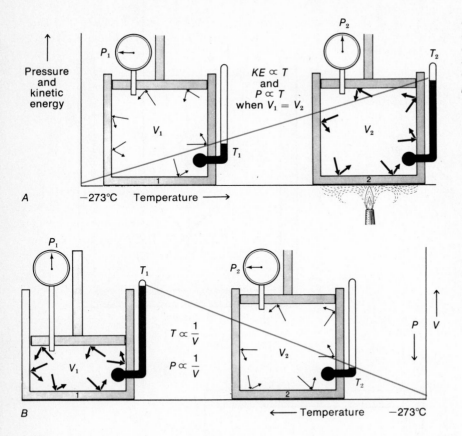

Pressure and kinetic energy

$KE \propto T$
and
$P \propto T$
when $V_1 = V_2$

−273°C   Temperature ⟶

A

$T \propto \dfrac{1}{V}$

$P \propto \dfrac{1}{V}$

⟵ Temperature   −273°C

B

**Figure 7-12** *The kinetic-molecular theory of matter. Represented by rebounding arrows, six particles collide with the walls of all the containers in the same period of time. A. The particles in container 2 strike with more force (longer and thicker arrows) and therefore possess a greater kinetic energy than those in container 1. B. A reduction in the volume of container 1 enclosing a constant concentration of gas causes a rise in gas temperature by increasing the force with which the particles bombard the inside walls (thick arrows). Conversely, increasing the volume of container 2 lowers gas temperature by dissipating some of the kinetic energy of the gas particles (thin arrows). The pressure and volume of a gas at −273°C is zero.*

it is at 0°C; at −1°C, gas pressure is always $\frac{1}{273}$ less than at 0°C.

The relation between the kinetic energy or pressure of a gas and its temperature is plotted on the superimposed graph in Fig. 7-12A. Note that the pressure line intersects zero at a temperature of −273°C. Thus, all gases should be at zero pressure at −273°C, a temperature called *absolute zero.*

Is this conclusion valid? The answer is both yes and no, in that the pressure of any gas continues to decrease by the constant amount of 1/273 for each degree Celsius drop in temperature, as − 273 degrees Celsius is approached. However, absolute zero has not yet been experimentally achieved. Furthermore, all gases tested change their state by liquefying at temperatures above absolute zero. Nevertheless, at absolute zero, the kinetic-molecular theory predicts that there would be a complete lack of am-

bient heat energy and a corresponding absence of thermal agitation. Without kinetic energy, gases cannot exert pressure.

What would happen to the temperature and pressure of a gas if its volume were suddenly changed? Suppose that a piston slides down the bore of a cylinder holding a gas, as in container 1, Fig. 7-12B. In this example, no attempt is made to hold temperature or pressure constant. As the volume of container 1 decreases, the kinetic energy of the enclosed gas particles increases. This is because the piston is moving toward the gas particles, thereby increasing the effective velocity of impact. The situation is similar to two cars which collide head on. If each car is traveling at 50 mi/hr at the moment of impact, then the effective impact velocity is 100 mi/hr. Thus, a reduction in volume means that the gaseous molecules will strike their enclosing walls with a greater vigor.

In the case of the gas particles in container 1, the increase in kinetic energy causes a corresponding increase in both pressure and temperature. This fact should become obvious when we recall that an increase in the kinetic energy of a gas is the same as an elevation in its temperature. When one increases, the other must also go up. Conversely, as the piston slides up the bore of the cylinder, increasing the volume of container 2, the temperature and pressure of the enclosed gas both decrease. Not only must gas particles travel a longer distance before striking a wall in container 2, compared with the situation in container 1, but some of the gas particles dissipate their kinetic energy as they strike the face of the receding piston. The loss of kinetic energy cools the gas.

From such experiments, we conclude that the temperature ($T$) or pressure ($P$) of a gas is inversely proportional to its volume ($V$). These relationships are expressed quantitatively as follows:

$$T \propto \frac{1}{V} \quad \text{and} \quad P \propto \frac{1}{V} \tag{7-4}$$

Notice that these expressions hold true only when $P$ and $T$ are changing simultaneously in response to changes in $V$.

Let us return once more to Fig. 7-12B. The superimposed graph plots change in temperature against volume. Note that the curve is a straight line which intersects the base line at a temperature of $-273°C$. This intersect is interpreted to mean that gases do not possess volume at absolute zero. Furthermore, the increase or decrease in volume as a function of temperature is the same as that for pressure. The volume of a gas therefore increases or decreases by $\frac{1}{273}$ for every degree Celsius rise or drop in temperature in a manner identical to the change in gas pressure with temperature. At absolute zero, gases would not only lack kinetic energy but would lack volume as well! Finally, observe that the slope of the line in $B$ is the exact opposite of the slope in $A$, reflecting the fact that $P$ and $T$ are inversely proportional to $V$.

**Boyle's law** Now, suppose that the temperature of a gas is held constant as its volume is changed. What, if anything, would happen to the gas pressure? The answer to this question is diagrammed in Fig. 7-13A. Note that the pressure of the enclosed gas increases as its volume decreases, and vice versa. Thus, the volume of a gas is inversely proportional to its pressure, provided that its temperature is held constant. Called Boyle's law, this relationship is expressed quantitatively as follows:

$$P \propto \frac{1}{V} \quad \text{under constant temperature} \tag{7-5}$$

**Charles' law** When the pressure of a gas is held constant, its volume must change when the temperature changes. Assume that the pistons in containers 1 and 2 of Fig. 7-13B are infinitely light and will rise or fall, depending upon the amount of kinetic energy of the gas particles which strike their faces. Under these conditions, an increase in the temperature of the gas also increases its volume, while a decrease in temperature has the opposite effect. In other words, the volume of a gas is proportional to its temperature, when pressure is held constant. The higher the temperature, the greater the volume, and vice versa. This relationship is called Charles' law and may be stated as follows:

$$V \propto T \quad \text{under constant pressure} \tag{7-6}$$

**The ideal gas law** Because the volume of a gas is proportional to its temperature ($V \propto T$, under constant pressure) and inversely proportional to its pressure ($V \propto 1/P$, under constant temperature), the quantitative expression for both these relationships becomes:

$$V \propto \frac{T}{P} \tag{7-7}$$

To remove the proportionality sign, add a constant $K$ to the right-hand side of the equation, thus:

$$V = K\frac{T}{P} \tag{7-8}$$

The value of the constant depends upon the mass of the gas and the units in which $V$, $T$, and $P$ are expressed. To solve for the constant $K$, multiply by $P$

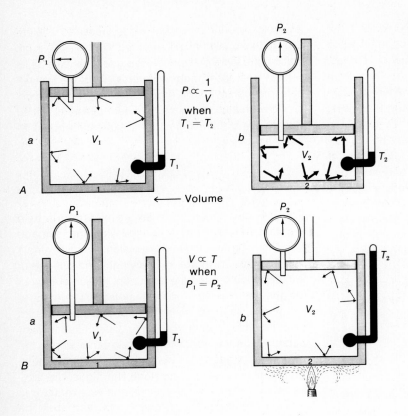

$$P \propto \frac{1}{V}$$
when
$$T_1 = T_2$$

← Volume

$$V \propto T$$
when
$$P_1 = P_2$$

**Figure 7-13** *The relationship between the pressure, volume, and temperature of gases.* A. *When the temperature is held constant, the pressure of a gas is inversely proportional to its volume.* B. *When pressure is held constant, the volume of a gas is proportional to its temperature.*

and divide by $T$ on both sides of the equation in the following manner:

$$(1) \quad PV = \frac{KT\not{P}}{\not{P}} \qquad (2) \quad \frac{PV}{T} = \frac{K\not{T}}{\not{T}} \qquad (3) \quad \frac{PV}{T} = K \qquad (7\text{-}9)$$

Expression 3 in Eq. (7-9) is referred to as the *ideal gas law*. The ideal gas law is interpreted as follows: The product of the pressure and volume of a gas divided by its temperature is always a constant value. In other words, although the individual values of $P$, $V$, and $T$ may change, the value for the total combination of these variables remains the same. If $K$ is to remain constant for a given gas, then an increase in $P$ requires a corresponding decrease in $V$ at constant $T$. Similarly, an increase in $T$ necessitates a corresponding increase in $V$ at constant $P$. These statements are the same as Boyle's law and Charles' law, respectively.

The ideal gas law describes the behavior of gases under ideal conditions. By this, we mean that certain assumptions about the nature of gases must be made before their behavior can be analyzed. The following are among the premises on which the ideal gas law is based: (1) The kinetic energy of a gas is proportional to its temperature; (2) the gas particles are relatively small, and the distances between them are relatively great; (3) although gas particles collide with each other, no kinetic energy is lost through molecular collisions; and (4) the gas particles do not exert forces on each other except during collisions. These assumptions break down when gas particles are large or when the density of a gas is increased through compression. Under these conditions, the mass of each particle exerts a slight attractive force on every other particle, thereby violating assumptions 2 and 4. In other words, the actual volume of a gas under extremely high pressure is not necessarily the same as that predicted by Boyle's law.

The basis of assumption 3 is probably not immediately obvious. Although energy can be converted from one form to another (see "The Laws of Thermodynamics," Chap. 1), as, for example, in the conversion of mechanical energy into heat energy, no transformation of energy is involved when one molecule strikes another. Brownian movement is a reflection of heat energy. Therefore, *molecular motion* is *heat!* Energy is transferred from one moving molecule to another when they collide with each other, but the total kinetic energy after collision is the same as that possessed prior to collision. Clearly, then, the kinetic energy of moving molecules cannot be reduced through collisions with other molecules.

***Heat and molecular motion*** Even though molecular motion is heat, not all ambient heat energy goes into the *translational* motion of molecules known as Brownian movement. Some part of the heat energy may be used in promoting the vibration, or *oscillation*, of molecules about their own axes. Another part may be expended in causing their *rotational* motion. The three types of motion characteristic of molecules are compared in Fig. 7-14. There is no net displacement of molecules as the result of oscillatory or rotatory motion. Since the degree of Brownian movement of a gas is related to its temperature, only that portion of the heat energy which causes an actual displacement of a molecule can increase its kinetic energy.

**Figure 7-14** *The three types of motion characteristic of gaseous molecules. A. Translational motion involves a net displacement of gaseous molecules and increases with temperature. B. and C. There is no net displacement of gas molecules or increase in gas temperature due to oscillatory or rotatory motions of gaseous molecules.*

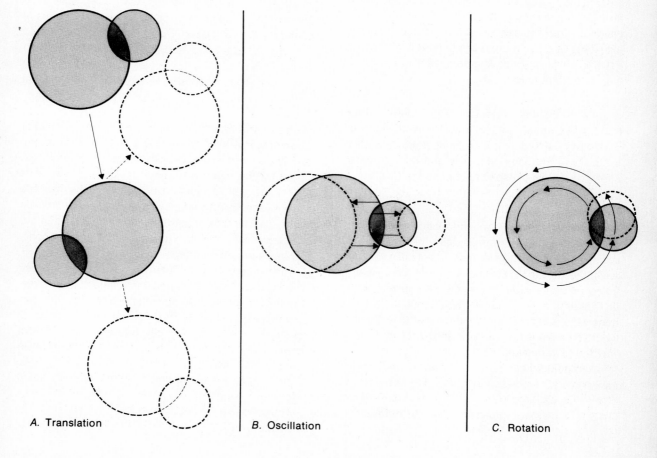

A. Translation      B. Oscillation      C. Rotation

Gases differ from each other quantitatively in the ways in which their molecules respond to heat energy. Some gases convert practically all the heat energy into the molecular energy of translational motion. Other gases may convert more of the heat energy into oscillatory and rotatory motions than into Brownian movement. It follows that the temperatures of different gases will vary when exposed to exactly the same amount of heat energy. Nevertheless, in each case, $PV/T = K$.

Perhaps we can better understand these relationships by considering an example. Suppose that two different gases are heated to the same temperature. Then

$$\frac{P_1 V_1}{T_1} = K_1 \quad \text{and} \quad \frac{P_2 V_2}{T_2} = K_2 \qquad (7\text{-}10)$$

Since $T_1 = T_2$ and $K_1 \propto K_2$, $P_1 V_1 \propto P_2 V_2$. This is just another way of stating Boyle's law. Thus, even though the pressures of gases do not all increase to the same extent with a given increase in temperature, the product of the pressure and volume of each gas will maintain a constant ratio.

**Unique properties of liquids** In general, liquids have structural properties which resemble those of solids and behavioral characteristics which are similar to those of gases. Viscosity and surface tension are two qualities of especial importance in a biologic context.

*Viscosity* Viscosity is defined as the resistance of a liquid to flow. Since liquids differ markedly in their rates of flow, their internal resistances must vary. Flow resistance is due to frictional forces which develop as liquid particles slide over each other in response to an applied force. The average amount of friction generated is in part proportional to both the density of the liquid and the nature and extent of the surface with which it is in contact.

When a liquid flows, its upper layers slide over its lower layers, much as individual cards in a new deck of playing cards behave when neatly stacked on a table and pushed with an evenly distributed force

applied from the side. In this example, the top card slides farthest and the bottom card slides least, because the weight of the pack increases from top to bottom. The "flow" of stacked cards under lateral pressure is illustrated on the left in Fig. 7-15A.

In a liquid, an increasing weight with depth causes a similar layering effect during flow. Suppose that a jar full of molasses is opened and placed on its side, as in Fig. 7-15A (right). The molasses will flow slowly out of the jar, the upper layers moving more rapidly than the lower layers. This phenomenon can readily be demonstrated by following the flow rate of bubbles at various depths in any syrup while the liquid is being poured from its container. Friction increases markedly from the interface between liquid and air to the interface between liquid and solid.

The behavior of solids and gases in response to an applied force is different from that of liquids. Solids do not flow, in contrast to liquids and gases. Solids are permanently deformed only when their tensile strengths are exceeded. Up to this point, the degree of "stretch" of a solid is proportional to the applied force. Gases, on the other hand, flow like liquids but are much less viscous. Therefore, under identical pressures, gases flow at greater rates than liquids.

*Surface and interfacial tensions* The layer of a liquid at the interface between it and some other medium is under tension. Caused by attracting forces exerted by the inner molecules of liquid, the force is commonly called a *surface tension* when the interface is between liquid and gas or an *interfacial tension* when the interface is between liquid and liquid or liquid and solid. This tension confers a unique cohesiveness on the outermost layer of a liquid.

The surface or interfacial tension tends to cause this layer to assume the least possible surface area, which is a sphere. Manifestations of surface or interfacial tension include the tendency of droplets or globules to assume spherical shapes, the formation of larger spheres when small oil globules coalesce in water following agitation, and the ability of a liquid to migrate up a glass capillary tube of extremely small bore.

The cohesiveness of the surface layer of a liquid is

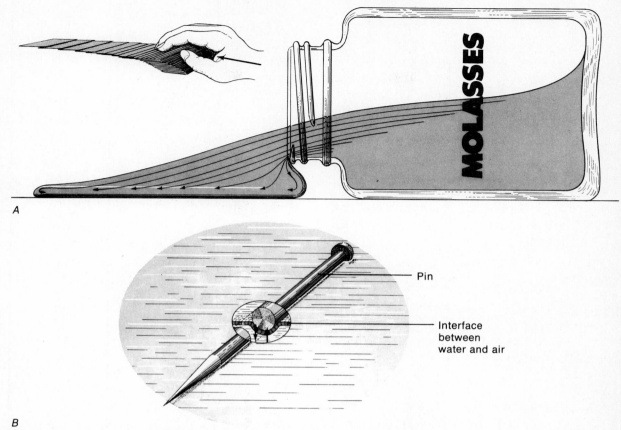

**Figure 7-15**  *Some properties unique to liquids. A. Fluid flow is analogous to the movement of stacked playing cards when pushed with an evenly distributed force applied from the side. B. Molasses moves slowly in sheets from an uncapped and opened jar, with the upper layers sliding over the lower layers because of a progressive decrease in frictional resistance. C. A pin carefully placed on the surface of water will not sink, so long as the surface of the pin is not made wet.*

exemplified by its ability to support a steel needle carefully placed on it. The surface layer acts as if it were membranous, stretching and finally rupturing in response to an applied force which gradually increases in magnitude. So long as the surface layer is not ruptured, or torn, however, the needle will not sink. This situation is illustrated in Fig. 7-15B. Notice that the needle is not made wet by the water but simply depresses the surface layer. Should the needle rupture the surface layer, it would become wet and would sink. In this event, the surface layer immediately

mends itself following the reestablishment of an undisturbed surface tension.

**Types of mixtures**  The classification of matter into three states—solids, liquids, and gases—is only the beginning of learning about how matter is organized. Since an understanding of the organization and behavior of matter will ultimately help us appreciate the unique qualities of protoplasm, we will next examine the types of mixtures which form when atoms and molecules are combined physically or chemically.

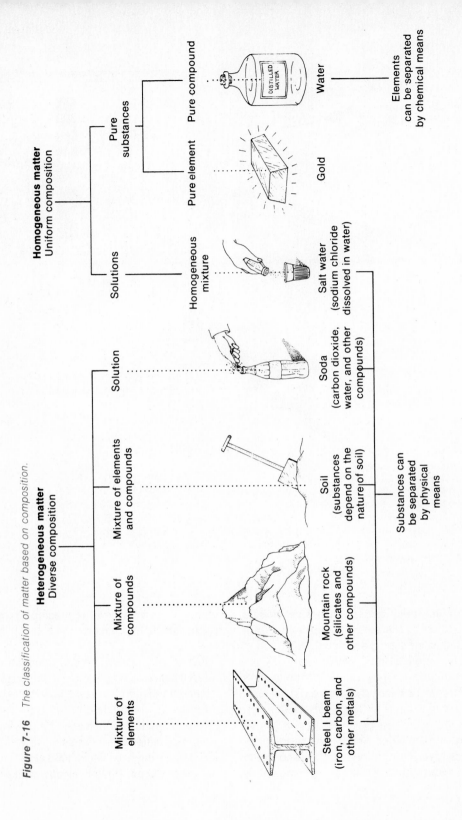

*Figure 7-16*   The classification of matter based on composition.

**Homogeneous matter**
Uniform composition

Pure substances

Solutions

Pure element

Pure compound

Homogeneous mixture

Gold

Water

Salt water
(sodium chloride
dissolved in water)

Elements
can be separated
by chemical means

**Heterogeneous matter**
Diverse composition

Mixture of
elements

Mixture of
compounds

Mixture of elements
and compounds

Solution

Steel I beam
(iron, carbon, and
other metals)

Mountain rock
(silicates and
other compounds)

Soil
(substances
depend on the
nature of soil)

Soda
(carbon dioxide,
water, and other
compounds)

Substances can
be separated
by physical
means

*Heterogeneous matter* Matter, whether solid, liquid, gaseous, or some combination of these three states, can be placed into one of two categories—heterogeneous matter or homogeneous matter.

Heterogeneous matter is of diverse composition. That is, the individual particles which make up heterogeneous matter are simply mixed together and may be easily separated from each other by mechanical or physical means. Heterogeneous matter, therefore, consists of mixtures of substances such as elements or compounds, or a combination of the two.

Examples of heterogeneous mixtures are shown in Fig. 7-16. The steel structural support is a mixture of elements including iron, carbon, and other metals which form an alloy stronger than pure iron. Since the individual metals melt at different temperatures, heating provides a simple physical method by which the components of steel may be separated. The jagged rocks of a mountain are formed of silicon compounds of various types. Soil is a mixture of compounds and trace elements. The uncapped bottle of soda is an example of a heterogeneous solution. Carbon dioxide, originally dissolved under pressure in the soda water, is evolved as the pressure is decreased. Since the evolution of gas separates it from solution, a decrease in pressure is one of several simple methods for promoting the release of dissolved gas. Can you think of any others?

*Homogeneous matter* By way of contrast, homogeneous matter is of uniform composition. Uniformity means either that the contents of the entire substance are identical in composition, or pure, or that the various substances are so well mixed together that they have become virtually indistinguishable. Thus, homogeneous matter consists of pure substances such as one element or compound as well as solutions in which particles are dissolved rather than dispersed.

Examples of homogeneous matter are given in Fig. 7-16. Prior to the removal of the cap from the soda bottle, the soda was a homogeneous solution in which the main ingredients were carbon dioxide dissolved in water. Recall that after opening the bottle, the soda becomes a heterogeneous solution.

Other common examples of homogeneous solutions include table salt or sugar dissolved in water. The water does not change visibly when either of these compounds is added. However, at the saturation point of the water, any additional salt or sugar will settle to the bottom without dissolving. In addition, the salt or sugar may be easily separated from the water by evaporating the water through boiling. Air is a homogeneous solution of different types of gas molecules, including nitrogen, oxygen, and carbon dioxide.

Moving from homogeneous solutions to pure substances, any pure element would constitute one category of pure substance. Finally, homogeneous matter may take the form of pure compounds, i.e., a group of molecules of the same kind not mixed with any other materials. Distilled water completely lacking in impurities is one of hundreds of thousands of pure compounds either identified following isolation from natural sources or prepared in chemical laboratories. Only through chemical means can the atoms of compounds be separated from each other.

Unless matter is pure, it is composed of more than one type of substance. When heterogeneous, such mixtures are of the following types:

1 Solids in solids
2 Solids in liquids
3 Liquids in solids
4 Solids in gases
5 Gases in solids
6 Liquids in liquids
7 Liquids in gases
8 Gases in liquids

As you will note, only the combination gases in gases has been omitted, because gas mixtures by definition are homogeneous.

The nature of colloids In physical makeup, protoplasm is a heterogeneous mixture of compounds in which solids and liquids are dispersed in a liquid dispersing medium. The dispersed particles in protoplasm include large molecules such as proteins, fats, and carbohydrates or clumps of molecules. Such particles cannot be dissolved in an aqueous, or water, solution and are therefore *insoluble*.

**Phase relationships**   Because protoplasm consists of dispersed particles in a dispersing medium, the insoluble particles are referred to as the *discontinuous phase* of the solution. Particles of the discontinuous phase are separated from each other by particles of the dispersing medium. The dispersing medium of protoplasm is water in which dissolved or soluble particles are found. The water molecules with their dissolved substances constitute the *continuous phase* of the solution, because their molecules establish widespread interconnections with, or are physically inseparable from, each other.

**Particle size**   The dispersed particles in protoplasm are extremely small, most of them measuring between one ten-millionth of a meter ($10^{-7}$ m, or 0.1 $\mu$m) and one-billionth of a meter ($10^{-10}$ m, or 0.0001$\mu$m). Particles in this size range, such as most proteins, are called *colloids*. Protoplasm is therefore a *colloidal solution*. When the dispersed particles are solid, the solution is called a *colloidal suspension*; the solution is an *emulsion* when the colloidal particles are liquid. Heterogeneous solutions in which the particles are coarser than 0.1 $\mu$m are termed *coarse suspensions.* A mixture of sand and water is a common example. Conversely, homogeneous solutions in which the particle size is less than 0.0001 $\mu$m, such as salt water or sugar water, are crystalloid solutions, or *true solutions*. While the particles of a coarse suspension settle out of solution as time progresses, colloidal particles are intimately mixed with their dispersing medium and display no such tendency.

**Interfacial phenomena**   Among the unique qualities of colloidal solutions, two are especially intriguing to biologists because of the insights they provide into the functional nature of protoplasm: (1) the unbelievably large surface area which all the individual colloidal particles collectively present to their dispersing medium, compared, for example, with the same mass of colloidal particles clumped together

and (2) the tendency of colloidal particles to remain suspended in solution indefinitely. Let us briefly examine the significance of each of these phenomena.

The tremendous surface area characteristic of colloids enables protoplasm to maintain a high degree of chemical activity. The myriad chemical reactions which normally occur in protoplasm are prerequisite to the maintenance of life. To appreciate the role of colloidal interfaces in this task, think of a cell as a manufacturing plant. The larger the floor space available for machinery, the greater the manufacturing capacity of the plant. In a cell, the "floor space" is represented by the surfaces of colloidal particles. Cellular products are various types of molecules synthesized from chemical raw materials.

However, the raw materials within a cell must be brought close enough together for chemical reactions to occur. Proximity is achieved when reactant particles become stuck together on the surface of another particle. When the "sticky" particles differ from the particle to which they are clinging, the process is called *adsorption*.

Adsorption occurs because of mutually attractive forces which exist between different solute particles. These forces are exactly the same as those which give rise to interfacial tensions, only here we are dealing with forces at the interface between solid particles rather than at the interface between liquids or between a liquid and a gas. Due to adsorption, particles in extremely low concentration in solution can be brought close enough together to interact chemically. The adsorption of particles on charcoal or on other materials with large surface areas is commonly used to remove impurities or to separate extremely dilute substances from solution. The low concentrations in which many raw materials are found within cells emphasize the importance of adsorption and its colloidal basis in promoting cellular metabolism.

Many colloidal particles are said to be *surface-active*, because they possess high interfacial tensions. Through the action of these forces, colloidal particles maintain a resistance to flow, or viscosity, even though they are dispersed in solution. Sometimes the interfacial tensions become large enough to cause a partial solidification of the entire solution. When these events occur in protoplasm, they are referred to as *phase reversals*.

The alternation of phases in protoplasm is diagramed in Fig. 7-17. When the dispersed particles (e.g., proteins) are separated from each other by their dispersing medium (i.e., water containing dissolved materials), the protoplasmic solution is more fluid than solid and is termed a *plasmasol*. If the dispersed particles link together because of an increase in their interfacial tensions, then they collectively become the continuous phase of the solution while the former dispersing medium becomes its discontinuous phase. This type of phase reversal gives the solution a structure not possessed by its sol state. Now, instead of being a water solution containing dispersed proteins, the solution consists of solidified protein containing pockets of water. This form of protoplasm is a semifluid solid called a *plasmagel*.

Phase reversals occur routinely in cells and may be precipitated by changes in one or more of the following factors: the electrostatic forces of attraction or repulsion between particles, the acidity or alkalinity of the solution, pressure and other mechanical forces acting on the solution, or the volume of solvent (water).

Even though protoplasm is a heterogeneous mixture of compounds, the dispersed particles do not settle out of solution. These particles tend to remain dispersed because of a number of influences which operate simultaneously: (1) Electrostatic forces of attraction and repulsion between charged particles keep them separated; (2) Brownian movement promotes the diffusion of dispersed particles throughout the solution; and (3) the fluxes set up through phase reversals continually redistribute protoplasmic constituents.

Each of the above factors serves to mix the components of protoplasm, tending to keep dispersed particles spread throughout their dispersing medium. For example, protoplasmic proteins and other charged particles are literally coated by a layer of water molecules which are oriented to them through forces of electrostatic attraction. Such particles are said to be *hydrated*. The water of hydration tends to prevent these colloidal particles from clumping together.

| A. Plasmasol | B. Plasmagel |

**Figure 7-17** *Phase changes in protoplasm. A. Solid particles (black) are dispersed in liquid (white), forming a plasmasol. B. In plasmagels, the solid particles come together to form fibrous strands enclosing pockets of water.*

Thus, protoplasm and other colloidal solutions are stable in part because attractive forces are exerted between the solvent and solute particles, binding the two phases together. We will explore the nature of these intermolecular forces in Chap. 10.

**The cell environment** Cells, of course, are made of protoplasm. Since many types of colloidal particles are dispersed in solution, protoplasm is basically a *polyphasic colloid*. Any theory which attempts to explain the living nature of protoplasm must of necessity deal with a mixture of different types of colloidal particles and the various physical principles which describe their behavior in nature. Let us see whether our knowledge of the physical nature of matter and of colloidal behavior in particular helps in understanding cellular dynamics.

*Physical properties* Cells have a granular structure and appear translucent. Does the presence of polyphasic colloids explain these qualities of protoplasm? The answer is a most emphatic yes! Both features are observed because of the characteristic and marked scattering which occurs when light rays are passed through any polyphasic colloidal system.

Many substances within protoplasm, e.g., granules and droplets, can be separated by centrifugation from the rest of the so-called ground substance without disturbing its essential nature. In centrifugation, the force of gravity is increased manyfold by the spinning action of a centrifuge. Particles separate from each other because of differences in their masses. That the ground substance maintains its integrity following centrifugation suggests that droplets and globules are dispersed rather than dissolved in the solution.

Protoplasm undergoes phase reversals, with viscosities ranging between a plasmasol and a plasmagel. Indeed, both phases are usually present in different parts of the same cell. Nevertheless, protoplasm seems to maintain a structural organization, even when granularity or structure is not apparent.

Colloidal behavior provides a sufficient explanation for both the fluidity of protoplasm and its semisolid structure. The resistance of protoplasm to flow is due to the surface-active properties of colloidal particles. The interfacial tensions associated with the surfaces of colloidal particles create a mutual attraction between them and their dispersing medium.

As dispersed particles diffuse through protoplasm, they move in an uneven manner reminiscent of objects tearing through successive barriers. The structural organization of protoplasm is believed to be disturbed by the temporary rupturing of connections between dispersed molecules when struck by large moving particles. The labile membranes held together by these attractive forces repair themselves immediately following such disturbances. The unbelievably large surface area of colloidal particles provides the necessary substrate on which these activities can occur.

Colloidal particles such as proteins ordinarily do not pass through cellular membranes, although other particles, including amino acids, fatty acids, glycerol, simple sugars, and ionized atoms, are capable of moving across the same membranes. The explanation of why one group of particles cannot pass through while another group may move in an unimpeded fashion has to do with their relative sizes. Colloidal particles are larger than the pores in cellular membranes and therefore do not diffuse. Crystalloid particles are smaller in diameter than the pore sizes of these membranous barriers and consequently pass easily.

Liquid droplets and globules in protoplasm tend to be spherical in shape. This phenomenon is caused by the attractive forces exerted by the inner molecules of the liquid droplet on molecules in its peripheral layer. The existence of an interfacial tension causes the droplet to assume a shape with the least amount of surface area, i.e., a sphere.

Proteins, the most abundant of the dispersed colloids in protoplasm, do not coalesce, or come together. Thus the stability of protoplasm is due to two factors: (1) Intracellular proteins are invariably electronegative and exert electrostatic forces of repulsion on each other, and (2) an oriented layer of water molecules surrounds the surface of each protein molecule, preventing them from clumping together.

The hydration of proteins and other charged particles in protoplasm has other important ramifications. Through protein hydration, for example, protoplasmic water is functionally separated into two groups— bound water and free water. The *bound water* is electrostatically attracted to the surfaces of charged particles and becomes part of their respective structures. *Free water*, on the other hand, is not bound to other particles and is therefore directly available for metabolic purposes.

Hydration also affects the rate of movement of ions. As more and more water molecules are added through electrostatic attraction to the surface of a positively charged ion, its rate of diffusion is progressively retarded. An ionized atom with an excess of one positive charge has one more proton in its nucleus than electrons outside the nucleus, or, more to the point, a deficit of one orbital electron. The electrostatic force of attraction between the negative portions of water molecules and the positively charged nucleus of an ionized atom decreases as the diameter of the atom increases. The decrease in attraction is due to an increase in the distance between electric charges of opposite sign.

The diameter of an atom is in part determined by the number of protons in its nucleus. The higher the atomic number, the greater the number of energy levels and, consequently, the larger the diameter of an atom. However, this relationship does not hold true for all elements.

Ionized atoms bearing negative charge have an

excess of one or more electrons, compared with the number of protons. The electron surplus is found in the outermost energy level of each ion. Thus, the larger negatively charged ions are capable of electrostatically attracting the positive ends of more water molecules, compared with ions of smaller diameter. Therefore, the relation between the diameter of a negatively charged ion and the degree of its hydration is the reverse of that of a positively charged ion.

Furthermore, the degree of hydration of intracellular proteins is directly affected by the specific ions electrostatically attracted to them. Recall that intracellular proteins bear a net negative charge and therefore attract ions of opposite sign. The amount of water bound to an ionized atom of positive charge is roughly proportional to its diameter. Therefore, the adsorption by proteins of positively charged ions of large diameter increases the hydration of the proteins to a greater degree than the adsorption of smaller ions of the same sign. In other words, the degree of protein swelling is related to the diameters of the adsorbed ions.

*Chemical regulatory mechanisms*  The sequence and timing of intracellular chemical reactions is accurately controlled. Although not all the chemical control mechanisms are well understood, we can confidently cite the following reasons for their existence: (1) If all the energy were released at one time from the fuels being degraded in every cell of our bodies, we would literally melt from the intensity of the heat evolved; (2) the chemical pathways of metabolism are controlled by the availability of, and the competition for, chemical raw materials transported into the cell, in addition to the requirements of the body for particular products; (3) enzymatic mediators of intracellular chemical reactions, as well as the presence of other essential molecules and ions, determine whether or not a reaction will occur and, in addition, establish its precise timing; and (4) the activity of the different enzymes and therefore the nature of available chemical pathways is under genetic control, i.e., chromosomal genes control enzymes.

That cells can carry out their complex chemical tasks is self-evident. How they do it is the subject of continuing and intense investigations by many scientists throughout the world. Let us consider some of the technical problems which cells must routinely solve to sustain their metabolism. First, thousands of different types of chemical reactions must be supported in the close quarters of every cell. Second, chemical raw materials in extremely low concentrations, perhaps even in trace amounts, must be brought close enough together with other appropriate raw materials to interact chemically. Third, intracellular chemical reactions must occur at body temperature and pressure rather than at the high temperatures and pressures ordinarily used by chemical industries. Here, as earlier, our knowledge of polyphasic colloidal systems will help us to understand how cells have surmounted these seemingly formidable obstacles.

The problem of keeping each of the chemical reactions isolated from the others in the crowded environment of a cell has been solved by nature in two ways: (1) Cellular membranes such as those of the endoplasmic reticulum, mitochondria, lysosomes, and the nucleus divide a cell into numerous intracellular compartments, each with unique characteristics, including the nature and types of chemical reactions supported, and (2) since proteins and other colloidal particles differentially adsorb specific particles, different chemical reactions may be supported on dispersed molecules adjacent to each other, even in the absence of intracellular partitions.

Adsorption is also behind the ability of a cell to obtain the appropriate raw materials for use in a particular chemical reaction. This is no mean feat, when we stop to consider that there is an infinitesimally large variety of raw materials within a cell and that many of them are in concentrations so low that a cellular sorting mechanism is absolutely essential. The situation is roughly analogous to a supermarket search for the last can of a particular product which has somehow been misplaced on the shelves — an almost impossible and, at the very least, extremely time-consuming task. A cell quickly "finds" the particular molecules it requires for chemical reactions through their adsorption to specific colloidal particles dispersed in solution. Adsorption plays a dual role here, in that the raw materials are not only concentrated on the surfaces of

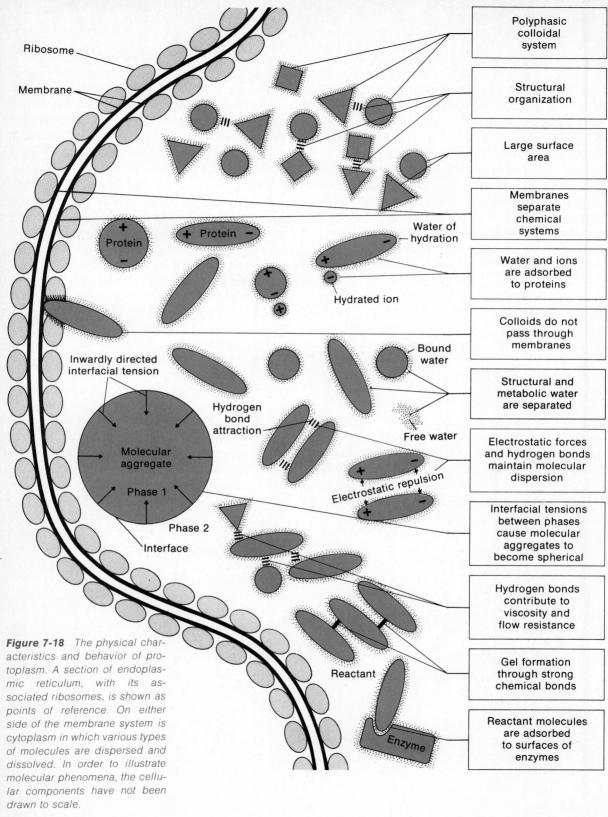

Ribosome

Membrane

Polyphasic colloidal system

Structural organization

Large surface area

Membranes separate chemical systems

Water of hydration

Protein +
Protein + −

Water and ions are adsorbed to proteins

Hydrated ion

Colloids do not pass through membranes

Bound water

Inwardly directed interfacial tension

Structural and metabolic water are separated

Hydrogen bond attraction

Free water

Molecular aggregate

Electrostatic forces and hydrogen bonds maintain molecular dispersion

Phase 1

Electrostatic repulsion

Phase 2

Interface

Interfacial tensions between phases cause molecular aggregates to become spherical

Hydrogen bonds contribute to viscosity and flow resistance

Gel formation through strong chemical bonds

Reactant

Reactant molecules are adsorbed to surfaces of enzymes

Enzyme

**Figure 7-18** *The physical characteristics and behavior of protoplasm. A section of endoplasmic reticulum, with its associated ribosomes, is shown as points of reference. On either side of the membrane system is cytoplasm in which various types of molecules are dispersed and dissolved. In order to illustrate molecular phenomena, the cellular components have not been drawn to scale.*

colloidal particles but are brought into close proximity with other molecules and ions with which they are capable of interacting chemically. Therefore, even those raw materials present in vanishingly small amounts within a cell are inevitably brought to sites of appropriate chemical activity.

Adsorption is also the cellular driving force which initiates and sustains chemical reactions at normal temperatures and pressures. The raw materials which engage in intracellular chemical reactions are called *reactant molecules,* or simply *reactants.* These reactants are adsorbed on the surfaces of enzymes. As you will remember, enzymes are protein molecules and therefore behave as colloids. The attractive forces exerted by an enzyme on its adsorbed reactant molecules decrease the amount of kinetic energy required to activate them chemically. Thus, enzymes lower the *energy of activation* of the reactants. Without enzymes and their surface-active forces of adsorption, much higher temperatures or pressures, or a combination of the two, would be required to support chemical reactions in protoplasm.

The physical properties and colloidal behavior of protoplasm are summarized in Fig. 7-18.

## COLLATERAL READING

Alder, B. J., and T. E. Wainwright, "Molecular Motions," *Scientific American,* October 1959. Computers are used to analyze the individual behavior of large numbers of particles during the transition from one state of matter to another.

Bernal, J. D., "The Structure of Liquids," *Scientific American,* August 1960. Liquids continually change their arrangements while maintaining a relatively high degree of organization. Bernal discusses this concept in relation to the general properties of liquids.

Buswell, Arthur M., and Worth H. Rodebush, "Water," *Scientific American,* April 1956. The behavior of water is analyzed by considering its varied molecular structures and arrangements.

Derjaguin, Boris V., "The Force Between Molecules," *Scientific American,* July 1960. Measurements of intermolecular forces indicate that they are entirely electromagnetic in nature. Derjaguin describes how the forces are measured and explains the mathematical expressions which define their magnitudes.

Feinberg, Gerald, "Ordinary Matter," *Scientific American,* May 1967. Feinberg believes that the details of particle physics are not directly applicable to the structure of bulk matter.

Gell-Mann, Murray, and E. P. Rosenbaum, "Elementary Particles," *Scientific American,* July 1957. Gell-Mann and Rosenbaum attempt to classify the myriad so-called "elementary particles" on the basis of certain physical characteristics and types of reaction.

Hofstadter, Robert, "The Atomic Nucleus," *Scientific American,* July 1956. The nucleus of an atom is compared optically to a "dirty" crystal ball.

Kendall, Henry W., and Wolfgang K. H. Panofsky, "The Structure of the Proton and the Neutron," *Scientific American,* June 1971. Experiments are described which demonstrate that nucleons have a complex internal structure consisting of pointlike entities which apparently are not mesons.

Schrödinger, Erwin, "What Is Matter?" *Scientific American,* September 1953. Matter and energy are discussed as matter waves which sometimes act as discrete particles and at other times behave as electromagnetic waves.

Wannier, Gregory H., "The Nature of Solids," *Scientific American,* December 1952. The functional division of solids into conductors and insulators is explained by analogy to a bottle consisting of separated sections connected by narrow tubes.

# CHAPTER 8
# matter's other face

Figuratively speaking, matter represents only one side of a coin. The other face of the coin is occupied by energy. They are intimately related because matter and energy are interchangeable. Under proper conditions, matter can be converted into energy and energy may be transformed into matter. The atomic age and the continuing threat of thermonuclear war are constant reminders of these facts.

We ordinarily talk about energy in terms of its functions or effects. It is common knowledge that energy supports life, causes motion, contributes to pollution, is responsible for the ebb and tide of the oceans, promotes mountain building, erosion, and sedimentation—in short, causes the world to go around. However, unlike matter, energy does not possess mass. What energy actually is and how it interacts with matter are more difficult to specify.

After examining the concept of energy, its forms and transformations, we will look once again at mole-cules and atoms, in order to appreciate the interaction of energy with matter. This knowledge will be especially important when studying the effects of ionizing radiation in the next chapter. Next, we will reconsider the relation between molecular kinetic energy and temperature. This discussion will return us full circle to matter in a biologic context, as we observe the thermal behavior of water and the physical basis of temperature regulation in human beings.

## THE NATURE OF ENERGY

Since energy lacks both mass and volume, it can be detected and measured only through its effects on matter. In other words, although an object may possess energy, it is only when that energy is put to work that we are aware of it. That you feel full of pep we must accept on faith. When you have washed the dishes, swept the floor, vacuumed the carpet, and dusted the furniture, all in one hour, then we must concede your claim.

Indeed, we ordinarily think of energy in terms of the ability to perform work. While work to some may suggest "blood, sweat, and tears," e.g., when studying science or other subjects taught in college, it is not until an expenditure of energy moves an object through a distance that the amount of work accomplished can be expressed physically. Pushing with all your might against an immovable brick wall can cause muscle fatigue in a short time, but no work will have been done as the result of the effort. Work is done only when a weight is moved by a force.

A classic example of the meaning of work is provided by the sport of weight lifting. The entire strength of the athlete is pitted against the weight to be lifted and the height through which it must be raised. The physical concept of work is illustrated in Fig. 8-1.

Thus we see that physical work depends on both the energy expended and the distance moved. The quantitative expression for the amount of work accomplished takes the following form:

$$W(\text{work}) = F(\text{force}) \times D(\text{distance}) \quad \text{or} \quad W = FD \quad (8\text{-}1)$$

When force is in pounds and distance is in feet, units of work are expressed in foot-pounds.

Application of Eq. (8-1) requires that the force be applied in the same direction as that in which the object moves. Of course, the direction of the applied force and that of the moving object are seldom exactly the same. Therefore, in practice, where more than one force acts on an object from different directions, as when attempting to fasten a diaper on a squirming baby, or where one force acts at an angle in relation to the direction of the object being moved, as when pulling a wagon by its handle, only the resultant force which expresses the magnitude acting in the same direction as the actual path of displacement is used to calculate the work done.

In determining the efficiency with which energy is expended, allowances must be made for losses due to frictional resistance. Assume that objects have been set in motion through the expenditure of energy.

Objects that passively glide through air such as gliders and sea birds are retarded by a force equal to that with which the air particles resist displacement or create counterforces such as turbulence and drag. Objects that move over solid or liquid surfaces, such as automobiles, people, and boats, are slowed up by a force equal to the sum of those generated by surface contact and air resistance. Objects which move

**Figure 8-1** *A comparison between the fruitless expenditure of energy and the utilization of energy to do work. A. Work is not done, in a physical sense, when someone is pushing against a brick wall or other immovable object, no matter how much energy is expended. B. A large amount of work is done by the weight lifter in moving the massive weights through the distance shown.*

**A.** No work done

**B.** Work done

entirely under water, such as fish and other aquatic animals, are opposed by a force equal to the viscous resistance of their fluid medium and the turbulence generated by their shape and the nature of their movements. All these resistive forces cause the dissipation of energy through the liberation of heat.

Although heat is thermal energy, it does not perform "useful" work when released as a consequence of friction. Therefore, to maintain the motion of an object which generates friction, an average amount of energy in excess of that required to overcome resistive forces must be continually supplied.

You will no doubt recognize that this statement is a consequence of the second law of thermodynamics, a concept which was introduced in Chap. 1. An object at rest is set in motion when sufficient energy is imparted to it but will ultimately stop moving as a result of energy losses through frictional resistance unless more energy is generated, just as your dormitory room, apartment, or home becomes more and more untidy as time elapses unless you put in the necessary and continuing effort to keep it clean and neat.

## FORMS OF ENERGY

Energy exists in two forms—kinetic and potential. *Kinetic energy* is the motive force which accomplishes work. *Potential energy* is "bottled up" and not actually being utilized at the moment. For example, potential energy is contained within the foods we eat. The kinetic energy obtained during the oxidation of foods powers muscle contractions and other activities which sustain life.

**Kinetic energy** We encountered the term kinetic energy in Chap. 7 where the effects of the thermal agitation of molecules were discussed. Recall, for example, the relation between the temperature of a gas and the average kinetic energy of its molecules. Matter is set in motion by kinetic energy.

The magnitude of the kinetic energy possessed by bulk matter in motion may be calculated when its mass and velocity are known. In a quantitative sense, kinetic energy is equal to the product of one-half the mass of the moving object and the square of its velocity. The formula is written as follows:

$$K(\text{kinetic}) \, E(\text{energy}) = \tfrac{1}{2} M(\text{mass}) \times V(\text{velocity})^2$$
$$\text{or } KE = \tfrac{1}{2} MV^2 \qquad (8\text{-}2)$$

When two objects are moving at the same velocity, one with double the mass of the other, the kinetic energy of the former is also double that of the latter. On the other hand, since kinetic energy also depends on the square of the velocity, an increase in the speed of an object causes a very large increase in its kinetic energy.

For example, when the speed of an automobile is increased from, say, 25 to 50 mi/hr, its kinetic energy is quadrupled ($KE \propto V^2$). If the speed of the car is increased from 25 to 75 mi/hr, its kinetic energy is increased ninefold. Should an accident occur, the destructive capacity of the machine is four times and nine times greater at 50 and 75 mi/hr, respectively, compared with that which could occur at 25 mi/hr. Contrast this with a situation where two cars, one weighing 2,000 lb and the other weighing 4,000 lb, are traveling at the same speed. Under these conditions, the heavier car possesses double the kinetic energy of the lighter one ($KE \propto \tfrac{1}{2} M$).

Not all the kinetic energy being released at any given time is used to do work, as we have already noted. A certain amount of this energy is required to overcome frictional losses of various types. Another portion may be "wasted" when no object is moved through a distance as the result of expending the energy. This does not negate the fact that the energy is kinetic; it simply means that the energy is not being used to accomplish "useful" work.

The efficiency with which kinetic energy does work is always less than 100 percent, because of the restrictions imposed by the second law of thermodynamics. Provided that we know the work accomplished and the total energy expenditure, the actual *efficiency* of the effort may be calculated in the following manner:

$$\text{Efficiency} = \frac{\text{work accomplished}}{\text{energy expenditure}} \qquad (8\text{-}3)$$

The above formula is just as applicable to human

movements as it is to inanimate machines. Since there is another useful way of expressing efficiency, we will return to this topic later when energy transformations are discussed.

Potential energy    Potential energy is the opposite of kinetic energy, in that it is stored rather than released. Since potential energy can be converted into kinetic energy, it has the capacity to accomplish work, although work cannot be done until an energy conversion takes place. Therefore the capacity to do work does not mean that work is actually being done. To appreciate the impact of this statement fully, consider yourself basking drowsily in the sun while recognizing in your unrepressed consciousness that numerous chores require your immediate attention. You have the capacity to do work because your muscles are capable of responding to the directives of your brain, even if you happen to be inert at the moment. That your muscles can contract on command is the outcome of a complicated series of intracellular chemical reactions. These oxidative reactions progressively liberate energy locked in the molecules formed by the digestive process. The chemical potential energy unlocked during the degradation of these molecules is ultimately released as kinetic energy of various types. The kinetic energy, in turn, is utilized to drive bodily processes, including muscle contraction, and is also evolved as heat. Some of the heat energy is used to maintain a constant body temperature, although most of the evolved heat serves no useful biologic function and is therefore wasted.

The world is replete with still other examples of matter possessing potential energy. Unconsumed fuels, water behind a dam, objects high up, unexploded gunpowder—all have potential energy because of the energy-yielding nature of their chemical compounds when burned or mixed or as the result of physical forces which have acted on them in particular ways. In the above examples, potential energy is converted into kinetic energy when fuel is burned through combustion, when water flows over a dam, when objects fall from higher to lower levels, or when a bullet is fired. A list of forms of matter which have potential energy and the specific requirements which must be met to convert potential energy into kinetic energy would be virtually endless. Can you think of other types of matter which contain potential energy and possible ways in which the energy can be liberated?

To possess potential energy, matter must have acquired it from other sources. The primary source of potential energy on earth is the sun. Solar energy maintains photosynthesis, the process by which green plants manufacture sugars and evolve oxygen in the presence of water and carbon dioxide. Green plants are used as food by many animals which, in turn, serve as food for still other animals. Without photosynthetic primary producers, herbivores and carnivores could not be supported. Similarly, because fossil fuels such as coal, oil, and natural gas as well as certain derived products such as gasoline are initially formed from plant remains kept under great pressures for geologically long periods of time, a lack of sunlight and, therefore, of photosynthesis would be reflected by an absence of these fuels.

Even though earth-bound energy is ultimately derived from the radiant energy of the sun, the relation between the potential energy of various objects and the sun may not be obvious. Consider explosives as an example. The nitroglycerin in certain dynamites as well as trinitrotoluene (TNT) are both highly explosive compounds. Each chemical is manufactured by an involved industrial process which requires the expenditure of considerable energy. The energy is necessary for their synthesis. The industrial energy can be traced to the combustion of fossil fuels and petroleum products derived from them. A portion of the energy generated industrially is incorporated as potential energy in each of the final products. The mechanical energy and heat energy released when these substances are ignited come from the potential energy which holds the atoms of the molecules together.

Interconversions of energy    The differences and conversions between potential and kinetic energy can probably best be explained by means of a specific example. Consider a child on a gym swing. The kinetic energy supplied by the child's efforts or through the efforts of someone else who is pushing initially causes the swing to move in the same direction as the applied force. Because the pivots of the swing remain in a fixed position, however, the swing arcs upward. All the kinetic energy imparted to the swing is supplied at the point of application of the propelling force. As the child in the swing is carried upward, kinetic energy is progressively dissipated by the following factors: (1) the mass of the child and swing acted on by the force of gravity, i.e., the weight being lifted and (2) the frictional resistance encountered because of both the total mass moving through the air and the rubbing that occurs between the chain or rope links supporting the swing and the pivotal hooks to which they attach. At some point during the upward sweep of the swing, all the original kinetic energy imparted to it will have been dissipated by the factors identified above. The velocity (speed in a given direction) of the swing decreases to zero.

In the period of time during which the swing and its occupant are losing kinetic energy, they are gaining an almost identical amount of potential energy. The gain in potential energy is the result of being lifted to a higher position in space. The amount of potential energy gained is always less than the amount of kinetic energy supplied because of the dissipation through frictional forces of some of the kinetic energy.

During the downward sweep of the swing, potential energy is progressively converted into kinetic energy, again with energy losses due to the factors mentioned above. Because the kinetic energy is greatest at the bottom of its arc, the swing is propelled in the opposite direction. The swing is said to have momentum. *Momentum* is the product of the mass of the moving body and its velocity; i.e., momentum = mass × velocity. Momentum has a directional component. By way of contrast, kinetic energy does not have a directional component and consequently informs us only about the amount of energy released.

Heat losses continue to decrease the forward movement of the swing. This inexorable consequence of the second law of thermodynamics means that the swing will move in smaller and smaller arcs until it ultimately stops, provided that no further kinetic energy is imparted to it.

However, when both useful and "nonuseful" energy are accounted for during each conversion of energy as the swing moves through space, we would find that the amount of potential energy gained at the top of the swing's excursion is exactly the same as the amount of kinetic energy used to move it through that distance. The relationship holds equally true for the progressive conversion of potential into kinetic energy during the downward excursion of the swing. The equivalence between the two forms of energy is a restatement of the first law of thermodynamics introduced in Chap. 1. That energy can be neither created nor destroyed, as the law of conservation of matter states, is substantiated by countless examples drawn from everyday experience.

If the applied force in the direction of motion and the distance traveled by the swing and its occupant are both known, then the amount of work accomplished may be calculated from the formula $W = FD$ (work = force × distance). The efficiency with which energy is expended can then be determined by dividing the work done by the applied force.

The efficiency of energy utilization may also be approached in a different way. Since usable energy is always less than 100 percent of the energy supplied, solving for the ratio between energy output and energy input will also provide a quantitative estimate of efficiency. The formula takes the following form:

$$\text{Efficiency} = \frac{\text{energy output}}{\text{energy input}} \qquad (8\text{-}4)$$

The relations between kinetic and potential energy and their interconversions are diagramed in Fig. 8-2.

**Units of measurement and comparison** To solve for the efficiency of any physical system in which energy is expended (or in which work is done), we must deal with units of energy. The amount of energy used or supplied is measured in calories or appropriate multiples of the calorie unit. A *calorie* is the amount of heat energy required to raise the temperature of one gram (1 Gm) of pure water one degree Celsius (1°C). Since the initial temperature of the water affects the value of a calorie, the water temperature is usually specified as having to rise from 14.5 to 15.5°C. In a biologic context, the energy which causes

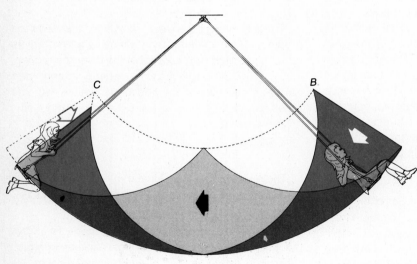

**Figure 8-2** *The relations between potential and kinetic energy while using a gym swing. A. The swing is set in motion by a push from a bystander. The kinetic energy (black arrows) is dissipated while potential energy (white arrows) increases during the upward excursion of the swing. B. During the downward excursion of the swing, the potential energy is progressively converted into kinetic energy. C. Its kinetic energy and momentum propel the swing in the opposite direction.*

this change is called a small calorie, and is abbreviated cal or c.

Ordinarily, a small calorie is too small a unit to work with. Therefore, a unit 1,000 times larger is used. This unit is called a kilocalorie (kcal) or a large calorie (sometimes abbreviated Cal or C). By definition, a *large calorie* is the amount of heat energy required to raise the temperature of one kilogram (1 kg) of water from 14.5 to 15.5°C. Large calories are preferentially used to express the energy released during the oxidation of various foodstuffs. We will have more to say on this subject in Chap. 26.

More heat energy is required to raise the temperature of 1 Gm of water 1°C than to raise the temperature of 1 Gm of any other substance the same amount. In other words, water will absorb a greater quantity of heat energy per unit of temperature increase than other materials of comparable mass.

When a calorie of heat energy is set equal to 1, the constant is referred to as the *specific heat*. The specific heat of water is 1 and is used as a reference point for comparing the heat-absorptive abilities of other materials. The specific heat of a substance is therefore the quantity of heat energy required to raise the temperature of 1 Gm of it 1°C. Compared with water, the specific heat of other materials is less than 1, because their temperatures rise more than 1°C for every calorie of heat energy added.

The tremendous heat-absorptive and heat-retentive capacities of water explain why the sand on a beach during midsummer is hot while the water next to the beach is still cool. The specific heat of sand is only about one-half that of water; per unit of heat energy added, the temperature of 1 Gm of sand rises approximately twice as fast as the same mass of water. Since the total mass of beach sand is ordinarily small in comparison to the total mass of water, the increase in the temperature of the sand is perceived more quickly.

The specific heat of water, one of its unique physical characteristics, makes water an admirable protoplasmic solvent. The human body is 70 to 85 percent water by volume. Chemical reactions which liberate or absorb heat occur in protoplasm at all times. In addition, the external temperature to which the body is exposed may fluctuate appreciably in a short time. Yet body temperature remains remarkably constant.

Numerous physical and physiologic mechanisms are primarily responsible for the maintenance of a constant body temperature, a subject discussd later in this chapter. However, the specific heat of water and its large volume in the body are added insurance against large-scale oscillations in body temperature. Such oscillations would otherwise occur in response to appreciable increases or decreases in metabolic rate, external temperature, or other factors which influence body temperature.

## THE INTERACTIONS OF ENERGY WITH MATTER

All types of energy may exist in the two forms, kinetic and potential. We have already mentioned several types such as mechanical energy, radiant energy including heat, and atomic energy. Additionally, we may cite electrical energy and chemical energy as other examples. Each type of energy may be transformed into any of the others, in accordance with the first law of thermodynamics.

**Mechanical energy**  Mechanical energy is exemplified by the motion of bulk matter. That a book can accidentally drop from your hands is due to the fact that you expended mechanical kinetic energy to lift the book, thereby conferring mechanical potential energy on it. When not adequately restrained, the book is dropped, because the force of gravity which pulls it toward the center of the earth overcomes the force holding the book back.

**Electrical energy**  Electrical energy is created by the movement of charged particles. The direction of movement is from a region in which the charged particles are in excess into a region in which they are deficient. Movement occurs because the particles electrostatically repel each other and are similarly attracted to the opposite charge borne by the particle-deficient region. Ordinarily, the charged particles are electrons or ionized atoms or molecules. In nature, electrons move most frequently through solids, whereas ionic movements are confined almost exclusively to liquids and gases. Ionic movements are therefore important in protoplasm.

The movement of electrons or other charged particles constitutes an *electric current,* and the materials through which they move are called *conductors*. The amount of friction which charged particles generate while moving through a conductor is referred to as the *resistance* of the conductor. Electrical resistance is analogous to the mechanical resistance generated by frictional forces when bulk matter is set in motion.

*Electrical potential energy* is a property of any conductor, simply because electrons or other charged particles are capable of moving through them. However, electrical potential energy cannot be stored like other types of energy, at least at the present state of our technology. Electrical energy must be generated through chemical reactions such as those which occur in a storage battery, a dry cell, and protoplasm or through large-scale transformations of other forms of energy into electrical energy, feats which are routinely accomplished by hydroelectric, fossil fuel, and thermonuclear power plants. Electrical potential energy is therefore a reality only when there is an imbalance in the quantity of electric charge between two points across which the charged particles are capable of moving.

The amount of electrical potential energy created by the existence of a charge difference between two points is expressed in *volts*, just as the distance through which matter is raised in acquiring mechanical potential energy is measured in feet or any other convenient unit of distance. Voltage is measured by passing electrons or other charged particles from the region in which they are in excess through the coil of an appropriately calibrated potentiometer, an instrument which measures potential differences. Voltage represents the energy with which the electrons are delivered. Since energy is involved, electrical work may be accomplished. We will discuss the concept of electrical work shortly. For now, we will simply note that the greater the electrical potential difference between two points a constant distance apart, the larger the voltage and the more the work which can be done.

A convenient and easily understood example of the concepts embodied by potential difference, electric current, and voltage is provided by an ordinary "D" battery. In this, as in other types of dry cells, chemical reactions are responsible for maintaining a potential difference between its negative and positive terminals. Even though the electron surplus is depleted through use, it is continually replenished as the result of chemical reactions.

The potential difference of the battery in our example is 1.5 volts. When a metallic conductor such as a copper wire contacts both poles, electrons move through the wire from the negative to the positive terminal. By tradition, however, current is said to flow in the opposite direction, i.e., from positive to negative. The voltage of the dry cell constitutes the "pressure" with which the electrons travel. Remember that the electrons flow because they are in superabundance at the negative pole and are poorly represented at the opposite pole. The movement is promoted by the electrostatic forces of repulsion between the densely packed electrons at the negative pole coupled with the electrostatic forces of attraction between the electrons and the electron-deficient, positively charged terminal.

The number of electrons which move is expressed in units called *coulombs* (sometimes abbreviated *coul*). Since one electron is equal to 0.000,-000,000,000,000,000,160 $(1.6 \times 10^{-19})$ coulomb, 1 coulomb represents a total charge of approximately 6,000,000,000,000,000,000 $(6 \times 10^{18})$ electrons. This number of free electrons is found in less than 1 cm$^3$ of copper, a fact responsible for the excellent electrical conductivity of this metal.

When the terminals of the D battery are connected by a copper wire, the quantity of charge which moves past a given point in a unit of time is a measure of the pressure exerted by the electrons, just as the volume of water moving past a given point in a unit of time is a measure of water pressure. The magnitude of the electric current indicates the rate of transfer of electrical energy.

One *ampere* of electric current is generated in a conductor when one coulomb of charge passes a given point in one second. The quantitative expression for the unit of electric current is as follows:

$$1 \text{ ampere} = \frac{1 \text{ coulomb}}{1 \text{ second}} \tag{8-5}$$

Since electricity is a type of energy, it is capable of accomplishing work. Indeed every electric appliance we use constitutes an example of the ability to convert electrical energy into useful work. Electrical energy is also an important participant in energy conversions within our bodies, especially during muscle contractions and nerve impulse conduction. Just as mechanical work depends on force and distance, the electrical equivalent of mechanical work depends on the quantity of electric charge (coulombs) used and the potential difference (volts) across which the total charge moves. The work done is expressed in joules.

One *joule* is equal to the work accomplished when one coulomb of charge moves through a potential difference of one volt: i.e.,

$$1 \text{ joule} = 1 \text{ coulomb} \times 1 \text{ volt} \qquad (8\text{-}6)$$

One foot-pound of work is equal to 1.36 joules. It is therefore possible to convert between mechanical and electrical work. Furthermore, because the mechanical equivalent of heat is 4.18 joules per calorie, a joule is a unit of both work and energy.

A *volt* is the electrical potential when one joule of work is done per coulomb of charge delivered. The quantitative expression is derived from Eq. (8-6) by dividing both sides by the unit, 1 coulomb, and rearranging to solve for voltage, thus:

$$\frac{1 \text{ joule}}{1 \text{ coulomb}} = \frac{\cancel{1 \text{ coulomb}}}{\cancel{1 \text{ coulomb}}} \times 1 \text{ volt} \quad \text{or} \quad 1 \text{ volt} = \frac{1 \text{ joule}}{1 \text{ coulomb}}$$
$$(8\text{-}7)$$

What does voltage mean in a practical sense? Consider the D battery in our example. The potential difference between its two terminals is 1.5 volts. If we were able to move a positively charged particle across an appropriate conductor from the negative to the positive terminal against the prevailing electrostatic forces, 1.5 joules of work will have been expended per coulomb of charge transported. Therefore, 1.5 joules of work per coulomb is obtained from the battery when it is properly connected in an appropriate electric circuit such as that inside a flashlight.

A battery of larger voltage will accomplish correspondingly more work per coulomb of charge, compared with the D battery. Because the potential difference of a 6-volt battery is four times that of a 1.5 volt battery, four times as many electrons will be delivered per unit time. In other words, the current provided by the 6-volt battery is four times as great as the current of the 1.5-volt battery.

However, as with all forms of energy, the efficiency with which electrical energy is utilized is never 100 percent. Some electrical energy is lost as heat during its transformation. Heat is liberated in overcoming the resistance of the conductor to the flow of electrons, as noted previously. Therefore, the magnitude of an electric current depends on two factors—the potential difference between the two points where the current is measured and the resistance of the conductor. The current increases as the potential difference increases; i.e., the current ($i$) is proportional to the voltage ($V$), or $i \propto V$. At the same time, the current decreases as the resistance increases; i.e., the current ($i$) is inversely proportional to the resistance ($R$), or $i \propto 1/R$. These two expressions are combined to remove the proportionality signs, thus:

$$\text{Current } (i) = \frac{\text{potential difference } (V)}{\text{resistance } (R)} \quad \text{or} \quad i = \frac{V}{R} \qquad (8\text{-}8)$$

Resistance is expressed in units called *ohms*. To solve for the *resistance*, Eq. (8-8) must be rearranged by dividing by $V$ and then inverting both sides, as follows:

$$\frac{\cancel{V}}{\cancel{V}R} = \frac{i}{V} \quad \text{or} \quad \frac{1}{R} = \frac{i}{V}$$
$$\text{Then} \quad \frac{R}{1} = \frac{V}{i} \quad \text{or} \quad R = \frac{V}{i} \qquad (8\text{-}9)$$

Equation (8-9) states that a conductor possesses one ohm of resistance if the potential difference is one volt when the current is one ampere.

The tingling sensations experienced or burns sustained when an electric current of sufficient strength passes through a human body which is grounded by contact with another conductor is a consequence of the relation between $R$ and $V$. As the voltage increases, so does the resistance. The nervous system is differentially affected. The reason these effects do not occur during the normal conduction of nerve impulses is that the voltages generated by nerve cells and, therefore, their resistances are extremely minute.

We will return to electrical concepts in Chap. 13, as we examine the process by which nerve impulses are propagated.

**Chemical energy**   Each molecule of every chemical compound known to us contains potential energy. The energy is locked in the connections, or bonds, which hold the atoms of a molecule together. To understand the nature of this energy, we must look again at atomic structure.

When atoms interact with each other chemically, i.e., when chemical reactions occur, orbital electrons in the outermost energy levels of the interacting atoms are displaced from positions they previously occupied. Depending on the identity of the chemical reactants, the new electronic arrangements possessed by the products of the reaction may either increase or decrease the potential energy of the affected electrons.

**Endergonic reactions**   An external source of energy such as heat, light, or mechanical or electrical energy must be supplied to reactants to increase the potential energy of the products. For example, if heat energy is added to sodium chloride (NaCl), sodium (Na) and chlorine ($Cl_2$) are formed. The reaction is written as follows:

$$2NaCl \xrightarrow{\text{heat}} 2Na + Cl_2 \qquad (8\text{-}10)$$

Reactions which require the addition of energy are called *endergonic*. During endergonic reactions, one or more electrons in the outermost energy levels of the atoms of the products acquire larger quantities of potential energy than they had in the same atoms before reacting. The acquisition of chemical potential energy involves a shift of such electrons to more distant points from the nucleus. To move an electron against the forces of electrostatic attraction of the nuclear protons demands an energy input. The process is really identical to the excitation of an atom by the jump of an electron from a lower to a higher energy level, only in endergonic reactions we are dealing with electrons in the outermost energy levels in atoms of molecular reactants.

In the human body, complex molecules such as proteins, fats, and carbohydrates are manufactured from amino acids, fatty acids and glycerol, simple sugars, and other less complex raw materials. These reactions occur in the presence of specific enzymes which lower the energy of activation of the reactants. Nevertheless, the intracellular syntheses of these chemicals still require energy inputs, just as the formation of sodium and chlorine from sodium chloride requires the addition of thermal energy to a test tube in which the salt is located. The biologic energy input is derived from the oxidation of foodstuffs in all the cells of the body.

**Exergonic reactions**   Infrared radiation, visible light, or other forms of kinetic energy are liberated during reactions in which the products attain a lower chemical potential energy than the reactants. The electronic displacements which occur in such reactions reduce the potential energy of affected orbital electrons. The released energy has one or more of the following effects: (1) The translational motion, or Brownian movement, of the molecular products is increased, causing an elevation in the temperature of the products as well as of the region which surrounds them; (2) other orbital electrons may become excited, causing the emission of other forms of radiant energy such as visible light as the electrons return to their unexcited state; and (3) heat evolution may be so rapid that gaseous products expand violently in an explosion which gives rise to large amounts of potentially destructive  mechanical energy.

Chemical reactions which release energy are referred to as *exergonic*. The burning of coal is a common example of an exergonic reaction. Here, carbon dioxide ($CO_2$) is formed, and heat energy is liberated during the oxidation of carbon (C) by oxygen ($O_2$). The reaction is written as follows:

$$C + O_2 \rightarrow CO_2 + \text{energy} \qquad (8\text{-}11)$$

The complete oxidation of simple sugars and other carbon compounds in the body forms carbon dioxide and water and yields a continual supply of heat and chemical energy required for metabolic and perpetuative functions. The chemical energy is trapped, transported to sites of utilization, and released when needed by special high-energy molecules called adenosine triphosphate (ATP). As we have already learned, ATP is the energy currency of the body.

*Stability of reactants* The amount of energy which must be added to make endergonic reactions occur or the quantity of energy liberated during exergonic reactions is, with few exceptions, an index of the stability of the chemical reactants. For example, chemicals which require the addition of large amounts of energy before they can interact with each other are more stable than chemicals which will react with little energy input.

Conversely, reactants which liberate little energy possess a greater stability than those which give up large amounts of energy. The great stability of carbon dioxide and water explains why these two chemicals are metabolic end products during the degradation of simple sugars in the presence of oxygen. While it is possible to obtain energy from both carbon dioxide and water, a large amount of energy must be added to do so.

The gradual loss of chemical potential energy during the many chemical reactions which result in the complete oxidation of intracellular carbon compounds is analogous to the progressive loss of mechanical potential energy incurred by a person unfortunate enough to slip and fall near the top of a flight of stairs. The fall is momentarily stopped by contact with each of the stairs on the way down. At the base of the stairs, movement abruptly ceases with a painful thud as the remaining potential energy is finally expended. During the tumble, the mechanical potential energy is completely transformed into kinetic energy of the falling body. The energy is dissipated through frictional losses as contact is made with each stair and, finally, with the floor at the foot of the staircase.

So it is during exergonic reactions. The breakdown of glucose into carbon dioxide and water is roughly comparable to the fall of a body down a staircase. At the top of the stairs, the body has a maximum of mechanical potential energy, just as glucose contains a specific quantity of chemical potential energy. The incremental loss of potential energy and the corresponding gain in kinetic energy as the body strikes each lower stair is analogous to the loss of chemical potential energy by glucose during the sequence of reactions in which it is completely degraded. The corresponding gain in kinetic energy is ultimately expended as heat as well as converted into other forms of energy. The fall ceases at the base of the stairs, just as the oxidation of glucose is complete when carbon dioxide and water are formed. No mechanical potential energy remains in the body heaped on the floor at the foot of the staircase; a minimum of chemical potential energy remains in the final end products of the degradation of glucose.

*Nuclear energy* This form of energy is emitted spontaneously from the nuclei of disintegrating atoms. Such atoms are unstable because they have either very large nuclei or an excess of neutrons. During disintegration, these radioactive atoms emit radiations such as alpha particles, beta particles, and gamma rays. These radiations contain an enormous amount of kinetic energy and may cause the ionization of other atoms with which they interact. The nature of nuclear radiations and their interactions with matter are fully explored in Chap. 9.

*Radiant energy* The types of radiant energy include radio waves, microwaves, infrared radiation, visible light, ultraviolet radiation, x-rays, and gamma rays. Visible light, of course, is probably the most commonly known form of radiant energy. Not withstanding their wavelike nature, all these radiations are assumed to travel in discrete bundles of energy called *photons*, each of which bears a specified amount of energy referred to as *quanta* (plural of *quantum*). Because the waves have *electrical* and *magnetic* components during their travel, radiant energy is also referred to as *electromagnetic radiation*. The electromagnetic nature of radiations, including their wavelike and particulate qualities, is discussed later in this and the following chapter.

The effects of the interaction of radiant energy with matter depend upon the characteristics of the radiation and of the object irradiated. Excluding thermal energy for the moment, let us review the nature of these interactions for some biologically important electromagnetic radiations.

*Visible light* The energies of visible light, such as white light from the sun, are differentially absorbed by specific types of matter. The *absorption* by matter of this form of radiant energy results in an increase in the degree of molecular oscillation, the excitation of atoms, or a combination of these effects. The excitation of atoms by visible light gives rise to photochemical reactions. The trapping of the radiant energy of visible light by chlorophyll, upon which the maintenance of all life depends, is a photochemical reaction of fundamental importance.

The energies of visible light which are not absorbed are reflected. White light, a blend of all the colors of the rainbow, can be separated during its passage through a glass prism into a spectrum of overlapping colors which include red, orange, yellow, green, blue, indigo, and violet. Thus, the energies of reflected light are perceived as specific colors. For example, because the energy of the light they reflect is interpreted as green, grass and leaves are green. All the rest of the energies of visible light to which grass and leaves are exposed are absorbed by them. The perception of colors will be discussed in conjunction with vision in Chap. 11.

*Ultraviolet, x-rays, and gamma rays*  The energy of ultraviolet radiation, x-rays, and gamma rays causes the excitation or ionization of atoms. Radiations which are capable of producing atomic ionization are called *ionizing radiations*. The interactions of ionizing radiations with matter in general and with the human body in particular are discussed in the next chapter. Our purpose in considering ionizing radiations here will be restricted to a comparison of the events which occur during excitation and ionization.

The *excitation* of an atom occurs when an orbital electron is displaced from an energy level close to the nucleus to an energy level farther removed from the nucleus. The process is illustrated in Fig. 8-3A.

Assume that an incident radiation, such as ultraviolet light, interacts with an orbital electron as it passes through an atom. Their interaction is analogous to a collision, as the result of which kinetic energy is transferred from the incident radiation to the orbital electron. In response to its excess kinetic energy, the affected electron "jumps" to a new position in a higher energy level. In the phosphorus atom shown in Fig. 8-3A, the affected electron moves from the K shell to the L or M energy level. The number of energy levels through which the electron is displaced during its outward journey depends on the amount of kinetic energy it acquires during the collision with the incident radiation; the greater the energy transfer, the larger the jump. The amount of energy transferred, in turn, is dependent upon the energy content of the incident radiation and the angle of its impact with the orbital electron. Since the more peripheral, or higher, energy levels serve as "stairs" which an orbital electron "climbs" during displacement, energy is thought

**Figure 8-3** *The interactions between various energetic radiations and an atom. The atom is of phosphorus. A. An orbital electron in the K energy level is displaced one or more energy levels outward following its interaction with the radiant energy of ultraviolet light. B. An orbital electron in the K energy level is completely ejected from an atom following its interaction with x-ray or gamma radiation.*

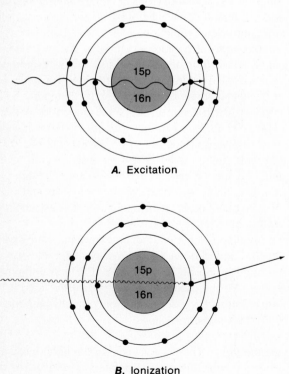

**A.** Excitation

**B.** Ionization

to be transferred in discrete bundles called *photons*. The quanta of energy acquired by the electron enable it to move through, or climb, one or more energy levels, or stairs. Here is an example where the particulate nature of electromagnetic radiation explains its behavior more satisfactorily than its wavelike qualities. Because of the atomic instability created when one or more orbital electrons jump from lower to higher energy levels, the affected atom is said to be excited.

Excited atoms are unstable by virtue of their excess energy. In the absence of further energy interactions, the atom immediately gives up this energy by refilling the vacant orbital position left by the displaced electron. The vacancy is filled by the movement of one or more electrons from higher to lower energy levels.The downward transition of electrons stops when all the original orbital positions are filled. During this time, an amount of energy equivalent to the excess acquired by the displaced electron is emitted as electromagnetic radiation. The *emission* of its excess energy brings about atomic stability by returning the atom to its ground state.

*Ionization* occurs when an orbital electron is ejected from an atom. Ionization in response to radiation is shown in Fig. 8-3*B*. Assume that an incident radiation, such as an x-ray or gamma ray photon, collides with an orbital electron in a manner similar to that described previously. However, instead of being displaced one or more energy levels outward by its excess energy, an affected electron is ejected completely from the atom. The loss of an electron leaves the atom with an excess of one nuclear proton above the remaining number of orbital electrons. Because the atom is transformed into an ion, in this case bearing one positive charge, the process is called *ionization*. Ionization differs from excitation in the amount of energy transferred to an orbital electron; only the most highly energetic electromagnetic radiations are capable of causing ionization. Ionization can also occur through atomic or molecular interactions, as we shall see in Chap. 9.

Ionization may also occur indirectly. For example, some of the more energetic ultraviolet rays can cause atomic ionization through a series of excitations which move an orbital electron progressively outward until it is pushed beyond the sphere of influence of the atom to which it belongs. Even x-rays and gamma rays may not cause ionization directly when their impact with orbital electrons is glancing rather than full or when other factors attenuate the amount of energy transferred to the electrons during collisions. In such cases, excitation occurs rather than ionization. However, since a displaced electron possesses an excess of energy, it takes correspondingly less energy to cause its ejection than the amount originally required. Therefore, if matter of which the atom is a constituent continues to be irradiated, there is an increasing probability that the remainder of the ionization energy will be supplied for the complete removal of the electron from the atom.

*Physical characteristics of electromagnetic radiations*   All types of radiant energy have important features in common, as we noted earlier. Their kinship is based on the following observations: (1) Radiant energies behave as waves, each of which is propagated through space as an electric and magnetic disturbance, and (2) in a vacuum, including the void of outer space, radiant energies travel with the speed of light.

In view of their striking resemblance to each other, the differences which set each type of radiant energy apart from the others are all the more remarkable. These differences have to do with their physical characteristics, which, in turn, affect the manner in which each interacts with matter. The basic distinction between the various electromagnetic radiations is in the range of their respective energies. To appreciate further why energy content is the basis for distinguishing between the various types of electromagnetic radiations, we must look at the relation between their frequencies and wavelengths.

The *frequency* of a radiation is defined as the number of energy waves in a "wave train" that pass a specific point in a unit of time, usually one second. As the frequency of an electromagnetic radiation increases, so does its energy content.

The *wavelength* of a radiation is the linear distance described by one cycle. Each cycle consists of one full oscillation of an electrical and a magnetic distur-

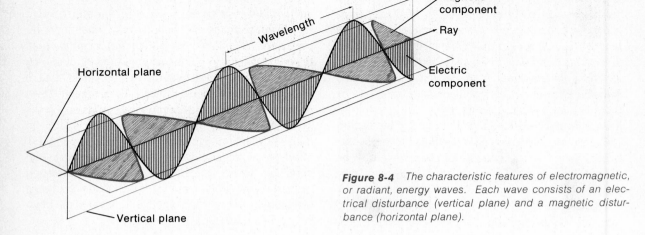

**Figure 8-4** *The characteristic features of electromagnetic, or radiant, energy waves. Each wave consists of an electrical disturbance (vertical plane) and a magnetic disturbance (horizontal plane).*

bance or field. The electric and magnetic fields of an electromagnetic radiation are depicted in Fig. 8-4.

The electrical disturbance (in color) oscillates in a vertical plane; the magnetic disturbance (in black) oscillates in a horizontal plane. One type of disturbance initiates, and is influenced by, the other. Both disturbances simultaneously increase to a maximum in one direction and then decrease to a minimum in the opposite direction, from their respective midpoints. Therefore, each wave consists of a disturbance which begins from a given point along the plane in which it is located, is propagated or displaced first in one direction and then in the opposite direction in the same plane, and, after traveling a certain distance established by its frequency, crosses a point parallel to that from which the disturbance originated. The linear distance between these two points is called one cycle, or wavelength. In Fig. 8-4, the wavelength is the linear distance from one peak to the next peak in the same direction.

Since all electromagnetic radiations travel at the same speed in a vacuum, their frequencies under ideal conditions are determined by the lengths of their respective waves. In other words, frequency is defined by wavelength because the speed of all electromagnetic radiations is constant. However, frequency and energy content are both *inversely proportional* to wavelength. Therefore, high-frequency,

high-energy electromagnetic radiations have extremely short wavelengths, and vice versa.

The relation between frequency, wavelength, and energy content can perhaps best be explained by an analogy: Let us imagine that we are observing the operation of a conveyor belt in a soft-drink bottling plant. Assume initially that the bottles are spaced 6 in. apart, i.e., that the distance from the center of one bottle to the center of the next in line is 6 in., and that the belt is moving at a rate of 1 ft/sec. At this speed an average of two bottles passes a given point in 1 sec. This is the frequency with which the bottles are moving, just as the frequency of an electromagnetic radiation is the number of energy waves per second. The 6-in. distance between the centers of adjacent bottles is analogous to the wavelength of an electromagnetic radiation. The speed of the conveyor belt is analogous to the speed of electromagnetic radiations, 186,000 mi/sec in a vacuum. Each filled bottle possesses a certain caloric value, just as each electromagnetic wave contains a particular photon energy. If one bottle is set equal to $n$ number of photons, instead of $n$ number of calories, a conveyor belt carrying two bottles per second contains a photon energy of $2n$.

Now, suppose that the bottles on this conveyor belt merge with similarly spaced bottles on another conveyor belt moving at the same speed in a manner such that the merged bottles are spaced 3 in. from center to center, i.e., half that of the original spacing. Assuming that the merged conveyor belt is moving at the same speed as before, the new frequency is four bottles per second, or double that of the original frequency. Since four bottles possess a photon energy of $4n$, the merged conveyor belt has twice the photon energy of either of the previous conveyor belts. Therefore, a decrease in the wavelength (or the distance between bottles) is responsible for an increase in both the frequency (or the number of bottles per second) and the photon energy (or total caloric value).

**The electromagnetic spectrum** Separated from each other by their frequencies, wavelengths, and photon energies, the radiant energies in the electromagnetic spectrum include radio waves, microwaves, infrared radiation, visible light, ultraviolet radiation, x-rays, and gamma rays. The qualitative relationships among these electromagnetic radiations are given in Fig. 8-5.

Radio waves have the longest wavelengths, the lowest frequencies, and the least energy content of any of the electromagnetic radiations mentioned above. More precisely, radio waves have wavelengths ranging from those of more than a mile to those of less than 10 ft, with frequencies lower than 10,000 ($10^4$) cps (cycles per second, also known as hertz, or Hz) to over 10,000,000 ($10^7$) cps (10 to 10,000 kilocycles per second, kc/s, or kHz). The energy content of radio waves ranges from values lower than

**Figure 8-5**   *The electromagnetic spectrum of radiant energies. Frequency and energy content increase in the progression from radio waves to gamma rays.*

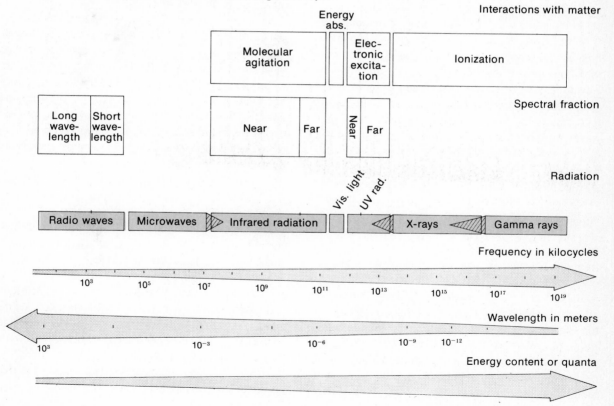

1/10,000,000 ($10^{-7}$) electron volt (eV) to more than 1/1,000,000 ($10^{-6}$) eV, where 1 eV is the quantity of energy gained by an electron accelerated to a potential difference of 1 volt, or 0.000,000,-000,000,000,000,160 ($1.6 \times 10^{-19}$) joule. These extremely long wavelengths and vanishingly small energy contents are distinctive features of radio waves.

At the other end of the electromagnetic spectrum are gamma rays bearing the shortest wavelengths, the highest frequencies, and the greatest photon energies. Gamma rays have wavelengths of less than one-billionth of a meter (0.000,000,000,001, or $10^{-12}$m). The frequencies of gamma rays range from approximately 10,000,000,000,000,000,000 ($10^{19}$) cps to over 10,000,000,000,000,000,000,000 ($10^{22}$) cps [10,000,-000,000,000,000 ($10^{16}$) kc/s or kHz to more than 10,000,000,000,000,000,000 ($10^{19}$) kc/s, or kHz]. The photon energies of gamma rays are usually greater than 1,000,000 ($10^{6}$) eV, where 1 million eV (1 MeV) are equal to 0.000,000,000,000,160 ($1.6 \times 10^{-13}$) joule. These enormous energies and fantastically short wavelengths are characteristic of gamma rays.

All the other types of electromagnetic radiations possess frequencies, wavelengths, and energy contents between those of radio waves and gamma waves. There is a tremendous decrease in wavelength and a corresponding increase in frequency and energy content in the progression from radio waves on the left to gamma rays on the right in Fig. 8-5, as suggested by the taperings and directions of the arrows.

Note, however, that the photon energies of some of the radiations overlap those of adjacent radiations in the spectrum. For example, high-frequency microwaves have an energy content identical to low-frequency infrared radiations. In addition, low-frequency x-rays and gamma rays possess photon energies similar to those of high-frequency ultraviolet radiations and x-rays, respectively. Since the energy content of an electromagnetic radiation determines the nature of its interactions with matter, high-frequency microwaves behave as thermal radiations capable of causing heating, high-frequency x-rays behave as ultraviolet radiations capable primarily of atomic excitation rather than ionization, and low-frequency gamma rays behave as soft, or mild, x-rays. The basic effects on matter of the biologically important electromagnetic radiations are given at the top of Figure 8-5.

*Gradations of interaction* The nature of the interaction between radiant energy and matter is determined by the energy content of the incident radiation, as we have already learned. To see what this means in a practical context, suppose it were possible to increase the energy of an object such as a steel bar ad infinitum through the continued application of thermal energy.

At first, only heat or infrared radiation is emitted from the surface of the steel bar. The radiation of heat occurs when the kinetic energy of the particles, in this case atoms, is transferred from the surface of the steel bar to the air. The radiation of heat is caused by the propagation of periodic electric and magnetic fields induced by the increased to-and-fro movements of the particles in the bar.

As the temperature of the steel bar rises, the bar not only becomes hotter but turns luminescent as well. This phenomenon occurs when the oscillatory motion of its atoms increases sufficiently. The increase in potential energy of the bar is matched by a corresponding increase in the intensity of the radiated energy. Thus, infrared radiation and visible light are emitted simultaneously. The first color radiated from the glowing bar is red. With the continued addition of heat energy, the color of the bar changes to yellow and, then, finally to white. During this time, of course, the bar becomes progressively hotter. The "white hot" glow is due to the fact that energies are being radiated in all parts of the visible spectrum in approximately equal amounts. Therefore, a greater range of energies is emitted in the visible light spectrum as the temperature of the bar continues to rise.

If the steel bar remained intact at much higher temperatures, it would glow with a bluish hue. At this point, all the visible radiations would be concentrated at the blue end of the spectrum. In other words, the potential energy of the bar would be great enough to give rise to a large number of the most highly energetic emissions possible in the visible portion of the electromagnetic spectrum. Little or no energy would be emitted in other, less energetic regions of the visible spectrum, although the intensity of heat radiation would continue to increase. However, long before this point is reached, the violent vibrations by the particles within the steel bar begin to tear free electrons away from the positions they occupy between the metallic atoms. Therefore, the process is not analogous to the ionization of orbital electrons from atoms.

When electrons which hold metallic atoms together are ripped away, the metal melts. Therefore, the steel bar in our example will melt when heated to a sufficiently high temperature. The metallic atoms are actually ions of positive charge. Adjacent ions are bound together by hordes of electrons. Since metals are good electric conductors, free electrons are readily available. The structural arrangement within a metal therefore consists of a mass of metallic ions separated by spaces occupied by clouds of electrons. Solids with this type of structure are called *metallic crystals*. Removal of electrons from between the metal ions causes the metal to lose its structural integrity. *Melting* of the steel bar is the visible effect of the loss of its crystalline structure.

*Vaporization* of the liquid metal begins when the kinetic energy of some of the metallic particles is great enough to overcome the attractive forces of neighboring particles holding them together. The oscillatory motion of these particles becomes so violent that they fly off into space from the surface of the liquid metal. The process continues until all the metallic particles acquire sufficient energy to vaporize.

**Heat energy**  Although there is a correlation between the temperature of matter and the amount of kinetic energy it possesses, the relationship may sometimes be deceptive. Consider, for example, a piping-hot cup of coffee and a large percolator full of the same beverage, only lukewarm rather than hot. More ice can actually be melted by the coffee in the percolator, compared with the coffee in the cup, even though the former is much hotter than the latter. This apparently anomalous result can be explained by comparing the total amount of thermal energy in each container. The total energy possessed by the coffee in

the cup will be found to be lower than that contained within the coffee in the percolator, despite the fact that its temperature is considerably higher.

The only other difference besides temperature between the coffee in the cup and that in the percolator is in their respective volumes. Therefore, we must conclude that it takes a larger amount of heat energy to bring the temperature of the coffee in the percolator to that of the coffee in the cup because there is a larger volume of coffee to be heated. This statement is really not different from the intuitive understanding that if 1 calorie raises the temperature of 1 Gm of water 1°C, raising the temperature of 1,000 times as much water by the same amount requires 1,000 times as many calories.

**Temperature**  The quantity of heat is expressed in calories, as we have already learned. The *intensity* of the heat is measured by the *temperature* of the heat source. Thermometers are instruments used to measure temperatures. The most commonly used thermometers are either bimetallic or liquid-in-glass. These thermometers are based on the principle that matter expands when heated and contracts when cooled.

*Bimetallic thermometers*, such as those found in ovens and other appliances in which high temperatures are monitored, consist of two different metals placed back to back at a given temperature. An increase in temperature causes the metals to expand, each at its own rate. Since the two strips of metal are fastened to each other, the bimetallic strip differentially bends in one direction. A drop in temperature causes the bimetallic strip to bend in the opposite direction. A temperature scale is calibrated by marking off the freezing and boiling points of pure water at 1 atm (atmosphere) and dividing the space between, below, and above these two points by the appropriate number of equally spaced divisions.

*Liquid-in-glass thermometers*, whether scientific, medical, or household, consist basically of a thin glass tube, at the base of which is a well containing a liquid such as mercury or some dye-colored compound. When heated, the liquid expands at a greater rate than the glass and therefore rises in the tube. When cooled, the liquid contracts more rapidly than

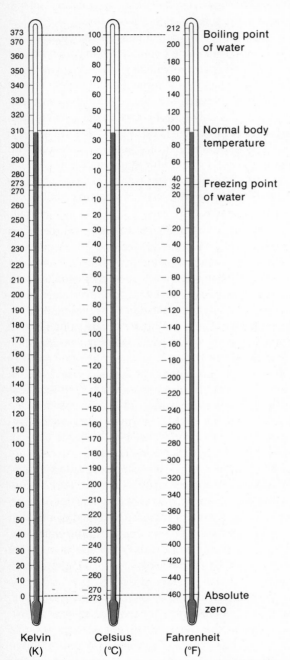

| Kelvin (K) | Celsius (°C) | Fahrenheit (°F) | |
|---|---|---|---|
| 373 370 | 100 | 212 200 | **Boiling point of water** |
| 360 | 90 80 | 180 | |
| 350 | 70 | 160 | |
| 340 330 | 60 50 | 140 120 | |
| 320 | 40 | | |
| 310 300 | 30 | 100 80 | **Normal body temperature** |
| 290 | 20 10 | 60 | |
| 280 273 270 | 0 | 40 32 20 | **Freezing point of water** |
| 260 | −10 −20 | 0 | |
| 250 240 | −30 −40 | −20 | |
| 230 | −50 | −40 −60 | |
| 220 | −60 −70 | −80 | |
| 210 200 | −80 | −100 | |
| 190 | −90 | −120 | |
| 180 170 | −100 −110 | −140 −160 | |
| 160 | −120 | −180 | |
| 150 140 | −130 −140 | −200 −220 | |
| 130 | −150 | −240 | |
| 120 110 | −160 −170 | −260 −280 | |
| 100 | −180 | −300 | |
| 90 80 | −190 −200 | −320 | |
| 70 | −210 | −340 | |
| 60 50 | −220 −230 | −360 −380 | |
| 40 | −240 | −400 | |
| 30 20 | −250 −260 | −420 −440 | |
| 10 0 | −270 −273 | −460 | **Absolute zero** |

**Figure 8-6** *A comparison of the kelvin (K), Celsius (C), and Fahrenheit (F) temperature scales.*

the glass, and its level in the tube drops. Such thermometers are calibrated by a method similar to that used for bimetallic thermometers.

Three different temperature scales are in use—kelvin, Celsius, and Fahrenheit (abbreviated K, C, and F, respectively). A thermometer calibrated in each of these scales is illustrated in Fig. 8-6. These temperature scales serve somewhat different purposes and constituencies.

The *kelvin scale* is used as a reference standard and is employed in extremely low temperature research and technology. Its lowest temperature, absolute zero, is set equal to zero.

The *Celsius scale* (formerly called centigrade) is preferentially used by scientists. The divisions on the Celsius scale are exactly the same distance apart as on the kelvin scale, only the freezing and boiling points of water on the Celsius scale are 0 and 100°, respectively, i.e., 100° apart. These two points establish absolute zero and body temperature at −273°C and 37°C, respectively.

By contrast, the *Fahrenheit scale* is deeply rooted in Western culture by tradition and is widely used in clinical, household, and industrial settings. The freezing and boiling points of water on the Fahrenheit scale are set equal to 32 and 212°, respectively. The difference between these two temperatures is 180°, or divisions.

Since 180 divisions on the Fahrenheit scale are equal to 100 divisions on either the Celsius or kelvin scale, each division on the Fahrenheit scale is fiveninths (5/9) those of the other scales (100/180). The freezing and boiling points on the Fahrenheit scale establish absolute zero and body temperature at −460 and 98.6°, respectively. The awkwardness of the Fahrenheit scale makes it unsuitable for use in most forms of scientific temperature measurement.

Since constants such as absolute zero, the freezing point, and the boiling point have their equivalents in each of the temperature scales discussed above, it is possible to convert temperatures read on one scale to either of the others. Interconversions between kelvin and Celsius scales are based on two facts: The distance between each division on these scales is identical, and 0 K is equal to −273°C. Therefore, converting from the Celsius to the kelvin scale is simply a matter of adding 273° to the temperature in question. The same number of degrees must be subtracted when converting a temperature reading from the kelvin to the Celsius scale. On the other hand, interconversions between the Celsius and Fahrenheit scales are based on the following considerations: Each division on the Fahrenheit scale corresponds to 5/9°C and 32°F is equal to 0°C. Therefore, a temperature reading is converted from the Fahrenheit to the Celsius scale by subtracting 32 from the reading and then multiplying the remainder by 5/9. The quantitative expression for this conversion takes the following form:

$$T_c = \tfrac{5}{9}(T_F - 32°)$$ (8-12)

Notice that this equation states that each Fahrenheit degree is equal to 5/9 of a Celsius degree and that 32 degrees must be subtracted when converting from the Fahrenheit to the Celsius scale. Conversely, a temperature reading is converted from the Celsius to the Fahrenheit scale by multiplying the reading by 9/5 and then adding 32 to the product. The formula is written as follows:

$$T_F = \tfrac{9}{5}(T_C + 32°)$$ (8-13)

Observe that this formula states that each Celsius degree is 9/5 of a Fahrenheit degree and that 32 degrees must be added when converting from the Celsius to the Fahrenheit scale.

*Changes of state*   The physical state of any type of matter can be changed by increasing or decreasing its temperature sufficiently. The transition gas → liquid → solid is achieved by subtracting heat energy and is ordinarily accompanied by a decrease in volume. The reverse progression solid → liquid → gas is accomplished by adding thermal energy and generally involves an increase in volume.

Although most substances change state in the orders given above, some materials vaporize directly from the solid state without going through a liquid state. The transition solid→gas is called *sublimation*. Sublimation occurs from the surfaces of all solids, al-

though the loss from this cause is usually negligible.

When heated, certain solids decompose before they melt. This effect is caused by a cleavage of the connections between the molecules of which these solids are composed and takes place before the connections are severed between the atoms which make up the individual molecules.

Any real gas, as opposed to the ideal gases discussed in the last chapter, can be liquefied with a sufficient increase in the applied pressure, provided the temperature remains below the point where both liquid and vapor states exist simultaneously. Each gas has a single pressure at any given temperature below this critical temperature where it can exist as a liquid, a combination of a liquid and vapor, or a vapor. Thus, pressure-induced changes of state are temperature-dependent. Conversely, pressure is intimately involved in determining the actual temperature at which changes of state occur for a given type of matter, as we shall see later.

Thus, whether a substance is solid, liquid, or gaseous depends on the prevailing temperature and pressure. The universal applicability of this principle is illustrated by the following anecdote: When a noted lawyer was asked, "How's your wife [or husband]," the reply was, "Compared with whom?" When a famous scientist was asked the same question, he replied, "Under what conditions of temperature and pressure?"

Let us now examine the relation between temperature, pressure, and changes of state. Since water is used as a reference against which other substances are compared, we will follow its changes of state. Water is also the solvent in protoplasm, a fact that emphasizes the importance of understanding its physical behavior. The changes of state of water at different temperatures are illustrated in Fig. 8-7.

In Fig. 8-7, the changes of state of pure water in a single beaker are followed over the time span $T_1$ to $T_9$. An open flame provides a continual source of heat energy to the beaker, beginning at $T_2$. Prior to the start of the experiment, a thermometer is inserted in the water within the beaker to measure changes in temperature. A large, snugly fitting container is secured

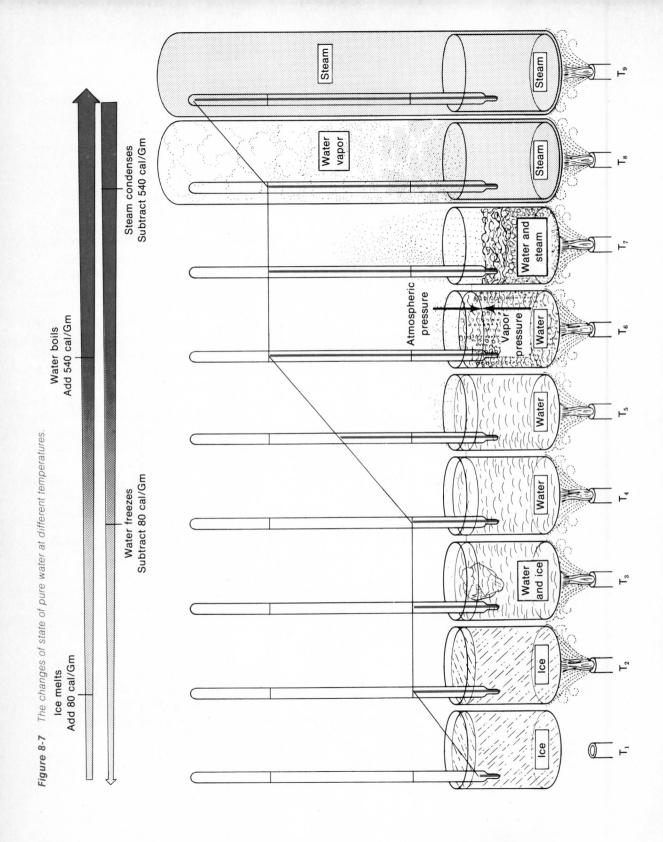

**Figure 8-7** The changes of state of pure water at different temperatures.

Ice melts
Add 80 cal/Gm

Water boils
Add 540 cal/Gm

Water freezes
Subtract 80 cal/Gm

Steam condenses
Subtract 540 cal/Gm

Atmospheric pressure

Vapor pressure

Ice

Ice

Water and ice

Water

Water

Water

Water and steam

Water vapor

Steam

Steam

Steam

$T_1$  $T_2$  $T_3$  $T_4$  $T_5$  $T_6$  $T_7$  $T_8$  $T_9$

around the mouth of the beaker at $T_8$ to trap water vapor. During the experiment, the air pressure is held constant at 1 atm.

The experiment begins at $T_1$. The water in the beaker is frozen solid. Since the beaker has just been removed from a freezer kept at a temperature below freezing, the temperature of the ice at $T_1$ is below 0°C.

The flame from a bunsen burner heats the beaker of ice, beginning at $T_2$. While the ice does not yet melt, its temperature rises relatively rapidly to the freezing point of water, 0°C. A relatively rapid rise in temperature occurs because less heat energy is required to raise the temperature of a given mass of ice by a specific amount than to raise the temperature of the same mass of liquid water by an identical amount. Immediately after the freezing point of water is attained, the water in the beaker still consists entirely of ice ($T_2$).

With the continued addition of thermal energy, the ice begins to melt ($T_3$). During this time, the water in the beaker consists of a mixture of ice and liquid water. Notice that the temperature remains at 0°C during the change of state from solid ice to liquid water. Immediately after all the ice has melted, the liquid water is still at its freezing point ($T_4$). Let us temporarily stop the experiment at this point to examine the events more closely.

Since there was no change of temperature during the time required to melt the ice completely once the freezing point of water was reached, where did the heat energy go? Remember that the attractive forces between densely packed molecules in a solid are quite large. As more and more heat energy is supplied at 0°C ($T_2$), some of the hydrogen bonds are broken between water molecules, causing an increase in the compactness of the water mass which will reach a maximum at 4°C. Less than 20 percent of the hydrogen bonds originally present in ice water are actually broken during the melting process. It is for this reason that liquid water more closely resembles a solid than a liquid. The result is the formation of polyhedral chains of liquid water molecules, around which are aggregated free water molecules.

However, even at 0°C, a few water molecules vaporize directly from the surface of the ice (dots above beaker at $T_3$). Such molecules have a sufficient amount of kinetic energy to evaporate without entering the liquid state first. Obviously, the presence of such energetic water molecules at 0°C is rare, and therefore the rate of sublimation is vanishingly small. Above 0°C, pure water is entirely in liquid form and therefore cannot undergo sublimation.

Thus, the heat absorbed by ice during melting is used to cleave intermolecular hydrogen bonds. All the ice in a mixture of ice and liquid water must melt before the temperature of the water can rise above 0°C, simply because all the energy provided while ice is present is used in breaking down its coordinated crystalline structure. Eighty calories of heat energy are required to melt 1 Gm of ice (80 kcal/kg). This constant is called the *heat of melting*. Conversely, if we liberate heat energy by chilling liquid water already at 0°C., i.e., when going from $T_5$ to $T_3$ by subtracting heat energy, 80 cal of heat energy must be removed to freeze each gram of water present. Called the *heat of fusion*, this constant has the same quantitative value as the heat of melting, only in the former case heat energy is removed from the freezing liquid water, whereas in the latter case heat energy is absorbed by the melting ice water. Each pure substance has its own heat of melting and fusion. Now let us return to the experiment in progress.

After all the ice has melted, a steady application of heat to the beaker containing liquid water at 0°C ($T_4$) will cause a gradual rise in temperature ($T_5$) to 100°C, the boiling point of water ($T_6$). During the period of time required to bring the liquid water from its freezing point to its boiling point, an increasing amount of water vapor or gaseous water (black dots) accumulates in the air above and surrounding the beaker.

Initially, water vapor forms when some of the liquid water molecules at the surface of the water acquire enough kinetic energy to be torn away from the attractive forces binding them to the surface. Remember that the temperature of the water is a reflection of the average amount of kinetic energy possessed by the water molecules. However, some molecules have more energy than others. Furthermore, those molecules below the surface of the water are bound more tightly to each other than those at the interface between water and air. Therefore, as the average translational motion of the water molecules increases during the addition of heat energy, some of the more energetic molecules at the surface are the first to fly off into space. Less energy is required to separate them from their neighbors, compared with those below the surface. The number of molecules of water vapor above the surface of the liquid steadily increases as the water temperature rises to 100°C. Bubbles of air are also forced out of solution during this time. Although the amount of liquid water left in the beaker decreases by a volume corresponding exactly to that of the lost water vapor and dissolved air, the drop in the level of the liquid water is hardly noticeable until boiling begins.

At $T_6$, the average kinetic energy of the liquid water is such that its temperature is 100°C. Now boiling begins as liquid water vaporizes below the interface layer. Gaseous water is less dense than the liquid water in which it is dissolved. Consequently, bubbles of expanding water vapor rise to, and are evaporated from, the surface. The force with which the water vapor is ripped from the liquid water and strikes the atmosphere is called the *vapor pressure*. At 100°C, the vapor pressure of pure water is equal to an air pressure of 14.7 lb/in.² (1 atm) pushing down on the surface of the water (vertical arrows at $T_6$). When the downward pressure of the atmosphere is balanced by the upward pressure of the water vapor, the tendency of the liquid water to vaporize is the same as the tendency of the gaseous water to condense.

If, however, the atmospheric pressure is decreased, e.g., by moving from a lower to a higher altitude above sea level, then boiling will begin at a correspondingly lower vapor pressure. Since the amount of water vapor above the surface of the water in the heated beaker is determined by the average kinetic energy of the water, vapor pressure varies with the temperature. The lower the final vapor pressure, the lower the temperature required to achieve it. For example, the boiling point of water at an altitude of 10,000 ft will be lower and will take less time to reach, compared with the boiling point of water at sea level, provided that the time rate of application of thermal energy and the water mass are the same in each case.

The converse is also true. To boil water at pressures above 1 atm, the vapor pressure and, therefore, the temperature of the water must increase accordingly. The disinfectant action of a bacterial autoclave and the rapid cooking in a home pressure cooker result from the fact that the average kinetic energy of liquid water is increased as its temperature rises above 100°C. After boiling begins, it follows that the penetrating powers of pressurized steam are considerably greater than of steam at 1 atm.

So long as the atmospheric pressure and the vapor pressure are balanced, boiling will progress if thermal energy is continuously applied. This condition was first observed in our experiment at $T_6$. The temperature remains at the boiling point ($T_7$) until all the liquid water has evaporated from the beaker ($T_8$). In other words, during the transition from liquid to gas, the temperature of water stays at the boiling point, just as its temperature remains at the freezing point during the transition from solid to liquid.

Therefore while boiling continues, all the heat energy added to liquid water is used in its vaporization. The additional thermal energy is expended in overcoming the attractive forces which liquid water molecules exert on each other. Approximately 540 cal of thermal energy is required to convert 1 Gm of liquid water into 1 Gm of gaseous water (540 kcal/kg). This constant is referred to as the *heat of vaporization*. Conversely, if heat is subtracted by cooling water vapor, i.e., when going from $T_8$ to $T_6$ by removing heat, the condensation of 1 Gm of liquid water from 1 Gm of gaseous water would require the liberation of 540 cal of heat energy. Called the *heat of condensation*, this constant is the same as the heat of vaporization, only in the former case heat is released from condensing water vapor, whereas in the latter case heat is absorbed by vaporizing liquid water. Each pure substance has its own heat of vaporization and condensation.

The fact that skin is more severely burned by contact with steam than by contact with an identical mass of liquid water at the same temperature is explained by the heat differential between these two states of water. Each gram of water vapor at 100°C contains 540 cal more of heat energy than a comparable amount of liquid water at the same temperature. This additional amount of heat energy is liberated by every gram of vapor which condenses on the skin.

After all the liquid water has vaporized ($T_8$), the continued application of heat will raise the temperature of the water vapor beyond 100°C ($T_9$), provided that the vapor is trapped in an enclosure ($T_8$ and $T_9$). Under these conditions, the additional heat energy accelerates the rate of Brownian movement of the gaseous water molecules while preventing an unrestricted outward expansion of the gas. Water vapor above the boiling point of water is called *steam*. If an enclosure to trap the water vapor were not provided at $T_8$, the temperature of the gaseous water would never rise above 100°C, because the increase in the average kinetic energy of the molecules would be used up in causing an infinite outward expansion of the gas.

## THE MAINTENANCE OF BODY TEMPERATURE

The human body is capable of maintaining a constant temperature, whether in the withering heat of a tropical rain forest or in the painfully frigid air of the Arctic. Temperature homeostasis is achieved by balancing heat production and heat losses.

The adjustments made to changes of temperature are both involuntary and voluntary. When exposed to cold, we shiver, our skin pales, and small bumps which give the appearance of gooseflesh may form on our arms and legs. We voluntarily attempt to get warmer by increasing the amount of body insulation with more and heavier clothing, the amount of heat supplied to the body by raising the external temperature, or the amount of heat produced by the body through physical activity or vigorous exercise.

When heat production exceeds heat loss, we perspire, or sweat, and our skin reddens. We voluntarily attempt to get cooler by decreasing the amount of body insulation by shedding our clothing, increasing the amount of evaporation from the surface of the skin by fanning ourselves, decreasing the amount of heat supplied to the body by lowering the external temperature, and decreasing the amount of heat produced by the body through a reduction in physical activity.

**Heat production** The sources of body heat are both external and internal. Heat is added to the body whenever the external temperature is higher than that of the body. The gaseous molecules of heated air bombard our bodies through Brownian movement, increasing the translational motions of molecules within our clothes and skin. Heat therefore flows from a region or body at a higher temperature to a region or body at a lower temperature. The process is identical to a cup which is warmed when hot coffee is poured into it. The coffee loses heat energy and becomes cooler. Equilibrium is reached when the coffee and the cup attain the same intermediate temperature.

Heat is also produced by the body as a consequence of metabolism. The oxidation of digested foodstuffs liberates energy, 55 percent or more of which is evolved as heat. Through muscular activity, such as tensing the muscles, physical exercise, or shivering, heat production may be temporarily increased at least fourfold.

The temperature of the body may also perceptibly increase following a heavy meal, especially one with a high protein content. A person may perspire after eating a steak with all the trimmings, because the degradation of protein requires a large expenditure of energy, most of which is lost as heat. Called *specific dynamic action*, this characteristic of food when metabolized is discussed in Chap. 26.

Finally, body heat may be raised as the result of specific pathologic conditions. For example, heat is one of the symptoms which appears during an inflammatory response, a body defense mechanism explained in Chap. 4.

**Heat losses** The human body is a source of heat energy. Most of the heat lost from the body is transferred physically to the surrounding air by a combination of three processes—conduction, convection, and radiation.

*Conduction* Conduction occurs when heat is transferred through physical contact from an object or region at a higher temperature to an object or region at a lower temperature. The highly agitated atoms or molecules in the hotter material collide with, and transfer a portion of their kinetic energy to, the more slowly moving particles in the colder material. As the kinetic energy of these particles increases, there is a corresponding rise in the temperature of this region. The particles which make up the newly warmed region, in turn, transfer part of their energy through collisions with adjacent particles which have a lower energy content, and so on. Thus, heat is conducted through contact between two objects, or portions of one object, at different temperatures. It follows that no heat is conducted when contact is made between objects at the same temperature.

Before heat can be transferred from the skin to the air, it must be brought to the surface from deeper, hotter regions of the body. A small portion of this heat energy is simply conducted from hotter tissues to cooler tissues until the surface of the body is reached.

Body heat is lost by conduction when some part of the body makes contact with another object at a cooler temperature. For example, heat is transferred by conduction from the hand to a glass when drinking cold water, from the feet to the ground when walking, and from the immersed body surface to water when bathing. About 2 percent of the total heat lost from the body occurs through conduction.

*Convection*  Convection occurs when thermal energy is transferred by the movement of a heated material, ordinarily a liquid or a gas, from one point to another. For example, blood is the medium of transport of most of the body heat from the deepest regions to the superficial layers. With each beat of the heart, a certain increment of the total heat energy in the blood is transported through the arteries to the surface of the body. The transport of heat in this manner is convective.

Heat is lost from the body by convection when air is warmed as it flows over the surface of the skin. Here, heat is transferred from a hot object to a circulating medium at a cooler temperature. The circulation of air may be promoted by a breeze or a body movement. As the air is heated, it becomes less dense through expansion. The void left by the heated current of air as it spreads outward from the surface of the skin is immediately occupied by more dense, cooler air. The replacement air is also warmed and expands, maintaining the convective loss of heat from the body.

In addition, heat is removed from the body during defecation, urination, and expiration. Feces, urine, and expired air are warm by virtue of having come from deep portions of the body. Since these materials carry heat energy from regions inside the body to points outside the body, the mode of heat transfer is at least analogous to convection, even though the feces are ordinarily not fluid in consistency and the heat content of expired air is due to evaporation.

Under normal conditions, less than 10 percent of body heat loss is ascribable to convection. In fact, in still air and at rest, the body does not lose heat through convection, because uniform thermal gradients are quickly established between the surface of the skin and the surrounding air.

*Radiation*  A large portion of the heat lost from the skin is emitted as periodic electric and magnetic disturbances caused by the radiation of thermal energy. This type of electromagnetic radiation is called *infrared*, as you will recall from our earlier discussions. Radiation is responsible for almost 60 percent of the heat transferred from the body to the external environment.

To appreciate further the operational differences between conduction, convection, and radiation, consider the following example: Imagine that you are in the vicinity of a spectacular fire. You may accidentally get burned if you brush against glowing embers or other objects charred by the flames. This, of course, is heat transfer by conduction. As a spectator just in back of the firemen in the front lines, most of the intense heat which you may experience is transferred convectively. If the heat of the flames is still experienced when your view of the fire is completely blocked by intervening buildings, then the heat is being transferred through radiation.

*Vaporization*  The physical methods of heat transfer described above account, however, for only two-thirds of the total heat lost from the body. Most of the remainder of the heat loss occurs during the evaporation, or vaporization, of water from the skin and lungs.

Recall that the heat of vaporization of water is 540 cal/Gm. However 540 cal evaporates considerably less than 1/2 oz of water. Since more than 1/2 qt of water (1 qt = 32 oz, on an average) is evaporated each day from the skin alone, the amount of heat energy expended during normal vaporization is appreciable.

Perspiration is produced by eccrine sweat glands in the skin. These glands are illustrated in Fig. 4-1. Even when perspiration cannot be felt or seen, the eccrine glands continue to lose water to the atmosphere through vaporization. Such sweating is called *insensible perspiration* and occurs at all times without our awareness. On the other hand, sweat which moistens or wets the skin is referred to as *sensible perspiration*, since we are aware of its presence. Two or more quarts of sweat an hour may be produced by a person perspiring profusely. The volume of sensible perspiration produced depends on many factors including the temperature and moisture content of the surrounding air, the body's rate of metabolism, and the person's emotional state.

Moisture is also present in expired air, as evidenced by the condensation of droplets which occurs when warm breath strikes a window pane chilled by frigid outdoor air. Water vapor is present in the airways of the lungs because fluids coat the surfaces of the airways. A certain amount of vaporization occurs at all temperatures below boiling, as we have already learned. At body temperature, the tendency of water molecules to both vaporize and condense is balanced at a vapor pressure of 47 mmHg. The evaporation of water from the surfaces of the airways cannot increase unless the average kinetic energy of the liquid water molecules exceeds 47 mmHg. In other words, an increase in the vapor pressure requires an increase in the temperature.

Compared with the moisture-laden air in the airways of the lungs, inspired air is usually relatively dry. However, this air is mixed with water vapor as it is sucked into the lungs. During expiration, a portion of the humidified air in the respiratory passageways is given up to the atmosphere. Actually, over 6 percent of the expired air is water vapor. Since thermal energy is consumed during vaporization, the loss of water vapor from the lungs reduces the amount of body heat.

**Thermoregulation** Because heat production must be balanced by heat losses, the human body has to produce, conserve, and lose heat in appropriate amounts in response to internal and external changes in temperature. These adjustments are under neural and hormonal control. However, our purpose here will be restricted to following the fate of body heat.

The mechanics of thermoregulation by the human body are illustrated in Fig. 8-8. Based on heat transfer and body temperature, the body can be divided into an inner *core* and an outer *shell*. Although the temperature of the shell varies from region to region, it is ordinarily lower than the temperature of the core. Most of the heat from the core is transferred to the shell convectively by the blood, although a small amount of heat is conducted directly to the surface of the shell.

Heat is conserved in the core of the extremities because of the nature of the blood circulation there. The arteries and veins ordinarily run in parallel. As circulating blood courses through the major arteries of the extremities, heat is continually dissipated to the surrounding tissues. However, much of this heat is subsequently transferred to the contiguous veins, ultimately to be returned to the heart. As a consequence, arterial blood reaching the hands and feet gives up less heat than would otherwise be the case. The heat gradients established in this manner preserve the temperature differential between the core and shell of each limb. In this way, heat conservation in the extremities maintains two heat gradients—one longitudinal and the other transverse. These gradients are identified by the tapered arrows in Fig. 8-8.

Control of the diameter of the blood vessels in the skin is of major importance in the conservation and loss of body heat. When the external temperature is appreciably below that of the skin, small arteries called *arterioles* in the skin automatically decrease in diameter. Large numbers of capillaries (minute blood vessels) supplied by the arterioles collapse when the

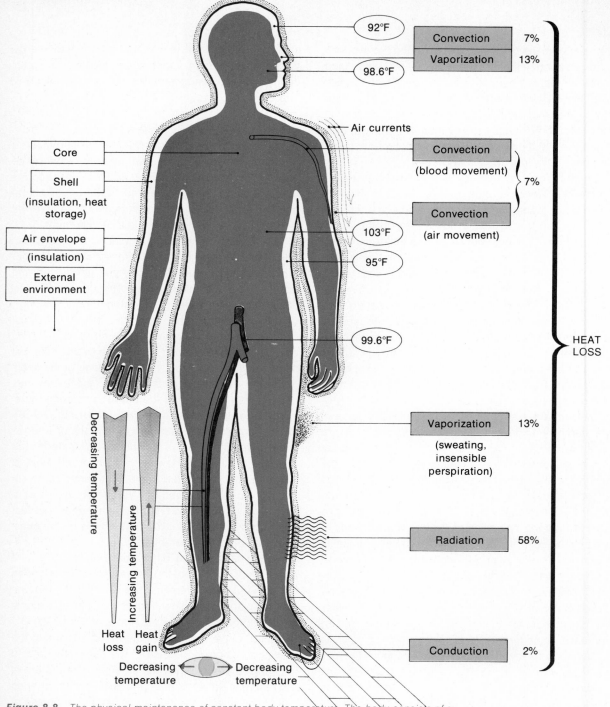

**Figure 8-8** *The physical maintenance of constant body temperature. The body consists of a central core and a peripheral shell. The shell insulates against heat loss and makes fine adjustments in temperature which hold core temperatures constant, despite wide variations in environmental temperature. Insulation is also provided by the air envelope which surrounds the shell.*

volume of circulating blood falls below a critical level. The skin blanches and may even turn bluish as the tissue supply of oxygen is diminished. Reduction in the amount of blood in the skin causes a corresponding decrease in the convective transfer of heat from the core to the shell.

The situation is reversed as the body temperature rises above its normal range. This condition causes an automatic increase in the diameters of the arterioles beneath the skin. The response increases the volume of circulating blood in the skin and promotes the convective transfer of heat from the core to the shell. The large blood volume in the skin during this time gives it a ruddy complexion.

If a rise in body temperature cannot be compensated by vasodilatation alone, then the body begins to perspire perceptibly. The eccrine glands are automatically stimulated to secrete sweat when the body temperature is excessive. The higher the body temperature, the more profuse the perspiration. If heat exposure persists for an extended period of time, e.g., following a move from a temperate to a tropical climate, the eccrine glands rapidly increase their secretory capacities.This adaptive response enables an individual acclimatized to heat to produce copious amounts of sweat. At the same time, the salt content ($Na^+$ and $Cl^-$) of the sweat decreases to conserve these important mineral ions. Therefore, while chronic exposure to warm air temperatures increases the volume of sweat produced, it also sets in motion hormonal mechanisms which conserve salt. People who are used to heat secrete proportionately less salt in their sweat than people who are not used to hot air temperatures.

Insulation against heat loss is provided by the shell and air envelope. The *air envelope* is the thin layer of air more or less trapped next to the surface of the skin. The shell consists of the skin, subcutaneous connective tissues, and associated fat depots.

Were it not for the fact that heat is transferred convectively from the core to the shell through the blood, all the heat from the core would have to be transferred through conduction. Conduction, of course, is a much less efficient method of heat transfer. This is one of the reasons why certain types of circulatory impairments may cause noninflammatory elevations of body tem-

perature. Thus, the volume of circulating blood in the cutaneous vessels adjusts the insulating capacity of the shell.

Fat has about four times the insulating value of other types of tissues. The presence of subcutaneous fat therefore enhances the insulating abilities of the shell. Since females possess proportionately more subcutaneous fat than males, women enjoy better heat insulation than men. It is not surprising that the professional skin divers who harvest shellfish from the cold depths of the ocean floor off the Korean coast, where this activity still takes place, are predominantly women.

The surface of the skin is at the interface between the shell and the air envelope and therefore has important insulating capacities. When the temperature drops abruptly, gooseflesh may appear because of the erection of hairs on the arms and legs. This automatic response increases the roughness and insulating effectiveness of the skin. In addition, body hairs trap air next to the skin, thereby reducing the rate of heat transfer through convection currents.

The air envelope communicates between the body and the rest of the external environment. This layer is considered as an extension of the body, because its temperatures are ordinarily appreciably higher than the rest of the outside air. Its insulating value to the body, while usually small, is measurable.

Body temperature variations   Under conditions of steady-state maintenance, the human body has a constant temperature of 98.6°F, or 37°C. However, this does not mean that all parts of the body are at the same temperature, or that any one part of the body is at the same temperature continually, or even that individuals matched in age, height, and sex have exactly the same temperature when taken at the same time under identical conditions. A temperature of 98.6°F is simply an average temperature beneath the tongue in a closed mouth. The temperature is still within the normal range if it deviates by about 1° to either side of the average value.

The normal value depends on the body region in which the temperature is taken. The rectal temperature is about 1° higher (99.6°F) than the oral temperature. The midregion of the body has a surface temperature in the vicinity of 95°F. The forehead is even cooler, with a temperature of slightly over 90°F. At the other extreme, the temperatures in the deepest portions of the body are over 100°F.

Body temperature also fluctuates in a cyclic manner, reaching a peak toward evening and a low point during sleep. Based on time periods of approximately 24 hr, these regular and entirely normal oscillations of body temperature may range beyond 2°F. Therefore, a normal body temperature taken during midday is not the same as that registered at midnight.

Body temperature also varies slightly from person to person, since the brain centers for temperature control (see Chap. 14) are not all at exactly the same "setting." In fact, what is amazing is that the average body temperature of 98.6°F is such a good rule of thumb. In the final analysis, the interpretation of a body temperature requires common sense as well as specialized knowledge.

When heat production exceeds heat loss, body temperature rises. If the body temperature exceeds 105°F, e.g., during a severe fever caused by a pyrogenic (fever-producing) bacterium, then the condition is referred to as *hyperthermia*. Conversely, when heat losses are greater than the total heat input of the body, body temperature falls. The condition is called *hypothermia* when body temperature is depressed below 94°F. Hypothermia may be due to heat losses sustained either during excessive urinations (polyuria) or through exposure.

## COLLATERAL READING

Oort, Abraham H., "The Energy Cycle of the Earth," in *The Biosphere,* San Francisco: W. H. Freeman and Company, 1970. The fate of solar energy is followed from the time it enters the atmosphere as sunlight until it is reradiated as heat.

Summers, Claude M., "The Conversion of Energy," *Scientific American*, September 1971. Methods are explored which may increase the efficiency with which energy is converted into useful forms by human beings.

Baker, Jeffrey J. W., and Garland E. Allen, *Matter, Energy and Life,* Reading, Mass.: Addison-Wesley Publishing Company, Inc., 1965. Chapter 1 reviews numerous concepts discussed in both this and the preceding chapter.

Hong, Suk Ki, and Hermann Rahn, "The Diving Women of Korea," *Scientific American*, May 1967. Fascinating descriptions of the habits of skin-diving women which enable them to withstand the icy rigors of the sea while harvesting shellfish from the ocean floor.

# CHAPTER 9
# IONIZING RADIATIONS AND MAN

Human beings are continually subjected to physical forces which interact with the molecules and atoms that make up their bodies. Natural or artificial in origin, these physical forces are particulate or electromagnetic, or have characteristics of both forms of radiation. Particulate radiations consist of various subatomic particles such as protons, neutrons, electrons, other subatomic nuclear fragments, and cosmic radiation which originates outside the earth's atmosphere. These particles move through space at extraordinarily high velocities. Electromagnetic radiations consist of wavelike particles without mass which travel in a vacuum at approximately 186,000 mi/sec, the speed of light. Among the biologically destructive portions of the electromagnetic spectrum are infrared, ultraviolet, x—, and gamma rays.

When the energies associated with these radiations interact with matter, including human tissues and cells, they may cause one of the following effects: (1) the physical oscillation of molecules, liberating heat, as in the case of infrared radiation; (2) the excitation of atoms by causing orbital electrons to move to higher energy levels and liberating visible or ultraviolet radiation when the excited electrons return to their stable positions, as in the case of most forms of ultraviolet radiation; (3) the ionization of atoms through the ejection of orbital electrons either directly, as in the case of particulate radiation, or indirectly, as in the case of electromagnetic radiation; or (4) the creation of instability within atomic nuclei which may be subsequently expressed through nuclear disintegration, as in the case of atomic bombardment with neutrons. Radiation with insufficient energy to cause the immediate ionization of an atom may still do so through continual bombardment which provides for the accumulation of enough excitation energy to finally cause the ejection of an orbital electron.

Ionizing radiations cause the ionization of atoms

with which they interact. From what has already been stated, we may conclude that such radiations are particulate or electromagnetic and cause ionization by interacting with orbital electrons either directly or indirectly.

Radiation-induced tissue damage results from the direct effects of the physical ionization of atoms and the chemical formation of highly reactive but electrically neutral molecules called *free radicals*. Their effects are exemplified by the shattering of chromosomes, the destruction of cellular organelles, and alterations in cell membrane permeability. Such changes cause disruptions of metabolism and perpetuation, resulting in mitotic inhibition, metabolic depression, genetic mutations, chromosomal aberrations, and carcinogenic (cancer-inducing) effects. The damage may be seen after a short time following irradiation as, for example, in "acute radiation syndrome" or by prenatal mortality and congenital defects among children of afflicted individuals. Some effects, however, may not be observed until after many years have elapsed, when mute testimony to prior radiation exposure is given through the appearance of a premature age-specific disease such as cancer.

The amount of damage is determined by factors associated with both the ionizing radiation and the portion of the body or specific tissue irradiated. The biologic damage incurred is, in part, a function of the specific type of incident radiation, its energy content before and after atomic impact, the number of secondary ionizations caused by ejected electrons which subsequently interact with other nearby atoms, and "the dose-rate effect," or rapidity with which a measured dose of radiation is actually delivered to the body. These physical factors are, in turn, influenced by various biologic characteristics. For example, certain tissues consisting of predominantly undifferentiated cells are more radiosensitive compared to those containing many highly specialized cells. Other biologic factors which influence the response to a given amount of radiation include age, sex, nutritional status, hormonal balance, blood oxygen tension, and body temperature. The damage wrought by radiation

exposure will be offset to some extent by the ability of the body to repair the cellular and tissue damage.

As a rough estimate, human beings are exposed to an average of about 150 milliroentgens (mr) of naturally occurring and artificially produced radiation each year. The difficulties of establishing safe minimum daily exposures to radiation are suggested by the many physical and biologic variables enumerated above. The fact that radiation damage may be either delayed or genetic militates against any dosage of radiation which is in fact "safe." Analyses of risks versus benefits for diagnostic, therapeutic, military, and industrial uses of radiation may ultimately turn out to be inappropriate methods for justifying potential or real environmental contamination by, or excessive individual exposure to, radiation. Standards for human exposure to radiation are continually being reexamined and remain an extremely controversial topic at the present time.

In this chapter, we will clarify and amplify the ideas introduced above. To appreciate the origins and characteristics of particulate and electromagnetic ionizing radiations, we will turn our attention first to an examination of the various kinds of radiations and the units which are used to measure their energy content. Then we shall consider their biologic effects and medical applications.

## RADIATIONS AND IONIZATION

Radiations are vehicles for the transfer of energy through matter or space. Many forms of radiation have no effect on atoms through which their energies pass. Other types of radiation provoke an imbalance of electric charge in atoms with which they interact. The electrical imbalance converts each affected atom into a pair of charged particles called ions. Radiations possessing these capabilities are referred to as *ionizing radiations*. Ionizing radiations have the potential for damaging cells and tissues.

In order to understand why a pair of charged particles results from the ionization of an atom, let us assume that an orbital electron is ejected from, or kicked out of, the atom to which it belongs by a particle or wave traveling at a great velocity and therefore possessing a large amount of energy. If we identify the atom prior to impact as A, then after collision, the loss of an orbital electron e⁻ from an atom results in an excess of one positive (+) charge in its nucleus. Therefore, the atom has become an ion which may be referred to symbolically as A⁺. This transformation is best understood when we recall that the atom prior to impact is electrically neutral, i.e., that it has the same number of protons in the nucleus as electrons in orbit outside the nucleus and that the loss of a negative charge results in an automatic surplus of one positive charge. Using the above symbols, the reaction may be summarized as follows:

$$A \xrightarrow{\text{ionizing radiation}} A^+ + e^- \qquad (9\text{-}1)$$

A free, fast-moving, and, therefore, energetic electron such as e⁻ is highly reactive and rapidly captured by another atom. Suppose that an atom B captures e⁻ to become a negatively charged ion B⁻, according to the following reaction:

$$B \xrightarrow{e^-} B^- \qquad (9\text{-}2)$$

Whether we consider A⁺ and e⁻ or A⁺ and B⁻ as the *primary ion pair* formed by the ionizing radiation, the fact remains that two particles of opposite charge have resulted from the interaction.

At this point, you may well be wondering what relevance atoms have in a practical sense to your body and to potential physical pathologic conditions associated with it. Bear in mind that atoms go into the makeup of molecules and that molecules constitute the largest chemical category of functional units within cells. In other words, biologic activity is a function of molecules, not atoms. As a result of atomic interactions, however, irradiated molecules may change chemically. The spectrum of possible chemical changes may range between oxidation and the formation of free radicals.

Although a free radical is electrically neutral, there is a single unpaired electron in its outer orbit. This solitary electron makes a free radical highly unstable,

extremely reactive, and unbelievably short-lived, existing in its free form for perhaps less than one ten-billionth of a second. Free radicals interact with other free radicals and molecules to form stable chemical products. However, biologically damaging chemical reactions can occur during the brief existence of free radicals.

The many kinds of chemical reactions triggered by the formation of ion pairs from irradiated atoms alter the biologic activity of affected molecules. Thus, enzymes may be broken down into smaller units, or degraded; separate strands of nucleic acids such as deoxyribonucleic acids (DNAs) in the chromosomes may link together or the chemical structure of the separate strands may be changed; and weak hydrogen bonds between portions of single organic molecules may be ruptured. In short, the unseen world of molecular machinery inside irradiated cells may be and frequently is drastically changed, compared with similar cells which have not been exposed to ionizing radiation above that of background.

These chemical changes induce structural and functional abnormalities in all parts of irradiated cells, but especially in their membranes, organelles, and chromosomes. Before we look more closely at these biologic alterations, we must gain an appreciation of the types of ionizing radiations which exist in nature and some of their more pertinent characteristics, such as their origins and interactions with matter.

## CHARACTERISTICS OF IONIZING RADIATIONS

The energies of ionizing radiations may be transferred as *particles* or as *waves*. Particles with sufficient energy to cause the ejection of orbital electrons from atoms with which they interact include subatomic particles such as protons and electrons, numerous nuclear emissions such as groupings of two protons and two neutrons called alpha particles and nuclear electrons called beta particles, and some forms of cosmic radiation. Wavelike radiations capable of causing atomic ionization include ultraviolet radiation, x-rays, and gamma rays. Since cosmic radia-

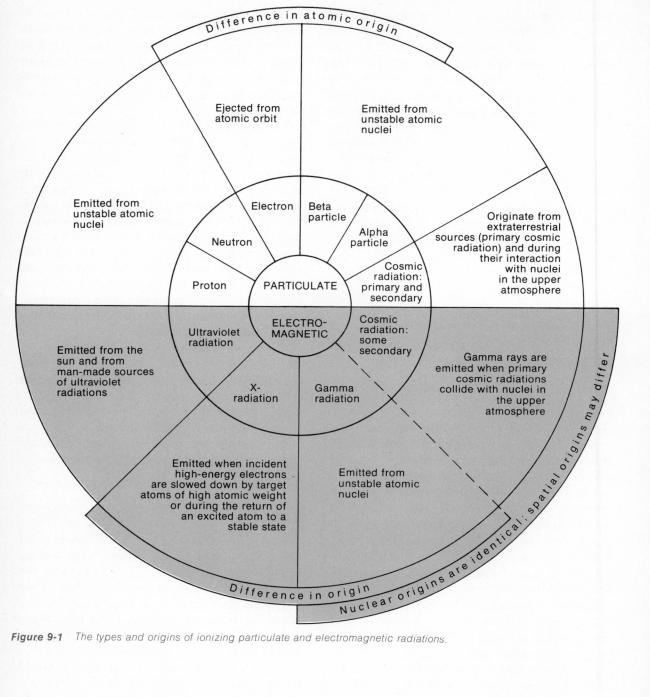

*Figure 9-1* *The types and origins of ionizing particulate and electromagnetic radiations.*

tion has a gamma ray component, certain cosmic rays are wavelike in nature.

**Particulate radiations**  The origins of particulate radiations are given in Fig. 9-1. From left to right, these particulate radiations include protons, neutrons, electrons, beta particles, and alpha particles.

*Protons*  Recall that protons are nuclear subatomic particles. Each possesses one positive charge and has a mass of 1.0073. Protons with high energy may be ejected during the disintegration of atomic nuclei following nuclear interactions with other ionizing particles such as neutrons. Protons constitute one type of particle found in cosmic radiation emanating from extragalactic (outside our galaxy) sources, although they may also originate within our own galaxy. Since these rays originate in outer space, they are referred to as *primary cosmic radiations*. Protons are also obtained from man-made sources for experimental, therapeutic, and other purposes.

A proton causes the ionization of an atom by impact with, and ejection of, an orbital electron. However, impact does not necessarily mean direct contact between the ionizing particle and an orbital electron. The energy of the ionizing particle, its charge, the degree of closeness of the interacting particles and their angle of impact, as well as other factors, will influence the outcome.

As mentioned previously, ionization results in the formation of an ion pair of opposite charge. In our example, the ion pair consists of the free electron and the remainder of the atom which is now positively charged because of the electron deficit. Ionization by impact is summarized in Fig. 9-2.

*Neutrons*  Neutrons, like protons, are subatomic particles within the nuclei of atoms. Lacking an electric charge, each has a mass of 1.0087. Therefore, compared with protons, neutrons have a slightly greater mass. Neutrons are emitted from artificially produced radioactive elements. Neutrons are one product, for example, of nuclear fission, or splitting, of uranium.

Because neutrons are uncharged, they ordinarily do not cause ionization by impact. Instead, they pass through atoms until one strikes an atomic nucleus. Then one of three things happens, depending on the energy of the incident neutron: nuclear recoil, nuclear absorption of the incident radiation, or nuclear fission. The subsequent nuclear radiations are capable of causing ionization by impact. Therefore, neutrons cause ionization indirectly or, more precisely, by secondary means.

In *nuclear recoil,* the nucleus becomes excited by virtue of its excess energy and may emit a neutron of lower energy as some but not all of the energy of the incident neutron is absorbed. When the nucleus returns to its stable, or ground, state, gamma radiation is emitted. The scattering of an incident neutron and the subsequent emission of gamma radiation by a recoiling nucleus are diagramed in Fig. 9-2.

*Nuclear absorption* occurs when the energy of the incident neutron completely disappears within a target nucleus. Complete absorption leads to the emission of a nuclear particle such as a proton. The absorption of a neutron by an atom changes its atomic mass number and represents an important method for the production of radioactive elements called *radioisotopes* for medical and experimental uses. Alpha particles, beta particles, and gamma rays are among the products of *nuclear fission,* or the disintegration of an atomic nucleus. These radiations may cause the ionization of other atoms by impact.

*Electrons*  *Orbital electrons* are subatomic particles found in discrete energy levels outside the nucleus of an atom, as you will recall from the last chapter. Each electron has a mass only 1/1,836 that of a proton and bears one negative electric charge. The origin of ionizing electrons is summarized in Fig. 9-1. Nuclear electrons, or beta particles, emitted by certain radioactive elements, will be discussed separately in the next section.

Electrons capable of causing ionization are ordinarily orbital electrons ejected by some form of ionizing radiation such as beta particles, themselves electrons, or by gamma rays. Let us assume that a high-

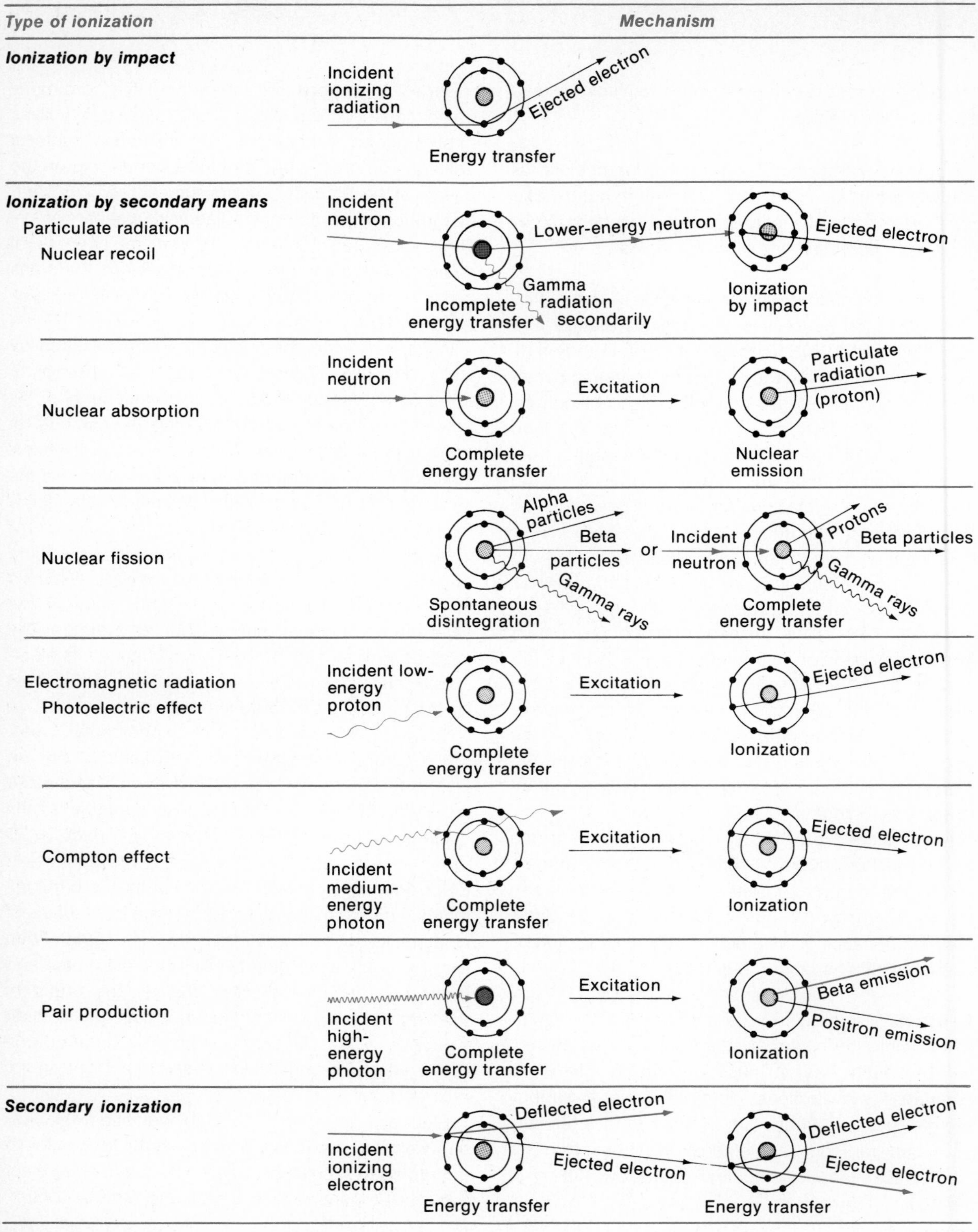

**Figure 9-2**   *The mechanisms by which radiations cause atomic ionization.*

velocity electron approaches an orbital electron during its passage through an atom. The orbital electron is repelled by the charge of the incident electron. The force of repulsion exerted on the orbital electron either causes it to shift to a higher energy level, in which case the atom is said to be excited, or forcibly ejects the orbital electron from the atom, in which case the atom is of course ionized.

When an orbital electron is ejected, it possesses a large portion of the original energy transferred to it. Its kinetic energy depends on the velocity of the incident electron and its angle of collision. The electron dislodged from the first atom may in turn collide or interact with an orbital electron of another atom. The result of this interaction is either the excitation or ionization of the second atom.

If the orbital electron of the second atom is ejected and has enough energy, then it too may cause the excitation or ionization of still another atom. Indeed, even the first ionizing electron will continue to cause excitation or ionization, provided that it still has sufficient energy. This process continues until the ejected electrons, through energy transfer reactions during collisions, are finally captured by, and incorporated within, other atoms.

Ionization does not occur directly when an incident radiation causes the emission of a beta particle. For ionization to occur directly, an ionizing radiation must eject an orbital electron by impact. When an incident radiation causes a nuclear emission which is capable of producing ionization by impact, the ionization mechanism is called *secondary ionization*. The events which occur in secondary ionization are summarized in Fig. 9-2.

Cellular and tissue damage to ionizing radiation is often a consequence of secondary ionizations induced indirectly by the original incident radiation. Both beta and gamma radiations cause large numbers of secondary ionizations. Secondary ionizations, therefore, compound the damage caused by irradiation.

*Beta particles* A beta particle arises during a change in the nature of a neutron when an atomic nucleus becomes unstable. Beta particles are iden-

tical to orbital electrons, except that they originate in the nucleus and are more energetic than orbital electrons following ejection from an atom. The origin of a beta particle is summarized in Fig. 9-1.

The formation of a beta particle is a relatively complex phenomenon. In an unstable and excited nucleus of a parent atom, there is an imbalance in the normal number of protons and neutrons. To achieve a more stable state, a neutron is transformed into a proton, a *neutrino*, and a beta particle. The proton formed in this transformation remains in the nucleus. The neutrino, a particle of negligible mass and no electric charge, is ejected along with the beta particle. If the atomic nucleus is still unstable following beta emission, then gamma rays are also emitted from the nucleus. This reaction serves to reduce any residual excess energy of the daughter nucleus, bringing the transformed atom to a stable, or ground, state. Beta emission is diagramed in Fig. 9-2.

The terms *parent atom* and *daughter nucleus* are used to emphasize the fact that the addition of a proton to a daughter nucleus increases its atomic number by one. Therefore, the daughter atom belongs to a different element, compared with the parent atom.

Beta emitters cause secondary ionizations in a manner identical to that of ejected orbital electrons. The cellular and tissue damage resulting from beta irradiation is dependent in part on the kinetic energy conferred on the beta particles and the nature of the tissue being penetrated. Because beta particles (and ejected orbital electrons) are small and possess only one electric charge, they may penetrate more deeply than other types of charged particles bearing the same amount of energy but a greater mass or charge. Beta particles may also give rise to gamma rays as they are precipitously decelerated during their passage through tissues. Therefore, some forms of beta radiation originating from sources outside the body are especially hazardous to peripheral tissues such as skin.

*Alpha particles* An alpha particle consists of two protons and two neutrons ejected in one bundle, or package, from the nucleus of an atom. An alpha particle is considerably heavier than other charged particles which induce ionizations. Its two protons confer a double positive charge.

An alpha particle is emitted by the nucleus of a parent atom during nuclear disintegration. The emission of an alpha particle may be followed by gamma ray emission from the daughter nucleus, for a reason similar to that following the emission of beta particles. The origin of alpha particles is summarized in Fig. 9-1. Alpha emission is diagramed in Fig. 9-2.

As alpha particles interact with nearby atoms, they may excite or eject orbital electrons through their strong attractive forces. Electrons which are ejected become a source of secondary ionization. Alpha particles continue to excite or ionize orbital electrons until their energies are dissipated. The rapid loss of kinetic energy by an alpha particle is followed by its capture of two electrons, converting it into an atom of helium. Therefore, an alpha particle is identical to the nucleus of a helium atom and is ultimately converted into an atom of this element. The structure of helium is reviewed in Chap. 7.

Their large mass causes alpha particles to dissipate energy quickly, since they move relatively slowly and are easily stopped by paper-thin surfaces. Because they cause a considerable number of ionizations in the extremely short distance of their travel through matter, including cells and tissues, the heaviness of an alpha particle is a two-edged sword. That is, while alpha emitters outside the body are not ordinarily considered to be hazardous, inhalation or ingestion of these particles is extremely serious. The damage due to internal exposure to an alpha emitter, while not extensive, is highly concentrated and restricted to specific cells and particular tissues close to the alpha particles. Ionizing radiations which dissipate their energies rapidly and cause the formation of many ion pairs in a short distance have a high *specific ionization.*

*Cosmic rays*   Primary cosmic rays are not actually rays but represent ionizing particles which enter the earth's atmosphere from extraterrestrial sources. The particles consist mainly of protons, alpha particles, and other atomic nuclei stripped of their orbital electrons. Primary cosmic rays are therefore positively charged particles. Although a few of these pristine cosmic radiations reach the earth's surface to cause atomic excitation or ionization, most of the highly energetic primary cosmic particles collide with the nuclei of atoms present in the upper atmosphere. As a consequence of these violent interactions, the atmospheric nuclei disintegrate, giving off particles called mesons, discussed in Chap. 7. Most mesons strike the earth's surface. Some mesons, however, instantaneously disintegrate into a shower of other particulate and wavelike radiations, including electrons, neutrinos, and gamma rays, before reaching the ground. The origins of cosmic rays are summarized in Fig. 9-1.

Primary cosmic rays and secondary cosmic particles and waves bombard our bodies continually but with no apparent effect. The significance of cosmic rays resides in the fact that they make up a small portion of the total natural background of ionizing radiation to which human beings and all other forms of life are exposed. The other component of natural background radiation comes from terrestrial sources such as rocks and water. While cosmic radiations are not considered hazardous at the earth's surface, they become potentially dangerous for space travelers who must pass through a concentrated layer (the Van Allen belt) of these rays in the outer reaches of the atmosphere.

**Electromagnetic radiations**   In contrast to particulate radiation, electromagnetic radiation is wavelike in nature. This form of radiation travels in a vacuum with the speed of light, has electric and magnetic field components while in motion, and of course travels through space and interacts with matter through energy transfer reactions.

Although wavelike in nature, electromagnetic radiations travel in discrete bundles of energy referred to as photons, mentioned earlier, instead of uniformly along an ever widening circle known as a wave front. Photons lack mass. Each photon, or "particle of energy," is designated one quantum. The total energy of one quantum is related directly to the frequency of the electromagnetic radiation of which it is a part. In other words, the greater the frequency of the individual photon, the greater its energy content.

Electromagnetic radiations cause ionization by secondary means following the absorption of their energies by irradiated atoms. Each excited atom responds to excess energy by emitting particulate radiation. The ejected orbital electrons are capable of producing secondary ionizations by impact.

Most of the energy imparted to ejected particles goes into the tremendous velocities with which they travel; only a small amount of the energy associated with electromagnetic photons is consumed in causing the actual removal of orbital electrons.

Ionizing electromagnetic radiations are absorbed by, and therefore transfer their energies to, atoms through one of three processes: the photoelectric effect, the Compton effect, and pair production. These processes are diagramed in Fig. 9-2 and are described below.

The *photoelectric effect* occurs when orbital electrons are ejected from certain atoms in response to the complete absorption of incident electromagnetic photons bearing appropriate energies. To produce a photoelectric effect, the photons must be of low energy content. The total absorption of incident photons results in the complete transfer of their energies to the affected atoms. Atoms which make good absorbers of low-energy electromagnetic photons have many orbital electrons, since the probability of reacting to an incident radiation increases with increasing electron density.

Because the number of negatively charged orbital electrons is balanced by the number of positively charged nuclear protons, absorbers with large numbers of orbital electrons possess high atomic weights. Therefore, low-energy electromagnetic radiations transfer their energies most efficiently to absorbers of high atomic weights. The excess energies induced in these atoms through the photoelectric effect cause each to expel one or more orbital electrons. Each ejected electron is called a *photoelectron* because a photon is responsible for dislodging it. The formation of a negative ion (an ejected photoelectron) is coupled with the simultaneous formation of a positively charged ion (the rest of the atom). The photoelectrons ejected through ionization by secondary means collide with orbital electrons of other atoms,

causing secondary ionizations. Secondary ionizations continue until the ionization energies of the ejected electrons are dissipated.

The *Compton effect* occurs when orbital electrons are ejected from almost any irradiated atom in response to the partial or incomplete absorption of incident electromagnetic photons of moderate energy content. Since the photons collide with orbital electrons at an angle, the travel paths of the orbital electrons after collision are correspondingly deflected. After a glancing impact with orbital electrons, the incident photons rebound with longer wavelengths. These secondary photons possess correspondingly lower energies than they had as ionizing photons prior to collision.

So long as the ionization energies of rebound photons remain sufficiently high, they may continue to cause ionizations by secondary means in other atoms. Finally, the secondary photons dissipate most of their energy. The remaining energy of each secondary photon is then absorbed by available atoms without causing further rebound.

While these events are occurring, photoelectrons ejected by secondary means promote secondary ionization as they interact with orbital electrons of other atoms. These electrons also dissipate their ionization energies through subsequent atomic collisions and are finally captured by available atoms.

*Pair production* occurs when an electron and its counterpart of opposite charge, a positron, are ejected from the nucleus of an atom of high atomic weight. These nuclear emissions are produced in response to nuclear displacement and subsequent recoil prompted by the complete absorption of an electromagnetic photon of high energy. In effect, some of the total energy of the photons is converted into the combined resting mass of a positron and electron. This transformation is referred to as an *energy-mass conversion*. The velocities imparted to the ejection of these particles from the irradiated atom are a reflection of the remainder of the energy originally transferred to the nucleus by the incident photon. Since high-energy electromagnetic radiations disappear within atoms, to be replaced by high-velocity electron-positron pairs, no rebound photons are formed as a consequence of pair production. However, the electron and positron may cause secondary ionizations by impacting with orbital electrons of other atoms.

Only three forms of electromagnetic radiation cause atomic excitation or ionization: ultraviolet radiation, x-rays, and gamma rays, including some secondary cosmic radiation. The sources and effects of these types of electromagnetic waves are summarized in Fig. 9-1.

*Ultraviolet radiations* Recall from the previous chapter that ultraviolet radiations have a greater energy content than the visible portion of the electromagnetic spectrum and a lower energy content than the x-ray portion of the same spectrum. Ultraviolet radiations closest in quality to those of visible light waves are referred to collectively as the *near-ultraviolet spectrum,* or band. Since their energies are only slightly greater than those of visible light waves, their effects are restricted to the excitation of atoms. Following excitation, the electrostatic force of attraction of the affected electron to the nuclear protons is lower than it was before electronic excitation. Therefore, less energy will be required to eject this electron, compared with the ionization energy which would ordinarily be necessary. Ejection is more likely because excitation energy has already been imparted to the affected orbital electron.

Ultraviolet radiations with frequencies sufficiently high to approximate those of x-rays are referred to collectively as the *far-ultraviolet spectrum,* or band. Since their energies are only slightly lower than those of x-rays, their effects include both excitation and ionization of target atoms. In fact, there is a gradual merger between far-ultraviolet radiations and lower-frequency x-rays. This fact has already been noted (refer back to Fig. 8-5). However, only radiations of the highest frequencies in the far-ultraviolet band are capable of causing atomic ionization. Therefore, virtually all ultraviolet radiations cause electronic excitation or potentiate atomic ionization by other incident ionizing radiations.

If an atom excited by ultraviolet radiation does not lose its displaced orbital electron, then the electron immediately moves back to its former energy level. During this process, electromagnetic energy equivalent to the original excitation energy is emitted by the atom. Under these circumstances, the energy of the emitted electromagnetic radiation may be in the ultraviolet or even in the far, visible portion of the electromagnetic spectrum.

The sources of ultraviolet radiation include the sun and various man-made devices such as certain incandescent and arc lamps. Ultraviolet radiations produced by these sources are stopped during their first millimeter of travel through skin. Their penetrating powers are therefore correspondingly poor. However, in sufficient quantities, these radiations cause chemical and physical changes in skin which give rise to symptoms such as erythema, or skin reddening, pain, and blistering. These symptoms of course are characteristic of sunburn. Ultraviolet radiations are also used to induce bacteriolysis, as pointed out in Chap. 3.

*X-rays* X-rays are located in the electromagnetic spectrum between ultraviolet and gamma radiations. They possess a range of energies greater than the former but lower than the latter. However, the frequency range of x-rays overlaps those of both ultraviolet and gamma radiations. Therefore, only a portion of the x-ray spectrum has properties distinctive enough to separate it from other, adjoining portions of the electromagnetic spectrum. The distinction between x-rays and gamma rays becomes even more blurred when we recognize that the ultimate difference between them is their source of origin. Gamma rays originate during the disintegration of atomic nuclei as diagramed in Fig. 9-2. X-rays, on the other hand, originate (1) when incident ionizing electrons are slowed down precipitously and deflected when close to, but not within, atomic nuclei or (2) when orbital electrons ejected by ionizing electrons are immediately replaced by other orbital electrons from more peripheral energy levels of the same atoms. These two reactions do not involve nuclear radiant emissions.

Because of their great penetrating abilities through matter, including lead shields, x-rays with frequencies close to those of gamma rays are termed *hard x-rays.* Those x-rays with frequencies close to the far-ultraviolet spectrum are labeled *soft x-rays,* since their penetrating powers are relatively poor by comparison.

X-rays are produced artificially when negatively charged electrons accelerated to high velocities in cathode-ray or x-ray tubes interact with atoms of a positively charged target metal known as an *anode.* The anode target is usually made up of an absorber of high atomic weight. Two types of x-radiation are emitted. One, called nonspecific, or general, x-radiation, consists of a whole spectrum of x-ray frequencies. *General x-radiation* is produced when high-speed electrons are suddenly slowed down and deflected as they approach the positively charged nuclei of appropriate target atoms. The other type, x-radiation characteristic of the element being bombarded, consists of a narrow frequency band of x-rays. *Characteristic x-radiation* is produced by target atoms following their ionization by incident electrons. During the production of characteristic x-radiation, an electron ejected from an inner energy level of a target atom is immediately replaced by an electron from an energy level farther removed from its nucleus. As the replacement electron undergoes transition from a higher to a lower energy level, an x-ray photon is emitted equivalent to the difference in energy content between the two energy levels. Since elements to which different target atoms belong vary in their number of energy levels and electrons, the energy content of the emitted x-rays is as unique as a fingerprint and can be distinguished by appropriate methods.

X-rays interact with matter, including cells and tissues, by one of the two processes described earlier—the photoelectric effect or the Compton effect. These electromagnetic phenomena cause the ionization of atoms by secondary means. The particulate radiations, especially electrons emitted following x-irradiation, provoke secondary ionizations of other atoms. Secondary ionizations produce most of the biologic damage associated with x-radiation.

X-rays have industrial, medical, and research applications. They enable the visualization for diagnostic purposes of opaque objects such as bones and fabricated industrial parts. X-rays are also used in certain types of cancer therapy to destroy tumors and other neoplasms (new pathologic growths). However, the diagnostic and therapeutic applications of x-rays are not without their dangers to health and safety. Stray x-radiation produced by an x-ray unit represents a potential hazard to technicians, physicians, nurses, engineers, and others who may be involved with its operation. Here, adequate safety precautions are necessary. Obviously, radiations which destroy microorganisms and cancerous cells are also capable of lethal or damaging effects in normal cells which they penetrate. Thus, therapeutic doses of x-rays unavoidably destroy normal tissues which surround cancerous growths. Sometimes the damage caused by therapeutic x-rays provokes symptoms worse than those associated with the tumor being treated.

Cellular damage is also a by-product of diagnostic x-rays. The justification for their use must always be in terms of an analysis of risks versus benefits. The analysis should take into account not only the potential benefits to an individual recipient but, in the case of annual chest x-rays for the detection of tuberculosis, to the entire population for which exposure to radiation is anticipated. In this type of public health program, a point of diminishing returns may be reached when the low frequency of disease detection does not justify the exposure of extremely large numbers of subjects to even the relatively low doses of radiation characteristic of diagnostic x-rays. This point is the subject of continuing and vigorous debate at the present time.

*Gamma rays* Gamma ray photons possess the highest frequencies, the shortest wavelengths, and the most energy of any of the radiations classified as electromagnetic. As mentioned earlier, considerable overlap exists in the energy content of x-ray photons and gamma ray photons. However, gamma rays ordinarily originate during the disintegration of the nuclei of unstable atoms. Such atoms are radioactive, and gamma rays represent one type of ionizing energy which may be emitted by their nuclei.

Only under certain conditions are gamma ray emissions a by-product of nuclear disintegration. When the nucleus of an atom breaks up, it emits particulate radiations such as alpha particles, beta particles, positrons, protons, neutrons, or neutrinos. When the emission of particulate radiation from the nucleus fails to rid itself completely of its excess energy, the remainder of the excess energy is emitted as gamma radiation. Gamma radiation is therefore a method by which atoms achieve nuclear stability.

Because of their physical resemblance to x-rays, gamma ray photons interact with matter in a fashion similar to that of x-ray photons. Gamma rays cause ionization by secondary means through one of the three processes by which ionizing electromagnetic radiations interact with matter. More specifically, low-, medium-, and high-energy gamma ray photons interact with matter through the photoelectric effect, the Compton effect, and pair production, respectively. As noted earlier, electrons or other radiations produced through these interactions cause secondary ionizations of other atoms with which they collide. The secondary ionizations are responsible for most of the biologic damage due to irradiation with gamma rays. Therefore, gamma rays and x-rays produce their biologic effects in an identical manner.

Strong gamma ray emitters such as the radioactive element cobalt 60 are used for experimental, therapeutic, industrial, and other peaceful purposes. On the other side of the coin, gamma ray emitters make up a substantial amount of the fallout of nuclear devices detonated in the atmosphere. In the final analysis, the uses to which atomic energy is put, just like the applications of any discovery, are the social responsibility of all humanity and not just of politicians and other decision makers.

*Cosmic rays*  As we mentioned earlier, primary cosmic rays consist mainly of protons, alpha particles, and a few other atomic nuclei which enter the earth's atmosphere from sources in outer space. Indeed, most of the secondary cosmic rays which result from the interactions of primary cosmic rays with the nuclei of atoms in the upper atmosphere are also particulate. These particulate radiations include mesons, electrons, and positrons. However, a small portion of the secondary cosmic radiation impinging on the earth's crust consists of electromagnetic photons emitted during the decay of mesons, the interaction of mesons with the nuclei of other atoms, or the braking action of high-speed electrons as they approach the nuclei of target atoms. In the latter case, the energy losses during braking are converted into x-ray emissions. Therefore, electrons which are constituents of secondary cosmic ray showers give rise to x-radiations in a manner identical to the general x-radiations produced by ionizing electrons in an x-ray tube.

The electromagnetic components of secondary cosmic rays are part of the natural background radiation to which everything living is exposed. For this reason, their biologic effects, if any, have not been adequately assessed. Of course, x-rays cause ionization by secondary means and indirectly induce secondary ionizations. These reactions must also take place in tissues irradiated by cosmic x-rays. Presumably the energy content of cosmic rays is low enough and tissue regeneration processes are great enough to prevent demonstrable radiation symptoms from developing.

On the other hand, most cosmic radiations which reach the earth's surface are so highly energetic that they traverse tissues without being slowed down or stopped. Therefore little if any damage usually results from their passage. This concept is readily understood when we recognize that the more rapidly an ionizing radiation is slowed down, the denser the medium through which it is passing and the greater the number of ion pairs produced per unit of its travel path. Thus, the most rapid dissipation of energy occurs near the end of the travel path because of the increased probability of interaction of the radiation with target atoms. The high specific ionizations of quickly decelerated radiations are then a consequence of their rapid dissipation of energy and the correspondingly shortened paths which they describe through matter.

## UNITS OF RADIATION MEASUREMENT

Because so many ionizing particulate and electromagnetic radiations exist, it is essential to express their potencies in such a manner that comparisons can readily be made between them. Clearly, the physical nature and energy content of an ionizing radiation are not the only important variables in its interactions with matter. The medium through which the radiation travels and the density of the absorber are other determinants of its ultimate effects. Yet we have no way of directly measuring the actual effects of an individual ionizing particle or wave on any of the atoms with which it interacts. All we can hope to do is express the average effect of a given type of ionizing radiation with a specified energy content traveling through or penetrating a particular medium, or absorber. All these qualifications are required because what we actually do in practice is observe the consequences of the interactions of a radiation with matter.

Basically, three features associated with ionizing radiation can be expressed with accuracy: (1) the average number of ion pairs produced throughout the total length of the travel path of an incident ionizing radiation in a given medium or, to be more precise, the average amount of energy transferred to matter in the same length of travel, since some portion of the total ionization energy is expended in causing the ejection of an electron and another portion is converted into its kinetic energy; (2) the average energy content of an incident radiation used as an index of the degree of its penetration through matter of a given kind; and (3) the degree of biologic damage associated with the exposure of a given tissue to measured amounts of an ionizing radiation. Our goal in the health sciences is to express with some degree of confidence the amount of damage to be expected in a given tissue following irradiation by a specified dosage of ionizing energy.

The *roentgen* (r) was the first unit to be used quantitatively to express the energy of ionizing radiations. One roentgen is defined as the energy content of x-rays or gamma rays required for the formation of $2.083 \times 10^9$ (2,083,000,000) ion pairs per cubic centimeter of air at standard temperature and pressure (0°C and 760 mmHg). Expressed on an energy basis, this value is 83 ergs/Gm. One joule (see Chap. 8) is equal to 10 million ($10^7$) ergs.

Notice three points about the definition of a roentgen: (1) A roentgen may be expressed in terms of the number of ion pairs produced or its average energy content; (2) without statistical modifications, the roentgen is useful for expressing the energy content of x-rays and gamma rays only; and (3) the amount of energy in 1 roentgen (1r) is valid only when measurements are made in air. The last two points impose major restrictions on the use of the roentgen as an all-encompassing unit of ionizing radiation.

To circumvent these limitations on the use of the roentgen unit, another unit known as the *roentgen equivalent physical* (rep) has been formulated. A rep is defined as the energy of particulate radiations absorbed by soft tissues or water equivalent to 1 r in air. Since the density of an absorber affects the specific ionization of an incident particulate radiation, soft tissues actually absorb 10 ergs/Gm more energy on an average than air irradiated under identical conditions. In other words, 1 r, or the energy absorption in air of 83 ergs/Gm, is 93 ergs/Gm when absorbed by soft tissues or water. Notice that the rep applies only to particulate radiations absorbed by soft tissues or water. Electromagnetic radiations cannot be measured with the rep. Another drawback is that the values of the rep obtained from the absorption of 1 r of energy are highly erroneous in dense tissues such as bone. These biologic limitations to the use of the rep are similar to those associated with the roentgen.

Clearly, another type of radiation unit, one applicable to all types of radiation and to all kinds of absorbers, including average tissues, is necessary. These requirements are met by a unit of ionizing radiation known as the *radiation absorbed dose* (rad). A rad is defined as the energy equivalent to one roentgen when any ionizing radiation is absorbed by any medium. Since average tissues (in contrast to soft tissues) will absorb 17 ergs/Gm more energy when exposed to ionizing radiation equivalent to 1 r in air, the energy content of 1 rad of any radiation in average tissues is equal to 100 ergs/Gm. Notice that a rad removes the restrictions imposed by the roentgen and the rep, in that the energy content of any ionizing radiation in any medium, including average tissues, can be compared to the effects of one roentgen of x-radiation or gamma radiation in air. However, the values are always in terms of average tissues and do not differentiate between variations in tissue density. A greater density shortens the path of an incident radiation and increases the rapidity with which it dissipates its energy. Thus, average values have a built-in error which increases tremendously as the density value of the tissue in question deviates in either direction from the average. The rad therefore requires further refinement before it can be applied with precision in expressing the absorbed dose of an ionizing radiation in a particular tissue.

The degree of precision necessary is accorded to the rad (and to the rep as well) with the use of appropriate conversion factors. These conversion factors provide a mechanism for taking into consideration the density of the irradiated tissue as well as the physical characteristics of the incident radiation. Thus, such corrections enable comparisons of the *relative biologic effectiveness* (RBE) of an ionizing radiation with that of a standard such as x-rays or gamma rays. In practice, the rad or the rep is multiplied by the RBE of the incident radiation or by any other appropriate conversion factor which attempts to correct for the relative inflexibility of the rad and rep units. The derived product, or value, is therefore indicative of the average energy of any ionizing radiation absorbed by a given tissue, compared with the energy of 1 r in air. Applicable to tissues only, this derived unit is referred to as a *roentgen equivalent man* (rem). In other words, a rem is equal to either rad · RBE or rep · RBE. For a given energy of ionizing radiation, bones possess a significantly higher rem than do other kinds of tissues. This concept is important when assessing radiation damage, since bone marrow and epiphyseal cartilaginous disks are two important structures profoundly affected by moderate or high doses of whole-body radiation.

The relationships between a roentgen, rep, rad, and rem are illustrated in Fig. 9-3.

The roentgen, rep, rad, and rem are all related to the number of ion pairs produced in a given medium by an ionizing radiation, converted into expressions of their average ionization energies or the average amount of energy transferred to matter due to their interactions. By way of contrast, another unit of radiation measurement called the *curie* (Ci) has nothing directly to do with the energy content of an ionizing radiation. A curie is a measure of the number of unstable nuclei of a radioactive element disintegrating each second. By convention, 1 Ci is set equal to $3.7 \times 10^{10}$ (37 billion) disintegrations per second (dps). Bear in mind that among the emissions of spontaneously disintegrating atoms are alpha particles, beta particles, and gamma rays. The ionizing energies of these emissions are obviously not the same. The only characteristic established by the unit curie is the number of nuclear dps in the atoms of a radioactive element. The curie is an extraordinarily large unit of nuclear energy. In medical practice, a unit 1,000 times smaller (1/1,000 Ci) is used. This unit is called a millicurie (mCi). One millicurie is equal to 37 million dps. In many research applications, a unit 1,000 times less than 1 mCi (1/1,000 mCi) or 1 million times less than a Ci (1/1,000,000 Ci) is preferred. Called a *microcurie* ($\mu$Ci), this unit is equal to 37,000 dps. The relationships between a curie, a millicurie, and a microcurie are shown in Fig. 9-3.

## THE BIOLOGIC EFFECTS OF IONIZING RADIATIONS

The damage inflicted by ionizing radiations in human beings may affect all levels of biologic organization from molecules to multiple organ systems. The available evidence indicates that *chromosomes* are the primary sites of damage when living cells are irradiated. The effects of the primary damage are compounded by alterations in the structures and functions of other cellular constituents.

Although enzymes can be changed structurally to the point of their inactivation if the energy of an incident radiation is high enough, the amount of energy required (up to 1 million or more rads) is rarely encountered outside experimental laboratories. Therefore, a change in enzymatic structure or a decrease in the concentration of particular enzymes in living cells after exposure to ionizing radiation does not necessarily imply a cause-and-effect relationship between them. In other words, the incident radiation probably affects another target molecule which in turn is reflected by an enzymatic change.

### INTRACELLULAR STRUCTURES AND FUNCTIONS

To understand what ionizing radiations can do biologically, we must concentrate on their effects at the

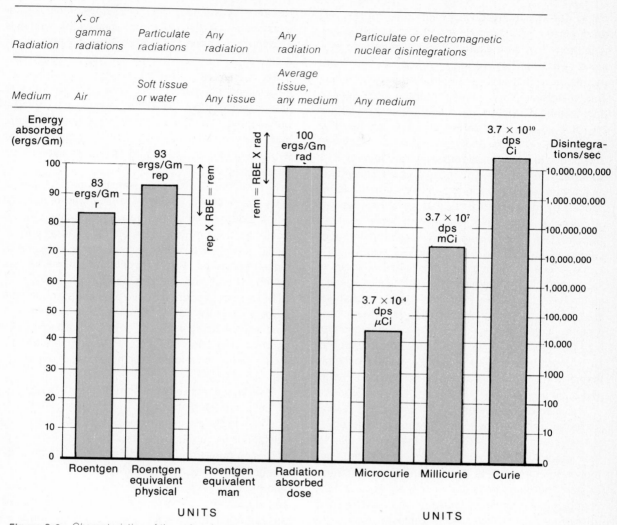

**Figure 9-3**  *Characteristics of the units of measurement of ionizing radiation. The left side of the graph compares the energy in ergs per gram absorbed by specified media when exposed to 1 r, rep, rad, and rem of particular types of radiation. The right side of the graph compares the number of nuclear disintegrations per second in a curie (Ci), millicurie (mCi), and microcurie (μCi).*

cellular level. As a result of the interactions of ionizing radiations with the atoms of cellular molecules, damage may be noted in three major types of structures: membranes, cytoplasmic organelles, and nuclei.

**Membranes and cytoplasmic organelles**  Ionizing radiations damage numerous types of cellular membranes or membrane systems of intracellular organelles including the plasma membrane, the lamellae of

the endoplasmic reticulum, and membranes associated with lysosomes and mitochondria. Membrane damage impairs the functions of these organelles and indirectly affects cellular integrity.

Moderate doses of ionizing radiation may rupture the plasma membrane as well as the membrane enclosing the enzymes contained within lysosomes. Rupture is an effect of the interaction of the ionization energy of a radiation on atoms of membrane molecules. The consequences of membrane rupture are determined by the nature of the structure of which the membrane is a part. When a plasma membrane ruptures, whatever its cause, extracellular and intracellular tissue fluids are permitted to intermingle. This disturbance in cellular integrity is a serious homeostatic disruption and may result in the death of the affected cell. Rupture of lysosomes, on the other hand, releases their enclosed hydrolytic enzymes. This action results in a degradation of cellular constituents such as proteins and nucleic acids. This mechanism normally helps in the elimination of debris caused by the presence of dead cells. However, its operation in living cells is generally lethal to them. Thus, rupture of the plasma or lysosome membranes ultimately leads to the same effect—cellular death.

Low doses of ionizing radiation cause changes in the permeability of the plasma membrane to specific electrolytes such as $Na^+$, $K^+$, and $Ca^{++}$. Such changes are not as drastic as those resulting from membrane rupture, although their physiologic consequences may be extensive. Since changes in the balance of electrolytes may result in changes of fluid balance, they give rise to disturbances in osmotic equilibrium and cellular irritability.

The membrane system of mitochondria is also profoundly affected by ionizing radiation. The physical effects include alterations in the permeability of the outer membrane of each irradiated mitochondrion as well as a disorganization of internal cristae. The direct physiologic consequence of these changes is a dramatic depression in the synthesis of adenosine triphosphate (ATP). Because ATP is the primary energy currency of cells, a decrease in the production of these molecules leads to a reduction in the overall rate of cellular metabolism.

In addition to these membrane disturbances, the lamellae of the endoplasmic reticulum may exhibit thickenings in response to incident ionizing radiation. The increased width and other structural alterations which develop may affect the movement of molecules between the cytoplasm and the cavity of the reticulum. These changes also contribute to alterations in cellular metabolism and may cause secondary effects in other initially unaffected cells.

**Nuclei** The irradiation of cells delays or inhibits mitosis or cellular division and depresses the synthesis of DNA (deoxyribonucleic acid). Probably the latter is a consequence of the former, although a cause-and-effect relationship is not clear-cut. In either case all cells, whether mitotic or intermitotic, show an immediate or subsequent delay in mitotic activity. The precise response is in part determined by the nature and energy content of the incident radiation as well as the type of cell irradiated and its mitotic stage at the time of irradiation.

In addition to a decrease in the manufacture of DNA, the synthesis of mRNA (messenger ribonucleic acid) is depressed in the nuclei of irradiated cells. A decrease in the mRNA content is reflected by a diminution in the size of the nuclear storage depots of these molecules. Such repositories of mRNA are called nucleoli and are discussed in Chap. 5.

The effects described above presumably arise from a decrease in the ability of nuclear DNA to serve as a template for the synthesis of mRNA, following the absorption of ionizing radiation. However, here too the evidence is presumptive and therefore not unequivocal.

The major effects of ionizing radiations in the nuclei of cells are chromosomal. The damage is due to changes in the chemical structure of DNA molecules (mutations—see Chap. 6) through alterations in, or losses of, purine and pyrimidine bases or to chromosomal aberrations through various types of fractures and cross linking. Both DNA changes and chromosomal aberrations alter or eliminate entirely a certain amount of molecular information ordinarily delivered to the cytoplasmic ribosomes by mRNA. The cytoplasm is therefore deprived of normal information necessary for the synthesis of molecules vitally involved in metabolism and perpetuation. The resulting metabolic dysfunctions set the stage for the establishment of vicious circles which culminate in the death of the irradiated cells and the induction of degenerative changes in other tissues and organs of the body.

Only an extremely small energy content of ionizing radiation is required for the induction of mutations. Indeed, there seems to be no minimum energy content of ionizing radiation below which mutations will not occur! Because a small component of natural background radiation comes from cosmic rays, even this radiation source contributes to the natural frequencies of mutations. These mutation frequencies have been established by genetic methods for various genes.

## SHORT-TERM RESPONSES

Overexposure to ionizing radiation may either follow a brief period of irradiation (acute exposure) or develop after receiving many increments of small doses of radiation over an extended period of time (chronic exposure). When the effects appear within the first 30 to 90 days after irradiation, the host response is said to be *short-term,* even if the response is ultimately fatal. A short-term response does not negate the possibility that various aspects of the symptomatology associated with sublethal radiation exposure may persist for months or even years. The expression "short-term" is used simply to distinguish certain symptoms which develop immediately or within a reasonably short time following irradiation from those which may appear only after a lapse of many months or years following the termination of radiation exposure.

The biologic effects of whole-body irradiation in man depend in part on the energy content of the incident radiation. The symptomatology of acute radiation sickness may be due to the indirect inactivation of enzyme systems, neural damage, intestinal lesions, or the destruction of the lymphoid tissues and red bone marrow. The latter tissues are of course responsible for the manufacture of the formed elements of blood.

Radiation deaths caused by the inactivation of enzymes require extremely high exposure levels; the *late effects* of radiation, discussed later in this chapter, are induced by relatively low doses. Other pathologic conditions mentioned above require exposure to ionizing radiations with energies between these two extremes. Thus, the nature of the radiation-induced symptoms is related in part to the energy content of the absorbed radiation.

Certain types of cells, tissues, and organs are little damaged by moderate or high doses of ionizing radiation; others are vulnerable to low doses. The former are referred to as *radioresistant* structures; the latter are termed *radiosensitive* structures.

With reference to the vulnerability of individual tissues and organs to ionizing radiation, we may state that (1) The nervous system and sensory structures are radioresistant; (2) the digestive tract and, to a greater extent, blood cell–forming tissues and organs are radiosensitive; (3) mitotically active tissues such as epiphyseal cartilaginous disks of long bones and the germinal cells in the testes and ovaries are extremely radiosensitive; and (4) the radiosensitivity of other cells is related directly to their lack of differentiation in that unspecialized cells will not tolerate as high a dose of ionizing radiation as specialized cells.

We are now in a position to review the more detailed effects of ionizing radiation on organ systems and body defense mechanisms. We shall emphasize the vicious circles which amplify the initial effects due to the ionization of atoms attached to cellular molecules. When we consider that the ionization of only 1 in every 10 million atoms, on an average, will prove lethal to the human body, the amplification effect produced at least in part by the establishment of vicious circles must be considerable.

Whole-body exposure to low or moderate doses of ionizing radiation may lead to some combination of the following symptoms:

1   A decrease in nonspecific and specific resistance and a corresponding increase in susceptibility to infection
2   Hemorrhaging and possible anemia
3   Stagnant hypoxia, or a depression in the amount of oxygen supplied to the tissues, due to a reduction in blood flow
4   A loss of both fluids and electrolytes
5   Anorexia, or loss of appetite
6   Sex cell and somatic, or body, cell mutations
7   Possible changes in behavior

A number of these symptoms lead to ischemia (tissue anemia) and necrosis (the death of tissues) either as a direct consequence of radiation damage or as a secondary effect of other changes at the cellular level. A combination of these symptoms increases the probability of secondary damage to more radioresistant tissues and organs of the body, thereby promoting the continuation of a vicious circle.

The vicious circle associated with radiation-induced damage is summarized in Fig. 9-4. The pathophysiology of these symptoms will now be discussed.

**Hemopoietic tissues and blood**  Among the most sensitive indicators of radiation damage are the concentrations of circulating lymphocytes and thrombocytes (blood platelets). Following exposure to ionizing radiation, both lymphocyte and thrombocyte concentrations are significantly depressed, conditions referred to as *lymphocytopenia* and *thrombocytopenia,* respectively. These effects are caused by damage to the lymphoid tissues and red bone marrow rather than by direct damage to circulating lymphocytes and thrombocytes. The concentrations of circulating granulocytes such as eosinophils, basophils, and neutrophils are similarly depressed, a condition called *granulocytopenia*. When the radiation dose is high, there is also a depression in the number of circulating erythrocytes, a condition known as *erythropenia*.

Lymphocytopenia and granulocytopenia depress

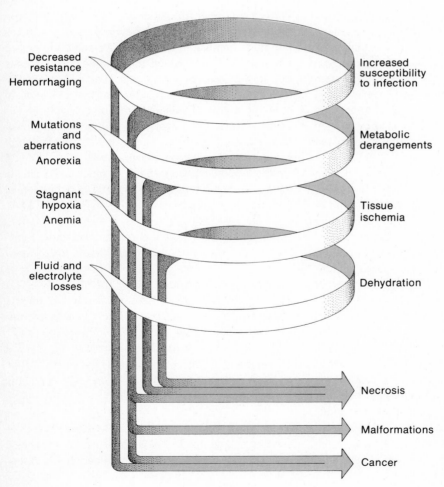

**Figure 9-4**  *The vicious circle of degenerative changes caused by radiation injury.*

Decreased resistance
Hemorrhaging

Increased susceptibility to infection

Mutations and aberrations
Anorexia

Metabolic derangements

Stagnant hypoxia
Anemia

Tissue ischemia

Fluid and electrolyte losses

Dehydration

Necrosis

Malformations

Cancer

the phagocytic abilities of the nonspecific cellular defense mechanism of the body, especially during inflammatory reactions. In addition, destruction of lymphoid tissues, including those of the thymus gland, directly or indirectly depresses the blood concentration of circulating antibodies, a condition known as *hypogammaglobulinemia.* Antibody induction, in which leukocytic stem cells become committed to the production of specific antibodies, is depressed when antigens are introduced after radiation exposure; antibody synthesis is delayed without affecting the actual induction process when antigenic entry occurs prior to radiation exposure. In either case, specific immunologic responses are impaired, the primary response being more radiosensitive than the secondary response. The inflammatory response and antigen-antibody interactions are reviewed in Chap. 4.

Thrombocytopenia contributes to hemorrhaging by prolonging the times required for intravascular clotting and clot retraction. Multiple hemorrhaging in the peripheral blood vessels results in purpura, or discoloration. The mechanism of clotting is discussed in Chap. 4; purpura is explained in Chap. 31.

Erythropenia is ordinarily not seen except in patients who have been exposed to high levels of ionizing radiation. Erythropenia leads to anemia and arterial hypoxia, i.e., a depression in the oxygen tension of arterial blood. Erythropenia is discussed in Chap. 31.

Hemopoiesis (the manufacture of erythrocytes) is a relatively radioresistant process. Apparently only erythroblasts (embryonic erythrocytes) are radiosensitive. In addition, part of the low red blood corpuscle (RBC) count results from the leakage of erythrocytes from the blood to the extracellular tissue spaces in response to a radiation-induced increase in capillary permeability and fragility. In any event, circulating erythrocytes are ordinarily not significantly affected by overexposure to ionizing radiation.

Skin  Most ionizing radiations to which human beings are exposed must penetrate the skin first. Only a small amount of radiation originates within the body after being inspired or ingested as contaminants of air, water, or food or after being administered for medical reasons. Therefore, the skin is a useful indicator of the degree of radiation exposure. The most obvious integumentary symptom following irradiation is a reddening called *erythema.* This symptom appears when the energy of the incident radiation is in excess of 30 rads (radiation absorbed dose). Since skin reddening in this instance is caused by ionizing radiation, the symptom is known as *radiation erythema.* Radiation erythema is due to dilatation of peripheral capillaries following the development of stagnant hypoxia.

A decrease in the blood supply to localized tissues is referred to as ischemia, as mentioned previously. Necrosis is a consequence of chronic ischemia. In an attempt to offset this vicious circle, the body responds by inducing a compensatory dilatation of blood vessels within the irradiated skin. This response increases the peripheral flow of blood and gives rise to the erythemic condition.

In addition to erythema, several other pathologic conditions characterize the response of skin to ionizing radiation: a sloughing off of skin known as *desquamation;* the loss of hair, a condition called *epilation,* and a lightening of the pigmentation of replacement hair; and an impairment of the process of tissue regeneration. Desquamation and epilation are reflections of the radiosensitivity of mitotically active tissues such as the stratum germinativum and papillae at the bases of hair follicles. These tissues are differentially affected by ionizing radiations. The stratum corneum and the nonliving portions of hair are radioresistant.

A lightening in the pigmentation of replacement hair and a disorganization of subcutaneous connective tissue following irradiation stems from the relative radiosensitivity of undifferentiated integumentary cells, including melanoblasts (embryonic pigment cells) and parenchyma (connective tissue cells).

An inability to heal flesh wounds follows from the effects of desquamation and connective tissue disorganization. In addition, a depression in the concentration of basophils, part of the granulocytopenia mentioned earlier, leads to a decrease in the number of fibroblasts which can be mobilized following skin damage. Under these circumstances, the available fibroblasts cannot secrete enough collagen to meet the demands for the synthesis of scar tissue to replace dead and dying tissue. Wound regeneration is therefore impaired. Wound regeneration is explained in detail in Chap. 4.

**Blood vessels** Aside from dilatation of peripheral capillaries, blood vessels respond in several other ways to irradiation. Radiation-induced damage to blood vessels is reflected by an increase in capillary permeability and the formation of occlusions, or blockages, in both arterioles and capillaries.

The permeability changes permit a leakage of blood serum, erythrocytes, and other formed elements through the walls of the capillaries to the extracellular tissue spaces. The effects of this leakage are profound and include fluid losses from the blood and a corresponding hemoconcentration and interstitial edema, a decrease in the number of circulating erythrocytes, and a depression in the concentration of blood proteins and other plasma constituents. These changes disrupt osmotic equilibria between the blood vascular and interstitial fluid compartments. Their effects are ultimately felt by tissue cells which must osmotically respond to imbalances of fluids and electrolytes.

Radiation-induced vascular occlusions (blockages) occur in a variety of ways. Basically, radiation occlusions form following the destruction or alteration of perivascular or intravascular cells. More specifically, such occlusions form as a result of (1) intravascular clotting subsequent to the destruction of radiosensitive endothelial cells, (2) the collapse of blood vessels due to interstitial pressures which create vascular constrictions, or (3) other factors which decrease the diameter of irradiated blood vessels. The occlusions produce localized patches of tissue in which the blood supply is decreased. When the ischemic condition persists, necrotic areas of degenerating tissues develop. The necrotic areas multiply and spread, ultimately enveloping the more radioresistant tissues of the body.

**Digestive system** Other major systemic effects of ionizing radiation appear in various portions of the digestive tract. Structural damage caused by radiation includes ulcerations in the mouth, stomach, and small intestines as well as the differential destruction of certain radiosensitive structures in the wall of the digestive tract. These structures include the chief cells of the gastric mucosa and the crypts of Lieberkühn in the duodenal wall. The functions of these structures are described in Chap. 26.

Radiation-induced changes in cellular permeability are responsible for decreasing the passive absorption of certain soluble end products of digestion. At the same time, automatic neural responses alter normal digestive reflexes. After whole-body exposure to moderate doses of ionizing radiation, there is an increase in the length of time food remains in the stomach as well as a decrease in the rate of segmental and mass movements of the intestines.

Accessory digestive structures such as the salivary glands, the liver, and the pancreas are relatively radioresistant. Nevertheless, high doses of radiation will cause a depression in the synthesis of various digestive enzymes which go into the makeup of pancreatic juice. In addition, overexposure to radiation causes the mobilization of fixed phagocytes within the liver. These specialized leukocytes neutralize foreign materials and cellular debris carried through the blood to the liver. In this instance, the elevated activity of the fixed phagocytes follows degenerative changes in the tissues of the small intestines. The contributions of fixed phagocytes to nonspecific body defense are discussed in Chap. 4.

The structural and functional changes described above have widespread ramifications. Ulcerative lesions are susceptible to microbial infection. The natural microbial flora and fauna within the intestines are also disturbed, giving rise to the production of both gas (flatus) and diarrhea. The fluid losses sustained because of diarrhea cause severe dehydration.

Radiation-induced modifications in digestive reflexes combine with alterations in chemical digestion to bring about a state of indigestion. The subjective feeling associated with indigestion contributes to a loss of appetite for food (anorexia).

**Skeletal system** Bone damage caused by radiation is imposed mainly on undifferentiated, rapidly growing, and mitotically active tissues such as epiphyseal cartilaginous disks. This form of damage is reflected by a stunting of growth. Other portions of bone are basically radioresistant. However, vascular damage sustained following high doses of radiation may involve bone marrow by decreasing the supply of nutrients and oxygen to, and the removal of waste products from, marrow cells. In this event, the marrow cells degenerate, resulting in one form of bone marrow disease. Abnormal bone growths or malformations may also develop following radiation injury. Presumably, disruptions occur in the normal patterns which govern the absorption of calcium from, and its deposition in, bone. Finally, the teeth may be affected by exposure of the face to ionizing radiations, causing transient lesions such as chalking. Skeletal disorders are considered further in Chap. 25.

**Reproductive system** Most portions of the male and female reproductive systems are radioresistant. However, the few components of each reproductive system which are radiosensitive make up in their effects for the relative lack of overall damage. The most vulnerable parts of the reproductive system are the formative sex cells. Such immature cells include spermatogonia, primary spermatocytes, and secondary spermatocytes in the seminiferous tubules of the testes as well as oogonia and primary oocytes in the follicles of the ovaries. Ionizing radiations, even in low doses, cause mutations and aberrations within these germinal cells. Such damage affects the genetic and chromosomal constitutions of the spermatozoa and ova. Mutations and aberrations are usually deleterious and give rise to one of the following consequences: (1) immediate lethality, abortion, or stillbirth by preventing conception, gestation, or prenatal development, or parturition (birth), respectively, or (2) hereditary diseases which are expressed either at birth or some time after birth. Hereditary diseases, irrespective of cause, are reviewed in Chap. 21.

Spermatogonia and oogonia are apparently more vulnerable to a given dose of ionizing radiation than primary spermatocytes and primary oocytes. In other words, cells in less advanced meiotic stages of gametic development are more radiosensitive than cells in later stages of development. Thus, a temporary period of sterility or at least a decrease in potential fecundity is an unavoidable consequence of gonadal exposure to ionizing radiation. In this instance, sterility results from the interactions of two physiologic mechanisms: Oogonia are inhibited from enlarging and initiating the first meiotic division for some time after irradiation, and, simultaneously, the number of mature sex cells is progressively depleted through attrition. These actions give rise to a gametic void which is not dispelled until maturational divisions gradually recommence. Meiosis is reviewed in Chap. 19. Other causes of sterility are discussed in Chap. 20.

In addition to the effects on formative reproductive cells described above, gonadal exposure to ionizing radiations damages cells responsible for the secretion of sex hormones. In the female, such cells are found in the stratum granulosum. These follicular cells are radiosensitive and consequently degenerate following irradiation. Since the granulosa cells secrete estrogen and progesterone, the resulting hormonal imbalances may produce various physiologic and behavioral abnormalities associated with the menstrual cycle and reproduction. To a lesser extent, damage similar to that sustained by granulosa cells is observed in Sertoli cells and interstitial cells of Leydig, two cell types found in the testes. Sertoli cells are nutritive, providing sustenance for the conversion of immature spermatids into mature spermatozoa. Leydig cells secrete testosterone, the primary male reproductive hormone. A radiation-induced degeneration of the Sertoli cells contributes to sterility by increasing the rate of depletion of spermatids. Destruction of interstitial cells causes a hyposecretion of testosterone.

**Nervous system and sensory structures** Neural components are among the most radioresistant structures in the body. The radioresistance of neurons and sensory receptors is a reflection of their high degree of cellular specialization. Structural alterations do not occur unless the dose of ionizing radiation is at least moderate.

*Cataract* is perhaps the best-known sensory abnormality due to radiation exposure. Opacities of the lenses of the eyes may gradually develop after exposing them to moderate doses of ionizing radiation. Lenticular opacities produced in this manner are termed *radiation cataracts*.

The penetration of the body by ionizing radiation

has been perceived immediately by certain persons. Such persons report pinpoint flashes of light within the eyes, even though they are closed. Called *phosphenes,* this phenomenon is consistent with the idea that stimulation of photoreceptive elements by inappropriate stimuli with sufficient energy may be detected under appropriate conditions. Other possible origins of phosphenes are considered in Chap. 12.

Nonvisual perception of ionizing radiation has also been reported. The strategy in such experiments is to train laboratory animals to avoid a particular location or a specific food associated with exposure to ionizing radiation. Since the animals are not aware of their radiation exposure, at least through their usual senses, avoidance must be indicative of the perception of radiation through other means. In addition, some aspects of conditioned-response learning may be altered by the absorption of ionizing radiation. Otherwise, however, learning and memory seem to be highly radioresistant.

**The general adaptation syndrome**  Because of the above effects, the body responds to radiation as it would to any nonspecific stress. The general adaptation syndrome (GAS) is triggered by the joint operation of the sympatheticoadrenal and pituitary-adrenal (hypophyseal-adrenal) axes. These mechanisms are explained in detail in Chap. 4.

## LATE EFFECTS

Nonlethal exposures to ionizing radiation shorten the natural life span, although the amount of shortening due to a given dose applied over a specified time is difficult to quantitate. Basically, a reduction in longevity due to irradiation may be caused by some combination of two interrelated events: (1) an acceleration of the aging process due to a progressive degeneration of tissue integrity and (2) the premature induction of age-specific diseases due in part to a lowering of the defense barriers of the body. A shortening of the natural life span is believed to result from the interaction of these two events. These ideas are expressed graphically in Fig. 9-5.

**Radiologic aging**  A radiation-induced acceleration of the aging process is in large part due to connective tissue and blood vascular alterations. Both parenchymal connective tissue cells and endothelial vascular cells are radiosensitive. Destruction of parenchymal cells results in their replacement by a more dense and extensive collagenous network known as *scar tissue*. The collagen molecules in scar tissue are extensively cross-bonded to each other. This feature is associated with the normal aging of collagen (see Chap. 4) and is in part responsible for a wrinkling in the overlying skin of older people. The increased thickness of scar tissue, in relation to its relatively unaltered blood supply, means that this replacement tissue will be chronically ischemic. In other words, a longer time will be required for the diffusion of oxygen from the capillaries to the scar tissue, compared with the diffusion rate of oxygen in the same location prior to radiation exposure. In addition, edema resulting from a radiation-induced increase in capillary permeability also decreases the efficiency of gaseous diffusion. The ischemic condition which develops is further compounded when the endothelial cells of nearby capillaries are destroyed. The cellular debris formed through this destruction as well as the subsequent appearance of intravascular clots plugs up affected capillaries. These effects cause stagnant hypoxia and thereby contribute to ischemia and necrosis.

Ionizing radiations also cause precocious involutional changes in the reproductive systems of both sexes. The absorption of high doses of radiation may be sufficient to cause permanent sterility, provided that the subject survives the overexposure. Following radiation-induced damage to cells of the stratum granulosum and interstitial cells of Leydig, a hyposecretion of sex hormones may cause other regressive changes in the reproductive system similar to those seen as a consequence of normal aging.

**Age-specific diseases**  Cancer and certain bone disorders (osteopathies) are among the age-specific diseases induced prematurely by ionizing radiation. That ionizing radiations are carcinogenic, i.e., cause cancer, is well documented. However, the mecha-

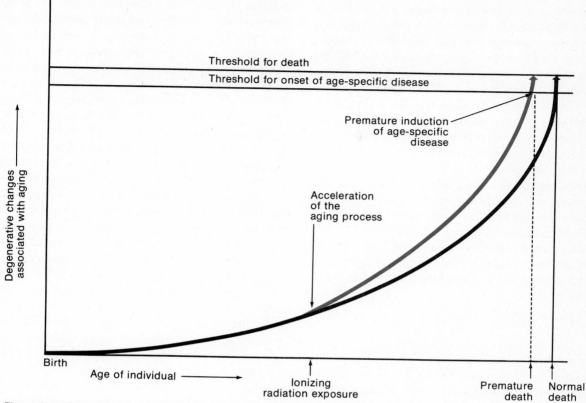

**Figure 9-5** *The probable way in which the natural life span is shortened by radiation exposure. Radiation causes an acceleration of the aging process and the premature induction of age-specific diseases, shifting the life-span curve to the left.*

nisms by which the energies of such radiations act on normal cells to cause their transformation into cancer cells are still in dispute. Cancer cells are not only structurally abnormal but seem to be released from the ordinary constraints imposed upon their neighboring but normal cells. Oblivious in a molecular sense to both their ordinary roles and the needs of the body, cancer cells undergo a series of rapid and uncontrolled mitotic divisions, usurping, or preempting, raw materials required for normal metabolism and perpetuation. The neoplasm increases in size until nearby normal tissues are smothered, pushed out of existence, or otherwise damaged.

The following theories have been proposed to explain the carcinogenic effects of ionizing radia-

tions: (1) the stimulation provided by the absorption of the energy of an ionizing radiation activates a genetically repressed carcinogenic virus already present in one or more irradiated cells; (2) a radiation-induced depression in body resistance and a corresponding increase in susceptibility to infection enable carcinogenic viruses outside the body to more easily attack tissue cells; (3) changes in the structure of DNA molecules in irradiated tissue cells give rise to somatic mutations which may be reflected by runaway cellular divisions of the mutated cells; or (4) radiation-induced chromosomal aberrations in somatic cells may lead to the same outcome. Although it is not yet possible to resolve which, if any, of these theories is actually correct, the activation or derepression of an already present but latent virus appears to be the most plausible explanation at the present time.

Numerous skeletal disorders found mainly in older people are also seen in greater-than-normal frequencies in irradiated subjects. Recall that the initial skeletal lesions caused by ionizing radiation are located in epiphyseal cartilaginous disks. When blood vessels within bone are damaged by ionizing radiation, the entire bone may gradually degenerate. However, this effect does not occur unless the radiation dose is high.

Thus, radiologic aging shortens life through a vicious circle of degenerative changes in irradiated tissues and promotes the early development of age-specific diseases. The precocious induction of diseases that are ordinarily age-specific feeds the fires that accelerate aging and may cause secondary complications which precipitate death prematurely.

## FACTORS WHICH ALTER RADIATION RESPONSES

The amount of damage sustained by the body following radiation exposure depends on numerous factors. Many of the determinants of the degree of damage have already been considered and consist basically of characteristics associated with both the ionizing radiation and the tissues absorbing the radiation. These variables include the kind of incident radiation, its total energy content, the time rate of application of the total dose, the density of the absorber tissues, their degrees of specialization, and the rate of mitotic activity. Many biologic, chemical, and physical factors which modify or influence the radiation response act through one or more of these mechanisms. Other factors exert their modifying influences through unrelated mechanisms. These modifying factors and their probable mechanisms of action are summarized in Fig. 9-6 and are discussed below.

Biologic factors These variables include the age, health, nutritional status, and hormonal balance of the irradiated subject as well as the specific portion and percentage of the body irradiated.

The very young, including embryos and fetuses, and the very old are radiosensitive, although for quite different reasons. Young people possess relatively undifferentiated tissues, compared with older persons. Therefore, radiation effects are seen earlier and are more pronounced in very young persons, compared with older age groups, when the radiation dose and its time course of application are held constant. At the other extreme, the radiosensitivity characteristic of older people is due in part to an inability to repair damaged tissues adequately and perhaps to a general lowering of body defenses against infection.

Animals in poor health are more radiosensitive than animals of the same age in excellent health. Apparently, good health promotes optimal metabolism. Optimal metabolism promotes the ability to maintain effective repair and defense mechanisms for a longer time than would otherwise be the case, despite the tissue damage and physiologic stress caused by exposure to ionizing radiation.

Nutritional status will also influence the biologic response to ionizing radiation. A diet lacking in protein, certain fats, and specific vitamins confers more potential radiosensitivity than one which is adequate in these nutrients. Apparently radiation-induced stress increases the demand for the synthesis of numerous molecules. These requirements can be met only with adequate supplies of essential amino acids, fatty acids, and vitamins. When the supply of these nutrients is depressed or depleted, the resulting metabolic dysfunctions lower defensive barriers of the body. The multiple roles of essential amino acids, fatty acids, and vitamins are described in Chap. 26.

The hormonal status of a subject may also modify the physiologic response to ionizing radiation. Depending on the concentration of particular hormones present in the blood, the response may be in the form of a decrease or an increase in radiosensitivity. In addition, males are somewhat more radiosensitive than females of the same age, when the radiation dose required for a particular effect is measured. The variation in radiosensitivity between the sexes is attributed to the different physiologic effects induced by androgens (testosterone) and estrogens. The actions of these hormones are reviewed in Chap. 19.

The portion of the body exposed to radiation or, conversely, the part shielded from radiation is an im-

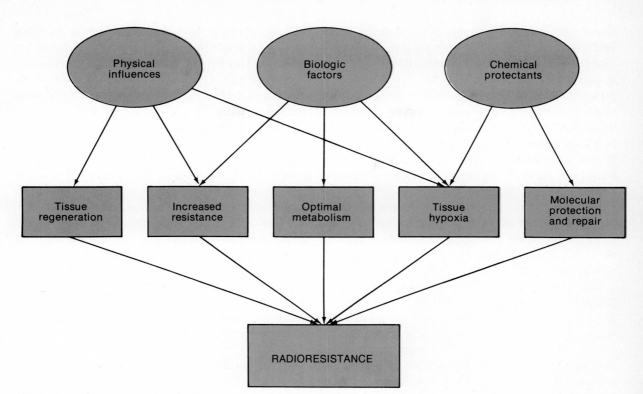

**Figure 9-6**   *Factors which contribute to radioresistance.*

portant determinant of the amount of damage which will subsequently appear. The protection provided by shielding certain portions of the body during radiation exposure is apparently related to an ability to maintain hemopoietic and immunogenic functions. Thus, shielding the thymus provides protection by sustaining the specific immunogenic functions of the body. In addition, shielding the spleen affords some radiologic protection through its ability to replenish the lymphoid tissues of the body with stem cells from which antibodies can be produced. Shielding the thighs during whole-body radiation exposure provides a small but measurable amount of radiologic protection because of the significant hemopoietic functions of the femurs (thigh bones).

**Physical influences**   The nature of the biologic response to ionizing radiation depends in part on a host of physical factors including the oxygen tension of the irradiated tissue cells, body temperature, time rate of application of the total dose, and efficiency with which energy is transferred by the ionizing radiation (the RBE—relative biologic effectiveness).

Tissue hypoxia which already prevails at the time of irradiation will provide some protection against the effects of ionizing radiation. Notice that this condition is different from the stagnant hypoxia which develops following the radiation-induced destruction of endothelial cells and the subsequent development of intravascular clots. How a depression in the oxygen concentration provides radiologic protection is not clear, but presumably the production of free radicals is reduced following irradiation during hypoxia. Recall that some portion of radiation damage is a consequence of the powerful oxidizing or reducing abilities of free radicals which form secondarily. Because of its high reactivity, available oxygen combines chemically with these free radicals. The result of this combination is the formation of other free radicals with powerful oxidizing abilities. Thus, a lack of oxygen during irradiation prevents the formation of oxidizing free radicals.

Hypoxia also plays a major role in the radiologic protection afforded by hypothermia (body temperatures lower than 94°F) and hyperthermia (body temperatures higher than 105°F). Hypothermia reduces the partial pressure of oxygen ($P_{O_2}$) in tissue cells, since the thermal effects which contribute to molecular agitation and, therefore, to gas pressure are reduced. Hyperthermia differentially reduces the oxygen supply to the spleen through direct vascular effects. The protection provided by splenic hypoxia enables this organ to repopulate the lymphoid tissues with leukocytic stem cells.

The rapidity with which the total radiation dose is applied to the body has an important bearing on the degree of damage sustained. As the exposure is made more rapidly, the damage produced by a given quantity of ionizing radiation becomes increasingly severe. The reason is to be found in the time required to repair damaged tissue. In acute radiation damage, the regenerative abilities of the body are swamped by the degree of tissue destruction. In chronic radiation damage, the rate of tissue repair keeps pace with, or is not too different from, the rate of cell death.

The RBE of a radiation is an expression of the efficiency with which it transfers energy to cells and tissues. This characteristic of ionizing radiation modifies its biologic effects because a specific tissue may increase or decrease the RBE. For example, a given volume of protoplasm may require only one hit, or strike, by an incident ionizing radiation before the appearance of structural or functional derangements. On the other hand, some types of protoplasm of identical volume may require multiple strikes before damage is detected. Consider a neutron which has a high efficiency of energy transfer, and, therefore, a high specific ionization. If a tissue irradiated by neutrons is damaged by relatively few strikes, then neutron radiation will cause considerable damage. In other words, neutron radiation is particularly destructive when the ionizations caused by neutrons are grouped close together and their travel paths are extremely short.

Conversely, when a tissue irradiated by neutrons requires multiple strikes before damage is visible, the effects of neutron radiation are much less severe.

Remember that multiple strikes are achieved only by ionizing radiations with a relatively low rate of energy transfer over the lengths of their travel paths. Since the distance of tissue penetration of such radiations is relatively great, the volume of tissue traversed is correspondingly large. The result is an increase in the probability of transferring energy at numerous but distant points of interaction with atoms along the lengths of the travel paths of the incident radiation.

Therefore, the degree of damage caused by any ionizing radiation is a function of the number of strikes a given volume of tissue must absorb before structural and functional changes are observed. In this context, RBE is itself a relative term. Neutron radiation, for example, has a high RBE in tissues which are damaged by only a few strikes but a low RBE in tissues damaged only by multiple strikes.

Chemical protectants    Numerous chemicals are capable of reducing the amount of radiation-induced damage, provided they are present in high concentration at the time of irradiation. However, these two requirements and the adverse side effects of the chemicals markedly impair their usefulness in a practical sense.

Chemical protectants seem to act by reducing the initial effects of the radiation in exposed tissues. Their mechanisms of protective action are still poorly understood, although several theories to account for their abilities have been suggested. Some chemical protectants are thought to act by somehow interfering with the production or action of free radicals formed as a secondary effect following exposure to ionizing radiation. Other chemical protectants may exert their beneficial effects by repairing some types of molecular damage caused by ionizing radiation. Molecular repair may require the replacement of an atom or an entire functional group or may involve the reversal of oxidation-reduction reactions initiated by the action of free radicals. In this instance, the production and action of free radicals is not directly affected. Finally, some chemicals may give radiologic protection by causing metabolic changes which confer radioresistance. Most frequently, such protectants act by inducing hypoxia, although interferences with the synthesis of DNA and corresponding delays in mitotic divisions or depressions in cellular metabolism may also result in protection.

## MEDICAL APPLICATIONS OF IONIZING RADIATIONS

Ionizing radiations are used in the diagnosis and treatment of various diseases. Most forms of radiodiagnosis and radiotherapy employ x-rays, although radioisotopes, especially gamma emitters, are also used extensively. The term *x-radiography* is restricted to the use of x-rays for these purposes. *Radioisotopic techniques* utilize particulate or electromagnetic radiations emitted from disintegrating nuclei of unstable or radioactive atoms of various elements.

Specific medical applications of ionizing radiations are summarized in Fig. 9-7. X-radiographic techniques are identified in the top half of the illustration. Radioisotopic techniques are given in the bottom half of the figure.

### RADIODIAGNOSIS

Ordinarily, the doses of ionizing radiation used in medical diagnoses are small enough to prevent short-term symptoms from developing. However, the role of ionizing radiations, even in low dosages, in shortening the natural life span and in inducing mutations and chromosomal aberrations emphasizes the caution and care which must be exercised when working with so-called harmless quantities of radiation in a clinical setting. In practice, this means good judgment, not timidity, when handling or using radiation emitters.

**X-radiography**  A photographic film penetrated by x-radiation is blackened. When an object such as a body part is placed between an x-ray tube and appropriate film and is then irradiated, the degrees of darkening of the film are related to the densities of the tissues through which the x-rays must pass. Soft tissues such as skin, connective tissue, and muscle permit most of the x-radiation to pass through without appreciable tissue absorption or scatter. Since most of the x-radiation passing through soft tissue reaches the film, those portions of the film in back of soft tissues will appear gray or dark following photographic development in a darkroom. Conversely, hard tissues such as bone absorb and scatter most of the incident x-radiation. Therefore, only a small portion of the energy of x-radiation penetrating bone actually reaches the film. Thus, those areas of the film in back of bone will appear light after chemical development of the film. The images produced in this manner are then interpreted by looking for abnormal shadows or peculiar shapes. When a fluorescent screen is used in place of x-ray film for the visualization of x-ray images, the technique is referred to as *fluoroscopy*. Both types of x-radiography are used routinely for the visualization of fractures, dental caries, skeletal abnormalities, and certain respiratory, digestive, and excretory pathologic conditions.

Other forms of x-radiography take advantage of the fact that the densities of certain structures can be changed by injecting air, pure oxygen, or special radiopaque substances into them. Such density changes enhance the contrast in certain neural, vascular, respiratory, excretory, and digestive structures and are therefore useful in identifying specific pathologic conditions. Among the x-ray techniques based on density changes in intraneural and perineural structures are intracranial pneumography, angiography, and myelography. Additional x-radiographic methods which enhance normal contrast are used in the diagnosis of diseases of the digestive tract, kidneys, and other portions of the excretory system.

In *intracranial pneumography,* cerebrospinal fluid is removed and replaced with air, oxygen, or an inert radiopaque chemical. Because air and oxygen are both less dense than cerebrospinal fluid, the shadow image of the subarachnoid spaces and brain ventricles will appear darker than normal. Since radiopaque materials are more dense than cerebrospinal fluid, the shadow images of intraneural and perineural

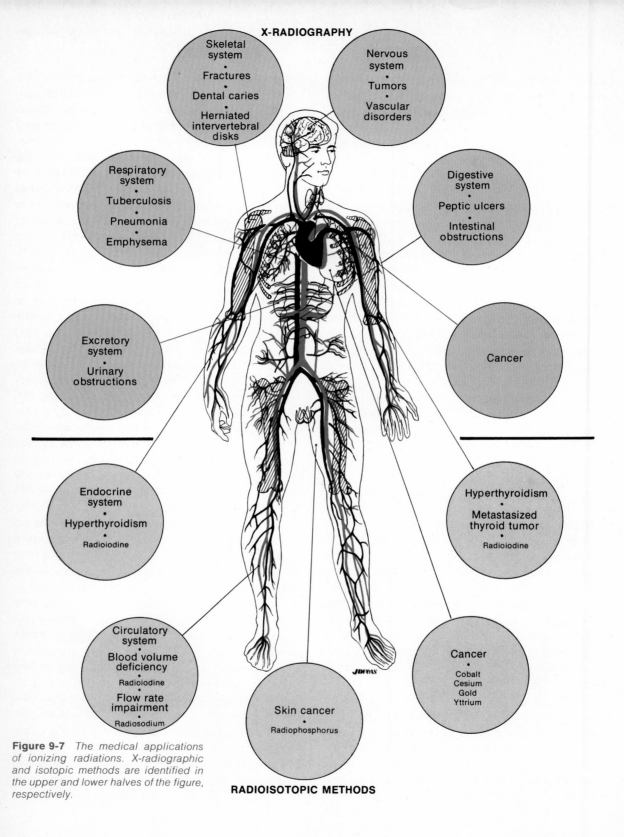

**X-RADIOGRAPHY**

Skeletal system
•
Fractures
•
Dental caries
•
Herniated intervertebral disks

Nervous system
•
Tumors
•
Vascular disorders

Respiratory system
•
Tuberculosis
•
Pneumonia
•
Emphysema

Digestive system
•
Peptic ulcers
•
Intestinal obstructions

Excretory system
•
Urinary obstructions

Cancer

Endocrine system
•
Hyperthyroidism
•
Radioiodine

Hyperthyroidism
•
Metastasized thyroid tumor
•
Radioiodine

Circulatory system
•
Blood volume deficiency
Radioiodine
•
Flow rate impairment
Radiosodium

Skin cancer
•
Radiophosphorus

Cancer
•
Cobalt
Cesium
Gold
Yttrium

**Figure 9-7** *The medical applications of ionizing radiations. X-radiographic and isotopic methods are identified in the upper and lower halves of the figure, respectively.*

**RADIOISOTOPIC METHODS**

252

structures injected with these substances will appear lighter than usual. The darkening or lightening enhances the contrast of these structures and accentuates the appearance in roentgenograms of pathologic conditions such as tumors and certain congenital brain disorders.

*Angiography* is a technique used for x-ray visualization of specific intracranial blood vessels following the introduction of a radiopaque substance. The opacity of the injected material lightens the x-ray image of the blood vessel in which damage is suspected, enabling the visualization of constrictions and other vascular abnormalities. Because certain tumors usually cause a decrease in the vascularization of the surrounding tissues, angiography is also useful in localizing certain intracranial tumors.

Barium is a radiopaque element used in the fluoroscopic visualization of certain disorders of the alimentary canal. To coat the inside wall of the esophagus, stomach, and small intestines, a barium compound is introduced orally; to coat the mucosa of the large intestine, barium is introduced by means of a special enema. Barium lightens the lining of the digestive tract in the fluoroscopic image produced by exposure to x-rays. This diagnostic procedure enhances the visualization of mucosal lesions, such as peptic ulcers in the stomach or duodenum, as well as obstructions in the intestines.

Barium and other similar radiopaque chemicals are also used for the x-ray visualization of the kidneys, renal pelvises, ureters, urinary bladder, and urethra. The opaque material is introduced either intravenously or via catheterization. Since the insides of the tubular structures of the excretory system are made opaque by this procedure, the urinary pathways will appear lighter than normal when visualized on film or with a fluoroscope. The enhancement of contrast is useful in the diagnosis of urinary obstructions and other excretory disorders.

**Radioisotopic methods**   Radioactive sources are also used diagnostically. However, the approaches employed in radiodiagnosis are different from those used in x-radiography. As we have already learned, x-radiography relies on the interpretations attached to shadow images, with or without enhancement of contrast, made on x-ray film or displayed on a fluoroscopic screen. Isotopic diagnosis, on the other hand, depends upon one of the following factors: (1) the differential uptake of a specific radioactive element by a particular tissue or organ or, alternatively, the differential distribution of a radioisotope introduced into the body; (2) the normal rate of metabolism and excretion of a radioactive element introduced into the body; or (3) the extreme sensitivity of the detecting and measuring equipment, enabling the external measurement of internal emitters in the millicurie and microcurie range of dps (disintegrations per second). Ordinarily, the individual disintegrations are converted electronically into an average number of counts per minute or into the specific number of counts in the unit of time during which measurements are made.

Numerous forms of medical diagnosis capitalize on these unique features of radioisotopes and the sensitivity with which they may be detected. Some examples follow.

Radiomercury, a gamma emitter, is sometimes introduced into intracranial blood vessels to enable their examination. The brain is scanned in both anteroposterior and superoinferior directions with appropriate detectors. A specific uptake of radiomercury within intracranial blood vessels is indicative of certain vascular abnormalities or tumors.

The differential uptake of a radioisotope is also the primary principle involved in the diagnosis of a hypersecretion of the thyroid gland known as *thyrotoxicosis*. In this type of diagnosis, radioiodine, an emitter of both beta particles and gamma ray photons, is administered orally. The radioiodine is incorporated into the natural hormone thyroxine in a manner identical to that of normal iodine. The thyroid gland is then scanned by one or more detectors at stated times after the start of the procedure. The counts per minute in each region scanned are integrated electronically and displayed in the form of an image of the distribution of radioiodine in the gland. The distribution pattern pinpoints the enlarged regions.

Other radiodiagnoses of thyrotoxicosis may be made in one of the following ways: (1) external detection over the thigh of protein-bound radioiodine in the blood, taking advantage of the fact that the blood concentration of thyroxine (natural) is elevated in thyrotoxicosis; (2) direct measurements of protein-bound radioiodine in blood samples, based on the same reasoning; and (3) the radioactivity of excreted urine over a given length of time following the start of the procedure, compared with the total amount of radioiodine administered. Fragments of thyroid tumor tissue which have metastasized, or spread, through the blood to other parts of the body may be localized with the use of radioiodine, provided that such abnormal tissues continue to differentially take up radioiodine. When these conditions are met, whole-body scanning will detect gamma emissions in localized regions containing the cancerous thyroid tissue. Therapy is then instituted in an effort to eliminate the metastasized tissue.

Measurements of total blood volume may also be done radioisotopically, when volume insufficiencies are suspected. The method depends upon the differential uptake by erythrocytes of radiophosphorus, a beta emitter. A known fraction of the total volume of radiophosphorus-labeled erythrocytes in a blood sample from the patient is reinjected. After a specified time has elapsed, another blood sample is taken from the patient, and the radioactivities in a given volume of the original blood sample and the new blood sample are counted electronically and compared. The total volume of blood may be calculated from the decrease in the radioactivity of the labeled sample after reinjection. Low blood volumes may be indicative of an excessive fluid loss.

Circulatory impairments in localized regions of the body may also be diagnosed by radioisotopic methods. The time required for radiosodium to travel from the point of injection to another prescribed point in the body is measured with the aid of detectors. The measured rate of blood flow between these two points is then compared with normal rates established for the region and distance in question. Depressions in flow rate are indicative of vascular abnormalities such as thrombosis, certain aneurysms, and varicosities. These and other vascular disorders are discussed in Chap. 31.

## RADIOTHERAPY

Whereas diagnostic doses of ionizing radiations are extremely small, therapeutic doses of the same radiations are usually massive. Since the object of radiotherapy is to destroy diseased tissues, it is not surprising to find that ionizing radiations are primarily used in the treatment of serious cancers which do not respond to chemotherapy or other forms of treatment. However, less serious skin cancers are also amenable to radiotherapy.

The fundamental problem which plagues this type of treatment is to hold the destruction of normal tissues at a level which does not provoke more serious symptoms than the outward manifestations of the cancer itself. In other words, since ionizing radiations destroy normal tissues as well as diseased tissues, the destruction of normal tissues must be kept to the minimum consistent with the continuation of the health and well-being possessed by the patient prior to treatment. As we shall see, this problem has been partially resolved by some rather unique methods. Nevertheless, the destruction of normal tissues by ionizing radiations imposes constraints on the efficacy of the technique in individual cases.

**X-ray therapy**   The treatment of cancers with x-rays has a fairly long medical history. Thousands of roentgens of radiation are ordinarily required for the destruction of neoplasms. The energy represented by this amount of radiation is lethal when absorbed by the whole body. However, only that part of the body in which the tumor is located is actually exposed to the radiation.

Although the x-rays are "focused" on the cancerous tissue by means of collimating slits, they must still pass through overlying but normal tissues. To reduce the damage to these tissues and at the same time maintain a collimated, or directed, stream of x-ray photons on the cancer, one of three techniques is employed: (1) The patient is rotated in a circle about a fixed axis, to distribute the ionization energy of the radiation over a larger area of normal tissue and thereby reduce the amount of radiation-induced damage at any one point in the overlying tissues; (2) the radiation source is rotated around the patient who remains in a fixed position; or (3) several radiation sources from different directions simultaneously irradiate the cancerous tissue. All these techniques are variations on the basic theme of preventing peripheral damage while delivering a concentrated dose to the diseased tissue beneath the skin.

In practice, the total radiation dose required to destroy a tumor is divided into increments or fractionated on the basis of the radiation tolerance of the individual patient. Thus, *fractionation* provides a schedule for the dose rate and total number of exposures.

Isotopic therapy  For deep-seated tumors, gamma emissions from cobalt 60 are preferred to x-rays, although the radioactive elements radium and cesium have also been used. Because gamma ray photons can be more energetic than x-ray photons, the use of high-energy gamma rays means that less ionization energy will be expended in the overlying, normal tissues and more energy per unit of dose will be transferred to the cancerous tissue. In virtually all other respects, however, cobalt therapy is similar to x-ray therapy.

In contrast to radiation exposure through a distance with the use of external emitters, other forms of radiotherapy employ skin applicators or internally implanted or introduced needles, capsules, or colloidal suspensions containing measured amounts of radioactive substances. Such devices include applicators impregnated with radiophosphorus, needles of radium (a gamma emitter), seeds of radon (an alpha emitter), and colloidal suspensions of both radioactive gold (a gamma emitter) and yttrium (a beta emitter). The use of these materials may be more advantageous than irradiation through a distance for the following reasons: (1) Because needles and seeds can be implanted next to, or within, cancerous growths, only millicurie amounts of radiation are required for their destruction; (2) colloidal suspensions of radioactive elements can be uniformly distributed within a benign or encapsulated tumor; and (3) these procedures are relatively uncomplicated and inexpensive. Thus where feasible to use, their efficiency, safety, and economy make these devices clinically attractive.

Finally, beta emitters such as radiophosphorus may be simply administered *in solution* to patients with metastasizing thyroid tumors. Assuming that the cancerous fragments still take up radioiodine, this element will differentially accumulate in the diseased tissues in far-flung parts of the body. Provided that the beta emissions of the radioiodine are intense enough, the cancerous tissues will be differentially destroyed by the localized accumulations of radiation.

## COLLATERAL READING

Andrews, Howard L., *Radiation Biophysics,* Englewood Cliffs, N.J.: Prentice-Hall, Inc., 1961. A textbook that presents a working knowledge of radiation fundamentals at an elementary level.

Benarde, Melvin A., *Our Precious Habitat,* New York: W. W. Norton & Company, Inc., 1970. Chapter 14, entitled "Ionizing Radiation," presents a lucid discussion of the hazards and benefits of ionizing energy sources.

Burbidge, Geoffrey, "The Origins of Cosmic Rays," *Scientific American,* August 1966. Based on mathematical calculations, Burbidge suggests that cosmic rays originate from radio sources outside our galaxy.

Casarett, Alison, *Radiation Biology,* Englewood Cliffs, N.J.: Prentice-Hall, Inc., 1968. This elementary and lucid book on radiobiology covers most of the field, in addition to an adequate amount of radiophysics and radiochemistry.

Cook, Earl, "Ionizing Radiation," in W. W. Murdoch (ed.), *Environment: Resources, Pollution and Society,* Stamford, Conn.: Sinauer Associates, Inc., 1971. Concluding that any buildup of long-lived isotopes in the environment will be almost irreversible and damaging, Cook criticizes present federal radiation standards.

Deering, R. A. "Ultraviolet Radiation and Nucleic Acid," *Scientific American,* December 1962. The cellular damage produced by ultraviolet radiation is traced to alterations in the pyrimidine bases thymine and cytosine of the DNA molecules in the chromosomes within the nucleus.

Hafen, Brent Q. (ed.), *Man—Health and Environment,* Minneapolis: Burgess Publishing Company, 1972. One essay—"Never Do Harm," by Karl Z. Morgan—is especially pertinent. Morgan compares risks versus benefits resulting from the use of various kinds of radiation equipment.

Hammond, Allen L., "Cancer Radiation Therapy: Potential for High Energy Particles," *Science,* 175:1230–1232, 1972. A review of the evidence suggesting that x-radiation and gamma radiation therapy for cancer may be advantageously replaced by irradiation with heavy particles such as neutrons, protons, and alpha particles.

Hollaender, Alexander, and George E. Stapleton, "Ionizing Radiation and the Cell," *Scientific American,* September 1959. The authors emphasize the distinction between the primary physical damage caused by radiation and the secondary biologic effects resulting from the molecular interactions of free radicals formed during ionization.

Puck, Theodore, "Radiation and the Human Cell," *Scientific American,* April 1960. The primary site of damage is located in the chromosomes within the nucleus.

Putnam, J. L., *Isotopes,* Baltimore: Penguin Books, Inc., 1960. This capsule paperback review emphasizes the uses of isotopes.

Wallace, Bruce, *Essays in Social Biology,* vol. II, *Genetics, Evolution, Race, Radiation Biology,* Englewood Cliffs, N.J.: Prentice-Hall, Inc., 1972. Section four, "Radiation Biology," is a succinct summary of some of the social implications of atomic radiation.

# CHAPTER 10
# WHAT makes YOU TICK

The ceaseless chemical activity in every cell and fluid of the body sustains metabolism and perpetuation. The two mechanisms by which biochemical reactions accomplish these objectives are *information transfer* and *energy flow*. Information transfer is necessary for the control of biologic activities and is presided over by various molecules such as DNAs (deoxyribonucleic acids) and RNAs (ribonucleic acids) in the nucleus and cytoplasm as well as by hormones produced in the ductless, or endocrine, glands of the body. Energy flow is required for the maintenance of cellular metabolism, upon which perpetuation depends, and is regulated primarily by enzymes and ATP (adenosine triphosphate).

All the biochemical reactions in your body result from an interaction between information transfer and energy flow. Whether you are well or ill is determined by the nature of the information transferred and the quantity of energy processed by available molecules. Without molecules, life could not exist.

But to deal with biologically important molecules and biochemical reactions in a meaningful way requires a background of basic chemistry. This chapter is intended to provide that necessary background.

## THE PERIODIC TABLE OF THE ELEMENTS

The periodic table is a tool. The functions of this tool are predictive. Among the predictions which the periodic table will enable us to make are the physical characteristics and chemical properties of individual elements, the proportions in which atoms combine chemically with each other, and the nature of the chemical bond between any two united atoms. Like all tools, the usefulness of the periodic table will depend upon its versatility, in this case as a predictive device, and the skill with which it is employed.

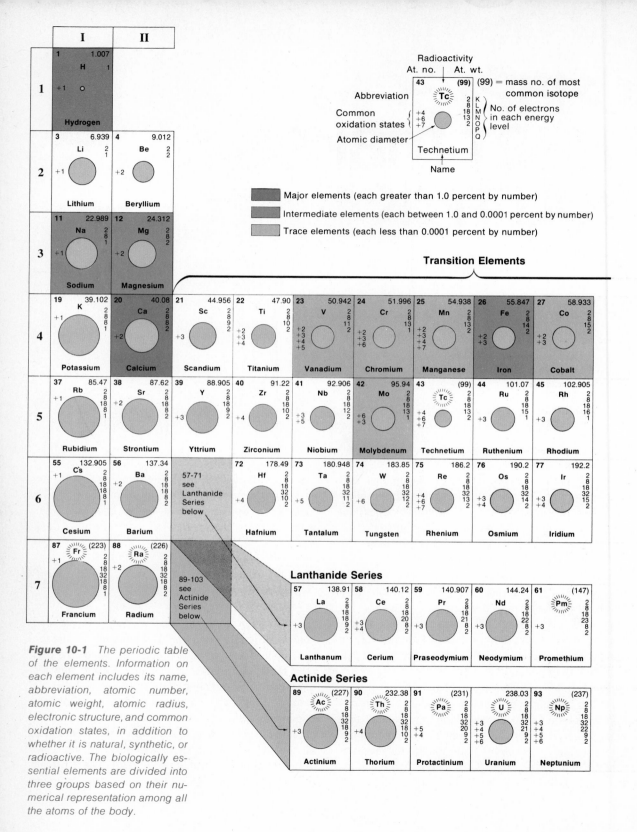

**Figure 10-1** *The periodic table of the elements. Information on each element includes its name, abbreviation, atomic number, atomic weight, atomic radius, electronic structure, and common oxidation states, in addition to whether it is natural, synthetic, or radioactive. The biologically essential elements are divided into three groups based on their numerical representation among all the atoms of the body.*

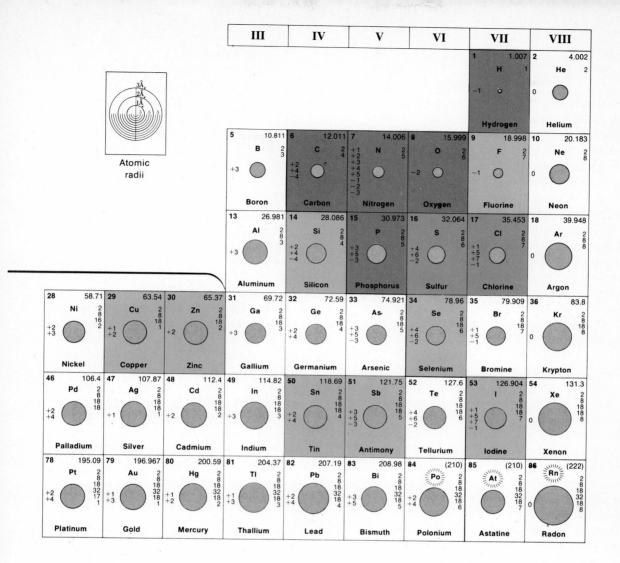

| | III | IV | V | VI | VII | VIII |
|---|---|---|---|---|---|---|
| | | | | | **1** 1.007 **H** 1 −1 o Hydrogen | **2** 4.002 **He** 2 0 Helium |
| | **5** 10.811 **B** 2 3 +3 Boron | **6** 12.011 **C** 2 4 +2 +4 −4 Carbon | **7** 14.006 **N** 2 5 +1 +2 +3 +4 +5 −1 −2 −3 Nitrogen | **8** 15.999 **O** 2 6 −2 Oxygen | **9** 18.998 **F** 2 7 −1 Fluorine | **10** 20.183 **Ne** 2 8 0 Neon |
| | **13** 26.981 **Al** 2 8 3 +3 Aluminum | **14** 28.086 **Si** 2 8 4 +2 +4 −4 Silicon | **15** 30.973 **P** 2 8 5 +3 +5 −3 Phosphorus | **16** 32.064 **S** 2 8 6 +4 +6 −2 Sulfur | **17** 35.453 **Cl** 2 8 7 +1 +5 +7 −1 Chlorine | **18** 39.948 **Ar** 2 8 8 0 Argon |

Atomic radii

| **28** 58.71 **Ni** 2 8 16 2 +2 +3 Nickel | **29** 63.54 **Cu** 2 8 18 1 +1 +2 Copper | **30** 65.37 **Zn** 2 8 18 2 +2 Zinc | **31** 69.72 **Ga** 2 8 18 3 +3 Gallium | **32** 72.59 **Ge** 2 8 18 4 +2 +4 Germanium | **33** 74.921 **As** 2 8 18 5 +3 +5 −3 Arsenic | **34** 78.96 **Se** 2 8 18 6 +4 +6 −2 Selenium | **35** 79.909 **Br** 2 8 18 7 +1 +5 −1 Bromine | **36** 83.8 **Kr** 2 8 18 8 0 Krypton |
|---|---|---|---|---|---|---|---|---|
| **46** 106.4 **Pd** 2 8 18 18 +2 +4 Palladium | **47** 107.87 **Ag** 2 8 18 18 1 +1 Silver | **48** 112.4 **Cd** 2 8 18 18 2 +2 Cadmium | **49** 114.82 **In** 2 8 18 18 3 +3 Indium | **50** 118.69 **Sn** 2 8 18 18 4 +2 +4 −3 Tin | **51** 121.75 **Sb** 2 8 18 18 5 +3 +5 −3 Antimony | **52** 127.6 **Te** 2 8 18 18 6 +4 +6 −2 Tellurium | **53** 126.904 **I** 2 8 18 18 7 +1 +5 +7 −1 Iodine | **54** 131.3 **Xe** 2 8 18 18 8 0 Xenon |
| **78** 195.09 **Pt** 2 8 18 32 17 1 +2 +4 Platinum | **79** 196.967 **Au** 2 8 18 32 18 1 +1 +3 Gold | **80** 200.59 **Hg** 2 8 18 32 18 2 +1 +2 Mercury | **81** 204.37 **Tl** 2 8 18 32 18 3 +1 +3 Thallium | **82** 207.19 **Pb** 2 8 18 32 18 4 +2 +4 Lead | **83** 208.98 **Bi** 2 8 18 32 18 5 +3 +5 Bismuth | **84** (210) **Po** 2 8 18 32 18 6 +2 +4 Polonium | **85** (210) **At** 2 8 18 32 18 7 0 Astatine | **86** (222) **Rn** 2 8 18 32 18 8 0 Radon |

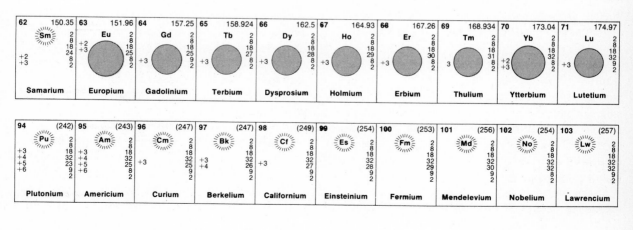

| **62** 150.35 **Sm** 2 8 18 24 8 2 +2 +3 Samarium | **63** 151.96 **Eu** 2 8 18 25 8 2 +2 +3 Europium | **64** 157.25 **Gd** 2 8 18 25 9 2 +3 Gadolinium | **65** 158.924 **Tb** 2 8 18 27 8 2 +3 Terbium | **66** 162.5 **Dy** 2 8 18 28 8 2 +3 Dysprosium | **67** 164.93 **Ho** 2 8 18 29 8 2 +3 Holmium | **68** 167.26 **Er** 2 8 18 30 8 2 +3 Erbium | **69** 168.934 **Tm** 2 8 18 31 8 2 3 Thulium | **70** 173.04 **Yb** 2 8 18 32 8 2 +2 +3 Ytterbium | **71** 174.97 **Lu** 2 8 18 32 9 2 +3 Lutetium |
|---|---|---|---|---|---|---|---|---|---|
| **94** (242) **Pu** 2 8 18 32 23 9 2 +3 +4 +5 +6 Plutonium | **95** (243) **Am** 2 8 18 32 25 8 2 +3 +4 +5 +6 Americium | **96** (247) **Cm** 2 8 18 32 25 9 2 +3 Curium | **97** (247) **Bk** 2 8 18 32 26 9 2 +3 +4 Berkelium | **98** (249) **Cf** 2 8 18 32 27 9 2 +3 Californium | **99** (254) **Es** 2 8 18 32 28 9 2 Einsteinium | **100** (253) **Fm** 2 8 18 32 29 9 2 Fermium | **101** (256) **Md** 2 8 18 32 30 9 2 Mendelevium | **102** (254) **No** 2 8 18 32 32 9 2 Nobelium | **103** (257) **Lw** 2 8 18 32 32 9 2 Lawrencium |

259

## ORGANIZATION

The 103 elements in the periodic table are grouped together on the basis of systematic and therefore fairly predictable differences in their physical characteristics and chemical behavior. Such differences, in turn, are dependent on the atomic number (at. no.), i.e., the number of protons in the nucleus. Since the atomic number of an element is the same as the number of electrons outside its nucleus, similarities and differences in physical and chemical properties among the elements can be established by grouping them together by the number of energy levels and the number of electrons in the outermost energy level. Periodic chemical similarities which identify family relationships and progressive differences in reactivity among diverse elements become apparent when they are arranged in this manner. Hence the name *periodic table.*

Figure 10-1 shows the periodic table of the elements. We will attempt to understand some of its basic features and to appreciate its utility, without trying to unravel the specific quantitative differences in properties among family members or the precise rules which describe the orbital positions of the electrons in each element. The table is displayed so that we might extract certain unifying principles that will enable us to predict the proportions in which elements will combine with each other chemically.

The elements in the periodic table are arranged in vertical groups and horizontal periods. The elements are ordered serially from left to right and from top to bottom by ascending atomic numbers. Atomic numbers are in the upper right-hand corner of each element.

**Groups** There are eight major *groups*, or families, numbered I to VIII. Between groups II and III is a large series of groupings collectively referred to as the *transition elements.* Transition elements are more or less related to each other in physical and chemical properties but are generally dissimilar to the elements of the major groups.

All the elements in a major group have the same number of electrons in their outermost energy levels and possess more or less similar chemical properties. Thus the number of electrons in the outermost shell of an element is identical to the number of the group to which it belongs. For example, all members of group I (at. nos. 1, 3, 11, 19, 37, 55, and 87) have one electron in their outermost shell. At the other extreme, the family of elements known as group VIII (at. nos. 2, 10, 18, 36, 54, and 86) have eight electrons in the outermost energy level. Helium (at. no. 2) is the only exception to this rule, since chemical stability is satisfied in the K energy level by two electrons.

In Fig. 10-1, the number of electrons in each of the energy levels is given by the vertically arranged numbers on the right-hand side of every element, the uppermost number representing the K energy level. The lowest number in a vertical grouping is the number of electrons in the most peripheral shell.

Compared with the major groups, systematic differences in the electronic structure among the transition elements are subtle, although they do exist. If you scan the electronic structures of the transition elements from 21 to 30, 39 to 48, or 72 to 80, you will note that most of the electrons are added to the next to the outermost shell. Only occasionally does the addition of electrons occur in the outermost energy level. This feature of transition elements confers a variability on their chemical behavior not possessed by elements of the major groups.

**Periods** There are seven horizontal periods. The period designations indicate the number of energy levels possessed by the elements which belong to them. For example, period 1 contains elements possessing only a K shell; period 2 embraces elements with two energy levels, K and L; etc. Since the atomic number and, therefore, the total number of orbital electrons increase in a stepwise fashion from element to element in the progression from left to right in any one period, there is a wide variation in the chemical properties of elements in the same period.

Elements are grouped in seven periods, as follows:

*1* Period 1 contains only two elements (at. nos. 1 and 2) because the maximum electron capacity of the K energy level is two. Hydrogen (at. no. 1) is listed in both groups I and VII because its structure fits the former group while its properties fit the latter group. Helium (at. no. 2) has the maximum number of electrons associated with the K energy level and therefore closes the first period.

2  Periods 2 and 3, with two and three energy levels, respectively, contain eight elements apiece. Period 2 holds elements with at. nos. 3 to 10. Neon (at. no. 10) has the maximum number of electrons contained in the L energy level and therefore closes the period. Elements with at. nos. 11 to 18 belong to period 3. Because argon (at. no. 18) contains a stable octet of electrons in its outermost (M) energy level, this element closes the third period.

3  Periods 4 and 5, with four and five energy levels, respectively, contain 18 elements apiece. Period 4 houses elements with at. nos. 19 to 36. Krypton (at. no. 36) closes the fourth period because it contains eight electrons in its outermost (N) energy level. Elements with at. nos. 37 to 54 are located in period 5. This period is closed by xenon (at. no. 54), because this element possesses a stable octet of electrons in its outermost (O) energy level.

4  Periods 6 and 7, with six and seven energy levels, respectively, contain 32 elements apiece. Period 6 contains elements with at. nos. 55 to 86. Elements of period 6 with at. nos. 57 to 71 are removed from their serial order by ascending atomic numbers and placed in a special column called the *lanthanide series* below the main body of the table. The lanthanide series is named for the element with which it begins, lanthanum. The elements in the lanthanide series resemble each other in various physical and chemical properties but are dissimilar in most respects to the transition elements to which they are related by atomic number. Radon (at. no. 86) closes period 6 because it contains a stable grouping of eight electrons in its outermost (P) energy level. Period 7 holds elements with at. nos. 87 to 103. Elements of period 7 with at. nos. 89 to 103 are placed in a special column called the *actinide series* for reasons similar to those which required the displacement of elements in period 6 belonging to the lanthanide series. Located below

the lanthanide series, the actinide series is named for actinium, the first element of the grouping. Observe that most of the elements in both series add electrons to the third from the outermost energy level. Also note that all the elements from radium (at. no. 88) through lawrencium (at. no. 103) have the maximum number of electrons (two) permitted in the Q energy level.

## ELEMENTS OF BIOLOGIC IMPORTANCE

Only 25 elements are thought to be essential to life. These biologically essential elements are identified in the periodic table (Fig. 10-1). Before we examine the essential elements more closely, let us review the reasons why the other elements have not become integral components of the body.

**The natural selection of vital elements**  We can only speculate about why nature has not found vital roles in life processes for most of the elements. The reasons, if we can state them as such, are varied. Some of the elements listed in Fig. 10-1, including all but two in the actinide series, are not naturally occurring, having been produced only in the laboratory. The synthetic elements are identified by parentheses around their atomic numbers. Numerous elements, a number of which overlap the synthetic group, are radioactive and create cellular damage within the body. Radioactive elements are identified by radiating lines surrounding their chemical abbreviations. The elements of group VIII, called *inert gases*, are highly stable by virtue of the octet of electrons in the outermost energy level of each. Their lack of reactivity at ordinary temperatures and pressures prevents them from chemically interacting with other atoms in the body. Certain heavy metals including platinum, gold, mercury, and lead are toxic to cells through their inhibition of enzyme activity. Other undesirable physical characteristics or chemical properties preclude essential biologic roles for many other elements. Nevertheless, there is no apparent reason why certain elements lack biologic functions, since most authorities agree they have the proper "credentials." Perhaps this is more an expression of our own ignorance than of reality, in that increasingly sophisticated tests in the future may yet reveal biologic roles for them.

**Vital elements and the periodic table**  Now, let us examine some of the relationships which exist among the biologically essential elements. In a nutritional sense, the elements vital to life are called *minerals*. If the minerals bear one or more electric charges by virtue of being in ionic form, they are also referred to as *electrolytes*.

The following generalizations about the essential elements are derived from an inspection of the periodic table in Fig. 10-1.

1  The essential elements are scattered among the first five periods but are completely excluded from groups III and VIII. Iodine (at. no. 53) is the heaviest essential element known to man. Four-fifths of the essential elements have atomic numbers below 31, a point which emphasizes the importance of the lighter elements to life.
2  Only six elements, hydrogen (H), carbon (C), nitrogen (N), oxygen (O), phosphorus (P), and calcium (Ca), account for most of the atoms in the body. In a biologic context, these are *major elements*. Each represents more than 1 percent of the total number of atoms in the body.
3  Another six elements, sodium (Na), magnesium (Mg), sulfur (S), chlorine (Cl), potassium (K), and iron (Fe), account for practically all the remaining atoms in the body. We may call them *intermediate elements*, in that each represents between 1 and 1/10,000 (one ten-thousandth, or $10^{-4}$) percent of the total number.
4  The remaining 13 essential elements combined account for only slightly more than 0.5 percent of the total number of atoms in the body. Since each of these elements constitutes less than $10^{-4}$ percent of the total number, they are collectively referred to as *trace elements*.

**The abundance of vital elements in the body**  Based on sheer bulk, the human body is composed primarily of four elements—hydrogen, carbon, nitrogen, and oxygen. The relative quantity of each of these elements, compared with the total number of atoms in, or total weight of, the body, is given in Fig. 10-2. The following generalizations are derived from a study of this figure:

1  The rank order by total number is hydrogen (63 percent), oxygen (about 25 percent), carbon (about 10 percent), and nitrogen (less than 2 percent). Thus, over three-fifths of the atoms of the body consist of hydrogen. The next most frequent atom, oxygen, represents a little over one-fifth of the atoms in the body.
2  The rank order by total weight is oxygen (65 percent), carbon (18 percent), hydrogen (10 percent), and nitrogen (3 percent). Therefore, almost two-thirds of the weight of the body consists of oxygen, even though hydrogen is the most numerous atom. The reason, of course, is to be found in the difference between their atomic weights. The atomic weight of oxygen is almost 16, whereas that of hydrogen is slightly over 1. Since each oxygen atom has about 16 times the mass of a hydrogen atom, in addition to having the greatest mass of the four major elements listed above, it is not surprising that 25 percent of the total number comes out to 65 percent of the total weight. Conversely, although hydrogen comprises almost 65 percent of the atoms of the body by number, the element only represents 10 percent of the body weight.

**The importance of vital elements**  The elements hydrogen, carbon, nitrogen, oxygen, phosphorus, sulfur, and calcium either are among the primary substituent atoms of biologically important molecules or bone or have metabolic roles in energy liberation. Ionized atoms including $Na^+$, $Mg^{++}$, $Cl^-$, $K^+$, and $Ca^{++}$ maintain the fluid and electrolyte balance between the various fluid compartments of the body, serve as activators for specific biochemical reactions, or sustain bioelectrical phenomena responsible for nerve impulse conduction and muscle contraction. The trace elements may serve variously as important components of enzymes, structural proteins, or other proteins with specific physiologic functions such as the transport or storage of oxygen.

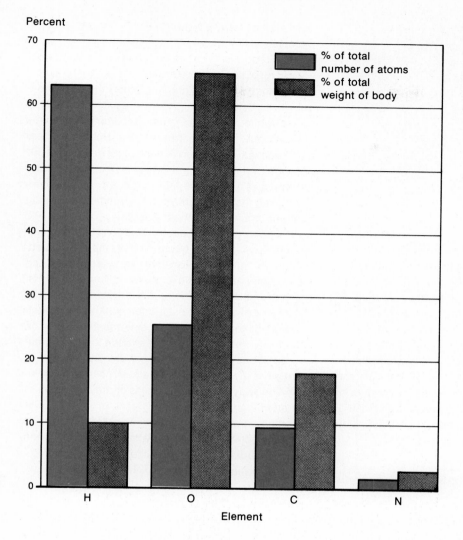

Percent

**Figure 10-2** *A comparison of the relative quantities of the four most common elements in the human body. The estimated percentages are based on both the total number of atoms in, and the total mass of, the body.*

Legend:
- % of total number of atoms
- % of total weight of body

(x-axis: Element — H, O, C, N)

## CHEMICAL INTERACTIONS BETWEEN ATOMS

A chemical interaction between two or more atoms occurs when their electron clouds overlap slightly. This phenomenon is illustrated conceptually in Fig. 10-3. Only the outermost portion of each electron cloud is actually involved in the interaction. The region of electronic overlap establishes electromagnetic forces of attraction between the atoms. Such atoms tend to remain together. Once the atoms are bound to each other, there is less of a tendency for

each to interact with other atoms, indicating that they are more stable or less reactive together than apart. The new unit is called a *molecule* when electrically neutral, i.e., when no net charge is present, or is referred to as an *ion*, or stable atomic grouping, when electrically charged. In either case, the new unit does not have properties which resemble those of its substituent atoms.

The above description, actually the essence of chemistry, raises numerous questions. As examples, consider the following: How do atoms get close

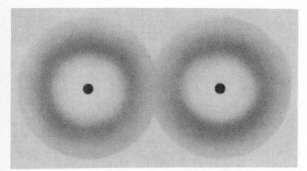

**Figure 10-3**   *The formation of a molecule by a chemical interaction between two atoms. The overlapping electron clouds create electromagnetic attractive forces called chemical bonds between the atoms.*

enough to each other to interact chemically? What is the nature of the electromagnetic forces of attraction which hold atoms together? What happens when atoms get so close together that there is a large degree of overlap between their electron clouds? Why are atoms more stable in molecules or ionic aggregations of atoms than they are individually? What factors determine whether, and in what proportions, atoms will combine chemically? Why are the physical characteristics and chemical properties of molecules or atomic groupings different from those of their substituent atoms? Let us attempt to obtain some insights into the answers to these questions.

## THE INITIATION OF CHEMICAL REACTIONS

Atoms may interact chemically when they bump into each other. Any factors which increase the probability of atomic collisions should therefore promote chemical reactions. Thus, the solution to the problem of how to get two or more atoms, each slightly larger than about 1/10,000,000,000 (one ten-billionth, or $10^{-10}$) meter, in contact with each other revolves around mechanisms which increase the probability of either collision or close physical proximity.

The second question regarding the nature of the attractive forces between atoms is more complex and will be answered in stages. Atoms are attracted to each other when one or more electrons are transferred from one atom to another or when one or more electrons from each interacting atom are shared either equally or unequally between them. Only electrons in the outermost energy level are involved.

When one or more orbital electrons in the outermost energy level of one atom are transferred to the outermost energy level of another atom, the first atom becomes positively charged as a result of its electron deficit while the other atom becomes negatively charged due to its electron excess. Both atoms are electrostatically attracted to each other because of the opposite charges which they bear. Once pairing occurs, the positive and negative charges are balanced. Thus, each molecule formed by a union of oppositely charged atoms is electrically neutral.

When one or more pairs of electrons in the outermost energy levels of interacting atoms are shared, the atoms are held together by electromagnetic forces generated by the opposite spins and orbital paths of the shared electrons. Depending upon the magnitude of the electrostatic force with which each interacting atom holds onto its outermost electrons or, conversely, the attractive force exerted by the nuclear charge of one atom on outermost electrons of one or more other atoms, the shared electrons may be positioned midway between the atoms or may be closer to one atom than to the others.

Actually, equal sharing and complete transfer of electrons between interacting atoms are two extremes of a continuum which reflects the magnitude of completing nuclear attractions for their outermost electrons. In other words, the principle behind electronic interactions in the outermost energy levels of atoms during chemical reactions is the same for all elements. The only difference is in the degree to which the electrons are shifted to one or the other of the interacting atoms. Atoms capable of interacting chemically always react with each other in a predictable manner, so that the degree of electron sharing or transfer can be specified precisely. We will return to this concept shortly.

## BALANCE BETWEEN NUCLEAR REPULSION AND ELECTRONIC ATTRACTION

When atoms get so close that more than their outermost energy levels overlap, what happens can be explained simply enough, although the theoretical foundations upon which the answer is based are more difficult to grasp and will not be considered here. Basically, at some point while the nuclei of two atoms approach each other, the electrostatic forces of repulsion between their positively charged nuclear protons will begin to exceed the electromagnetic forces of attraction created by the electronic interactions between their outermost energy levels. In other words, as atoms come together, they are first mutually attracted to each other by their electronic interactions and then are mutually repelled by nuclear forces if their nuclei approach too closely.

Therefore, individual reactants will react chemically only when provided with just enough kinetic energy to permit electronic transfer or sharing between the outermost energy levels while preventing nuclear interactions between protons. Since not all atoms in a population of atoms possess the same amount of kinetic energy, we can only deal with the probability of chemical interaction as evidenced by their rate of reaction. Increasing the average kinetic energy in a population of reactants increases the number of atoms which have the requisite amount of energy to interact chemically.

## THE ATTAINMENT OF CHEMICAL STABILITY

The stability of molecules and atomic aggregations is based on a reduction in the tendency of united atoms to transfer their outermost electrons to, or share them with, other nearby atoms. For atoms of elements which contain electrons only in the K energy level, i.e., hydrogen and helium, stability is achieved when this energy level is filled to its maximum capacity of two electrons. Reference to the periodic table (Fig. 10-1) shows that helium, with two orbital electrons, fulfills this specification. We would expect, and indeed find, that this element is chemically inert. Hydrogen, on the other hand, is highly reactive because it contains only one electron, rather than two, in the K energy level. In most reactions in which hydrogen participates, it gives up its solitary electron to another atom, thereby becoming a positively charged proton or hydrogen ion. In those instances where hydrogen shares its

electron with one from another atom during chemical combination, it becomes as inert as an atom of helium is ordinarily.

Atoms of elements with electrons in more than one energy level, i.e., those in periods 2 to 7 of the periodic table, exhibit maximum chemical stability when they have eight electrons in their outermost energy levels. When their electrons are shared unequally rather than transferred outright from one atom to another, they approximate this condition as closely as possible. In any event, no element may contain more than eight electrons in its outermost energy level, since this arrangement represents the lowest energy state of an element.

Inspection of the periodic table reveals that five of the elements in group VIII, neon, argon, krypton, xenon, and radon, already possess eight electrons in their outermost energy levels. Called inert gases as mentioned earlier, these elements have little tendency to interact chemically either with any of the other elements or among themselves.

## CHEMICAL BONDING MECHANISMS

Whether elements gain, lose, or share electrons during chemical combination revolves around how tenaciously their atoms hold on to their outermost electrons or, conversely, how strongly they attract the outermost electrons of other nearby atoms. Called *ionization energy*, the forces required to remove one of the outermost orbital electrons from an atom is an index of its electronic tenacity. The ease with which an atom gains electrons is a measure of its electronic attractance called *electronegativity*.

**The effect of the nuclear charge**   Both the ionization energy potential and the electronegativity of an atom are related to the effective nuclear charge which electrostatically holds electrons in or attracts them to its outermost energy level. The *effective nuclear charge* of an atom is exactly the same as the number of electrons in its outermost shell. This relationship is due to the partial screening of the nuclear charge from electrons in the outer energy level by orbital electrons belonging to the inner energy levels. The magnitude of the screening effect is identical to the number of electrons in all but the outermost energy level. The outermost electrons as well as electrons from other atoms are exposed to the unscreened protons.

During chemical interactions, atoms with low ionization energy potentials and electronegativities tend to lose electrons from their outermost energy levels. Atoms with high ionization energy potentials and correspondingly large electronegativities tend to gain electrons from the outermost energy levels of other atoms. Atoms between these two extremes ordinarily share electrons.

**The effect of the atomic radius** The atomic radius influences the strength of the effective nuclear charge, i.e., the *binding power* with which an atom holds electrons in its outermost energy level. The magnitude of the nuclear attraction for any orbital electron is inversely proportional to the square of the distance between them. The relationship can be expressed as follows:

$$F\text{(force)} \propto \frac{1}{D} \text{ (distance)}_2 \quad \text{or} \quad F \propto \frac{1}{D^2} \quad (10\text{-}1)$$

For example, the nuclear force exerted on the outermost electrons by elements in the same major group will decrease by fourfold and ninefold, respectively, for a doubling and tripling of the atomic radius.

In practice, the strength of the nuclear charge on the outermost electrons is the resultant of two opposing influences: (1) an increase in the radius of atoms as the number of energy levels increases and (2) a decrease in the distance between energy levels as their number increases. The first factor decreases while the second factor increases the effective nuclear charge on the outermost electrons.

The increasing compactness of atoms with six or seven energy levels, compared with those with one or two shells, is attributed to their greater electronic densities and correspondingly larger magnitudes of nuclear charge. Therefore, it does not necessarily follow that atoms with high atomic numbers are larger than atoms with low atomic numbers. This fact can be easily confirmed by comparing the relative sizes of the elements in the periodic table (Fig. 10-1).

**The predictive value of the periodic table** As we have already learned, the number of electrons in the outermost energy level increases from left to right across the periodic table, i.e., from group I to group VIII. At the same time, the atomic radius gradually decreases in each period from group I to group VII and then markedly increases in group VIII. Notice also that the atomic radii of elements in the same group progressively increase from period 1 to period 7. It follows that group I elements in periods 2 to 7 are slightly larger in diameter than the group VIII elements which end the preceding periods.

Now we are in a position to consider a few simple rules which predict whether atoms of the elements in the periodic table will donate, accept, or share electrons during chemical combination and in what proportions they will combine.

1 During chemical combination, elements in groups I to III tend to lose their outermost electrons to strongly electronegative elements. The tendency to lose electrons decreases from group I to group VIII and from period 7 to period 1. The number of electrons lost by these elements is equal to the number of electrons in their outermost energy levels, leaving them with a new outer electronic configuration identical to the inert gas which closes the period preceding them.

2 Elements in groups V to VII tend to gain electrons from elements with low ionization energy potentials. The tendency to gain electrons increases from group V to group VII and from period 6 to period 2. The number of electrons gained by these elements brings their outermost electron number to eight, so that the electronic configuration is identical to the inert gas which closes the period to which each belongs.

3 All the elements in group IV, many of the elements in groups III and V, elements close to each other in the periodic table generally, and atoms of the same element tend to share, rather than transfer, electrons during chemical combination. Whether electrons are shared equally or unequally by interacting atoms depends on the difference in their electronegativities. In any event, each of the united

atoms approximates a stable octet of outermost electrons. Identical atoms share their outermost electrons equally because there is no difference in electronegativity between them. The greater the disparity in the electronegativities of united atoms, the more unequally their outermost electrons are shared. Electrons shared unequally between atoms spend more time in orbit around, and therefore are closer to, the more electronegative atom.

To illustrate the correspondence between the number of electrons lost by atoms in groups I and II during chemical interaction and the number of electrons in their outermost energy levels, consider sodium (Na) and calcium (Ca) as representative examples. Na belongs to group I and contains one outermost electron. When Na combines chemically, it donates its outermost electron to a strongly electronegative atom. Using dots to represent the electrons in the outermost energy level, we can express the chemical change in Na as follows:

$$\cdot Na - e^- \rightarrow Na^+ \tag{10-2}$$

In other words, when an atom of Na loses its outermost electron during chemical union with another atom, it becomes an ion bearing one positive charge. Because Na$^+$ does not possess electrons in its M, or outermost, energy level, no dots are shown around it. However, Na$^+$ is stable since the L energy level (which has now become the outermost shell) is filled to its maximum capacity of eight electrons. Observe that Na$^+$ bears two electrons in the K shell and eight electrons in the L shell. In this respect, Na$^+$ resembles Ne, the inert gas which ends period 2.

Ca, on the other hand, has two electrons in its outermost energy level and, of course, belongs to group II. When Ca donates its two outermost electrons during chemical interaction with strongly electronegative atoms, it becomes an ion with a double positive charge, i.e., Ca$^{++}$. Recall that the loss of two electrons from any atom leaves it with an excess of two nuclear protons over the number of orbital electrons. The overall chemical change in Ca can be shown in the following manner:

$$:Ca - 2e^- \rightarrow Ca^{++} \tag{10-3}$$

Notice that Ca$^{++}$ is chemically unreactive because the L shell with eight electrons becomes the new outermost energy level. Therefore, the electronic configuration of Ca$^{++}$ is also the same as that of neon (Ne), the inert gas closing the period preceding it.

The correspondence between the number of electrons gained by a strongly electronegative atom and the number of electrons required to bring its outer electron number to eight will be shown with the use of a common example, chlorine (Cl). Belonging to group VII, Cl possesses seven outermost electrons. During chemical combination, Cl accepts one electron from an atom with a low ionization energy potential. The following equation shows the chemical change undergone by Cl:

$$:\overset{..}{\underset{..}{Cl}}\cdot + e^- \rightarrow :\overset{..}{\underset{..}{Cl}}:^- \tag{10-4}$$

Equation (10-4) states that Cl accepts one electron when it unites with another atom, becoming an ion bearing one negative charge. The gain of one electron by Cl$^-$ brings its outermost electron number to eight. The electronic configuration of Cl$^-$ is identical to that of argon (Ar), the inert gas which closes the period to which Cl belongs. Thus, Cl$^-$ is stable, or chemically unreactive.

The electronic interactions between chemically united atoms are *chemical bonds.* Two types of chemical bonding are recognized: (1) *ionic,* or *electrovalent,* in which one or more electrons are transferred from one atom to another and (2) *covalent,* in which one or more pairs of electrons are shared between atoms.

*Ionic bonds*   Since elements in group I give up one electron and elements in group VII accept one electron, atoms from each of these groups combine in equal proportions, i.e., in a ratio of 1:1 when in close physical proximity. For example, when Na unites chemically with Cl through the formation of ionic bonds, the overall chemical change is summarized as follows:

$$\cdot Na + :\overset{..}{\underset{..}{Cl}}\cdot \rightarrow Na^+ :\overset{..}{\underset{..}{Cl}}:^- \tag{10-5}$$

The compound formed in Eq. (10-5) is called sodium chloride. The ionic products of the chemical reaction, $Na^+$ and $Cl^-$, are electrostatically attracted to each other because they bear opposite charges of identical magnitude, forming a solid crystal. Both $Na^+$ and $Cl^-$ in $Na^+Cl^-$ have a stable octet of electrons in their outermost energy levels.

Because group II elements donate two electrons while group VII elements accept only one electron, ionic bonds will bind one atom from group II to two identical atoms in group VII. For example, when Ca unites with Cl, the overall chemical change is expressed thus:

$$:Ca + 2 \, :\overset{..}{\underset{..}{Cl}} \cdot \; \rightarrow \; Ca^{++} \; :\overset{..}{\underset{..}{Cl}}: \; \bar{}_2 \qquad (10\text{-}6)$$

Equation (10-6) states that one Ca reacts with two Cl, forming the ionic compound calcium chloride, or $Ca^{++}Cl_2^-$. The product $Ca^{++}Cl_2^-$ consists of $Ca^{++}$

which is electrostatically attracted to $2Cl^-$. In other words, each $Cl^-$ has accepted one electron from Ca, thereby resulting in the formation of $Ca^{++}$. Each of the three ionized atoms in $Ca^{++}Cl_2^-$ has achieved a stable outer electron configuration. Thus, the proportion in which atoms from group VII combine with atoms from group II is 2:1.

Figure 10-4 summarizes the electronic changes which result from ionic bonding between atoms. Observe that a chemical bond is established through electron transfer and that electrostatic forces of attraction hold the atoms together.

*Covalent bonds* Covalent bonds are formed between atoms with identical or only moderately diverging electronegativities. In covalent bonding, one or more pairs of electrons are shared by the united atoms. The electron pairs shared by each atom

**Figure 10-4** *The electronic interactions between atoms with widely diverging electronegativities. The chemical reactions are shown by symbols below the interacting atoms. The atomic diameters of all the atoms are drawn to the same scale.*

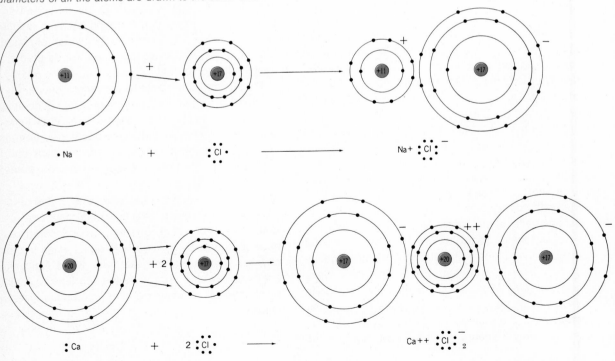

provide both of them with stable octets of electrons in their outermost energy levels. This type of bonding is characteristic of biologically important molecules.

When two atoms of the same element combine chemically to form a molecule, the electrons are shared equally. In other words, each of the shared electrons is as likely to be found in orbit around one of the atoms as the other. Equal sharing, of course, occurs when there is no difference in electronegativity between the atoms. This form of electronic interaction is called *nonpolar* covalent bonding.

Numerous gaseous molecules are formed by covalent bonding between two identical atoms. Such molecules include hydrogen ($H_2$), chlorine ($Cl_2$) and other halogens, oxygen ($O_2$), and nitrogen ($N_2$). Let us consider each of these examples, using dots to keep track of the electronic interactions between the outermost energy levels of the united atoms.

Atomic H (hydrogen) can resemble the electronic configuration of He (helium) when its solitary electron and an outermost electron from another atom are shared. Thus, when two H unite by equally sharing one pair of electrons, $H_2$ is formed. The formation of $H_2$ is written as follows:

$$H + \cdot H \rightarrow H \colon H \qquad (10\text{-}7)$$

The two dots between the H atoms represent the shared pair of electrons. Each pair of shared electrons represents one chemical bond. Therefore, the two H atoms in molecular H are linked together by one covalent chemical bond.

When two Cl (chlorine) atoms interact chemically, $Cl_2$ is generated. Each Cl resembles the electronic configuration of Ar (argon) after adding one electron to its outermost energy level through covalent bonding. The chemical interaction between two Cl is summarized in Eq. (10-8):

$$:\!\overset{\cdot\cdot}{\underset{\cdot\cdot}{Cl}}\!\cdot \; + \; :\!\overset{\cdot\cdot}{\underset{\cdot\cdot}{Cl}}\!\cdot \; \rightarrow \; :\!\overset{\cdot\cdot}{\underset{\cdot\cdot}{Cl}}\!:\!\overset{\cdot\cdot}{\underset{\cdot\cdot}{Cl}}\!: \qquad (10\text{-}8)$$

Neither Cl in $Cl_2$ is electrically charged, because the pair of electrons which unite them is shared equally.

Atomic O (oxygen) is two electrons shy of the stable electronic configuration possessed by Ne (neon). Therefore, when two O (oxygen) atoms unite to form $O_2$, each atom of the pair must share two electrons. In other words, stability requirements are met when two O (oxygen) atoms are united through the sharing of two pairs of electrons. Equation (10-9) summarizes the overall chemical change:

$$\overset{\cdot\cdot}{\underset{\cdot\cdot}{O}}\!: \; + \; :\!\overset{\cdot\cdot}{O} \; \rightarrow \; :\!\overset{\cdot\cdot}{O}\!:\!:\!\overset{\cdot\cdot}{O}\!: \qquad (10\text{-}9)$$

The two pairs of dots between the O atoms represent the two pairs of shared electrons. In other words, two covalent bonds unite the O atoms in molecular O.

As a member of group V, N (nitrogen) is deficient by three electrons of the stable electronic configuration possessed by Ne (neon). However, stability may be achieved if each of two N can share three of its outermost electrons with the other. In other words, $N_2$ forms when three pairs of electrons are equally shared by two N. Equation (10-10) summarizes the reaction:

$$:\!\overset{\cdot\;\;\cdot}{N}\!: \; + \; :\!\overset{\cdot\;\;\cdot}{N}\!: \; \rightarrow \; :\!N\!:\!:\!:\!N\!: \qquad (10\text{-}10)$$

The three pairs of dots between each N represent the three pairs of shared electrons. Thus, three covalent bonds unite the N atoms in molecular N.

Figure 10-5 summarizes the electronic changes which result from covalent bonding between identical atoms.

Whenever covalent bonds are formed between atoms of different elements, the electrons are shared unequally. Under these conditions, the shared electrons are electrostatically attracted to a greater degree by the more electronegative atom. Because the shared electrons are differentially pulled more closely to the atom with the higher electronegativity, they also spend more time in orbit around it than around the less electronegative atom.

This form of electronic interaction between different atoms is called *polar* covalent bonding. Thus, if united atoms cannot attain stable octets of electrons in their outermost energy levels through ionic bonding or nonpolar covalent bonding, they can at least approximate this condition through polar covalent bonding. In other words, polar covalent bonding is a compromise between nonpolar covalent bonding at one extreme and ionic bonding at the other extreme.

Let us look at a few examples of molecules formed by the establishment of polar covalent bonds between atoms. First, consider the chemical union be-

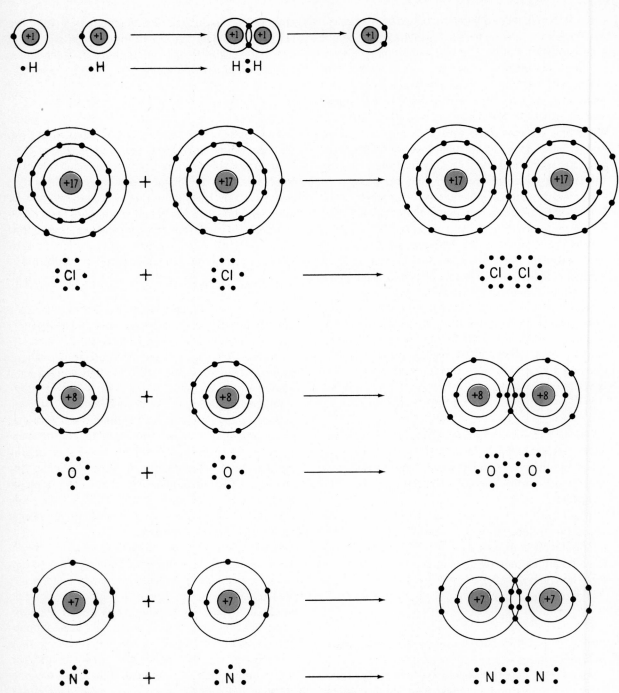

**Figure 10-5** *The electronic interactions between identical atoms. The chemical reactions are shown by symbols below the interacting atoms. The atomic diameters of all the atoms are drawn to the same scale.*

tween hydrogen (H) and chlorine (Cl) to form hydrogen chloride, or hydrochloric acid (HCl). The stability requirements of both elements can be met by sharing one pair of electrons. However, Cl is considerably more electronegative than H. Therefore, the shared pair of electrons spend proportionately more time in orbit around Cl, compared with the amount of time in orbit around H. When these two atoms unite, their electronic interactions can be expressed as follows:

$$H^+ \; : \overset{..}{\underset{..}{Cl}} : \; ^- \tag{10-11}$$

Even though HCl is not an ionic compound like $Na^+Cl^-$, the H part of the molecule is slightly positively charged, and the Cl portion of the molecule is correspondingly negatively charged.

Next, observe the nature of the electronic interactions between the atoms which form a molecule of water ($H_2O$). A $H_2O$ molecule is formed from two atoms of H and one atom of O. The stability requirements of both elements can be met when O shares two of its electrons, one with each of two H. Similarly, each H becomes unreactive chemically by sharing its electron with one electron from O. However, O is appreciably more electronegative than either H. Therefore, the two pairs of electrons which unite the three atoms spend proportionately more time in orbit around O, compared with the amount of time in orbit around either H, i.e., the electron pairs are closer to O than to either H. Polar covalent bonding in a $H_2O$ molecule is summarized in the following manner:

$$\begin{array}{c} H^+ \\ H^+ \; : \overset{..}{\underset{..}{O}} : \; _= \end{array} \tag{10-12}$$

Because the electron from each H is pulled somewhat away from its nuclear proton, both H atoms in $H_2O$ have a positive charge. Conversely, since the electron from each H is differentially attracted to O, this atom has a charge of −2. Of course, the molecule is still electrically neutral; i.e., the positive charge on each H is cancelled out by the two negative charges on O.

Now, consider the chemical union between one atom of carbon (C) and two atoms of oxygen (O), forming a molecule of carbon dioxide ($CO_2$). Since C is a member of group IV and O belongs to group VI, the difference in electronegativity between these two elements, while considerable, is not sufficient to cause them to interact through ionic bonding. Instead, electron pairs are shared through covalent bonding. Because O is the more electronegative of the two elements, the shared electrons are pulled more closely to either O than to C.

C is capable of sharing four electrons, whereas O is capable of sharing two electrons. When two O combine with C, the outer energy level of each atom attains the equivalent of eight electrons. Even though the electrons are shared rather than transferred outright, the electronic configuration of each atom resembles that of Ne (neon), the inert gas closing the period to which these atoms belong.

These points can be summarized by writing the structure of $CO_2$ in the following way:

$$: \overset{.}{\underset{.}{O}} :: C :: \overset{.}{\underset{.}{O}} : \tag{10-13}$$

The two pairs of dots binding each O to C represent two pairs of electrons or two chemical bonds. In other words, C shares four electrons, one pair from each O, while each O shares two electrons from C.

The electronic arrangement in $CO_2$ confers a symmetry of charge distribution which prevents polarization of the molecule. Thus, even though the electrons are shared unequally, there is no separation of electric charge within $CO_2$ as there is in $H_2O$.

As you remember, a $H_2O$ molecule is effectively separated into two oppositely charged regions, one containing two H with a double positive charge and the other containing O with a double negative charge. We can emphasize the charge separation in a $H_2O$ molecule in the following manner:

| | |
|---|---|
| $H^+ \quad H^+$ | positively charged end |
| $: \overset{..}{\underset{..}{O}} : \; _=$ | negatively charged end   (10-14) |

The asymmetric geometry results in a discrete separation of charge and is the reason $H_2O$ is a *polar solvent*.

By contrast, $CO_2$ is structurally symmetric. The carbon atom is flanked squarely on each side by O, so that there is no apparent angle between them. The geometric relationship among the atoms in $CO_2$ prevents a separation of electric charge within the molecule. Carbon dioxide is therefore a *nonpolar molecule*, even though its covalent bonds are distinctly polar.

C and the other group IV elements have the capability of combining chemically with as many as four atoms. C has the ability to form compounds which consist of long chains of C atoms to which H, O, and N are linked. Excluding $CO_2$ and its derivatives, compounds of C are called *organic compounds*. Organic compounds are responsible for metabolism and perpetuation. Since C is one of the most important elements in protoplasm, let us focus more closely on its combining powers.

Although this event does not occur in protoplasm, C can combine covalently with four H, forming a molecule called methane ($CH_4$). This union is possible because C can share four electrons and each of four H can share one electron apiece. While there is very little difference in the electronegativities of these two elements, the electron-attracting ability of C is slightly greater than that of H. Thus, the five atoms in $CH_4$ are linked by four somewhat unequally shared electron pairs, i.e., by four polar covalent bonds.

The polar covalent bonding between the five atoms of $CH_4$ is shown below:

$$
\begin{array}{c}
\text{H} \\
\text{H} :\!\overset{..}{\underset{..}{\text{C}}}\!: \text{H} \\
\text{H}
\end{array}
\qquad (10\text{-}15)
$$

Notice that the stability requirements of all five atoms are satisfied in that each H achieves the equivalent of two outermost electrons while C attains the equivalent of a stable octet of outermost electrons.

Because H is a very reactive atom, it readily combines with a C atom. As a consequence of this reactivity, C is united more frequently with H than with any other element. Molecules constructed entirely of C and H atoms are called *hydrocarbons*. An entire branch of chemistry has been built around a study of their behavior. As we will find out later, *carbohydrates* are nothing more than long-chain C compounds in which the C atoms are hydrated; i.e., the general

formula for a carbohydrate is $C(H_2O)_n$ where $n =$ the number of C atoms in the chain.

Carbon combines with four atoms of Cl for basically the same reasons it unites with H. The molecule which results from polar covalent bonding between these atoms is called *carbon tetrachloride* ($CCl_4$). Cl (chlorine) is somewhat more electronegative than C. Thus, the difference in electronegativity between Cl and C is slightly greater than that between C and H. With these facts in mind, attempt to diagram the electronic interactions between the five atoms of $CCl_4$. How do the stability requirements of $CCl_4$ differ from those of $CH_4$? Where are the four pairs of shared electrons located in $CCl_4$, compared with their positions in $CH_4$?

Figure 10-6 summarizes the electronic changes which result from polar covalent bonding between different atoms in various molecules.

*Oxidation states* The rules discussed above are only qualitative. They do not tell us how high the ionization energy potential must be before it becomes impossible to remove the outermost electrons from an atom through chemical interaction or how great the difference in electronegativity between atoms must be before the more electronegative atom will attract, rather than share, electrons during chemical combination. It is also extremely difficult to determine the actual electronic interactions which have occurred during the formation of compounds composed of atoms of more than two elements, such as sulfuric acid ($H_2^+SO_4^{--}$) or ammonium chloride ($NH_4^+Cl^-$). To reach a new level of simplicity when dealing with the electronic interactions among atoms in complex molecules, i.e., molecules composed of more than two or three atoms, it is necessary to recast some of our earlier formulations.

Let us return for a moment to ionic bonding. An atom which gives up one or more electrons is said to be *oxidized*. Conversely, an atom which gains one or more electrons is said to be *reduced*. Both oxidation and reduction take place in the same chemical reaction. The electron donor is called a *reductant*; the electron acceptor is referred to as an *oxidant*. Since oxidation requires the removal of electrons and reduction demands the acquisition of electrons,

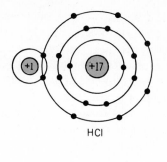

HCl

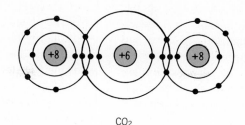

$CO_2$

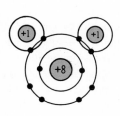

$H_2O$

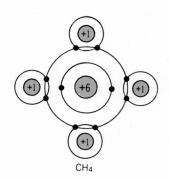

$CH_4$

**Figure 10-6** *The electronic interactions between atoms with slightly or moderately diverging electronegativities. The molecules are identified by symbols below the interacting atoms. The atomic diameters of all the atoms are drawn to the same scale.*

chemical interactions in which these reciprocal events occur are called *oxidation-reduction*, or *redox*, reactions.

Now, suppose we extend the term oxidation to chemical reactions which result from polar covalent bonding. In other words, when one pair of shared electrons is closer to atom B than to atom A because B is more electronegative than A, let us assume that B has gained one electron from A or that A has lost one electron to B. While the assumption is not strictly correct, it enables us to deal with the electronic interactions among atoms in complex molecules in a fashion not otherwise possible. First, we will examine the rules permitted by the assumption, and then we will see why they are so expedient.

The following rules define the *oxidation numbers*, or states, of the elements:

1  All elements in their natural states have the oxidation number zero; i.e., they have neither lost nor gained electrons.
2  The loss of one or more electrons from the outer energy level of an atom results in an increase in its oxidation number when it is zero or already positive or a decrease in its oxidation number when already negative. The magnitude of the increase or decrease corresponds to the number of electrons lost.
3  The gain of one or more electrons in the outer energy level of an atom results in a decrease in its oxidation number when it is zero or already positive or an increase in its oxidation number when already negative. The magnitude of the decrease or increase corresponds to the number of electrons gained.
4  Group VIII elements have an elemental oxidation number of zero.
5  Since molecules are electrically neutral, the algebraic sum of the oxidation numbers of the individual atoms in a molecule must equal zero.

The practical application of these rules is relatively straightforward. Consider, for example, the compound magnesium sulfate, $Mg^{++}SO_4^{--}$. There are six atoms in the molecule, and we wish to identify the oxidation numbers of each. Since Mg (magnesium) is in group

II, it donates its two outermost electrons during chemical combination. Therefore, the oxidation number of Mg in $Mg^{++}SO_4^{--}$ is $+2$; i.e., Mg has donated its two outermost electrons to the rest of the molecule, becoming $Mg^{++}$ in the process. Because the sulfate group accepts two electrons from Mg, it bears a double negative charge. The sulfate group is therefore abbreviated $SO_4^{--}$. In other words, $Mg^{++}$ is electrostatically attracted to $SO_4^{--}$ as the result of ionic bonding. Furthermore, the oxidation number for each oxygen atom in $SO_4^{--}$ must be $-2$, since O (oxygen) usually accepts two electrons in achieving stability during chemical interaction. Thus, the sum of the oxidation numbers of the four O atoms in $SO_4^{--}$ is $-8$.

We are now in a position to derive the oxidation number of the S (sulfur) atom, bearing in mind that the algebraic sum of all the oxidation numbers must equal zero. In other words, knowing that the oxidation number of $Mg^{++}$ is $+2$ and that of $O_4$ is $-8$, what oxidation number must be assigned to S in order to fulfill this rule? The answer, of course, is $+6$, since $(+)2 + (+)6 + (-)8 = (+)8 + (-)8 = 0$. The formula and oxidation numbers for $Mg^{++}SO_4^{--}$ are shown below:

$$\overset{+2}{Mg^{++}} \quad \overset{+6 \; -8}{SO_4^{=}} \tag{10-16}$$

In aqueous solution, $Mg^{++}SO_4^{--}$ dissociates into $Mg^{++}$ and $SO_4^{--}$. The S and O atoms in $SO_4^{--}$ remain together in a stable atomic grouping. We may justifiably conclude that the chemical bonds between S and each O are covalent rather than ionic. Indeed, if ionic bonds were responsible for the chemical union between S and O, the atoms in $SO_4^{--}$ would also have dissociated in $H_2O$ (water). This, of course, does not happen.

We may begin to draw several inferences, based on our knowledge of the electronic interactions among the atoms in $Mg^{++}SO_4^{--}$. First, $Mg^{++}SO_4^{--}$ and other ionic compounds with stable atomic groupings such as $H_3^+PO_4^{3-}$ and $NH_4^+Cl^-$ are formed through both covalent and ionic bonding. Covalent bonds link the atoms of the ionic aggregate together. An ionic bond joins the oppositely charged ions to each other. Second, the actual dissociation of $Mg^{++}$ from $SO_4^{--}$ in

aqueous solution is not a chemical reaction, because electronic interactions between these ions do not occur during the ionization process. The ions separate because the force of electrostatic attraction between them is weakened. Therefore, *molecular dissociation* into ions represents a *physical* rather than a chemical change. Third, the tendency of a molecule to dissociate into ions in aqueous solution must be related directly to the strength of the ionic or polar covalent bond between the oppositely charged portions of the molecule.

If we did not know any of the rules governing oxidation states, we could still predict the nature of the electronic interactions between the S (sulfur) and O (oxygen) atoms. Recall that electronic stability in the outermost energy level of any atom is achieved when it contains eight electrons (or as close to eight as possible when polar covalent bonding is involved). Since S and O both belong to group VI, each of the atoms in $SO_4^{--}$ must acquire the equivalent of two electrons through polar covalent bonding to attain chemical stability.

Using dots to indicate the outermost electrons in these atoms and ignoring the difference in electronegativity between S and O, we can show that the only electronic configuration of $SO_4^{--}$ which meets these specifications is as follows:

$$\begin{matrix} & :\ddot{O}: & ^= \\ :\ddot{O}&:\!\!\overset{..}{S}\!\!:&\ddot{O}: \\ & :\ddot{O}: & \end{matrix} \tag{10-17}$$

In other words, the ionic aggregation of atoms is stable only when each atom is surrounded by eight electrons. Having acquired two electrons from Mg through ionic bonding prior to dissociation, S is able to share two of its eight electrons with each oxygen atom. Thus, the stability requirements of $SO_4^{--}$ are satisfied, while its double negative charge is explained by the acquisition of two electrons from Mg. Obviously, we can avoid these rather cumbersome manipulations simply by following the rules which describe how the oxidation states of elements are determined.

Now, consider a chemical reaction in aqueous solution between $Mg^{++}SO_4^{--}$ (magnesium sulfate) and

Na (sodium). Na is a more active electron donor than Mg, since less ionization energy is required to remove an electron from Na than from Mg. Therefore, each atom of Na will donate an electron to Mg, thereby becoming oxidized. At the same time, $Mg^{++}$ will accept an electron from each of two Na, becoming reduced in the process. In other words, Na changes from elemental to ionic form as $Mg^{++}$ undergoes the reverse transformation. The following equation expresses the chemical changes which occur:

$$2Na + Mg^{++} + SO_4^{=} \rightarrow Mg + 2Na^+ + SO_4^{=} \qquad (10\text{-}18)$$

Since $SO_4^{--}$ does not undergo chemical change during the reaction, we can ignore it and focus our attention on the electronic interactions between Na and $Mg^{++}$.

The oxidation number of Na increases from zero to +2 while that of $Mg^{++}$ decreases from +2 to zero. In other words, Na is oxidized as $Mg^{++}$ is reduced. Therefore Na displaces $Mg^{++}$, forming a new ionic compound, $Na_2^+SO_4^{--}$. This compound would be left behind following evaporation of the $H_2O$.

A redox reaction in which one element in ionic form is displaced from a compound by another, more active element in atomic form is called a *displacement reaction*. Redox reactions in general are important in the energy metabolism of cells. Displacement reactions in particular are fundamental to an understanding of chemical interactions in protoplasm and body fluids.

To guide you in verifying the actual electronic interactions among atoms in molecules and stable atomic groupings, common oxidation states are given for most elements in the periodic table (Fig. 10-1). The oxidation numbers are located in the center on the left-hand side of the elements.

The effect of chemical union on the properties of the combined atoms Electronic interactions change the diameters of those atoms that interact. Atoms which donate electrons generally become smaller while atoms which accept electrons characteristically become larger. Changes in atomic radii also affect the ability of united atoms to engage in electron transfer or sharing interactions with other nearby atoms.

At the same time, the electromagnetic forces of attraction and repulsion among adjacent atoms in any molecule establish predictable spatial relationships between them. The physical locations of the atoms in a molecule are expressed through their bond angles. Bond angles between atoms confer a three-dimensional shape to molecules. *Shape* is of paramount importance in determining molecular properties.

Of course, the differences outlined above are not the only ones which physically separate molecules from atoms. Since molecules consist of united atoms, the mass of a molecule is equal to the sum of the masses of the individual atoms. Mass affects solubility and other physical characteristics. The size of a molecule is related to the number of its substituent atoms and its overall shape. Size influences the diffusibility of molecules. While atoms and molecules are by definition electrically neutral, the nature of the electromagnetic interactions between substituent atoms in a molecule may result in a distinct separation of positively and negatively charged atoms, thereby conferring polarity on the molecule. The behavior of the polar solvent $H_2O$ makes life possible. These physical features of molecules obviously cannot be duplicated by any individual atom.

## INTERMOLECULAR BONDS

Several types of bonds link molecules to each other. We shall review two types—hydrogen bonding and van der Waals' forces. Although weak in magnitude, compared with ionic and covalent chemical bonds, they are vital to the structures and functions of biologically important molecules.

Hydrogen bonds Hydrogen bonds may be established when oppositely charged parts of either the same molecule or of neighboring molecules come close to each other. In hydrogen bonding, one of the participating polar groups is hydrogen, while the other participating polar group is generally oxygen or nitrogen. The polarity of these atoms is of course de-

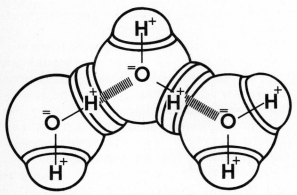

**Figure 10-7** *Hydrogen (H) bonds between water (H₂O) molecules. The intermolecular bonds are identified with hatched lines.*

termined by the relative electronegativities of the atoms with which they are combined chemically.

Water ($H_2O$) is a common example of a polar molecule which forms hydrogen bonds with other polar molecules. Hydrogen bonding between several water molecules is shown in Fig. 10-7. Notice that a hydrogen atom from one molecule is electrostatically attracted to the oxygen atom of an adjacent molecule. The hatched lines represent hydrogen bonds. However, since an oxygen atom in a water molecule has a double negative charge while each hydrogen atom bears a single positive charge, as many as two hydrogen atoms from one or two other molecules can form hydrogen bonds with it. Therefore, water molecules may be linked to each other in several different ways, depending on the number and origin of the hydrogen atoms linked to the oxygen atoms.

There are two types of hydrogen bonds—weak and strong. *Weak hydrogen bonds* are entirely due to electrostatic forces of attraction between polar groups. No displacements in the outermost electrons of the interacting atoms occur during their formation. Therefore, weak hydrogen bonds are physical in nature rather than the result of chemical interaction.

Only a small amount of energy is needed to rupture a weak hydrogen bond. For example, to change water from its solid to its liquid state requires the addition of 80 cal/Gm. The change of state is accomplished when about 20 percent of the hydrogen bonds are actually ruptured. The rest of the hydrogen bonds must be ruptured before water can boil.

By way of contrast, *strong hydrogen bonds* are partly electrostatic, partly covalent. The amount of energy required to rupture a strong hydrogen bond is about double that needed to break a weak hydrogen bond but still almost 10 times less than that consumed in cleaving chemical bonds.

Strong hydrogen bonds are important determinants of the shape of protein molecules and the way in which one protein relates structurally to another protein. Hydrogen bonds linking various points along a protein molecule can mold its shape into a fold, loop, spiral, or some other configuration, just as glue can fuse different parts of one sheet of paper when they are brought in contact. Such molding stabilizes the structure of a protein molecule by maintaining it in a particular configuration. Hydrogen bonds, again acting as glue, are in part responsible for linking different chains of one protein molecule together. The individual chains consist of molecules called *polypeptides*, which are somewhat smaller than proteins. Indeed, the chains would be separate molecules were it not for the fact that hydrogen bonds and chemical bonds between special atoms attach them to the rest of the macromolecule. Finally, hydrogen bonding "glues" separate nucleic acid molecules together (see Chap. 6).

Van der Waals' forces   Van der Waals' forces are weak attractive forces which exist between atoms, ions, or molecules in any physical state when they are very close together. The interatomic distance at which the greatest van der Waals' force is generated is that which is just great enough to prevent chemical interaction. These forces play prominent roles in holding specific portions of adjacent protein molecules together in a semisolid mass such as that of protoplasm. However, they are much weaker than ionic or covalent bonds and can easily be ruptured by heat.

Since proteins and other organic molecules are nonpolar, what is the explanation for their mutual attraction? Actually the answer to this question is based on our concept of a polar molecule.

Although there is no distinct separation of charge in nonpolar molecules, this rule is valid only when the charge distribution of one molecule is followed over a sufficiently long period of time. In other words, the statement is statistical in nature. At any one instant, a nonpolar molecule actually has a separation of charge, because the shared electrons between substituent atoms are asymmetrically distributed or skewed toward one end of the molecule; at the next instant, the asymmetry may be in the opposite direction; etc. This is why the charge symmetry of a nonpolar molecule is an average condition.

Thus, when nonpolar molecules approach each other, the random variations in their charge distribution at the time of their physical proximity become fixed. When this happens, the molecules are held together electrostatically because of an alignment between oppositely charged portions of adjacent molecules.

## THE LANGUAGE OF CHEMISTRY

Because of the clinical importance of chemistry in the health-related professions, we should develop an ability to summarize chemical reactions succinctly and to state chemical concentrations precisely. In this unit, therefore, we will concentrate on expressing chemical reactions, balancing chemical equations, and understanding some commonly used units of concentration.

## CHEMICAL EQUATIONS

Equations summarize with symbols what happens during the course of chemical reactions. In other words, a chemical equation is a shorthand method of stating the input, output, and direction of the reaction as well as the conditions required to initiate or sustain it.

For example, consider cellular respiration, the process by which cells obtain energy from the oxidation of sugar. The overall reaction, summarizing about 30 individual reactions, is written as follows:

$$C_6H_{12}O_6 + 6O_2 \xrightarrow{\text{enzymes, vitamins, minerals}} 6CO_2 + 6H_2O + \text{energy} \quad (10\text{-}19)$$

Reactants                    Products

We could have said that glucose ($C_6H_{12}O_6$) and oxy-

gen (O) interact chemically in the presence of certain enzymes, vitamins, and minerals, forming carbon dioxide ($CO_2$) and water ($H_2O$), in addition to releasing energy. However, the equation tells us specifically how much of each reactant combines and how much of each product forms. In the above reaction, for example, one molecule, or part, of $C_6H_{12}O_6$ combines with six molecules, or parts, of $O_2$, yielding six molecules, or parts, of $CO_2$, six molecules, or parts, of $H_2O$, plus energy. Therefore, the number in front of the formula for each chemical represents its quantity, while each subscript represents the number of atoms of the element immediately preceding it.

Balancing equations    Notice that the number of atoms of each element on both sides of the equation is equal. There are six C atoms in both $C_6H_{12}O_6$ and $CO_2$ (six molecules of $CO_2$ × one atom of C in each molecule), twelve H in both $C_6H_{12}O_6$ and $H_2O$ (six molecules of $H_2O$ × two H per molecule), and eighteen O on either side of the equation (six O in $C_6H_{12}O_6$ and twelve O in $6O_2$ on the left-hand side, and twelve O in $6CO_2$ and six O in $6H_2O$ on the right-hand side).

The number of each type of atom on both sides of an equation must be equal, because matter can be neither created nor destroyed, at least under ordinary conditions of chemical reaction. The law of conservation of matter demands that chemical equations be balanced. The atomic bookkeeping required in balancing equations is analogous to the electronic bookkeeping involved with the use of dots to represent the outermost level of electrons in interacting atoms. Both systems allow us to keep track of particles during combination.

Most of the equations with which we will be dealing can be balanced through simple visual inspection, following an analysis of the relative electronegativities of the participating atoms. To take a simple reaction where we may not know the quantity of reactants and products, consider the formation of water through the chemical union of oxygen and hydrogen. Both gases exist as diatomic molecules, i.e., molecules consisting of two identical atoms under natural conditions. Furthermore, we know that the stability requirements of hydrogen and oxygen in water ($H_2O$) are satisfied when two atoms of hydrogen (H) are chemically united with one atom of oxygen (O). Alternatively, we can reason that the oxidation number of hydrogen atoms in water must be +2, since that of oxygen is −2 and the molecule must be electrically neutral.

Our first inclination is to write the reaction as follows:

$$H_2 + O_2 \rightarrow H_2O \qquad (10\text{-}20)$$

However, notice that there is one more atom on the left-hand side than on the right-hand side of the equation. We can balance the number of oxygen atoms on both sides by placing a 2 in front of $H_2O$, thus:

$$H_2 + O_2 \rightarrow 2H_2O \qquad (10\text{-}21)$$

Balancing oxygen atoms in this manner also doubles the number of hydrogen atoms on the right-hand side. To bring the hydrogen atoms back into balance on both sides of the equation, another molecule of hydrogen must be added on the left-hand side. The balanced equation is therefore

$$2H_2 + O_2 \rightarrow 2H_2O \qquad (10\text{-}22)$$

In other words, to form water from the chemical union of hydrogen and oxygen, it is necessary to maintain a ratio of two parts of molecular hydrogen to one part of molecular oxygen under appropriate conditions of chemical reaction. For every two molecules, or parts, of hydrogen and every molecule, or part, of oxygen which unite, two molecules, or parts, of water are generated. When chemical equilibrium is reached, as evidenced by a lack of change in the concentrations of the reactants and product, any excess $H_2$ or $O_2$ simply remains in an uncombined state.

## WEIGHING AND COUNTING ATOMS AND MOLECULES

In Chap. 7, we learned that (1) the atomic mass of an atom of any element is equal to the number of protons and neutrons in its nucleus; (2) the atomic weight of any element is the average of the individual atomic masses of its various isotopes, based on their relative abundances; and (3) atomic weights are relative values, since they are based on the atomic mass of the most common isotope of carbon, $C_{12}$. The atomic weight or average atomic mass of each element is given in the periodic table (Fig. 10-1). For expediency, we will round off atomic weights to the nearest whole numbers in the following discussion.

Other atomic measures  The relative weight of an element, compared with $C_{12}$, can be converted into a measurable, or real, weight by weighing out enough of the element to equal its atomic weight in grams. The atomic weight of an element in grams is called its *gram atomic weight* (Gm at. wt.). Thus, the gram atomic weight of hydrogen is 1 Gm, that of carbon is 12 Gm, that of calcium is 40 Gm, etc.

One gram atomic weight of any element always contains the same number of atoms, $6.023 \times 10^{23}$. For expediency, the number is simplified, or rounded, to $6 \times 10^{23}$. This number is unbelievably large and is actually 6 with 23 zeros strung to the right of it. Since $6 \times 10^{23}$ atoms are found in 1 Gm at. wt., 1 Gm of hydrogen, 12 Gm of carbon, and 40 Gm of calcium all contain the same number of atoms. This quantity of any element is called a *mole*. Although a mole is based on the number of atoms of an element, it is also equal to its atomic weight in grams. The equivalence is based on practicality, for it is possible to weigh out the number of grams of an element equal to its atomic weight, whereas it is impossible to count $6 \times 10^{23}$ individual atoms.

Molecular measures  The principles on which the above measures are based are just as applicable to molecules as they are to atoms. This should not be surprising since molecules consist of two or more similar or dissimilar atoms. The molecular counterparts of atomic weight and gram atomic weight are *molecular weight* (mol. wt.) and *gram molecular weight* (Gm mol. wt.). Let us explore their similarities and differences.

The molecular weight of a molecule is the sum of the atomic weights of its substituent atoms. For example, to determine the molecular weight of glucose, $C_6H_{12}O_6$, we must multiply the atomic weight of each type of atom by the number of atoms of the same type in the molecule and obtain the sum of the products. The molecular weight of glucose is calculated below:

| Atom | No. of atoms | | At. wt. | | |
|------|------|------|------|------|------|
| C | 6 | × | 12 | = | 72 |
| H | 12 | × | 1 | = | 12 |
| O | 6 | × | 16 | = | 96 |
| | | | | | 180 |

The molecular weight of glucose is 180. This number represents the sum of the atomic weights of 6 atoms of carbon, 12 atoms of hydrogen, and 6 atoms of oxygen. Notice that the value is relative, since it is based on the average atomic mass of carbon.

The number of grams of a compound equal to the weight of one of its molecules is called its gram molecular weight (Gm mol. wt.). The gram molecular weight of glucose is 180 Gm; that of carbon dioxide is 44 Gm (12 + 16 + 16); and that of water is 18 Gm (1 + 1 + 16).

If 1 Gm at. wt. contains $6 \times 10^{23}$ atoms, then 1 Gm mol. wt. contains the same number of molecules. In other words, there are $6 \times 10^{23}$ molecules in 180 Gm of glucose, 44 Gm of carbon dioxide, or 18 Gm of water. This number of molecules is called a mole, just as $6 \times 10^{23}$ atoms is a mole. Therefore, 1 mole of any chemical substance contains $6 \times 10^{23}$ particles. Whether they are atoms, ions, or molecules does not matter. Since the number of particles is obviously related to concentration, the number of moles of a chemical expresses its concentration.

Table 10-1 summarizes and compares the various units described above.

Expressing concentrations in aqueous solution Moles are especially useful measures of the concentrations of chemicals in solution, i.e., for expressing the number of solute particles per unit volume of solvent. When enough solvent is added to 1 mole of any solute to bring the total volume (solute + solvent) to 1 liter, the concentration of the solution is 1 *molar* (*M*). A 1 *M* aqueous solution of any chemical contains $6 \times 10^{23}$ solute particles in 1 liter of solution. In addition, a 1 *M* solution contains 1 atomic, ionic, or molecular weight of solute particles per liter of solution. For example, a 1 *M* glucose solution contains 180 Gm of glucose dissolved in a total volume of 1 liter. There are two ways to change the molarity of a solution: (1) by increasing or decreasing the total volume of the solution without changing the concentration of solute and (2) by increasing or decreasing the solute concentration while holding the total volume of solution at 1 liter.

Often, the concentration of a chemical is expressed in *percent*, rather than by its molarity. Suppose we are given a 1.8 percent glucose solution and wish to know its molarity. Percentage concentrations are usually expressed in either Gm/100 ml or mg/100 ml. In a 1.8 percent glucose solution, there are 1.8 Gm/100 ml (1,800 mg/100 ml) or 10 times as much, 18 Gm (18,000 mg) in 1,000 ml (1 liter) of solution. Since a 1 *M* glucose solution contains 180 Gm of glucose and the above solution has only 18 Gm of solute, or just one-tenth as much, the 1.8 percent glucose solution is 0.1 *M*. This means that 1 liter of a 1.8 percent glucose solution contains $0.1 \times 6 \times 10^{23}$ molecules, i.e., only

**Table 10-1**  Ways in which the abundance of atoms and molecules are expressed

| Atomic mass units (amu) | Weight | Number of particles |
|------|------|------|
| At. wt. = protons + neutrons | Gm at. wt. = at. wt. (Gm) | Mole = $6.023 \times 10^{23}$ atoms |
| Mol. wt. = sum of at. wts. of all atoms in molecule | Gm mol. wt. = mol. wt. (Gm) | Mole = $6.023 \times 10^{23}$ molecules |

one-tenth the number of molecules contained in a mole.

We can formalize the relationship between percentage concentration and molarity by the following formula:

$$\text{Molarity} = \frac{\text{Gm/100 ml} \times 10}{\text{Gm atomic, ionic, or molecular weight}} \quad (10\text{-}23)$$

In other words, to obtain the molar concentration of an atomic, ionic, or molecular solute when its percentage concentration in Gm/100 ml of solution is known, multiply the latter by 10 and divide by the gram atomic, ionic, or molecular weight of the solute. When the percentage concentration is in mg/100 ml, the concentration is expressed in millimoles (mM).

Although a useful measure of concentration, the molarity must be appropriately corrected when compounds dissociate into ions in aqueous solution. Since ionization reactions are a common phenomenon in the body, the ability to express the activity of ionic solutions is vital. Units other than molarity are necessary under the following conditions: (1) when an ionic compound, e.g., $Na^+Cl^-$, or a strongly polar covalent compound is added to an aqueous solution, since 1 mole of $Na^+Cl^-$ dissociates into 1 mole of $Na^+$ and 1 mole of $Cl^-$; (2) when oppositely charged but electrically imbalanced ions, e.g., $Ca^{++}$ and $Cl^-$, are present in solution, since each $Ca^{++}$ can combine with $2Cl^-$; and (3) when acids, e.g., HCl or $H_2^+SO_4^{--}$, and bases, e.g., $Na^+OH^-$ or $Ca^{++}(OH)_2^-$, capable of donating or accepting different numbers of protons are present in the same solution, since there is a difference in the number of $H^+$ given up by HCl and $H_2^+SO_4^{--}$ during dissociation or in the number of $H^+$ accepted by $Na^+OH^-$ and $Ca^{++}(OH)_2^-$ following dissociation.

In the first case, 2 moles of solute particles result from the addition of 1 mole of $Na^+Cl^-$ to the solvent. Following the ionization of $Na^+Cl^-$, the solution contains $2 \times 6 \times 10^{23}$ ions rather than $6 \times 10^{23}$ molecules. The actual concentration of solute particles is therefore twice the stated molarity. If the molarity of $Na^+Cl^-$ is 0.1 M, then the concentration of the solution after adding the salt is 0.2 M.

In the second case, the combining powers of $Cl^-$ and $Ca^{++}$ are different because of their numerical imbalance of electric charge. When this happens, the combining powers of solute particles are expressed with units called *equivalents*.

For example, let us express the combining power of $Ca^{++}$ in blood plasma. First we must know that its concentration in plasma is approximately 0.0025 M. Since the combining power of an ion is related directly to the number of electric charges borne by it, i.e., to its oxidation number, the combining power of $Ca^{++}$ is 2. We can express the combining power of $Ca^{++}$ by multiplying its molarity, 0.0025 M, by its oxidation number, 2. The product 0.005 is the number of equivalents of $Ca^{++}$ in 1 liter of plasma. Since 0.005, or 5/1,000 (five-thousandths), of an equivalent is the same as 5 *milliequivalents* (mEq), the combining power is expressed in milliequivalents. Thus, 5 mEq of $Ca^{++}$ per liter of plasma will combine with 5 mEq of any other ion bearing a double negative charge or with 10 mEq of ions with a single negative charge.

In the third case, virtually all $H_2^+SO_4^{--}$ molecules ionize into $2H^+ + SO_4^{--}$, while most $Ca^{++}(OH)_2^-$ molecules ionize into $2OH^- + Ca^{++}$. In neutralization reactions between acids and bases in aqueous solution, it is important to keep track of the number of $H^+$ which go into solution upon the addition of an acid or the number of $H^+$ in solution which combine with $OH^-$ following the addition of a base. When these concentrations are expressed in gram equivalents (gEq), the concentration of the solution is expressed in *normality* (N). A 1 N (1 normal) solution contains 1 gEq of $H^+$ or the capability of combining with 1 gEq of $H^+$.

For example, consider the strong acid $H_2SO_4$. How many gram equivalents are there in a 1 M solution of $H_2SO_4$, and what is its normality? A 1 M solution of any compound contains 1 Gm mol. wt. of solute per liter of total solution. Since each $H_2SO_4$ molecule gives up $2H^+$, there are 2 Gm ionic weights of $H^+$ in a 1 M $H_2SO_4$ solution. In other words, an aqueous solution containing 1 Gm mol. wt. of $H_2SO_4$ is the equivalent of a 2 M solution in terms of $H^+$. This is the same as stating that a 1 M $H_2SO_4$ solution contains 2 gEq of $H^+$, or is 2 N.

Similarly, an aqueous solution containing 1 Gm mol. wt. of $Ca(OH)_2$ is the equivalent of a 2 $M$ solution in terms of the number of $H^+$ which can be electrostatically attracted to $OH^-$ following ionization of the base. This is comparable to saying that a 1 $M$ Ca·$(OH)_2$ solution can combine with 2 gEq of $H^+$, or is 2 $N$.

However, the situation is quantitatively different for HCl, $Na^+OH^-$, or any other acid or base which gives up or accepts one $H^+$ per molecule. A solution containing 1 Gm mol. wt. of HCl contains only 1 gEq of $H^+$, whereas a solution with an identical concentration of $Na^+OH^-$ can combine with 1 gEq of $H^+$. Therefore, a 1 $M$ solution of each of these compounds is also 1 $N$.

The relation between molarity and normality can probably best be illuminated by looking at several reactions between acids and bases with various gram equivalent values. For example, in the reaction

$$HCl + Na^+OH^- \rightarrow Na^+ + Cl^- + H_2O \qquad (10\text{-}24)$$

1 mole of HCl and 1 mole of $Na^+OH^-$ interact, resulting in the formation of 1 mole of $H_2O$ and leaving 1 mole of $Na^+$ and $Cl^-$ in solution. In other words, 1 Gm mol. wt. of HCl is neutralized by 1 Gm mol. wt. of $Na^+OH^-$, since the number of gram equivalents of $H^+$ given up by HCl is matched by the capacity of $OH^-$ from $Na^+OH^-$ to accept the gram equivalents of $H^+$.

Compare this with the interaction between $H_2^+SO_4^{--}$ and $Na^+OH^-$. The reaction is written as follows:

$$H_2^+SO_4^= + 2Na^+OH^- \rightarrow 2H_2O + 2Na^+ + SO_4^= \qquad (10\text{-}25)$$

Since each molecule of $H_2^+SO_4^{--}$ gives up $2H^+$ while each molecule of $Na^+OH^-$ accepts $1H^+$, neutralization is complete when 2 moles or Gm mol. wt. of $Na^+OH^-$ interact with 1 mole or Gm mol. wt. of $H_2^+SO_4^{--}$. In other words, the 2 gEq of $H^+$ given up by 1 mole of $H_2SO_4$ are neutralized completely only when 2 moles of NaOH are present. This means that the number of gram equivalents of $H^+$ with which $OH^-$ from NaOH can combine is only half that provided by the same quantity of $H_2^+SO_4^{--}$.

Following neutralization, 2 moles of $H_2O$ are formed, leaving 2 moles of $Na^+$ and 1 mole of $SO_4^{--}$ in solution. Upon evaporation of the water, 1 mole of the salt sodium sulfate ($Na_2^+SO_4^{--}$) can be recovered as a solid.

## REACTION CHARACTERISTICS

Each intracellular reaction, once initiated, determines the direction of an entire biochemical pathway by fixing the rate of the next reaction in the sequence. The determinants of direction and rate are fundamental to an understanding of chemical reactions, whether in the body or not. In this section, we will examine these reaction characteristics.

### CHEMICAL EQUILIBRIUM

A chemical reaction will eventually cease if there is no change in the conditions under which it is occurring. When this point is reached, there is no further change in the concentration of either the reactants or the products, even though chemical interactions continue to take place. This condition is referred to as *chemical equilibrium.*

**The dynamics of equilibrium** Let us look quantitatively at a sample chemical reaction to see what factors are responsible for the attainment of equilibrium. Assume that we are able to count the number of molecules of reactants and products at various times during the following chemical reaction:

$$A + B \rightleftharpoons C + D \qquad (10\text{-}26)$$
Reactants    Products

The significance of the difference in the lengths of the two arrows between the reactants and the products will become obvious as we follow the course of the chemical reaction.

For clarity, we shall initiate the reaction by placing 10 molecules of A and 10 molecules of B in a beaker of water and mixing them thoroughly. Products C and D are not initially present in the solution into which the reactants are placed. The conditions of the reaction are held constant with the exception, of course, of the chemical changes brought about following the introduction of the reactants.

The reaction is initiated at $T_0$. The rate of the reaction is measured at consecutive and equally spaced time intervals $T_1$ to $T_3$. The rate of disappearance of the reactants and the rate of appearance of the products are traced in Fig. 10-8.

As we have already learned, the rate of disappearance of the reactants is initially quite rapid but decreases as time elapses. In the reaction we are following, the number of reactants decreases from 20 to 10 to 6 from $T_0$ to $T_2$, respectively, while the number of products increases from zero to 10 to 14 in the same time span. Therefore, we note that the rate of decrease of the reactants is proportional to the rate of increase of the products. There is no change in the concentration of either reactants or products at $T_3$, compared with their quantities at $T_2$. Thus, chemical equilibrium is established at $T_2$. Observe that 6 reactant molecules are left and 14 product molecules have formed by the time chemical equilibrium is reached. Can we explain these findings on the bases of known mechanisms of chemical interaction?

Taking note of the difference in the number of reactant and product molecules at chemical equilibrium,

we can suggest that the activation of reactant molecules A and B requires a lower energy input than that of product molecules C and D. In other words, A and B interact chemically more readily than do C and D because the activation energy required by the former is lower than that required by the latter. Let us see how this explanation helps to interpret when the point of chemical equilibrium is reached.

At $T_0$, there are no products. Therefore, only the reactants can combine chemically. However, as products C and D accumulate, they also begin to combine chemically with each other, although at a slower rate than A and B. Therefore at $T_1$, the reaction is still predominantly from left to right, even though a few of the product molecules have become reactants in their own right.

The rapid depletion of reactant molecules A and B decreases their density in solution and correspondingly reduces the opportunity for collisions which will result in chemical interaction.

At $T_2$, the frequency of interaction between A and B is identical to the frequency of interaction between C and D; i.e., the lengths of the arrows in the two reac-

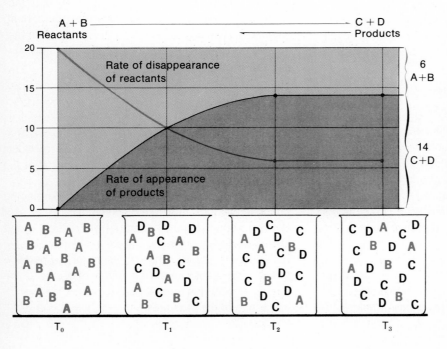

**Figure 10-8** *The achievement of chemical equilibrium in the reaction system $A + B \rightleftharpoons C + D$. The chemical reaction is followed in the same beaker at four consecutive and equal time intervals, $T_0$ to $T_3$. The letters in the solution within the beaker at each time interval represent individual molecules of reactants A and B (in color) and products C and D (in black). The rate of disappearance of the reactants and the rate of appearance of the products are followed graphically.*

tions, A + B → C + D and C + D → A + B, are equal. This is the point of chemical equilibrium, and, in this reaction, it requires a far greater density of C and D than of A and B (14 C and D molecules as opposed to 6 A and B molecules). Stated in other terms, at chemical equilibrium, 7C + 7D × higher activation energy = 3A + 3B × lower activation energy.

The reason for the differential in the lengths of the arrows in the overall equation $A + B \rightleftharpoons C + D$ should now be apparent. The equilibrium point of the reaction is far to the right. This means that most of the reactants will be depleted when chemical equilibrium is reached or, alternatively, that there will be more products than reactants at equilibrium. The equilibrium point depends on the nature of the reactants and the products as well as on the conditions of the reaction.

Irreversibility    One factor of extreme biologic importance is responsible for making many intracellular reactions irreversible, even though all reactions are theoretically reversible. To accomplish irreversibility, nature frequently removes the products as fast as they form. To see how this mechanism operates, let us attempt to drive the reaction $A + B \rightleftharpoons C + D$ further to the right; i.e., our goal is to increase the amount of product formed or decrease the amount of reactant left at chemical equilibrium. Ideally, we wish to establish an irreversible reaction A + B → C + D.

What we can do to promote irreversibility is best described with the aid of Fig. 10-8. If some or all of products C and D could be removed as fast as they form at $T_1$, the remaining molecules in the beaker would consist of nothing but A and B. Following removal of the products, the only direction in which the reaction can go is from left to right. If this procedure is continued indefinitely, eventually all the reactant molecules will disappear. Therefore, the continual removal of the products shifts the reaction as far to the right as possible, making it irreversible.

During biochemical reactions, we commonly find that the products of one reaction become the reactants of the next reaction in the sequence. This drives all the reactions in a particular chemical pathway to the right, tending to make them irreversible.

## THE MODIFICATION OF REACTION RATES

The balance in diametrically opposed reaction rates at chemical equilibrium suggests that factors which affect the rate of reaction can also alter the point of chemical equilibrium. However, this conjecture breaks down somewhat in practice. The reason is not hard to grasp. It is difficult to adjust the conditions of a reaction in a way which will cause a change in the rate in one direction without causing a change of the same relative magnitude in the opposite direction. Nevertheless, it is sometimes possible to adjust conditions to differentially increase or decrease the rate of reaction in one direction, thereby shifting the point of chemical equilibrium. In other words, some factors which influence the rate of a chemical reaction can also shift the equilibrium point.

Only a few factors are capable of changing the rate of chemical reactions. These factors are the temperature, pressure, concentration, and particle size of the reactants as well as the presence and concentration of catalysts such as enzymes. The mechanisms by which these factors increase reaction rates are diagramed in Fig. 10-9.

The effect of temperature    An elevation in temperature increases the rate of a chemical reaction, and vice versa. As a rule of thumb, a 10°C (Celsius) rise in temperature doubles the reaction rate. Since molecular kinetic energy is heat, as we learned in Chap. 8, a rise in the temperature of a reaction is actually an increase in the average velocity of the reactant molecules. When all other conditions are held constant, a rise in the rate of molecular movement increases the probability of molecular collisions. However, doubling the temperature does not come close to doubling the average velocity of reactant molecules. Therefore, the rate of a chemical reaction must depend upon more than improving the chances that molecules will bump into each other.

The discrepancy between reaction rate and molecular velocity is explained by a disproportionately rapid rise in the number of molecules which are activated as the temperature increases. Activated molecules have acquired enough potential energy to become unstable. The gain in potential energy in-

**Figure 10-9** *Factors which affect the rate of chemical reactions. Asterisks indicate activated molecules. A. The effect of temperature. B. The effect of pressure. C. The effect of reactant concentration. D. The effect of reactant size. E. The effect of enzymes.*

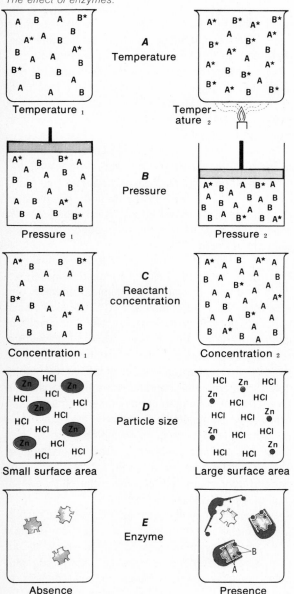

creases their oscillatory and rotatory motions but not their translational motion. Recall from Chap. 8 that only translational motion is measured by temperature. Molecular instability is reflected by internal rearrangements of certain atoms. Only activated molecules can engage in chemical reactions. The rapid increase in the number of activated molecules accounts for a doubling of the reaction rate when the temperature is doubled.

This concept is illustrated in Fig. 10-9A. At temperature$_1$, only 4 of the 20 molecules are activated. After heating to temperature$_2$, 14 of the 20 molecules are in an activated state. Since molecular collisions are more probable and the number of activated molecules is much greater at temperature$_2$ than at temperature$_1$, the average rate of reaction will be considerably higher at the former than at the latter temperature.

**The effect of pressure** An increase in the pressure decreases the volume of space in which reactant molecules can interact chemically. Since volume is compressed, the density of the reactants increases. As their density rises, so does the probability of molecular collisions.

This effect is shown in Fig. 10-9B. At pressure$_1$, the 20 reactant molecules are randomly distributed throughout the volume of the container. Four of the molecules in the container are activated, but they are relatively far apart. At pressure$_2$, on the other hand, the activated molecules are closer to each other. Thus, chemical interactions between them do not require as much time, compared with the reaction rate at pressure$_1$.

**The effect of reactant concentration** We have already seen that the removal of products, by reducing the reaction rate from right to left, shifts the equilibrium point further to the right. Should we expect a similar result from an increase in the concentration of reactants?

The answer is conceptualized in Fig. 10-9C. There are 10 molecules of A and 10 molecules of B at concentration$_1$. Two of each type of molecule are activated. The only change at concentration$_2$ is a doubling of the number of A molecules. Therefore at concentration$_2$, there are 20 A molecules, of which 4 are activated and 10 B molecules, of which 2 are activated. Since these 30 molecules are crowded into the space formerly occupied by 20 molecules, their density and the probability of collisions between activated molecules are increased.

Doubling the concentration of A doubles its chances of colliding with B. Thus, the probability of collisions between activated molecules also increases 100 percent. Since A combines with B in a ratio of 1:1, the amount of product formed by 20 A molecules and 10 B molecules is identical with that formed by 10 A and 10 B molecules. In other words, if the chemical reaction between A and B at concentration$_2$ is irreversible, then 10 A molecules will remain uncombined after completion of the reaction. To summarize, when the concentration of one of the reactants is increased, the rate of the reaction is increased but the equilibrium point remains unaffected.

Doubling the concentration of both reactants will quadruple the reaction rate, since the probability of collision has doubled for each of the two reactants. During the same period, the concentration of products formed and the probability of their chemical interaction also quadruple. The net result of these equal but diametrically opposed increases in reaction rate is to increase the concentration of products without shifting the equilibrium point.

**The effect of particle size**  The relative size of the reactant molecules also influences the rate of reaction. Decreasing the particle size increases the surface area of the particle, thereby exposing a greater number of individual atoms to potential collisions with other particles.

The relation between size and chemical activity is depicted in Fig. 10-9D. Imagine zinc (Zn) particles immersed in a solution of hydrochloric acid (HCl). When the Zn particles are large, i.e., exhibit a relatively small surface area per unit mass, the reaction to form $ZnCl_2$ is slow. Conversely, small Zn particles have a relatively large surface area, exposing many atoms to the action of the acid. As a consequence, many molecules of $ZnCl_2$ are formed.

This principle explains numerous phenomena which would be difficult to interpret otherwise. For example, many chemical reactions are promoted by mixing, because the surface area of the reactants is increased. Methods used to increase the chemical activity of solids include slicing, grinding, and pulverizing.

Ionic compounds such as $Na^+Cl^-$ dissociate immediately in aqueous solution, whereas polar covalent compounds such as HCl require a longer period to ionize. The difference in their rates of ionization is in part related to their relative sizes. $Na^+$ and $Cl^-$ are already in ionic form and therefore present a relatively greater surface area than HCl which is a molecule.

**The effect of enzymes**  Depending on the conditions of a reaction, enzymes can either increase or decrease the reaction rate. Since the products of one biochemical reaction are usually the reactants of the next reaction in a metabolic pathway, we ordinarily say that enzymes increase the rate of biochemical reactions without mentioning their potential to accomplish the opposite task.

The mechanism by which enzymes increase the reaction rate is diagramed in Fig. 10-9E. At the left is an aqueous solution such as protoplasm containing a trace amount of a reactant. Although chemical activity of a low level may occur when a few of the molecules acquire enough potential energy to have one or more of their chemical bonds broken, few of these molecules do so at the temperatures and pressures which prevail in the body.

At the right in Fig. 10-9E, an enzyme with a shape which can conform to the active parts of the reactant is added to the aqueous solution. Each enzyme molecule adsorbs (see Chap. 7) a molecule of reactant to its surface by combining with specific portions of the reactant. There are two kinds of combining sites on an enzyme, active sites (A) and catalytic sites (B). Both types of sites must conform exactly to the shape of the reactant molecule. Once adsorbed, the reactant easily undergoes a structural change which results in its splitting or breakdown into products. Since a certain amount of potential energy is ordinarily required to degrade the reactant molecule, the enzyme must *lower the activation energy* of the reactant. How enzymes accomplish this feat is not known with certainty.

The formation of products through the enzymatic degradation of reactants involves the displacement of electrons in the outermost energy levels of particular atoms in the reactant molecules. Therefore, the changes undergone by the reactants are chemical rather than physical. However, the products do not remain adsorbed to the enzymes. Instead, the products are set free in solution. After the release of the products, the enzymes revert to the shapes they had when uncombined. The situation is somewhat analogous to gloves when not worn on the hands. Thus, after each enzyme-mediated reaction is complete, enzymes are again free to adsorb reactants for which they are specific.

Rate versus time    The reaction rate can be expressed in terms of the time rate of disappearance of reactant molecules or the percentage of reactant molecules left after each unit of time. In either case, curves of the general type shown in Fig. 10-10 are obtained for the same reactants under different conditions.

The lower curve is characteristic of fast reactions, whereas the upper curve is characteristic of slow reactions. We have already followed the dynamics which lead to chemical equilibrium. Here we are concerned only with differential rates due to varying the conditions of a reaction. Notice that the reaction described by the upper curve has not yet reached equilibrium during the period in which it is followed. If the equilibrium point of the chemical reaction is not affected by the factors identified in the figure, then the upper curve can be expected to unite with the horizontal portion of the lower curve at some future time. If, however, chemical equilibrium is shifted by the imposed conditions, the two curves will never unite.

To better visualize the second case, let us imagine that the end of the upper curve on the right-hand side of the graph is the equilibrium point of the chemical reaction under a given set of conditions. From this point on the graph, draw a horizontal line to the left until it intersects the vertical scale. Notice that the amount of reactant which has disappeared is slightly over 50 percent. Thus, a similar amount of reactant still remains when chemical equilibrium is attained. This is far different from the percentage of reactant left at the equilibrium point of the lower curve.

## AQUEOUS SOLUTION

Water ($H_2O$) is the solvent of an aqueous solution. Many compounds dissolve in $H_2O$ while many others do not. This section is devoted to the various types of interactions between $H_2O$ and introduced solutes.

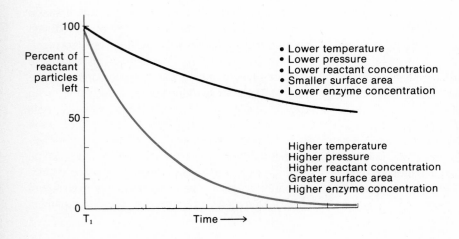

*Figure 10-10* *The time rate of disappearance of reactants under various conditions. The reaction rate is slow when the temperature, pressure, and reactant and enzyme concentrations are low or when the size of the reactants is large (upper curve). The reaction rate is rapid when these conditions are reversed (lower curve).*

## IONIZATION AND SOLUTION

Each molecule of $H_2O$ is capable of electrostatically attracting, and being attracted to, charged solute particles. The reason is that $H_2O$ is a polar solvent. Indeed, $H_2O$ is the most polar solvent available. However, $H_2O$ acts in different ways on solute molecules, depending on whether they are ionic, polar, or nonpolar.

**Ionic compounds in aqueous solution**   If they are mixed with $H_2O$, ionic compounds break up, or *dissociate*, into positively and negatively charged ions. The attractive forces exerted by $H_2O$ on the charged portions of ionic compounds give rise to charged particles called *electrolytes*. This process is of fundamental importance in the aqueous medium of protoplasm.

Therefore, we shall focus our inquiry first on what happens to an ionic compound such as common table salt when mixed with $H_2O$. Called sodium chloride ($Na^+Cl^-$), table salt is built up of alternating ions of sodium ($Na^+$) and chloride ($Cl^-$) which form a crystalline solid. Because of the attractive forces exerted by their opposite electric charges, the $Na^+$ and $Cl^-$ are held together physically. In other words, although $Na^+Cl^-$ is a crystalline solid, its substituent atoms are already in ionic form.

When $Na^+Cl^-$ is placed in aqueous solution, each of the salt crystals dissolves or disappears. Practically instantaneous, the reaction is familiar to everyone. What really happens can be explained by reference to Fig. 10-11.

Notice that some $H_2O$ molecules begin to orient electrostatically to the $Na^+$ and $Cl^-$ on the surface of the crystal, because these ions are exposed directly to the solvent action of $H_2O$. For example, the positive portions of two $H_2O$ molecules are aligned to the $Cl^-$ on the left-hand side of the crystal in the drawing. Similarly, the negative portions of two $H_2O$ molecules are oriented to the $Na^+$ on the right-hand side of the crystal. The other peripheral $Na^+$ and $Cl^-$ are also being surrounded by oriented $H_2O$ molecules, but , for simplicity, only the fates of $Cl^-$ and $Na^+$ defined by dashed circles will be followed. Bear in mind, however, that what happens to them also happens to the rest

of the ions in the salt crystal and that the entire ionization process is essentially instantaneous.

As more and more $H_2O$ molecules surround the outermost $Na^+$ and $Cl^-$ of the crystal, the electrostatic forces holding adjacent $Na^+$ and $Cl^-$ together are increasingly diminished and finally disrupted completely. Therefore, as $Cl^-$ is surrounded by the positive ends of an increasing number of $H_2O$ molecules, a point is reached when the attractive forces pulling $Cl^-$ away from the crystal exceed the forces holding $Cl^-$ to the crystal. The crystal loses the tug-of-war with $H_2O$, so that $Cl^-$ is literally torn from it and set free in solution. Of course, $Cl^-$ is still surrounded by $H_2O$ molecules oriented through electrostatic forces of attraction, as shown in the upper left-hand corner of the illustration. Removal of $Cl^-$ exposes the next $Na^+$ within the crystal to the solvent action of $H_2O$.

The same fate awaits $Na^+$. The negative ends of an increasing number of $H_2O$ molecules are responsible for overcoming the attractive forces holding the ion to the crystal. After being set free in solution, $Na^+$ is still surrounded by oriented $H_2O$ molecules, as shown in the lower right-hand corner of the illustration.

The ionization process is continued until every $Na^+$ and $Cl^-$ in the original crystal are set free in solution. Each free ion is of course surrounded by a layer of oriented $H_2O$ molecules. Bound electrostatically to the ions which they surround, bound $H_2O$ is referred to as *water of hydration*. Ions wearing a coat of oriented $H_2O$ molecules are called *hydrated ions*.

The ionization of ionic compounds in aqueous solution does not represent a chemical change. Remember that chemical changes involve electronic interactions between atoms. The weakening of electrostatic forces of attraction between $Na^+$ and $Cl^-$ in a salt crystal due to the solvent action of $H_2O$ does not affect electrons in the outermost energy levels of either ion. In other words, the reaction

$$Na^+Cl^- + \cancel{H_2O} \rightarrow Na^+ + Cl^- + \cancel{H_2O} \qquad (10\text{-}27)$$

must represent a physical rather than a chemical change. That $Na^+Cl^-$ has gone into solution or is completely soluble in $H_2O$ can be verified by the recrystallization of salt when $H_2O$ is completely evaporated from an aqueous solution containing nothing but $Na^+$ and $Cl^-$.

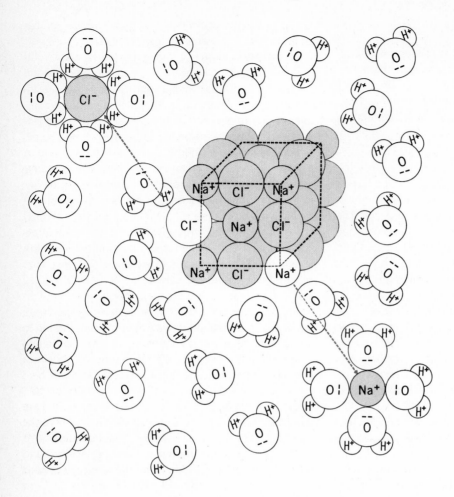

**Figure 10-11** *The ionization of Na⁺Cl⁻ in H₂O. The electrostatic forces of attraction exerted by the H₂O molecules overcome the forces holding the ions to the crystal, permitting them to enter solution.*

**Acids and bases in aqueous solution**  The ionization of acids and bases in $H_2O$ depends on another mechanism. The mechanism by which a strongly polar covalent compound such as hydrochloric acid (HCl) dissociates will serve as a model. Since the shared pair of electrons which hold the atoms of a HCl molecule together are considerably closer to Cl than to H, the molecule has the equivalent of a positive end represented by the nucleus of a H atom and a negative end represented by Cl with a de facto excess of one electron. When HCl is mixed with $H_2O$, the double negative charge of an oxygen atom belonging to each $H_2O$ molecule wins a tug-of-war with the single nega-

tive charge of each HCl molecule for possession of the H nuclei or protons. Left in solution are Cl⁻ surrounded by water of hydration.

Notice that this ionization mechanism, although different from that described for ionic compounds, still does not involve new electronic interactions between atoms. The shared pair of electrons, one of which is contributed by H, is already close to Cl in HCl before the H nucleus is ripped from each molecule. On the other hand, considerable energy is released when HCl is mixed with $H_2O$. The release of energy suggests that the reaction is chemical rather than physical. We may conclude that the reaction between strongly polar covalent compounds and $H_2O$ can be considered either physical or chemical, depending on what aspects of the interaction are being stressed.

**Polar compounds in aqueous solution**  Some other types of molecules do not ionize in $H_2O$. What does happen revolves around the polarity or lack of polarity exhibited by the compound. Polar molecules have a distinct separation of charge, so that one end of the molecule is positive and the other end of the molecule is negative.

Alcohol is one example of a polar molecule. However, only the alcohols with a small number of carbon (C) atoms in their structures are soluble in $H_2O$. When methyl or ethyl alcohol ($CH_3OH$ or $CH_3CH_2OH$) or other short-chain alcohols are mixed with $H_2O$, they dissolve but not through the process of ionization. The solubility of alcohol in $H_2O$ is explained by the electrostatic attraction between the $H^+$ portion of $H_2O$ and the $OH^-$ portion of the alcohol. Since the C portion of an alcohol is insoluble, an increase in the C chain length of an alcohol also increases its degree of insolubility.

The solubility of other types of polar molecules is explained by the electrostatic attraction between oppositely charged portions of the solvent and solute. That is, the positive and negative ends of water molecules are aligned to the negative and positive ends, respectively, of the polar solute molecules. We can conceptualize this arrangement in the following manner:

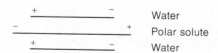

The shorter horizontal lines represent water molecules. The longer horizontal line represents a molecule of polar solute. The water molecules are oriented to the polar solute by electric charges of opposite sign.

**Nonpolar compounds in aqueous solution**  Nonpolar molecules with polar covalent bonds also bear equal but opposite charges. However, the charges are symmetrically distributed along each molecule so that there is no distinct separation of charge. When nonpolar molecules are mixed with $H_2O$, they do not dissolve and are therefore called *insoluble*. Many types of C compounds are nonpolar.

**pH**

Chemicals which give up hydrogen ions ($H^+$) in aqueous solution are called *acids*. Recall that hydrogen ions are the nuclei of hydrogen atoms and are therefore protons. Thus, *proton donors* are acids. On the other hand, chemicals which take up hydrogen ions from the solution are referred to as *bases*. Usually these chemicals contain hydroxyl ions ($OH^-$). It follows that bases are *proton acceptors*. In other words, whether a substance is an acid or a base depends on its ability to either give up or accept protons.

**The dissociation of pure water**  Pure water, itself, is slightly dissociated. The particles which form from the dissociation of a water molecule are a hydrogen ion ($H^+$) and a hydroxyl ion ($OH^-$). The $OH^-$ bears a negative charge because of the electron left behind by $H^+$. Only 0.0000001 mole (1/10,000,000, or $10^{-7}$ mole) of $H^+$ is present in 1 liter of water. Since we are dealing with ionized hydrogen atoms in solution, the concentration of the entire solution is identical to this value, or $10^{-7}$ normal ($N$). In other words, $10^{-7}$ $N = 10^{-7}$ $M = 10^{-7}$ Gm. Furthermore, because each dissociated $H_2O$ molecule gives rise to one $H^+$ and one $OH^-$, 1 liter of pure water also contains $10^{-7}$ mole, or Gm, of $OH^-$. Thus, pure water contains the same number of $H^+$ and $OH^-$ per liter, i.e., $10^{-7}$ $N$.

The value of the hydrogen ion concentration [$H^+$] in pure water fixes the value of the hydroxyl ion concentration [$OH^-$]. The quantitative expression of this concept is written as follows:

$$[H^+]\,[OH^-] = K \qquad (10\text{-}28)$$

Equation (10-28) states that the product of the concentrations of the hydrogen and hydroxyl ions in an aqueous solution is equal to a constant. An increase in the [$H^+$] requires a corresponding decrease in the [$OH^-$] and vice versa, since their product is always the same.

These principles provide us with a basis for

measuring the acidity or alkalinity of any aqueous solution. Remember that the ability of pure water to give up protons, as reflected by its $[H^+]$, is exactly the same as its ability to accept protons because of the presence of the same number of $OH^-$. Pure water is therefore neither acidic nor basic; instead, it is neutral. Because pure water is $10^{-7} N$, the strength of any acid or base, i.e., the number of gram equivalents of $H^+$ or $OH^-$ per liter of aqueous solution, can be compared to that of pure water.

**The expression of acidity** Now, suppose that an acid, or proton donor, is added to pure water until the $[H^+]$ per liter of aqueous solution is $10^{-6} N$. The $[H^+]$ changes from 1/10,000,000 mole/liter to 1/1,000,000 mole/liter, a tenfold increase. At the same time, the $[OH^-]$ changes from 1/10,000,000 mole/liter to 1/100,000,000 mole/liter, a tenfold decrease, so that $[H^+]$ $[OH^-] = K$.

If more acid is added until the concentration of the solution is $10^{-5} N$, the $[H^+]$ undergoes an additional tenfold increase, i.e., from 1/1,000,000 mole/liter to 1/100,000 mole/liter, while the $[OH^-]$ undergoes an additional tenfold decrease, i.e., from 1/100,000,000 mole/liter to 1/1,000,000,000 mole/liter. Compared with pure water, a $10^{-5} N$ solution contains 100 times as many $H^+$ and 100 times fewer $OH^-$ per unit volume, since each unit increase in normality ($N$), i.e., from $10^{-7}$ to $10^{-6}$ and from $10^{-6}$ to $10^{-5}$, represents a tenfold increase in $[H^+]$ and a corresponding decrease in $[OH^-]$.

Assume that enough acid is finally added to the solution to bring its $[H^+]$ from the original value of $10^{-7}$ mole/liter (1/10,000,000 mole/liter) to $10^{-0}$ mole/liter (1/1 = 1 mole/liter). To increase the acidity of pure water to 1 mole of $H^+$ per liter of solution requires a 10 millionfold increase in $[H^+]$. At the same time, the $[OH^-]$ of the solution decreases from its original value of $10^{-7}$ mole/liter (1/10,000,000 mole/liter) to $10^{-14}$ mole/liter (1/100,000,000,000,000, or one hundred-trillionth of a mole/liter), a 10 millionfold decrease.

**The expression of alkaanity** Conversely, if a base or proton acceptor is added to pure water until the $[H^+]$

per liter of aqueous solution is reduced to $10^{-8} N$, the $[H^+]$ changes from 1/10,000,000 mole/liter to 1/100,000,000 mole/liter, a tenfold decrease. At the same time, the $[OH^-]$ changes from 1/10,000,000 mole/liter to 1/1,000,000 mole/liter, a tenfold increase. This increase is required to maintain a constant product between $[H^+]$ and $[OH^-]$.

If more base is added until the concentration is $10^{-9} N$, the $[H^+]$ undergoes an additional tenfold decrease, i.e., from 1/100,000,000 mole/liter to 1/1,000,000,000 mole/liter, while the $[OH^-]$ undergoes an additional tenfold increase, i.e., from 1/1,000,000 mole/liter to 1/100,000 mole/liter. Compared with pure water, a $10^{-9} N$ solution contains 100 times fewer $H^+$ and 100 times more $OH^-$ per unit volume, because of the two tenfold decreases and increases in $[H^+]$ and $[OH^-]$, respectively.

Assume that enough base is finally added to the solution to reduce its $[H^+]$ from the original value of $10^{-7}$ mole/liter to $10^{-14}$ mole/liter. To decrease the acidity of pure water by this amount requires a 10 millionfold decrease in $[H^+]$. At the same time, the $[OH^-]$ of the solution increases from its original value of $10^{-7}$ mole/liter to $10^0$ mole/liter, a 10 millionfold increase.

**The pH scale** The principles we have just reviewed are embodied in a pH scale which gives a quantitative measure of the acidity or alkalinity of any aqueous solution. The midpoint of the scale is fixed by the $[H^+]$ of pure water, $10^{-7} N$. For our purposes in dealing with the practical aspects of expressing $[H^+]$, the negative exponent is transformed into a positive whole number. Thus, the pH of pure water is 7. This, of course, means that the $[H^+]$ is $10^{-7} N$ but eliminates the awkwardness of using negative numbers with many zeros.

The extremes of the pH scale are fixed by the expression $[H^+]$ $[OH^-] = K$. Recall that the strongest acids have a $[H^+]$ and $[OH^-]$ of $10^0$ mole/liter and $10^{-14}$ mole/liter, respectively, and that the strongest bases possess a $[H^+]$ and $[OH^-]$ of $10^{-14}$ mole/liter and $10^0$ mole/liter, respectively. Therefore, in terms of $[H^+]$, the strongest acids have a pH of 0 ($10^0$ mole/liter = 1/1 mole/liter = 1 mole/liter), and the strongest bases possess a pH of 14 ($10^{-14}$ mole/liter).

The pH scale and the corresponding [H⁺] and [OH⁻] in moles per liter are given in Fig. 10-12.

While the pH scale ranges from 0 to 14, there are acids whose [H⁺] is greater than $10^0$ mole/liter (1 N) and bases whose [H⁺] is less than $10^{-14}$ mole/liter ($10^{-14}$ N). However, these facts do not detract from the usefulness of the pH scale, since most acids and bases, including all those of biologic importance, have a [H⁺] within its range of measurement.

**The measurement of pH**  The actual measurement of pH can be done colorimetrically or electrometrically. *Colorimetric measurements* rely on a predict-

**Figure 10-12**  *The pH scale. The scale ranges between zero and 14. Each step represents a tenfold difference in [H⁺] from the next in line. A pH of 7 is called neutrality. A decrease in pH is caused by an increase in [H⁺] (colored bars) and is reflected by a corresponding decrease in [OH⁻] (black bars). An increase in pH is caused by the opposite conditions. The ranges of some physiologically important pH values are given above the graph.*

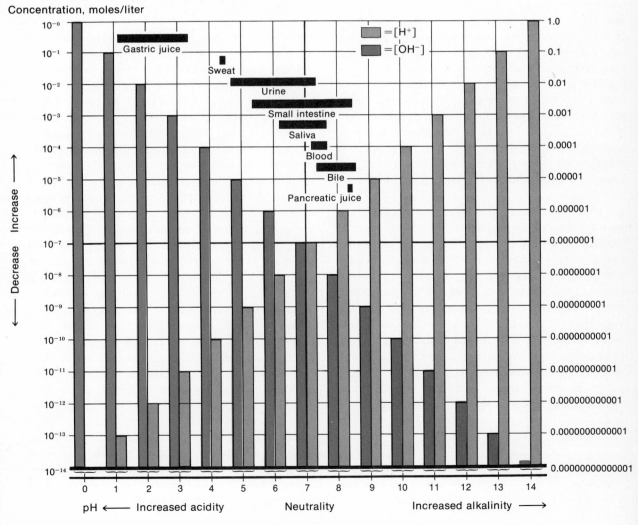

able change in the color of litmus paper or an indicator dye when placed in an acid or alkaline solution. Colorimetric pH readings are relatively imprecise.

*Electrometric measurements* of pH require a pH meter. This instrument is a potentiometer which converts measurements of differences in voltage caused by variations in $[H^+]$ directly into pH readings on a properly calibrated dial. The virtue of electrometric pH readings is their accuracy.

**pH and body functions** A few representative pH values from regions or natural products of the human body are displayed in Fig. 10-12. Notice that gastric juice is extremely acid, bile is the most strongly alkaline component listed, and blood is slightly alkaline. In addition, observe that the normal pH values of the small intestine, urine, and gastric juice are highly variable while those of blood, sweat, and pancreatic juice have very narrow ranges. Of the biologically important pH values shown, only those of urine, the contents of the small intestine, and saliva normally vary between acid and alkaline values. Most of the pH values of the body fall between 6 and 8. These are relatively narrow limits, compared with the range of the entire pH scale, since there is only a hundredfold difference in $[H^+]$ and $[OH^-]$ in the pH range 6 to 8.

The importance of $[H^+]$ to the normal functioning of the body is probably best exemplified by the maintenance of a constant blood pH. The pH of arterial blood normally ranges between 7.35 and 7.45. A long-term (other than very temporary) increase in the pH of arterial blood beyond 7.45 gives rise to a pathologic condition called alkalosis, or alkalemia. Conversely, a long-lasting decrease in the pH of arterial blood below 7.3 results in an abnormal condition referred to as acidosis, or acidemia. The severity of the symptoms depends on how divergent the arterial blood pH is from its normal range. The causes and effects of alkalemia and acidemia are considered in detail in Chap. 31.

## ACID-BASE REACTIONS

In the previous discussion, we indicated that $H_2O$ has a vanishingly small but measurable tendency to

ionize into $H^+$ and $OH^-$. Since $H^+$ represents a proton and is extremely small, compared with the size of atoms, it is highly unlikely that protons remain as isolated particles in aqueous solution. Rather, by virtue of the positive charge borne by a free proton, each proton is electrostatically attracted to the negative end of a $H_2O$ molecule. Using dots to indicate the outermost electrons of the atoms in $H_2O$, its ionization can be conceptualized as follows:

$$\overset{=}{\underset{H_+}{:\overset{..}{\underset{..}{O}}:}}\ H^+ +\ \overset{=}{\underset{H_+}{:\overset{..}{O}:}}\ H^+ \rightarrow H^+\ \overset{=}{\underset{H_+}{:\overset{..}{O}:}}\ H^+ +\ \overset{=}{:\overset{..}{O}:}\ H^+ \quad (10\text{-}29)$$

Eliminating the dots and expressing the net charge on each ion, the reaction is reduced to:

$$H_2O + H_2O \rightleftharpoons H_3O^+ + OH^- \quad (10\text{-}30)$$

$H_3O^+$ is called a *hydronium ion* and is really nothing more than a hydrated $H^+$.

These equations inform us of certain points which the earlier formulation, based on the dissociation of $H_2O$ into $H^+$ and $OH^-$, does not. First, note that $H_2O$ serves as a proton donor, or acid, as well as a proton acceptor, or base. Second, observe that the free proton donated by a $H_2O$ molecule is electrostatically attracted to one of the two unshared pairs of electrons which surround the oxygen atom of another $H_2O$ molecule. No electronic displacements are involved in the formation of $H_3O^+$ and $OH^-$, even though the electrostatically attracted proton establishes the equivalent of a polar covalent bond with the oxygen atom.

Since $H_2O$ has such a slight tendency to ionize into $H_3O^+$ and $OH^-$, these ions alternatively exhibit a strong tendency to form water when they are present in solution. In other words, a proton has a much greater probability of being attracted to $OH^-$ from $H_3O^+$ than it has of being attracted to one $H_2O$ from another $H_2O$ during collisions between these molecules. To show these ideas, the ionization reaction of Eq. (10-30) should be written in reverse, i.e.,

$$H_3O^+ + OH^- \rightarrow H_2O + H_2O \quad (10\text{-}31)$$

Equation (10-31) informs us that the equilibrium point is all the way to the right, so that every time $H_3O^+$ forms, it gives up its proton to $OH^-$, forming $H_2O$. This, in turn, means that $H_3O^+$ and $OH^-$ are a stronger acid and base, respectively, than $H_2O$.

We already know that $H_2O$ can act as either a proton donor or acceptor. Therefore, when $H_3O^+$ donates a proton to $OH^-$, the $H_2O$ formed becomes a potential acid. Conversely, the $H_2O$ which remains after $H_3O^+$ dissociates is a potential base. In other words, $H_2O$ has a much lower tendency to donate a proton, compared with $H_3O^+$, or to accept a proton, compared with $OH^-$. Using the ionization equation for $H_2O$ in Eq. (10-31), we can identify the acids and bases on each side in the following manner:

$$H_3O^+ + OH^- \rightarrow H_2O + H_2O \qquad (10\text{-}32)$$
Acid$_1$  Base$_2$   Acid$_2$  Base$_1$

Applying the above terminology, let us review the nature of the changes which occur during an acid-base reaction:

1   When acid$_1$ donates a proton, it becomes potential base$_1$.
2   When base$_2$ accepts a proton, it becomes potential acid$_2$.
3   Acid$_1$ is stronger than acid$_2$.
4   Base$_2$ is stronger than base$_1$.

Since pure $H_2O$ has a pH of 7 and is therefore neutral, an interaction between an acid and a base in aqueous solution is called a *neutralization reaction*. Neutralization is complete when all the available $H^+$ and $OH^-$ are combined as $H_2O$. Neutralization is incomplete when one of these ions is in greater abundance than the other. That is, the pH will be less than 7 when $H^+$ is in excess or more than 7 when $OH^-$ is in excess.

Let us see how these principles are applied to situations in which $H_2O$ contains an acid, a base, or a combination of both.

**Strong acids in aqueous solution**   First, consider the addition of HCl (hydrochloric acid) to $H_2O$. HCl is a strong acid because the shared pair of electrons between H and Cl is pulled much closer to Cl, the more electronegative of the two atoms. The strongly polar covalent bonding means that the electrostatic force of attraction between $H^+$ and $Cl^-$ can easily be broken when the negative portion of a water molecule comes in close proximity. In fact, virtually all the pro-

tons of HCl are donated in aqueous solution. Using dots to indicate the outermost electrons of the atoms in HCl and $H_2O$, the reaction can be conceptualized as follows:

$$\overset{-}{:}\overset{\cdot\cdot}{\underset{\cdot\cdot}{Cl}}\overset{\cdot\cdot}{:}\ H^+ +\ \overset{=}{\underset{H^+}{:}}\overset{\cdot\cdot}{\underset{\cdot\cdot}{O}}\overset{\cdot\cdot}{:}\ H^+ \rightarrow\ {}^+H\ \overset{\cdot\cdot}{\underset{H^+}{:}}\overset{\cdot\cdot}{\underset{\cdot\cdot}{O}}\overset{\cdot\cdot}{:}\ H^+ +\ \overset{\cdot\cdot}{:}\overset{\cdot\cdot}{\underset{\cdot\cdot}{Cl}}\overset{\cdot\cdot}{:}\ ^- \qquad (10\text{-}33)$$

Eliminating the dots and expressing the net charge on each of the molecules and ions, the reaction is reduced to the following equation:

$$HCl + H_2O \rightleftharpoons H_3O^+ + Cl^- \qquad (10\text{-}34)$$
Acid$_1$  Base$_2$       Acid$_2$  Base$_1$

The above reaction is essentially one between a very strong acid (HCl) and a very weak base ($H_2O$). Although HCl has little difficulty donating protons, $Cl^-$ has little tendency to accept protons, compared with $H_2O$. In other words, since base$_2$ is stronger than base$_1$, HCl tends to ionize virtually completely in aqueous solution.

**Strong bases in aqueous solution**   Now, consider the addition of sodium hydroxide ($Na^+OH^-$) to $H_2O$. Sodium hydroxide is a strong base because of the ionic bond between Na and O. Therefore $Na^+$ is bound electrostatically to $OH^-$. Notice that bases are ionic compounds while acids are polar covalent compounds.

When $Na^+OH^-$ is mixed with $H_2O$, both ions are pulled apart by oppositely charged portions of $H_2O$ molecules. In other words, $Na^+$ is surrounded by the negatively charged portions of some $H_2O$ molecules, while $OH^-$ is enveloped by the positively charged regions of other $H_2O$ molecules. Using dots to indicate the outermost electrons in $Na^+OH^-$, the ionization of $Na^+OH^-$ in $H_2O$ can be summarized in the following manner:

$$Na^+\ \overset{\cdot\cdot}{:}\overset{\cdot\cdot}{\underset{\cdot\cdot}{O}}\overset{\cdot\cdot}{:}\ H^- + H_2O \rightarrow Na^+ +\ \overset{\cdot\cdot}{:}\overset{\cdot\cdot}{\underset{\cdot\cdot}{O}}\overset{\cdot\cdot}{:}\ H^- + H_2O \qquad (10\text{-}35)$$

Observe that $Na^+OH^-$ is already ionic before entering solution. Because the solvent $H_2O$ does not change chemically during the reaction, the only change of any consequence involves the ability of the oppositely charged ions in $Na^+Cl^-$ to dissolve almost completely in the polar solvent $H_2O$.

**Neutralization reactions between strong acids and bases** Next, let us add equal amounts of HCl and $Na^+OH^-$ to $H_2O$. Each molecule of HCl and $Na^+OH^-$ gives rise to one $H^+$ and $OH^-$, respectively. Since the addition of HCl to $H_2O$ forms $H_3O^+$ and $Cl^-$ while the addition of $Na^+OH^-$ to $H_2O$ gives rise to $Na^+$ and $OH^-$, the overall reaction can be summarized as follows:

$$H_3O^+ + \cancel{Cl^-} + \cancel{Na^+} + OH^- \rightarrow H_2O + H_2O + \cancel{Na^+} + \cancel{Cl^-} \tag{10-36}$$

Because $Na^+$ and $Cl^-$ are common to both sides of the equation, we need not concern ourselves with these terms. Indeed, since crystalline $Na^+Cl^-$ is a salt, we may generalize by stating that the salt concentration does not change during a neutralization reaction. In other words, salt does not form as a result of the interaction between a strong acid and a strong base, since it is present before the reaction is initiated.

When condensed, Eq. (10-36) is identical to the ionization equation for $H_2O$ in Eq. (10-31). Thus, when a strong acid and a strong base are mixed together in $H_2O$, what really happens is that $H_3O^+$ donates protons to $OH^-$, forming $H_2O$.

**Weak acids and bases in aqueous solution** The strengths of different acids at the same concentration vary widely. Are the principles we have developed applicable to weak acids and weak bases?

To answer this question, we will look at a common example, the ionization reaction of acetic acid ($HC_2H_3O_2$, abbreviated HAc) in $H_2O$. Only about 1 percent of HAc molecules give up their protons in aqueous solution because of the relatively large ionization energy required to remove $H^+$ from $C_2H_3O_2^-$ (acetate ion). In other words, the covalent bond between H and C is only weakly polar.

The following reaction shows what happens when HAc is added to $H_2O$:

$$HAc + H_2O \rightleftharpoons H_3O^+ + Ac^- \tag{10-37}$$
$$\text{Acid}_1 \quad \text{Base}_2 \qquad \text{Acid}_2 \quad \text{Base}_1$$

One in approximately one hundred HAc molecules ionizes, forming $H^+$ and $Ac^-$. Because HAc is a weak acid, losing a proton only with difficulty, the potential base $Ac^-$ left after ionization of HAc must have a strong tendency to accept protons. The proton-accepting abilities of $Ac^-$ replenish a large portion of the HAc molecules which have undergone ionization.

To be an effective proton acceptor, $Ac^-$ must be a stronger base than $H_2O$. Otherwise, HAc would be a strong rather than a weak acid. Therefore, the tendency of $H_3O^+$ to donate protons is greater than that of HAc; i.e., $H_3O^+$ is a stronger acid than HAc.

To summarize, the equilibrium point of the reaction is shifted far to the left for two reasons: (1) $Ac^-$ accepts protons much more readily than HAc donates them and (2) $H_3O^+$ donates protons more readily than $H_2O$ accepts them.

**Neutralization reactions between weak acids and strong bases** What happens when a strong base is added to a weak acid in aqueous solution? Assume that $Na^+OH^-$ is added to HAc in $H_2O$. The following equation shows the results of the reaction:

$$HAc + Na^+OH^- \rightarrow Na^+ + Ac^- + H_2O \tag{10-38}$$
$$\text{Acid}_1 \quad \text{Base}_2 \qquad\qquad \text{Base}_1 \quad \text{Acid}_2$$

Note that $OH^-$ does not remain free in solution even though $Na^+OH^-$ ionizes almost completely. The reason becomes obvious when we compare the relative strengths of the bases $OH^-$ and $Ac^-$. Both are capable of accepting protons, but $OH^-$ is much stronger than $Ac^-$. Since $OH^-$ readily combines with free protons given up by HAc, the proton deficit causes more HAc to ionize. In other words, when $OH^-$ combines physically with $H^+$, the effect is equivalent to an increase in the concentration of un-ionized HAc. Thus, $OH^-$ is effectively removed from solution as $H_2O$ by promoting the further ionization of HAc, leaving behind $Na^+ + Ac^-$. These two ions form a salt known as sodium acetate which would remain as a crystalline solid if $H_2O$ were evaporated completely.

To summarize, the reaction between a weak acid and a strong base gives rise to $H_2O$, thereby removing $OH^-$ and leaving in solution ions composed of the salt of the weak acid.

**Buffering** What happens when a strong acid such as HCl is added to HAc, $Na^+$, and $Ac^-$ in $H_2O$? Since HAc is only slightly ionized, its $H^+$ contribution is negligible compared with that from HCl; i.e., HCl is a strong acid while HAc is a weak acid. On the other hand, $Na^+$ and $Ac^-$ are already in ionic form in solution. Therefore, we need only concern ourselves with the interaction between $Na^+$, $Ac^-$, and $HCl$. The reaction is shown below:

$$\cancel{Na^+} + Ac^- + HCl \rightarrow HAc + \cancel{Na^+} + \cancel{Cl^-} \qquad (10\text{-}39)$$
$$\text{Base}_1 \; \text{Acid}_2 \qquad \text{Acid}_1 \qquad \text{Base}_2$$

Because $Ac^-$ is a much stronger base than $Cl^-$, protons donated by $HCl$ are readily accepted by $Ac^-$. This interaction results in the formation of $HAc$, leaving $Na^+$ and $Cl^-$ free in solution. Since $acid_1$ is much weaker than $acid_2$, $H^+$ is removed from solution. Thus, when a strong acid is added to the salt of a weak acid in $H_2O$, the weak acid is formed while the salt of the strong acid remains in ionic form.

The important feature about the interaction between a strong acid or a strong base and a weak acid (or a weak base) plus its salt [Eq. (10-39)] is the lack of a significant change in either the $[OH^-]$ or $[H^+]$ of the solution. This fact is reflected by a lack of a change in the pH of the solution.

A group of chemicals which stabilize pH by tying up free $H^+$ or $OH^-$ is called a *buffer system*. Weak acids or weak bases and their salts admirably qualify as buffer systems. Because salts in ionic form are always available in protoplasm and body fluids, weak acids or bases are referred to simply as *buffers*. Biologically important buffers include carbonates, phosphates, and proteinates. Buffers have vital roles in maintaining a constant pH in the blood, other fluid compartments, and tissue cells of the body.

## COLLATERAL READING

Buswell, Arthur M., and Worth H. Rodebush, "Water," *Scientific American*, April 1956. The behavior of water is analyzed by considering its varied molecular structures and arrangements.

Frieden, Earl, "The Chemical Elements of Life," *Scientific American*, July 1972. A review of the bases for the selection of elements essential to life, the suggested roles of trace elements, and the characteristics of some essential elements recently added to the list.

Kendrew, John C., "The Three-Dimensional Structure of a Protein Molecule," *Scientific American*, December 1961. Vividly conveys the difficulties of trying to decipher the shape of the protein molecule myoglobin, bringing the reader to almost the same degree of physical exhaustion that must have been experienced during the actual determinations.

Kennelly, Rosemary, and Raymond E. Neal, *Chemistry: With Selected Principles of Physics*, 2d ed., New York: McGraw-Hill Book Company, 1971. Emphasizes the application of physical principles in clinical situations.

Lambert, Joseph B., "The Shapes of Organic Molecules," *Scientific American*, January 1970. A discussion of the dynamics involved when different three-dimensional shapes are assumed by certain molecules, the atoms of which rotate freely around one or more chemical bonds.

Sackheim, George I., and Ronald M. Schultz, *Chemistry for the Health Sciences*, New York: The Macmillan Company, 1969. In a book keyed to students preparing for an allied health profession, the authors emphasize the more practical aspects of chemistry.

Schmitt, Francis O., "Giant Molecules in Cells and Tissues," *Scientific American*, September 1957. Schmitt suggests that individual protein and nucleic acid molecules are separate units in their inactive state and giant, highly organized conglomerates of individual units after activation.

# Part Three
# Integration

# CHAPTER 11
# THE SENSES

The senses are our windows of awareness of both the world "out there" and the internal environment of our own bodies. From the moment we are born (and probably a good deal earlier than that) until the day we die, we are constantly bombarded with sensory stimuli from our external and internal environments. Our knowledge, attitudes, the way we react under a given set of conditions—indeed, our basic personalities—are molded by our perceptual experiences (perception is defined below).

Yet almost everyone can agree on the names of the primary colors, can identify a musical instrument by the quality of the sounds it produces, can distinguish which way is up, can describe the texture of a material by its feel, the quality of an odor by its smell, or the taste of a foodstuff through its unique stimulation of the taste sense.

That we agree on the basic nature of the world "out there" is not by accident but by design. The designs are those built into our senses and exemplified by

their innate capacities to provide us with more or less accurate representations of what is going on in the world. Although pain, for example, is a learned concept and therefore abstract, the experience of being stuck by a pin is as real to a newborn baby as it is to an adult. That some people have an extremely high tolerance for pain (can bear pain with less discomfort than other people can) does not negate the fact that the sensory experience is real; it only emphasizes the fact that there is a difference between reception and perception.

To appreciate the distinction, we must develop a few working definitions. A *receptor* converts, or transduces, the energies of specific types of stimuli into the electrical energy of nerve impulses (see Chap. 13) that are then conducted to the central nervous system, which consists of the brain and spinal cord (see Chap. 14). Therefore, reception occurs when a stimulus initiates changes which lead to stimulus transduction in a receptor and the propagation of nerve impulses. Each receptor, whether composed of a free nerve ending or a multicellular grouping of specialized end organs, responds only to a specific type of stimulus or range of stimuli.

In a physical sense, your central nervous system is far removed from receptors. Before reaching the spinal cord or brain, the stimuli in their original form may cause functional interactions among a population of receptors which change their informational content.

The transduced signals are conducted along numerous neurons in a chain toward the central nervous system. Each of the neurons synapses, or comes in close physical proximity, with one or many neurons along the way. Sometimes, one or a few neurons may diverge to synapse with many other neurons; other times many neurons may converge to form synaptic connections with one or a few neurons. Convergence and divergence alter the informational content of the electric signals. In addition, the entire electric signal or a specific component of it may be enhanced, inhibited, or otherwise filtered by specialized neurons which influence information processing at particular synaptic junctions.

In short, before ever reaching the central nervous system, the quality of an electric signal is changed. This type of information processing contributes to perception by selecting certain characteristics of the original stimuli and filtering out other components.

Now to tackle a more difficult definition, that of perception. *Perception* occurs when a response is elicited by a stimulus. Since we tend to ignore irrelevant stimuli to which we are exposed, the lack of a response may itself be a response to a given stimulus. Although perception is commonly presumed to be mediated at a conscious level, it may also occur entirely subconsciously. That your footsteps quicken when you see a fast-moving car bearing down on you is of course a conscious response to a stimulus perceived to be dangerous. However, the dilatation of your pupils when you first look at a good-looking and scantily clad person of the opposite sex indicates that your sexual perceptions are perfectly normal, in spite of the fact that your pupillary response is automatic and, therefore, involuntary. Both responses are examples of perception. Notice that, in each of these cases, the sensory information must have been com-

pared in the brain to previous experiences of a similar nature in order to elicit appropriate responses such as a quickened stride or sexual appreciation.

Although perception requires some degree of integration by the central nervous system, the response may be integrated in the spinal cord rather than in the brain. Reflex withdrawal from a hot object with which the body has just made contact does not involve, and indeed occurs too rapidly for the intervention of, conscious awareness.

Withdrawal from a stimulus requires its prior perception accompanied by a sensation. The ability to judge, while traveling behind a truck on a two-lane highway, that the passing distance is adequate and then take appropriate action based on the judgment is a consequence of integrating perceptive experiences at both the conscious and subconscious levels. Thus, perception is not a property unique to the central nervous system and in fact occurs every time there is a change in the informational content of a neural electric signal.

We must never lose sight of the fact that the senses influence, and are influenced by, the rest of the body. In other words, sensory structures do not operate as independent entities. Perception to some degree is influenced by our basic personality and the attitudes which we hold. That we screen out irrelevant stimuli and pay attention to only those stimuli which are perceived to be important emphasizes the adaptive value of central integration.

On the other hand, certain senses are absolutely more important than others. When, for example, we are faced with conflicting information from our visual and tactile senses, we will respond preferentially to the visual cues and transform the information generated by the touch sense to conform to our visual perceptions, even when the tactile information is correct and the visual is not!

Human beings are endowed with at least 10 different senses. Senses for which there are definite receptors include vision, or sight; audition, or hearing; gustation, or taste; olfaction, or smell; touch, or the tactile sense; pressure; pain; cold and hot, or the temperature senses; and proprioception, or the sense of position. Because vision, audition, gustation, and olfaction are highly specialized and ultrasensitive sensory abilities associated with the head, they are traditionally referred to as the *special senses*. The so-called "five senses" of human beings include the special senses plus tactile sensations. Obviously human beings are better endowed with sensory abilities than the traditionalist view would have us believe!

Many attempts have been made over the years to classify sensory abilities. Classification schemes have been based on anatomic location, functional characteristics, or some combination of these criteria. Because each sensory ability is unique, none of these schemes is entirely satisfactory. Therefore, for our purposes, we will simply divide the senses into the special senses and other senses.

In this chapter, we shall review the human senses, emphasizing the correlation between structure and function. Sensory disorders are discussed in Chap. 12.

## THE SPECIAL SENSES

### VISION

Of all our senses, special or otherwise, we probably depend most upon our ability to see. Vision not only enables us to see but contributes importantly to our perceptions of equilibrium, motion, and touch. Finally, the appearance of the eyes is an important nonverbal mode of human communication.

**The structure of the eye** The gross (general) structure of the eye is illustrated in Fig. 11-1. The diagram is of a horizontal section through the region of the optic nerve of the left eye. The sectioned eyeball is viewed from above.

Basically, the eyeball consists of three coats which form its wall, a crystalline lens and its supporting structures, and specialized intraocular (within the

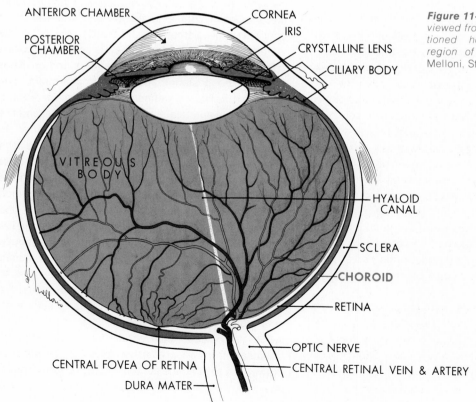

ANTERIOR CHAMBER
POSTERIOR CHAMBER
CORNEA
IRIS
CRYSTALLINE LENS
CILIARY BODY
VITREOUS BODY
HYALOID CANAL
SCLERA
CHOROID
RETINA
OPTIC NERVE
CENTRAL RETINAL VEIN & ARTERY
CENTRAL FOVEA OF RETINA
DURA MATER

**Figure 11-1** *The left eyeball viewed from above. The eye is sectioned horizontally through the region of the optic nerve. (From Melloni, Stone, and Hurd.)*

eye) fluid. The coats of the eyeball, from outside inward, are fibrous, vascular, and neural.

*Fibrous coat*    The fibrous coat is the outermost layer of the eyeball. This coat is divided into two regions—the *sclera* and the *cornea.*

Excluding the most anterior portion of the eyeball, the fibrous coat consists of the sclera. The sclera is a capsule which helps to maintain the spherical shape of the eyeball and provides attachment points for the extrinsic eyeball muscles.

The cornea is a transparent, anterior extension of the sclera. The sclerocorneal junction is marked by a perceptible notch which causes a bulge on the anterior surface of the eyeball. The cornea is the window of our visual world and the first part of the eye struck by the radiant energy of incident light rays. The outer surface of the cornea is covered by a thin, transparent mucous membrane called the *conjunctiva.* The conjunctiva continues around the inner surfaces of the eyelids.

*Vascular coat*    This coat constitutes the middle layer of the eyeball. The vascular coat is divided into three structures—the *choroid, ciliary body,* and *iris.*

The choroid contains many blood vessels. The extensive plexus of choroid vessels supplies the eye with nutrients and $O_2$, removes waste products of metabolism, including $CO_2$, and is the ultimate source of intraocular fluid. The blood vessels of the choroid are supplied and drained by the central artery and vein of the retina. These vessels are branches of the *ophthalmic artery* and *vein* and course through the center of the optic nerve. The *hyaloid canal* is a remnant of an artery which once ran from the beginning of the optic nerve to the posterior surface of the lens. Since the hyaloid canal is in the region of the *optic disk,* or blind spot, which defines the exit of the optic nerve fibers from the eyeball, this canal is never perceived visually.

Supplied by fibers of the *oculomotor nerve* (IIId cranial nerve—see Chap. 16), the ciliary body is the thickened anterior extension of the choroid. There are two portions to the ciliary body—the *ciliary process* and the *ciliary muscle.*

The thickest part of the ciliary body is composed of the ciliary muscle. The fibers of the ciliary muscle are arranged in three directions—meridional, radial, and circular. The ciliary muscle indirectly controls the thickness of the lens, as we will see later.

The ciliary process is composed of fingerlike projections which completely encircle the crystalline lens. The lens is supported in place by means of *suspensory ligaments* (zonular fibers) which attach to the ciliary process. The choroid vessels in the ciliary process give rise to an intraocular fluid known as the *aqueous humor.* This fluid provides sustenance for the cornea and lens, neither of which has a blood supply, and serves as one of the refracting media of the eye during the passage to the retina of incident light rays. The aqueous humor is drained into venous sinuses at the sclerocorneal junction.

The *iris* is a muscular diaphragm which arises from the anterior end of the ciliary body. Projecting inwardly, the pigmented iris is immediately in front of the crystalline lens. The spherical opening in the center of the iris is called the *pupil.* The iris regulates the amount of light entering the crystalline lens by controlling the diameter of the pupil.

Because of the position of the iris and the lens, two chambers are formed inside the anterior portion of the eyeball—the posterior and anterior chambers. The *posterior chamber* is defined by the posterior surface of the iris, the anterolateral surface of the lens, and the suspensory ligaments. The *anterior chamber* is bounded by the anterior surfaces of the iris and lens and the posterior surface of the cornea. Both chambers communicate with each other through the pupillary opening, and both contain aqueous humor.

After secretion from the ciliary process, the aqueous humor diffuses in the space between the iris and lens to the pupil, through which this intraocular fluid enters the anterior chamber. The aqueous humor also diffuses into the large cavity behind the lens, to be incorporated into a gelatinous material called the *vitreous body* which fills its interior. The vitreous body contributes to the refraction of incident light rays before they reach the retina. In addition, the intraocular pressure created in part by the vitreous body helps to mold the contours of the eyeball. The intraocular pressure is maintained at about 30 mmHg. Maintenance of a constant intraocular pressure is essential for the normal focusing of images on the retina.

*Neural coat* Consisting of the *retina,* this coat constitutes the innermost layer of the eyeball. Notice that the retina thickens slightly from anterior to posterior. Derived embryonically from brain ectoderm, the retina contains 10 discrete (separate) layers.

The *pigment layer* of the retina contains a single layer of cells which have a heavy brown pigment. This layer prevents light from being reflected backward after having stimulated the photoreceptive rods and cones.

The retinal layer next to the pigment layer contains the photoreceptors. The ends of the photoreceptive retinal cells called the *rods* and *cones* actually project into the pigment cells. Pigments in the rods and cones bleach in the presence of light, causing stimulation of the receptor cell membranes and the propagation of nerve impulses. To reach the rods and cones, light must pass through the conjunctiva, cornea, aqueous humor in the anterior chamber, pupil, lens, vitreous body, and several outer layers of the retina.

There are about 130 million rods and cones in each retina. Only about 5 million of these photoreceptors in each eye are cones.

Providing information on form, contrast, and movement, the rods detect gradations in light intensity and are not involved in color vision. The heaviest concentration of rods is on the periphery of the retina. This is the region of the retina most sensitive to vision in the dark (*scotopic vision*). The number of rods progressively decreases toward the central region of the retina.

Possessing one of three types of pigments for the determination of colors, the cones are responsible for color vision. Cones are also differentially used for visual acuity, or sharpness of vision. Forming a sphere about 1 mm$^2$ in area, the greatest concentration of cones in the retina is displaced laterally from the optic disk, or blind spot. Colored yellow, this sphere of cones is called the *macula lutea.* The yellow carotenoid pigment of the macula helps to correct for

chromatic aberration, i.e., irregularities of color in the visual field.

There is a slight depression in the center of the macula called the *fovea centralis.* Containing nothing but a single layer of cones about six times thinner than in the rest of the macula, the fovea is the point of most acute vision on the retina (*photopic vision*) and the point to which the light rays converge when a visual image is in focus. The lateral displacement of the fovea is believed to enhance depth perception, or stereoscopic vision (*stereopsis*).

The rest of the neuronal layers of the retina are concerned with information processing of the light energy transduced by the photoreceptive elements. The photoreceptors synapse with *horizontal cells* which run parallel to the length of the retina and with *bipolar cells* which course toward the apical border of the retina. The horizontal cells are affected by the average luminance, or light intensity, in the regions surrounding the immediate visual field and are responsible for sharper contrast. The other ends of the bipolar cells, in turn, synapse with *amacrine cells,* which run in parallel with the horizontal cells as well as with *ganglion cells* that reach the apical margin of the retina. The amacrine cells are affected by spatial or temporal changes in the surrounding regions.

The axons of the ganglion cells leave the eyeball at the region of the optic disk and form the *optic nerve.* The information conducted by the ganglion cells through the optic nerve depends on the nature of the interactions between the photoreceptors, horizontal cells, and amacrine cells. The whitish and spherical optic disk at the junction between the retina and the optic nerve is devoid of photoreceptive elements.

*The crystalline lens* The *lens* is a biconvex transparent disk about $\frac{3}{4}$ in. in diameter. The lens consists of concentrically arranged elastic fibers of protein covered by a thin capsule. The lens has the greatest refractive powers of any of the refractive media of the eye and functions in accommodation, or the ability to focus on both near and far objects. With age, the elasticity of the lens decreases in a predictable manner. Because elasticity is required for accommodation, the closest distance to the eyes at which an object remains in focus recedes with age. The progressive loss of the ability to focus on near objects is called *recession of the near point*.

## The physiology of vision

***Refraction and the visual image***  A light ray is *refracted* when it is bent while crossing the interface between two media of different densities. The greater the density of the medium, the higher the *index of refraction*. The index of refraction of air is set equal to 1.00. When light passes from a less dense to a more dense medium, it is bent toward an imaginary line perpendicular to the surface of the entering medium. Under the opposite set of conditions, light is bent away from the perpendicular. These principles are illustrated diagrammatically in Fig. 11-2.

Light rays entering the eye from air pass through four refractive media on the way to the retina:

1  The *cornea*. The cornea has a refractive index of 1.38. Of any of the refractive media of the eye, the cornea bends light the greatest amount. Since light rays enter the cornea from air, via the conjunctiva, the rays are bent toward the perpendicular.

2  The *aqueous humor*. The intraocular fluid in the anterior chamber has a refractive index of 1.33. Thus, light passing from the cornea through the aqueous humor is bent slightly away from the perpendicular.

3  The *crystalline lens*. The lens, while having the highest refractive index of any of the refracting media of the eye, is only slightly more dense than the aqueous humor. Thus, light entering the lens from the aqueous humor is bent slightly toward the perpendicular. However, since the degree of convexity of the lens can be adjusted, the lens is re-

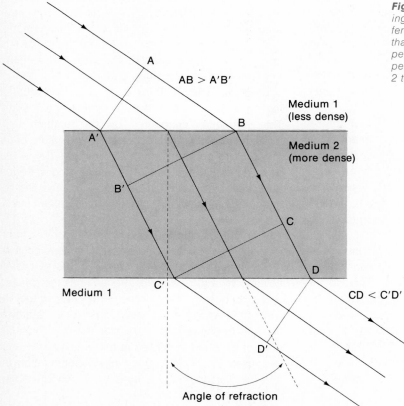

**Figure 11-2**  *The refraction of light passing the interface between two media of different densities. Medium 2 is more dense than medium 1 and bends light toward the perpendicular. Light is bent away from the perpendicular when passing from medium 2 to medium 1.*

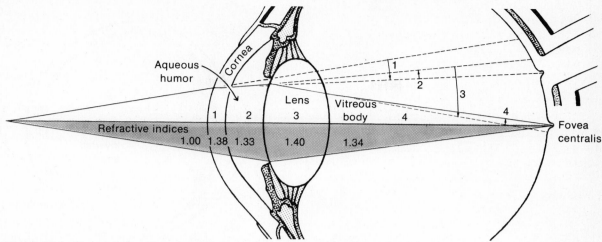

**Figure 11-3** *The refraction of an incident light ray passing through the eye on the way to the retina. As shown above the horizontal line, light is bent toward the perpendicular by the cornea (1) and the lens (3) and away from the perpendicular by the aqueous humor (2) and the vitreous body (4). The path of the light ray below the horizontal line is an algebraic summation of the degree of refraction.*

sponsible for focusing the image on the retina. In other words, the refractive powers of the lens are due much more to changes in its shape than to differences between its refractive index and that of the aqueous humor.

4 The *vitreous body.* The intraocular fluid behind the lens has a refractive index of 1.34. Since the refractive index of the vitreous body is slightly less than that of the crystalline lens, light entering the former from the latter is bent slightly away from the perpendicular. When focused properly, incident light rays fall on the fovea centralis.

The bending of a light ray by the refractive media of the eye is diagramed in Fig. 11-3. The path of the light ray below the horizontal line which bisects the lens represents an algebraic summation of the effects of the individual indices of refraction.

The image of an object on the retina is inverted and reversed. Similar in physical principle to the formation of an image by a camera lens, the mechanism by which the eye forms an image is shown in Fig. 11-4. The object being viewed, an upward-pointing arrow, is far from the eye, so that all the incident light rays

are parallel to each other. The light ray which passes through the exact center of the lens describes a path which is perpendicular to the surfaces of both the cornea and the lens. This particular light ray remains unrefracted as it passes through all the refracting media of the eye and falls precisely on the fovea centralis. The other light rays enter the cornea and lens at various angles to the perpendicular and are therefore refracted. The paths of two of these light rays are traced, one at either end of the object being viewed. These two light rays cross at a point between the lens and the retina. The distance between the midpoint of the lens and this intersection is the *focal length* of the lens. From the point of convergence, the light rays diverge to strike the retina on opposite sides of the fovea centralis. Thus, the image of the object projected on the retina is upside down and reversed.

To the question of how we learn to correct for the inverted and reversed image produced by the eye, the answer is quite simple: we don't! For example, whether an odor molecule is right side up or upside down makes little difference, so long as it stimulates hair cells in the olfactory epithelium. Likewise, so long

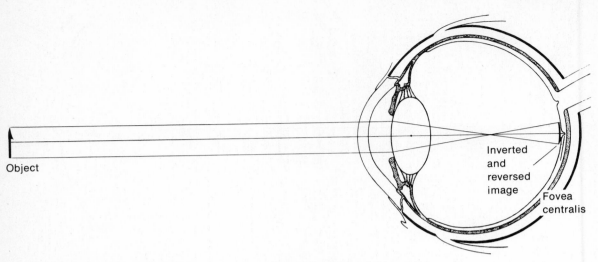

Object

Inverted
and
reversed
image

Fovea
centralis

**Figure 11-4** *Diagram of image formation by the eye. The path of three light rays is traced from a far object in the visual field. The image formed on the retina is upside down and backward.*

as the neural organization processing visual information is consistent, our perceptions of the visual world do not rely on which way the image is projected on the retina.

***Accommodation to near and far objects*** *Accommodation* refers to the ability to focus clearly on objects that are either near to or far from the eyes. This ability depends upon the elasticity of the lenses and the operation of the ciliary muscle of the vascular coat of each eyeball.

When the ciliary muscle contracts, the choroid of the vascular coat is pulled anteriorly and inward, releasing the tension which the ciliary process exerts on the suspensory ligaments which support the lens. As a consequence, the lens decreases in length and increases in thickness. As the convexity of the lens becomes more and more circular, its ability to bend light rays, referred to as the *refracting power* of the lens, progressively increases. An increase in refracting power is necessary when viewing objects close to the eyes.

On the other hand, a thin elongated lens with a low refracting power is required for far vision. The refracting power of the lens is decreased by relaxing the ciliary muscle. As the ciliary muscle relaxes, the choroid is drawn posteriorly, increasing the tension exerted on the suspensory ligaments and, therefore, the lens.

***Control of light intensity*** The iris is similar in operation to the diaphragm of a camera lens and controls the diameter of the pupil. This action is accomplished by the operation of the *circular* and *radial muscle fibers* of the iris. When the circular muscle fibers contract, the pupillary diameter is decreased, thereby reducing the amount of light entering the eye. These relationships are reversed when the radial muscle fibers contract. The muscular portion of the iris is innervated by fibers of the oculomotor nerve (IIId cranial nerve).

***Visual pigments and the transduction process*** There are two types of visual pigment—photopsins and rhodopsin. Three types of *photopsin* are found in cones. When bleached, these pigments transduce (convert) the wavelengths of incident light rays into nerve impulses which are interpreted centrally as colors. Color vision is discussed in the next unit in this chapter.

*Rhodopsin* is the visual pigment found in rods. When rhodopsin is bleached upon exposure to light, the stimulated rods transduce the energies of incident light rays into nerve impulses which are interpreted in terms of gradations of light intensity lacking color.

However, both the photopsins and rhodopsin consist of the chemical *retinene* combined with specific visual pigments. Our remarks below will be restricted to the events which occur when rhodopsin is bleached.

The rhodopsin in rods is so sensitive that some of it can be bleached when exposed to just 1 photon of light bearing 1 quantum of energy. The bleaching of rhodopsin causes it to split into retinene and the visual pigment *scotopsin*. The reaction is summarized as follows:

$$\text{Rhodopsin} \quad \underset{\text{dark}}{\overset{\text{light energy}}{\rightleftharpoons}} \quad \text{retinene} + \text{scotopsin} \qquad (11\text{-}1)$$

The reaction is reversible in the dark or at diminished light intensities, provided that scotopsin and retinene are available as raw materials. This reversible reaction is rapid. The bleaching of rhodopsin by the radiant energy of visible light is transduced by the stimulated rods into nerve impulses.

Following the bleaching of rhodopsin, the product retinene is converted enzymatically into *vitamin A*. This reaction is summarized as follows:

$$\text{Retinene} \quad \overset{\text{enzyme}}{\rightleftharpoons} \quad \text{vitamin A} \qquad (11\text{-}2)$$

Although the reaction in Eq. (11-2) is also reversible, it is relatively slow, compared with that summarized by Eq. (11-1). The fact that retinene and vitamin A are interconvertible emphasizes the importance of this vitamin in vision mediated by the rods (see Chap. 26).

*Color pigments and vision*  We are still in the Dark Ages with regard to how colors are perceived. The predominant theory, one which has a biologic basis, is called the *three-pigment theory of color vision*. We shall cite the salient features of this theory and then consider some recent evidence which is hard to explain by the use of the theory.

The cones are responsible for detecting different wavelengths of visible light which are then interpreted centrally as specific colors. There are defi-

nitely three different types of cone based on the specific kind of photopsin pigment contained within: (1) *red-detecting cones,* the responsiveness of which is greatest for light at a wavelength of 575 nanometers (nm, formerly called millimicrons, or m$\mu$); (2) *green-detecting cones,* the responsiveness of which is greatest for light at a wavelength of 535 nm; and (3) *blue-detecting cones,* the responsiveness of which is greatest for light at a wavelength of 430 nm. Each of the three types of color pigments will also absorb light at wavelengths to either side of the peak absorbency values. However, the degree of absorbency progressively diminishes, the farther incident waves diverge from the peak values characteristic of the pigments of the stimulated cones. The shoulders of the absorbency curves for each of the three color pigments overlap to some extent. Thus, light of one wavelength may stimulate two or even all three color-detecting pigments.

To appreciate how the theory operates, we must review some elementary principles of color. Red, green, and blue are primary colors. Any other color in the visible spectrum can be synthesized by blending two or all of the primary colors in various proportions and intensities. For example, white can be blended by mixing the colors red, green, and blue in approximately equal proportions. Black is perceived in the absence of light, i.e., when all light rays are absorbed and none is reflected from an object in the visual field.

According to the three-pigment theory of color vision, the colors we see are determined by the number of red-, green-, and blue-detecting cones stimulated and the intensities of their respective stimulation. Their responses, in turn, are dependent upon the wavelengths and intensities represented in the incident light waves.

Some astonishing experiments done by Edwin Land will undoubtedly require some revision of this theory. Land and his coworkers photographed colored objects with a special camera containing two lenses. One lens was covered by a red filter, the other by a green filter. Thus, one filter differentially admitted light of long wavelengths; the other, light of short wavelengths. Each lens focused the incident light rays on separate frames of ordinary black-and-white film. Black-and-white positive transparencies were made of each negative. Next, the black-and-white positives were projected onto a screen by a special dual lens projector. The red-exposed transparency was projected through a lens with a red filter. The green-exposed transparency was projected through an unfiltered lens. The fused image on the screen appeared in full, vivid, and authentic colors!

Based on these and other intriguing experiments, Land suggests that the perception of color is based on the presence in the visual field of long and short wavelengths of light and does not depend upon the specific wavelengths of the incident light rays at each point in the visual field. Apparently, to see in full color simply requires the visual stimulation of cones by any two wavelengths of light in the visible spectrum. How this theory can be reconciled with the demonstrated presence of three different color pigments in the cones has yet to be worked out.

**Visual pathways in the brain** The visual pathways in the brain are traced in Fig. 11-5. The figure is of the inferior (basal) aspect of the brain with the eyes in situ (in their natural position).

Based on location, the retina of each eye is divided into two regions: (1) the *nasal retina,* located on the medial (toward the midline) half of the eyeball and (2) the *temporal retina* located on the lateral (toward the side) half of the eyeball. Notice that incident light rays from the temporal region of the visual field fall on the nasal retina and that optic nerve fibers from this region of the retina cross over, or *decussate,* to the opposite side of the brain at the *optic chiasm.* The optic chiasm is located immediately anterior to the hypophysis, or pituitary gland, at the base of the diencephalic region of the brain. Furthermore, observe that incident light rays from the nasal region of the visual field fall on the temporal retina and that optic nerve fibers from this region of the retina remain on

**Figure 11-5** *The visual pathways from the eyes to the visual cortex of the brain.* (From Melloni, Stone, and Hurd.)

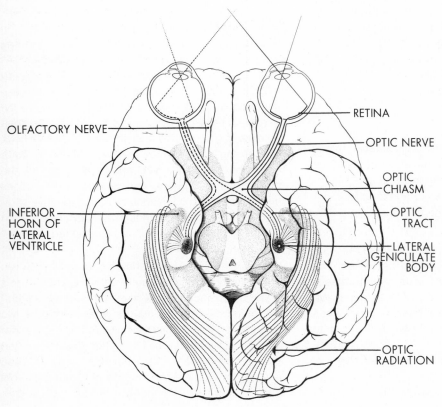

OLFACTORY NERVE

RETINA

OPTIC NERVE

OPTIC CHIASM

OPTIC TRACT

INFERIOR HORN OF LATERAL VENTRICLE

LATERAL GENICULATE BODY

OPTIC RADIATION

the same side of the brain after passing through the optic chiasm.

Following partial decussation of the optic nerve fibers at the optic chiasm, the fibers of the optic nerves enter the brain, where they are collectively referred to as *optic tracts*. The fibers of each optic tract synapse with neurons in the *lateral geniculate body* of the thalamus.

The lateral geniculate bodies enhance contrast, providing a sharper image than that originating on the retina, and analyze differences in the same image from both retinas. The latter function plays a prominent role in binocular vision. These interpretations are based on both the structural organization of the lateral geniculate bodies and the electrophysiologic responses of some of their individual neurons to various types of retinal stimulation.

A number of fibers from each optic tract make synaptic contact with neurons in the *pretectal area* of the midbrain. The axons of the pretectal neurons indirectly synapse with neurons of the oculomotor nerve (IIId cranial nerve). These autonomic axons indirectly innervate the iris muscle of each eye. This complex pathway is responsible for mediating the *light reflex* by causing pupillary constriction upon exposure to bright light. While reflex in nature, the response is far from simple in terms of the number of neurons involved.

The fibers of some neurons originating in each lateral geniculate body make synaptic connections with neurons in the *superior colliculus* of the midbrain. Through complex pathways which ultimately innervate the extrinsic eyeball muscles and the muscles of the neck, the superior colliculus is responsible for reflexly coordinating movements of the eyes and neck when moving objects must be kept in the visual field.

However, the main visual pathways from the retina lead through the optic nerves to the optic chiasm and from there via the optic tracts to the lateral geniculate bodies of the thalamus. The last part of the visual pathways of the brain consists of fibers which radiate from the lateral geniculate bodies to the *visual area* in the *occipital cortex* of the *cerebrum* (see Chap. 14), forming discrete layers or columns.

The visual cortex lies at either side of the calcarine fissure (calcarine sulcus) in the occipital region of the cerebrum. The largest and most posterior portion of the occipital cortex receives fibers that carry visual information originating from the fovea centralis of each eye. Thus, a large area of the visual cortex is occupied with the processing of visual information from the fovea.

The columnar organization of the visual cortex and electrophysiologic responses from its neurons indicate that this region of the brain is concerned with analyzing form, shape, and movement. Each column in the visual cortex is responsive to only a specific type of contour. The position of the image on the retina does not seem to be crucial to this analysis. Indeed, since the neural organization of the visual cortex is present at birth, human beings and other animals are predisposed to respond to specific forms, shapes, contours, and movements without prior learning!

Fibers from the visual cortex project to the adjacent *visual association area*. Here, the processed visual information is presumably compared with prior experiences of a similar nature. Other parts of the brain integrate the visual information with other sensory information, so that neural directives can be issued for an appropriate response based on the entire sensory input.

**Accessory structures of the eye**  Several structures which surround various portions of the eyeball are required for, or ensure, its normal functioning. These structures include the extrinsic eyeball muscles, the lacrimal apparatus, and the eyelids.

*The extrinsic eyeball muscles*  Movements of the eyeballs are accomplished by six paired, striated muscles which insert at strategic points on the scleras. The muscles consist of the medial, lateral, inferior, and superior rectus muscles and the superior and inferior oblique muscles. The superior oblique and lateral rectus muscles are innervated by the *trochlear nerve* (IVth cranial nerve) and the *abducens nerve* (VIth cranial nerve), respectively. The rest of the eyeball muscles are innervated by the oculomotor nerve (IIId cranial nerve). The insertion points of the extrinsic muscles on the eyeball correspond to their

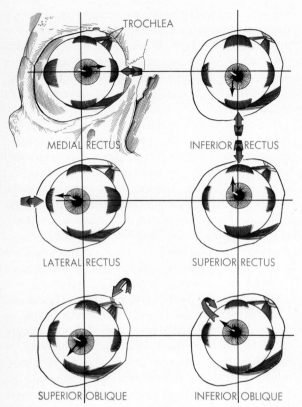

**Figure 11-6** *The extrinsic eyeball muscles and the movements they accomplish. The thick arrows outside the eyeballs indicate the directions they move in when the respective muscles contract. The thin arrows originating from the irises indicate the directions in which the pupils move.* (From Melloni, Stone, and Hurd.)

first names. Their last names refer to the directions of the muscles relative to the normal position of the eyeball.

The natural positions of these muscles and the movements they perform are shown in Fig. 11-6. Notice that the muscles partially anchor the eyeball in place. Also observe the pulleylike *trochlea,* which changes the direction of the force applied to the eyeball by the superior oblique muscle.

**The lacrimal apparatus** The paired *lacrimal glands* produce tears, a watery solution containing salts such as $Na^+Cl^-$ and sodium bicarbonate ($Na^+HCO_3^-$), as well as enzymes. Tears continually wash the conjunctiva covering the cornea of the eyeball. This action cleanses the eyeball of windblown particulates, kills many types of bacteria by the presence of a bacteriolytic enzyme, and prevents the conjunctiva and cornea from drying. The moistening effect reduces the possibility of subsequent infection. The lacrimal apparatus is illustrated in Fig. 11-7.

The size of an almond, the lacrimal gland is located in the lacrimal fossa (depression) in the lateral roof of the orbit. The gland is divided into a large superior (upper) and a small inferior (lower) division. Its secretion is drained by a small number of *excretory ductules,* which empty into a chamber formed by the conjunctiva in the superolateral region of the eye.

The tears wash continually across the eye in the direction of the arrows and collect in a notch called the *lacrimal lake* at the medial angle of the eye. From here, the tears drain into two small orifices referred to as *lacrimal puncta.* Each lacrimal punctum lies just above or below the lacrimal lake. The lacrimal puncta lead into small canals called *lacrimal canaliculi* which communicate with the *lacrimal sac.* This sac is located in the lacrimal fossa at the side of the nose. By means of the *nasolacrimal duct,* the lacrimal sac communicates with the nasal cavity lateral to and

**Figure 11-7** *The lacrimal apparatus.* (From Melloni, Stone, and Hurd.)

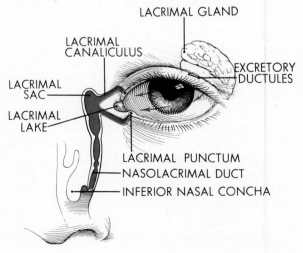

below the inferior nasal concha. The tears which accumulate in the lacrimal sac are transported passively through the nasolacrimal duct to the nasal cavity. The rate of production of tears is related to their rate of evaporation, the presence of conjunctival irritants, certain chemicals called *secretogogues* in the blood, and specific emotional states.

*The eyelids*   There are two eyelids in front of each eye—a superior and an inferior eyelid. The eyelids and eyelashes on their free margins are basically protective in function. The eyelashes prevent larger particulates in the air from lodging on the conjunctiva. The eyelids close when sleeping, as much to keep the conjunctiva and cornea moist as to screen out light or provide a period of rest for the eyes. Blinking helps the tears keep the conjunctiva moistened. The eyelids partially or completely close when exposed to either intensely bright light or particulates blown in the face by gusts of wind.

The outer portion of each eyelid is formed of skin. The muscular portion of each eyelid is shown in Fig. 24-11. The muscle is called the *orbicularis oculi pars palpebralis*. Arranged as a sphincter around each eyeball, this muscle closes the eyelids. The conjunctiva which lines the inside surface of the eyelids is continuous with that which covers the outside surface of the cornea. The inner borders of the eyelids contain numerous *tarsal glands* which secrete an oily substance that prevents the eyelids from adhering to each other.

## AUDITION AND EQUILIBRIUM

The senses of hearing and balance are associated with the ears.

**The structure of the ear**   The ear is divided into three regions, as follows:

1   The *external ear*. This portion of the ear consists of the *pinna* and *external auditory meatus*. The pinna is the fleshy portion of the ear. Consisting of cartilage enclosed by skin, the pinna collects sound waves which are conducted through the air within the external auditory meatus to the tympanic membrane, or eardrum, of the middle ear. *Ceruminous glands* which line the wall of the external auditory meatus secrete *cerumen*, or earwax. Cerumen and hairs which surround the external orifice of the meatus tend to prevent particulates from entering the deeper portions of the ear.

2   The *middle ear*. This portion of the ear consists of the *tympanic membrane, auditory ossicles*, and auditory tube (*eustachian tube*). Forming the distal border of the middle ear, the tympanic membrane oscillates in resonance with the frequencies of sound waves which strike its outer surface. A chain of three small bones called auditory ossicles connects the eardrum with the inner ear. The *malleus*, or hammer, attaches to the inner surface of the tympanic membrane and articulates with the *incus*, or anvil. The incus, in turn, articulates with the *stapes*, or stirrup. The footplate of the stapes articulates with the *fenestra vestibuli*, or oval window, of the cochlea, a portion of the inner ear. The auditory ossicles conduct the energies of sound waves from the oscillating eardrum to perilymphatic fluid within the cochlea (see below). The reduction in the size of the membrane from eardrum to oval window and the physical arrangement of the auditory ossicles combine to greatly amplify the sound waves which set the perilymphatic fluid into oscillatory motion. Opening into the nasopharynx, the eustachian tube equalizes air pressures between the middle ear cavity and the pharynx, tending to prevent pressure-induced damage to the tympanic membrane.

3   The *inner ear*. This portion of the ear consists of the *cochlea, vestibule*, and three *semicircular canals*. The energies of sound waves are conducted through the perilymphatic fluid of the cochlea to *auditory receptor cells*. Changes in the force of gravity caused by physical displacements of the head and the rest of the body result in shifts of the endolymphatic fluid and its particulates. The fluid and particulate flow stimulate hair cells in specialized areas within the vestibule and the semicircular ducts, providing information on position in space.

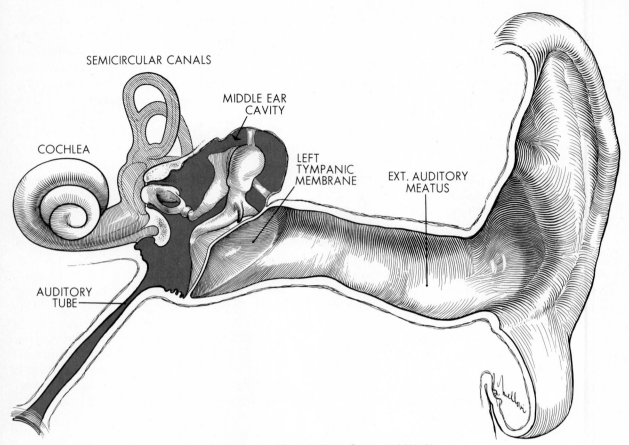

**Figure 11-8** *The structures associated with hearing.* (From Melloni, Stone, and Hurd.)

The structures of the external, middle, and inner ears are illustrated in Fig. 11-8. A portion of the wall of the middle ear and eustachian tube has been removed to reveal the auditory ossicles and the continuity between the middle ear cavity and the eustachian tube.

The middle and inner ears are exposed in situ in Fig. 11-9. The illustration is of the upper aspect of the right *temporal bone*. This bone fits into the space on the right side of the skull shown in Fig. 23-7. The superior wall of the petrous portion of the temporal bone has been scraped away, revealing the auditory ossicles in the middle ear cavity as well as the cochlea, vestibule, and semicircular canals of the inner ear.

**Figure 11-9** *The organs of hearing and equilibrium exposed from superior aspect in the petrous portion of the temporal bone.* (From Melloni, Stone, and Hurd.)

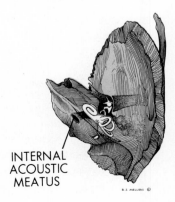

INTERNAL
ACOUSTIC
MEATUS

The inner ear is enlarged in Fig. 11-10 to show some of its finer structural details. The outer, lighter portion of the inner ear consists of cancellous bony tissue called the *bony labyrinth,* or *osseous labyrinth.* Within the interstices of the bony labyrinth is a fluid called *perilymph.* The inner, darker region of the inner ear consists of nonbony tissue called the *membranous labyrinth. Endolymphatic fluid* is contained within the membranous labyrinth. In the cochlear portion of the inner ear, the membranous labyrinth is called the *cochlear duct.* The *organ of Corti,* or spiral organ of hearing, is located in the cochlear duct. Notice that the cochlear duct describes $2\frac{1}{2}$ turns in the cochlea.

The middle region of the inner ear is called the *vestibule.* The oval window, identified by a dark oval outline in Fig. 11-10, is on the lateral aspect of the vestibule. During bone conduction by the auditory ossicles, the footplate of the stapes rocks back and forth against the oval window, setting into motion the perilymphatic fluid within the bony labyrinth of the cochlea. The nearby *fenestra cochleae,* or round window, permits the dissipation of the sound energies after fluid conduction through the cochlea is completed. Thus, the round window prevents the continued stimulation of the organ of Corti in the absence of sound waves and also prevents an excessive build-up in pressure within the cochlea.

The *saccule* and *utricle* are the membranous labyrinths in the vestibular region. Detecting physical displacements of the head when their specialized hair cells are stimulated by movements of endolymphatic fluid and calcium carbonate ($CaCO_3$) particles, these membranous labyrinths indirectly communicate with each other via a system of ducts, as suggested in Fig. 11-10.

The right-hand portion of the inner ear in Fig. 11-10 consists of three semicircular canals which enclose a

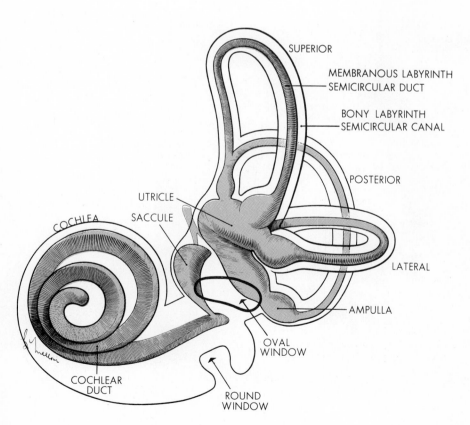

**Figure 11-10** *The inner ear.* (From Melloni, Stone, and Hurd.)

SUPERIOR

MEMBRANOUS LABYRINTH
SEMICIRCULAR DUCT

BONY LABYRINTH
SEMICIRCULAR CANAL

POSTERIOR

UTRICLE

SACCULE

LATERAL

COCHLEA

AMPULLA

OVAL WINDOW

COCHLEAR DUCT

ROUND WINDOW

corresponding number of *semicircular ducts.* The canals are portions of the bony labyrinth and contain perilymph, whereas the ducts are parts of the membranous labyrinth and enclose endolymph. The superior, posterior, and lateral semicircular canals lie in the coronal, sagittal, and horizontal planes, respectively, when the head is in an erect position. The *ampullae* at the bases of the semicircular ducts possess hair cells. The displacement of the hair cells by movements of the endolymph provides information on the position of the body in space.

The physiology of hearing   Amplification is a function of the middle ear. However, the initial analysis of the qualities of sound waves is accomplished by the cochlea of the inner ear. To appreciate this aspect of hearing, we must examine the fine structure on the inside of the cochlea.

A cross section through a portion of the cochlea is shown in the upper part of Fig. 11-11. The upper and lower chambers are called the *scala vestibuli* and *scala tympani,* respectively. Both chambers are enclosed within the bony labyrinth of the cochlea and contain perilymph. The oval window is at the entrance of the scala vestibuli. The round window is at the exit of the scala tympani. These two chambers communicate with each other by means of a small duct at the apex of the cochlea. The cochlear duct lies between the scala vestibuli and the scala tympani. Containing endolymph, the cochlear duct is bounded above by the *vestibular membrane* and below by the *basilar membrane.*

An enlargement of a portion of the cochlear duct is shown in Fig. 11-11. The organ of Corti lies in the endolymph of the cochlear duct, and is supported by the basilar membrane. The organ of Corti consists of supporting cells which lie on the basilar membrane and specialized hair cells which lie above it. The

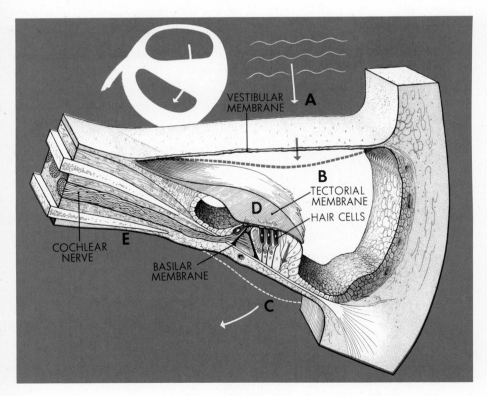

**Figure 11-11**  *Cross section through a portion of the cochlea showing the cochlear duct and the organ of Corti.* (From Melloni, Stone, and Hurd.)

VESTIBULAR MEMBRANE

A

B

TECTORIAL MEMBRANE

HAIR CELLS

D

COCHLEAR NERVE

E

BASILAR MEMBRANE

C

apical portion of each hair cell bears a group of specialized cilia embedded in the gelatinous substance of the overlying *tectorial membrane.* Dendritic fibers of the *cochlear nerve,* a division of the acoustic nerve (VIIIth cranial nerve, also called the auditory nerve), contact the basal portions of the hair cells. The hair cells transduce the mechanical stimuli resulting from the energies of sound waves into nerve impulses which are conducted along the cochlear nerve fibers to the brain.

There are about 16,000 hair cells in the organ of Corti of each cochlear duct. Based on position, the hair cells are divided into two groups: (1) an inner grouping of approximately 4,000 hair cells and (2) an outer grouping containing the remainder of the hair cells.

Hair cells are stimulated when a portion of the basilar membrane is displaced. The transduction process is the end result of a chain of events initiated when the perilymphatic fluid in the scala vestibuli is set in motion in response to the rocking action of the stapes against the oval window. The energies of sound waves conducted through the perilymph deform the vestibular membrane (*A*) and those portions of the basilar membrane (*C*) resonant with the frequencies of the incident sound waves. Deformation of the basilar membrane causes a relative displacement of the overlying tectorial membrane (*D*), resulting in the stimulation of the hair cells.

The *intensity,* or loudness, of a sound is in part determined by the rate of oscillation of resonant portions of the basilar membrane. The greater the rate of vibration, the greater the frequency of nerve impulses generated by the hair cells. In other words, the magnitude of a stimulus is related physiologically to the frequency of nerve impulse propagation. However, because of the design characteristics of the cochlea, human beings are much more sensitive to sounds of intermediate frequencies than to sounds at either end of the audible frequency spectrum. For example, when sounds of different frequencies are matched in intensity, sounds at a frequency of 1,000 cps (cycles per second, hertz, or Hz) sound louder than those at 100 cps.

The *frequency,* or pitch, of a sound is established

when a portion of the basilar membrane is set into oscillatory motions which are resonant with the frequency of the sound waves. Only resonant *transverse fibers* on the basilar membrane respond to specific frequencies. The basilar membrane contains about 25,000 transverse fibers of different lengths. The shortest fibers are at the base of the cochlear duct and respond to the highest frequencies of the audible spectrum. The longest fibers are at the apex of the cochlear duct and respond to the lowest audible frequencies. Fibers graded in length between these two extremes resonate in response to intermediate sound frequencies. The audible spectrum ranges from 20 to 20,000 or more cps during childhood. Both hearing range and sensitivity become more restricted with age.

The *quality* of a sound is established by the number of different frequencies represented and by each of their respective intensities. Each frequency sets a different transverse fiber into resonant oscillations. The number of oscillations of each resonating transverse fiber establishes the intensities of the various frequencies of a complex sound.

Further analysis of sounds and the integration of auditory information occurs in the *auditory cortex* in the temporal lobe of the cerebrum. Auditory information is conducted to the auditory cortex by cochlear nerve fibers and intervening neurons of the auditory pathway in the medulla and thalamus.

The *direction* from which a sound comes can often be established by turning the head until the sound seems loudest. This point will obviously be at right angles to one of the two ears. This method of sound localization is sometimes unreliable because of its ambiguity. True *binaural hearing,* the counterpart of binocular vision, requires an analysis in the auditory cortex of intensity differences of the same sound after its transduction by hair cells in both organs of Corti.

**The physiology of equilibrium** Equilibrium receptors in the saccule and utricle are located within specialized structures called *maculae.* Each macula consists of two types of hair cells interspersed among supporting cells. The hair cells of the macula are the receptors which transduce mechanical stimuli resulting from changes in the position of the head into nerve impulses. The apical portions of the hair cells contain cilia which project into a thin, transparent layer called the *otolithic membrane.* The otolithic membrane contains crystals of calcium carbonate ($CaCO_3$) and special proteins which move in response to positional changes of the head. The fluid and particulate movements within the otolithic membrane displace the cilia of the hair cells. This mechanical stimulation is transduced into nerve impulses which are conducted to the brain by fibers of the *vestibular division* of the *acoustic nerve* (vestibulocochleas, auditory, or VIIIth cranial nerve). The connections with the surfaces of the hair cells.

Equilibrium receptors in the ampullae at the bases of the semicircular ducts are located within specialized structures called *cristae.* The composition of each crista is basically similar to that of a macula. However, the specialized cilia on the apical portions of the hair cells in each crista project into a thick, elongated gelatinous mass called a *cupula.* Movements of the endolymphatic fluid inside the semicircular ducts displace the cupulae and the hair cell cilia embedded within. Initiated by movements of the head in any direction, these mechanical stimuli are transduced by the hair cells into nerve impulses. As in

the case of the macular hair cells, the hair cells of the cristae are supplied with dendritic fibers of the vestibular nerve.

## GUSTATION

The receptors for taste are located primarily on the back of the tongue but are also distributed on the soft palate of the roof of the mouth and in the pharyngeal region. The taste receptors are specialized groupings of cells called *taste buds.* The taste buds are located in slight depressions around the bases of small projections called *papillae.* The roughness of the surface of the tongue is due to the presence of these papillae.

Based on shape, the papillae are classified as follows:

1 *Vallate papillae.* These papillae are large and spherical. There are only a small number of vallate papillae located along a V-shaped line on the posterior third of the tongue. The apex of the V points toward the pharynx. These papillae and the posterior third of the tongue in general are innervated by the *glossopharyngeal nerve* (IXth cranial nerve).
2 *Fungiform papillae.* These papillae are smaller than the vallate papillae and are shaped like a mushroom, with a thin elongated base and an expanded end portion. The fungiform papillae are distributed over the anterior two-thirds of the tongue, although the margins have a somewhat heavier concentration than the more centrally located regions. These papillae and the anterior two-thirds of the tongue in general are innervated by the *facial nerve* (VIIth cranial nerve).
3 *Filiform papillae.* Slender in shape, these papillae are the smallest of the three kinds of papilla. The filiform papillae are interspersed among the fungiform papillae and are also innervated by the facial nerve.

The papillae on the superior aspect of the tongue are shown in Fig. 11-12.

Although we can distinguish a great variety of

**Figure 11-12** *The superior aspect of the tongue showing the three types of papillae and the regions in which the various modalities of taste are discriminated.* (From Melloni, Stone, and Hurd.)

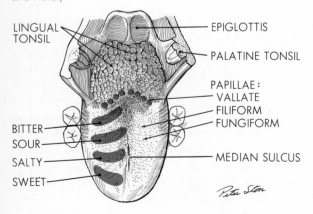

LINGUAL TONSIL — EPIGLOTTIS — PALATINE TONSIL — PAPILLAE: VALLATE FILIFORM FUNGIFORM BITTER SOUR SALTY SWEET — MEDIAN SULCUS

tastes, there are only four *modalities* of taste sensation—*salty, sweet, sour,* or acid, and *bitter*. The wide range of taste sensations which we experience is due to the stimulation in various proportions of taste buds which mediate these basic qualities of taste and the possible simultaneous sensory stimulation of olfactory (odor) receptors. Indeed, some so-called tastes, such as that of chocolate, are mediated exclusively by olfactory reception!

While all taste buds appear to be similar in structure, the surface of the tongue is divided into functional regions in each of which one of the modalities of taste predominates. For example, sweet tastes are detected best on the tip of the tongue, salty and sour along its lateral aspects, and bitter sensations on its posterior border. These regional specializations along the surface of the tongue are identified in Fig. 11-12. To appreciate the full flavor of a food or bever-

age, it is well known that it must be swished around the entire surface of the tongue, thereby stimulating the widest variety of taste buds possible.

Molecules which stimulate taste buds must be dissolved in an aqueous medium. If the ingested materials are not already in liquid form, then the saliva acts as a solvent for their dissolution. When the tongue is dry, taste sensations are depressed. This knowledge can be put to practical use the next time you feel the need to take an aspirin. To avoid the sour taste sensations of an aspirin allowed to dissolve in the mouth, simply make sure the tongue is dry and rapidly move the aspirin to the pharynx with the tongue before washing the aspirin down with water.

Gustatory (taste) nerve fibers of the facial and glossopharyngeal nerves enter the medulla, where they form synaptic connections with *secondary neurons*. These secondary neurons synapse, in turn, with neurons in the thalamic region of the brain. Gustatory information is then carried by *thalamic projection fibers* to the *parietal cortex* of the cerebrum for integration.

## OLFACTION

Olfaction refers to the sense of smell. This sense is probably the most poorly developed and least understood of the major senses of man.

Olfactory receptor cells form part of the epithelial lining in the upper part of the nasal cavity, above the superior conchae of the ethmoid bone. The olfactory receptors are modified bipolar neurons interspersed in the mucosa among columnar supporting, or sustentacular, cells. The apical surfaces of the *olfactory receptors* bear tufts of cilia called *olfactory hairs*. The surface of the mucosa, including the olfactory receptors, is kept moist by special secretions produced by olfactory glands in the submucosa. Moisture is necessary for the dissolution of olfactory molecules on the receptor surface. A diagrammatic section through a portion of the olfactory epithelium and surrounding tissue is shown in Fig. 11-13.

Olfactory reception and transduction occur when odor molecules drawn through the nose or mouth with inspired air are dissolved on, and interact with,

**Figure 11-13** *Section through a portion of the olfactory epithelium in the superior part of the nasal cavity. Axons of the olfactory receptors pass through the cribriform plate and synapse with secondary neurons called* mitral cells *whose axons contribute to the olfactory tract.* (From Langley, Telford, and Christensen.)

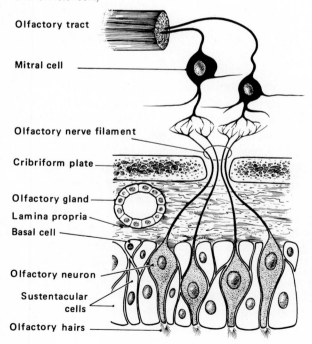

Olfactory tract

Mitral cell

Olfactory nerve filament

Cribriform plate

Olfactory gland
Lamina propria
Basal cell

Olfactory neuron

Sustentacular cells

Olfactory hairs

membrane enzymes of specific olfactory receptors. The *strength* of an odor is determined by the number of olfactory receptors stimulated. The stronger the odor, the greater the number of stimulating molecules and the greater the frequency of nerve impulses conducted away from the olfactory receptors by their axons.

Nevertheless, the presence of one odor molecule is sufficient to excite an olfactory receptor. Although olfactory receptors have extremely low response thresholds, they adapt quickly to the presence of an odor by decreasing the frequency of nerve impulses propagated.

The *quality* of an odor is apparently determined by the geometry of the stimulating molecule. Based on similarities in molecular structure and receptor excitation, odorous molecules have been classified according to the following primary odor qualities: camphoraceous, musky, floral, pepperminty, ethereal, pungent, and putrid. Apparently most odors are compound, in that they consist of several of the primary odors combined in various proportions. Although the olfactory powers of most human beings are relatively poor, the trained human nose can distinguish thousands of different odors.

Refer again to Fig. 11-13. Notice that numerous axons from olfactory receptors converge in the submucosa before passing through each orifice of the cribriform plate of the ethmoid bone. As a result of convergence, about 20 fibers emerge from the superior aspect of the cribriform plate. The axons of these fibers converge synaptically with the dendrites of *mitral cells*. Mitral cells are neurons which originate in the *olfactory bulb*. The grouped axons from the mitral cells give rise to the *olfactory tract*. Fibers of the olfactory tract terminate in the *olfactory lobe* of the brain. The *olfactory nerve* (Ist cranial nerve) is considered to be in part a fiber tract of the brain and thus includes not only the axons of the olfactory receptors but the olfactory bulb and tract as well. From the olfactory lobes, the axonal fibers of other neurons carry partially processed olfactory information to other parts of the cerebrum for integration with other stimuli and to the thalamic region of the brain for possible response at a subconscious level.

## OTHER SENSES

### TOUCH AND PRESSURE

The sense of *touch* (*tactile sensation*) is mediated by several specialized end organs in the skin. These receptors include *free nerve endings* which wrap around the base of hair follicles and *Meissner's corpuscles* in the dermal papillae just beneath the stratum germinativum of the epidermis. These touch-receptive structures are illustrated in Fig. 11-14*A* and *B*. Both sensory structures adapt rapidly to mechanical stimulation. The free nerve endings associated with hair follicles are stimulated only during displacements of the hairs.

The sense of *pressure* is also mediated by several specialized skin end organs. One of these sensory structures, called a *pacinian corpuscle,* is shown in Fig. 11-14*E*. Scattered around the interface between the dermis and subcutaneous connective tissue, pacinian corpuscles are responsive to deep pressure. Pacinian corpuscles adapt about as rapidly as Meissner's corpuscles. Presumably, free nerve endings in the epidermis are sensitive to pressure as well as to pain.

### PAIN

The sense of *pain* is mediated by free nerve endings distributed throughout the body. A free nerve ending on the surface of the skin is illustrated in Fig. 11-14*A*. The quality of the pain is determined by the nature and location of the pain-eliciting disturbance as well as by previous experience and cultural influences. While all people have a similar threshold for pain, the overt response of different subjects to painful stimuli of the same magnitude may be highly variable. Pain is adaptive and alerts its sufferer to the possibility of potential injury or the existence of real injury.

Many types of stimuli are capable of eliciting pain. The stabbing, throbbing, and burning pains caused by physical injuries are probably due to the release of certain chemicals such as *histamine* or *bradykinin* from dead and dying cells. Interestingly, such pain may last for only a short period following physical

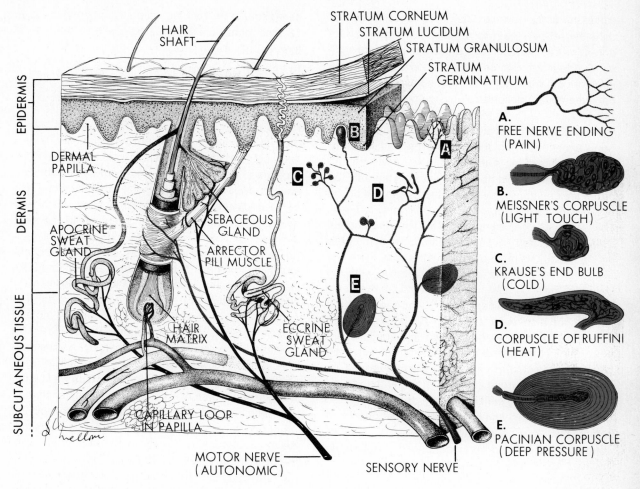

**Figure 11-14** *Some types of sensory nerve endings in the skin.* (From Melloni, Stone, and Hurd.)

trauma. The pain experienced during muscle spasms and heart attacks is most likely due to a significant decrease in the blood supply and, therefore, the amount of $O_2$ supplied to the affected tissues. This condition is referred to as *ischemia*. Pain always accompanies ischemia. Presumably, other causes of pain, such as excessive pressure, gas distension, excessively hot or cold stimuli, and chemical intoxication, act through these same mechanisms. In other words, whatever the origin, pain becomes evident when cells begin to die from unnatural causes such

as an irritative or destructive lesion or ischemia due to local hypoxia (a decrease in $O_2$).

Perception of the presence of pain is mediated by the *thalamic region* of the brain. However, the thalamus does not localize the pain to a specific part of the body. This function is relegated to the *postcentral gyrus,* or *somesthetic cortex,* of the cerebrum, to which fibers project that carry pain information from the thalamus. Presumably, the neural integration of painful stimuli in the cerebral cortex can attenuate or inhibit the continued perception of pain following an insult to the body. However, the nerve fibers which mediate the sense of pain do not seem to adapt or adapt only slowly to the continued presence of painful stimuli. Furthermore, cortical intervention in the experience of pain does not imply that human beings cannot remember or describe a physically painful experience.

*Referred pain* originates from an internal structure but is experienced on or near the surface of the body. The segments of the spinal cord to which visceral afferent (sensory) nerve fibers project from the afflicted organ are determined by the segments from which the organ was derived during embryonic development. Thus, when free nerve endings in the wall of a visceral organ are excited by painful stimuli, the stimuli are conducted to this related region of the cord. Here, the axons of the visceral afferents synapse with the dendrites of secondary neurons which also receive the axons of sensory nerve fibers that originate in a specific location on the surface of the body. Because pain receptor neurons are diffuse in visceral structures, the localization of the pain by the somesthetic cortex is projected to the body surface supplied by the secondary neurons in the cord, rather than to the visceral organ from which the pain originates.

Abnormalities in which pain is the most prominent symptom such as neuralgia, neuritis, and polyneuritis are discussed in Chap. 16. Hyperalgesia is discussed in Chap. 12. Headache is reviewed in Chap. 15.

## TEMPERATURE

The temperature sense is mediated by the end bulbs of Krause (Fig. 11-14C) and the corpuscles of Ruffini (Fig. 11-14D). *Krause's end bulbs* are receptive to *cold* temperatures or, more precisely, to a sudden decrease in temperature. *Ruffini's corpuscles* are sensitive to *heat*. Both these end organs are located in the dermis of the skin. However, the end bulbs of Krause are more numerous and superficial than the corpuscles of Ruffini.

Free nerve endings are also responsive to extremely cold or hot stimuli. The mechanism by which pain is experienced during continued stimulation by either cold or hot running water is explained in Chap. 12. Most people experience pain when the temperature reaches 45°C.

## PROPRIOCEPTION

Proprioceptors are specialized end organs which provide information on the position of the body or a body part and its rate of displacement in space. Such receptors transduce the mechanical stimulation caused by stretch or physical displacement into nerve impulses. Aside from some specialized end organs which subserve other functions in the skin, such as pacinian corpuscles, corpuscles of Ruffini, and the highly specialized equilibrium receptors in the inner ear (see "The Physiology of Equilibrium," earlier in this chapter), which operate in this way, stretch receptors are also located in muscles, tendons, and ligaments. Such proprioceptors include *muscle spindles* and *Golgi tendon organs*.

These highly specialized sensory structures provide automatic feedback to the cerebellum (see Chap. 14) and perhaps to the cerebrum as well on the position of a region of the body and its rate of change with time. The sensory information provided by the proprioceptors is utilized by the cerebellum to bring the motor performance of the body parts in line with the central directives issued by the motor area of the cerebral cortex. A proprioceptor provides information about any disparity between instructions and performance, so that appropriate adjustments can be made to bring the latter in line with the former.

A physical analogy to proprioceptive feedback with which most of us are familiar is the power steering mechanism of an automobile. When the front wheels are held in a particular alignment during travel, a bump which causes momentary misalignment of the wheels is corrected by the automatic action of the power steering unit. Of course, the intervention of an automatic correcting device causes a certain amount of sluggishness when new directives are issued, whether by the driver behind the steering wheel or by the brain, but the responsiveness is obviously adequate for either type of performance.

## COLLATERAL READING

Amoore, John E., James W. Johnston, Jr., and Martin Rubin, "The Stereochemical Theory of Odor," *Scientific American,* February 1964. The authors put forth a theory of odor discrimination based on the specificity of olfactory receptors and the geometry of odor molecules.

Botelko, Stella Y., "Tears and the Lacrimal Gland," *Scientific American,* October 1964. Describing the anatomy and physiology of the lacrimal gland and duct system which surrounds each eye, Botelko raises important questions about the mechanisms that regulate the flow of tears.

Bower, T. G. R., "The Visual World of Infants," *Scientific American,* December 1966. Bower suggests that infants can detect most of the visual information that adults can, although infants cannot integrate as much of the visual information received.

Case, James, *Sensory Mechanisms,* New York: The Macmillan Company, 1966. A good review of sensory biology for students who wish to extend their knowledge beyond the basics.

Fender, Derek H., "Control Mechanisms of the Eye," *Scientific American,* July 1964. Experiments are described which suggest that eye-tracking movements are regulated by feedback based on the degree to which the image of the object being tracked is displaced from the two foveae.

Gesell, Arnold, "Infant Vision," *Scientific American,* February 1950. Gesell cites evidence indicating that an infant can discern small objects with his or her eyes by 20 weeks of age but cannot grasp these objects manually until double this age. The first visual observations made by an infant are accomplished by fixating only one of the two eyes.

Haagen-Smit, A. J., "Smell and Taste," *Scientific American,* March 1952. A review of the basic types and physiology of taste and smell, emphasizing relationships between molecular structure and perception.

Hubel, David H., "The Visual Cortex of the Brain," *Scientific American,* November 1963. Hubel's findings indicate that lines and contours become the most important stimuli at this level of neural integration.

Land, Edwin H., "Experiments in Color Vision," *Scientific American,* May 1959. Land critically examines a fundamental conflict between the firmly entrenched three-pigment theory of color vision and the theory of natural images based on the relative balance of long and short wavelengths over the entire visual field.

Livingston, W. K., "What Is Pain?" *Scientific American,* March 1953. Based on his surgical experience, Livingston believes that "dying is merely the closing event in a sequential loss of function which accompanies brain depression" and is not painful.

MacNichol, Edward F., Jr., "Three-Pigment Color Vision," *Scientific American,* December 1964. Experiments by MacNichol and others indicate that perception of color is initiated in the retina by three different types of cone receptor, each with either a red-, green-, or blue-detecting pigment. The information is encoded into two-color, on-off signals by color-sensitive retinal ganglion cells and processed at higher neural levels.

Melzack, Ronald, "The Perception of Pain," *Scientific American,* February 1961. According to Melzack, pain is a reflection of the plasticity of the central nervous system and the physiologic and psychological status of the subject at any given time.

Pettigrew, John D., "The Neurophysiology of Binocular Vision," *Scientific American,* August 1972. Often a single point of an object being viewed does not lie in corresponding regions of both retinas, a condition referred to as image disparity. Pettigrew has identified several kinds of simple and complex disparity-specific neurons in the visual cortex of the cat. Binocular inhibition of the image formed by one eye by the complex interplay of disparity-specific neurons in the visual cortex may explain binocular fusion phenomena.

Rock, Irving, and Charles S. Harris, "Vision and Touch," *Scientific American,* May 1967. Experiments are described which show that when the sense of touch conveys information that disagrees with the sense of vision, the visual information determines actual perception.

Rosenzweig, Mark R., "Auditory Localization," *Scientific American,* October 1961. A historical review of experiments which attempt to determine how human beings, using only the auditory sense, determine the source of a sound.

Sperry, R. W., "The Eye and the Brain," *Scientific American,* May 1956. From interesting experiments in which the optic nerves of newts were severed and allowed to regenerate after the eyes were rotated through a fixed number of degrees, Sperry concludes that many features of visual perception such as location in space, pattern organization, and perception of motion are innate and apparently not acquired through experience.

Thomas, E. Llewellyn, "Movements of the Eye," *Scientific American,* August 1968. From the results of experimental research, Thomas concludes that many fixation movements never reach the level of consciousness, even though appropriate motor responses are elicited by those portions of the visual fields upon which the eyes fixate.

Von Bekesy, Georg, "The Ear," *Scientific American,* August 1957. A general review of the anatomy of the ear and the physiology and pathology of hearing.

Werblin, Frank S., "The Control of Sensitivity in the Retina," *Scientific American,* January 1973. A discussion of the nature of the retinal interactions among nerve cells. These enable the retina to form high-contrast, high-acuity images over a wide range of light intensities.

Wilentz, Joan Steen, *The Senses of Man,* New York: Apollo Editions, Inc., and New York: Thomas Y. Crowell Company, 1971. Written for the layman, this engaging paperback contains a wealth of information on all the senses with which human beings are endowed.

# CHAPTER 12
# SENSORY DISORDERS

Sensory disorders affect reception and perception. Therefore, sensory defects may be a malfunction in the operation of one or more receptors or sense organs, in the conduction and transmission of nerve impulses along peripheral nerve pathways from sensory structures, or of integration of the sensory information within the central nervous system.

More specifically, sensory dysfunctions may take the following forms: (1) complete absence of, marked depression in, or other alterations involving sensation from a localized region of the body or from one or more specialized sense organs due to damage of sensory structures, peripheral nerves, or central nervous system structures; (2) aberrations of, or distortions in, sensory stimuli due to ambiguous sensory information, structural defects of sense organs, or abnormal secretions, obstructions, and pressures; (3) loss of muscle control of sense organs such as the eyes and ears due to peripheral nerve disorders or the aging process; and (4) perceptual difficulties

which prevent normal sensory integration due to psychogenic, hereditary, or idiopathic factors. Idiopathic factors are those of unknown origin.

This chapter reviews the causes and effects of the more common and well-known sensory disorders.

## PERCEPTUAL ILLUSIONS AND CONFUSIONS

An *illusion* is a misinterpretation or misrepresentation of a sensory experience. Many illusions occur because there is insufficient sensory information upon which to base an unequivocal interpretation of the real experience. Several examples of visual illusions are given in Fig. 12-1.

When several interpretations fit or explain a sensory experience, we tend to have a difficult time deciding which to accept and so oscillate periodically between them. This phenomenon is probably best exemplified by the "flip-flopping" image perceived while staring

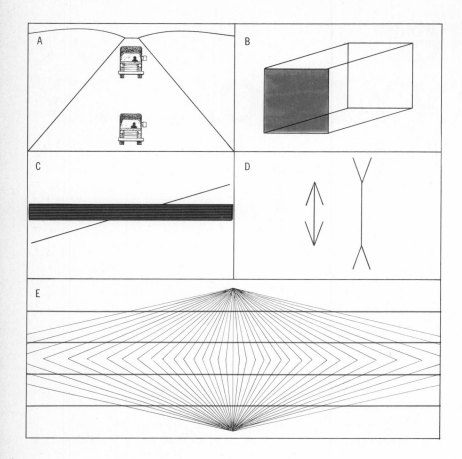

**Figure 12-1** *Some examples of visual illusion. A. Although the trucks are identical in size, the truck at the top looks larger because of depth perspective. B. The cube appears to "flip-flop" during steady visual inspection because the brain cannot decide whether the colored face is nearest to, or farthest from, the observer. C. When a continuous line is broken by an object of some thickness, the brain tends to interpret their broken ends as parallel to, rather than continuous with, each other D. Although the vertical lines are of the same dimension, the diametrically opposed arrangement of the V's on their ends affects perception of their lengths. E. Although all the horizontal lines are straight, the two top lines appear bent upward at their ends because of convergence of the obliquely arranged lines, and the two bottom lines appear bent downward through divergence of the same lines.*

at the edges of the transparent cube in Fig. 12-1. Other visual illusions distort size or positional relationships. Traditional wisdom accumulated from our previous experience is our usual guide but can be misleading when we rely upon it intuitively. Although we should do more than judge on the basis of our immediate sensory impressions, an analysis which goes beyond intuition is often not feasible. We do not usually look for a symbolic or abstract interpretation of what seem to be straightforward sensory experiences. Therefore, either our intuitive impression differs from the real experience, or we have difficulty deciding which of several impressions to select.

Alternatively, when sensory experience is ambiguous or incomplete because extraneous stimuli detract from that which we are concentrating on, as when coughing in an audience obscures a portion of a sentence spoken by the speaker, we tend to interpolate, or fill in, the missing words subconsciously. This ability results in a misrepresentation of the real sensory experience, which was coughing and not audible words. The most extreme forms of this type of illusion are noted in subjects who suffer from visual or auditory *scotomas*, nonfunctional portions of the retina or the basilar membrane which lead to abnormal blind spots or differential frequency deafness, respectively. In many such cases, the missing portions of the visual image and the absent range of sound frequencies are automatically supplied during integration of the sensory input by the central nervous system. In other words, the subjects may be totally unaware of their sensory deficiencies, because the brain supplies the missing elements of the visual or auditory experience.

The *phantom pain,* a type of referred pain, of a recent amputee is another case in point. The non-visual sensory impression is that of a limb or a portion of a limb which is not there. Since real sensory stimuli arise from the cut nerve endings in the stump which remains, the referred pain is obviously not a hallucination. Phantom pain has a definite physiologic basis, and the inaccuracy of perception is not due to the ambiguity or incompleteness of the sensory stimuli.

Not all illusions are visual and auditory; stimuli mediated by any sense may be inaccurately perceived. We know much more about visual and auditory illusions, compared with other types of illusions, simply because the eye and the ear have been more intensively studied than other sensory structures. We are also more aware of the uses and limitations of these sense organs in interpreting the world around us, compared with those of other sensory structures.

However, an "inaccurate" perception of a sensory stimulus may not be due to an illusion at all. Consider for a moment the sensation of temperature. Cold running water or a bag of ice cubes on the wrists is initially perceived as cold, but if stimulation is continued over an extended period of time, the sensation changes to that of hot or burning pain. This condition results because very cold stimuli excite pain receptors rather than cold receptors. At the other end of the temperature spectrum, when hot running water stimulates the wrists, the sensation of heat may change with continued stimulation into cold and pain sensations. In this instance, stimulation of cold receptors and pain receptors is initiated as stimulation of hot receptors ceases.

In these examples, sensation changes even though the stimuli remain constant (although the skin becomes colder or hotter with time). Because different types of receptors are being activated at different times and under particular circumstances, changes in perception with time are real although they may be thought to be illusory. In other words, the perceptual ambiguities are not a function of the sensory stimulus but, instead, reside in a comparison between our knowledge of the sensory stimulus and our actual perceptual experience. This fact does not prevent us from commonly thinking, albeit incorrectly, about these phenomena as types of illusions.

## COMMON DISTURBANCES

The foregoing illustrations are not really sensory disturbances or disorders, although the outcome is an inaccurate perception of real stimuli and therefore maladaptive. Other sensory disturbances fall into this same gray area, with one important difference: Such sensory disturbances, whether mild or moderate, originate in the sense organ or receptor neuron, or in the peripheral nerve pathways which supply these structures. Among these sensory disorders are visual alterations, disturbances of audition and balance, and unique somatic, or body, sensations.

### VISUAL DISTURBANCES

Common visual disturbances originate in the eye and include light and dark adaptation, negative and positive afterimages, and phosphenes (seeing stars).

Light and dark adaptation *Light adaptation* is a transient dimness of vision due to an overstimulation of the rods and cones of the retina by bright light. Recall that the visual pigments of the rods and cones (rhodopsin and photopsin) are broken down into opsins and retinene when stimulated by light. Retinene, in turn, is converted into vitamin A. Since the conversion of the chemical intermediates into rhodopsin requires a longer time than the breakdown of rhodopsin into these same intermediates, the result is a bleaching of the image due to a lack of contrast between its light and dark portions. Most of us are undoubtedly familiar with this experience from having attempted to look directly at the sun or following prolonged exposure to a landscape painted white by a freshly fallen snow. Both these experiences are potentially dangerous, the former because of the damaging effects of ultraviolet rays to the eyes and the latter because severe bleaching may cause functional blindness and possible irreversible damage to the rods and cones.

*Dark adaptation,* on the other hand, is due to a transient increase in visual sensitivity to light. The phenomenon occurs because of an excessive accumulation of rhodopsin, owing to prior understimulation of the rods. Under these conditions, vitamin A is converted into retinene, from which rhodopsin is synthesized at a rate in excess of its degradation. The increased amount of visual pigments in the rods makes them extremely sensitive, even in dim light.

The subjective impression obtained from dark-adapted eyes during stimulation by bright light is of bleached images similar to that obtained from light-adapted eyes, but for a different reason. The dark portions of images produced by dark-adapted eyes are relatively bright because of the sensitivity of the photoreceptors is unusually great. Therefore, the entire image appears washed out until photoreceptor sensitivity decreases sufficiently.

**Negative and positive afterimages** A *negative afterimage* is a persistence of a visual image after cessation of photic, or light, stimulation. The colors of the image are complementary to those of the object or visual field. In effect, what is perceived is a color reversal, or an impression similar to that of a photographic negative, only colored. To experience this phenomenon, look with fixed eyes for a reasonable length of time at a brightly lighted visual field containing contrasting regions of light and dark. During this period of observation, there is a rapid breakdown of visual pigments in the rods and cones stimulated by light from the bright portions of the visual field and a less rapid degradation of the pigments in photoreceptors excited by light from the darker regions. Following light adaptation, quickly gaze at a well-lighted white wall or sheet. Under these circumstances, a complementary image of the sharply contrasting visual field first perceived should persist for a short time.

The explanations for negative afterimages are to be found in light adaptation, image persistence, and pigment complementarity. Following the first period of visual observation, the eyes are light-adapted. During the second period of visual observation, the originally less stimulated photoreceptors are relatively more sensitive than those rods and cones in which most of the visual pigment has disappeared. The awareness of a transient persistence of a previous visual image possessing sharp contrasts of shading is increased by looking at a uniform and brightly lighted visual field. Color reversal is due to the brief, differential activity of the initially understimulated and therefore more sensitive cones following the first period of observation. Complementary colors are perceived because these cones contain pigments which are complementary to those of the previously light-adapted cones (see Chap. 11 for explanations of color vision).

A *positive afterimage* is similar to a negative afterimage, except that the perceived image is identical in color to the original object and visual field. To experience this phenomenon, permit the eyes to become dark-adapted by remaining in a darkened room for a reasonable period of time. Under these conditions, the eyes become hypersensitive to light stimulation because of an excessive accumulation of visual pigments in the photoreceptors. Following dark-adaptation, expose the eyes momentarily to a bright light. Then extinguish the light and observe the nature of the afterimage during the first moments in darkness. The afterimage is positive, i.e., of the same colors as those possessed by the object.

Positive afterimages are more simple to explain than negative afterimages. Obviously, dark adaptation is essential to increase the sensitivity of the rods and cones. The brief flash of light causes a differential degradation of visual pigments among the photoreceptors. During the darkness which immediately follows, the original image and colors persist because the rods and cones initially stimulated are still being stimulated chemically.

Both negative and positive afterimages are based on events which occur within the rods and cones following light and dark adaptation, respectively. Under these conditions, perception is an artifact of image persistence due to the biochemical events occurring in the stimulated photoreceptors. Here, inaccurate perception is due to the peculiarities of photoreceptor function rather than to factors which create illusions.

**Phosphenes (seeing stars)**  Phosphenes are visual patterns not related to the real visual field. Phosphenes move and are usually colored in pastel shades of orange, yellow, green, and blue. The variegated patterns and color combinations seem infinite, and each such experience is unique. Sometimes points of light or bright lines move across an otherwise darkened visual field.

The perception of phosphenes requires mechanical or certain types of electromagnetic stimulation of either the photoreceptors or the optic nerve fibers. To experience phosphenes, close both eyes and rub the eyelids by exerting moderate pressure with the fingers on the medial aspects of the orbits.

The mechanical pressure created by rubbing the eyes is probably the most common cause of phosphenes. Phosphenes may also be elicited in the following ways: (1) by intense photic stimulation, as from a photographic flashbulb which is fired from close range and head on; (2) by particles such as alpha particles and protons which occasionally penetrate the atmosphere at high velocity to interact directly with photoreceptive elements; and (3) through visual deprivation, which promotes an increase in the amount of spontaneous endogenous activity in the optic nerve fibers.

Thus, perception of phosphenes requires stimulation of the photoreceptors or other neural elements by nonvisual forms of energy. The stimulation is real, albeit artifactual, in that the rods and cones are converting inappropriate stimulus energies into nerve impulses. In any event, the perceptual experience is irrelevant.

## DISTURBANCES OF AUDITION AND BALANCE

Transient abnormalities of hearing and equilibrium may originate in any portion of the ear or associated sensory and neural structures. Such disturbances include clogged ears, tinnitus (ringing or hissing in the ears), motion sickness, dizziness, and vertigo.

**Clogged ears**  Ears become clogged when the external auditory meatus becomes filled or impacted by water or cerumen (wax), respectively. The pressure of such materials against the outer surface of the tympanic membrane, or eardrum, prevents this structure from oscillating freely in response to sound pressure waves. Thus, a reduced amount of energy is transferred by the middle ear ossicles within the tympanic cavity to the perilymphatic fluids inside the oval window of the cochlea. The end result is a reduction in auditory acuity; sounds seem muted, or muffled. Sometimes, the pressure of impacted cerumen or water in the ears may elicit wooshing, rushing, and popping sounds, an experience which is extremely annoying. Abrupt changes of altitude or of atmospheric pressure may also induce these sensations.

**Tinnitus**  When perceived sounds are subjective and not based on sound pressure waves generated outside of the body, the disorder is called *tinnitus*. The nature and localization of the perceived sounds are determined by the portions of the ear or neural components in which the affection is localized. More specifically, the abnormality may affect the outer, middle, or inner ear; the acoustic nerve (VIIIth cranial nerve); or central structures such as the auditory cortex of the cerebrum. Most of the time, tinnitus is simply a transient disturbance due to obstructions within the external auditory meatus. Sometimes, tinnitus is a more serious, longer-lasting disability due to an inflammation of the middle ear (otitis media) or to ankylosis, or immobility, of the stapes through the formation of cancellous bone around the oval window (otosclerosis). More rarely, tinnitus is evoked by lesions of the VIIIth cranial nerve (see Chap. 16) or by some psychogenic factor such as hysteria which involves the auditory sense (see "Conversion Hysteria" in Chap. 15).

Tinnitus is often experienced by subjects with Ménière's syndrome. This disorder is discussed later in the chapter.

**Dizziness and vertigo**  *Dizziness* is a feeling of light-headedness or giddiness accompanied by spinning sensations. Dizziness may be brought about in the following ways: (1) steady rotation of the body about a vertical axis; (2) abrupt changes in the force of gravity, as when rapidly accelerating, decelerating, or arising from a prone position; (3) double vision, or diplopia, due to lesions of the VIth cranial nerve; and (4) a depression in either the volume of blood or the amount of oxygen supplied to the brain.

Sudden or repetitive changes of position or gravitational forces and loss of motor control of the eyes have several basic effects which provoke dizziness: (1) Rotational and oscillating movements cause the endolymphatic fluids within the semicircular ducts to move continually; (2) abrupt gravitational changes deprive the brain of necessary nutrients and oxygen; and (3) cranial nerve disorders which affect the extrinsic eyeball muscles disrupt important feedback information from the eyes about position in space.

Although *vertigo* is also a disturbance in the sense of balance, its causes differ somewhat from those of dizziness. Whereas most forms of dizziness are in response to real stimuli detected by equilibrium receptors within the inner ear, vertigo is usually due to various pathologic conditions of the eyes, ears, semicircular canals, brain, blood, blood vessels, or other parts of the body. The subjective experience is either that of moving around a stationary world or of the world moving around the subject. Severe vertigo may cause the body to be flung to the ground. However, in a clinical sense, the distinction between dizziness and vertigo tends to become blurred, because symptoms are known more often than causes.

Common symptoms associated with both dizziness and vertigo include nausea and vomiting. Gastric disturbances may precede some cases of vertigo, but nausea and vomiting are ordinarily consequences rather than causes of disequilibrium.

Disturbances in the sense of equilibrium excite the vomiting center in the brainstem directly (see Chap. 14). In addition, digestive functions are easily disrupted by disturbances in the sense of balance. Nausea and vomiting are considered further in Chap. 27.

Dizziness and vertigo may also lead to pallor and faintness. Both these symptoms are due to a depression in cerebral circulation, although psychogenic factors may also be involved. Fainting, or *syncope,* may follow a severe bout of dizziness or vertigo.

Treatment and prevention of dizziness and vertigo involve a reduction in symptoms, the prevention of physical injury, and the elimination of the causative factors responsible for the abnormal sensations. When a depression in the cerebral circulation or psychogenic factors which result in the same effect are at fault, placing the subject in a sitting or lying-head-down position may reestablish normal circulation and thereby relieve the symptoms and prevent fainting. Eye movements which counteract the sensory information carried by the vestibular nerve during rapid rotational movements also decrease the intensity of the subjective experience of disequilibrium. Dramamine, the proprietary name for dimenhydrinate, or other drugs with similar properties may be used to counteract dizziness or vertigo as well as nausea and vomiting. Toxins and microbial pathogens which may be responsible for various types of vertigo can be controlled with antitoxins, antibiotics, and other forms of chemotherapy.

Motion sickness (seasickness)   Dizziness, nausea, vomiting, and other symptoms induced during turbulent or rough travel through the air, on land, or at sea comprise a syndrome known as *motion sickness.* As in the case of most forms of dizziness, motion sickness is precipitated in susceptible individuals by erratic oscillations of the endolymphatic fluids within the semicircular ducts.

Treatment of motion sickness is preventive, symptomatic, and supportive. Both chemoprophylaxis (drugs used for prevention) and drug therapy rely upon the use of Dramamine, antihistamines, and sedatives. *Antihistamines* counteract the effects of histamines, inflammatory chemicals released by dead and dying cells. *Sedatives* depress neural excitability, thereby minimizing the effects of disequilibrium. Motion sickness may also be reduced by abstaining from alcohol prior to an anticipated rough journey. The severity of vomiting may be lessened by eating lightly. Maintaining visual reference to the horizon as a ship rocks and rolls in an angry sea is also useful in allaying motion sickness. Symptomatic relief from this condition may sometimes be obtained by lying down. Supportive therapy is directed toward the prevention of dehydration during a severe bout of vomiting triggered by motion sickness. Hydration is usually achieved by drinking copious quantities of water or by the provision of intravenous fluids, where water ingestion cannot be tolerated.

## DISTURBANCES IN SOMATIC SENSATIONS

Sensory dysfunctions associated with the skin include itching and paresthesias. Closely related to itching is another somatic sensation commonly known as tickling.

Itching and tickling   *Itching* is a sensation caused by mild irritation of the skin. This symptom is especially prevalent in fungal and other forms of ectoparasitic infestations or endoparasitic infections but, of course, may be caused by many other factors. The various types of irritative itch are named for the parasites which induce this symptom, the portions of the skin frequented by the parasites, the parts of the environment from which the parasites are contracted, or the occupations in which such irritations are an especially frequent hazard.

Itching is probably caused by stimulation of special pain or touch receptor neurons. In either case, the dendrites of these neurons terminate externally on the skin. The available evidence does not yet permit us to select one pathway of stimulation over the other. All we are justified in concluding is that the stimulation causes irritation which is then interpreted as itching.

*Tickling,* on the other hand, is a pleasant sensation elicited by light touch on sensitive regions of the skin. The pathway of stimulation therefore seems to be initiated by excitation of receptors which differentially respond to gentle tactile stimuli. The population of these receptors must be greater in sensitive regions than in less sensitive portions of the skin.

The response to tickling is reflex laughter, sometimes bordering on hysteria. The susceptibility, or sensitivity, to tickling stimuli seems to vary from time to time in the same subject. In other words, sometimes an individual is hypersensitive to the same strength of stimulation that at other times has no discernible effect. Interestingly, a subject cannot elicit reflex laughter through self-stimulation of the appropriate type. Presumably, some neural feedback mechanism inhibits the reflex when self-initiated.

Paresthesias and related phenomena   *Paresthesias* are abnormal sensations referred to localized regions of the skin. These abnormalities are due to irritative or mechanical lesions of sensory portions of peripheral nerves. The peculiar sensations usually take the form of numbness (pins and needles), tingling, or prickling. The direct stimulation of a sensory nerve trunk through compression, as from bumping the elbow or sleeping on the wrists, causes neural excitation. The stimulation is interpreted as if the sensory receptors supplied by peripheral neurons of the nerve trunk have actually been stimulated. Therefore, when a sensory nerve trunk is pinched, bruised, or otherwise damaged, the nerve potentials induced at the site of injury (injury potentials) are sustained for some time but are referred back to sensory structures on the periphery of the body. The same principle is responsible for paresthesias perceived when lesions occur in the sensory nuclei of cranial nerves, e.g., from a severe blow to the head (concussion) or in the posterior roots of spinal nerves, e.g., from a slipped, or herniated, intervertebral disk.

Sometimes the experience is one of a heightened sensitivity to normal stimuli rather than abnormal stimulation by endogenous factors. This form of abnormal sensation is called *hyperesthesia,* when the tactile sense is primarily involved. If responsiveness to pain is exacerbated, then the phenomenon is referred to as *hyperalgesia.* In both cases, elevated sensitivity to a normal sensory input may be due to physiologic events which occur during either transduction of the stimulus energies or chemical transmission of the encoded sensory information across a synapse.

In more serious lesions of the peripheral nerves,

there may be a depression in one or more somatic sensations (*hypesthesia*) or total lack of sensation (*anesthesia*) from localized regions of the skin. Because adjacent pain receptors may be supplied by different peripheral nerves, impairment of pain sensation is not as great as that of touch. However, the opposite is true in posterior (sensory) root lesions. Here, considerable overlap in the nerve supply to touch receptors exists between adjacent spinal nerve roots.

Paresthesias and anesthesia due to peripheral nerve lesions are discussed further in Chap. 16.

## SERIOUS DISORDERS

The more serious sensory disorders are associated with vision and audition, simply because these two senses are more important to our everyday functioning than any of the other senses with which we are equipped. Although not as frequent in the general population, such diseases are more damaging physically, more debilitating functionally, and longer-lasting, compared with the disturbances mentioned heretofore.

### VISUAL DISEASES

Visual diseases are related to the following causes: (1) dysfunctions of the retina or other structural components of the eyeball, (2) lesions of cranial nerves associated with either the eyes (IId cranial nerve) or the extrinsic or intrinsic muscles of the eyeballs (IIId, IVth, and VIth cranial nerves), or (3) central nervous system disorders and mental diseases which involve visual perception. Included among these diseases of vision are blindness, glaucoma, strabismus, diplopia, and amblyopia.

Blindness    Strictly speaking, blindness is a loss of vision. The loss may be partial or total. In the former case, shadow images or gradations of light and dark are still detected; in the latter case, no image is perceived. However, the term blindness may be qualified by modifiers such as "color" (protanopia and deuteranopia; see Chap. 21), "night" (see Chap. 27), or other terms which refer either directly or indirectly to a specific type of visual deficit, its anatomic localization, or its causation. There are also localized regions of blindness on the retina called *visual scotomas.*

Blindness is caused by one of the following factors:

1  Physical trauma
2  Extreme bleaching of the visual pigments in the rods and cones
3  Severe deficiencies of vitamin A, a chemical necessary for regeneration of rhodopsin
4  Opacities of the lens or capsule
5  Increase in intraocular pressure, causing atrophy of the optic nerve
6  Neurologic diseases affecting the retina, optic nerve, midbrain, or visual cortex
7  Congenital factors such as hereditary defects, developmental anomalies, and congenital gonorrhea (see Chap. 21)
8  Psychogenic factors such as hysteria (see Chap. 15)

Glaucoma    Glaucoma is one cause of blindness. It results from an acute or chronic increase in the pressure of the intraocular fluids, causing the vitreous body to press against the inside wall of the eyeball. The condition is illustrated in Fig. 12-2.

The accumulation of fluid is due to one of the following factors: (1) a blockage of the venous sinuses between the sclera and cornea which ordinarily drain the aqueous humor, (2) a chronic increase in the blood pressure of the arteries which supply the eyeball, or (3) a decrease in the colloid osmotic pressure of intraocular blood or, alternatively, an increase in the permeability of the capillary walls in the vascular coat of the eyeball.

The increased pressure on the inside of the eye has one or more of the following effects: (1) enlargement of the eyeball, (2) hardening of the eyeball, (3) atrophy of the optic nerve, and (4) degeneration of the rods and cones of the retina. The eyeball may become as hard as a rock as the disease progresses. Severe eye pains, corneal insensitivity, and pupillary dilatation may accompany the partial or complete loss of vision. Most of these effects presumably result from alterations in the circulation of blood within the eye due to the abnormal increase in intraocular pressure.

Strabismus    Strabismus, or squint, is due to a pathologic deviation of one or both eyes. The deviation takes one of the following forms: (1) Both eyes turn inward so that their visual axes cross; (2) both eyes turn outward so that their visual axes diverge; or (3) one eye deviates from the other in either a vertical or horizontal plane. The deviation may cause the affected eye to remain in a fixed position or move in the same altered position, relative to the other eye. The abnormality may be displayed continuously or intermittently, or the condition may alternate between both eyes.

Strabismus is caused by either refractive disorders which impair accommodation or lesions of the IIId, IVth, and VIth cranial nerves. These nerves control the extrinsic muscles of the eyeball.

The treatment of strabismus depends on its causation. Accommodative strabismus is treated by securing proper refraction with the use of eye glasses. Paralytic strabismus may require the strengthening and

**Figure 12-2**  *The cause of glaucoma. Glaucoma is caused by a chronic increase in the intraocular pressure. The fluid pressure pushes against the inside walls of the eyeball (arrows).*

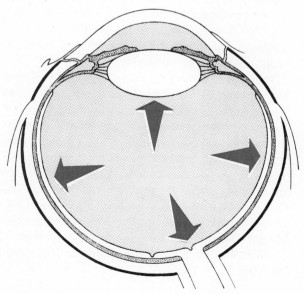

retraining of the extrinsic muscles of the affected eye or eyes. Ideally, any pathogenic agent responsible for the disorder should be eliminated.

Diplopia    When the visual axes of the two eyes are not parallel to each other, the result is a double image or a marked blurring of the perceived image. This phenomenon is called *diplopia* and, as we have just noted, can be caused by strabismus. The effect may be perceived in only one eye or may be a function of both eyes operating together.

The causes of diplopia are similar to those of strabismus. Where one eye is involved, the abnormality usually stems from an inability to accommodate adequately. When both eyes are affected, the disorder can generally be traced to cranial nerve lesions which paralyze one or more extrinsic eyeball muscles. Sometimes, central nervous system disorders may be responsible for diplopia.

Amblyopia    Amblyopia is a dimness of vision due to a reduction in visual acuity. The abnormality is caused by the direct action of certain toxic and irritative chemicals such as methyl, or wood, alcohol and quinine on the optic nerve or retina. Sometimes, amblyopia is caused by mechanical or microbial irritation of the eyes, e.g., rubbing the eyes with dirty fingers. Reduced visual acuity may also result from strong light adaptation due to the excessive bleaching of the visual pigments of the rods and cones. Because amblyopia appears during severe alcoholism, toxemia, and uremia, the disorder is symptomatic of chemical intoxication.

## AUDITORY DISEASES

The more serious disorders of hearing may affect any portion of the ear, acoustic nerve, or auditory cortex. Among such diseases are deafness, Ménière's syndrome, otitis, and otosclerosis. Central nervous system disorders which result in deafness are discussed in Chap. 15.

Deafness    Deafness is a partial or complete loss of hearing. The difficulty may be due to defects in conduction mechanisms, lesions in nervous structures, or personality disorders without demonstrable organic damage.

Conduction deafness is due to one or more of the following factors: (1) obstructions such as cerumen in the external auditory meatus; (2) perforations of the tympanic membrane due to sound concussion or mechanical injuries; (3) ankylosis of the ear ossicles, immobilizing the stapes against the oval window (*otosclerosis*—see below); and (4) damage to a portion of the basilar membrane giving rise to *aural scotomas*.

Neural deafness, on the other hand, results from some combination of the following abnormalities: (1) damage to the hair cells due to physical trauma or microbial infection, (2) lesions of the acoustic nerve (VIIIth cranial nerve, see Chap. 16), and (3) damage to acoustic nerve tracts in the brain or the auditory cortex of the cerebrum.

Finally, functional deafness may be caused by mental disorders, such as hysteria, which block central integration.

Deafness is either hereditary or acquired. Otosclerosis has a hereditary basis, at least in some patients, and the incidence of this disease is much higher in females than in males. Acquired deafness may be due to anomalies that arise during prenatal development, injuries incurred at the time of birth, or diseases and age-specific changes which occur during postnatal development. Most causes of deafness are a consequence of the aging process, the average incidence of this disorder rising dramatically to a peak in the sixth decade of life.

The techniques used to restore hearing to the deaf are determined by the nature of the deafness exhibited. Where bone conduction is depressed, hearing aids are useful in amplifying the energy of the sound waves. To bypass the spongy bone which forms around the oval window in otosclerosis, an artificial opening into the labyrinth is made to receive the stapes. This surgical procedure is called *fenestration.* Ruptured eardrums can also be repaired surgically. Impacted obstructions in the outer ear can be dislodged by dissolution in dilute hydrogen peroxide followed by irrigation of the external auditory meatus with water under pressure. Functional deafness is managed with psychiatric treatment. When structural damage cannot be repaired, as in neural deafness, lipreading can partially remedy the disability.

Ménière's syndrome   This syndrome is a chronic, progressive, and intermittent disease marked by severe vertigo, tinnitus, and an increasing loss of hearing. Nausea, vomiting, and profuse perspiration frequently accompany the attacks. The vertigo is usually characterized by the sensation that the world is spinning around the subject, although the reverse of this situation sometimes occurs. Often, attacks are serious enough to throw the subject to the ground. Rapid to-and-fro eye movements called *nystagmus* (see Chap. 16) are performed reflexly during such bouts. As the disease progresses, tinnitus and deafness persist, even with complete remission of vertigo.

The cause of the disease is unknown. Some patients with Ménière's syndrome show a dilation of the endolymphatic ducts, presumably as the result of edema. Therefore, some investigators believe that this disease may follow certain inflammatory conditions within the middle ear. However, in other patients the disease develops without demonstrable damage to, or inflammation of, either the ear or the nervous system. Since the disorder is age-specific and more frequently found in men than in women, a hereditary predisposition may be involved.

The treatment of Ménière's syndrome is prophylactic, symptomatic, and remedial. Prophylaxis is intended to prevent the patient from physical injury during bouts of vertigo and simply consists of lying down at the first signs of physical imbalance. The symptoms of vertigo may be reduced by drug therapy. When conventional treatment is ineffective and the disorder is severe, surgical sectioning of the vestibular nerve trunk or destruction of the labyrinth may be attempted, to eliminate the feelings of disequilibrium. There is no effective method at present of ameliorating the accompanying tinnitus and nerve deafness.

Otitis   An inflammation of the ear is called *otitis*. A common form of this inflammatory condition is called *otitis media* and affects structures in the middle ear. When infection results in the immobility of one or more ear ossicles, the disorder is referred to as *otosclerosis*. Otosclerosis causes progressive deafness due to the formation of cancellous bone around the oval window. This pathologic process results in ankylosis of the stapes.

Otitis may be caused by one of the following factors: (1) sudden changes of air pressure which block the eustachian tubes; (2) the development of abnormal growths such as boils within the external auditory meatus; (3) inflammation of the mastoid cells within the temporal bone, resulting in mastoiditis; or (4) the presence of microbial pathogens in any portion of the ear.

## COLLATERAL READING

Brindley, G. S., "Afterimages," *Scientific American,* October 1963. Studies of afterimages suggest that different products are liberated by cone pigments struck by light, giving rise to different rates of diffusion or reflecting different cone sensitivities.

Day, R. H., "Visual Spatial Illusions: A General Explanation," *Science,* 175: 1335–1340, 1972. According to Day, illusions appear when stimuli which preserve constancy, such as information about distance, bearing, lateral tilt, and movement of the observer with respect to the object, are normally operative, while the retinal image of the object is not varied.

Gregory, R. L., *The Intelligent Eye,* New York: McGraw-Hill Book Company, 1970. Equipped with stereo pictures and special viewing glasses, this intriguing paperback develops the idea that human perception is based more on symbolism than on intuition.

Kalmus, Hans, "Inherited Sense Defects," *Scientific American,* May 1952. A broad, general discussion of color blindness, tone deafness, and several other inherited sensory abnormalities that raises some interesting questions with regard to the possible relationships between specific diseases and diet.

Oster, Gerald, "Phosphenes," *Scientific American,* February 1970. Oster describes the different ways in which phosphenes may be elicited and contributes to our understanding of the functional organization of visual pathways.

# CHAPTER 13
# THE CHARACTERISTICS OF NEURAL TISSUES

The biologic processing of information is initiated when receptors transduce stimuli, i.e., convert their energies into the electrical energy of nerve impulses (see Chap. 11). The transduction process causes the generation of nerve impulses which are conducted toward the brain or spinal cord along fibrous extensions of nerve cells, or neurons, and transmitted across junctions called synapses between successive neurons. The encoded information contained in the nerve impulses is further processed in the central nervous system (CNS). Following neural integration of the sensory input, the information may be ignored, stored, or acted upon. When action is initiated, encoded information is conducted along neurons as nerve impulses and transmitted across synapses leading from the brain or spinal cord to effectors such as muscles or glands. The responses may be reflex or volitional, excitatory or inhibitory. The muscular responses are monitored by other receptors which

provide feedback information to the CNS so that any necessary corrections can be made to achieve or maintain homeostasis.

The enormous complexity of the nervous system may be appreciated by considering the following:

1   The nervous system is made up of many thousands of millions of neurons and countless other nonneural cells which affect neural functions.
2   Each neuron receives synaptic connections from about 100 neurons and makes approximately as many connections with other neurons.
3   A staggering number of nerve impulses, estimated at roughly 3 billion, are propagated each second.
4   Neurons secrete chemicals called neurohumors which serve as either transmitter substances at synaptic junctions or hormones which enter the bloodstream.
5   Nervous and endocrine systems reciprocally in-

teract with each other, giving rise to a higher order of control than can be achieved by either system acting separately.

6 Nerve impulses cause chemical changes in the structures of certain proteins in neurons, changes believed to be related to information storage of the type necessary for learning and memory.

Our physiologic knowledge of neuronal functioning is limited in such fundamental areas as synaptic transmission; information storage and retrieval including learning, remembering, and forgetting; motivation; the roles of heredity and experience in determining neural inflexibility and plasticity; and chemical mechanisms by which brain neurons interact with each other. These complexities often force us to talk about the nervous system in generalities based more on educated guesses than on well-established facts and principles.

In this chapter, we will explore those qualities of neural tissue which endow it with its unique abilities of intercellular communication. The functional anatomy of the nervous system will be discussed in the next chapter.

## THE ORGANIZATION OF NEURAL TISSUE

Neural tissue consists of neurons, or nerve cells, nonneural cells called *neuroglia* associated with the neurons and nearby structures, as well as blood vessels and connective tissue. These are the tissues which form the substance and coverings of the brain, spinal cord, and peripheral nerves.

### NEUROGLIA

Neuroglia collectively represents a variety of cells closely associated with neurons. Based on location in the nervous system, there are two types of glia—central and peripheral.

Central glia   These are nonnervous cells associated with neurons in the CNS. Glial cells comprise almost

half of the brain by volume. The following types of central glia are recognized:

1   *Astrocytes* (astroglia). The largest of the central glial cells, astrocytes are star-shaped and possess numerous fibrous processes. Astrocytes are closely associated with nerve cell bodies and either the walls of blood vessels or the meningeal coverings of the brain and spinal cord. In addition to nourishing and supporting the neurons to which they are attached, these cells are believed to play important roles in information storage, or memory.

2   *Oligodendrocytes* (oligodendroglia). Intermediate in size, these cells possess fewer fibrous processes than do astrocytes. Lying next to the dendrites, cell bodies, and axons of central neurons, oligodendrocytes secrete myelin sheaths just as neurilemma cells do for peripheral neurons. Thus, these cells provide neural insulation, thereby increasing the rapidity of nerve impulse conduction in the CNS.

3   *Microcytes* (microglia). These are the smallest and most highly variable of the central glial cells and have few fibrous processes. Related to phagocytes, microglial cells are ameboid and function phagocytically during microbial infections of neurons. They may also provide neural support.

4   *Ependymal cells* (ependyma). Lining the ventricles of the brain and the central canal of the spinal cord, these ciliated cells are closely associated with the walls of choroid blood capillaries. The ependyma participates in the production of cerebrospinal fluid (CSF) by forming part of the differentially permeable membrane through which the cerebrospinal fluid must pass as it enters the ventricular lumina from the choroid vascular plexuses. The formation of cerebrospinal fluid is discussed in Chap. 14.

Several types of central glia are illustrated in Fig. 13-1.

Peripheral glia These nonnervous cells are associated with neurons in the peripheral nervous sys-

A. Astrocyte     B. Oligodendrocyte     C. Microcyte

tem. The following types of peripheral glia are recognized:

1   *Neurilemma* (Schwann cells). These cells are the peripheral counterparts of the central oligodendrocytes. Surrounding peripheral axons, the neurilemma is responsible for the formation of myelin sheaths. Therefore, Schwann cells form the insulation of peripheral nerve fibers. Several neurilemma cells are shown in situ (in their natural place) in Fig. 13-2.
2   *Satellite cells.* These cells surround the nerve cell bodies and axons of peripheral neurons and probably subserve nutritive functions. A satellite glial cell is shown in situ in Fig. 13-2.

## THE NEURON

Neurons, or nerve cells, are the structural units of the nervous system. These cells are the most highly specialized and longest cells in the body. Most neurons are functionally specialized for the conduction and transmission of nerve impulses. Some neurons secrete hormones called neurohumors, as mentioned earlier, rather than propagate nerve impulses. All neurons are involved in their own energy metabolism and protein synthesis.

**Structure-function correlations**   Each neuron contains a cell body and one or more fibrous processes which may range up to several feet in length. The fibrous extensions communicate functionally with receptors, effectors, or other neurons. Portions of the cell body and fibrous processes are invested by glial cells.

The description which follows is with reference to a *multipolar* neuron, a type of neuron commonly found in the brain and spinal cord. Other types of neurons will be compared with the multipolar neuron after its structure is explained.

A multipolar neuron contains three regions—dendrites, cell body, and axon. The gross structure of a multipolar neuron is diagramed in the upper portion of Fig. 13-2. Resembling the twigs of a bush, the *dendrites* are short, highly branching fibers continuous with the cell body of the neuron. The *cell body* contains a nucleus bearing a well-defined nucleolus. The dark-staining structures in the cytoplasmic portion of the cell body are called *Nissl bodies*. Nissl bodies are the functional counterparts of the endoplasmic reticulum of other types of cells and participate in protein synthesis. The long fibrous process which projects from the cell body on the side opposite the dendrites is called an *axon*. The axon is surrounded by an inner and outer wrapping referred to, respectively, as the *myelin sheath* and *neurilemma.* The myelin sheath and neurilemma are accessory structures and not part of the neuron. The myelin sheath and neurilemma disappear periodically along the length of the axon, forming constrictions called *nodes of Ranvier.* The terminal portion of the axon loses its investment of myelin sheath and neurilemma and branches extensively. Collateral branches (not shown in Fig. 13-2) may also project laterally from points along the length of an axon. Slender tubules called *neurofibrils* course through the substance of the dendrites, cell body, and axon.

The fine structure of an axon and its complex coverings are shown in the lower half of Fig. 13-2. The cell membrane of the axon is called the *axolemma.* Notice that the axolemma is exposed at each node of Ranvier. A bundle of tightly packed, longitudinally arranged neurofibrils, randomly distributed mitochondria, and a cytoplasmic fluid called *axoplasm* are within the axon. The myelin sheath is a whitish-appearing insulation formed from concentric layers of fatty material called *myelin.* The myelin sheath is laid down by the neurilemma, or Schwann cells. The neurilemma also functions in the regenera-

**Figure 13-2**  *The structure of a multipolar neuron.* A. *An entire neuron.* B. *Enlarged view of a section of a myelinated axon in the peripheral nervous system.* (From Melloni, Stone, and Hurd.)

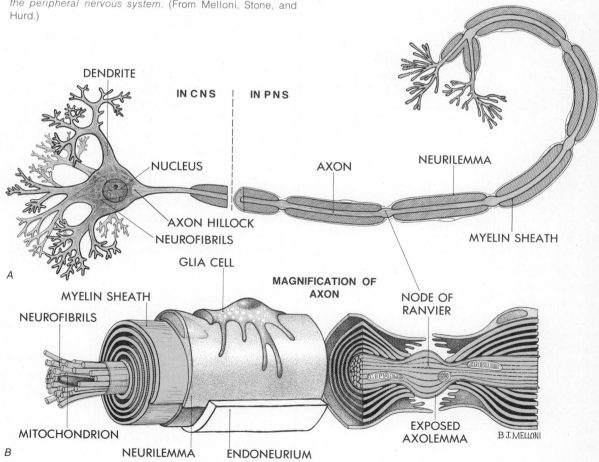

tion of axons which have been damaged. As the neurilemma is thrown into concentric folds, it deposits myelin between adjacent portions of its multiply overlapping cell membrane, giving rise to a number of concentric alternations between myelin sheath and neurilemma. The outermost concentric investment of the axon contains the nucleus and main portion of each Schwann cell. Each *internodal region* of the axon, i.e., the region between two consecutive nodes of Ranvier, is invested by a Schwann cell. Thus, there are as many Schwann cells as there are internodal regions along the length of the axon. An oligodendrocyte is shown straddling the outer surface of the neurilemma. The entire unit is covered by a thin connective tissue sheath called *endoneurium.*

The high degree of structural organization of a neuron reflects its functional specializations. Based on functional characteristics, a neuron can be divided into the following regions: *reception,* or generation, *propagation, conduction,* and *transmission.* Because the dendrites and cell body of a neuron receive the specialized terminal portions of axons from other neurons through synaptic connections (see below), these portions of a neuron receive encoded electric signals.

Sensory neurons transduce specific changes in the external environment. Transduction gives rise to *generator potentials,* a step which must occur if nerve impulses are to be conducted to the CNS from receptor neurons. Several types of receptor neurons are shown in Fig. 11-14.

The propagation of a nerve impulse is initiated in a region of the cell body called the *axon hillock.* Once initiated, nerve impulses are conducted from the axon hillock along the length of the axon to its naked terminal branches. Upon arrival in the terminal branches of an axon, nerve impulses somehow stimulate the secretion from presynaptic axonal vesicles of one of several types of transmitter substance. These chemicals act during the transmission of nerve impulses from the axonal branches of the active neuron to the dendrites or cell body of the next neuron in the chain or to the cell membranes of individual skeletal muscle fibers, smooth muscle fibers, or gland cells.

Thus, nerve impulse conduction is a function of the axon, whereas synaptic transmission is accomplished by actions initiated in the terminal branches of the axon. The electrical changes responsible for nerve impulse conduction are regulated by the axolemma of the axon.

The insulation afforded by the presence of a myelin sheath prevents electrical leakage to nearby axons and enables rapid conduction. Indeed, the rate of impulse conduction is roughly related to the amount of myelin which surrounds an axon. Axons, or nerve fibers, range from unmyelinated to heavily myelinated. Completely unmyelinated nerve fibers conduct nerve impulses at rates usually of less than 0.5 m/sec (meter per second). Heavily myelinated nerve fibers conduct impulses at more than 100 m/sec. Other nerve fibers conduct at intermediate rates, depending upon the degree of myelinization.

The nodes of Ranvier contribute to an increase in the velocity of nerve impulse conduction along an axon. Because of the nature of axonal electric conduction, the uninsulated axonal regions at the nodes of Ranvier permit nerve impulses to travel from node to node without actually traversing the axolemma in the internodal regions. Because the nerve impulse literally jumps from one node to the next, this type of conduction is referred to as *saltatorial* (leaping).

Bear in mind that neurons do not exist in isolation. Rather, they are found in series, forming neuronal chains that lead toward, through, or away from the CNS. The dendrites and cell body of a neuron receive the terminal axonal branches of other neurons. The axonal branches of the same neuron terminate synaptically on either the dendrites or the cell bodies of still other neurons, as shown in Fig. 13-3.

Thus, nerve impulses are transmitted synaptically from the axons of one neuron to the dendrites or cell bodies of other neurons and not in reverse. Called *polarity,* this one-way flow of encoded information across a synapse is due to the functional organization of a synapse. When an axon of an isolated neuron is adequately stimulated mechanically or electrically, a nerve impulse is conducted in both directions from the point of stimulation. In other words, while each neuron in vivo (in the living organism) is potentially capable of conduction in both directions, synapses maintain the polarity of the nervous system. The polarity of the nervous system is obviously essential to the smooth flow of information from receptors to the CNS and from the CNS to effectors.

The multipolar neuron of the CNS described earlier is classified as such on the basis of the number of fibrous processes it possesses—in this case, many. Other types of neurons based on fiber number are as follows: (1) *Bipolar* neurons, each of which consists of a cell body separating a single dendrite from a single axon, are found in the retina of the eye and in the acoustic ganglion of the inner ear, and (2) *pseudounipolar* (unipolar) neurons, each of which appears to consist of a single fiber from which the cell body extends, are characteristically associated with

**Figure 13-3** *Multiple synapses formed between neurons. Convergence and divergence are illustrated. Some of the synapses are axodendritic; others are axosomatic.*

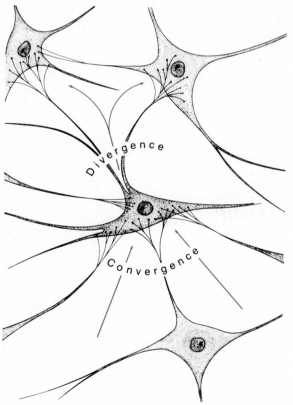

peripheral ganglia. These and several other neuronal types are shown in Fig. 13-4.

In practice, this classification breaks down because of the wide variability in neuronal architecture and in the number of dendrites, axons, and collateral fibers possessed by individual neurons in various parts of the central and peripheral nervous systems. For example, multipolar *pyramidal cells* in the motor cortex of the cerebrum have cell bodies with a pyramidal shape. Multipolar *Purkinje cells* in the cerebellum have a dense, gigantic, treelike grouping of dendrites. Indeed, there are so many exceptions to the simplistic classification given above that it is used more for convenience than to provide true insights into the relation between neuronal structure and function.

Other neuronal classification schemes are based on the following factors: (1) the direction of nerve impulse conduction, with reference to the CNS, and (2) the speed of nerve impulse conduction. For our purposes, the first scheme, the classification of neurons by their functional location, will probably prove most useful. According to this system, neurons are divided into three groups: (1) *sensory,* or *afferent,* neurons which conduct nerve impulses toward or to higher levels within the CNS; (2) *motor,* or *efferent,* neurons which conduct nerve impulses away from or to lower levels within the CNS; and (3) *internuncial neurons,* or *interneurons,* which connect between sensory and motor neurons within the CNS.

**Nerve impulse conduction** Nerve impulses are encoded electric signals conducted along the entire length of a neuron. It is by means of nerve impulses that receptors supply sensory information to the CNS and that the brain and spinal cord integrate the sensory input and initiate and coordinate the motor activities of the body.

If nerve impulses are electric signals, what information do they encode? Within the normal physiologic range of responsiveness, the number of nerve impulses conducted in a given time along the length of a neuron is related to the magnitude or strength of the stimulus. For example, when pressure is exerted against the skin, the intensity of the pressure is trans-

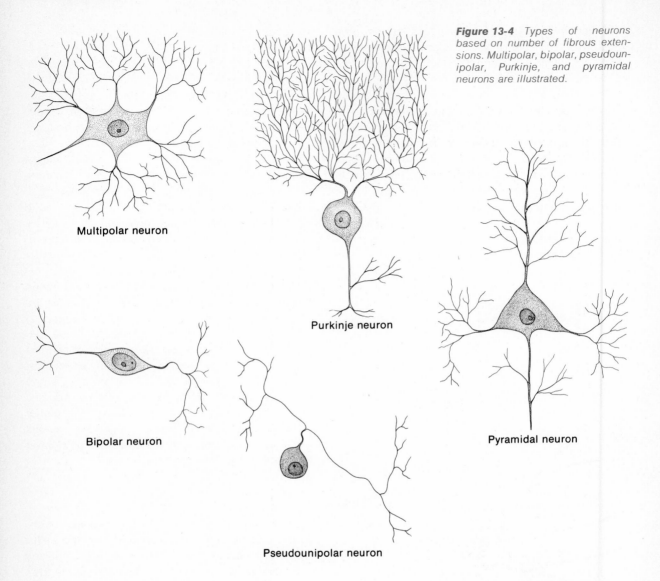

**Multipolar neuron**

**Purkinje neuron**

**Bipolar neuron**

**Pyramidal neuron**

**Pseudounipolar neuron**

**Figure 13-4** *Types of neurons based on number of fibrous extensions. Multipolar, bipolar, pseudounipolar, Purkinje, and pyramidal neurons are illustrated.*

duced by pacinian corpuscles (see Chap. 11) into a given number of nerve impulses per unit of time. If the pressure is increased, there is a corresponding increase in the number of nerve impulses conducted in the same unit of time by a nerve fiber from the receptor.

However, nerve impulses conducted by neurons in one pathway are not qualitatively different from those conducted by neurons of other sensory or motor pathways of the body. Since only the origins and destina-

tions of neural pathways differ, the sensations or responses elicited by the nerve impulses are determined by the nature of the receptors and synapses in the pathway, the portion of the brain in which the electric signals are interpreted, and the nature of the effector to which the efferent signals are conducted.

To understand the mechanisms by which nerve impulses are initiated and conducted along a nerve fiber by an adequate stimulus, we must look more closely at the ionic events which occur in the region of the nerve cell membrane. Although we will focus our attention on the axolemma of an axon, it should be understood that all fibrous portions of the nerve cell membrane, including dendrites, are conductile.

*Membrane potentials*   The plasmalemma, or cell membrane, of a neuron and the sarcolemma of a muscle cell are polarized electrically. With the use of appropriate equipment for transducing, processing, and displaying these minute voltages, the magnitude of the electrical potential and its polarity can be accurately measured. In the case of an axolemma, the

electrical potential difference across the membrane is approximately 70 mV (millivolts). The axoplasm is electronegative to the extracellular tissue fluid. Thus, the outside of the axon is electropositive, with reference to the axoplasm in its interior. The polarity of an axolemma and the nature of the equipment required for measurements of bioelectric potentials are suggested in Fig. 13-5.

The value of 70 mV is the *membrane potential* of an axolemma at "rest," i.e., not conducting nerve impulses. Because electrical measurements of membrane potentials are made with the use of an in-

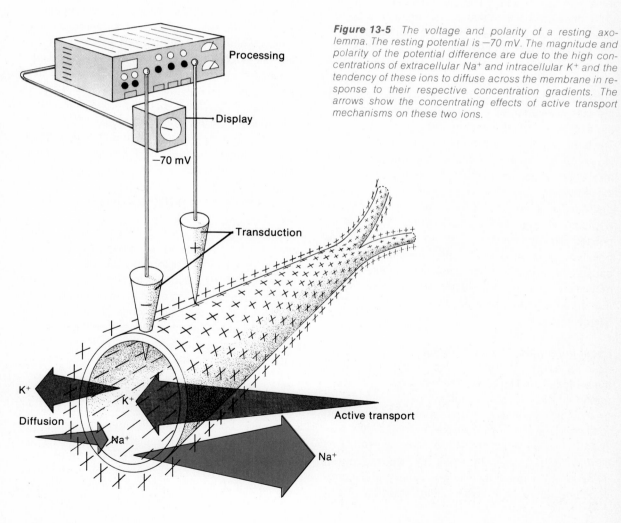

**Figure 13-5**   *The voltage and polarity of a resting axolemma. The resting potential is −70 mV. The magnitude and polarity of the potential difference are due to the high concentrations of extracellular Na⁺ and intracellular K⁺ and the tendency of these ions to diffuse across the membrane in response to their respective concentration gradients. The arrows show the concentrating effects of active transport mechanisms on these two ions.*

Processing

Display

−70 mV

Transduction

K⁺

K⁺

Diffusion

Na⁺

Active transport

Na⁺

tracellular recording microelectrode and an extracellular reference microelectrode, the membrane potential of a resting nerve fiber is written with a minus sign by convention, e.g., $-70$ mV. The minus sign simply means that the axoplasm is electronegative to the extracellular tissue fluid.

The membrane potential is maintained by the skewed distribution of specific ions on both sides of the axolemma. The ions are sodium ($Na^+$), potassium ($K^+$), and chloride ($Cl^-$). Other electrolytes, while present, are negligible in concentration compared with these three ions. The concentration of extracellular $Na^+$ is 10 times greater than that within the axoplasm, while the intracellular concentration of $K^+$ is 30 times greater than that in the extracellular tissue space. The axolemma is permeable to all three ions, $Na^+$ being the least and $Cl^-$ the most freely diffusible.

The polarity of the axolemma is determined by the direction of movement of $Na^+$, $K^+$, and $Cl^-$ at equilibrium. The tendency of extracellular $Na^+$ to leak into the axon is offset by an active transport mechanism in the axolemma which expends energy to expel $Na^+$. The tendency of intracellular $K^+$ to leak out of the axon is neutralized by the same chemical pump. The net membrane potential created by the movement of these two ions is $-70$ mV at equilibrium. The skewed distribution of $Na^+$ and $K^+$ across the axolemma is depicted in Fig. 13-5. This "resting" potential is caused almost entirely by the flux of $K^+$, since this positive ion is more freely diffusible than $Na^+$. Because the equilibrium potential due solely to the movement of $Cl^-$ is also $-70$ mV, we can safely ignore it and concentrate on the positively charged electrolytes $Na^+$ and $K^+$.

*Local potentials* Stimuli potentially capable of initiating a nerve impulse increase the permeability of the axolemma to $Na^+$. Since the extracellular concentration of $Na^+$ is significantly greater than its intracellular concentration, the stimulus causes an influx of extracellular $Na^+$. As $Na^+$ moves into the axoplasm, the resting potential of the axolemma is slightly reduced, say, to $-69.5$ mV. Caused by an increase in intracellular electropositivity and corresponding decrease in extracellular electropositivity,

this reduction in membrane potential results in partial depolarization. However, each neuron has its own threshold before complete depolarization and the conduction of a nerve impulse will be initiated. Let us assume that the threshold for conduction of a nerve impulse by an axolemma is $-60$ mV. Excluding receptor neurons, the stimulus which causes partial depolarization originates from another neuron. Thus, the arrival in rapid succession of many presynaptic nerve impulses will be required before a postsynaptic action potential can be propagated.

The change in membrane potential will remain below threshold if only one or a few nerve impulses are responsible for the alteration in postsynaptic membrane permeability. These partial depolarizations are called *local potentials.* A subthreshold excitatory potential is conceptualized at the left in Fig. 13-6. Although the potentials may be conducted a short distance along the length of a nerve fiber, they diminish in magnitude and velocity and quickly disappear. Thus, local potentials are said to be *propagated with decrement*; i.e., they progressively disappear. The decremental nature of a local potential is explained by the still relatively great impermeability of the axolemma to extracellular $Na^+$ and the ease with which the "sodium pump" restores the equilibrium potential of $Na^+$ following a slight imbalance.

Subthreshold stimuli do not change the $Na^+$ permeability of the axolemma sufficiently to reach the neuronal threshold value for complete depolarization. Therefore, the magnitude of the local potential can vary between the resting membrane potential and the critical threshold. Although complete depolarization does not occur, subthreshold stimuli facilitate the propagation of a nerve impulse by increasing the probability that the critical threshold will be reached.

*Action potentials* When the threshold value for the propagation of a nerve impulse is attained, the axolemma becomes completely depolarized. A propagated nerve impulse is called an *action potential,* or "spike." An action potential is conceptualized at the right in Fig. 13-6. Once the threshold for complete depolarization is reached, the potential difference

**Potential difference across axolemma**

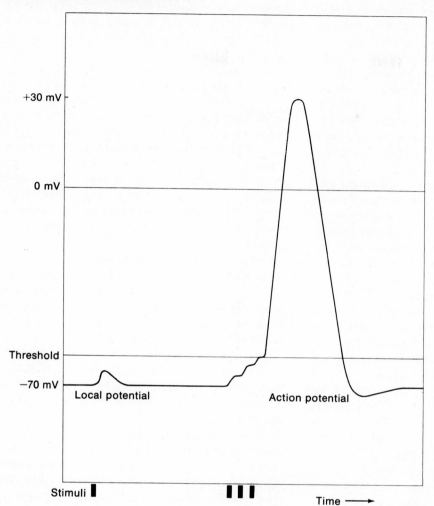

*Figure 13-6* *The degree of de-polarization undergone during the passage of a local potential and an action potential. The vertical arrows below the graph show when the excitatory stimuli arrive at the neuron.*

across the axolemma changes abruptly from −70 to +30 mV and then reverts back to the resting potential at a slightly less rapid rate. An action potential travels at a constant velocity along the entire length of the axon, from its site of initiation.

An action potential differs from a local potential in the following ways: (1) It is self-propagating, i.e., is conducted at a constant velocity along the entire length of an axon, and (2) it maintains a constant magnitude, i.e., is conducted without decrement.

Because a neuron will respond to a stimulus by ei-

ther propagating or not propagating a nerve impulse, the neuron response is said to be "all-or-none." However, we should reemphasize that, in the absence of an action potential, a neuron still responds to subthreshold stimuli with a graded series of decremental local potentials. *All-or-none* refers to whether or not a nerve impulse is propagated; nothing more is implied.

Now, let us look more closely at the ionic events responsible for depolarization and repolarization. Recall that the membrane potential of the axolemma not actually conducting a nerve impulse is $-70$ mV. As the threshold for conduction of an action potential is reached in a stimulated region of the axolemma, its permeability to $Na^+$ becomes so great that extracellular $Na^+$ diffuses into the axoplasm in large concentrations. The greatly increased permeability of the axolemma to $Na^+$ causes its complete depolarization and a reversal in the sign of the charge across it. The massive influx of $Na^+$ is responsible for the charge reversal. The inward diffusion of $Na^+$ stops when the membrane potential measures approximately $+30$ mV.

At this point, the same region of axolemma described above becomes impermeable to $Na^+$. Following $Na^+$ invasion, intracellular $K^+$ is actively transported from the axoplasm to the extracellular tissue fluid at a rate about 30 times faster than by diffusion. The expulsion of intracellular $K^+$ repolarizes the axolemma by reducing intracellular electropositivity and correspondingly increasing extracellular electropositivity. The egress of $K^+$ continues until the resting membrane potential of $-70$ mV is restored. Thus, repolarization restores the original electrical potential across the membrane and results in a second reversal of electrical polarity of the membrane.

Even though the original electrical polarity of the axolemma is reestablished, there is still an ionic imbalance due to the presence of excessive concentrations of intracellular $Na^+$ and extracellular $K^+$. The ionic imbalance is rectified through the reactivation of the Na (sodium) pump which actively expels $Na^+$ from, and reconcentrates $K^+$ in, the axoplasm. This step is called *restoration of ionic equilibrium*.

A nerve impulse lasts for about 1 msec (1 millisecond, or one one-thousandth of a second). During depolarization and a portion of repolarization, the neuron propagating a nerve impulse is completely refractory to another stimulus, irrespective of its magnitude. This functional period is referred to as the *absolute refractory period*. During the remainder of repolarization, another action potential can be initiated, provided that the new stimulus is greater in magnitude than that which caused the propagation of the previous nerve impulse. This functional period is called the *relative refractory period*. The maximum number of nerve im-

pulses propagated each second by one neuron is determined by the length of its refractory period and the physical characteristics of the nerve fiber. Neurons seldom propagate nerve impulses at their maximum theoretical rates.

The reason why an action potential is self-propagating is explained by the progressive increase in the permeability to $Na^+$ of the axolemma in the region immediately adjacent to that in which a nerve impulse exists. The increase in permeability is the result of an ionic flux between contiguous regions of opposite polarity. Threshold for complete depolarization in this adjacent region is inevitably reached because of the induction of a large influx of extracellular $Na^+$. This phenomenon causes the nerve impulse to move from point to point along the axolemma.

A rough analogy can be made between progressive axonal depolarization and a series of dominoes. Once the first domino is toppled, the rest topple in sequence. Similarly, once complete depolarization occurs at one point on the axolemma, the rest of the membrane is depolarized sequentially.

The electrical events which occur during the conduction of an action potential can be explained with the aid of Fig. 13-7. This figure traces the electrical changes across the axolemma during the propagation of a nerve impulse. The action potential is followed in the upper portion of the figure as it moves along the length of an axon. Polarity changes of the axolemma are traced in the lower portion of the figure. Voltages are measured with a potentiometer attached to an intracellular recording microelectrode and an extracellular reference microelectrode.

In the drawings at the left, the propagated nerve impulse has not yet reached the recording electrode. Thus, the membrane potential in this region is still $-70$ mV. However, the disparity in polarity between the portion of the axolemma in which the nerve impulse exists and the immediately adjacent resting region causes an influx of $Na^+$ in the latter. The inward diffusion of $Na^+$ precipitates complete depolarization of the membrane. In the center drawings, the propagated nerve impulse reaches the region in which the recording electrode is located. The membrane po-

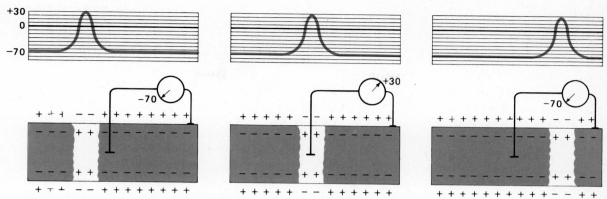

**Figure 13-7** *The propagation of a nerve impulse. An action potential (upper portion of figure) is conducted from left to right along an axon. The potential differences across the axolemma are measured by a potentiometer (lower portion of figure). (Left) Since the action potential has not yet reached the recording electrode, the potential difference across the membrane is −70 mV. (Center) The complete reversal in charge across the membrane as the action potential passes the region of the recording electrode results in a potential difference of +30 mV. (Right) When the action potential leaves the region of the recording electrode, the membrane potential returns to its resting state of −70 mV. (From Langley, Telford, and Christensen.)*

tential, as measured by the potentiometer, changes abruptly from −70 mV to +30 mV while the membrane in the region previously occupied by the nerve impulse has repolarized. In the drawings at the right, the nerve impulse has left the region of the recording electrode by causing depolarization in the immediately adjacent resting region. The secondary expulsion of intracellular K+ in the region of the recording electrode causes repolarization of the axolemma at this point. Thus, the resting potential of −70 mV is reestablished. This self-propagating process is repeated from point to point until the entire nerve fiber is traversed by the action potential.

*Saltatorial conduction* Nerve impulses propagated along myelinated fibers jump from one node of Ranvier to the next. As noted earlier, this action is referred to as *saltation;* it is conceptualized in Fig. 13-8. Saltation is caused by the myelin sheath which, where present, prevents an influx of Na+. The efficiency of insulation depends upon the thickness of the myelin sheath. Since the axolemma is exposed directly to the extracellular tissue fluid only at the nodes of Ranvier, these are the only places where depolari-

zation of the nerve fiber can occur. Thus, a series of ionic "sinks" is established at the nodes of Ranvier.

The ionic fluxes which result in progressive depolarization bypass the axolemma in the internodal regions. Instead, Na+ flows through the extracellular tissue space outside the neurilemma in one direction, through the sink provided by the node of Ranvier in the region of the spike, and then in the opposite direction through the axoplasm. Since the exposed axolemma in the adjacent node of Ranvier bears the opposite polarity, an ionic flux is established between these two nodes, causing depolarization of the second, etc. Because the action potential jumps from node to node, bypassing the myelin insulation in the internodal regions, the velocity of nerve impulse conduction is related directly to the degree of myelinization and the lengths of the internodal regions.

## THE SYNAPSE AND SYNAPTIC TRANSMISSION

A *synapse* is a functional connection established between two neurons or between a neuron and an effector. A synapse is formed when a specialized expansion called a *terminal bouton* at the end of one

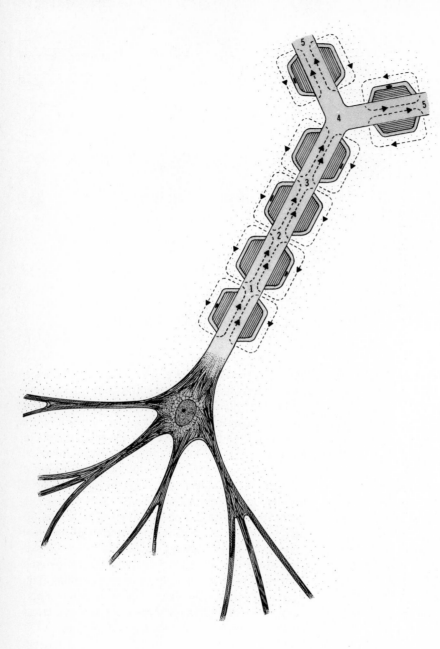

**Figure 13-8** *Saltatorial conduction of a nerve impulse along a myelinated axon. The ionic electric current flows from one node of Ranvier to the next, successively depolarizing each region of exposed axolemma. The arrows show the direction of the ionic fluxes. The numbers indicate the sequence of depolarization.*

axonal branch of a neuron lies extremely close to, but not in direct contact with, the cell body or a dendrite of another neuron or the cell membrane of a muscle fiber or gland cell. A synaptic junction between two neurons is diagramed in Fig. 13-9. Synapses formed between neurons and muscle fibers are shown in Fig. 13-10.

The distance separating the terminal bouton from another neuron or other innervated cell is approximately 1/1,000,000 in. (one-millionth of an inch). This distance of separation is called a *synaptic cleft*. For a nerve impulse to travel beyond the axon of a neuron, the encoded information contained in the impulse must somehow cross the synaptic cleft.

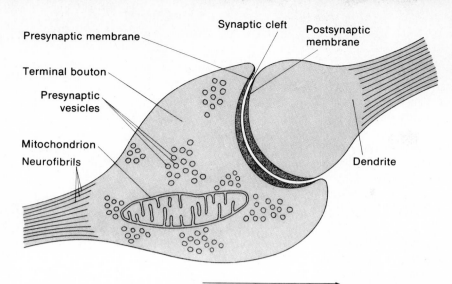

Presynaptic membrane

Terminal bouton

Presynaptic vesicles

Mitochondrion

Neurofibrils

Synaptic cleft

Postsynaptic membrane

Dendrite

Direction of nerve impulse transmission

**Figure 13-9** *A synaptic junction between two neurons. This axodendritic synapse is between the terminal bouton of an axonal branch of a presynaptic neuron and the dendritic branch of a postsynaptic neuron.*

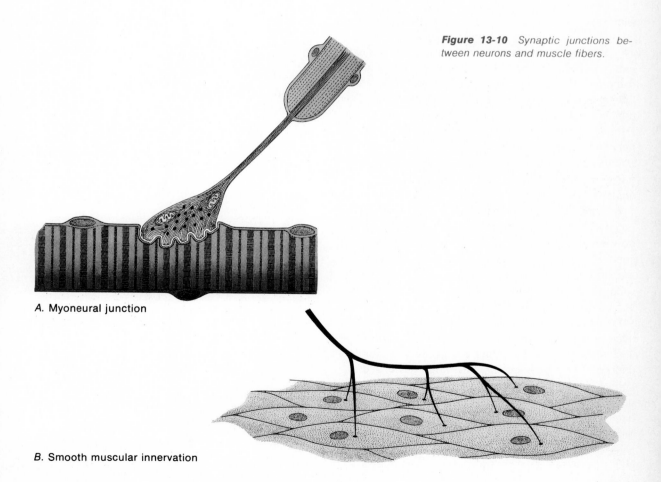

**Figure 13-10** *Synaptic junctions between neurons and muscle fibers.*

*A.* Myoneural junction

*B.* Smooth muscular innervation

Several types of synapse are recognized: (1) an *axodendritic synapse*, established when an axonal branch of one neuron terminates synaptically on a dendrite of another neuron; (2) an *axosomatic synapse*, established when an axonal branch of one neuron terminates synaptically on the cell body of another neuron; and (3) a *myoneural junction*, established when an axonal branch of a neuron terminates synaptically on the cell membrane of a muscle fiber (see Chap. 24). When a nerve impulse travels from one neuron to another across a synapse, the neuron transmitting the encoded information is called a *presynaptic neuron*, while that receiving the encoded information is referred to as a *postsynaptic neuron*.

Synapses are found in unmyelinated gray matter within the brain and spinal cord and in neural structures called ganglia outside the CNS. Synapses within the CNS occur when a bundle of nerve fibers terminates synaptically on a mass of nerve cell bodies which form anatomically discrete structures called *nuclei, centers,* or *ganglia*. The same situation exists in the peripheral nervous system, except that the anatomically discrete structures in which the synaptic connections are made are called *ganglia* exclusively. Peripheral nervous system neurons which conduct nerve impulses to a ganglion are called *preganglionic* neurons; those which conduct impulses from a ganglion to an effector are called *postganglionic* neurons.

We still do not fully understand how a nerve impulse is transmitted across a synapse. However, we do know that chemicals called *transmitter substances* secreted from the terminal boutons of the axonal tips are important to synaptic transmission. The most widely accepted theory of how synaptic transmission occurs is chemical in nature.

Transmission occurs when a chemical is released into the synaptic cleft from one or more specialized vesicles located in the terminal boutons of a neuron conducting nerve impulses. Presumably, the secretion of a transmitter substance from these vesicles occurs when the nerve impulses conducted along the axon reach the vesicles. However, the actual mechanism which causes transmitter release is obscure.

Potential difference across axolemma

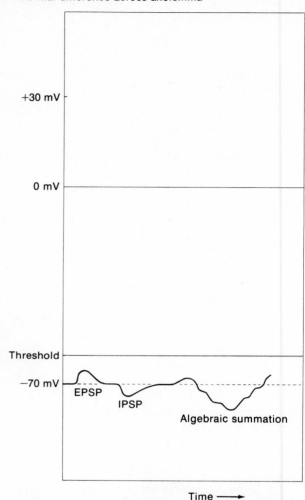

**Figure 13-11** *The effects of excitatory and inhibitory stimuli on the resting potential of an axolemma. (Left) An EPSP produced by a subthreshold excitatory stimulus. (Center) An IPSP caused by an inhibitory stimulus. (Right) The instantaneous value of the membrane potential is determined by algebraically summating the total excitatory and inhibitory inputs. There is a preponderance of inhibitory input at the right.*

Transmitters are either *excitatory* or *inhibitory;* i.e., they will either facilitate depolarization of the postsynaptic membrane and, therefore, the propagation of a nerve impulse or cause hyperpolarization of the postsynaptic membrane and, therefore, inhibit the propagation of a nerve impulse.

The chemical theory suggests that the transmitter substance crosses the synaptic cleft and affects the permeability of the postsynaptic membrane, i.e., the membrane of the postsynaptic cell body or dendrite. How the transmitter alters the postsynaptic membrane is not known with certainty. Excitatory transmitters may combine with specific binding sites on the postsynaptic membrane, possibly "unclogging" wide channels in the membrane through which extracellular $Na^+$ can pass to the intracellular environment of the postsynaptic neuron. An increased membrane permeability to $Na^+$ would of course facilitate depolarization of the neuronal membrane. Presumably, inhibitory transmitters are bound to other sites on the postsynaptic membrane. Their action may result in the unclogging of small channels through which intracellular $K^+$ can pass without permitting the ingress of extracellular $Na^+$. This action would increase the membrane potential beyond its normal range, thereby resulting in *hyperpolarization* and the inhibition of nerve impulse conduction. Because the terminal boutons of many neurons may converge on one postsynaptic neuron, its net response is determined by an algebraic summation of the influences exerted by the kinds and quantities of transmitter released by the presynaptic neurons.

Transmitters that facilitate depolarization are called *excitatory transmitters* and give rise to *excitatory postsynaptic potentials* (EPSP). Transmitters that promote hyperpolarization are called *inhibitory transmitters* and give rise to *inhibitory postsynaptic potentials* (IPSP). These potentials are compared in Fig. 13-11.

Because the response of the postsynaptic neuron is an algebraic summation of the influences exerted by available excitatory and inhibitory transmitters, it may take many more than the minimum number of nerve impulses before the net amount of excitatory transmitter is sufficient to transform the postsynaptic membrane potential into an action potential. This dynamic equilibrium tends to preserve the resting potential of the postsynaptic membrane, thereby preventing neuronal discharges to random fluctuations in membrane potential.

Acetylcholine (ACh) and norepinephrine are among the better-known transmitter substances. Their relations to, and roles in, the CNS and autonomic nervous system are described in the next chapter. Other transmitter substances include $\gamma$-aminobutyric acid (GABA), glutamine, serotonin, and dopamine. Recent research suggests that the neurotransmitter dopamine may act to cause hyperpolarization following the formation of cyclic adenosine monophosphate (cyclic AMP). Since numerous hormones which circulate in blood act through the formation of cyclic AMP (see Chap. 17), it may well be that at least certain neurotransmitters act in a similar manner.

## COLLATERAL READING

Baker, Peter F., "The Nerve Axon," *Scientific American,* March 1966. By removing the axoplasm from giant axons of squids and perfusing the empty axons with substitute solutions, Baker was able to study the dynamics of axonal membrane permeability under specified conditions.

Deutsch, J. Anthony, "The Cholinergic Synapse and the Site of Memory," *Science,* 174:788-794, 1971. Deutsch recounts experiments on memory retention after treatment with drugs which affect synapses.

Eccles, Sir John, "The Synapse," *Scientific American,* January 1965. A classic description of what is believed to happen ionically during postsynaptic excitation and inhibition.

Katz, Bernhard, "The Nerve Impulse," *Scientific American*, November 1952. An interestingly written general review of the events which cause the propagation of a nerve impulse and the anatomic features of nerve fibers which aid nerve impulse conduction.

Katz, Bernhard, "How Cells Communicate," *Scientific American,* September 1961. A review of the nature of neuronal excitability.

Katz, Bernhard, "Quantal Mechanism of Neural Transmitter Release," *Science,* 173:123–126, 1971. In an article intended for an advanced audience, Katz discusses the possible mechanism by which a nerve impulse causes the release of acetylcholine at a synaptic junction.

Marx, Jean L., "Cyclic AMP in Brain: Role in Synaptic Transmission," *Science,* 178:1188–1190, 1972. A review of recent evidence suggesting that neurohumors released from brain neurons cause the formation of cyclic AMP in postsynaptic neurons.

Nachmansohn, David, "Proteins in Excitable Membranes," *Science,* 168:1059–1066, 1970. A reinterpretation of critical experiments leads Nachmansohn to conclude that both axonal conduction and synaptic transmission are electrical in nature and basically the same.

# CHAPTER 14
# THE nervous SYSTEM

Basically, the nervous system maintains homeostasis by continually adjusting the diverse functions of the body to environmental changes. More specifically, the nervous system has the following functions: (1) to provide a continuous flow of information on the states of the internal and external environments; (2) to coordinate, integrate, and store encoded sensory information; (3) to initiate and monitor responses based on the processed information; and (4) to continually recode the stored information so that it becomes more consistent with the results of new experiences.

The integrative actions of the nervous system are enhanced by its ability to interact with the endocrine (ductless gland) system. For example, hormones from the hypothalamus or adrenal medullae are secreted in response to appropriate neural stimulation. That behavior and, therefore, neural mechanisms are profoundly influenced by the presence of reproductive hormones is also abundantly clear. However, hormonal effects are relatively slow, usually being measured in minutes, hours, or days, while neural effects are rapid, being elicited in a fraction of a second.

Regionally, the nervous system is divided into two parts—central and peripheral. The *central nervous system* (CNS) includes the *brain* and the *spinal cord*. The *peripheral nervous system* encompasses the *peripheral nerves* and the *autonomic nervous system*. The peripheral nerves communicate between the CNS, receptors, and effectors. There are 12 pairs of *cranial nerves* associated with the brain and 31 pairs of *spinal nerves* which enter and leave the spinal cord. The autonomic nervous system communicates indirectly with the CNS. The autonomic nervous system is divided into two divisions—the *sympathetic division*, associated with the thoracolumbar body regions, and the *parasympathetic division*, associated with the craniosacral body regions. It is with these components that neural control of the body is accomplished.

Although the nervous system is divided into regions

for convenience, the various regions are interconnected. Thus, while neural properties from region to region may vary quantitatively, we do not find qualitative differences in operational characteristics among them.

Homeostatic disruptions of neural control at any point between receptor and effector may give rise to neural disorders and the establishment of new steady states. Central nervous system diseases are discussed in the next chapter. Peripheral nervous system disorders are reviewed in Chap. 16. In this chapter, we will concentrate on the normal mechanisms by which the nervous system controls physiology and behavior.

## THE CENTRAL NERVOUS SYSTEM

As already mentioned, the CNS consists of the brain and the spinal cord. The basic components of the CNS are diagramed in Fig. 14-1A.

### THE BRAIN

Goethe called it "the flower at the top of the human plant." A glance at Fig. 14-8 should confirm the magnificence of this poetic judgment.

The brain lies in the *cranial cavity* of the skull and is continuous with the spinal cord at the foramen magnum. Three obvious regions can be distinguished in an adult brain—the cerebrum, cerebellum, and brainstem. The gross structure of the brain is illustrated in Fig. 14-1B.

Weighing between 3 and 4 lb in an adult, the brain originates embryonically from the upper portion of a straight tube derived from neural ectoderm. This tube is found in the posterior midline of the embryonic body, running lengthwise. The earliest differentiation of the brain during prenatal development occurs when the upper portion of the neural tube in the presumptive head region forms three vesicles called the *prosencephalon, mesencephalon,* and *rhombencephalon.* These vesicles will give rise, respectively, to the forebrain, midbrain, and hindbrain. As prenatal development proceeds, the prosencephalon sub-

divides into the *telencephalon,* which gives rise to the cerebrum, and the *diencephalon,* from which the thalamic region of the brain is derived; the rhombencephalon subdivides into the *metencephalon,* from which the cerebellum and pons develop, and the *myelencephalon,* which becomes the adult medulla oblongata. The brainstem is usually considered to include the diencephalon, midbrain, pons, and medulla oblongata.

Although the brain will be discussed in terms of its individual anatomic regions, functional portions of the brain usually overlap two or more structurally separate parts. Therefore, it will frequently be necessary to discuss several regions simultaneously to gain an appreciation of the control mechanisms of the brain.

The overlapping functions of the brain include:

1  Qualitative and quantitative interpretations of encoded sensory information initially transduced by senory receptors throughout the body
2  Information storage and retrieval, which permit learning and abstract thought
3  Integration of the entire sensory input which reaches this level
4  The initiation of skeletal muscular or viscercal responses, either consciously or subconsciously
5  Coordination of the organ systems of the body, thereby contributing to homeostasis

Not only does the brain interact with the endocrine system, but it secretes numerous hormones of its own.

The cerebrum    Derived from the embryonic telencephalon, the cerebrum is the anterior part of the forebrain. Since the cerebrum and thalamus originate from the prosencephalon, they are extensively interconnected. As can be seen in Fig. 14-1B, the cerebrum is the largest portion of the human brain. The cerebrum completely envelops the diencephalon and midbrain and partially surrounds the cerebellum and pons. The cerebrum is obviously Goethe's "flower."

The cerebrum consists of right and left *hemispheres* connected by a massive transverse commissural tract called the *corpus callosum.* The outermost

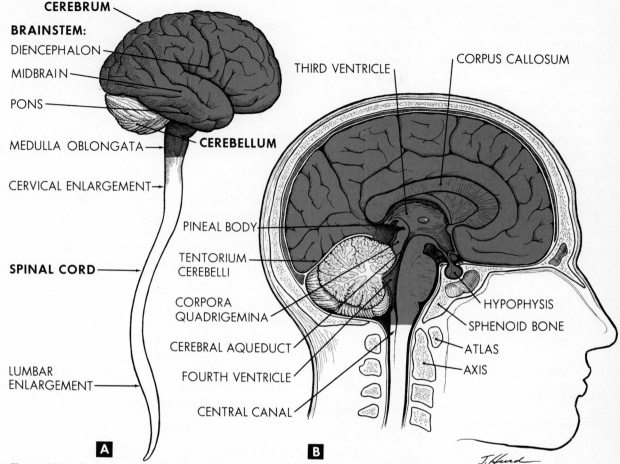

**CEREBRUM**

**BRAINSTEM:**
DIENCEPHALON
MIDBRAIN
PONS
MEDULLA OBLONGATA — **CEREBELLUM**
CERVICAL ENLARGEMENT

PINEAL BODY
**SPINAL CORD** —
TENTORIUM CEREBELLI
CORPORA QUADRIGEMINA
CEREBRAL AQUEDUCT
LUMBAR ENLARGEMENT
FOURTH VENTRICLE
CENTRAL CANAL

THIRD VENTRICLE          CORPUS CALLOSUM

HYPOPHYSIS
SPHENOID BONE
ATLAS
AXIS

**A**          **B**

**Figure 14-1**  *The gross structure of the central nervous system. A. The brain and spinal cord. B. The brain.* (From Melloni, Stone, and Hurd.)

cerebral tissue consists of *gray matter* called the *cerebral cortex*. Consisting largely of layers of nerve cell bodies, the cerebral cortex ranges between 2 and 4 mm in thickness. The subcortical tissue is composed of *white matter* called the *cerebral medulla*. The medulla consists primarily of fiber tracts which (1) link the cerebrum with other regions of the brain, especially with the thalamus by way of the internal capsule (see Fig. 14-8); (2) join the two cerebral hemispheres together by means of the corpus callosum; and (3) provide functional connections between different regions in each hemisphere. Also found in the

subcortical region are isolated masses of gray matter called *basal ganglia*. Because the basal ganglia are coordination nuclei which are part of a larger functional control system of the brain, they will be discussed after all the anatomic brain regions have been considered individually.

*The cerebral cortex*  The cortical portion of the cerebrum is highly convoluted, or folded. The elevated regions are called gyrations, or *gyri* (plural of *gyrus*). The depressions which define each gyrus are referred to as *sulci* (plural of *sulcus*). Deep sulci are usually called *fissures,* although this convention is not always followed. Providing a surface area of over 20 m², the cerebral convolutions permit a large amount of nervous tissue to be packed into the restricted volume of the cranial cavity.

Each cerebral hemisphere is divided into five lobes—*frontal, parietal, occipital, temporal,* and the *insula.* The first four lobes mentioned are external and are named for the cranial bones which cover them. The insula is an internal lobe not visible externally. The external lobes of the left cerebral cortex are illustrated in Fig. 14-2*A.* The internal lobe of the cerebral cortex is shown in Fig. 14-2*B.*

The cerebral hemispheres and the lobes in each hemisphere are separated, at least in part, by fissures, or deep sulci. These fissures are identified as follows:

(1) The *longitudinal fissure* occupied by the falx cerebri (see "Meninges," later in this chapter) separates the right and left cerebral hemispheres; (2) the *central sulcus* (fissure of Rolando) separates the frontal and parietal lobes; (3) the *parietooccipital fissure* incompletely separates the parietal and occipital lobes; and (4) the lateral *Sylvian fissure* completely separates the temporal lobe from the frontal lobe and incompletely separates the temporal lobe from the parietal lobe. The internal insula is formed from folds of the cerebrum on the inside aspect of the Sylvian fissure. The fissures and sulci which divide the cerebrum into lobes are shown in Fig. 14-2*A.* The longitudinal fissure between cerebral hemispheres is illustrated in Fig. 14-8.

In respect to function, the cerebral cortex has three

**Figure 14-2** *The gross structure of the cerebral cortex. A. Lateral aspect of the left cerebral hemisphere. B. A portion of the left cerebral hemisphere has been removed to reveal the insula. (From Melloni, Stone, and Hurd.)*

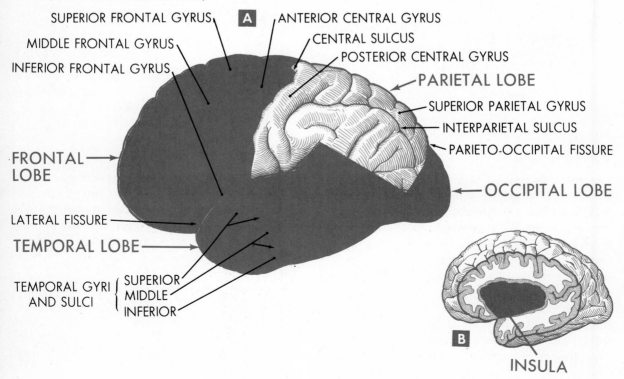

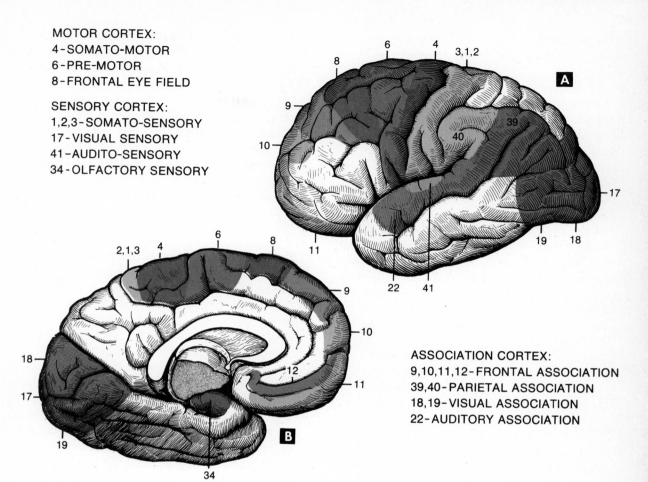

MOTOR CORTEX:
4-SOMATO-MOTOR
6-PRE-MOTOR
8-FRONTAL EYE FIELD

SENSORY CORTEX:
1,2,3-SOMATO-SENSORY
17-VISUAL SENSORY
41-AUDITO-SENSORY
34-OLFACTORY SENSORY

ASSOCIATION CORTEX:
9,10,11,12-FRONTAL ASSOCIATION
39,40-PARIETAL ASSOCIATION
18,19-VISUAL ASSOCIATION
22-AUDITORY ASSOCIATION

**Figure 14-3** *Functional specializations within the cerebral cortex. A. Functional areas on the periphery of the left cerebral cortex, viewed from a lateral aspect. B. Functional areas along the midsagittal plane of the left cerebral cortex, viewed from medial aspect.* (From Melloni, Stone, and Hurd.)

divisions—sensory, motor, and association. The *sensory cortex* processes sensory information. The *motor cortex* initiates and coordinates motor activities. The *association cortex* is not directly involved in either sensory or motor functions and presumably performs information storage and retrieval, thereby providing the substrate for interpreting and integrating sensory input, as well as for learning, memory, and abstract thought.

The distribution of the three functional regions of the cerebral cortex is shown from lateral aspect and

in midlongitudinal section in Fig. 14-3A and B, respectively. By convention, each functional area is identified with a *Brodmann number*. Assignments of the Brodmann numbers are based on differences in cellular structure in various cortical regions, the results of animal experiments in which specific regions are either damaged or removed, and observations both in human beings and animals of the responses elicited by direct electrical stimulation of specific portions of the cortex. Brodmann numbers are identified in Fig. 14-3A and B.

The following sensory and sensory association regions are recognized:

1  The *somesthetic cortex* (areas 1 to 3). Since this brain convolution lies immediately posterior to the central sulcus, it is called the *postcentral gyrus.* Located in the anterior portion of the parietal lobe, the somesthetic cortex processes sensory information on touch, cutaneous and deep pressure, pain, proprioception, and temperature, and localizes the stimuli to the specific parts of the body being stimulated. Sensory information from the toes is projected to the superior aspect of the somesthetic cortex, while sensory input from the head is projected to its lower lateral aspect. Other parts of the body are represented by intermediate regions. Thus, the postcentral gyrus is a rough upside-down map of the regions of the body. Because this functional organization reflects the lateralization of nerve fibers from tactile, pressure, pain, kinesthetic, and temperature receptors as the fibers enter various ascending spinal tracts at successively higher levels of the cord, the somesthetic cortex is said to be *somatotopically organized.* The amount of gray matter devoted to each region of the body is related to the number of sensory receptors in each region. The largest areas of the somesthetic cortex process sensory information from the tongue, lips, hand (including the fingers), and foot. These regions of the body are among the most richly supplied with sensory receptors. Information received by the somesthetic cortex is immediately relayed to the somesthetic association area (area 40). It is here that the sensory information is interpreted, presumably by comparing it with the results of previous experiences of a similar nature. The integrated information is used to adjust the activities of the motor cortex.

2  The *visual cortex* (areas 17 and 18). Located in the occipital lobe, this region continues the processing of visual information begun in the retinas. The cortical processing of visual images is discussed in Chap. 11. After processing by the visual cortex, the integrated information is relayed to the visual association area (area 19), where it is interpreted.

3  The *auditory cortex* (auditory projection area, or area 41). Located on the superior aspect of the temporal lobe, this region continues the processing of auditory information begun in the organ of Corti in the inner ear. Following integration, the processed information is relayed to the auditory association area (area 22), where it is interpreted.

4  The *olfactory cortex* (area 34). Located on the anteromedial aspect of the temporal lobe, this region integrates olfactory information initially transduced by the hair cells in the nasal epithelium.

The following motor areas are recognized:

1  The *motor cortex* (area 4). Since this brain convolution lies immediately anterior to the central sulcus, it is called the *precentral gyrus.* Located in the posterior portion of the frontal lobe, the motor cortex controls and coordinates the movements of skeletal muscles by way of the pyramidal and extrapyramidal tracts. Large neurons called pyramidal cells originate in the motor cortex. Their axons contribute to the pyramidal tracts. The somatotopic organization of the motor cortex is similar to that of the somesthetic cortex; i.e., it is a rough inverted map of the body regions. The amount of gray matter devoted to each region of the body is related to the number of skeletal muscles found in each region and the degree of control required by the individual muscles. The largest areas of the motor cortex coordinate skeletal muscle movements of the lips, hand (including the fingers), and foot. About one-third of the total area of the motor cortex controls the muscles of speech.

2  The *premotor area* (areas 6 and 8). Located in the frontal lobe immediately anterior to the motor area, the region called area 6 gives a higher level of motor coordination than can be effected by the motor cortex alone. Acting through the motor cortex, the premotor cortex is responsible for the initiation of patterned and precision movements acquired through motor practice. An anterior portion of the premotor cortex (area 8) controls eye movements which enable voluntary visual fixation.

3  The *motor speech area* (Broca's area). Located in the frontal lobe on the superior border of the later-

al fissure, this region (area 44) controls patterned activities of the muscles of speech.

Vast amounts of cerebral cortex consist of association areas which are not involved with the direct interpretation of sensory information transduced by the various types of receptors. Such unassigned areas are especially prevalent in the frontal and temporal lobes. These areas of association cortex are described as follows:

1   The *prefrontal association areas* (areas 9 to 12). Located in the region of the frontal lobe anterior to the premotor area, these areas play predominant roles in the performance of abstract thought, the establishment of memory and intelligence, the execution of goal-directed mental activities, and the inhibition of socially unacceptable behavior. If there is a discrete location for conscience, then it must be assigned to the prefrontal lobes.
2   The *temporal association area.* Extending throughout the inferior portion of the temporal lobe and into the region surrounding the superior end of the Sylvian fissure, this huge area is primarily responsible for the storage of complex memories which must be consulted before coherent responses can be made to the total sensory input. Cerebral dominance in speech and motor coordination by the left temporal lobe is exhibited by over 90 percent of all human beings. However, recent research indicates that the hemisphere which will be dominant depends on the nature of the ongoing activity. For example, although the left cerebral hemisphere usually exerts dominance in speech, the right cerebral hemisphere seems to be preferentially consulted during activities which require perception of the environment.

Now we are ready to put the functions of the cerebrum all together. The sensory input of each type of receptor is processed by the appropriate sensory cortex. The regions of association cortex which surround the sensory areas give meaning to the various sensory experiences. Information processing continues as the integrated electrical signals are projected to the prefrontal association areas for deliberation and reflection and to the temporal association area, where the entire sensory input is sorted out and its significance is analyzed. This information is then relayed to a cortical center in the parietal lobe just above the Sylvian fissure, where decisions on the best course of action are made. Once the most appropriate response is decided upon, the information is projected to both the premotor and motor areas. These areas initiate the motor directives for the control and coordination of the skeletal muscles. Based on feedback from sensory receptors distributed over the entire body, the directives of the motor cortex are modulated, or adjusted, by the somesthetic cortex and the cerebellum (see below). Of course, when time is of the essence, one or more intermediate steps between reception and the initiation of a motor response are bypassed.

Facilitated by extensive synaptic connections with neurons which originate in the reticular formation (see "The Reticular Formation," later in this chapter), neurons of the cortex are spontaneously active, even during deep sleep. It is possible to record the electrical activity from neurons in the outermost layer of the cerebral cortex with special scalp electrodes attached to appropriate recording equipment. The record thus obtained is called an *electroencephalogram,* or EEG. Because of the relation between the reticular formation and the electrical activity of the cerebral cortex, the status of the reticular formation in large part determines the nature and characteristics of the brain waves. These characteristics include the frequency, or number of waves per second; amplitude, or average voltage; regularity, or rhythmicity; and shape, or appearance, of consecutive waves. An analysis of brain waves is used in the diagnosis of several kinds of brain disorder. The application of the EEG to the diagnosis of epilepsy is described in Chap. 15. Normal and abnormal EEGs are presented in Fig. 15-2.

Although at least four characteristic types of brain wave have been described, the most common types recorded from adults under normal conditions are called *alpha waves* and *beta waves.* Elicited mainly during intervals of quiet rest, alpha waves have a frequency of 8 to 12 waves per second and measure about 50 $\mu$V (microvolts; 1 $\mu$V = one-millionth of a volt). Usually facilitated by closing the eyes while resting, alpha waves are regularly spaced and synchronous. The rhythmicity of alpha waves suggests that all the active cortical neurons are depolarizing simultaneously. Apparently, alpha waves are predominant when the reticular formation exerts little influence on the cerebral cortex.

Beta waves, on the other hand, appear during periods of mental concentration. Presumably, the facilitating effects of the reticular formation on the cerebral cortex at this time induce asynchronous, or irregular, waves with higher frequencies and lower amplitudes than those of alpha waves. There is a close correlation between the depth of mental concentration and the frequency of beta waves. These waves range from 14 to more than 30 waves per second in frequency and between 5 and 10 $\mu$V in amplitude.

*The cerebral medulla* The white core of the cerebrum is called the cerebral medulla. Fiber tracts and specialized nuclei make up its substance.

The great commissure between the hemispheres, as mentioned earlier, is called the corpus callosum. Complete severance of this connective structure in experimental animals and in human beings following natural accidents or neurosurgery gives rise to two separate "brains" in one individual. Tasks learned by one cerebral hemisphere are not transferred to the other. Each hemisphere must separately learn the same lesson. Although both hemispheres in an individual with a "split brain" initially compete with each other for cerebral dominance, cerebral competition is usually resolved, and dominance is finally established.

These and other observations suggest that each hemisphere is fully equipped to serve as a functional "brain." Their cooperation under normal conditions is a marvelous example of natural redundancy of the type that engineers build into highly complex technological systems so that, in the event of failure in one component, another can immediately take over. Obviously, two cerebral hemispheres also tremendously increase the information processing capacities of the human brain.

The diencephalon The diencephalon is the hind part of the forebrain and connects the cerebrum with the midbrain. This region surrounds the *third ventricle*. The *pineal gland,* or pineal body, is located on its dorsal (back) surface, and the *hypophysis,* or pituitary gland, projects from its ventral (front) surface. The approximate location of the diencephalon is shown in Fig. 14-1B. The diencephalon contains the *thalamus* and *hypothalamus*. Several other diencephalic regions also surround the thalamus. Each of these regions contains numerous nuclei which have important sensory and motor functions.

*The thalamus* The thalamus is located directly above the midbrain. The thalamus receives sensory nerve fibers from all the lower regions of the brain and receives motor nerve fibers from the cerebral cortex. The thalamus also maintains interconnections with the hypothalamus, thereby influencing both somatic and visceral responses.

Of the many nuclei in the thalamus, we will mention only two—the *medial* and *lateral geniculate bodies*. The medial and lateral geniculate bodies are parts of the auditory and visual pathways, respectively, in the brain. The role of the lateral geniculate body in vision is discussed in Chap. 11.

Certain aspects of perception are believed to occur in the thalamus. Presumably, the thalamus calls our attention to the presence of tactile, painful, and cold and hot stimuli but does not localize the sensations to a particular region of the body. Localization is accomplished in the somesthetic cortex, to which thalamic fibers project. Because rapid responses to potentially harmful stimuli can be initiated without relaying the information to the somesthetic cortex, this division of labor between the thalamus and cortex is obviously adaptive.

*The hypothalamus* This portion of the diencephalon lies below the thalamus and above the infundibulum of the hypophysis. The numerous nuclei of the hypothalamus participate in diverse activities which affect visceral motor functions and behavioral states. The hypothalamus receives fibers from the limbic cortex (see "The Limbic System," below), maintains neural connections with the thalamus and midbrain, and communicates neurally with the pars nervosa (posterior lobe) of the hypophysis and by means of hormones with the adenohypophysis (see Chap. 17).

Functions of the hypothalamus include:

1  Control of *reproduction* by regulating the secretion of adenohypophyseal gonadotropins and secreting oxytocin directly by way of the pars nervosa

2  Control of body *metabolism* by regulating the secretion of adenohypophyseal somatotropic hormone (STH), thyroid-stimulating hormone (TSH), and adrenocorticotropic hormone (ACTH)

3  Control of *water balance* through the secretion of antidiuretic hormone (ADH) and the corresponding stimulation of the hypothalamic thirst center

4  Regulation of *food intake* by the reciprocal interactions of lateral feeding and medial satiety centers

5  Subcortical control of both sympathetic and parasympathetic divisions of the *autonomic nervous system*

6  Maintenance of a *constant body temperature* by measuring the internal temperature and autonomically regulating the degree of peripheral vasoconstriction or dilatation

7  Regulation of the *cardiovascular system* by increasing or decreasing the rate of the heartbeat and the blood pressure

8  Regulation of mechanical and chemical *digestion* by promoting or inhibiting peristalsis and digestive secretions

9  Regulation of *affective states* (current emotional states) by influencing the degree and kind of appetitive (goal-seeking) and consummatory (goal-achieved) behavior

**The midbrain**   Linking the forebrain and hindbrain by means of various fiber tracts and containing numerous coordinating nuclei, this region lies immediately cephalad (toward the head), in relation to the pons. Basically, the midbrain contains three regions—a dorsally located tectum, a middle tegmentum, and ventrally positioned cerebral peduncles.

The *tectum* contains two pairs of bodies, or nuclei, collectively called the *corpora quadrigemina*. The corpora quadrigemina are divided into paired *superior* and *inferior colliculi*. The superior colliculi control certain visual reflexes (see Chap. 11), while the inferior colliculi mediate specific auditory reflexes. Running through the tectum is the cerebral aqueduct of Sylvius, which connects the third and fourth ventricles of the brain.

The *tegmentum* contains the nuclei of the IIId and IVth cranial nerves and the red nucleus (*nucleus ruber*). Neurons of the rubrospinal tract (see "Spinal Tracts," later in the chapter) originate in the nucleus ruber. The cranial projections of numerous ascending spinal tracts, including the medial and lateral lemnisci (plural of lemniscus) and ascending tracts such as the spinothalamic and spinotectal tracts, also course through the tegmentum.

The *cerebral peduncles,* large fiber tracts on the inferior surfaces of the cerebral hemispheres, approach each other as they pass through the midbrain. The descending corticospinal, or pyramidal, tracts (see "Spinal Tracts," later in the chapter) occupy the basal portion of the cerebral peduncles.

**The cerebellum**   Most of the cerebellum is located posterior to the medulla oblongata. The posterior region of the cerebrum is separated from the cerebellum by a tentlike membranous partition called the *tentorium cerebelli* formed by the dura mater of the meninges. The position of the cerebellum is illustrated in Fig. 14-1*B*. An enlargement of the brain in the region of the cerebellum is shown in Fig. 14-4.

The cerebellum is the second largest part of the brain and bears a superficial resemblance to the cerebrum. The outside of the cerebellum is laced with transverse sulci, and the inside is highly convoluted. Indeed, per unit of volume, the cerebellum is more convoluted than the cerebrum. Containing an outer *cortex* of *gray matter* and an inner *medulla* of *white matter,* the cerebellum is divided into three structural lobes: a midlongitudinal *vermis* appearing externally like a worm turned on itself and internally like a tree with numerous branches, and two enlarged *lateral hemispheres*. The treelike internal configuration and the ancient belief that it was the seat of the soul led to the designation of the vermis as the "arbor vitae," or tree of life.

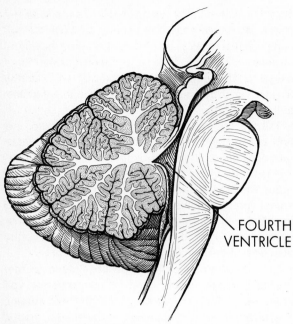

**FOURTH VENTRICLE**

*Figure 14-4* *The gross structure of the cerebellum. The cerebellum and the rest of the brain associated with it are sectioned along the midsagittal line.* (From Melloni, Stone, and Hurd.)

Many *fiber tracts* run through the substance of the cerebellum. The fibers of these tracts enter and leave the cerebellum through three paired stalks, or bundles of tracts, called *peduncles.* The superior cerebellar peduncles project to the cerebral cortex by way of the thalamus. The middle cerebellar peduncles project to the cerebellum from the cerebral cortex by way of the pons. The inferior cerebellar peduncles project to the cerebellum from the spinal cord by way of the medulla oblongata. The relations of the cerebellar peduncles to the various spinal tracts are discussed later in this chapter.

The cerebellum coordinates skeletal muscle activities by adjusting the tone of skeletal muscles. The muscular adjustments effected by the cerebellum are based on (1) proprioceptive feedback information supplied by muscle spindles and Golgi tendon organs on the position of body parts and their rate of displacement with time; (2) sensory information on equilibrium provided by the utricle, saccule, and

semicircular canals of the inner ear; (3) other sensory information such as touch and vision; and (4) information supplied by the precentral gyrus of the cerebral cortex by way of the pons. In other words, the cerebellum brings the posture and movements of the body regions in line with the motor directives issued by the cerebral cortex. The unique sensory and motor functions of the cerebellum facilitate or inhibit the tone of functional groupings of skeletal muscles, thereby refining the gross movements initiated by the motor cortex and relieving it of moment-by-moment responsibility for the execution of routine, or patterned, motor activities such as sitting, standing, walking, or running.

**The pons** The pons is an expanded portion of the hindbrain on the anterolateral walls of the medulla oblongata. Its position is illustrated in Fig. 14-1. The pons contains the nuclei of the Vth (trigeminal), VIth (abducens), VIIth (facial), and VIIIth (acoustic) cranial nerves; a pneumotaxic center which influences the respiratory center in the medulla oblongata; a transverse commissure which connects its two halves; longitudinal fiber tracts from the cerebrum; and the middle cerebellar peduncle. The pons therefore functions in various aspects of *reflex control.*

**The medulla oblongata** Because it is an expansion at the end of the hindbrain, the medulla oblongata is sometimes referred to as the *bulb.* The medulla is continuous with the spinal cord. The fourth ventricle is contained within its substance. The position of the medulla is illustrated in Fig. 14-1. The medulla contains the nuclei of cranial nerves IX (glossopharyngeal), X (vagus), XI (accessory), and XII (hypoglossal) and reflex centers for heart rate, respiration, vasomotor tone, winking, sneezing, coughing, vomiting, and gastrointestinal activity. In addition, all the ascending and descending fiber tracts of the spinal cord either pass through or terminate synaptically in the medulla. Many of the fiber tracts decussate in the medulla.

The medulla oblongata can be divided functionally into three regions: (1) a posterior division in which are found nuclei of ascending spinal tracts known as the fasciculus gracilis and fasciculus cuneatus, and the inferior cerebellar peduncle; (2) a middle division, consisting of the medial lemniscus and the olivary body containing numerous olivary nuclei; and (3) an anterior division containing the pyramidal, or corticospinal, tracts. These structures are either associated with, or are, spinal tracts and will be discussed later in this chapter.

Because of its strategic location between the brain and spinal cord and its many reflex control centers coordinating homeostatic activities throughout the body, the medulla is a vital conduction and integrating region of the brain. The presence of the heart rate and respiratory centers makes the medulla oblongata absolutely essential to the continued existence of life.

Functional systems  In several functional systems within the brain, two or more anatomic regions overlap. These systems include the limbic system, the basal ganglia, and the reticular formation.

The limbic system  Embracing the ventromedial aspects of the frontal lobes and the deep portions of the temporal lobes of the cerebral cortex, as well as various portions of the thalamus, hypothalamus, and certain surrounding regions in the diencephalon, this system is involved in both the excitation and inhibition of motivated and emotional behavior. The limbic system maintains multiple synaptic connections with the frontal cortex, with the reticular formation in the brainstem, and with neurons within the system itself. The limbic system participates in the mediation of such diverse activities as attention, "flight-or-fight" behavior, reward ("pleasure") versus punishment (avoidance), docility versus rage, as well as appetitive and consummatory behaviors associated with food and sex.

The basal ganglia  These ganglia, or nuclei, are masses of gray matter consisting of the cell bodies of neurons deep in the medulla, or white matter, of the cerebrum. Located below the lateral ventricles and medial to the insula, the basal ganglia surround the

thalamic region of the diencephalon. The four paired nuclei which comprise the basal ganglia are the caudate, lentiform, amygdaloid, and claustrum. Each lentiform nucleus, in turn, is divided into two smaller nuclei—the putamen and globus pallidus. Together with part of the internal capsule, the caudate and lentiform nuclei are referred to as the *corpus striatum.*

In addition to integrating sensory information on the way to the cerebral cortex, these nuclei maintain multiple synaptic connections among each other; receive the axons from many neurons which originate in the motor cortex of the cerebrum; maintain multiple synaptic connections with the *substantia nigra,* a nucleus which runs from the midbrain to the hypothalamus; and make synaptic connections with the red nucleus (nucleus ruber) of the midbrain and the reticular formation of the brainstem. Thus, a great motor pathway is established from the motor cortex through the basal and brainstem ganglia to the reticular formation. This pathway influences diverse skeletal muscle responses conducted over the extrapyramidal spinal tracts. The normal functioning of the extrapyramidal tracts depends in turn upon intact pyramidal tracts.

The basal ganglia coordinate and integrate posture, movement, and locomotion by (1) maintaining posture during patterned, automatic movements in which the center of gravity of the body is continually shifting, e.g., during standing, walking, and running; (2) adjusting the tone of muscles participating in fine, intentional movements such as threading a needle; and (3) integrating the actions of different functional groupings of muscles during patterned movements so that the execution of a complex activity occurs smoothly, e.g., playing a musical instrument or typewriting.

The reticular formation  Running through the core, or center, of the brainstem is a network of nerve fibers which monitors and integrates all the sensory and motor messages passing to and from the brain. Neurons of the reticular formation form multiple synaptic connections among themselves; project to all parts of the cerebral cortex, the cerebellum, and lower levels of the brain; and receive collateral branches from cranial nerves and descending spinal tracts. The reticular formation contains the heart rate and respiratory centers of the medulla oblongata and gives rise to the upper motor neurons of the reticulospinal tract, a descending spinal tract which coordinates extensor reflex activities.

The reticular formation, frequently called the *reticular activating system* (RAS), is also implicated in waking up and falling asleep. How this interlacing and extensively branched network sounds the physiologic alarm which rouses the cerebrum to a state of alert wakefulness or induces drowsiness and sleep states is the subject of the following discussion.

The RAS is an accelerator of the cerebrum. While it is active, an individual is in a wakeful state. Because the status of the RAS affects the frequency of the EEG produced by the cerebral cortex, it will be useful to follow changes in the EEG during the transition from wakefulness to deep sleep. During *wakefulness,* the EEG consists predominantly of beta waves with a frequency of approximately 13 to 30 waves per second.

After drowsiness sets in and following closure of the eyes, a subject enters a period of *light sleep.* Light sleep is characterized by a relatively slow EEG of about 4 to 6 waves per second. The frequency is somewhat slower than that of alpha waves, which appear right after the eyes close under relaxed conditions. While light sleep is in progress, the skeletal muscles maintain their tone, and the eyeballs are still. Some evidence indicates that a nucleus in the lower portion of the brainstem induces light sleep by secreting an increasing concentration of the neurohumor serotonin. According to this theory, serotonin exerts a braking effect on the RAS which in turn results in light sleep.

Light sleep gradually blends into *deep sleep* when the subject remains undisturbed. In contrast to light sleep, deep sleep is characterized by a return to a fast EEG of 14 to 16 waves per second. This stage is often referred to as *paradoxical sleep* for this reason. The increased EEG frequencies are apparently a reflection of attentiveness to vague and disjointed thoughts generated during dream states. The muscle tone maintained during light sleep is lost during deep

sleep. However, frequently the eyes can be seen to move from side to side or up and down behind the closed lids. The rapid eye movements (REMs) are correlated with periods of dreaming, and the eyes are believed to be following internally generated visual images. These characteristic eye movements have given deep, or paradoxical, sleep still another name—REM sleep. Apparently, the induction of deep sleep depends upon the secretion of norepinephrine, a neurohumor secreted by a specific nucleus in the pons. Thus, norepinephrine acts as a second brake, further inhibiting the actions of the RAS.

While sleep lasts, deep and light sleep alternate with each other periodically. One sleep cycle lasts approximately 1 hr, on an average. Initially, deep sleep lasts perhaps 15 to 20 min, but the length of deep sleep usually increases as its depth decreases with each succeeding cycle. Finally, deep sleep cannot be sustained, and the RAS arouses the cerebral cortex to a wakeful state.

**Blood circulation** The brain has an exceedingly high metabolic rate and receives a correspondingly large volume of blood. Approximately 20 percent of the total blood and, therefore, one-fifth of the total nutrients and oxygen in blood are sent to, and used by, this organ. Energy metabolism in the brain is almost totally dependent upon the aerobic degradation of carbohydrates. Because of its high rate of metabolism and reliance on glucose as an energy source, the brain enters a state of unconsciousness within 10 sec of oxygen deprivation, even though it contains almost a liter of blood at any one time. Irrevocable brain damage results from oxygen deprivation for 5 to 10 min.

*The arterial blood supply* The arterial circulation at the base of the brain is unique in several respects. First, there are two separate arterial blood supplies to the brain. Second, although these two arterial trees anastomose, or interconnect, with each other, for the most part they remain functionally separate circuits. Third, and finally, several arteriolar branches give rise to plexuses associated with the walls of the brain ventricles. Cerebrospinal fluid (CSF) is formed from these plexuses.

Blood is transported to the brain through the paired *vertebral* and *internal carotid arteries*. These two arterial systems communicate through anastomoses which form the *circle of Willis*. This confluence of the two systems surrounds the optic chiasm in the inferomedial region of the brain. The vascular components of the circle of Willis include the anterior cerebral arteries from the internal carotid arteries; the posterior cerebral arteries from the basilar artery, an extension of the vertebral arteries; an anterior communicating artery between the anterior cerebral arteries; and posterior communicating arteries between the posterior cerebral and internal carotid arteries. The circle of Willis and the rest of the arterial blood supply at the base of the brain are illustrated in Fig. 14-5.

The circle of Willis provides a mechanism for equalizing pressure differentials which may develop as the result of an obstruction or other damage to one or more vessels in either system. Cranial vascular disorders involving the base of the brain are considered in detail in Chap. 31.

Several of the arteries at the base of the brain, after penetrating deeply into its substance, form specialized capillary plexuses in the walls of the brain ventricles. Choroid plexuses of the lateral ventricles are formed by branches of the anterior choroid arteries, while similar plexuses associated with the third ventricle are formed from branches that originate from the superior cerebellar arteries.

Each vessel in a choroid plexus is packed tightly together and lined externally by ependymal cells.

**Figure 14-5**  *The arterial supply of the brain. The brain is viewed from inferior aspect.* (From Melloni, Stone, and Hurd.)

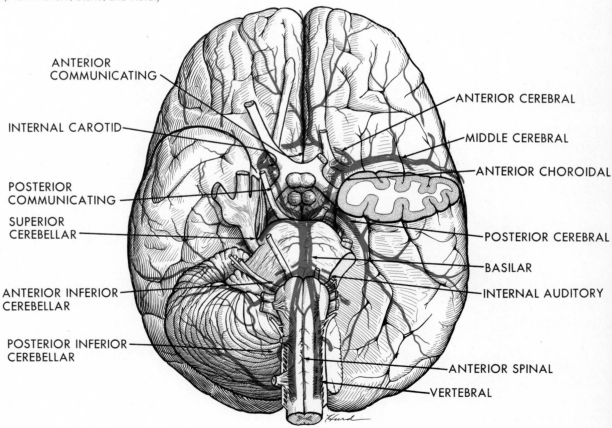

Thus, little interstitial tissue is associated with these vascular beds. The compound membrane formed by the vascular wall and the ependyma, coupled with the lack of extracellular tissue space, gives rise to a differentially permeable membrane. CSF is both actively transported and filtered through this membrane from the blood within the choroid vessels to the lumina of the ventricles. The differential permeability due to the arrangement of the compound membranes of the choroid plexuses is the reason why the plexuses are frequently called the *blood-brain barrier*. The formation of CSF will be described later.

*The venous drainage* The venous drainage associated with the brain is adapted to receive CSF as well as blood. Two anatomically different but connected structures drain venous blood from the brain: (1) venous sinuses, located in the dura mater of the meninges, and (2) venous vascular plexuses, located in the superficial aspects of the cranium. The major venous sinuses of the brain are shown in Fig. 14-6. Blood entering the venous sinuses comes from blood vessels associated with the pia mater. Since the venous sinuses communicate with plexuses of veins on the periphery, some of the venous blood leaves the

cranial region by this route. However, most of the blood is ultimately drained from the venous sinuses by the internal jugular veins.

THE SPINAL CORD

**Structural organization** The spinal cord is that portion of the CNS (central nervous system) located in the vertebral or spinal column. Measuring about 18 in. in length, the spinal cord extends from the foramen magnum, where it is continuous with the medulla oblongata of the brain, to the *conus medullaris,* in line with the body of the second lumbar vertebra.

The position of the spinal cord within the body is illustrated in Fig. 14-7. In this view from the posterior aspect, the spinous processes of the vertebrae and the posterior portion of the meningeal covering of the cord have been removed to expose the cord and the origins of the spinal nerves.

The divergence between the caudal terminations of the spinal cord and vertebral column causes the lumbar and sacral spinal nerves to be drawn caudad (downward), forming a brushlike configuration of spinal nerves called the *cauda equina* ("horse's tail," lit-

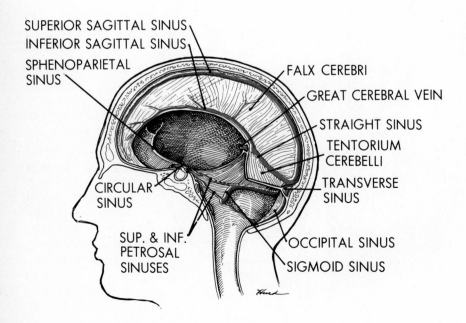

SUPERIOR SAGITTAL SINUS
INFERIOR SAGITTAL SINUS
SPHENOPARIETAL SINUS
FALX CEREBRI
GREAT CEREBRAL VEIN
STRAIGHT SINUS
TENTORIUM CEREBELLI
TRANSVERSE SINUS
CIRCULAR SINUS
SUP. & INF. PETROSAL SINUSES
OCCIPITAL SINUS
SIGMOID SINUS

**Figure 14-6** *The venous drainage of the brain.* (From Melloni, Stone, and Hurd.)

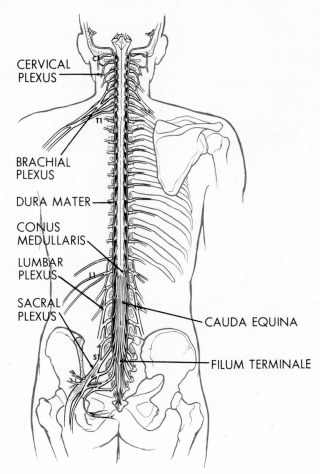

CERVICAL PLEXUS

BRACHIAL PLEXUS

DURA MATER

CONUS MEDULLARIS

LUMBAR PLEXUS

SACRAL PLEXUS

CAUDA EQUINA

FILUM TERMINALE

*Figure 14-7* *The relation of the spinal cord and nerves to the vertebral column. The roof of the vertebral column has been removed to expose the spinal cord and nerve roots. (From Melloni, Stone, and Hurd.)*

erally). Each of the spinal nerves represented in the cauda equina still exits through the meninges and intervertebral foramina of the vertebral column in correct serial order based on their points of origin in the cord. Notice that the sacral spinal nerves originate in the terminal portion of the spinal cord associated with the level of the first lumbar vertebra (L1). Thus, the conus medullaris is also called the *sacral cord,* even though it is located in the lumbar region of the vertebral column. The lumbar spinal nerves, which

comprise the lateral aspects of the cauda equina, are associated with the lumbar cord in line with the eleventh and twelfth thoracic vertebrae (T11 and T12). In other words, regional differentiations of the cord are established by the regions supplied by the spinal nerves found there and not necessarily by the anatomic position of the cord in the vertebral column.

The divergence between the terminations of the cord and vertebral column is attributed to differential growth rates between these two regions during prenatal development. The vertebral column grows at a more rapid rate than the spinal cord, resulting in a relative decrease in the length of the spinal cord compared with that of the vertebral column. As the spinal column increases in length, in relation to the spinal cord, the lumbar and sacral nerves elongate caudally, thereby maintaining their proper anatomic positions with respect to the lumbar and sacral vertebrae.

The cauda equina is in a large cistern, or chamber, filled with CSF. The meninges which surround the cauda equina terminate in the region of the second sacral vertebra, or S2. The caudal extension of the innermost meningeal membrane attaches to the first coccygeal vertebra. This structure is called the *filum terminale.* The outermost meningeal membrane surrounds the filum terminale.

The relations of the spinal cord to the brain are shown in Fig. 14-8. Representative cross sections through various regions of the brainstem and spinal cord are given in perspective to illustrate their continuity. The spinal cord sections, from top to bottom, are in regions associated with cervical, thoracic, lumbar, and sacral spinal nerves. The spinal cord is composed of neural tissue called *white matter* which surrounds an inner H-shaped mass of neural tissue referred to as *gray matter.* Observe that the relative amounts of gray and white matter vary from region to region but that, in general, the spinal cord increases in thickness as it approaches the brain. This is because more and more nerve fibers are found in both ascending and descending tracts within the spinal cord at successively higher levels. Finally, notice that there are two localized enlargements of the spinal cord which correspond to regions which form large nerve plexuses supplying the upper and lower

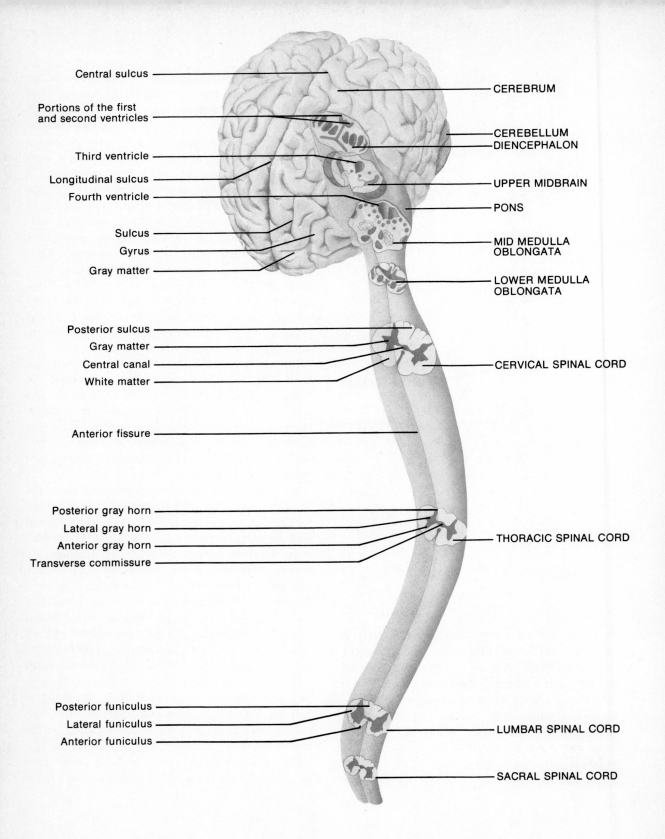

Central sulcus

Portions of the first
and second ventricles

Third ventricle

Longitudinal sulcus

Fourth ventricle

Sulcus

Gyrus

Gray matter

CEREBRUM

CEREBELLUM
DIENCEPHALON

UPPER MIDBRAIN

PONS

MID MEDULLA
OBLONGATA

LOWER MEDULLA
OBLONGATA

Posterior sulcus

Gray matter

Central canal

White matter

CERVICAL SPINAL CORD

Anterior fissure

Posterior gray horn

Lateral gray horn

Anterior gray horn

Transverse commissure

THORACIC SPINAL CORD

Posterior funiculus

Lateral funiculus

Anterior funiculus

LUMBAR SPINAL CORD

SACRAL SPINAL CORD

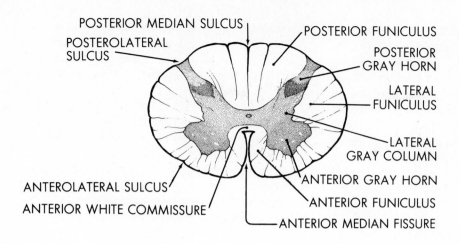

POSTERIOR MEDIAN SULCUS
POSTEROLATERAL SULCUS
POSTERIOR FUNICULUS
POSTERIOR GRAY HORN
LATERAL FUNICULUS
LATERAL GRAY COLUMN
ANTERIOR GRAY HORN
ANTERIOR FUNICULUS
ANTERIOR MEDIAN FISSURE
ANTEROLATERAL SULCUS
ANTERIOR WHITE COMMISSURE

**Figure 14-9** *Cross section of the spinal cord in the cervical region.* (From Melloni, Stone, and Hurd.)

extremities. The upper plexus, or *brachial plexus,* extends from the fifth cervical to the first thoracic spinal cord segments (C5 to T1) and supplies the upper extremities. The lower plexus, or *lumbosacral plexus,* extends from the third lumbar to the second sacral spinal cord segments (L3 to S2) and supplies the lower extremities.

An enlargement of a cross section through the cervical region of the spinal cord is shown in Fig. 14-9. The posterior aspect of the cord, closest to the spinous process of the vertebra which encloses it, is located at the top of the drawing. The midsagittal line on the posterior aspect of the cord contains a shallow groove called the *posterior median sulcus.* The midsagittal line on the anterior aspect of the cord possesses a deep furrow referred to as the *anterior median fissure.* The *posterolateral* and *anterolateral sulci* are superficial indentations which mark the entrances and exits, respectively, of spinal nerve roots. The sulci and the fissure run the length of the spinal cord.

The H-shaped gray matter of the spinal cord contains the cell bodies of sensory and motor neurons, as well as nerve fibers which cross from one side of the cord to the other. The gray matter on each side of the cord is divided into posterior, lateral, and anterior columns, or horns. The *posterior horns* receive senso-

ry nerve fibers. Autonomic and somatic motor neurons originate in the *lateral* and *anterior horns,* respectively. Internuncial neurons make synaptic connections between sensory and motor neurons. The transverse bridge which connects the gray matter on opposite sides is called the *gray commissure.* The central canal in the center of the gray commissure divides it into two regions—the posterior and anterior gray commissures. Filled with CSF, the *central canal* is continuous with the fourth ventricle of the medulla oblongata and terminates at the conus medullaris.

White matter is primarily composed of myelinated nerve fibers which run lengthwise in the cord. The regions of white matter are established by the horns of gray matter. The white matter lying between the posterior horns is called the *posterior funiculus.* The *lateral funiculi* are masses of white matter lying between the posterior and lateral horns on each side of the cord. The anterior horns demarcate a region of white matter referred to as the *anterior funiculus.* The transverse bridge of white matter associated with the anterior funiculus is called the *anterior white commissure.* Ascending and descending spinal nerve tracts course through the substance of the funiculi. The spinal tracts will be described below.

The spinal cord has the following functions: (1) the

**Figure 14-8** *The brain and spinal cord, showing representative cross sections in perspective at various levels. The sections, from top to bottom, are in the diencephalon, the upper midbrain, the midmedulla, the lower medulla, and the cervical, thoracic, lumbar, and sacral segments of the spinal cord.*

mediation of spinal reflexes by integrating sensory information and initiating motor directives without the intervention of the brain, (2) the coordination of skeletal muscles during the execution of complex reflexes involving many segments of the spinal cord, (3) the conduction of sensory information from the receptors of the body to the brain, and (4) the conduction of motor directives from the brain to peripheral nerves which innervate the effectors of the body. Thus, the spinal cord serves in both reflex and volitional activities by providing channels of communication between receptors, effectors, and the brain.

**Spinal tracts**  The funiculi of the spinal cord contain spinal tracts which carry encoded electrical messages to and from the brain. All the spinal tracts are paired, since the spinal cord is bilaterally symmetric. Each spinal tract, or *fasciculus,* is composed of a bundle of myelinated nerve fibers that have a common origin, conduction pathway, and termination. Fiber tracts which carry sensory, or afferent, nerve impulses to the brain are called *ascending tracts.* Fiber tracts which conduct motor, or efferent, nerve impulses away from the brain are called *descending tracts.* Each "cable" of a spinal tract consists of a chain of at least two successive neurons linked synaptically. Most spinal tracts are somatotopically organized; i.e., fibers which enter or leave the spinal cord at any level are the most laterally or medially located in the fasciculus at that point. Information processing of the encoded electrical signals occurs at each of the synaptic connections along the spinal tracts.

**Ascending tracts**  The origins of ascending tracts are occupied by neurons whose cell bodies are located in posterior root ganglia. Called *first-order neurons,* these nerve cells link receptors with the spinal cord. Successive neurons in the fiber tract are called *second-order neurons, third-order neurons,* etc. Second-order neurons begin in the spinal cord and terminate in the brain, while third-order neurons begin at a lower level of the brain and terminate at a higher level.

The major ascending spinal tracts include the fasciculus gracilis, fasciculus cuneatus, lateral spinothalamic tract, anterior spinothalamic tract, posterior spinocerebellar tract, and anterior spinocerebellar tract. These tracts are illustrated in Fig. 14-10A to F.

1  *Fasciculus gracilis.* Located in the posteromedial aspect of the spinal cord, this tract provides information from the sacral, lumbar, and lower six thoracic segments on proprioception from muscle spindles and Golgi tendon organs, deep pressure from pacinian corpuscles, and discriminative or light touch from Meissner's corpuscles. The axons of the first-order neurons from the above regions enter the posterior horns, run through the fasciculus gracilis on the same (ipsilateral) side of the cord, and terminate synaptically in the nucleus gracilis of the lower medulla oblongata. The fibers of the fasciculus gracilis are somatotopically organized, so that they are layered more and more laterally as they are added at successively higher levels. Fibers of the second-order neurons decussate in the lower medulla oblongata, ascend in the medial lemniscus, and terminate synaptically in the thalamus. Third-order thalamic neurons project through the internal capsule to the postcentral gyrus (somesthetic cortex). Because of medullary decussation, the somesthetic cortex receives touch, pressure, pain, and proprioceptive information from the opposite (contralateral) side of the body.

2  *Fasciculus cuneatus.* This tract is similar in function and somatotopic organization to the fasciculus gracilis. Located immediately lateral to the fasciculus gracilis on each side of the cord, the fasciculus cuneatus is made up of first-order fibers which enter the posterior horn from the upper six thoracic segments and the entire cervical region. Axons of the first-order neurons ascend in the posterior funiculus of the cord. The axons terminate synaptically in the nucleus cuneatus of the lower medulla. Fibers of the second-order neurons decussate immediately, following a pathway similar to that of the fasciculus gracilis.

3  *Lateral spinothalamic tract.* This tract provides information on pain from free nerve endings and

*Figure 14-10* The major ascending spinal tracts of man. (Continued on pages 370 and 371.)

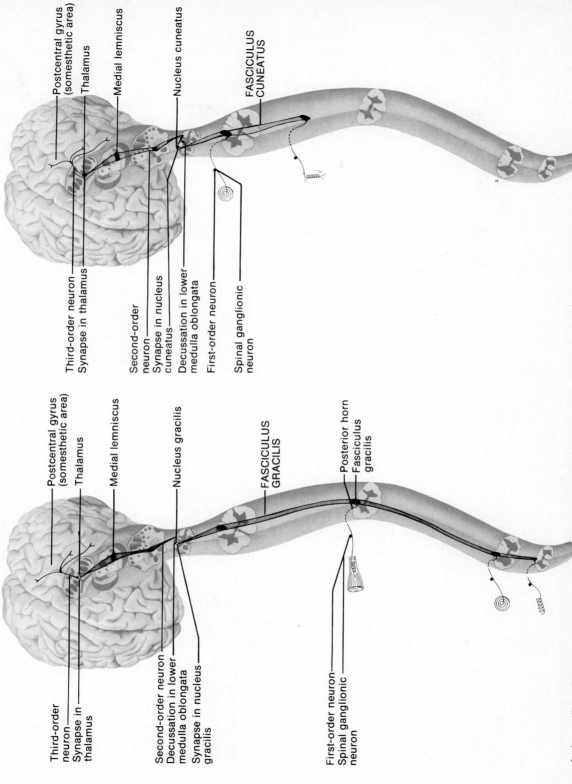

Postcentral gyrus (somesthetic area)
Thalamus
Medial lemniscus
Nucleus cuneatus
FASCICULUS CUNEATUS

Third-order neuron
Synapse in thalamus
Second-order neuron
Synapse in nucleus cuneatus
Decussation in lower medulla oblongata
First-order neuron
Spinal ganglionic neuron

Postcentral gyrus (somesthetic area)
Thalamus
Medial lemniscus
Nucleus gracilis
FASCICULUS GRACILIS
Posterior horn
Fasciculus gracilis

Third-order neuron
Synapse in thalamus
Second-order neuron
Decussation in lower medulla oblongata
Synapse in nucleus gracilis
First-order neuron
Spinal ganglionic neuron

*A.* Ascending—Fasciculus gracilis: touch, pressure, and proprioception

*B.* Ascending—Fasciculus cuneatus: touch, pressure, and proprioception

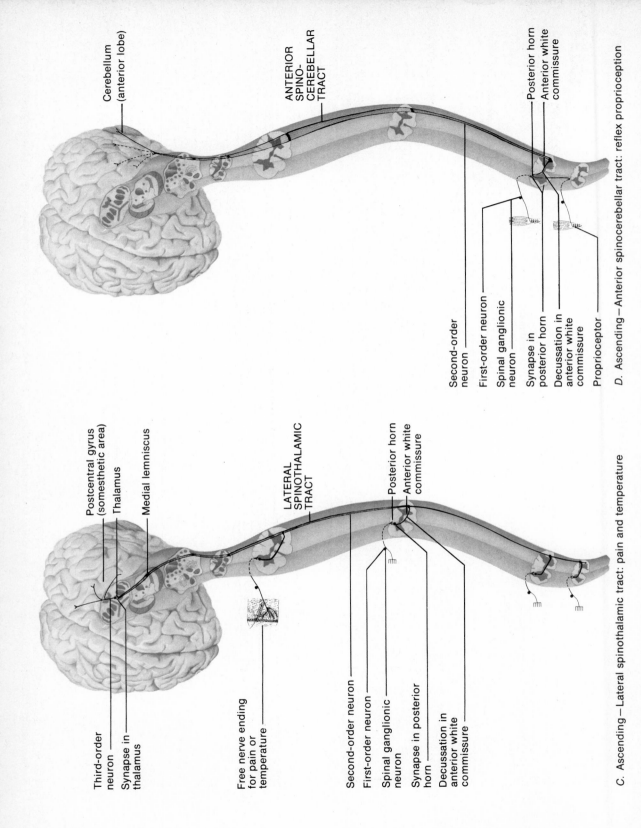

Cerebellum (anterior lobe)

ANTERIOR SPINO-CEREBELLAR TRACT

Posterior horn

Anterior white commissure

Second-order neuron

First-order neuron

Spinal ganglionic neuron

Synapse in posterior horn

Decussation in anterior white commissure

Proprioceptor

D. Ascending—Anterior spinocerebellar tract: reflex proprioception

Postcentral gyrus (somesthetic area)

Thalamus

Medial lemniscus

Third-order neuron

Synapse in thalamus

LATERAL SPINOTHALAMIC TRACT

Posterior horn

Anterior white commissure

Free nerve ending for pain or temperature

Second-order neuron

First-order neuron

Spinal ganglionic neuron

Synapse in posterior horn

Decussation in anterior white commissure

C. Ascending—Lateral spinothalamic tract: pain and temperature

370

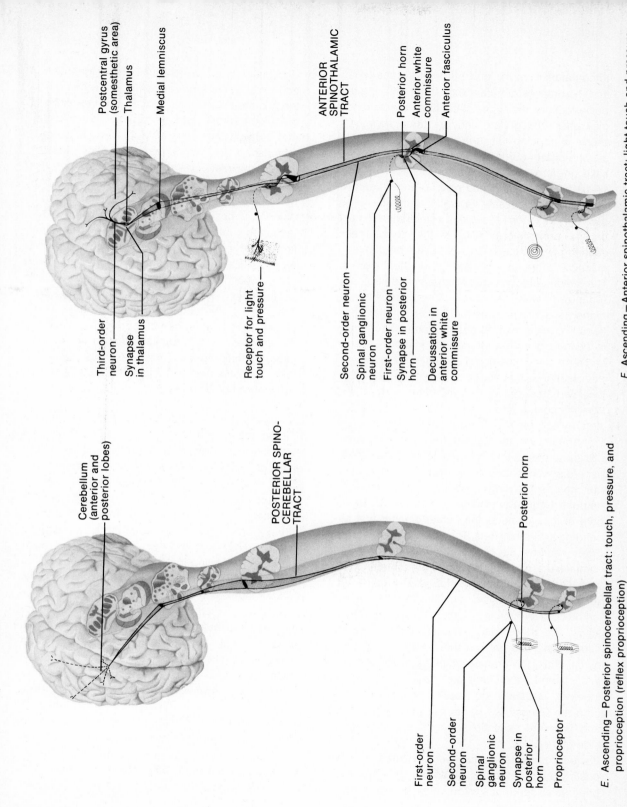

Postcentral gyrus (somesthetic area)

Thalamus

Medial lemniscus

Third-order neuron

Synapse in thalamus

Receptor for light touch and pressure

ANTERIOR SPINOTHALAMIC TRACT

Posterior horn

Anterior white commissure

Anterior fasciculus

Second-order neuron

Spinal ganglionic neuron

First-order neuron

Synapse in posterior horn

Decussation in anterior white commissure

*F.* Ascending—Anterior spinothalamic tract: light touch and pressure

Cerebellum (anterior and posterior lobes)

POSTERIOR SPINO-CEREBELLAR TRACT

Posterior horn

First-order neuron

Second-order neuron

Spinal ganglionic neuron

Synapse in posterior horn

Proprioceptor

*E.* Ascending—Posterior spinocerebellar tract: touch, pressure, and proprioception (reflex proprioception)

371

temperature from the end bulbs of Krause (cold) and the brushes of Ruffini (hot). After entering the posterior horn, the axons of the first-order neurons may pass through as many as three segments before synapsing with the nerve cell body of a second-order neuron in the posterior horn. For the sake of clarity, this has not been shown in Fig. 14-10C. The axons of the second-order neurons immediately decussate through the anterior white commissure to form the lateral spinothalamic tract on the contralateral side of the cord. These second-order axons pass through the medial lemniscus and terminate synaptically in the thalamus. The lateral spinothalamic tract is somatotopically organized, with the addition of successively higher levels to progressively more medial positions. The axons of third-order neurons which arise in the thalamus pass through the internal capsule and terminate in the postcentral gyrus. Because of decussation within the cord, the somesthetic cortex integrates pain and temperature sensations from the contralateral side of the body.

4   *Anterior spinothalamic tract.* Located in the anterior funiculus throughout the cord, this tract provides information on light touch from Meissner's corpuscles and deep pressure from pacinian corpuscles. The axons of the first-order neurons terminate synaptically in the posterior horn of the segments they enter or after first ascending to higher levels. The axons of the second-order neurons decussate in the anterior white commissure to form the anterior spinothalamic tract on the contralateral side of the cord. The second-order axons ascend through the medial lemniscus, terminating synaptically in the thalamus. The third-order thalamic neurons project through the internal capsule to the postcentral gyrus. Thus, the cranial locations of the anterior and lateral spinothalamic tracts are close together. Because the fibers decussate in the cord, the somesthetic cortex integrates light touch and pressure sensations from the contralateral side of the body.

5   *Posterior spinocerebellar tract.* Located in the posterolateral funiculus, this tract provides proprioceptive feedback information from the lower extremities to the cerebellum. This information is conveyed by the axons of the first-order neurons to the posterior horn in the lower cord. The axons of the second-order neurons ascend on the ipsilateral side as the posterior spinocerebellar tract, terminating in the cerebellum after passing through the inferior cerebellar peduncle. Thus, this tract is uncrossed and consists of a chain of only two, rather than three, successive neurons.

6   *Anterior spinocerebellar tract.* Located in the anterolateral funiculus, this tract has a function similar to the posterior spinocerebellar tract. The axons of the first-order neurons enter the lower cord, ascend, and terminate synaptically in the posterior horn within the thoracic or lumbar region. The axons of the second-order neurons decussate in the anterior white commissure and ascend as the anterior spinocerebellar tract on the contralateral side of the cord. These axons terminate in the cerebellum after passing through the superior cerebellar peduncle.

**Descending tracts**   Each descending tract consists of upper motor neurons and lower motor neurons which share a common origin and termination. The nerve cell bodies of the *upper motor neurons* are located in the brain. Their axons terminate synaptically on either association (internuncial) neurons, which then synapse with lower motor neurons, or directly on lower motor neurons, which exit the spinal cord from the anterior horn. Thus, upper motor neurons are entirely within the CNS. Since the axon of an anterior horn neuron leaves the cord through the anterior root to innervate, or supply, an effector, it is called a *lower motor neuron.* In other words, lower motor neurons originate in the cord but terminate elsewhere. Many internuncial and other inhibitory and excitatory neurons synapse with each anterior horn neuron.

The pathway of a lower motor neuron is often called the *final common path.* The coded electrical information carried by the final common path is fixed, and no further information processing is possible.

The major descending spinal tracts include the lateral corticospinal tract, anterior corticospinal tract, rubrospinal tract, vestibulospinal tracts, reticulospinal tracts, and tectospinal tract. The lateral and anterior corticospinal tracts are commonly referred to as the crossed and uncrossed *pyramidal tracts,* respectively. The rest of the descending tracts are collectively called the *extrapyramidal tracts.* These tracts are illustrated in Fig. 14-11*A* to *F*.

1 *Lateral corticospinal tract* (crossed pyramidal tract). This great tract controls the actions of voluntary muscles, especially those associated with the upper and lower extremities, by coordinating muscle agonists, antagonists, and synergists in executing gross and fine movements. The pyramidal cells which constitute the upper motor neurons originate in the precentral gyrus (motor area 4), descending through the internal capsule and midbrain to the medulla oblongata. About 90 percent of the pyramidal fibers decussate in the medullary pyramids to form the lateral corticospinal tract in the contralateral cord. A small percentage remain uncrossed, contributing to the lateral corticospinal tract on the ipsilateral side of the cord. The remainder of the pyramidal fibers give rise to the anterior corticospinal tract (see below). The pyramidal fibers of the lateral corticospinal tract either synapse directly with lower motor neurons, the cell bodies of which are located in the anterior horn, or synapse with internuncial neurons which in turn make synaptic connections with lower motor neurons. Because of pyramidal decussation, the precentral gyrus primarily controls the skeletal musculature on the contralateral side of the body.

2 *Anterior corticospinal tract* (uncrossed pyramidal tract). About 7 percent of the pyramidal fibers which remain uncrossed as they descend through the medulla give rise to the anterior corticospinal tract. The pyramidal fibers of this tract decussate in either the cervical or upper thoracic region of the cord and synapse either directly or indirectly with lower motor neurons which originate in the anterior horn in the region of decussation. These lower motor neurons innervate the skeletal musculature of the trunk.

3 *Rubrospinal tract.* Located in the lateral funiculus, this tract makes adjustments in the tone of skeletal muscles based on motor directives from the precentral gyrus and proprioceptive feedback information processed by the cerebellum. The cell bodies of the upper motor neurons are located in the red nucleus (nucleus ruber) of the midbrain. Uncrossed corticorubral fibers from motor area 4 and crossed cerebellorubral fibers from the cerebellum make synaptic connections in the nucleus ruber. The fibers of the upper motor neurons decussate in the midbrain, descending in the tegmentum to enter the contralateral cord as the rubrospinal tract. Most rubrospinal fibers terminate synaptically on the cell bodies of lower motor neurons in the anterior horn of the cervical region. Because of decussation in the midbrain, the rubrospinal tract adjusts the tone of skeletal muscles on the contralateral side of the body.

4 *Vestibulospinal tracts.* These tracts help maintain posture and equilibrium by adjusting the tone of skeletal muscles to the proprioceptive information transduced by hair cells in the cristae and maculae of the membranous labyrinths. The upper motor neurons originate in the vestibular nuclei below the fourth ventricle in the pontine region of the brain. The vestibular nuclei receive uncrossed vestibular fibers from the membranous labyrinth and cerebellovestibular fibers from the cerebellum. The fibers of the upper motor neurons descend in the lateral and anterior funiculi of the ipsilateral cord as the lateral and medial vestibulospinal tracts, respectively. Vestibulospinal fibers of the lateral tract synapse indirectly with lower motor neurons that, for the most part, innervate skeletal muscles in the upper and lower extremities. The lateral vestibulospinal tract promotes extensor muscle tone. Vestibulospinal fibers of the medial tract synapse by way of internuncial neurons (interneurons) with lower motor neurons primarily in the thoracic segments of the spinal cord. The medial vestibulospinal tract inhibits extensor muscle tone in this region of the body.

5 *Reticulospinal tracts.* These tracts coordinate extensor reflex activities. The upper motor neurons originate in the reticular nuclei of the pons and medulla. Uncrossed corticoreticular fibers from

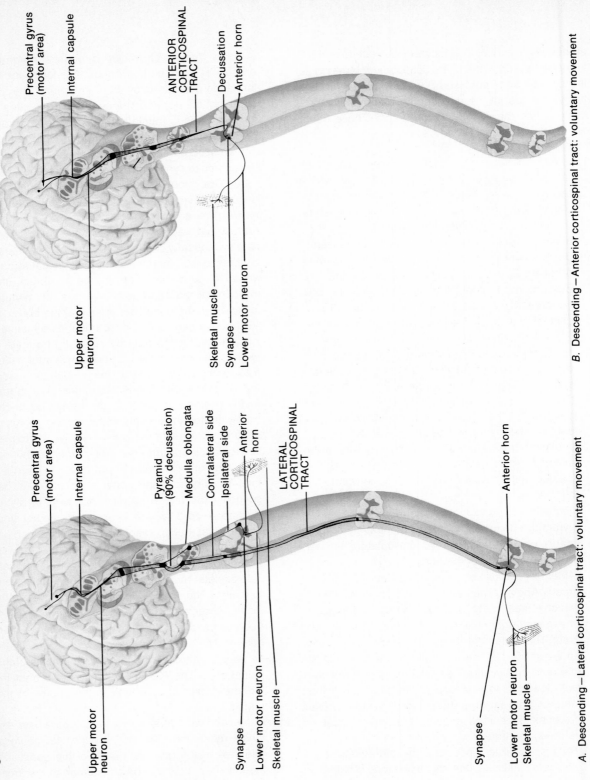

**Figure 14-11** The major descending spinal tracts of man. (Continued on pages 375 and 376.)

Precentral gyrus
(motor area)

Internal capsule

Upper motor
neuron

Pyramid
(90% decussation)

Medulla oblongata

Contralateral side

Ipsilateral side

Anterior
horn

LATERAL
CORTICOSPINAL
TRACT

Synapse

Lower motor neuron

Skeletal muscle

Anterior horn

Synapse

Lower motor neuron
Skeletal muscle

A. Descending—Lateral corticospinal tract: voluntary movement

Precentral gyrus
(motor area)

Internal capsule

ANTERIOR
CORTICOSPINAL
TRACT

Decussation

Anterior horn

Upper motor
neuron

Skeletal muscle

Synapse

Lower motor neuron

B. Descending—Anterior corticospinal tract: voluntary movement

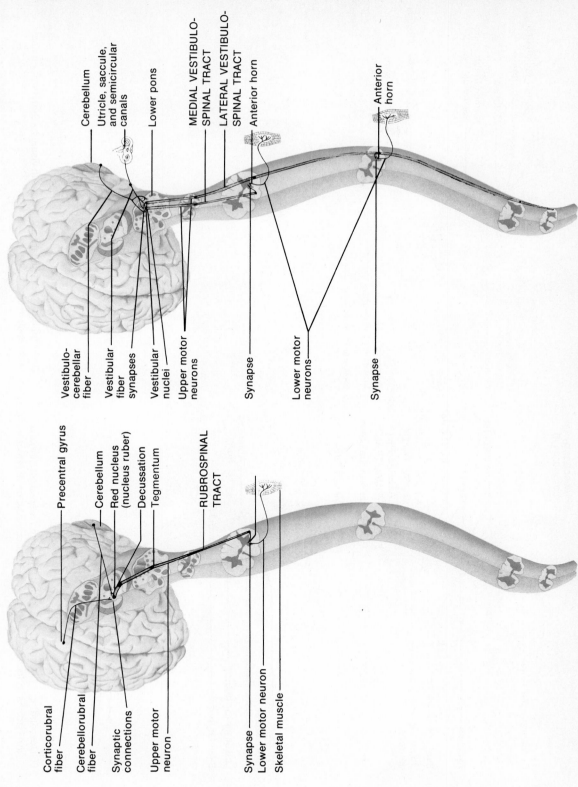

Cerebellum

Utricle, saccule, and semicircular canals

Lower pons

MEDIAL VESTIBULO-SPINAL TRACT

LATERAL VESTIBULO-SPINAL TRACT

Anterior horn

Anterior horn

Vestibulo-cerebellar fiber

Vestibular fiber synapses

Vestibular nuclei

Upper motor neurons

Synapse

Lower motor neurons

Synapse

Precentral gyrus

Cerebellum

Red nucleus (nucleus ruber)

Decussation

Tegmentum

RUBROSPINAL TRACT

Corticorubral fiber

Cerebellorubral fiber

Synaptic connections

Upper motor neuron

Synapse

Lower motor neuron

Skeletal muscle

C. Descending—Rubrospinal tract: adjustments between proprioceptive feedback and motor area directives    D. Descending—Vestibulospinal tracts: maintenance of equilibrium and posture through correct muscle tonus

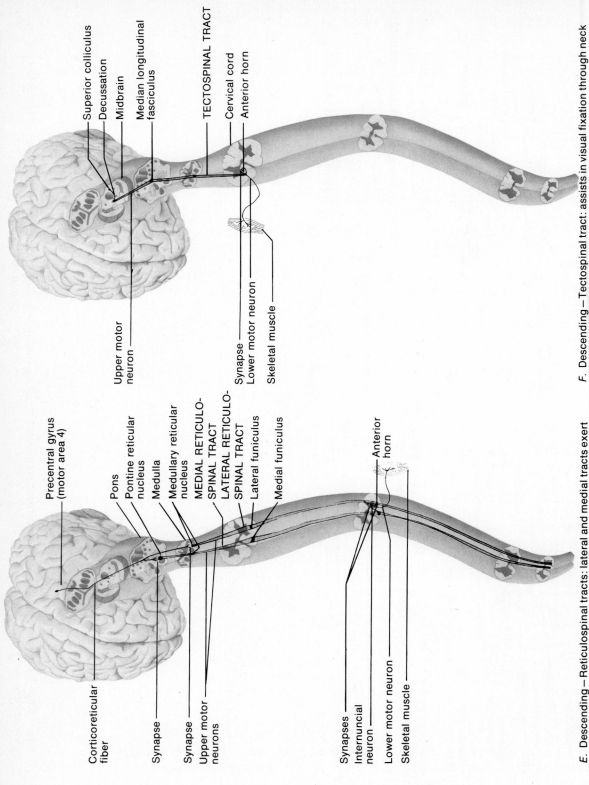

Superior colliculus
Decussation
Midbrain
Median longitudinal fasciculus
TECTOSPINAL TRACT
Cervical cord
Anterior horn

Upper motor neuron

Synapse
Lower motor neuron
Skeletal muscle

Precentral gyrus (motor area 4)

Pons
Pontine reticular nucleus
Medulla
Medullary reticular nucleus
MEDIAL RETICULO-SPINAL TRACT
LATERAL RETICULO-SPINAL TRACT
Lateral funiculus
Medial funiculus

Anterior horn

Corticoreticular fiber

Synapse

Synapse
Upper motor neurons

Synapses
Internuncial neuron
Lower motor neuron
Skeletal muscle

*E. Descending* — Reticulospinal tracts: lateral and medial tracts exert inhibitory and facilitatory influences, respectively, on extensor reflexes

*F. Descending* — Tectospinal tract: assists in visual fixation through neck reflex movements

the precentral gyrus make synaptic connections with the reticular nuclei. Fibers from the pontine and medullary reticular nuclei descend in the lateral and anterior funiculi of the ipsilateral cord as the lateral and anterior reticulospinal tracts. These fibers synapse indirectly with lower motor neurons in the anterior horn. The lateral and anterior reticulospinal tracts exert inhibitory and excitatory influences, respectively, on extensor reflexes.

6  *Tectospinal tract.* This tract assists in visual fixation through neck reflex movements. The upper motor neurons originate in the superior colliculi of the midbrain. Following decussation in the midbrain, the tectospinal fibers descend through the median longitudinal fasciculus to become the tectospinal tract. The tectospinal fibers synapse indirectly with lower motor neurons in the anterior horn of the cervical spinal cord. Because of decussation in the midbrain, the tectospinal tract reflexly controls the cervical musculature on the contralateral side of the body.

## SPECIALIZED STRUCTURES

**The meninges**   The brain and the spinal cord are covered on the outside by three protective membranes collectively referred to as *meninges*. From outside inward, the meninges are called the *dura mater, arachnoid,* and *pia mater.* The cranial meninges are illustrated in Fig. 14-12A. The spinal meninges are shown in Fig. 14-12B.

The dura mater, the outermost of the meninges, is a thick fibrous membrane. The cranial dura mater is merged with the endosteum, or inferior portion, of the cranial bones. At various locations in the cranium, the dura mater is split into a double membrane by triangular sinuses which collect both venous blood and CSF. The dura mater below the venous sinuses usually penetrates deeply into the sulci of the brain, forming partitions which separate paired lobes or adjacent regions.

In Fig. 14-12A, the meninges and associated regions are shown in a partial coronal section, thereby exposing a part of the left and right cerebral hemispheres. The dura mater which fills the longitudinal fissure between hemispheres is called the *falx cerebri.* The venous sinus in the dura mater immediately superior to the falx cerebri is named the *superior sagittal sinus.* The spinal dura mater is continuous with the cranial dura mater, although the former is separate from the inside surfaces of the surrounding vertebrae.

The arachnoid, a clear thin membrane named for its spiderlike surface configurations, is sandwiched between the dura mater and pia mater. Except in the region of the falx cerebri, the arachnoid does not follow the dura mater into the sulci. The space between the dura mater and arachnoid is called the *subdural space.* More potential than actual, this space contains a thin layer of serous fluid which prevents the membranes from adhering to each other. The space between the arachnoid and pia mater is referred to as the *subarachnoid space.* The subarachnoid space contains CSF in the interstices formed by many fine trabeculae, in addition to pial blood vessels. In the region of the venous sinuses, the arachnoid forms periodic fingerlike projections called *arachnoid villi* which penetrate these blood-filled spaces. Thus, CSF in the subarachnoid spaces within the villi can move by diffusion through the arachnoid membrane into the blood within the venous sinuses.

The spinal arachnoid is continuous with the cranial arachnoid, although it lacks specialized arachnoid villi. However, CSF fills the spinal subarachnoid space, just as it does the cranial subarachnoid space.

The pia mater lies next to the neural tissue. This delicate membrane sends extensions into the fissures and sulci in the cranial region. The spinal pia mater sends out lateral specializations called *dentate ligaments* which separate the posterior and anterior roots of the spinal nerves.

**The cerebrospinal fluid**   CSF is a special fluid of the body found inside and outside the brain and the spinal cord. The fluid is divided between two compartments which are continuous with each other: (1) an inner compartment consisting of the ventricles of the brain and the central canal of the spinal cord and (2) an outer compartment consisting of the cranial and spinal subarachnoid spaces. CSF is formed from blood serum in specialized capillaries associated

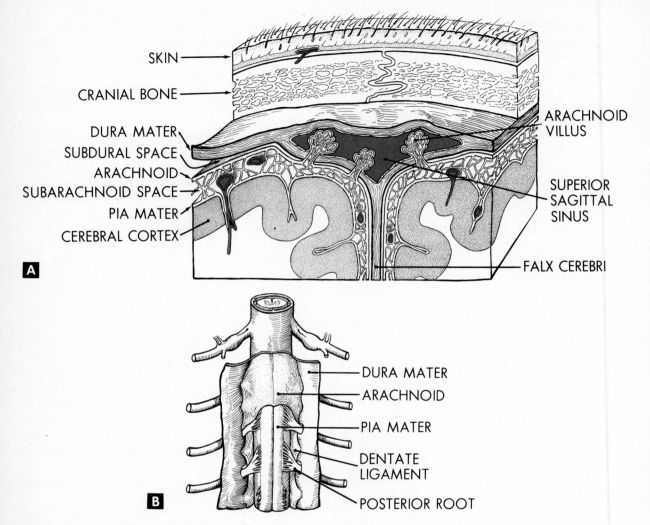

SKIN

CRANIAL BONE

DURA MATER
SUBDURAL SPACE
ARACHNOID
SUBARACHNOID SPACE
PIA MATER
CEREBRAL CORTEX

ARACHNOID VILLUS

SUPERIOR SAGITTAL SINUS

FALX CEREBRI

**A**

DURA MATER
ARACHNOID
PIA MATER
DENTATE LIGAMENT
POSTERIOR ROOT

**B**

*Figure 14-12* *The relation of the meninges to the central nervous system. A. The meningeal coverings of a portion of the brain from posterior aspect. B. The meningeal coverings of a portion of the spinal cord.*

with the brain ventricles, circulates in a definite pattern throughout the inner and outer compartments, and is ultimately returned through the arachnoid villi to the venous blood in the dural sinuses. The functions of CSF include protection of the CNS by absorbing mechanical shocks transmitted through the body and participation in the metabolic functions of the brain and spinal cord through the transport of nutrients and waste products. Also, CSF formation

bars the entry into the CNS of potentially harmful materials in the blood, as we will see.

To appreciate the pattern of circulation of CSF through the brain, we must look more closely at the interconnected network of brain cavities called *ventricles*. The brain ventricles are diagramed in Fig. 14-13A. Each cerebral hemisphere contains a lateral ventricle. Each *lateral ventricle* connects with the *third ventricle* in the midbrain by means of an *inter-*

ventricular foramen of Monro. The third centricle communicates with the *fourth ventricle* between the cerebellum and pons by the *cerebral aqueduct of Sylvius*. At the end of the fourth ventricle are three foramina—a median *foramen of Magendie* and two lateral *foramina of Luschka*. These three foramina communicate with the subarachnoid spaces between the brain and the spinal cord. The terminal portion of the fourth ventricle is continuous with the central canal of the cord.

CSF is formed by both the diffusion of blood serum and the active transport of specific plasma components through the walls of specialized groups of blood capillaries called *choroid plexuses* located in the ventricular walls. One group of these capillaries is located in the walls of the third and lateral ventricles, while another group is found in the walls of the fourth ventricle. The choroid plexuses are illustrated in a midsagittal section of the brain in Fig. 14-13B. The

compound structure formed by the endothelial walls of the choroid blood vessels and the ependymal epithelial lining of the ventricles is functionally a differentially permeable membrane called the blood-brain barrier, mentioned earlier. Ordinarily, materials in the blood must breach the blood-brain barrier to enter the brain tissues.

Since most CSF forms from the choroid plexus associated with the lateral and third ventricles, CSF moves by hydrostatic pressure from the forebrain to the midbrain through the ventricles. From the fourth ventricle, CSF enters the central canal of the spinal cord or an enlarged subarachnoid space called the *cerebellomedullary cistern* at the junction of the brain and spinal cord. The pressure of the fluid causes it to circulate either in the subarachnoid spaces of the spinal cord or through cisterns associated with various portions of the brain. While circulating, CSF bathes the brain tissues. Ultimately, the CSF reaches the

**Figure 14-13** *The formation and circulation of cerebrospinal fluid (CSF).* A. *Brain ventricles.* B. *CSF circulates through the ventricles and subarachnoid spaces in the direction of the arrows.* (From Melloni, Stone, and Hurd.)

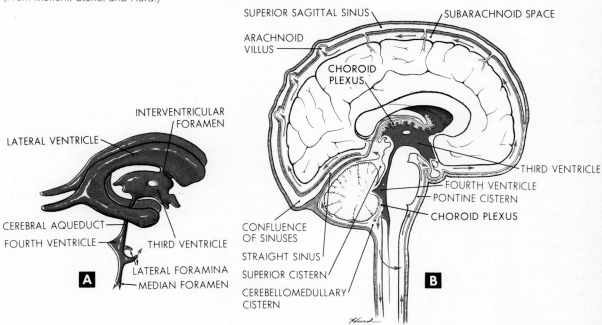

subarachnoid spaces associated with the cranial dura mater and diffuses through arachnoid villi into the dural sinuses.

The circulatory pathways of CSF are diagrammed in Fig. 14-13*B*. The arrows indicate the direction of fluid flow.

Somewhat similar to lymph in composition, CSF is ordinarily aseptic. On an average, there are 200 ml of CSF in the body. An excessive increase in CSF caused by obstructions or other factors is called hydrocephaly. Hydrocephaly is reviewed in Chap. 15.

## THE PERIPHERAL NERVOUS SYSTEM

The peripheral nervous system includes the cranial nerves, spinal nerves, and autonomic nervous system. The peripheral nervous system links receptors and effectors to the CNS (central nervous system) and provides pathways for the execution and coordination of motor responses, following sensory stimulation and central integration. The responses may be either reflex or volitional.

## THE REFLEX ARC AND REFLEXES

If the structural unit of the nervous system is the neuron, then its functional unit is called a *reflex arc*. A reflex arc consists of a receptor; an afferent, or sensory, neuron which conducts the encoded information transduced by a receptor to the CNS (either the brain or the spinal cord); and an efferent, or motor, neuron which conducts the integrated neural directives of the CNS to an effector such as a muscle or gland. The response of the effector may be excitatory or inhibitory. Usually, but not always, the afferent and efferent neurons are linked together functionally in the CNS by a connecting, or internuncial, neuron (interneuron).

A reflex arc is diagrammed in Fig. 14-14. Assume that a sensory ending is adequately stimulated. The encoded sensory information is conducted by the axon of the afferent neuron through a spinal nerve to the posterior horn of the spinal cord, where the axon terminates synaptically on the cell body of an internuncial neuron (interneuron). Notice that the cell body of the afferent neuron is located in the posterior root

ganglion. The axon of the interneuron makes synaptic connections with an efferent neuron located in the anterior horn of the cord. Thus, the termination of the afferent neuron and the origin of the efferent neuron are both within the CNS. The integrated motor information is conducted by the axon of the efferent neuron to a myoneural junction of a skeletal muscle where a response to the stimuli is elicited. Notice that a reflex is mediated below the level of consciousness, i.e., a response to the stimulus is reflected back to the periphery from the spinal cord without conscious intervention. Hence, the term *reflex*. Reflexes are relatively simple, innate patterns of behavioral response which are exhibited without prior learning. In addition to skeletal muscles, reflexes can affect smooth muscles or glands. In these cases, the efferent neurons originate in the lateral horns of the cord.

The reflex arc described above consists of a three-neuron chain in which the afferent and efferent neurons enter and leave the same side of the cord. Called a *three-neuron ipsilateral reflex arc,* it mediates flexion reflexes such as rapid withdrawal of a portion of the body in contact with a hot or cold object. However, the three-neuron ipsilateral reflex arc is only one of several common types. The following types of reflex arc are variations on the basic theme:

1  *Two-neuron ipsilateral reflex arcs.* The axons of the afferent neurons in these reflex arcs synapse directly with efferent neurons. Such arcs commonly mediate stretch reflexes. A well-known example is the knee jerk, which is elicited in response to sudden stretch of the patellar tendon of the quadriceps femoris muscles.
2  *Three-neuron contralateral reflex arcs.* The internuncial neurons linking afferent and efferent neurons in these reflex arcs cross over to the opposite side of the cord. Therefore, the response elicited occurs on the side of the body opposite that which is stimulated. Contralateral responses are exhibited during crossed-extension reflexes. For example, the body remains balanced during ipsilateral flexion because of contralateral extension.
3  *Long spinal reflex arcs.* The axons of the afferent

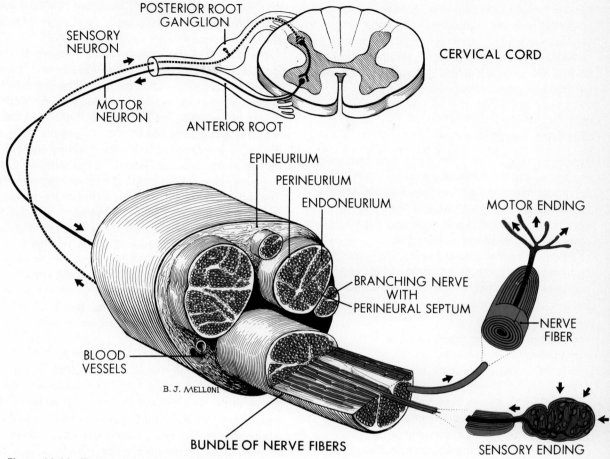

**Figure 14-14** *The structure of a peripheral nerve.*

neurons in these intersegmental reflex arcs form ascending and descending branches which synapse with internuncial neurons located in higher levels of the cord. The internuncial neurons may or may not decussate before terminating synaptically on motor neurons which originate in the anterior horns. Since one sensory fiber may make synaptic connections with numerous internuncial neurons (divergence), the response to the stimulus is necessarily complex. Long spinal reflex arcs are exemplified during rapid avoidance movements involving the arms and legs and the execution of patterned motor activities such as walking, running, and swimming.

Reflexes, as opposed to reflex arcs, are classified either by the region of the CNS with which they are associated, e.g., cranial or spinal reflexes, or by the distribution of the effectors in the body. In the latter classification, three categories of reflex are commonly recognized: (1) superficial reflexes associated with the skin and mucous membranes such as blinking the eye when the cornea is irritated (corneal reflex), sneezing when the nasal mucosa is irritated (sneeze reflex), or plantar flexion of the toes when the sole of the foot is touched gently (plantar reflex); (2) deep reflexes associated with the skeletal muscles such as extension at the knee (knee jerk) or flexion at the elbow (biceps reflex) when the tendon of the biceps femoris or biceps brachii, respectively, is suddenly stretched; and (3) visceral reflexes associated with visceral structures such as pupillary constriction when the retina is illuminated (light reflex) or contractions of the urinary bladder upon appropriate parasympathetic stimulation (urination reflex).

## THE CRANIAL NERVES

Twelve pairs of cranial nerves originate from various regions along the base of the brain. These nerves carry information to the brain from the special senses, including the sense of balance, innervate muscles and glands of the face and neck, and supply certain other visceral structures in the trunk.

The roots of the cranial nerves are identified by numbers at the base of the brain in Fig. 14-15. The numbers and names of the cranial nerves are as follows:

cranial nerve I—the olfactory nerve
cranial nerve II—the optic nerve
cranial nerve III—the oculomotor nerve
cranial nerve IV—the trochlear nerve
cranial nerve V—the trigeminal nerve
cranial nerve VI—the abducens nerve
cranial nerve VII—the facial nerve
cranial nerve VIII—the acoustic (auditory) nerve
cranial nerve IX—the glossopharyngeal nerve
cranial nerve X—the vagus nerve
cranial nerve XI—the spinal accessory nerve
cranial nerve XII—the hypoglossal nerve

The assignments of number are based on serial location from anterior to posterior. The names in most cases are descriptive of their functions.

Cranial nerves are sensory, motor, or mixed (containing both sensory and motor fibers). Cranial nerves I, II, and VIII are purely *sensory*. Cranial nerves III, IV, VI, XI, and XII are purely *motor*. The remainder of the cranial nerves, i.e., V, VII, IX, and X, are *mixed*. This classification arbitrarily ignores the possibility that each of the mixed and motor nerves may in fact contain sensory nerve fibers originating from muscle spindles of the muscles they innervate. Nevertheless, the convenience of this classification outweighs its possible shortcomings.

The origins of the cranial nerves depend on the location of the cell bodies of the sensory and motor nerve fibers which make up the nerve. The cell bodies of efferent fibers are located in nuclei deep in the substance of the brain. The cell bodies of afferent fibers, with few exceptions, are located in ganglia outside the brain. The exceptions are the olfactory and optic nerves, in which the cell bodies of the sensory fibers originate in the nasal mucosa and retina, respectively. The reason for the anomalous origins of the sensory nerve cell bodies of cranial nerves I and II is due to the fact that these nerves are really fiber tracts of the brain. Thus, based on their anatomic relation to the brain, cranial nerves I and II are extensions of the forebrain, cranial nerves III and IV originate in the midbrain, cranial nerves V to VIII enter in the pontine region of the hindbrain, and the last four cranial nerves originate in the medullary region of the hindbrain.

In addition, cranial nerves III, VII, IX, and X collectively make up the cranial portion of the *parasympathetic* division of the autonomic nervous system. Some of the motor fibers of these nerves terminate synaptically in autonomic ganglia close to various structures in the face, neck, and upper thoracic region. These preganglionic fibers participate in the automatic control of certain structures in these regions. The autonomic nervous system will be described later in this chapter.

The functions and common neuropathies resulting from irritative or destructive lesions of the various cranial nerves are discussed in detail in Chap. 16.

## THE SPINAL NERVES

A spinal nerve consists of a myelinated bundle of sensory and motor nerve fibers, along with their accompanying blood vessels and connective tissue sheaths, which enter and exit at the same segment of the spinal cord. The organization of a spinal nerve is shown in Fig. 14-16.

In a strict sense, most spinal nerves are only a few

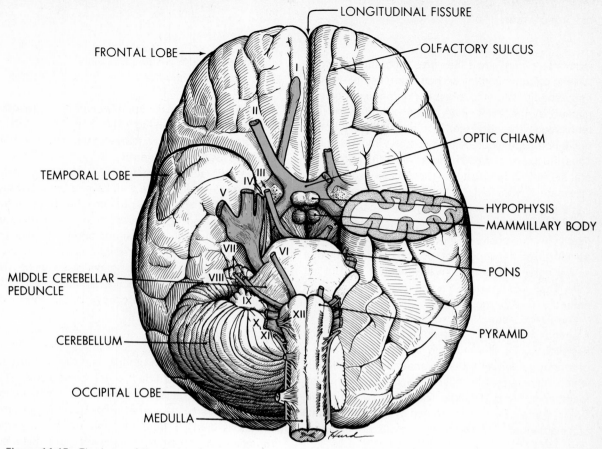

**Figure 14-15** *The base of the brain, showing the roots of the cranial nerves. The cranial nerves are identified with roman numerals. (From Melloni, Stone, and Hurd.)*

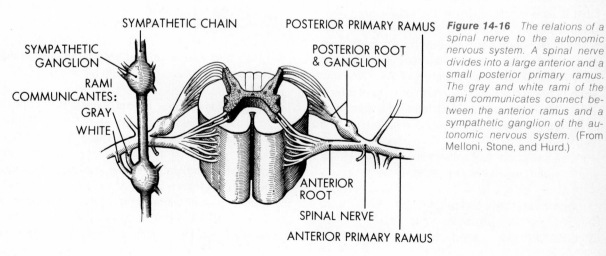

**Figure 14-16** *The relations of a spinal nerve to the autonomic nervous system. A spinal nerve divides into a large anterior and a small posterior primary ramus. The gray and white rami of the rami communicates connect between the anterior ramus and a sympathetic ganglion of the autonomic nervous system. (From Melloni, Stone, and Hurd.)*

millimeters long, since they give off branches which have separate identities on both their medial and lateral aspects. The medial aspect of a spinal nerve is marked by the confluence of the posterior and anterior roots. The nerve cell bodies of sensory spinal nerve fibers are located in the posterior root ganglion. Thus, the axons of these neurons enter the posterolateral sulcus via the posterior root. These axons either terminate synaptically in the posterior horn of the same segment or course through the white matter of one or more spinal cord segments to terminate synaptically in the posterior horn at a higher level of the cord. The nerve cell bodies of motor neurons of a spinal nerve are located in the anterior horn of the cord. Their axons exit at the anterolateral sulcus and enter the anterior root.

The lateral aspect of a spinal nerve bifurcates, or branches, into a small posterior primary ramus which supplies the muscles and skin on the posterior aspect of a body region and a large anterior primary ramus which innervates the muscles and skin in the remainder of the body region supplied by the nerve. Spinal nerve plexuses are formed by the coalescence of the anterior primary rami of numerous spinal nerves (see below).

The internal organization of a spinal nerve is shown in Fig. 14-16. The nerve fibers are grouped together in bundles called *fasciculi*. Each fasciculus is wrapped together by a connective tissue sheath referred to as *perineurium*. Septal partitions which project internally from the perineurium divide the fasciculus into numerous smaller *lobules*. The fasciculi which comprise a spinal nerve are held together by a connective tissue sheath known as *epineurium*. Arterioles and venules run within the substance of the epineurium. Each of the nerve fibers is in turn ensheathed by *endoneurium*. The internal organization of a spinal nerve is analogous to that of a skeletal muscle (see Chap. 24).

There are 31 pairs of spinal nerves, divided among the following regions of the vertebral column:

1   Cervical (8 pairs called C1 to C8)
2   Thoracic (12 pairs called T1 to T12)
3   Lumbar (5 pairs called L1 to L5)
4   Sacral (5 pairs called S1 to S5)
5   Coccygeal (1 unpaired nerve)

The relationship of the spinal nerves to the vertebral column is illustrated in Fig. 14-7. Spinal nerves C1 to C7 are named for the vertebrae in front of which they pass. C8 passes between the seventh cervical and the first thoracic vertebrae. The rest of the spinal nerves are named for the vertebrae in back of which they pass. The fibers of each spinal nerve, excluding C1 and the coccygeal nerve, enter and exit at the spinal cord through a lateral opening in the vertebral column. Called an *intervertebral foramen*, this opening is formed by the articulation between two adjacent vertebrae. The coccygeal nerve passes inferiorly to terminate on the coccyx, or tail bone.

### THE SPINAL PLEXUSES

The spinal nerves form groupings called plexuses in several locations outside the vertebral column. Their origins are associated with the cervical and lumbar enlargements of the spinal cord. Each spinal nerve plexus is formed from the anterior primary rami (branches) of numerous spinal nerves. The major spinal plexuses are identified in Fig. 14-7 and include:

1   The *cervical plexus,* formed from spinal nerves C1 to C4
2   The *brachial plexus,* formed from spinal nerves C5 to C8 and a portion of T1
3   The *lumbar plexus,* formed from a portion of spinal nerve T12, all of L1 to L3, and a portion of L4
4   The *sacral plexus,* formed from a portion of spinal nerve L4, all of L5 and S1, and a portion of S2 and S3
5   The *sacrococcygeal plexus,* formed from a portion of spinal nerves S2 and S3, all of S4 and S5, and the unpaired coccygeal nerve

Except where indicated above, the thoracic spinal nerves do not form a plexus.

Numerous peripheral nerve branches originate from each spinal plexus. These peripheral nerves are usually named for the blood vessel which they accompany. The peripheral nerves innervate the skeletal muscles and supply muscle spindles, Golgi tendon organs, and sensory receptors of the skin.

The spinal nerve plexuses, their peripheral nerve branches, and the common neuropathies which result from irritative or destructive lesions in these plexuses are considered in detail in Chap. 16.

## THE AUTONOMIC NERVOUS SYSTEM

The autonomic nervous system is essentially an automatically or subconsciously controlled grouping of motor neurons and ganglia which help to maintain homeostasis by effecting routine adjustments in the tone of smooth muscles and the secretory output of glands throughout the body. Each autonomic nerve consists of a chain of two neurons separated by a ganglion. The preganglionic neuron originates in the CNS and terminates synaptically with one or more cell bodies of postganglionic neurons located in autonomic ganglia outside the CNS. The axons of the postganglionic neurons innervate smooth muscle or glandular tissues in various visceral structures.

Based on location and function, there are two divisions of the autonomic nervous system: (1) the parasympathetic division originating in the craniosacral regions of the body and (2) the sympathetic division originating in the thoracolumbar regions of the body. Usually each visceral organ and gland is innervated by postganglionic neurons of both divisions. However, certain internal structures lack a parasympathetic innervation, as we will note later.

It is difficult to make a general statement about the specific effects of each division that will hold true for every visceral structure innervated. Both divisions exert their effects at the molecular level by influencing cellular metabolism. Basically, the function of the sympathetic division is to mobilize the metabolic resources of the body to meet stressful situations and to inhibit autonomic functions not directly related to these activities. The general adaptation syndrome (GAS), preparing an individual for "flight or fight" through the action of the pituitary-adrenal (adrenohypophyseal) axis (see Chap. 4), is a good example of the profound effects exerted by the sympathetic

division. The role of the parasympathetic division is diametrically opposed to that of the sympathetic division, restoring and conserving body resources and permitting "housekeeping" and perpetuative activities such as urination and penile erection.

When a visceral structure has a double innervation, i.e., receives the terminal branches of axons belonging to postganglionic neurons of both sympathetic and parasympathetic divisions, the response of the organ or gland represents an algebraic summation of the degree of stimulation provided by each division. Depending upon the visceral structure, each system may increase or decrease, accelerate or decelerate various aspects of its metabolism. Thus, the algebraically summated stimulation provided through double innervation to a visceral structure may result in some degree of excitation or inhibition, or in no change from the level of previous activity.

When a visceral structure has a single innervation, e.g., blood vessels which are innervated only by sympathetic autonomic fibers, the frequency of nerve impulse propagation via the autonomic pathway determines the visceral response. In the example of a blood vessel cited above, the more nerve impulses conducted per unit of time by the sympathetic fibers, the greater the degree of vasoconstriction. Thus, in the absence of sympathetic stimulation, the blood vessel would be maximally dilated. In other words, the lack of sympathetic stimulation is tantamount functionally to parasympathetic stimulation.

The preganglionic fibers of both the sympathetic and parasympathetic divisions of the autonomic nervous system originate in the CNS and are influenced by various regions of the brain. With special training using *operant conditioning* techniques and direct feedback on the status of visceral structures, it is possible to learn to voluntarily control the pupillary reflex, rate of heartbeat, degree of intestinal motility, and other ordinarily autonomic functions. In operant conditioning, reinforcement (e.g., a reward or the avoidance of pain) is offered for a desired aspect of ongoing behavior in order to increase the probability of its reappearance. Demonstrations that many, if not all, autonomic responses can be controlled voluntarily following the application of special learning techniques indicate that the autonomic nervous system is really not qualitatively different from other types of peripheral nerve. Gone forever is a deeply rooted belief that the autonomic nervous system is a "vegetative" system beyond voluntary control or immune to conscious regulation.

**The parasympathetic division** This division originates in the brain and sacral region of the spinal cord; hence, its association with the craniosacral portions of the body. More specifically, the preganglionic neurons originate in the nuclei of cranial nerves III (oculomotor), VII (facial), IX (glossopharyngeal), and X (vagus), as well as in the lateral horn of spinal cord segments S2 to S4. The axons of the preganglionic neurons are long and synapse with the cell bodies of postganglionic neurons located in terminal ganglia near or within the visceral structures innervated by the autonomic nerves. The axons of the postganglionic neurons are short and terminate in smooth muscle or glandular tissue of a visceral organ or gland. Because of the short postganglionic pathway and the relative lack of multiple synaptic connections in the terminal ganglia, the effects of parasympathetic stimulation are mostly local in nature.

Parasympathetic innervation by the vagus nerve (Xth cranial nerve) is extensive and includes the heart, trachea, bronchi, lungs, esophagus, stomach, liver, pancreas, small intestine, and large intestine. Visceral structures which lack a parasympathetic innervation include most blood vessels, adrenal medullae, the arrector pili muscles associated with the body hairs, and sweat glands.

*Acetylcholine* (ACh) is secreted by the axonal tips of both the preganglionic and postganglionic neurons belonging to the parasympathetic division. When ACh is secreted by the postsynaptic fibers of an autonomic nerve, the pathway is referred to as *cholinergic*. Thus, parasympathetic nerve pathways are cholinergic.

The structural and functional organization of a parasympathetic autonomic nerve is shown at the left in Fig. 14-17.

**The sympathetic division** This division originates in the thoracic and lumbar regions of the spinal cord, hence, its association with the thoracolumbar portions

of the body. More specifically, the preganglionic neurons originate in the lateral horn of spinal cord segments T1 to T12 and L1 to L3. Following their passage through the anterior root and white ramus, the short axons of the preganglionic neurons may terminate synaptically in one of three ways: (1) on the cell bodies of numerous postganglionic neurons located in vertebral (sympathetic) ganglia next to their points of origin, (2) on the cell bodies of numerous postganglionic neurons located in sympathetic ganglia at levels either above or below their points of origin, or (3) on the cell bodies of numerous postganglionic neurons located in prevertebral (collateral) ganglia some distance removed from the vertebral ganglia through which the fibers pass without synapsing. Thus, sympathetic preganglionic fibers form multiple synapses in adjacent or distant vertebral ganglia or in prevertebral ganglia which form nerve plexuses with multiple synaptic connections. The prevertebral ganglia are associated with the aorta in the regions of the celiac, superior mesenteric, and inferior mesenteric arteries. Each of these ganglia is named for the blood vessel with which it is associated. These ganglia give rise to large nerve plexuses which innervate numerous structures in the abdominal and pelvic regions of the body.

The vertebral, or sympathetic, ganglia are located against the posterior thoracic and abdominal walls, at either side of the vertebral column. Resembling beads on a string, each sympathetic ganglionic chain consists of 2 cervical, 10 thoracic, and 3 or 4 lumbar ganglia, all of which are connected by nerve fibers. Several sympathetic ganglia are illustrated in Fig. 14-16.

The postganglionic neurons of the sympathetic division therefore arise in either vertebral or prevertebral ganglia. The relatively long axons of the postganglionic neurons terminate in smooth muscle or glandular tissue in a visceral organ or gland. Because of the multiple synaptic connections made by the axons of the preganglionic neurons and the extensive nerve plexuses formed from the prevertebral ganglia, widespread and profound effects may result from minimal stimulation of one sympathetic nerve pathway.

While ACh is secreted by the axonal tips of the

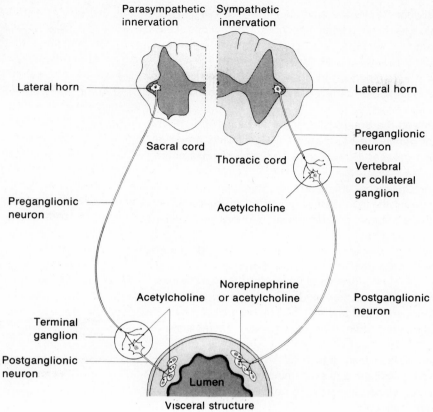

Parasympathetic innervation

Sympathetic innervation

Lateral horn

Lateral horn

Sacral cord

Preganglionic neuron

Thoracic cord

Vertebral or collateral ganglion

Preganglionic neuron

Acetylcholine

Acetylcholine

Norepinephrine or acetylcholine

Postganglionic neuron

Terminal ganglion

Postganglionic neuron

Lumen

Visceral structure

**Figure 14-17** *A comparison of the visceral innervation by the parasympathetic and sympathetic divisions of the autonomic nervous system. (Left) Parasympathetic innervation. (Right) Sympathetic innervation.*

preganglionic neurons and may also be secreted by some postsynaptic fibers such as those which innervate sweat glands and intramuscular blood vessels, many postganglionic axons of the sympathetic division secrete a transmitter substance called *norepinephrine*. This chemical is identical to one of the catecholamines secreted by the medullae of the adrenal glands (see Chap. 4). Sympathetic pathways in which norepinephrine is secreted by the postsynaptic fibers are called *adrenergic.*

The organization of a sympathetic autonomic nerve is shown at the right in Fig. 14-17.

## COLLATERAL READING

Benzinger, T. H., "The Human Thermostat," *Scientific American,* January 1961. Benzinger concludes that the hypothalamus measures as well as regulates body temperature and is therefore a major sensory organ of the body.

Fisher, Alan E., "Chemical Stimulation of the Brain," *Scientific American,* June 1964. Experiments are described which demonstrate that thirst is elicited in the rat by the introduction of acetylcholine into various portions of the limbic system of the brain.

French, J. D., "The Reticular Formation," *Scientific American,* May 1957. A general review of experiments which have led to our present concepts of the actions of the RAS (reticular activating system).

Gazzanega, Michael S., "One Brain–Two Minds?" *American Scientist,* 360:311–318, 1972. A review of experiments and observations which indicate that accidental or surgical sectioning of the corpus callosum gives rise to two independent "brains" in the same individual.

Geschwind, Norman, "Language and the Brain," *Scientific American,* April 1972. Through a historical review of human cases of aphasia, Geschwind provides some insights into the relations between Broca's area and other regions of the cerebral cortex.

Gordon, Barbara, "The Superior Colliculus of the Brain," *Scientific American,* December 1972. Gordon examines the role of this region of the midbrain in receiving and processing sensory information from visual, auditory, and tactile inputs.

Jouvet, Michel, "The States of Sleep," *Scientific American,* February 1967. From experiments with cats, Jouvet concludes that (1) light sleep is caused by the secretion of serotonin from the central portion of the pons; (2) deep sleep is caused by the secretion of norepinephrine from laterally located nuclei in the pons; (3) both chemicals in their turn inhibit the RAS; and (4) arousal of the RAS from a state of sleep is accomplished when these accumulated chemicals are degraded enzymatically.

Kimura, Doreen, "The Asymmetry of the Human Brain," *Scientific American,* March 1973. Although the left cerebral hemisphere plays a dominant role in speech, the opposite hemisphere exhibits dominance in perception of the environment, according to the results of Kimura's research.

Luria, A. R., "The Functional Organization of the Brain," *Scientific American,* March 1970. From an examination of human patients with various types of brain injury, Luria concludes that (1) similar behavioral processes may not be integrated in the same region of the cerebral cortex; (2) apparently unrelated behavioral processes may be integrated in the same region of the cortex when they share in common the same perceptual ability, e.g., spatial visualization; and (3) habituation changes the organization of neural control so that cortical damage to motor areas may not disturb a particular motor response when elicited without thought or analysis.

Noback, Charles R., and Robert J. Demarest, *The Nervous System: Introduction and Review,* New York: McGraw-Hill Book Company, 1972. This excellent paperback provides a concise and comprehensive review of the central nervous system.

Olds, James, "Pleasure Centers in the Brain," *Scientific American,* October 1956. Olds reviews his early work on the mapping of hypothalamic areas involved in emotions.

Pribram, Karl H., "The Neurophysiology of Remembering," *Scientific American,* January 1969. From extensive experimentation on monkeys, Pribram concludes that the association cortex plays a major role in organizing memory built on diverse encoded and recoded sensory data.

Rosenzweig, Mark R., Edward L. Bennett, and Marian Cleeves Diamond, "Brain Changes in Response to Experience," *Scientific American,* February 1972. The authors conclude that an enriched environment causes an increase in (1) the number of synaptic junctions in the dendrites of pyramidal cells in the motor cortex and (2) the size of the postsynaptic membrane at synaptic junctions.

Snider, Ray S., "The Cerebellum," *Scientific American,* August 1958. A discussion of the structure and functions of the cerebellum.

# CHAPTER 15
# CENTRAL NERVOUS SYSTEM DISORDERS

Disorders of the central nervous system (CNS) are caused by structural or functional abnormalities of the brain, the spinal cord, and the nuclei or roots of the peripheral nerves. The disorders are due to one of the following factors:

1 Physical trauma, including birth injuries
2 Microbial pathogens which damage neural tissues
3 Degeneration of neural tissues due to aging, hereditary defects, or other factors
4 Metabolic abnormalities, including severe nutritional deficiencies
5 Toxicity due to the effects of drugs or other chemicals
6 Vascular diseases which cause brain hypoxia
7 Tumors which cause increased intracranial pressures and subsequent necrosis of nerve tissues
8 Congenital defects such as hereditary diseases and developmental anomalies which involve the brain and spinal cord
9 Psychological trauma or other factors which give rise to mental diseases

CNS disorders are either *organic* or *functional*. When physical, physiologic, or biochemical damage of the CNS is demonstrable, the disease is organic. Mental diseases affect the brain, especially the cerebrum, and cause disturbances of intellect and personality. Either organic or functional in cause, mental diseases are divided into mild personality disturbances called *psychoneuroses* and more serious behavioral aberrations called *psychoses*. Contact with reality is temporarily or permanently lost in a psychosis but not in a neurosis. *Psychosomatic diseases* originate within the CNS, although their pathologic effects may be expressed elsewhere in the body.

This chapter reviews the causes and symptoms of the more common and well-known CNS disorders.

## BRAIN DISORDERS

Disorders of the brain may involve any portion of the neural or nonneural tissues enclosed within the cranium. Because the brain is a highly specialized mass of neural tissue subserving integrative, regulatory, and conceptual functions, brain lesions cause a disintegration of some combination of sensory, motor, intellectual, and personality processes.

## GENERAL PATHOLOGIC CONDITIONS

Our knowledge of the effects of brain lesions is derived from observations on the effects of specific natural brain injuries in human patients; behavioral and physiologic observations following neurosurgery which requires incision in, or removal of, specific parts of the brain in human patients; and animal experimentation in which alterations of behavior and physiology are observed following electrical and chemical stimulation or destruction of specific portions of the brain.

The cerebrum   Lesions of the cerebrum may involve gross portions of one or both hemispheres or may be restricted to specific regions (see Chap. 14) of sensory, motor, or association cortex, or to a functional cerebral system in which several regions overlap.

Sensory areas of the cerebrum include the somesthetic cortex (areas 1, 2, and 3), visual cortex (area 17), auditory area (area 41), and olfactory and gustatory areas.

The somesthetic cortex precisely identifies those portions of the body from which sensations originate. Lesions of the somesthetic cortex may cause peculiar somatic sensations such as numbness on the side of the body opposite that in which the lesion is located (contralateral paresthesia) or can lead to a depression in the sensitivity to pain (hypalgesia) or to all sensations (hypesthesia).

The visual cortex is involved in the integration of

sensory information originating from the photoreceptors of the retina. Lesions of the visual cortex may cause visual hallucinations or visual field defects on the side of the body opposite that in which the lesion is located.

The auditory area integrates the encoded sensory information transduced by the phonoreceptive hair cells. Lesions of the auditory area cause some degree of hearing impairment and may give rise to tinnitus (see Chap. 12).

The olfactory area is concerned with the integration of sensory information originating from olfactory, or distance, chemoreceptors. Lesions of the olfactory area may cause olfactory hallucinations, such as are sometimes experienced just before an epileptic seizure, or can completely block the ability to smell (anosmia; see Chap. 16)

The gustatory, or taste, area integrates sensory information originating from gustatory chemoreceptors. Lesions of the gustatory area usually cause a depression in the ability to taste (hypogeusia; see Chap. 16), although damage restricted solely to the gustatory area is rarely seen.

Motor areas of the cerebrum include the motor cortex (area 4) and the premotor area (area 6).

The motor cortex exerts contralateral control over the skeletal muscles of the body. Lesions of this region can cause muscular convulsions and related symptoms or can give rise to a flaccid paralysis.

The premotor cortex regulates the activity of the motor cortex and lower brain centers, processing and storing information for the execution of patterned activities involving many muscles. Lesions of this region cause the disintegration of patterned movements in specific parts of the body. Since experimental removal of the premotor cortex in monkeys induces forced grasping, this region may also exert inhibitory effects on certain patterned movements controlled elsewhere in the brain.

Association areas are found in all portions of the cerebrum. Such regions include the prefrontal portions of the frontal lobes as well as the auditory association (area 42), visual association (areas 18 and 19), and somesthetic association areas (areas 5 and 7).

The association areas compare present sensory

sensations with memories of previous experiences and initiate the most appropriate responses to them. These regions are thought to be the seats of intellectual processes including concept formation, higher-order learning and recall abilities, and personality modification.

*Mental deterioration* is most apparent in damage to the frontal lobes. Lesions of areas 9, 10, and 11 reduce the ability to concentrate on ongoing tasks. When the dominant motor speech area (area 44) is damaged (left side in right-handed subjects), the patient may not be able to speak in a coordinated fashion, even though the muscles of speech are functional. This condition is called *motor aphasia*. On the other hand, lesions of area 39 of the left temporal lobe in right-handed subjects cause *sensory aphasia*. In this abnormality, auditory sensations including spoken words are detected, transduced, conducted, and transmitted to the brain normally, although the brain is incapable of taking appropriate action.

Parietal lobe damage is usually reflected by a deterioration in the sense of touch and the integration of words which are read. Because cutaneous sensations are not integrated, lesions of areas 5 and 7 of the parietal lobe eliminate the ability to recognize objects by touch alone. This condition is known as *astereognosis*. In addition, damage to areas 39 and 40 of the same lobe depresses the ability to comprehend written words, even though the eyes and optic nerves are functioning normally. This abnormality is referred to as *word blindness*.

Damage to the rhinencephalon may result in *hyperactivity* and *intellectual deficits*. Restlessness may be a consequence of the release of the inhibitory influence of the intact rhinencephalon on the hypothalamus. Memory may also deteriorate, especially the recall of recent events.

**The diencephalon** The diencephalon is a major switchboard for the relay between the cerebrum and the midbrain of both sensory and motor information. The diencephalon also controls numerous visceral functions. The thalamic portion of the diencephalon participates in the integration of a number of sensations including touch, pain, and temperature. The sensory modality is identified but not the region of the body being stimulated. The hypothalamic part of the diencephalon has centers for the regulation of numerous activities, including body temperature; food intake and water balance; synthesis of hormones stored in the posterior hypophysis and control of the release of hormones synthesized in the anterior hypophysis; regulation of sleep, arousal, and emotional states; and general control of both sympathetic and parasympathetic divisions of the autonomic nervous system.

Lesions of the thalamus result in a partial or complete loss of touch, pain, and temperature sensations from that side of the body mediated by the affected region. This condition is called *hemianesthesia*. Partial losses usually alter the quality of incoming somatic sensations, so that they are perceived as obnoxious or painful. Thalamic damage may also induce changes in muscle tone on the opposite side of the body, because the thalamus participates in both the motor control and sensory feedback regulation of skeletal muscle tone.

Hypothalamic lesions cause some combination of the following effects:

1 *Hypothermia*, when the posterior hypothalamus is differentially damaged and *hyperthermia,* when the anterior hypothalamus is similarly affected
2 *Obesity* and related disturbances in the metabolism of lipids
3 *Diabetes insipidus*, due to an inability to synthesize antidiuretic hormone (ADH) following destruction of the supraoptic nuclei or the tracts connecting these structures with the posterior lobe of the hypophysis (see Chap. 17)
4 *Reproductive and other hormonal abnormalities,* due to the failure to secrete hypothalamic releasing and inhibiting factors (see Chap. 17)
5 A sleepy or drowsy condition called *somnolence* which lasts for a prolonged period of time
6 Profound *alterations of emotional states,* ranging from sexual pleasure to violent aggression

**The midbrain** The midbrain connects the cerebellum and the pons with the diencephalon and the cerebral hemispheres. The midbrain participates in sensory and motor activities by means of the fiber tracts which course through it. In addition, cranial nerve nuclei located in the midbrain carry out directives for the motor coordination of the extrinsic eyeball muscles controlled by the oculomotor and trochlear cranial nerves. Finally, this portion of the brain is assigned numerous integrative tasks related to the tone of skeletal muscles.

Lesions in several parts of the midbrain can cause *muscular disabilities of the eyeballs*. Destruction of the nuclei of the IIId and IVth cranial nerves causes unilateral or bilateral paralysis of the superior, inferior, and medial rectus muscles and the inferior oblique muscle (oculomotor nerve), as well as of the superior oblique muscle (trochlear nerve). In addition, damage to the superior colliculi of the corpora quadrigemina prevents the eyes from moving upward.

*Skeletal muscle movements* are also affected in various ways by midbrain lesions. Involuntary movements such as tremors (parkinsonism), erratic muscular twitching (chorea), and slow, undulating movements (athetosis) or muscular rigidity occur upon the destruction of the reticular formation, the substantia nigra, or the red nucleus. Damage to the inferolateral portion of the midbrain (cerebral peduncles) results in tonic spasms on the side of the body opposite that of the lesion (spastic hemiplegia). Lesions in other parts of the midbrain elicit slow, involuntary, or otherwise abnormal movements, especially in the upper and lower extremities.

**The cerebellum**   The cerebellum consists of three functional regions—archicerebellum, paleocerebellum, and neocerebellum. These portions of the cerebellum are identified in Fig. 15-1. The *archicerebellum* helps to maintain equilibrium by integrating proprioceptive, visual, and vestibular feedback information regarding spatial orientation. The *paleocerebellum* contributes to the control of the postural muscles by causing tonic changes through stretch-reflex mechanisms. The *neocerebellum* coordinates the activities of the skeletal muscles, controlling the execution of motor tasks by damping voluntary movements at precisely the right time and by calculating the position in space of various parts of the body ahead of time. These regulatory functions ensure the successful completion of motor tasks which have already been initiated.

Lesions of the archicerebellum result in an incoordination of the skeletal muscles called *ataxia*. The subject staggers about, much like a drunkard. Muscular incoordination is especially prevalent in the thoracic region of the body.

Damage to the paleocerebellum usually causes an abnormal increase in the stretch reflexes of postural muscles. Such damage results in various *postural disabilities*.

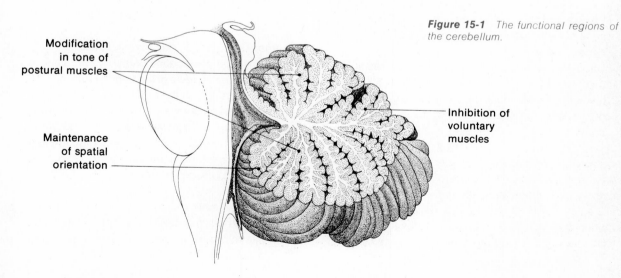

**Figure 15-1**   *The functional regions of the cerebellum.*

Modification
in tone of
postural muscles

Maintenance
of spatial
orientation

Inhibition of
voluntary
muscles

Destruction of the neocerebellum leads to a deterioration in the ability to perform motor tasks requiring measured responses or intense concentration. Motor incoordination takes one of the following forms: (1) an inability to determine the right amount of force required to execute a particular movement, resulting in an overshoot or awkwardness (*dysmetria*); (2) rapid, perceptible, and involuntary movements of the muscles of the fingers only when consciously concentrating on the performance of a task requiring fine finger dexterity (*intention tremors*); and (3) an inability to perform gross or fine movements with precision because of a loss of damping and predictive functions associated with the neocerebellum (*ataxia*).

**The pons**   The pons connects the cerebellum with the midbrain and higher levels superiorly and with the medulla inferiorly. The pons contains the nuclei of origin of the Vth to VIIIth cranial nerves and the pneumotaxic center, which modifies the rate of respiration but not its basic pattern.

Pontine lesions may give rise to some combination of the following abnormalities:

1   Paralysis of the muscles of chewing and mastication and anesthesia in the facial region due to damage to the nucleus of the trigeminal nerve
2   Partial or complete paralysis of the lateral rectus muscle of the eyeball because of destruction of the nucleus of the abducens nerve
3   Paralysis of the muscles of the scalp, face, and neck through lesions of the nucleus of the facial nerve
4   Deafness, tinnitus, and vertigo as the result of damage to the acoustic (auditory) nerve nucleus
5   Involuntary alterations in the rate of respiration due to damage which involves the pneumotaxic center

**The medulla oblongata**   The medulla connects the spinal cord with all other parts of the brain. All ascending and descending tracts must pass through this region. Many of these tracts decussate here. In addition, the medulla houses the nuclei of the last four cranial nerves. Furthermore, the region contains several centers of vital importance, including centers for heart rate, vasoconstriction and vasodilatation, and respiration. Finally, the medulla possesses reflex centers for swallowing and vomiting, in addition to sharing the regulation of postural reflexes with other brain regions.

Damage to the medulla oblongata may result in one of the following consequences:

1   Immediate death because of the involvement of either the heart rate center or the respiratory center
2   Sensory and motor impairments mainly in the pharyngeal, laryngeal, and cervical regions due to lesions of the nuclei of the IXth to the XIIth cranial nerves
3   Extreme *hypotension*, or low blood pressure, due to involvement of the vasoconstrictor center, or *hypertension* (high blood pressure) because of damage to the vasodilator center
4   Spastic contractions of the extensor muscles of the arms and legs due to medullary lesions which release the cerebral cortex from inhibitory influences (*decerebrate rigidity*)
5   Lack of control of the antigravity muscles of the legs and loss of the neck-righting reflex

The latter reflex maintains a normal alignment between the head and the rest of the body.

**Functional brain systems**   The functional systems of the brain incorporate more than one anatomic region.

The *limbic system* includes the hypothalamus and other portions of the diencephalon and cerebrum, especially the inferior aspects of the frontal lobes. These structures are involved in the regulation of behavioral states. Among the behaviors controlled by the limbic system are motivation, sexual receptivity, and emotional state. The limbic cortex is an association area for the integration of behavioral reponses, in conjunction with other portions of the cerebrum and the hypothalamus.

Lesions of the limbic system cause restlessness or a release from social inhibitions. The first of these behavioral alterations, *hyperkinesis*, results from involvement of the inferoposterior portions of the frontal lobes. A *lack of social consciousness* follows involvement of the cingulate gyrus, located above the corpus callosum.

Although its validity has been questioned, the *extrapyramidal system* is considered to consist of all the efferent pathways, excluding the pyramidal tracts, leading from the motor cortex to various segments of the spinal cord. After integration in both the forebrain and midbrain, the motor information from the extrapyramidal system is carried primarily through the reticulospinal tracts in the lateral and anterior gray horns of the spinal cord. The extrapyramidal system is concerned with posture, movement, and autonomic functions.

Lesions of the extrapyramidal system result in one of the following abnormalities: (1) *parkinsonism* due to involvement of one of the basal ganglia called the globus pallidus of the cerebrum or the substantia nigra and reticular substance of the midbrain; (2) *athetosis* due to involvement of the globus pallidus or the lentiform nucleus (caudate nucleus and the putamen) of the corpus striatum; (3) *chorea* due to the involvement of the corpus striatum; and (4) *contralateral spastic paralysis* due to involvement of the internal capsule of the cerebrum.

The *reticular activating system* (RAS) extends from the brainstem through the midbrain and thalamic and hypothalamic portions of the diencephalon, making synaptic connections with all parts of the cerebral cortex. The RAS controls the state and quality of consciousness, including the degree of wakefulness or sleepiness and the ability to think coherently. That part of the system associated with the brainstem is thought to be responsible for the maintenance of normal wakefulness, although the activity of the motor cortex can also sustain this function.

Lesions of the RAS can result in *anesthesia* or *coma*. The pathologic hyperexcitability of neurons within the brainstem is responsible for several types of *epilepsy*. Epilepsy is discussed in detail later in this chapter.

The brain ventricles are part of a larger system which nourishes the brain and the spinal cord with cerebrospinal fluid (CSF). The hydraulic effects of CSF also protect the CNS against mechanical shock.

Because CSF circulates in a closed system, any increase in its volume is usually reflected by an increase in CSF pressure. Increase in the volume of CSF is normally brought about in one of two ways: (1) by an increase in the pressure or volume of blood within the choroid plexuses and (2) by a decrease in the venous drainage of CSF from the head.

Pathologic conditions arise when excessive intracranial pressures build up because of an accumulation of CSF. One of these conditions, *hydrocephaly*, is usually caused by an obstruction of the aqueduct of Sylvius or the foramina of Luschka and Magendie. This disease is discussed later in the chapter.

## SPECIFIC BRAIN DISORDERS

**Cortical damage**    Specific syndromes caused by cortical damage include mental retardation and cerebral palsy.

*Mental retardation*    It is probably easier to say what mental retardation, or mental deficiency, is not than to say what it is. It is not a mental disease. Mental diseases are psychoneuroses and psychoses. Mental retardation is actually a diverse grouping of disorders which have in common the fact that intellectual functioning is significantly lower than normal.

Intelligence quotients (IQs) are used to define the quantitative levels of intellect. There are three levels of mental deficiency, compared with the normal IQ range of 90 to 109: (1) *morons*, with an IQ between 50 and 70, equivalent in intellectual capacity to normal children between 7 and 12 years of age; (2) *imbeciles*, with an IQ between 25 and 50, equivalent in intellectual capacity to normal children between 3 and 7 years of age; and (3) *idiots*, with an IQ between 0 and 25, equivalent in intellectual capacity to normal children up to 3 years of age. Of these three levels, only morons have no demonstrable brain damage.

The validity of the various types of IQ test has been severely criticized in recent years. Such tests are more often than not culturally bound and emphasize verbal and intellectual skills rather than social interactions and capabilities within the ethnic or racial group to which the individual belongs. Test scores and interpretations based on them should therefore not be made in a vacuum, without the benefit of other diagnostic aids.

The symptoms of mental retardation usually start at an early age. Excluding morons, the degree of disability is ordinarily determined by the amount and location of damage to the cerebrum. Nevertheless, developing social maladjustments may increase the apparent severity of the disorder. Mentally retarded subjects are unable to conceptualize, in addition to being easily distracted, restless, and erratic. Such individuals are therefore compulsive and unpredictable in behavior and unable to concentrate on one particular task for a reasonable length of time. In short, the profile presented is one of continual appetitive behavior with little rational basis.

Besides exhibiting some combination of these behavioral aberrations, subjects suffering from mental retardation frequently show physical deformities and personality disturbances. Malformations are indicative of congenital defects. Mental disease, as opposed to mental retardation, develops in parallel with the degree of social maladjustment.

There are many causes of mental retardation. High on the list are congenital defects such as hereditary diseases and developmental anomalies as well as prematurity and birth injuries. Prenatal infections are a frequent cause of neural anomalies. Hypoxia and subdural hematoma are often associated with mental retardation resulting from prematurity and birth injuries, respectively. Hypoxia develops because of the small size of the lungs in a premature baby. Obstetric manipulations may cause subdural hematoma during a difficult delivery. Encephalitis is probably the most common cause of postnatal mental retardation. However, other causes of mental deficiency include protein and thyroid hormone deficiencies during early childhood, conditions which result in kwashiorkor (see Chap. 27) and cretinism (see Chap. 18), respectively, or cerebral hemorrhage, a condition which may precipitate a cerebrovascular accident (see Chap. 31).

The treatment of subjects with mental retardation is determined by its cause and severity. Mental retardation induced by certain types of metabolic dysfunctions may be corrected, at least partially, by the addition of protein or thyroid hormone. Most forms of encephalitis are amenable to antimicrobial therapy. Much of the treatment of a mental retardate, however, is designed to assist this form of brain-damaged patient in becoming independent in the activities of daily living and gainful employment through institutionalized care or special home programs.

*Cerebral palsy* While mental retardation may be one result of cerebral palsy, these two categories of cerebral disorder are mutually exclusive. As in the case of mental retardation, however, cerebral palsy is a conglomeration of disorders. All forms of cerebral palsy have in common the fact that there is some type of muscular paralysis or dysfunction early in childhood. In addition, there is usually some form of sensory involvement, and these deficits may ultimately overshadow the motor disabilities. Indeed, the term *palsy* is derived from this combination of sensory and motor symptoms.

The symptoms of cerebral palsy are therefore a reflection of sensory, motor, and integrative dysfunctions. Spastic paralysis on one side of the body (*hemiplegia*) is the most frequent motor symptom. Usually the upper extremity is more severely involved than the lower one. Sometimes athetosis or ataxia is observed, instead of spastic paralysis. Among the common sensory deficits is a loss of proprioceptive abilities. This sensory defect is usually coupled with ataxia. Visual abnormalities occur when the occipital lobe is damaged.

The causes of cerebral palsy are similar to those of mental retardation. Most cases of cerebral palsy are caused by birth injuries and prematurity. Next in frequency are physical trauma and microbial infection during postnatal development.

Midbrain, diencephalic, and deep cerebral damage  Among the disorders caused by damage to these regions, especially the substantia nigra and the basal ganglia, are parkinsonism, chorea, and athetosis.

*Parkinsonism (paralysis agitans)* This degenerative disease causes muscle tremors, progressive deterioration of muscle functions, and unique pathologic patterns of muscle movement, without impairing intellectual abilities. Exhibited most frequently by persons in their fifth or sixth decade of life, the disorder is believed to be due to a dysfunction in the metabolism of brain amines.

Brain damage in patients with parkinsonism is restricted to circumscribed portions of the midbrain and forebrain. Among the structures which are found to differentially degenerate are the substantia nigra of the midbrain and the globus pallidus, a component of the basal ganglia.

The symptoms of parkinsonism are predominantly motor. However, paresthesias and personality disturbances may develop as the disease runs its course. The characteristic motor abnormalities include:

1  Paroxysmal (periodic) or continuous tremors of the wrists, hands, fingers, and head which become progressively worse with time and may ultimately involve the entire body
2  So-called pill-rolling tremors caused by oscillatory movements of the thumb and index finger over each other
3  A partial flexing and rigidity of the arms, knees, and thumbs, especially when standing
4  A shuffling gait which may spontaneously become faster during movement, with the arms stiffly at the sides and the head and thorax bent forward slightly
5  A progressive loss of strength, so that movements and speech become more feeble with time
6  A blank facial expression
7  A progressive muscular incoordination, so that it is difficult or impossible to execute motor tasks requiring the participation of numerous groups of muscles

Not all these symptoms necessarily appear, either simultaneously or at all, in any one patient. In other words, the syndrome is variable and depends on numerous physiologic and psychological variables such as degree of fatigue and emotional instability.

The treatment of parkinsonism is directed toward symptomatic relief, partial restoration and maintenance of function, and psychological support. The amelioration of symptoms has been approached (1) with the use of drugs such as sedatives to reduce awareness, antihistamines to combat the effects of histamines, antispasmodic drugs to reduce tremors, and L-dopa (dihydroxyphenylalanine) to replace the lost dopamine ordinarily secreted by axons originating in the substantia nigra and required for neurohumoral stimulation of the basal ganglia; and (2) by surgically destroying the globus pallidus or the subthalamus in an attempt to reduce rigidity and tremor. Muscular spasticity can also be minimized through a well-planned program of active and passive exercises and massage of the affected body regions. Psychological support is also important, because a depressed emotional state can exacerbate the symptoms of parkinsonism.

*Chorea (Sydenham's chorea)* In contrast to the tremors and rigidity of parkinsonism, the predominant motor signs of Sydenham's chorea consist of constant, rapid, erratic, or arrhythmic and often grotesque movements of the face, thorax, arms, and legs following a period of extreme restlessness. Completely involuntary in nature, these contractions and relaxations are superimposed on normal motor patterns, causing almost complete motor incoordination. The loss of control is especially dangerous to afflicted subjects because of their heightened vulnerability to injury during this time. The physical exertion suffered during a bout of the disease leads to severe muscle fatigue and hypotonia (lack of muscle tone). The result is a vicious circle which, if left untended, may lead to physical collapse. This type of chorea is an affliction of children and adolescents.

The cause of Sydenham's chorea is not clear. Often, symptoms of the disease are related to the presence of an active case of rheumatic fever. Apparently, the corpus striatum is somehow affected. The disease runs its course in one or two months, although relapses have been reported.

The treatment of chorea emphasizes prophylaxis against injury, relief of symptoms, and physical and mental rest. Protection from physical harm through bed rest and sedation are standard treatments during severe bouts. Drug therapy may attempt to increase physiologic resistance with the use of ACTH or cortisol. Rest also has definite therapeutic value.

*Athetosis* Athetosis is characterized by slow, repeated, involuntary, and arrhythmic movements primarily of the digits and extremities. The effect is that of writhing motions or undulations in these portions of the body. During a bout of athetosis the muscles are hypertonic and often spastic, and voluntary movements are difficult to perform. The affliction is most frequently seen in children. Brain damage in patients who exhibit athetosis is generally localized in the globus pallidus or in this structure plus other portions of the corpus striatum complex.

The physiologic disruptions which cause parkinsonian, choreiform, and athetoid movements are incompletely understood. However, we do know that damage to the globus pallidus prevents the execution of motor tasks requiring a high degree of coordination in the distal portions of the extremities, apparently through an inability to provide the proper degree of muscle tone. We also know that destruction of other portions of the corpus striatum results in a loss of the ability to subconsciously regulate gross movements of the voluntary musculature. Normally, since the globus pallidus and the rest of the corpus striatum complex act in part through a feedback loop to the motor area of the cerebral cortex, loss of this feedback regulatory mechanism may prevent accurate voluntary control of muscular activity. In addition, lesions of the basal ganglia release the inhibitory control over muscle contractions exerted by the reticular area of the brainstem and thereby induce hyperactivity and spasticity, or rigidity.

Cerebellar damage and motor incoordination    The cerebellum acts as a feedback mechanism for the maintenance, control, and coordination of automatic movements of skeletal muscles. Destruction of the cerebellum leads to an inability to control precisely the position of the body in space and to predict accurately the counterforces required to modulate ongoing muscle contractions. The result is a series of movements which are jerky and poorly executed. Locomotor movements become awkward and ungainly. Walking is accomplished stiffly, with the legs spread apart. Each leg moves erratically, overshooting its mark to strike the ground or floor with more force than anticipated. Often accompanied by vertigo and nystagmus (see Chap. 12), the gait is reeling or staggering. As noted earlier, this form of muscular incoordination is referred to as ataxia.

Cerebellar damage may also result in other types of motor incoordination. These abnormalities may take one of the following forms: (1) *decomposition movements*, in which the entire sequence of the motor pattern is chopped up into jerky segments; (2) *intention tremors*, in which concentration at a motor task causes alternations of fine contractions and relaxations of the muscles being used whereas automatic motor tasks lack such tremors; and (3) *adiadochokinesis*, in which repetitive movements controlled by antagonistic muscles, such as continued alternations between pronation and supination, cannot be performed without profound muscular incoordination.

Damage to functional systems and structures    Several disorders involve more than one anatomic region of the brain or are defined more by the nature of the functional loss than the extent of the structural damage. Among such abnormalities are decerebrate rigidity, epilepsy, and hydrocephaly.

*Decerebrate rigidity* Decerebrate rigidity arises when the cerebral cortex and the basal ganglia are structurally and functionally separated from the inhibitory portion of the reticular area of the brainstem. Such a separation most frequently occurs in destructive lesions between the superior colliculus of the midbrain and the vestibular nuclei of the pons, although it may also occur in lesions of higher brain regions. This form of injury removes the inhibitory influence on postural reflexes of the inhibitory center in the reticular area of the brainstem from the facilitatory effects of the cerebrum. Without this brake on the normal degree of muscle tone to be maintained, the postural muscles, especially the extensors of the upper and lower extremities, are thrown into tonic spasms. The response is mediated primarily by motor neurons associated with the muscle spindles, rather than the neurons which innervate the muscle fibers directly. Therefore, decerebrate rigidity results from an overexaggeration of the stretch reflex.

**Epilepsy**    Epilepsy is characterized by a temporary loss of consciousness as well as by localized seizures or generalized convulsions. The neural activity and functions of the brain are altered during attacks. These modifications are reflected in sensory, motor, and psychogenic abnormalities.

Basically, four types of epilepsy are recognized: (1) *grand mal* epilepsy, the so-called major form, characterized by hallucinatory auras, an extended loss of consciousness of several minutes' duration, tonic rigidity during seizures, clonic spasms during recovery, and muscle pains and stupor following an attack; (2) *petit mal* epilepsy, the so-called minor form, characterized by fleeting, unrecognized losses of consciousness which superficially resemble daydreaming, a depression of both muscle tone and motor activity during an attack, and resumption of normal functions following an attack; (3) *Jacksonian focal* epilepsy, characterized by convulsions restricted to a particular group of muscles or side of the body; and (4) *psychomotor seizures*, characterized by a loss of recent memory and behavioral aberrations which mask, or appear instead of, classic convulsions. Psychomotor seizures are particularly insidious, since criminal or antisocial acts may be perpetrated during such attacks.

Characteristic changes in brain waves, seen on electroencephalograms (EEGs), occur during attacks of epilepsy. Electroencephalographic changes are demonstrated by comparing brain wave frequencies (waves per second) and amplitudes (height of individual waves) as well as waveform patterns (shape of individual waves) with normal EEGs from the same region of the cerebral cortex. Ordinarily, EEGs are recorded simultaneously and displayed separately from many predetermined regions of the cerebrum. However, recordings from specific locations may be singled out for special study. Two such comparisons

are illustrated in Fig. 15-2. Normal brain waves are compared with those obtained during petit mal epilepsy (upper) and grand mal epilepsy (lower). Recorded from the left temporal–left occipital regions of the cerebrum, the electroencephalogram of the subject with petit mal epilepsy shows a typical repeated *doublet* of "spike domes," or "spike waves," with an amplitude of about $250\mu V$ (microvolts) and a frequency of approximately 3 doublets per second. The EEGs of the subject exhibiting grand mal epilepsy were recorded from the right temporal–right occipital regions of the cerebrum and consist of highly irregular waveforms of variable frequency and amplitude. The EEGs of patients experiencing other forms of epilepsy are also unique and therefore diagnostic.

**Figure 15-2**    *A comparison of EEGs between normal subjects and subjects with petit mal and grand mal epilepsy.*

Recorded from left temporal–
left occipital lobes

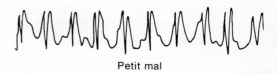

Normal

Petit mal

Recorded from right temporal–
right occipital lobes

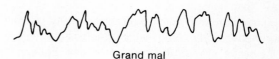

Normal

Grand mal

Epilepsy is caused by a variety of factors. Basically, these factors are traceable to organic brain damage or are not easily traceable, i.e., are idiopathic. Apparently, there is a marked hereditary predisposition toward epilepsy. Any brain injury in predisposed individuals may initiate the disease. The cause of epilepsy is also related to the age at which the disease is first exhibited. Epilepsy which appears during infancy or childhood is usually caused by hereditary disorders, developmental anomalies, birth injuries, postnatal physical trauma, microbial infections of the CNS (central nervous system), or high fever. During adolescence and the adult stages of development, the causes of epilepsy change, with physical trauma, neoplasms, and cranial arteriosclerosis the primary precipitating factors.

Whether caused by organic brain damage or idiopathic factors, the localization of altered brain function is ascribed to either the RAS (reticular activating system) of the brainstem, the thalamic portion of the diencephalon, or sensory, motor, or association portions of the cerebral cortex. During an attack of either grand mal or petit mal epilepsy, there is a profound increase in the excitability of neurons in the RAS of the brainstem. This hyperexcitability spreads in both directions from the point of origin, activating the cerebral cortex as well as skeletal and smooth muscles and giving rise to characteristic sensory and motor symptoms. Diencephalic involvement leads to autonomic responses including involuntary urination and defecation. The perception of visual, auditory, odorous, and other types of auras preceding attacks is indicative of localization in a corresponding sensory area of the cerebral cortex. Psychomotor seizures are believed to be referable to lesions in the frontal lobes, and the focal damage of Jacksonian epilepsy is in some portion of the cerebral cortex.

The treatment of epilepsy emphasizes the amelioration of symptoms, although protection against physical injuries or recurrence of attacks and control of the basic causal factors are also important objectives. Symptomatic relief is approached through chemotherapy with anticonvulsants such as Dilantin and phenobarbital, drugs which depress the activity of the motor area of the cerebral cortex. Epileptics should not perform tasks which are potentially dangerous to themselves or to others because of the possibility of attention lapses. The threat of more severe physical injury during an attack may be lessened by positioning the head to the side and slightly downward. Such positioning facilitates breathing, drains froth from the mouth, and prevents the tongue from being bitten or swallowed.

The probability of recurrence of attacks may be reduced (1) by diet therapy which emphasizes fasting or ingestion of ketogenic foods, both of which increase the acidity of the blood, and by abstaining from alcohol to avoid direct irritation of cranial blood vessels; and (2) by maintaining an optimal metabolism through periodic exercise, regularity, and adequate rest. Where a discrete causative factor such as a brain tumor is responsible for epilepsy, its surgical removal should be considered seriously. However, only a minority of cases is amenable to surgical intervention.

*Hydrocephaly*  Hydrocephaly is a condition in which the head becomes enlarged because of an increase in intracranial pressure due to an accumulation of CSF (cerebrospinal fluid). This abnormality is frequently exhibited at birth or soon thereafter. The fluid accumulation ordinarily is intraventricular. The increased fluid pressure causes a dilation of the brain ventricles and a corresponding atrophy of the periventricular neural tissues. As the intracranial pressure increases, the scalp is stretched, causing a decrease in the thickness of the flat bones, an increase in the distance of their separation at suture joints, and a bulging of the fontanelles. Because of the large size of the fourth ventricle, the occipital area of the head often becomes pyramidal in shape, causing a relative decrease in facial dimensions. The ensuing brain damage may result in mental retardation, spastic hemiplegia, blindness, or other neural abnormalities.

CSF accumulates in the cranial region because of a blockage of its circulatory path through the brain ventricles. In other words, the foramina of Monro, Luschka, and Magendie or the aqueduct of Sylvius may be obstructed, malformed, or absent. Defective drainage of CSF is usually traceable to congenital abnormalities such as brain anomalies or meningitis, or to postnatal disorders such as physical trauma and tumors in the cranial region.

Hydrocephaly is a serious abnormality which usually requires extreme measures such as surgery for the reduction or relief of symptoms. If the obstructions to the circulation of the cranial CSF cannot be removed, then two surgical strategies are still possible: (1) Decrease the amount of CSF formed in the cranial region, or (2) increase the amount of CSF drained from the same region. The first strategy is accomplished by reducing the extent of the choroid plexus associated with the lateral ventricles through cauterization. Since this choroid plexus is responsible for producing most of the CSF, reduction in its extent should result in a corresponding decrease in the volume of fluid produced. The second strategy requires the establishment of artificial drainage channels between the brain ventricles or the subarachnoid space and certain veins. The use of artificial channels removes the obstacles to the venous drainage of CSF, thereby reestablishing normal intracranial pressures.

Other degenerative diseases Of the numerous CNS disorders which result in progressive brain degeneration, three are especially important in a conceptual sense. These degenerative diseases are parkinsonism; senile degeneration, or senility; and multiple sclerosis. Parkinsonism has already been reviewed. The other two diseases are subjects of the following discussions.

Senile degeneration (senility)  Senility is an age-specific disease which is highly variable in its causes and symptomatology. Indeed, many people live to a ripe old age without exhibiting the debilities associated with the aging process that some other persons show prematurely around age 40. When the behavioral aberrations of the elderly are seriously maladaptive in a social context, the disease is called senile dementia, or senile psychosis.

Certain changes linked with the aging process are predictable enough to be labeled "normal." Nevertheless, the signs and symptoms of aging are so widely variable, when comparing different individuals of the same chronological age, that the term "normal" as used above may in fact turn out to be a misnomer or at least an inadequate frame of reference for purposes of comparison.

In any event, the consistent functional changes as they are viewed at the present time include a decrease in the speed with which reflexes are performed, a depression in visual and auditory acuity, lowered physical stamina and corresponding increase in the frequency of tiredness, as well as a reduction in the ability to remember recent events or the names of acquaintances. In other words, there is a decline in physical energy, sensory functions, and memory. These losses are observed to some degree in all people starting sometime between the fourth and sixth decade of life.

This general decrease in physical and mental abilities is often exacerbated by psychogenic factors. The "golden" years following retirement from work can be tarnished by frustration, intolerance, self-pity, and depression. The retired person is often no longer gainfully employed and may have to rely upon family, relatives, or others for some or all financial support. Various physical disabilities or personality disturbances may necessitate a certain amount of physical dependence on, and psychological support from, others. This need may come at a time when such help may be difficult to obtain. Usually, the children are grown, live away from home, and have their own lives to lead. Help, even when offered, frequently reinforces the idea of personal inadequacy after a lifetime of relative independence. The psychological conflicts between these feelings may be expressed as resentment, suspicion, and an excessive amount of inflexibility toward new ideas and changing times. These efforts to preserve self-respect serve to deepen the cleft between the generations, making psychological adjustments even more difficult.

More serious personality disturbances occur when psychopathologic behavior or pathophysiologic changes which directly or indirectly affect the brain are superimposed upon the more predictable age-specific changes described above. When intolerance and suspicion are transformed into agitation and delusions or when a loss of recent memory leads to confusion and transient losses of contact with reality, the boundaries of unacceptable social adjustment have been crossed. Hallucinations, where present, are indicative of cerebral involvement. This form of senility is psychotic. If cerebral aneurysms or arteriosclerotic changes occur during this time, then the degree of psychosis may even be more severe. Symptoms similar to parkinsonism or epileptic seizures may appear when the basal ganglia are involved. Restlessness, confusion, incoherence, and motor incoordination are longer-lasting, recur more frequently, and generally require more highly specialized treatment and care, compared with the symptoms of senile degeneration with no psychotic overtones.

Because the changes associated with senility, like aging itself, are usually gradual, probably no single causative factor is responsible for all its signs and symptoms. However, one leading theory suggests that senility is the result of chronic and progressive hypoxia of all the tissues, but especially of the cerebrum. Hypoxia would explain the depression in overall metabolism as exemplified by a general decrease in motor abilities. Since localized hypoxia gives rise to tissue ischemia and subsequent necrotic regions, a progressive cerebral hypoxia would explain the decrease in the number of intracranial nerve fibers and the increase in the number of atrophic neurons and nerve cell nuclei in older brains, compared with the number of fibers, neurons, and cell nuclei which exhibit these abnormalities in younger brains.

An increase in the amount of collagenous connective tissue and neuroglia with age may contribute to an increase in neural ischemia by adding to the time required for oxygen to diffuse from the blood to the neural tissues. Arteriosclerosis at the base of the brain may further reduce the amount of oxygen and decrease the amount of nutrients supplied to the cerebrum. Changes in capillary permeability with age increase the degree of hypoxia. These pathologic changes may set up a vicious circle of oxygen and nutrient deprivation which leads to further degeneration and incapacitation. Portions or all of the cerebrum in older patients with cranial arteriosclerosis

may soften and disintegrate. This condition is referred to as *encephalomalacia*. Focal lesions are surrounded by hypoxic but living cells which are believed to be responsible for many of the physical and mental disabilities suffered in this form of senile psychosis.

When the entire cerebrum is involved in encephalomalacia, its size shrinks, the thickness of the cortex decreases, and the sulci become wider and deeper. Often, the ventricles of the brain become dilated because of an increase in the volume of intracranial CSF. The basal ganglia also decrease in size and undergo degenerative cellular changes. Usually, damage to this portion of the extrapyramidal system is the origin of a form of senile motor incoordination which resembles parkinsonism, as mentioned previously. Epileptic seizures which, as we have noted, sometimes occur during senile degeneration are probably the result of an interaction between general brain damage and a genetic predisposition toward epilepsy. Epileptiform seizures may also be caused by a degeneration of one of the basal ganglia called the *amygdaloid nucleus*. This nucleus is located at the tail end of the caudate nucleus.

That chronic brain hypoxia is behind the development of senility is supported by the remission of symptoms frequently obtained by exposing senile, arteriosclerotic patients to pressures several times that of the atmosphere in *hyperbaric* (greater than barometric pressure) *chambers* for appropriate periods of time. This form of therapy increases the amount of oxygen dissolved physically in the plasma by several percent. The small increase in oxygen tension is sometimes sufficient to relieve the hypoxic condition and thereby reduce the severity of senile symptoms.

*Multiple sclerosis* Multiple sclerosis (sometimes referred to as MS) is a degenerative syndrome of the CNS with diverse causes. It is first exhibited sometime between the second and fifth decade of life. Although the symptoms can be sudden in onset, they are usually gradual and become progressively more severe with time. Generally, there are cycles in which remission of symptoms alternates with their recurrence. In general, each cycle results in more debilitation than that caused by the previous cycle and leads to a correspondingly shortened life span.

The symptomatology includes sensory, motor, and behavioral abnormalities. More specifically, one or more of the following disorders are observed: (1) amblyopia, nystagmus, and pains behind the eyeballs (retrobulbar neuritis); (2) numbness or pins-and-needles sensations (paresthesias) in localized regions of skin innervated by affected nerve pathways; (3) intention tremors, spastic paralysis, or slurred speech; and (4) hyperkinesis and euphoric states. Each of these symptoms is indicative of the involvement of a specific region of the brain or spinal cord.

The CNS lesions in multiple sclerosis are usually, but not always, multiple in number. Called *plaques*, these lesions are illustrated in the cross section of the diseased brain in Fig. 15-3. Several plaques are indicated by arrows. The plaques are variable in size, ranging between 1 and 15 mm (millimeters), and are more frequently located in the white matter than in the gray matter. The affinity of these plaques for the white matter is explained by the differential degeneration of the myelin sheaths of nerve fibers and a subsequent increase in the amount of neuroglia. This excessive deposition of neuroglia is called *gliosis*. Gliosis is comparable to the formation of scar tissue generally, except that this type of tissue is found within the CNS. The *demyelinization* of axons in the brain and spinal cord is correlated with the appearance of increased amounts of lipase in the blood and alterations in the metabolism of lipids within the CNS.

As the disease progresses, body resistance decreases, and there is a correspondingly greater susceptibility to respiratory and urinary infections. The decreased life span in multiple sclerosis patients is attributable in part to the effects of secondary infection and the involvement of the medullary area of the brainstem.

The treatment of multiple sclerosis emphasizes symptomatic relief, maintenance of existing functions, and psychological support. The usual treatment modes consist of adequate rest, the avoidance of temperature fluctuations which may exacerbate symptoms, physical and occupational therapy for the

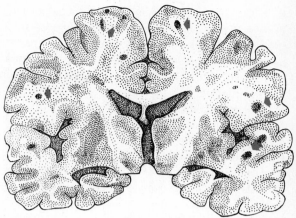

**Figure 15-3** *The appearance of plaques in the brain of a patient with multiple sclerosis.*

reduction of spasticity, and psychotherapy to engender positive attitudes. All these treatments are prescribed with one goal in mind: to enable the patient to live as normally as possible within the constraints imposed by the disability.

Types of physical trauma   Basically, there are three types of exogenous physical injury which can occur to the brain—concussion, contusion, and laceration.

A cerebral *concussion* is a mild form of head injury due to a sharp blow or jarring. Although not resulting in demonstrable pathologic changes in the involved neurons, cerebral concussions usually result in temporary unconsciousness. Other symptoms induced by concussions often include headache, paralysis, and amnesia. Concussions are produced by either direct blows to the head or the transmitted force arising from a fall on the base of the spine.

A cerebral *contusion* is a bruising of the brain. Since the scalp is not usually broken in this form of physical injury, it is difficult to distinguish unequivocally between a contusion and other forms of brain trauma. Ordinarily more severe than a concussion, however, this form of brain injury is characterized by a longer period of unconsciousness, observable pathologic changes in the involved neurons, and the cardinal signs of inflammation. A slight cerebral edema, due to an increase in the volume of cranial CSF, may be noted. The bruising may be responsible for hemorrhaging of superficial cerebral blood vessels. The pressure of extravascular blood against nearby neurons causes necrotic changes of varying degrees of severity. Other common symptoms of contusion include paralysis on the opposite side of the body (contralateral hemiplegia) and the inability to speak (aphasia).

A brain *laceration* is a serious form of injury in which there is a tearing of brain substance. The skull may or may not be broken by the force which causes the tear, but a scalp wound is usually visible. The brain tissue tears either at the point at which the greatest force is applied directly or on the opposite side of the head through the effects of transmitted energy. In addition to headache, paralysis, and amnesia, symptoms of laceration include extended periods of unconsciousness and widespread subdural or intracerebral hemorrhaging and the subsequent development of necrotic foci. The CSF becomes bloody because of an accumulation of pooled blood within the meninges. Called *subdural hematoma,* this condition creates excessive pressures against the neurons of the brain. Such pressures, in turn, cause encephalomalacia, a condition mentioned previously.

The severity of damage caused by physical trauma to the brain depends upon the magnitude of the applied force, the region of the brain affected, and the size of the lesion produced. Lesions of the cerebrum result in amnesia. Involvement of the basal ganglia gives rise to parkinsonism, chorea, or athetosis. Injury to the RAS causes stupor or coma. Medullary damage may precipitate immediate death. In all these examples, the symptoms are related directly to the functions of the damaged brain regions. Because headache is a form of referred pain in the cranium, trauma to any part of the cranium may provoke this symptom. Headache is discussed below.

The treatment of brain injuries caused by physical trauma is determined by the magnitude of the injury and the location of the involved region. Bed rest reduces the possibility of further physical injury and promotes healing. Cold compresses followed by massage of the cranial region may be soothing. Sedatives are useful when hyperexcitability is pronounced. Psychological support is required to counteract the effects of mental depression and confusion. Firm bandaging of the head reduces the effects of cerebral edema. Surgical intervention may be required when severe lacerations cause intradural or extradural bleeding.

Types of headache    A headache, or *cephalalgia*, is a form of pain referred to the periphery of the cranial region. Although headaches are commonplace, some types are symptomatic of more deep-seated disorders. Most headaches are intracranial in origin; some originate in extracranial locations. Many headaches are referred to the periphery of the head along branches of either the cervical plexus or the trigeminal nerve. Other headaches do not seem to follow the distribution of a single cranial nerve and are therefore more diffuse.

There is no standard classification of headaches, although probable causation is emphasized. Based on causal factors, the most common headaches include those induced by tension, nasal vasomotor reactions, cranial vascular responses, prolonged muscular contractions in the cranial and cervical regions, and facial inflammation. The approximate peripheral locations of these types of headache are identified in Fig. 15-4.

*Tension headaches* represent the most common form of headache. Caused by an interaction between psychological stress and fatigue, tension headaches are not referred to a particular part of the cranium. The experience is of dull pain throughout the periphery of the head. The pains may be increased by worry or frustration and can be dissipated by massage.

*Nasal vasomotor headaches* are also induced by psychological stress, but the effects are seen in the nasal portion of the facial region. Accompanying pains in this region is a nasal discomfort which may be expressed as either an excessive secretion of a watery fluid from the nose or a difficulty in breathing due to nasal congestion. Such attacks are not correlated with microbial infections or the presence of allergens. Since the response includes pain, edema, and redness due to irritation, the symptoms in some ways mimic those caused by nasal inflammation.

*Vascular headaches*, whether migrainous or non-

**Headache Type**

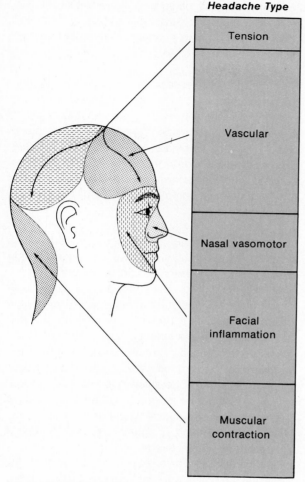

*Figure 15-4* The cranial regions affected by different types of headache.

of the cranium. In migraine, the recurrent pains are excruciating and may cause prostration. In addition, migraine attacks are usually preceded and accompanied by sensory and motor symptoms such as blurred vision, nausea, vomiting, and anorexia (loss of appetite). Nonmigrainous vascular headaches are not as severe as migraines and are usually not accompanied by visual and digestive abnormalities.

*Muscle contraction headaches* are caused by a chronic tightening of the muscles of the face, scalp, neck, and shoulders due to nervous tension. This form of headache is frequently referred to the region below the occiput in the inferoposterior aspect of the cranium. The pains are highly variable in frequency, intensity, and duration.

*Facial inflammation headaches* are referred to the facial region and, to this extent, are superficially similar to those associated with nasal vasomotor responses. However, headaches attributed to inflammations of the face are not restricted to the nasal region. Such headaches may be caused by inflammatory reactions of the eyes, ears, paranasal sinuses, teeth, or throat.

The above list by no means exhausts the possible causes of headache. Indeed, the causative agents of headache include any factors which excite pain receptors in the facial, cranial, or cervical regions; any factors which overexcite, irritate, inflame, or exert mechanical pressure against portions of the CNS or the peripheral nerves; any factors which irritate, inflame, or distend arteries in the cranial region; and psychogenic factors which exert their effects in one of the above-stated ways or through functional pathways, i.e., with no demonstrable organic basis. Therefore, it is not surprising to find headache as a symptom in the following disorders:

1 Emotional disturbances
2 Microbial infections which cause septicemia and fever
3 Physical trauma to the head or spine by concussion, contusion, or laceration
4 Vascular disorders such as cranial aneurysms and essential hypertension
5 Metabolic disturbances resulting from nutritional

migrainous, are caused by an initial vasoconstriction followed by a secondary vasodilatation of the cranial arteries. The spasmodic contractions of these blood vessels are usually, but not always, induced by tension. For some reason not well understood, the walls of the affected vessels suddenly become weak and distended. The pulsatile pressures of the blood against the inside of the already stretched and flaccid vascular walls is interpreted by the migraine sufferer as a lancinating, or stabbing, pain in the frontal region

deficiencies, hormonal imbalances, or drug reactions

6 Neuropathies such as neuralgia, neuritis, or polyneuritis (see Chap. 16)

7 Visual disorders due to errors of refraction because of eye strain and corresponding muscle fatigue

8 Dental caries

9 Intracranial tumors, or neoplasms

Headaches are treated symptomatically. This form of therapy frequently ignores the cause of the symptom. "Take two aspirin tablets and go to bed" is standard medical advice for a headache. Since most headaches are due to nervous tension, however, this treatment often works. Such advice, furthermore, is consistent with the two most common symptomatic approaches to the treatment of headache—rest to decrease anxiety and the resultant physical tension, and aspirin to reduce fever and pain through its antipyretic and analgesic actions, respectively.

The mechanism by which aspirin accomplishes these feats is not well understood, but some recent findings are at least partially suggestive. Aspirin has been found to inhibit the synthesis of prostaglandins, a group of specialized hormones normally released during numerous physiologic processes, including inflammatory reactions and cold stress. Since the presence of prostaglandins is associated with the onset of pain, a depression in their synthesis results in a decrease in the intensity of the perceived pain.

Other tension-reducing therapeutic practices for headache include massage and heat applications to reduce muscular spasticity and improve circulation, the application of evaporating solutions to the affected region, and the use of cold compresses to numb peripheral pain receptors. These forms of therapy are especially useful in relieving the symptoms of headaches caused by tension and muscular contraction. Chemotherapy is also practiced with drugs other than aspirin. Vasoconstrictive drugs are frequently prescribed for migraine in an attempt to counteract the effects of cranial arterial dilatation. Antispasmodics, or muscle relaxants, are sometimes used to control muscle contraction headaches. Saline laxatives are administered to create dehydration, thereby indirectly reducing the severity of hypertensive headaches. Psychotherapy is suggested for migraines and other severe, chronic headaches attributed to psychogenic factors.

**Psychosomatic disorders** When a personality disturbance contributes to or precipitates a pathologic condition anywhere in the body, the resulting disorder is termed *psychosomatic* ("mind-body"). Psychosomatic disorders originate within the CNS but are expressed elsewhere in the body. Among the more common and well-known psychosomatic disorders are peptic ulcers, chronic diarrhea (colitis), and essential hypertension.

*Peptic ulcers* Peptic ulcers are disorders in which there are open, bleeding sores or lesions of the mucosa in either the stomach or the first portion of the small intestine. An ulcerative lesion in a patient with peptic ulcers is illustrated in Fig. 15-5. Since peptic ulcers are gastric or duodenal in location, the disease is digestive in a strict sense. However, the condition is often caused by a chronic hyperactivity of the vagus nerve (Xth cranial nerve, part of the parasympathetic division of the autonomic nervous system), which may in turn be triggered by personality disturbances such as suppressed longings for acceptance, warmth, and affection, or aggressive, violent feelings expressed as anger or rage.

Since peptic ulcer involves a specific portion of the digestive tract, this disease usually causes some combination of the following signs and symptoms: (1) functional disruptions of digestive functions through the production of flatus (gas) and eructation (belching), as well as by causing heartburn, nausea, vomiting, diarrhea, and anorexia; (2) physical evidence of organic damage such as the appearance of blood in vomitus and the feces; and (3) subjective symptoms such as gnawing pains in the epigastric region which appear several hours following the completion of most meals. Disruptions of digestive processes are experienced in a wide variety of organic and functional disorders and are to be expected in digestive diseases such as peptic ulcers. Internal hemorrhaging is a consequence of mucosal ulcerations that rupture submucosal blood vessels.

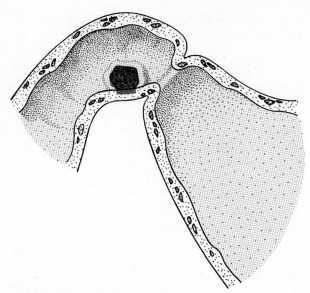

**Figure 15-5**  *A case of peptic ulcer in the proximal duodenum.*

The characteristic pains associated with this disease are due to the stimulation of visceral afferent pain receptors at the foci of physical damage.

The pathophysiology of peptic ulcers is explained by a chronic hyperexcitability of the parasympathetic division of the autonomic nervous system initiated and sustained by the action of the limbic system of the brain. Through a branch of the vagus nerve to the stomach, this excessive autonomic activity increases gastric motility and promotes a continual secretion of both pepsinogen (see Chap. 26) and hydrochloric acid (HCl) from the chief and parietal cells, respectively, of the gastric mucosa, even in the absence of food. The hyperacidity eventually causes a decrease in the thickness of the protective layer of mucus which lines the inside wall of the stomach. The sustained secretions of pepsinogen and HCl then begin to dissolve portions of the mucosa, causing ulcerations. Duodenal ulcers are more prevalent than gastric ulcers, simply because the duodenal mucosa does not normally have as thick a covering layer of mucus, compared with the gastric mucosa.

The basic problem in treating peptic ulcers is to reduce or eliminate the chronic hyperacidity within the stomach. Two techniques are especially useful:

diet therapy and psychotherapy. The methods and goals of dietary control for patients with peptic ulcers are (1) a high-protein diet and the administration of alkaline substances to serve as buffers for the binding of excess hydrogen ions; (2) more frequent meals of smaller volume to counterbalance the potentially damaging effects of chronic hyperacidity on the gastric and duodenal mucosae; (3) high-fat diets to decrease the volume of HCl secreted by reducing the need to activate pepsinogen; and (4) the use of soft, bland foods to reduce the degree of irritation at ulcerative foci on the gastric and duodenal mucosae.

Psychotherapy is a more basic tool designed to expose and eliminate the psychogenic basis of the pathologic condition. The goals of psychotherapy are long-range, compared with the ameliorative functions of diet therapy. Bringing repressed feelings of longing or violence out into the open in a way that is meaningful to the patient is the first step toward a complete cure.

Intractable cases sometimes respond to the surgical severation of the vagus nerve branch to the stomach. Vagectomy eliminates the parasympathetic innervation of the stomach, thereby preventing an increase in the secretions of pepsinogen and HCl ordinarily evoked by hyperactivity in this part of the autonomic nervous system.

***Chronic diarrhea (colitis)***   The psychosomatic form of colitis causes spasms of the smooth muscle wall of the colon. Spastic colitis is brought about by a chronic hyperactivity of that portion of the parasympathetic division of the autonomic nervous system which innervates this region of the large intestine. The excessive autonomic activity may be provoked by obsessive anxiety, guilt feelings, or other psychogenic factors. Therefore, the neural bases of spastic colitis and peptic ulcers are similar.

The pathophysiology of spastic colitis also resembles that of peptic ulcers. The motility and secretory activity of the colon are increased in response to excessive parasympathetic stimulation. Copious amounts of mucus are produced by the mucous glands in this region. The spasmodic contractions of the colon give rise to intermittent bouts of pains in the midabdominal region. These colicky pains are the most prevalent symptom of colitis. The homeostatic disruption of normal digestive functions causes an initial period of constipation which is later followed by chronic diarrhea. (Constipation and diarrhea are explained in Chap. 27.) Examination of the watery feces during the diarrheal stage of the syndrome reveals large quantities of mucus and sloughed-off portions of the affected mucosa. The lesions which develop and secondary microbial infection may cause ulcerations or perforations of the colon, further complicating the symptomatology and treatment of the disorder.

In view of the similar neural bases for and pathophysiology of colitis and ulcers, it is not surprising to find that their treatments are also closely allied. For example, diet therapy for spastic colitis emphasizes bland, high-protein, easily digestible foods containing a minimum of roughage to reduce irritation to, and motility of, the colic wall. In addition, psychotherapy is required for an understanding of the root causes of the condition.

*Essential hypertension* In "essential" hypertension, there is no demonstrable organic basis for the disorder. This form of hypertension is therefore attributed to psychogenic factors which lead to hyperactivity in branches of the sympathetic division of the autonomic nervous system which innervate the blood vessels. The personality disturbances responsible for such changes are poorly understood, but studies of patients suffering from this disorder reveal personality characteristics which revolve around benign, submissive, and dependent attitudes superimposed on a personality substrate of seething rage which is neither fully expressed nor repressed. In other words, the anger is recognized by the subject but is not well regulated or managed. Most cases of hypertension are of this variety. Hypertension is discussed further in Chap. 31.

Psychoneuroses Psychoneuroses are mild personality disturbances which cannot be correlated with organic damage. Individuals with neurotic personalities which express themselves in clinically identifiable ways are, nevertheless, tolerated by society even if some of their actions are abnormal, antisocial, or abrasive. Since contact with reality is maintained, institutionalization is not required. Although numerous forms of psychoneurosis have been described, we will direct our attention toward three of the more well-known types: neurasthenia, anxiety neurosis, and conversion hysteria. As these disorders are discussed, bear in mind that they are the discernible products of neurotic personalities and are not to be considered as distinct diseases.

*Neurasthenia* Neurasthenia incorporates a group of ill-defined neuroses in which the most characteristic symptom is a chronic feeling of physical and mental exhaustion. Strength ebbs quickly when vigorous activity is attempted, and the muscles tremble. Mental alertness is depressed, a condition which may be exacerbated by chronic insomnia. An increase in the activity of the sympathetic division of the autonomic nervous system causes tachycardia (rapid heartbeat), polyuria, and perspiration. These factors and the precipitating causes of the disorder interact to bring about an irritable feeling, vague physical complaints, and general nervousness. Although usually irrational, fear may become pervasive. Sometimes the factors which precipitate the disorder are severe enough to elicit hysterical symptoms. Conversion hysteria is discussed later in this chapter.

The antecedent of neurasthenia is chronic anxiety. Such apprehension develops out of a sharp conflict between current actions or status and expected performance or goals. Psychological defenses are erected by the subject to maintain equilibrium in the face of the conflict.

Equilibrium may falter when defenses weaken or intensify significantly. Such changes occur in response to external influences which directly or indirectly threaten defensive barriers. Neurasthenia is one way in which this neurotic breakdown may be expressed.

Treatment of neurasthenia is based on symptomatic relief and treatment of the basic causes. Diseases such as anemia and nutritional deficiencies should be ruled out or treated, if necessary. Psychotherapy is indicated when neurasthenia causes significant functional incapacitation.

*Anxiety neurosis* The causes and symptomatology of anxiety neurosis are similar to those of neurasthenia. However, the predominant symptoms are expressed as bouts of fear or delusions with physiologic overtones such as cardiovascular, respiratory, digestive, and muscular abnormalities. Even facial expression is often frozen in fright. The digestive upsets may lead to nausea, vomiting, or diarrhea. Hyperventilatory efforts, the results of diffuse excitement, can cause dyspnea, or gasping, and corresponding feelings of suffocation and dizziness. Tachycardia and angina pectoris (chest pain) are frequent symptoms in such patients. Hypertension, profuse sweating, and polyuria are some of the other autonomic responses set in motion by the neurosis. The skin feels clammy, the throat tight, and the head aches. Muscle tremors are also common.

Anxiety neurosis, like neurasthenia, is one possible outcome of a neurotic breakdown. A longing or fear is completely repressed because of its psychologically threatening nature. Feelings of inadequacy and profound frustration are engendered. Attacks of irrational terror or delusions take the place of the repressed feelings when the defensive barriers of the patient cannot completely cope with mounting anxiety. A feeling of profound depression may accompany the fearful demeanor. That defensive barriers are still functioning is attested to by the lack of a complete personality collapse and the absence of a personality regression to an earlier time in life.

Chemotherapy and psychotherapy are used in the treatment of patients with anxiety neurosis. Organic disorders such as chronic alcoholism should, of course, be ruled out. Chemotherapy may involve the use of antidepressants, sedatives, or tranquilizers, depending upon the patient's needs and physiologic status. Psychotherapy and other forms of counseling are essential for either the reestablishment of mental equilibrium or the attainment of a more appropriate level of social adjustment.

*Conversion hysteria* Conversion hysteria is a functional disability precipitated by physical injury or psychogenic (mental) trauma. Hysterical symptoms are frequently caused by a crisis superimposed on an underlying neurotic conflict. The crisis is resolved by developing either physical or mental disablements which relieve the patient of further responsibilities or tensions related to the neurotic conflict. The rapid development of the syndrome is not under conscious control, although the subject does subconsciously arrange the nature of the desired response. In other words, the conflict is suddenly converted by the subject into appropriate hysterical symptoms. Indeed, the term conversion hysteria is based on this assumption.

The symptoms of conversion hysteria are legion. Functional disabilities which have been developed to resolve neurotic conflicts are of the following types: sensory disorders such as blindness, deafness, anesthesia, and paresthesias; motor abnormalities such as spastic or flaccid paralysis, tremors, tonic or clonic spasms, aphonia (loss of speech), and hysterical fits resembling seizures; and mental illnesses such as amnesia for the period encompassed by a precipitating crisis. The sensory and motor disabilities are more often complete than partial.

The functional symptoms usually do not correspond to those produced by known lesions of the nervous and muscular systems. Functional paralysis of the hands, for example, is not accompanied by paralysis of the forearm. Functional paralysis in one extremity (monoplegia) does not eliminate deep tendon reflexes in it. Atrophy and degeneration are not consequences of functional paralysis. Neurologic (nerve) and myographic (muscle) examinations indicate that the sensory organs and receptor neurons, peripheral nerves, as well as the nervous and muscular systems are undamaged and functioning in spite of the apparent sensory and motor deficits.

The question of the diversity of structures which may be singled out for conversion of the neurotic conflict, i.e., why one patient suddenly becomes functionally blind while another patient becomes functionally paralyzed, etc., cannot be answered unequivocally. In numerous cases, a physical injury which initially disables finally heals, even though the symptoms still persist. Possibly, the violence associated with the initial injury or other nonviolent forms of neurotic conflict may promote reflex activity which perpetuates the former disability.

Sometimes a neurotic conflict is resolved by suddenly exacerbating the symptoms of a chronic disorder. Long-standing myopia turns into blindness, a decreased hearing acuity is transformed into deafness, or a chronic limp is converted into paralysis of a lower extremity.

Frequently, the functionally affected structures bear some symbolic relation to the crisis which precipitated the hysterical symptoms. The parts of the body which were performing work at the time of an explosion that causes a concussion, for example, may become functionally paralyzed.

Finally, it is not uncommon to find that the functionally disabled parts of the body are related to the occupation of the patient. The singer whose voice is suddenly lost, the soldier whose muscles uncontrollably tremble, the athlete whose legs are suddenly paralyzed, and the truck driver who is suddenly blinded obviously cannot perform work-related tasks.

As with so many other psychogenic disorders, the treatment of conversion hysteria involves the relief of symptoms and the elimination of the basic causes of the syndrome. Such individuals should be removed from the stresses which precipitated the crises that culminated in hysterical symptoms. This recommendation is easy to follow where the precipitating factors are well defined, but the symbolism between the hysterical symptoms and the precipitating crisis may be subtle or elusive. Helpful measures include sedatives to reduce hyperemotional states and massage to improve muscle tone and stimulate circulation. A reduction in the degree of external mental stimulation is also desirable. The patient must be protected from possible physical injury which can result from the functional disability. Psychotherapy may reveal and thereby remove the sources of neurotic conflict repressed from the consciousness of the patient.

## SPINAL CORD DISORDERS

Disorders of the spinal cord are usually due to physical trauma or microbial infection. Since the spinal cord is located within the vertebral column, injuries of the cord are identified according to the affected region, i.e., cervical, thoracic, lumbar, or sacral. Because complex reflexes requiring coordinated movements of numerous groupings of muscles are controlled by a number of segments of the spinal cord, segmental damage eliminates patterned movements regulated by the affected segments. The effects of more localized lesions are determined by the regions and segments of the cord in which the injuries are located as well as by the specific fiber tracts involved.

The symptoms are categorized as sensory, motor, or mixed. More specifically, the precise location of a lesion can be related to the following deficits: (1) a partial or complete, localized loss of sensations involving touch, pain, temperature, and proprioception; (2) a partial or complete paralysis of specific skeletal muscles and changes in deep reflexes; or (3) some combination of these sensory and motor symptoms.

## LOCALIZED LESIONS

Lesions may be localized in specific ascending or descending tracts in particular segments of the spinal cord. Such lesions are usually caused by physical trauma, microbial infection, or a tumor. Physical injuries are primarily due to spinal contusions or excessive compressional forces.

Ascending pathways    The major afferent tracts of the spinal cord include the fasciculus gracilis and fasciculus cuneatus, the lateral and anterior spinothalamic tracts, and the posterior and anterior spinocerebellar tracts.

The *fasciculus gracilis* arises from spinal ganglia associated with the sacral, lumbar, and lower thoracic segments of the spinal cord. Fibers of this fasciculus make synaptic connections with neurons of the nucleus gracilis in the medulla oblongata. After decussating in the medulla, fibers from the nucleus gracilis ascend within the brain to terminate in the thalamic area of the diencephalon. Recall that the thalamus communicates with the somesthetic cortex.

Because the fasciculus gracilis is involved in tactile discrimination, damage to this tract causes *astereognosis*, the inability to distinguish objects solely by touch. This sensory deficit appears in body regions below the level of the lesion.

The *fasciculus cuneatus* arises from spinal ganglia associated with the upper thoracic and cervical segments of the spinal cord. Fibers of this fasciculus make synaptic connections with neurons of the nucleus cuneatus in the medulla oblongata. Fibers from the nucleus cuneatus run parallel to those from the nucleus gracilis, decussating in the medulla, terminating in the thalamus, and indirectly communicating with the somesthetic cortex.

Because the fasciculus cuneatus is primarily concerned with proprioception (awareness of position), its destruction results in a loss of sensory feedback information on the degree of stretch of muscles, tendons, and joints in the thoracic region and upper extremities. This sensory deficit impairs the control of muscular movements. In addition, there is a depression in the ability to localize the part of the body from which deep pain and pressure stimuli are arising.

The *lateral* and *anterior spinothalamic tracts* arise from neurons located in the posterior gray horns of the spinal cord. Many fibers of these tracts immediately decussate in the white matter and ascend in their respective funiculi. Both tracts make synaptic connections in the thalamus. Since the lateral spinothalamic tract on one side of the cord carries information on pain, heat, and cold from the contralateral side of the body, unilateral lesions of this pathway eliminate these sensory abilities in contralateral regions (hemianesthesia) distal to the injury.

Fibers of the *posterior* and *anterior spinocerebellar tracts* arise from neurons in the spinal ganglia. The primary neurons of both tracts make synaptic connections with secondary neurons in the same segment of the spinal cord through which entry is made. The fibers of the secondary neurons of the posterior and anterior spinocerebellar tracts form the inferior and superior cerebellar peduncles, respectively, within the medulla oblongata. Both peduncles terminate in the cerebellum. Because the spinocerebellar tracts are concerned with kinesthetic (movement) sensa-

tions from all parts of the body, lesions of these tracts lead to a loss of proprioceptive feedback from muscles, tendons, and ligaments and a depression in the ability to execute deep reflexes.

Descending pathways   The efferent tracts of the spinal cord are either pyramidal or extrapyramidal. The pyramidal pathways include the lateral and anterior corticospinal tracts. The extrapyramidal pathways incorporate the lateral and anterior reticulospinal tracts as well as the vestibulospinal, rubrospinal, tectospinal, and olivospinal tracts.

The pyramidal tracts are direct motor pathways from the brain to the spinal cord. The upper motor neurons of the *lateral* (crossed) and *anterior* (uncrossed) *corticospinal tracts* originate in the motor area (precentral gyrus) of the cerebral cortex. After decussating in the medulla oblongata, the heavily myelinated fibers of the crossed pyramidal tract descend to all segments of the spinal cord where synaptic connections are made with lower motor neurons originating in the anterior horns. Most of the fibers of the uncrossed pyramidal tracts remain on the same side of the body when they exit from the medulla. The myelinated fibers of these tracts also make synaptic connections with lower motor neurons originating in the anterior horns of the cord. Because both tracts are involved in the initiation and control of gross and fine voluntary movements, their destruction results in either contralateral or ipsilateral spastic paralysis distal to (below) the point of the lesion.

The extrapyramidal system of tracts is an indirect motor pathway from the brain to the spinal cord. Motor information to be conducted along this pathway is integrated both directly and indirectly. Direct integration is accomplished by extrapyramidal portions of the motor cortex, the basal ganglia, and the thalamus, as well as the substantia nigra and surrounding regions of the midbrain. Indirect regulation is exerted by the reticular formation of the brainstem acting in conjunction with information originating in the cerebellum, basal ganglia, and motor cortex.

The *lateral* and *anterior reticulospinal tracts* arise from reticular substance associated with the midbrain and the rest of the brainstem. After coursing through the lateral funiculi, fibers of the upper motor neurons of the lateral reticulospinal tract terminate in the lateral horns. Conversely, fibers of the upper motor neurons of the anterior reticulospinal tract course through the anterior funiculi to make synaptic connections with lower motor neurons originating in the anterior horns. Functions of these tracts include inhibition (lateral reticulospinal tract) and facilitation (anterior reticulospinal tract) of stretch reflexes.

The *vestibulospinal tract* arises from upper motor neurons which collectively make up the vestibular nuclei of the pons and the medulla oblongata. The vestibular nuclei make synaptic connections with both the vestibular branches of the acoustic (auditory) nerve and the cerebellum. Fibers from the vestibular nuclei descend through the medulla and cord without decussating, making connections at various segments with lower motor neurons in the anterior horns. The functions of this tract include the control of the postural muscles and the maintenance of equilibrium.

The *rubrospinal tract* arises from upper motor neurons which collectively compose the red nucleus of the midbrain. Functionally similar to the vestibular nuclei, the red nucleus also makes synaptic connections with the vestibular nerve and the cerebellum. Fibers from the red nucleus decussate in the medulla oblongata, descend the spinal cord on the contralateral side, and make synaptic connections with lower motor neurons originating in the anterior horns of the thoracic cord. This tract regulates thoracic postural reflexes.

The *tectospinal tract* arises from upper motor neurons which collectively make up the superior and inferior colliculi of the corpora quadrigemina of the midbrain. The superior and inferior colliculi communicate synaptically with the visual and auditory areas, respectively, of the cerebral cortex. Fibers from the corpora quadrigemina follow a pathway similar to that of the rubrospinal tract, decussating in the medulla, descending in the cervical cord, and synapsing with lower motor neurons in its anterior horn. The functions of this tract include the control of postural reflexes of the head and neck.

The *olivospinal tract* arises from upper motor neurons which collectively compose the olivary nucleus of the medulla oblongata. The olivary nucleus communicates synaptically with the cerebellum. Fibers from the olivary nucleus descend the cord to synapse with lower motor neurons originating in the anterior horns. The functions of this tract include the control of postural reflexes and voluntary movements.

Because integration of motor information conducted by the extrapyramidal tracts occurs in diverse regions of the brain, lesions of the extrapyramidal system may result in one or more of the following symptoms distal to the point of injury: a loss of voluntary movements and the appearance of spastic paralysis; an inability to maintain posture and equilibrium; a loss of deep tendon and visceral autonomic reflexes; and the appearance of involuntary movements resembling parkinsonism, chorea, or athetosis due to the involvement of the basal ganglia and substantia nigra (see "Specific Brain Disorders," earlier in this chapter).

## PHYSICAL TRAUMA

The spinal cord may be injured by physical forces directed at the cervical, thoracic, lumbar, or sacral regions. The classification of damage is identical to that used for the cerebrum and includes concussion, contusion, and laceration. However, concussion of the spinal cord is rare because of the protection afforded by the vertebral column. Similarly, the violence of a force capable of causing laceration of the lumbar spinal cord is sufficient to precipitate instantaneous death.

**Concussion, contusion, compression, and laceration** As in the case of cerebral concussion, *spinal concussion* is due to a severe blow and results in a mild, temporary loss of sensory and motor functions. Swelling and hemorrhaging may occur at the site of trauma. Usually, but not always, the injury is reversible and does not leave a lesion.

*Contusions*, or bruises, of the spinal cord generally occur as a consequence of fracture of the vertebral column. Subdural hemorrhaging often accompanies this form of physical trauma and usually results in a bloody CSF (cerebrospinal fluid). The loss of sensory and motor functions is more severe and longer-lasting than in spinal concussion.

*Fracture* of the vertebral column may also cause *compression* of the spinal cord. This form of injury is frequently serious, causing degenerative changes in the damaged neural tissues and adhesions in the surrounding meninges. The sensory and motor deficits are often complete distal to (below) the lesion. The severity of the compressional forces causes meningeal vascular hemorrhaging and the subsequent appearance of blood in the CSF.

*Whiplash* injuries also cause compression of the spinal cord. The cervical spine is compressed first in one and then in the other direction. Such injuries are induced in most cases by rear-end accidents in motor vehicles. The compressional injuries result in severe cervical pains, especially when attempting to turn the head. Additional symptoms may include a prolonged headache, gastrointestinal upsets, paresthesias, and nervousness. The cervical curvature is often temporarily straightened or reversed by the force of the whiplash injury.

*Spinal lacerations* are tears in the substance of the spinal cord. These are among the most severe cord injuries caused by physical trauma. Such damage results from gunshot and stab wounds, fractures of the vertebral column, or other violent forces which pull, rip, or snap the neural tissues. Partial lesions cause damage to localized regions of the cord. In complete lesions, the spinal cord is severed at the site of trauma.

Broken back (complete severation) and spinal shock   When the spinal cord is *transected*, the back is said to be "broken." The immediate effects are paralysis, anesthesia, and loss of visceral reflexes in regions of the body distal to the point of severation. This condition is transient, however, lasting between 1 and 2 weeks. As spinal shock dissipates during the first month following its onset, there is an involuntary and simultaneous exaggeration of many flexion, extension, and visceral autonomic reflexes. This phenomenon is referred to as the *mass reflex*. Spontaneously, or in response to appropriate stimuli, the arms flex, perspiration becomes profuse, and urination and defecation occur.

Presumably, spinal shock is precipitated when the corticospinal tracts are interrupted. The widespread responses during the mass reflex in the recovery phase are therefore due to the removal of inhibitory influences on spinal reflex activity.

## COLLATERAL READING

Chusid, Joseph G., and Joseph J. McDonald, *Correlative Neuroanatomy and Functional Neurology*, Los Altos, Calif.: Lange Medical Publications, 1964. Detailed but somewhat advanced information on the whole gamut of neural disorders.

Dean, Geoffrey, "The Multiple Sclerosis Problem," *Scientific American*, July 1970. A review of the pathology, history, and present geographic distribution of multiple sclerosis and its similarity to certain other neural disorders.

Gazzaniga, Michael S., "The Split Brain in Man," *Scientific American*, August 1966. A review of the effects of surgically separating the two cerebral hemispheres by hemisecting the corpus collosum.

Gruenberg, Ernest M., "The Epidemiology of Mental Disease," *Scientific American*, March 1954. Contains a fascinating historical description of what was apparently a communicable neurologic disease with aberrant behavioral overtones called the "dancing mania" which spread throughout Europe between the thirteenth and sixteenth centuries.

Hammond, Allen L., "Aspirin: New Perspective on Everyman's Medicine," *Science*, 174:48, 1971. A review of some implications of current research findings linking the pharmacologic actions of aspirin with prostaglandin inhibition.

Pitts, Ferris N., Jr., "The Biochemistry of Anxiety," *Scientific American*, February 1969. Pitts demonstrates that the symptoms of anxiety neurosis can be induced by infusions of lactate, a normal metabolite which usually increases in the blood of patients during attacks of anxiety.

Satinoff, Evelyn, "Salicylate: Action on Normal Body Temperature in Rats," *Science*, 176:532–533, 1972. Satinoff demonstrates that sodium salicylate lowers the rectal temperature of normal rats exposed to cold temperatures. This finding is consistent with the hypothesis that salicylates inhibit the release of prostaglandins, natural products ordinarily released when an animal is exposed to cold stress.

White, Robert W., *The Abnormal Personality: A Textbook*, 2d ed., New York: The Ronald Press Company, 1956. Several chapters are of especial usefulness as supplementary reading: chap. 11, "Psychosomatic Disorders"; chap. 12, "Effects of Injuries and Abnormal Conditions in the Brain"; and chap. 13, "Common Symptom Syndromes of Cerebral Disorder."

# CHAPTER 16
# PERIPHERAL nerve DISORDERS

Peripheral nerve disorders result from impairments in the functioning of the cranial and spinal nerves and their various branches. Diseases of the autonomic nervous system are included here for convenience. Neural impairments of the peripheral nervous system are caused by the following factors:

1 Lesions or injuries due to microbial infection or physical trauma
2 Vascular abnormalities such as hemorrhages and aneurysms
3 Drug effects such as irritation by alcohol, excitation by strychnine, and structural disruption through the release of nitrogen gas dissolved under pressure
4 Degenerative diseases caused by infections or congenital defects
5 Neoplasms (new growths) which place pressure on peripheral nerve trunks

6 Chronic psychogenic factors which facilitate or inhibit autonomic reflexes
7 Congenital defects such as hereditary diseases and developmental anomalies which affect the peripheral nerves, nerve roots, or central nuclei

Disorders of the peripheral nerves can affect sensory, motor, or metabolic functions, or may influence all these processes. Sensory involvement is reflected by a depression in, or complete loss of, sensory abilities. Motor effects lead to a weakness in, or partial or complete paralysis of, skeletal muscles. Metabolic disturbances are brought about directly by a loss of neural control or indirectly by a decreased blood supply, frequently causing changes in the condition of the skin and its accessory structures such as nails and hair.

Derangements of cranial and spinal nerves may be accompanied by pains and peculiar sensations from the skin. Loss of autonomic control is correlated with impairments in the operation of smooth muscles, glands, and autonomic reflexes.

This chapter explores the causes and symptoms of specific peripheral and autonomic nerve disorders.

## CHARACTERISTICS OF NEUROPATHY

### TYPES

A number of conditions are descriptive of the kind and degree of peripheral nerve involvement. These conditions are neuralgia, neuritis, and polyneuritis.

Neuralgia    Neuralgia is a condition characterized by sudden, sharp, and usually short-lived pains along the course of one or more cranial or spinal nerves. The pains are generally paroxysmal, recurring periodically, and often consist of cutting, or stabbing, sensations in the facial or thoracic regions. Because vasomotor responses resulting in edema and skin reddening often accompany the pains, the major effect resembles an inflammation of the affected region.

The origins of neuralgic pains are diverse, but pressures which directly or indirectly affect a nerve are predominant causal factors. Sometimes, stimulation of a hypersensitized region by cold or drugs, or simply local movements, can trigger a paroxysm of neuralgia. In most instances, the affected nerve does not undergo pathologic changes. Occasionally, however, neuralgia is accompanied by degeneration of the affected neurons.

Neuritis    Neuritis is potentially more serious than neuralgia, consisting of an inflammation or degeneration of one or more peripheral nerves.

The following are among the causes of neuritis: (1) physical trauma of a nerve induced by mechanical factors more severe than simple pressure, such as compression or contusion; (2) various microbial infections of a structural component of nerve fibers, such as the axoplasm, neurilemma, myelin sheath, as well as perineural tissues; (3) chemical intoxications of nerve cell bodies by alcohol, heavy metals, or other noxious chemicals; and (4) metabolic dysfunctions of neurons due to either peripheral vascular disturbances which cause ischemia and necrosis, insulin deficiency which results in diabetes mellitus, or severe malnutrition due to chronic gastrointestinal disorders or specific nutritional defects such as thiamine deficiency.

The symptoms of neuritis are determined by the nature of the causative agent, the degree of neural involvement, and the specific nerve or nerves affected. Neuritic symptoms include (1) sensory loss, especially of touch (anesthesia) along the length of the affected nerve; (2) motor deficits such as flaccid paralysis and the possibility of muscular atrophy and contractures, as well as the loss of reflex activity in the affected region; (3) metabolic dysfunctions which may lead to nerve and perineural tissue degeneration; and (4) irritative effects including tenderness, pains, and peculiar peripheral sensations (paresthesias) such as numbness or tingling along the length of the inflamed nerve. The pains may not necessarily parallel the distribution of the affected nerve, since pain endings of adjacent nerves overlap to a considerable extent.

Polyneuritis    Polyneuritis is a condition characterized by extensive sensory and motor abnormalities caused by peripheral nerve involvement. Usually many nerves are affected.

Basically, the causes and symptoms of polyneuritis are similar to those of neuritis. Differences are mostly quantitative rather than qualitative. The following are examples of these differences: Hereditary diseases represent one of the etiologic factors; numerous nerves are involved; anesthesia is present in the affected region; and extreme muscular weakness is found in the distal portions of the extremities as reflected in wristdrop or foot drop and other neuromuscular abnormalities (see Chap. 25).

### TREATMENT

The treatment of a neuropathy is dictated by the nature and degree of the disability. Treatment objectives include the elimination of the pathogenic agent, relief of symptoms, promotion of healing, rehabilitation, and prevention of both more serious damage and the recurrence of symptoms. These goals are reached by employing some combination of the following practices: (1) elimination or neutralization of any microbial infection or toxicant which may have caused the disorder; (2) reduction of symptoms with analgesics to arrest pain, devices to minimize pressure on affected body parts, and splints to prevent contractures; (3) encouragement of healing through bed rest, massage, physical comfort, and a diet high in calories and vitamins; and (4) physical rehabilitation by applying the techniques of physical and occupational therapy to increase the range of passive and active motion.

## CRANIAL NERVE DISORDERS

These disorders are the result of lesions to one or more cranial nerves. The cranial nerves are described in Chap. 14 and illustrated in Fig. 14-15.

### CRANIAL NERVE I

The *olfactory nerve* (Ist cranial nerve) subserves the sense of smell. Actually a fiber tract of the brain, the nerve is entirely sensory in function.

Dysfunctions of the olfactory nerve are caused by any factors which irritate, inflame, or otherwise damage the mucosa lining the nasal cavity, the nerve fibers of the olfactory tract, or the nerve cells in the inferior region of the frontal lobe of the cerebrum. More specifically, olfactory impairments result from infections (rhinitis) of the olfactory receptor cells of the nasal mucous membrane; skull fractures involving the cribriform plate of the ethmoid bone through which olfactory nerve fibers must pass on their way to the brain; or vascular, inflammatory, neoplastic, and congenital disorders which result in lesions of the brain superior and anterior to the pituitary gland.

The most prominent symptom of olfactory nerve involvement is anosmia, the loss of the sense of smell. In lieu of anosmia, the sense of smell may be altered, giving rise to false, misleading, or peculiarly sharpened olfactory sensations.

### CRANIAL NERVE II

The *optic nerve* (IId cranial nerve) subserves the sense of vision. Like the olfactory nerve, the optic nerve is a sensory fiber tract of the brain.

Abnormalities associated with the optic nerve affect the retina, optic tract, or visual cortex of the cerebrum. More specifically, impairments in the function of the optic nerve are the result of (1) inflammation of the rods and cones of the retina due to infections and vascular disorders, (2) neuritis of the optic bulb or tract due to infections, or (3) an increase in intraocular pressure due to glaucoma or in intracranial pressure due to tumors and cerebrovascular accidents.

Retinitis and optic neuritis may cause a degeneration of the retina and optic nerve, respectively. Optic atrophy, in turn, results in a decrease in visual acuity (amblyopia). In addition, damage to the retina or optic nerve frequently results in the presence in the visual field of abnormal blind spots called scotomas, as mentioned earlier. Circumferential, partial, unilateral, or even total blindness (amaurosis) may also occur in response to lesions of the retina, optic nerve, and visual cortex, without apparent eye damage.

### CRANIAL NERVES III, IV, AND VI

The *oculomotor nerve* (IIId cranial nerve) is entirely motor, supplying fibers to the superior rectus, inferior rectus, medial rectus, and inferior oblique extrinsic eyeball muscles and to the pupillary sphincter and ciliary intrinsic eyeball muscles. The *trochlear nerve* (IVth cranial nerve) and *abducens nerve* (VIth cranial nerve) supply motor fibers to the superior oblique and the lateral rectus extrinsic eyeball muscles, respectively.

Disorders (see Chap. 12) related to involvement of the oculomotor, trochlear, and abducens nerves result primarily in abnormalities in the movements of the eyeballs. Impairment in the operation of the extrinsic muscles of the eye leads to one or more of the following disorders: (1) muscular weakness affecting vision and balance, giving rise to paresthesias and headaches; (2) extrinsic muscular paralysis (ophthalmoplegia), causing a deviation in the alignment of one or both eyes (strabismus) as well as double vision (diplopia); and (3) involuntary, repetitive movements of the eyes in one or more planes in space (nystagmus). Because parasympathetic fibers of the oculomotor nerve supply the ciliary and pupillary sphincter muscles within the eye, lesions of this nerve may also cause almost imperceptible alternations of pupillary constriction and dilatation under conditions of constant illumination. Upper lid drop, or ptosis, due to paralysis of the levator palpebrae muscle, is also noted when the oculomotor nerve is damaged. Other symptoms of impairment to the oculomotor, trochlear, and abducens nerves include compensatory adjustments of the head in attempts to correct for diplopia, diminution in the pupillary reflex, restrictions in the range of movements exhibited by the eyeballs, and vertigo due to visual disorientation.

The disorders enumerated above have diverse causes including physical trauma such as fractures of the sphenoid and frontal bones as well as the petrous portion of the temporal bone, vascular disturbances such as aneurysms of the internal carotid artery or cerebrovascular accidents involving other arteries at the base of the brain, and microbial infections such as cerebral meningitis or neurosyphilis.

## CRANIAL NERVE V

The *trigeminal nerve* (Vth cranial nerve) mediates the following functions:

1 Facial sensations originating from the superoanterior and lateral aspects of the head, by means of the ophthalmic division
2 Oral sensations originating from the upper part of the oral cavity, by means of the maxillary division
3 Oral sensations originating from the lower part of the oral cavity and sensations associated with the external ear, both by means of the mandibular division

4 Proprioceptive feedback information from the muscles of mastication, by means of fibers associated with the maxillary and mandibular divisions
5 Motor control of the muscles of mastication and certain muscles associated with hearing and swallowing, by means of fibers which run parallel to the mandibular division

The trigeminal nerve is therefore a mixed nerve (sensory and motor).

Abnormalities associated with the trigeminal nerve affect sensations of touch, pain, and temperature from various portions of the face and related regions of the head; the ability to chew or masticate food, because of paralysis of the masseter, temporal, internal ptyergoid, and external ptyergoid muscles; the ability to hear, because of paralysis of the tensor tympani muscle of the tympanic membrane; and the ability to perform specific facial reflexes such as the jaw jerk, sneezing, and eye blink (corneal reflex). In addition, severe neuralgic or neuritic pains, anesthesia, and paresthesias may accompany trigeminal nerve involvement.

One of the well-known clinical syndromes associated with lesions of the trigeminal nerve is called *tic douloureux*, or *trifacial neuralgia*. The disorder is characterized by severe, extremely transient, paroxysmal pains along the distribution of one or more branches of this cranial nerve. Autonomic activity which promotes an increased rate of secretion of the lacrimal glands and nasal mucosa usually accompanies a paroxysm of tic douloureux. The disability is unilateral (appearing on only one side of the face).

The causes of tic douloureux are not known with certainty, but infections of the teeth and paranasal sinuses are implicated. Aside from microbial infections which affect it, the trigeminal nerve may be damaged by physical trauma such as skull fractures of the petrous portion of the temporal bone which involve the mandibular division, fractures of the maxillary bone which involve the maxillary division, or fractures of the frontal and temporal bones which involve the ophthalmic division; vascular disorders such as aneurysms of the circle of Willis or cerebrothromboses at the base of the brain; and brain tumors and other diseases which affect either the cerebral aqueduct and pons by involving motor roots or the midbrain by involving sensory roots.

## CRANIAL NERVE VII

The *facial nerve* (VIIth cranial nerve) mediates taste sensations from the anterior two-thirds of the tongue; proprioceptive feedback information from various facial muscles; motor control of the muscles of facial expression and the tongue; and secretory regulation of the lacrimal, submandibular, and sublingual glands. The facial nerve is therefore a mixed nerve.

Lesions of the facial nerve result in both sensory and motor disabilities in the facial, oral, and cervical regions. The following disabilities may be seen: (1) impairment in the salty, sweet, and sour modalities of taste (hypogeusia) due to lesions of peripheral branches of sensory fibers; (2) altered or impaired auditory abilities due to lesions which involve the stapedius muscle of the stapes or which spread along the facial nerve to involve the tympanic membrane and the acoustic (VIIIth cranial) nerve (see below); (3) flaccid or spastic muscle paralysis in the scalp, face, and neck which affect the ability to voluntarily control facial expression and specific movements of the eyes and mouth, or puckering the lips due to lesions of motor fibers which innervate these structures; and (4) autonomic responses such as tearing, or lacrimation, and depression in the secretion of saliva due to lesions of parasympathetic fibers associated with the facial nerve. In addition to facial palsies (loss of sensation and movement), involvement of the facial nerve may induce neuralgia and other neurologic symptoms and signs. Sometimes along with the facial nerve, the abducens and acoustic nerves are also damaged because of their close anatomic relation to the facial nerve. In these cases, additional symptoms may include abnormalities in the movement of the eyeballs or the sense of hearing.

One of the well-known clinical syndromes associated with lesions of the facial nerve is called *Bell's palsy*, or peripheral facial paralysis. The disorder is characterized by an involuntary, upward-rolling movement of the eyeball on the affected side as the eyelids close. This characteristic sign is combined with some combination of the disabilities described above.

Bell's palsy and other types of abnormalities involving the facial nerve are caused by fractures of the petrous portion of the temporal bone which involve motor fibers of the facial nerve; diseases of the pons such as vascular disturbances, infectious disorders, or tumors which involve the nucleus of the facial nerve; and other factors which have neuromuscular effects in the facial region such as prolonged exposure of the face to extremely low temperatures or infections of the middle ear which spread to the peripheral nerves.

## CRANIAL NERVE VIII

The *acoustic nerve* (VIIIth cranial nerve) is entirely sensory, conducting afferent nerve impulses from the following structures: the sense of balance from the semicircular canals, utricle, and saccule by means of the vestibular nerve, and the sense of hearing from the organ of Corti within the cochlea by means of the cochlear nerve.

Lesions of the acoustic nerve result in sensory disabilities with respect to hearing and equilibrium. The symptoms indicative of *cochlear nerve* involvement are nerve deafness due to an interruption of the neural pathway from the cochlea; deafness to selective frequencies of sound (aural scotomas) due to differential damage of the hair cells of the organ of Corti; and noises within the ears (tinnitus) caused by endogenous factors. Symptoms of *vestibular nerve* damage include vertigo, caused by pathogenic factors which affect the middle or inner ear, and to-and-fro repetitive movements (slow "to" and rapid "fro" movements) of the eyes (nystagmus), induced by functional disruption in the normal feedback mechanism between the semicircular canals and the eyes.

One of the well-known clinical syndromes associated with lesions to both the cochlear and the vestibular nerves, called Ménière's syndrome, is discussed in Chap. 12.

Lesions of the cochlea, semicircular canals, and peripheral nerve fibers and roots may be caused by inflammation of the middle or inner ear, fractures involving the temporal bone, and degenerative diseases such as multiple sclerosis. Damage to the vestibular and cochlear nuclei in the pons may result from congenital defects, microbial infections, brain tumors, degenerative diseases, or psychogenic factors.

## CRANIAL NERVES IX, X, XI, AND XII

The glossopharyngeal and vagus nerves (IXth and Xth cranial nerves, respectively) are both mixed, while the spinal accessory (XIth cranial) and hypoglossal (XIIth cranial) nerves are both motor. The last four cranial nerves have the following functions:

1  Mediation of the sense of taste from the posterior third of the tongue (IXth cranial nerve)
2  Mediation of sensations associated with the pharynx (IXth)
3  Mediation of sensations associated with the external ear and visceral structures in the thoracic and abdominal regions (Xth)
4  Motor control of specific pharyngeal muscles and secretory regulation of the parotid gland (IXth)
5  Motor control of pharyngeal, laryngeal, thoracic, and abdominal muscles (Xth)
6  Motor control of the sternocleidomastoid muscle (XIth) and the tongue (XIIth)

Because of their anatomic proximity, the last four cranial nerves are frequently damaged simultaneously. Sometimes only one, two, or three of these cranial nerves are involved. Occasionally, portions of other cranial nerves, such as the oculomotor and trigeminal, are also affected.

Lesions of the IXth and Xth cranial nerves may result in the following *sensory* disabilities: loss of bitter taste sensations (IXth) and anesthesia of the pharynx (IXth and Xth) and larynx (Xth).

The following *motor* dysfunctions are due to damage of the IXth to XIIth cranial nerves:

1  Increased salivation due to excessive parasympathetic stimulation of the parotid gland (IXth)
2  Impairment in, or loss of, speech due to paralysis of the vocal cords (Xth)
3  Difficulty in swallowing due to impairments of the pharyngeal muscles (Xth)
4  Difficulty in nodding or shaking the head due to paralysis and possible atrophy of the sternocleidomastoid muscle (XIth); the presence of a drooping or squared shoulder and a displaced scapula due to paralysis and subsequent degeneration of the trapezius muscle (XIth)
5  Spastic or flaccid paralysis of the tongue, with or without degeneration (XIIth)

Wry neck, or torticollis, a disorder caused by involvement of the XIth cranial nerve, is shown in Fig. 25-4.

These sensory and motor disabilities lead to numerous disturbances in reflex acts and autonomic functions. Among the automatic activities affected are the gag reflex; deglutition, or the swallowing reflex; salivation, or the salivary reflex; the cardiac reflex; and the respiratory, or Hering-Breuer, reflex.

Probably the most frequent cause of damage to one or more of the last four cranial nerves are fractures of the base of the skull which involve the temporal and occipital bones. Such fractures cause lesions to the roots of these cranial nerves as they emerge from the skull. Damage to their nuclei can result from brain tumors and infectious diseases, e.g., neurosyphilis. Dislocations of the upper cervical vertebrae may injure motor fibers of the XIth and XIIth cranial nerves. In addition, irritation or inflammation of the IXth to XIIth cranial nerves can be brought about by vascular disorders including aneurysms and thromboses; chemicals such as heavy metals, alcohol, and carbon monoxide; or microbes which cause neuritic symptoms.

## SPINAL NERVE DISORDERS

These disorders are the result of lesions to one or more spinal nerves. The anatomic organization of the spinal nerves is described in Chap. 14.

The spinal nerves are arranged in functional and structural groupings called plexuses. For purposes of the following discussion, we will recognize the following specialized regions: cervical plexus, brachial plexus, thoracic nerves, lumbar plexus, sacral plexus, and sacrococcygeal plexus. The spinal nerve plexuses are illustrated in Fig. 14-7.

## CERVICAL PLEXUS

The cervical plexus is made up of the first four pairs of cervical nerves. The nerves supply sensory fibers to the skin surrounding the ear, the neck and posterior portion of the shoulders, the superolateral aspect of the arm, the dorsal segments of the chest, and the thoracic peritoneum and diaphragm. Motor fibers of the cervical plexus innervate skeletal muscles in these regions. Of major importance as branches of the cervical plexus are the *phrenic nerves*, which exit the spinal cord through the intervertebral foramina associated with the anterior roots of the third and fourth cervical nerves as well as the fifth cervical nerve of the adjacent brachial plexus. The phrenic nerves innervate their respective sides of the diaphragm, controlling contractions of this compound muscle.

Lesions of the cervical plexus usually involve one or both phrenic nerves. Therefore, the most important functional disability of cervical plexus injuries is partial or complete paralysis of the diaphragm. The loss of diaphragmatic respiration leads to extreme dyspnea (difficult breathing) upon the slightest physical exertion. The tremendous increase in the respiratory work load is reflected by an increase in activity of the thoracic and abdominal muscles, causing a profound heaving of the shoulders, chest, and abdomen. Neuralgic pains in the anterior region of the neck and upper chest may accompany the dyspneic form of breathing. Phrenic nerve involvement also results in hiccuping (see Chap. 29), since this nerve initiates the hiccuping reflex.

Physical trauma and microbial infection are probably responsible for most lesions of the cervical plexus. Mechanical injuries can result from whiplash, dislocations, or fractures of the cervical vertebrae, or from tumors in the neck region. Irritation or inflamma-

tion may also follow other diseases which affect the central nervous system (CNS).

## BRACHIAL PLEXUS

The brachial plexus is composed of the last four pairs of cervical nerves and the first pair of thoracic nerves. These nerves innervate skeletal muscles associated with the lateral aspect of the upper chest, shoulder, scapula, upper arm, forearm, and hand. The sensory distribution of the brachial plexus parallels its motor supply.

The brachial plexus is large and ramifying, made up of the musculocutaneous, axillary, radial, median, ulnar, and numerous lesser nerves. The *radial nerve* is the largest of the nerves which come from the brachial plexus and the most vulnerable of the peripheral nerves to injury. Consisting of both superficial and deep portions which supply the forearm, the radial nerve takes its origin from branches of the sixth to eighth cervical nerves and the first pair of thoracic nerves.

Any factor which damages the cervical spinal cord, the anterior roots of nerves belonging to the brachial plexus, or the peripheral branches of these nerves causes paralysis of the skeletal muscles supplied by the affected nerves. Damage may result from pulling, pinching, pressing, compressing, or tearing the nerves. Brachial plexus damage is a common birth injury and, therefore, an affliction not uncommonly seen in early childhood. Upper plexus lesions are frequently caused by tears, occurring during birth, of the fifth and sixth cervical nerves. These lesions impair movements of the arm and forearm. The affected muscles may subsequently atrophy. Lesions of the brachial plexus are also caused by skeletal injuries such as dislocations or fractures of the shoulder, neck, arm, and forearm, especially of the humerus, radius, ulna, and olecranon process; abnormal pulling such as in wrenching the arm and excessive compression such as in sleeping on an arm or hand, carrying heavy weights on the head, or violent blows in these regions; injuries resulting from gunshot or stab wounds which cause nerve lacerations; abnormal mechanical pressures induced endogenously by aneurysms or tumors; and irritative lesions produced by chronic intoxications or microbial pathogens which involve the CNS.

The syndromes produced by lesions of the brachial plexus are highly variable, because of the size and complexity of the grouping of spinal nerves and their various branches. Muscular weakness or partial or complete paralysis and atrophy of the skeletal muscles supplied by the damaged nerves are the most obvious motor symptoms. Wristdrop, a condition caused by lesions of the radial nerve, is shown in Fig. 25-6. Sensory impairments are not as noticeable because of overlaps between sensory nerve endings of adjacent spinal nerves. However, localized anesthesias are a common corollary of brachial plexus involvement. Metabolic disturbances which follow loss of a nerve supply lead to skin abnormalities in the affected region such as drying and chapping, coldness and discoloration, or keratinization and ulcerations. Bleeding sores are due to an impairment of tissue regeneration following injury. The fingernails can become brittle and cracked. When blood vessels are also affected, the skin may appear cyanotic (bluish) or edematous (swollen). In addition, neuralgic, neuritic, or polyneuritic pains and paresthesias may be experienced along the distribution of the involved nerves.

## THORACIC NERVES

The 12 pairs of thoracic nerves emerge from the spinal cord through intervertebral foramina associated with the 12 thoracic vertebrae. Most fibers of the first pair of thoracic nerves contribute to the brachial plexus, with only a small component merging with the second pair. The twelfth pair of thoracic nerves contributes fibers to the lumbar plexus. Thoracic nerves provide sensory and motor fibers to the skin and skeletal muscles, respectively, of the mid- and lower back, the lateral aspect of the chest, and the anterior region of the abdomen.

Since the intercostal and abdominal muscles are innervated by branches of the thoracic nerves, respiration, defecation, and some other reflexes in which the musculature in the thoracic and abdominal walls participate may be somewhat impaired by thoracic nerve lesions. However, the effects are not usually dramatic and may be partially compensated for by supplemental movements of uninvolved muscles. Neuritic pains sometimes accompany the motor loss.

The causes of thoracic nerve damage are similar to those which involve other groupings of spinal nerves and include mechanical pressure, physical trauma, microbial infection, and chronic intoxication.

## LUMBAR PLEXUS

Components of the lumbar plexus include a portion of the twelfth pair of thoracic nerves, all of the first three pairs of lumbar nerves, and a small contribution from the fourth lumbar pair of spinal nerves. The *femoral* and *obturator nerves* are major branches of the lumbar plexus. The femoral nerve is the largest peripheral nerve belonging to this plexus. Sensory fibers which belong to these spinal nerves supply the skin in the pubic and genital regions as well as the skin of the thigh and medial aspect of the shank of the leg and foot. Motor fibers of the lumbar plexus innervate skeletal muscles of the lower extremities.

The most marked symptoms of involvement of the lumbar plexus are weakness or paralysis and possible atrophy of muscles of the thigh and shank of the leg. Movements affected include flexion and extension, abduction and adduction, knee jerk, and crossing the legs. These disabilities are sometimes accompanied by neuritic pains, paresthesias, and anesthesia in the affected regions.

These disablements are usually induced by factors which place extreme pressure on the femoral nerve or other branches of the lumbar plexus. Such mechanical pressures may result from an enlarged uterus during pregnancy, abdominal tumors, fractures of the femur, obesity, or even tight clothing. Since the lumbar vertebrae are the thickest and strongest vertebrae of the vertebral column, direct damage to the lumbar spinal cord and its nerve roots is either immediately fatal or causes complete paralysis of that portion of the body distal to the point of injury.

## SACRAL PLEXUS

The sacral plexus is composed of a portion of the fourth pair and all of the fifth pair of lumbar nerves, in addition to the first three pairs of sacral nerves. The largest peripheral nerve in the body, the *sciatic nerve*, belongs to this plexus. The sciatic nerve is compound, dividing into the common peroneal and tibial nerves near the distal end of the femur. The sacral plexus supplies the buttocks, the hamstring muscles of the thigh, and significant portions of the shank of the leg and the foot.

Lesions of the sacral plexus may cause motor deficits ranging from paralysis of the gluteal and hamstring muscles to complete muscular paralysis of the leg and foot. The motor disabilities are reflected by losses of the Achilles tendon reflex and plantar reflex. Sometimes isolated damage to the common peroneal or the tibial nerve leads to an inability to support the foot, a condition known as foot drop, or to an involuntary spreading and flexion of the toes, a symptom referred to as claw foot.

Sensory deficits consist of anesthesia, paresthesias, and possible neuritic or polyneuritic pains in the regions supplied by the involved nerves. The sciatic nerve is particularly susceptible to factors which cause irritation and inflammation. The intense, sometimes burning pains characteristic of sciatic nerve involvement are part of a syndrome called *sciatica*. The pains are localized along the posterior and medial aspects of the thigh. Variable in quality, frequency, and extent, these pains are described as *causalgic*. Changes in the barometric pressure, as at the approach of an impending storm, frequently exacerbate a chronic sciatic condition.

Involvement of branches of the sacral plexus occasionally causes metabolic changes in the skin overlying affected regions. Such effects include a skin which is cold and clammy, loss of integumentary hair (hypotrichosis), and an increased brittleness and cracking of the toenails. In addition, possible vasomotor responses may result in edema and cyanosis.

One of the most frequent causes of damage to some portion of the sacral plexus is a rupture, or herniation, of an intervertebral disk which compresses either the sacral cord or roots of the last two lumbar spinal nerves. The herniation of the intervertebral disk is illustrated in Fig. 25-2. Mechanical compression may also result from pelvic tumors and fractures, from dislocations and fractures involving the long bones of the legs, or even from kneeling or sitting with legs crossed for extended periods of time. Toxic chemicals such as alcohol and heavy metals or microbial pathogens which infect the CNS can lead to toxic or infectious neuritis of the sacral plexus. Lacerations of the peripheral nerves of this plexus may result from automobile accidents or gunshot and stab wounds which involve the lower extremities.

## SACROCOCCYGEAL PLEXUS

The sacrococcygeal plexus is derived from portions of the second, third, and fourth pairs of sacral nerves, all of the fifth sacral nerve pair, as well as the sole nerve associated with the coccyx. The fourth pair of sacral nerves forms the major contribution to this plexus. Branches from the plexus supply the external anal sphincter, the penis and scrotum or the clitoris and labia majora, other structures in the perineal region, and the coccyx.

Involvement of the sacrococcygeal plexus results in impairment of the defecatory reflex. Involuntary defecation is induced through the loss of voluntary control of the external anal sphincter. Spasmodic contractions of the anal sphincter create a desire to defecate which is mostly frustrated when attempted. Urogenital pains frequently accompany this type of motor disability. Neuralgic pains emanate from the coccygeal region, especially in females, when the last two pairs of sacral nerves and the coccygeal nerve are damaged.

Because the second to fifth pairs of sacral nerves are a significant part of the sacrococcygeal plexus, lesions of this plexus are usually due to damage of the sacral plexus. In other words, rarely is there neural involvement in the urogenital, perineal, and coccygeal regions without motor and sensory disturbances in the buttocks and legs.

## COLLATERAL READING

Chusid, Joseph G., and Joseph J. McDonald, *Correlative Neuroanatomy and Functional Neurology,* 12th ed., Los Altos, Calif.: Lange Medical Publications, 1964. Chapters 4, 5, and 6 review disorders of the cranial nerves, the spinal nerves, and the autonomic nervous system, respectively.

# CHAPTER 17
# THE ENDOCRINE SYSTEM

The endocrine system is a loosely knit and widely distributed grouping of cells, tissues, and organs which synthesize and secrete hormones. *Hormones* are natural products of the body released from glands into the blood without passing through ducts. For this reason, endocrine glands are commonly referred to as *ductless glands*.

Endocrine secretions elicit both short- and long-term physiologic responses in specific organs, called *target organs*, that are sensitive to their presence. Often the response is restricted to a specific type of tissue or cell within the target organ, although sometimes the response involves most body tissues. To the extent that a hormonally mediated response is highly specific, it at least superficially resembles a neural reflex arc.

The endocrine system also interacts with the nervous system; i.e., hormones affect neural functions, and vice versa. A few hormones secreted as a consequence of direct neural stimulation elicit physio-

logic responses as rapidly as the nervous system itself. Most of us at one time or another have experienced a rapid, pounding heart, and perhaps thin, shallow breathing, perspiration, and heightened alertness following an emergency maneuver such as avoiding an automobile accident. Caused by the release of epinephrine when the sympathetic autonomic nervous system is stimulated, the physiologic reaction is immediate, forceful, and obviously adaptive.

By way of contrast, the menstrual cycle, pregnancy, and lactation are periodic, relatively long-lasting events which are largely regulated by the complex interplay between numerous hormones and their target organs. The action of hormones and their functional resemblance to reflex arcs lead us to the conclusion that both endocrine and nervous systems integrate the diverse functions of the body and that their actions are in many respects complementary.

In contrast to the fixed path of a reflex arc, however,

hormones are transported through the blood to their target organs. Since numerous hormones are always present in the blood, the excitation or inhibition of an endocrine gland simply increases or decreases the intensity of one or more physiologic responses already being exhibited. You may recognize that this characteristic of hormones is analogous to the mediating role of enzymes in chemical reactions (see Chap. 10).

Since hormones are molecules, their effects are molecular. The story of how hormones exert their actions is rapidly unfolding at the present time. Current research indicates that the molecular events during memory storage, inflammation, and hormonal action may well have important common denominators. In other words, it seems that many messages translated by cells in response to the stimulation provided by hormones, neurons, or certain other natural products of the body may follow a common molecular pattern. These exciting findings suggest mechanisms by which the interactions between, and regulatory functions of, the nervous and endocrine systems are accomplished.

In this chapter, we will examine the organization and regulation of the endocrine system as well as the general characteristics and mechanisms of action of hormones. The normal physiologic responses of individual hormones are discussed in the chapters relevant to their actions. Responses of the body to abnormal hyper- and hyposecretions of the various hormones are considered in detail in the next chapter.

## ORGANIZATION OF THE ENDOCRINE SYSTEM

The endocrine system consists of at least 13 glands which secrete hormones from specialized tissues or cells. The locations in the body of the various endocrine glands and the hormones they secrete are identified in Fig. 17-1. The list is expected to grow as a result of the application of more sophisticated physiologic and biochemical techniques to the detection and identification of hormones.

Notice that most endocrine glands produce more than one hormone. Usually each hormone is pro-

duced by a specialized type of tissue or cell within the gland, although there may be exceptions to this rule. The synthesis of more than one hormone in a cell, if it does occur, requires a physical separation of synthetic pathways. As we learned in Chap. 7, this separation can be accomplished by membranes which divide a cell into different functional compartments or by the differential adsorption of specific raw materials to particular enzymes in the same cellular compartment.

Although hormones have profound effects on every physiologic process in the body, physiologic mediation by hormones is especially prominent in reproduction and development, digestion and energy metabolism, electrolyte and fluid balance, and body defense, i.e., in various aspects of metabolism and reproduction.

In many cases, the locations of the endocrine glands in the body are related to their functions. The following are examples of this location-function relationship. Brain hormones are produced in direct response to appropriate brain stimulation. Pancreatic hormones affect the stores of liver glycogen. Hormones of the digestive tract regulate chemical digestion within its lumen. Gonadal hormones are responsible for the maturation of the testes or ovaries and related reproductive structures. Placental hormones influence the ovaries and, therefore, the uterus.

Many hormones act by synergizing or antagonizing the effects of another hormone. Such hormones belong to the same functional constellation. The following examples identify the major functional constellations. Parathyroid hormone, calcitonin, and calciferol from the parathyroid gland, thyroid gland, and skin, respectively, are all mediators of $Ca^{++}$ and $HPO_4^{--}$ levels in blood and bone. ADH from the hypothalamus (by way of the hypophysis) and aldosterone from the adrenal cortices are both important in the maintenance of the fluid balance of the body. A specific releasing factor from the hypothalamus, ACTH from the hypophysis (pituitary gland), cortisol from the adrenal cortices, and insulin and glucagon from the pancreas are the primary mediators of carbohydrate metabolism and energy metabolism in general. The rate of metabolism is regulated by two

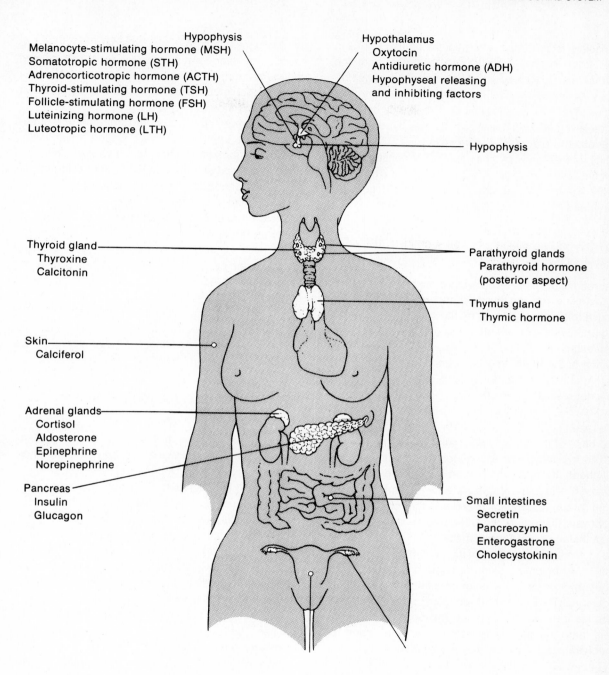

**Hypophysis**
Melanocyte-stimulating hormone (MSH)
Somatotropic hormone (STH)
Adrenocorticotropic hormone (ACTH)
Thyroid-stimulating hormone (TSH)
Follicle-stimulating hormone (FSH)
Luteinizing hormone (LH)
Luteotropic hormone (LTH)

**Hypothalamus**
Oxytocin
Antidiuretic hormone (ADH)
Hypophyseal releasing
and inhibiting factors

Hypophysis

Thyroid gland
Thyroxine
Calcitonin

Parathyroid glands
Parathyroid hormone
(posterior aspect)

Thymus gland
Thymic hormone

Skin
Calciferol

Adrenal glands
Cortisol
Aldosterone
Epinephrine
Norepinephrine

Pancreas
Insulin
Glucagon

Small intestines
Secretin
Pancreozymin
Enterogastrone
Cholecystokinin

**Figure 17-1** *The locations and natural products of endocrine glands. Since the female body is illustrated, the testes are not shown.*

releasing factors from the hypothalamus, TSH and STH from the hypophysis, and thyroxine from the thyroid gland. Reproduction is controlled either directly or indirectly by hormones from the hypothalamus, hypophysis, and gonads. Pregnancy and lactation are mediated by hypothalamic, hypophyseal, ovarian, and placental hormones.

Notice that many of these functional constellations involve the hypothalamus and hypophysis. The regulatory roles of these brain structures will now be explained.

## HYPOPHYSEAL MEDIATION OF HORMONAL SECRETIONS

Secretions of all the reproductive hormones as well as of thyroxine from the thyroid gland and the adrenocorticosteroids are regulated by the hypophysis. Weighing about one-thousandth of a pound, the *hypophysis*, or pituitary gland, lies at the base of the diencephalic region of the brain, inferior to the hypothalamus and immediately posterior to the optic chiasm. It is nestled in a depression of the sphenoid bone called the sella turcica, because of its resemblance to a Turkish saddle (see Fig. 23-7). The location of the hypophysis at the base of the brain is illustrated in Fig. 14-1.

The hypophysis is a complex gland with multiple functions. Its posterior portion, called the *pars nervosa* or *neurohypophysis*, is derived embryonically from brain ectoderm. Therefore, the posterior hypophysis, including the infundibular stalk from the hypothalamus, is a part of the brain. On the other hand, the anterior portion of the hypophysis is derived from embryonic skin ectoderm. Called the *adenohypophysis*, the anterior pituitary gland does not consist of neural tissue. This derivational difference between the anterior and posterior hypophysis has profound implications for the types of mechanisms by which the hypothalamus regulates hypophyseal functions, as we shall see. The adenohypophysis also includes a specialized region called the *pars intermedia*. As its name implies, the pars intermedia lies between the anterior and posterior hypophysis.

The adenohypophysis is highly glandular. Excluding the pars intermedia, six different hormones are secreted by the adenohypophysis, each from its own specialized cell type. The target organs and effects of these adenohypophyseal hormones are identified on the left-hand side of Fig. 17-2.

Only one hormone is secreted from the pars intermedia of the adenohypophysis. Called *melanocyte-stimulating hormone* (MSH), or *intermedin*, this hormone in superabundance causes a darkening of the skin. Since changes in skin pigmentation are unimportant in man, in contrast to many lower animals, the physiologic role, if any, of MSH in human beings is poorly understood.

The two hormones released from the *pars nervosa*, on the other hand, are actually synthesized in the hypothalamus. Apparently, the pars nervosa serves only as a storage depot for hypothalamic hormones and can be removed without appreciably affecting the functions mediated by the hormones stored there. The target organs and effects of these hormones are identified on the right-hand side of Fig. 17-2.

## HYPOTHALAMIC REGULATION OF HYPOPHYSEAL FUNCTIONS

The two hypothalamic mechanisms which regulate the production and release of hypophyseal hormones are diagramed in Fig. 17-3.

Basically, the secretion of adenohypophyseal hormones depends upon the presence of hormonal factors synthesized in the hypothalamus (refer to Fig. 17-3B). The hypothalamic hormones are produced in response to appropriate neural or hormonal stimulation. Each of the six hormonal factors secreted by the hypothalamus is synthesized in its own specialized neurosecretory cells. Instead of conducting nerve impulses, these cells release hormones from their axonal tips. The hormonal factors enter the blood within a capillary plexus inside the infundibulum and are transported by a special grouping of venules known as the *hypophyseal portal system* to a venous plexus surrounding nests of secretory cells inside the adenohypophysis.

After leaving the blood within the anterior hypophysis, most of the hypothalamic hormonal factors

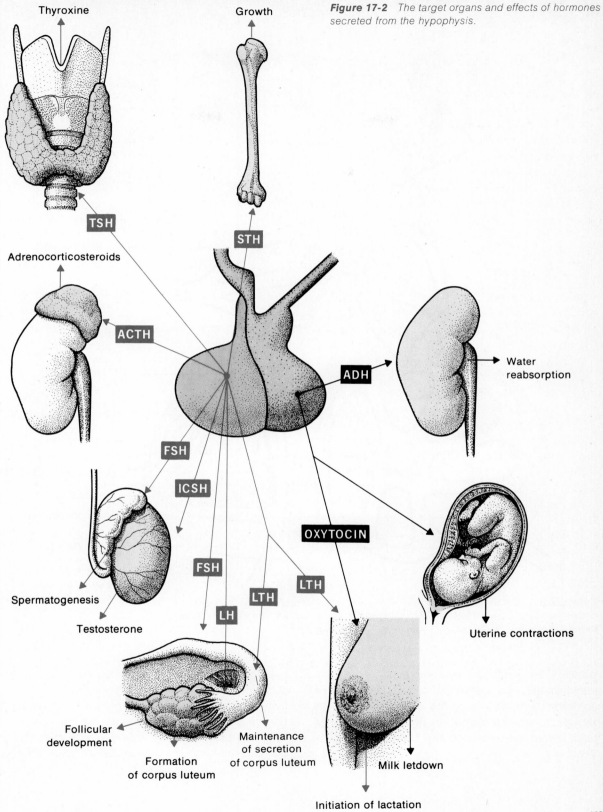

Thyroxine

Growth

**Figure 17-2** *The target organs and effects of hormones secreted from the hypophysis.*

TSH

STH

Adrenocorticosteroids

ACTH

ADH

Water reabsorption

FSH

ICSH

OXYTOCIN

FSH

LTH

LTH

Spermatogenesis

LH

Testosterone

Uterine contractions

Follicular development

Maintenance of secretion of corpus luteum

Milk letdown

Formation of corpus luteum

Initiation of lactation

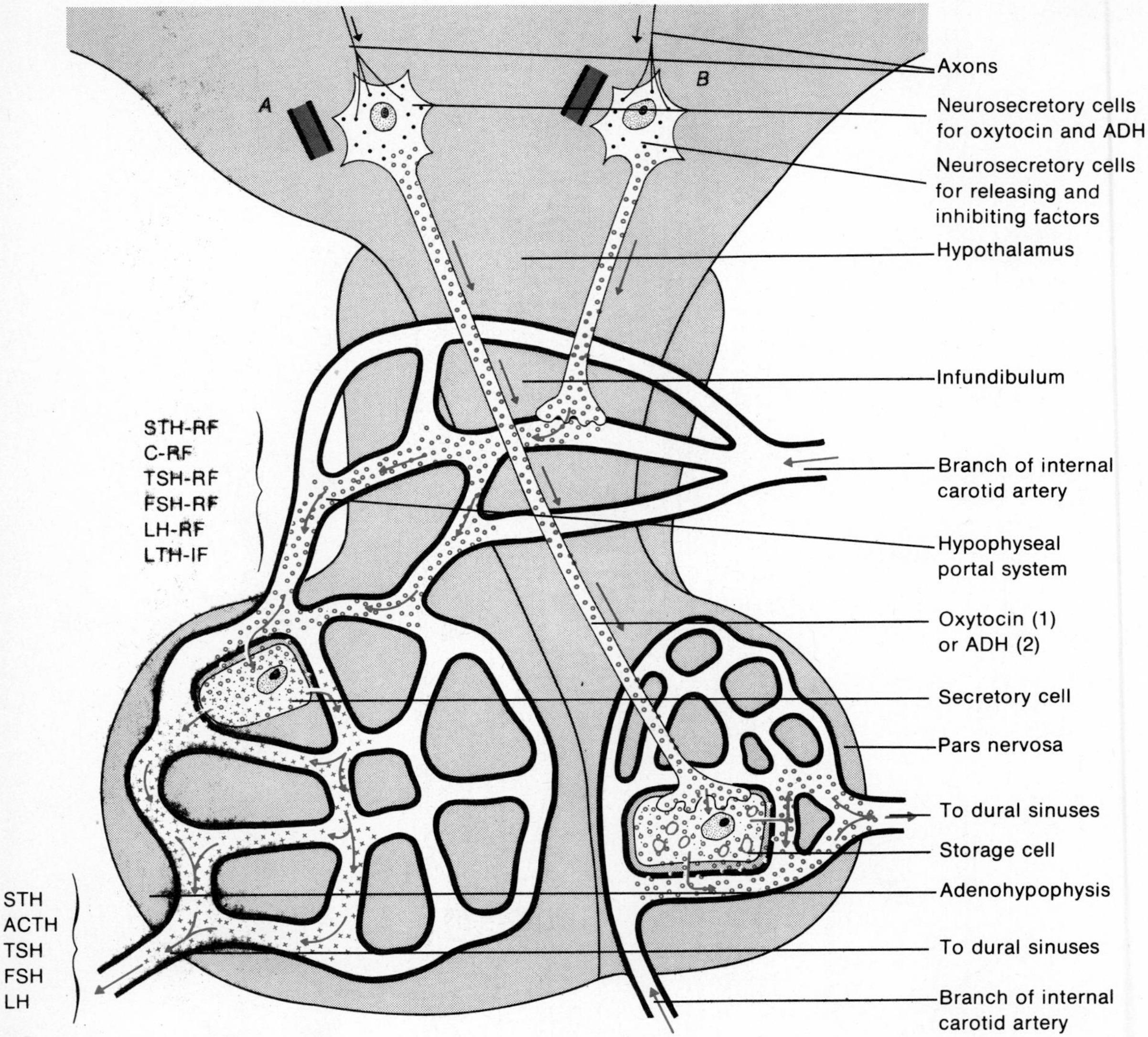

**Figure 17-3** *Hypothalamic involvement in the production of hypophyseal hormones.* A. *The mechanism by which hormones are secreted from the pars nervosa.* B. *The mechanism by which adenohypophyseal hormones are secreted. RF = releasing factor; IF = inhibiting factor.*

stimulate the synthesis and secretion of specific adenohypophyseal hormones. Following their synthesis, these hormones enter the surrounding venous plexus and are transported to dural sinuses in the head before reaching the heart and the general circulation.

All but one of the hypothalamic hormonal factors stimulate the release of adenohypophyseal hormones. These hypothalamic hormones are called *releasing factors*. The exception to the rule is a hypothalamic *inhibiting factor* which specifically prevents the synthesis of luteotropic hormone (LTH);

i.e., in the absence of this inhibiting factor, LTH is secreted by the adenohypophysis.

As we noted earlier, hormones released from the pars nervosa are actually secreted by separate clusters of neurosecretory cells originating in the hypothalamus (see Fig. 17-3A). The axons of these highly modified neurons terminate around clusters of storage cells inside the pars nervosa in close proximity to a network of capillaries. Even when unstimulated, these neurosecretory cells synthesize small quantities of specific hormones which enter the storage cells of the pars nervosa at a low rate.

When excited by appropriate neural or hormonal stimulation, on the other hand, the neurosecretory cells synthesize large quantities of their hormonal products. These hypothalamic hormones either enter the posterior hypophyseal blood directly or pass through the storage cells and into the blood. This blood is transported to dural sinuses in the head before reaching the heart and the general circulation, following a circulatory pathway similar to that of the adenohypophyseal hormones.

Observe that releasing or inhibiting factors play no part in the secretion of hormones from the pars nervosa. Indeed, it is worth reemphasizing that the role of the pars nervosa in hormonal secretion is entirely passive, revolving around the storage and release of hormones synthesized by hypothalamic neurosecretory cells.

## HORMONES

### CHEMICAL NATURE

Basically, hormones are divided into two chemical classes: (1) proteins (more than 50 amino acid residues), derivatives of proteins such as polypeptides or peptides (from 2 to 50 amino acid residues), and derivatives of amino acids; and (2) a class of lipids called *steroids*. The structures of most of the hormones are shown in Figs. 18-1 through 18-10.

The basic molecular skeleton of any steroid is given in Fig. 17-4. The hormonal derivatives of this heterocyclic (multiringed) hydrocarbon molecule include (1) the adrenocorticosteroids such as the glucocorticoid cortisol and its derivatives as well as the mineralocorticoid aldosterone; (2) the male sex hormones, or androgens, of which the most common is testosterone; (3) the female sex hormones including the estrogens, of which the most common is $17\beta$-estradiol, as well as progesterone; and (4) calciferol, a hormone formed following the ultraviolet irradiation of 17-dehydrocholesterol in the skin. Notice that each grouping of steroidal hormones such as the glucocorticoids, the androgens, and the estrogens contains numerous chemically related compounds, but only one of the compounds in each group is in greatest abundance within the body and, therefore, is most important physiologically.

The reason for the similarities in the chemical structures between the gonadal hormones and the corticosteroids has to do with the embryonic origin of the glands in which they are synthesized. This topic is discussed in Chap. 19.

Several hormones are derivatives of individual amino acids. These hormones include thyroxine from the thyroid gland and epinephrine and norepinephrine from the adrenal medullae. All three hormones are actually derivatives of the amino acid tyrosine. However, the catecholamines epinephrine and norepinephrine are also synthesized from the amino acid phenylalanine or from a related compound called dopamine.

Apparently all the hypothalamic hormones consist of polypeptides containing relatively small numbers of amino acids. Hormones thus far synthesized from the hypothalamus include oxytocin and ADH as well as two of the releasing factors, thyroid-stimulating hormone releasing factor (TSH releasing factor) and luteinizing hormone releasing factor (LH releasing factor).

**Figure 17-4** *The basic molecular skeleton of steroidal hormones. The skeleton is a heterocyclic hydrocarbon.*

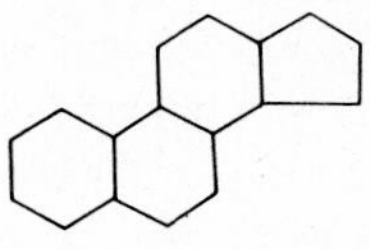

Both hypothalamic hormones released from the pars nervosa of the hypophysis are peptides containing nine amino acid residues. Each of these hormones differs in composition by only two of the nine amino acid residues in their chains. In other words, oxytocin and ADH are similar structurally.

TSH releasing factor also consists of a chain of nine amino acid residues, although its amino acid composition differs considerably from either oxytocin or ADH. LH releasing factor, on the other hand, is a tripeptide, i.e., a peptide consisting of three amino acid residues.

In addition to the hormones of the hypothalamus, other hormonal polypeptides include (1) the digestive hormones secretin, enterogastrone, cholecystokinin, and pancreozymin (the last two hormones are probably identical); (2) MSH from the pars intermedia; (3) the metabolic hormone glucagon from the pancreas; (4) several hormones involved with $Ca^{++}$ and $HPO_4^=$ regulation, including parathyroid hormone and calcitonin; and (5) ACTH from the adenohypophysis.

The following hormones consist of proteins: (1) most of the hormones from the adenophypophysis, including STH, FSH, LTH, and TSH; (2) chorionic gonadotropins; and (3) the metabolic hormone insulin from the pancreas.

## MECHANISM OF ACTION

Until recently, how hormones accomplish their regulatory feats remained a mystery. Now the mystery is gradually being replaced by an elegant story of the interplay between biologically active molecules that promises to open up new vistas in biology and medicine. At the present time, all we have are bits and pieces of the story, some of which will undoubtedly have to be modified as we enlarge our specific knowledge of the biochemical pathways which underlie the physiologic responses evoked by hormones.

Our story begins as a hormone molecule leaves the blood by diffusing through the walls of capillaries The hormone binds to a specific *receptor site* on the outside surface of a cell membrane. Each hormone binds to its own specific receptor site. Therefore, the specificity of the primary response elicited by the hormone is determined by the number of different types of cells which have identical receptor sites. After binding to the cell membrane, the work of the hormone is completed!

The hormone somehow activates an enzyme called adenylate cyclase inside the cell membrane. Probably, the activation of this membrane enzyme is a consequence of a conformational, or positional, change in its molecular structure. Once adenylate cyclase is activated, this enzyme catalyzes the conversion of adenosine triphosphate (ATP) into adenosine 3',5'-monophosphate (cyclic AMP). The structure of this functionally unique molecule is shown in Fig. 17-5. Cyclic AMP enters the cytoplasm and initiates chemical reactions which culminate in the characteristic physiologic response of the cell to the hormone bound to its membrane.

From this brief description, it is apparent that there is a complementary relationship between a hormone and cyclic AMP. If a hormone is the first chemical messenger in the molecular scenario described above, then cyclic AMP is the second chemical messenger. The situation is analogous to a baseball "battery" consisting of a pitcher and catcher. The pitcher delivers the specific pitches but the catcher calls the shots. In our analogy, the hormone is the pitcher and cyclic AMP is the catcher.

Once inside the cell, cyclic AMP may elicit a characteristic physiologic response to the membrane-bound hormone in one or more of three somewhat overlapping ways—by altering gene activity, enzyme activity, or membrane permeability. Let us explore each of these mechanisms in turn.

Cyclic AMP in some instances enters the cell nucleus and somehow causes the derepression of specific genes. Since each type of cell in the body contains the same genes but is different in structure and function from the others, many genes in any specific cell are nonfunctional. Apparently, other genes repress the potential expression of these genes. Cyclic AMP may somehow remove the influence of a specific repressor gene. In any event, in response to the presence of cyclic AMP, new messenger ribonucleic acid molecules (mRNAs) are manufactured by that portion of the chromosome in which the derepressed gene is located.

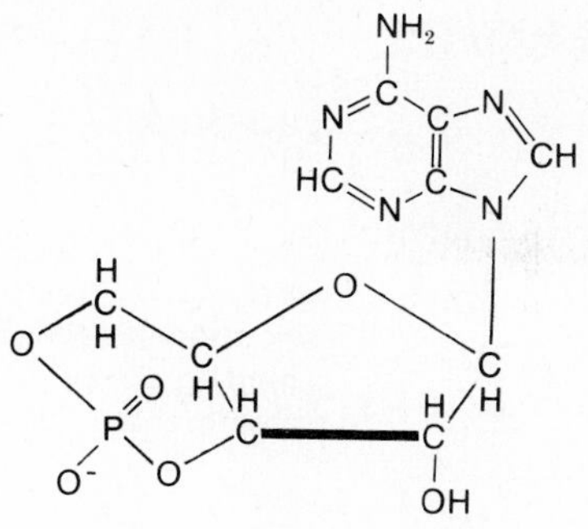

**Figure 17-5** *The structure of cyclic AMP. This secondary messenger is known chemically as adenosine 3',5-monophosphate.*

Recall from Chap. 6 that mRNAs carry nuclear instructions to the ribosomes in the cytoplasm for the synthesis of specific proteins. These proteins may be enzymes which mediate specific chemical reactions, hormones which are subsequently secreted into the bloodstream, or other natural products with unique types of biologic activity.

In other instances, cyclic AMP causes the activation of intracellular enzymes which are already present. This action obviates the necessity of derepressing a gene to accomplish the same purpose. However, it is likely, in some cases at least, that cyclic AMP activates an intracellular enzyme at the same time that it derepresses a specific gene to increase the intracellular quantity of the same enzyme. The former action would elicit an immediate physiologic response, and the latter action would provide a mechanism for sustaining the response over the longer term.

A third and common pathway by which the presence of cyclic AMP is translated into a physiologic response is through changes in the permeability of the cell membrane to which the hormone is bound. Membrane permeability may be increased or decreased in response to cyclic AMP. Ordinarily, permeability changes are the result of an excitation or inhibition of an active transport system in the membrane, although physical changes in membrane permeability are also likely.

The general mechanisms by which hormones act are shown in Fig. 17-6. Although numerous hormones

seem to exert their effects through the activation of cyclic AMP in the manner described above, we are not yet in a position of being able to generalize this mechanism to all hormones. However, enough hormones act through this second chemical messenger to establish the cyclic AMP mechanism as a general working model for hormonally mediated responses.

## FEEDBACK AS A REGULATOR OF SECRETION

In a healthy person, the blood concentrations of hormones in the same functional constellation are homeostatically balanced. This type of dynamic equilibrium is dependent upon the ability of an endocrine gland to somehow detect and respond appropriately to the magnitude of the physiologic response elicited in its target organ.

Recall that the response of a target organ to a bound hormone is to increase or decrease the synthesis and secretion of a hormone, enzyme, or some other natural product of the body. Therefore, the physiologic responsiveness of the target organ must somehow be detected chemically by the endocrine gland. Ordinarily, this is accomplished by monitoring the blood concentrations of various constituents released or conserved by the target cells.

Often, an increase in the physiologic responsiveness of a target organ is followed by a decrease in the secretory activity of the endocrine gland which mediates its responsiveness, and vice versa. This type of reciprocity between an endocrine gland and its target organ is patterned after a regulatory control principle called *negative feedback*.

Negative feedback in the regulation of hormonal secretion is diagramed in Fig. 17-7. Let us assume we are following the interaction between two endocrine glands in the same functional constellation. In response to appropriate stimuli, the first endocrine gland secretes an increasing quantity of a particular hormone. The hormone is transported in the blood throughout the body but binds specifically to receptor sites on the membranes of its target cells.

Following the formation of cyclic AMP in our example, the target cells respond by increasing the syn-

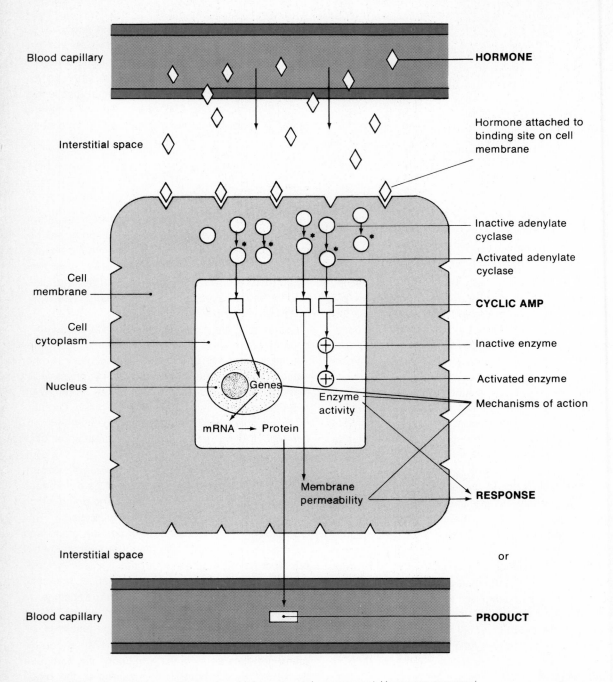

Blood capillary — **HORMONE**

Interstitial space

Hormone attached to binding site on cell membrane

Cell membrane

Cell cytoplasm

Inactive adenylate cyclase

Activated adenylate cyclase

**CYCLIC AMP**

Inactive enzyme

Activated enzyme

Nucleus

Genes

Enzyme activity

Mechanisms of action

mRNA → Protein

Membrane permeability

**RESPONSE**

Interstitial space

or

Blood capillary

**PRODUCT**

**Figure 17-6** *The general mechanism by which numerous hormones act. Hormones represent the first chemical messengers. Cyclic AMP is the second chemical messenger. Cyclic AMP causes one or more of the following intracellular effects: (1) activation of certain genes; (2) alteration of cellular membrane permeability; and (3) alteration of the activity of specific intracellular enzymes.*

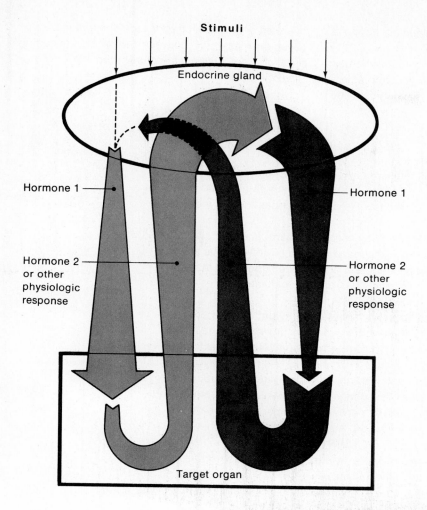

Stimuli

Endocrine gland

Hormone 1

Hormone 1

Hormone 2
or other
physiologic
response

Hormone 2
or other
physiologic
response

Target organ

**Figure 17-7** *Negative feedback in the regulation of hormonal secretion. The color arrows represent increased physiologic responses; the black arrows represent decreased physiologic responses.*

thesis and secretion of their hormone. As the blood concentrations of this hormone increase, the first endocrine gland responds by diminishing its secretory activity. This decreased activity in the first endocrine gland as a consequence of increased activity in its target organ is an example of negative feedback.

As the blood concentrations of the first hormone decline, the secretory activity of the target organ also decreases. This depression in the physiologic responsiveness of the target organ in turn becomes a stimulus to increase the secretory activity of the first endocrine gland. Again, this inverse responsiveness is an example of negative feedback.

By this type of reciprocal modulation of physiologic responsiveness, both the endocrine gland and its target organ maintain a dynamic, homeostatic equilibrium. Only pronounced fluctuations in neural or other hormonal stimuli can cause shifts in the finely attuned, physiologic balance between the two glands in our example, or more generally, between an endocrine gland and its target organ.

However, negative feedback is not the only mechanism which modulates the activity of endocrine glands and their target organs. In some cases, a target organ cannot respond to the presence of one hormone without the simultaneous presence of another hormone; in other words, the second hormone seems to permit the first hormone to act.

There are still other types of hormone-target organ interrelationships. However, by now you appreciate that whether, and to what extent, a hormone will be secreted or will function after secretion depends upon a complex interplay between the blood concentrations of other hormones, natural products, and raw materials.

## COLLATERAL READING

Collier, H. O. J., "Kinins," *Scientific American*, August 1962. Collier summarizes what is known about this class of localized hormones which participates in the early phases of the inflammatory response. He also describes the polypeptide structures of two kinins.

Davidson, Eric H., "Hormones and Genes," *Scientific American*, June 1965. Written before the discovery of cyclic AMP by Sutherland and his coworkers, this article summarizes the experimental evidence for and against the hypothesis that hormones, without actually entering their target cells, activate particular genes within them.

Guillemin, Roger, and Roger Burgus, "The Hormones of the Hypothalamus," *Scientific American*, November 1972. The authors discuss the implications of the recent synthesis of two hypothalamic releasing factors—thyrotropin releasing factor and luteinizing hormone releasing factor.

Melven, P. V., "Interaction Between Endocrine and Nervous Systems," *BioScience,* 20(10): 595–601, 1970. Describes the CNS-pituitary-ovarian axis as an operational example of reciprocal effects between the nervous and endocrine systems.

Pastan, Ira, "Cyclic AMP," *Scientific American*, August 1972. Pastan reviews the work which led to the discovery of cyclic AMP and discusses the potential implications of deficits or excesses of cyclic AMP in various disease states.

Sutherland, Earl W., "Studies on the Mechanism of Hormone Action," *Science*. 177:401–408, 1972. Written by the investigator who won the Nobel Prize in Physiology and Medicine in 1972 for his discovery of cyclic AMP, this article reviews the history and potential significance of his discovery.

Zuckerman, Sir Solly, "Hormones," *Scientific American*, March 1957. Although outdated in some respects, this article is an excellent general review of the natural products of the endocrine system.

# CHAPTER 18
# endocrine
# DISORDERS

The hormones secreted by endocrine glands have long-range behavioral and physiologic effects. As we learned in the last chapter, these effects are integrative, serving to maintain homeostatic conditions with regard to various aspects of metabolism and perpetuation.

However, the functional equilibria which exist from one endocrine gland to another and between the endocrine glands in each functional constellation may be disturbed by a variety of factors. These influences either depress glandular secretion or elicit a superabundance of a particular hormone. The former condition is referred to as *hyposecretion* of the hormone and *hypofunction* of the gland; the latter situation is called *hypersecretion* and *hyperfunction*, respectively. Because abnormal hypofunction and hyperfunction are obviously not homeostatic, such hormonal imbalances result in disease.

*Abnormal hyposecretion* is brought about in the following general ways:

1 Through congenital defects resulting from hereditary factors or developmental anomalies
2 Through dietary deficiencies of substances which are essential components of hormonal molecules
3 Through a lack of environmental factors which are necessary for hormonal activation
4 Through nutritional deficiencies which indirectly depress glandular metabolism
5 Through glandular inflammation due to physical trauma or infection
6 Through the surgical removal of all or part of a diseased endocrine gland
7 Through the administration of specific hormonal extracts or the excessive production of certain hormones, both of which may cause the disappearance of secretory cells in unrelated endocrine glands as a side reaction

The following mechanisms are responsible for *abnormal hypersecretion*: (1) feedback mechanisms which promote hypersecretion of one endocrine gland in a functional constellation following depression in hormonal output of another gland in the same constellation, and (2) pathologic hypertrophy or hyperplasia (tumor) of an endocrine gland.

The above lists of mechanisms by which abnormal hypo- and hypersecretions originate are interesting for several reasons. First, we note that there are considerably more pathways for the development of hypofunction than there are for hyperfunction. Therefore, we can anticipate the existence of a larger number of clinical syndromes due to hypofunction, compared to the number caused by hyperfunction. Second, an abnormal hypersecretion from one endocrine gland can cause a simultaneous and equally abnormal hyposecretion from another endocrine gland in the same functional constellation, and vice versa. Thus, a reciprocal imbalance may result from a disruption in the normal functional equilibrium between the two glands. Third, and finally, the above lists allow us to place the causes of endocrine imbalance into one of three major categories—congenital factors, exogenous influences, and pharmacologic interactions.

Since the various abnormalities caused by hormonal imbalances are reviewed below, we will pay especial attention to the general mechanisms by which they originate. The endocrine disorders are discussed according to the gland from which the hormones are secreted. Only those hormones which are known to contribute to or cause disease through their excess or deficiency are considered.

## THE HYPOPHYSIS (PITUITARY GLAND)

Recall from Chap. 17 that the hypophysis is divided into two lobes: (1) the neurohypophysis, pars nervosa, or posterior lobe; and (2) the adenohypophysis, or anterior lobe, including the pars intermedia. Each lobe produces hormones which, when in excess or deficit, cause abnormalities.

### THE NEUROHYPOPHYSIS OR PARS NERVOSA (POSTERIOR LOBE)

Two hormones are secreted by the pars nervosa—oxytocin and vasopressin, or antidiuretic hormone (ADH). The structures of oxytocin and ADH are illustrated in Fig. 18-1A and B, respectively.

Oxytocin   The following general effects are caused by oxytocin: (1) contraction of the myometrium of the uterus during the birth process; and (2) milk letdown, i.e., its expression from the mammary glands in response to sucking.

A *depression* in the secretion of oxytocin in the last trimester of pregnancy may result in birth difficulties by delaying uterine contractions. Such difficulties are usually circumvented by providing an exogenous source of oxytocin or a synthetic derivative of the hormone. Milk production is also decreased, which may impair nursing.

Although not well documented, an *excess* of oxytocin may cause abortion of a fetus by initiating vigorous uterine contractions prematurely. Hypersecretion

**Figure 18-1**  *The molecular structures of the hormones secreted from the pars nervosa. A. Oxytocin. B. Antidiuretic hormone (ADH). The amino acid residues are identified in the key.*

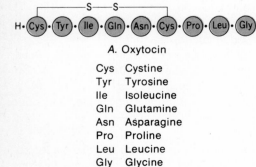

### A. Oxytocin

| | |
|---|---|
| Cys | Cystine |
| Tyr | Tyrosine |
| Ile | Isoleucine |
| Gln | Glutamine |
| Asn | Asparagine |
| Pro | Proline |
| Leu | Leucine |
| Gly | Glycine |

### B. Antidiuretic hormone (ADH)

| | |
|---|---|
| Cys | Cystine |
| Tyr | Tyrosine |
| Phe | Phenylalanine |
| Gln | Glutamine |
| Asn | Asparagine |
| Pro | Proline |
| Arg | Arginine |
| Gly | Glycine |

of oxytocin may also cause mammary gland discomfort, milk leakage, and feeding difficulties during nursing. These abnormalities of lactation arise from vigorous contractions of cells which compress the alveoli forcibly and expel the milk through the mammary ducts to the outside.

Vasopressin, or antidiuretic hormone (ADH)  ADH is secreted when the particle concentration of the extracellular tissue fluid increases. The increased blood concentration of this hormone causes water to be reabsorbed into the blood from the formative urine. The opposite situation prevails when the interstitial tissue fluid surrounding the nephrons becomes too dilute. In the former case, the urine becomes concentrated; in the latter case, the urine becomes watery.

A *deficit* of ADH causes a copious loss of water from the body via the urine. When the urine is voluminous and pale, the condition is referred to as *polyuria*. When polyuria is caused by a lack of ADH, the entire syndrome of symptoms is called *diabetes insipidus*. Other symptoms of diabetes insipidus include (1) unquenchable thirst because of excessive fluid loss and corresponding hemoconcentration (concentration of the blood), and (2) hypothermia due to an unusually large heat loss because of voluminous urine. Among the causes of hyposecretion are hypophyseal tumors, physical trauma to the head, cerebrovascular diseases which impair circulation in the inferior diencephalic region, and microbial infections.

## PARS INTERMEDIA (INTERMEDIATE LOBE)

Recall from Chap. 17 that the hormone secreted by the middle portion of the pituitary gland is called *intermedin,* or *melanocyte-stimulating hormone* (MSH). The structure of MSH is shown in Fig. 18-2.

Intermedin (melanocyte-stimulating hormone)  Intermedin stimulates specialized cells called *melanocytes* located between the epidermis and dermis to form a pigment called *melanin*, which darkens the skin. A *deficiency* of intermedin causes the skin to pale. By itself, MSH is medically unimportant. Its significance lies in its supression or production initiated

by excesses or deficiencies of other hormones or in its lack due to hypophyseal insufficiency in general. Such a lack may result from physical trauma, microbial infection, or cerebral vascular disease.

Adrenocortical insufficiency causes Addison's disease. One of the symptoms of this disorder is a darkening of the skin. The increased pigmentation is brought about by a compensatory hypersecretion of MSH from the pars intermedia. Simultaneously, the adenohypophysis is stimulated to produce adrenocorticotropic hormone (ACTH). ACTH promotes the production of adrenocorticosteroids and also contributes to the melanism seen in Addison's disease.

## THE ADENOHYPOPHYSIS (ANTERIOR PITUITARY)

The hormones secreted by the adenohypophysis include (1) growth, or somatotropic, hormone (STH); (2) adrenocorticotropin, or adrenocorticotropic hormone (ACTH); (3) thyroid-stimulating hormone (TSH); (4) luteinizing hormone (LH), or interstitial cell-stimulating hormone (ICSH); and (5) prolactin, or luteotropic hormone (LTH). The structure of ACTH is illustrated in Fig. 18-3.

Growth hormone (STH)  Somatotropin is secreted in response to increased blood concentrations of somatotropin releasing factor from the hypothalamus. STH causes cellular hypertrophy and hyperplasia through the following mechanisms: (1) promoting protein synthesis by enhancing amino acid uptake and RNA activity by cells; (2) promoting fatty acid oxidation through the degradation of fats; and (3) depressing both glucose uptake by cells and carbohydrate utilization by the body.

*Dwarfism* is caused by a *deficit* of growth hormone while bones are still growing in length; in other words, dwarfism can only occur during childhood and early adolescence. The stunting of growth may result in a height only half the normal potential, although the body features are usually in proportion to each other.

Other causes of dwarfism include dietary deficiencies, renal insufficiency, skeletal disorders, and hereditary defects. If the blood concentration of STH is high or normal in a particular case of dwarfism, then

H·(Ala)·(Glu)·(Lys)·(Lys)·(Asp)·(Glu)·(Gly)·(Pro)·(Tyr)·(Arg)·(Met)·(Glu)·(His)·(Phe)·(Arg)·(Try)·(Gly)·(Ser)·(Pro)·(Pro)·(Lys)·(Asp)·OH

**Melanocyte-stimulating hormone (MSH)**

| | | | |
|---|---|---|---|
| Ala | Alanine | Arg | Arginine |
| Glu | Glutamic acid | Met | Methionine |
| Lys | Lysine | His | Histidine |
| Asp | Aspartic acid | Phe | Phenylalanine |
| Gly | Glycine | Try | Tryptophan |
| Pro | Proline | Ser | Serine |
| Tyr | Tyrosine | | |

**Figure 18-2** *The molecular structure of melanocyte-stimulating hormone (MSH). The amino acid residues are identified in the key.*

the disorder is probably not the result of adenohypophyseal insufficiency.

An *excess* of STH causes *gigantism* if the epiphyseal cartilaginous disks are still present to permit growth in the length of long bones, or causes *acromegaly* after full skeletal growth in length has been achieved.

Gigantism gives rise to an individual who is extremely tall in physical stature, and acromegaly leads to an overgrowth in bones most often associated with the hands, feet, and face. Whether resulting in gigantism or acromegaly, hypersecretion of STH is usually caused by a tumor or abnormal proliferation of specialized cells of the adenohypophysis.

Individuals suffering from gigantism or acromegaly caused by an excess of STH usually suffer from ketosis and mild diabetes mellitus. The presence of ketone bodies is due to the increased metabolism of fat and correspondingly large concentrations of acetoacetic acid in the blood. This effect of STH is called *ketogenic*. The *hyperglycemia*, or glucosemia, associated with sugar diabetes is the result of an inability to metabolize carbohydrates, an effect of STH called *diabetogenic*.

**Figure 18-3** *The molecular structure of adrenocorticotropic hormone (ACTH). The amino acid residues are identified in the key.*

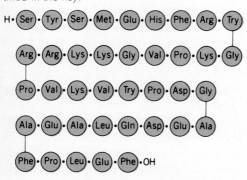

H·(Ser)·(Tyr)·(Ser)·(Met)·(Glu)·(His)·(Phe)·(Arg)·(Try)
(Arg)·(Arg)·(Lys)·(Lys)·(Gly)·(Val)·(Pro)·(Lys)·(Gly)
(Pro)·(Val)·(Lys)·(Val)·(Try)·(Pro)·(Asp)·(Gly)
(Ala)·(Glu)·(Ala)·(Leu)·(Gln)·(Asp)·(Glu)·(Ala)
(Phe)·(Pro)·(Leu)·(Glu)·(Phe)·OH

**Adrenocorticotropic hormone (ACTH)**

| | | | |
|---|---|---|---|
| Ser | Serine | Lys | Lysine |
| Tyr | Tyrosine | Gly | Glycine |
| Met | Methionine | Val | Valine |
| Glu | Glutamic acid | Pro | Proline |
| His | Histidine | Asp | Aspartic acid |
| Phe | Phenylalanine | Ala | Alanine |
| Arg | Arginine | Leu | Leucine |
| Try | Tryptophan | Gln | Glutamine |

**Adrenocorticotropic hormone (ACTH)** Recall from Chap. 4 that ACTH is secreted in response to increasing blood concentrations of corticotropin releasing factor from the hypothalamus. ACTH stimulates the adrenal cortex to secrete adrenocorticosteroids, especially cortisol. An increased concentration of cortisol shuts down production of ACTH through a negative feedback loop (see Chap. 17).

Because the general adaptation syndrome (GAS) revolves around the pituitary-adrenal axis, *hyposecretion* of ACTH results in hypofunction of the adrenal cortex and a correspondingly lowered resistance to stress. Lack of secretory stimulation may also cause degeneration of the specialized cells of the adrenal cortex which manufacture adrenocorticosteroids.

An *excessive output* of ACTH induces hypersecretion of the adrenocorticosteroids, causing some of all of the symptoms of *Cushing's disease*. The characteristics of this disease are explained later in this chapter under "Hypersecretion of Cortisol." If adrenalectomy is performed on such a patient to reduce the blood concentration of cortisol or to remove an adrenal tumor, the compensatory hypersecretion of ACTH in conjunction with an increased synthesis of MSH causes a deepening in skin pigmentation characteristic of Addison's disease. Addison's disease is attributed primarily to a hyposecretion of cortisol and is reviewed later in the chapter.

Thyroid-stimulating hormone (TSH) TSH is secreted in response to increasing blood concentrations of thyrotropin releasing factor from the hypothalamus. TSH stimulates the thyroid gland to secrete thyroxine. An elevated blood concentration of thyroxine reduces the secretion of TSH.

Although the above mechanism regulates the secretion of thyroxine in health, there is some question as to whether the same feedback loop is applicable during disease. For example, in thyrotoxicosis, thyroid secretion is excessive even though TSH secretion is depressed. Furthermore, in hypothyroidism due to thyroid malfunction, hypersecretion of TSH does little to restore normal production of thyroxine.

Even assuming that the feedback loop between the adenohypophysis and the thyroid gland is identical in health and disease, hyposecretion of TSH results in hypofunction of the thyroid. Lack of the hormone thyroxine leads to simple goiter, cretinism, or myxedema and a corresponding depression in the basal metabolic rate (BMR). These diseases will be explained when thyroxine is reviewed later in this chapter.

An excess of TSH may induce hyperthyroidism or exophthalmic goiter, as well as an elevation in BMR. Hyperthyroidism will also be discussed below.

Follicle-stimulating hormone (FSH) FSH is secreted in response to increasing blood concentrations of FSH releasing factor from the hypothalamus at puberty. FSH stimulates the development of follicles in the ovaries and initiates menstrual cycles. Along with an optimum balance of luteinizing hormone (LH), FSH is also responsible for ovulation. In the male, FSH initiates the conversion of primordial germ cells into spermatids, although mature spermatozoa are not formed unless testosterone is also present.

*Hypogonadism* may result from a *deficit* of FSH at a time when ovarian follicles should normally be forming. Since the proliferative stage of the menstrual cycle and ovulation are partly controlled by the blood concentration of FSH, a lack of this hormone causes abnormalities in both menstrual timing and flow. For example, hypogonadism can lead to absent cycles and excessive bleeding.

Luteinizing hormone (LH) LH is secreted in response to increasing blood concentrations of LH releasing factor from the hypothalamus. LH has the following effects: (1) the promotion of follicular growth and secretion; (2) the induction of ovulation, an action which occurs only when FSH is present in critical blood concentrations; and (3) the conversion of a graafian follicle into a corpus luteum.

A lack of LH may result in an absence of ovulation because of an imbalance in the concentrations of FSH and LH in the blood.

Interstitial cell–stimulating hormone (ICSH) The counterpart of female LH is male ICSH. ICSH is secreted in response to increasing blood concentrations of ICSH releasing factor from the hypothalamus at puberty. ICSH causes the production of testosterone by the interstitial cells of Leydig of the testes.

A *depression* in ICSH secretion may lead to *hypogonadism,* or hypofunction of the testes. A chronic lack of stimulation to the interstitial cells may cause their eventual degeneration.

Luteotropic hormone (LTH) LTH is secreted in response to decreasing blood concentrations of LTH inhibiting factor from the hypothalamus. LTH has the following effects: (1) promotion of the secretion of progesterone and maintenance of the secretion of estrogens by the corpus luteum, provided that LH concentrations are diminished, and (2) stimulation of milk production in the mammary glands, provided that the blood concentrations of estrogens and progesterone are sufficiently diminished in the later stages of pregnancy.

Because a minimum blood concentration of LTH is required for the continued secretion of progesterone, a *deficit* of LTH may lead to a lack of progesterone and the initiation of *atresia* (follicular degeneration), or an inability to accomplish implantation or development following fertilization.

A deficit of LTH following birth depresses the secretion of milk from the mammary glands. Instead, a fluid called *colostrum* continues to be secreted in extremely reduced amounts by functionally immature mammary glands. Colostrum differs from milk in that it lacks fat.

## OTHER ENDOCRINE GLANDS

### THE THYROID GLAND

The thyroid gland produces two hormones: thyroxine and calcitonin. The structures of these hormones are shown in Fig. 18-4*A* and *B*, respectively.

Thyroxine    Thyroxine is liberated in response to increasing blood concentrations of thyrotropin releasing factor (TRF) from the hypothalamus. TRF causes an increased secretion of thyroid-stimulating hormone (TSH) from the adenohypophysis. Thyroxine increases the rate of metabolism of all the cells of the body. These effects include increases in the following factors:

1   Cardiovascular functions such as heartbeat, mean systemic pressure, and blood volume
2   The rate and depth of respiration
3   Neural excitation as reflected in hyperexcitability
4   Digestive functions such as secretion of digestive juices, absorption of soluble foodstuffs, and motility of the digestive tract
5   Muscular tone or contractility as reflected by an increased strength and the appearance of tremors

A goiter is an enlargement of the thyroid gland. Simple goiter is due to a lack of iodine, depressing or preventing the synthesis of active thyroxine by the thyroid. In response to the deficit of active thyroxine, an increased blood concentration of TRF causes a corresponding elevation in the concentration of TSH. The thyroid gland reacts to a hypersecretion of TSH by undergoing compensatory hypertrophy and secreting copious amounts of inactive hormone.

Thyroid enlargement causes a protrusion, or swelling, at the front of the neck. At an earlier time in history, as in the painting of the *Madonna* by Leonardo da Vinci, and in certain primitive but contemporary societies, a goitrous neck was or still is considered a sign of physical beauty.

Goiters are endemic to geographic regions where natural supplies of iodine are lacking. Such regions are referred to as "goiter belts." One such belt includes the Great Lakes region in the United States. The natural deficiency of iodine in otherwise arable soil is remedied by adding iodine to table salt. Iodized salt is used throughout the world to supply iodine and prevent simple goiter.

*Cretinism* is a condition resulting from hypothyroidism during infancy or childhood. Cretins are distinguished by some degree of mental retardation and a stunting of growth. These effects are brought about by a reduction in BMR during stages critical to the normal growth and development of the nervous and skeletal systems. Other systems of the body are also adversely affected, but to a lesser extent.

Treatment of cretinism requires its early detection and the prompt administration of a thyroid extract containing natural thyroxine or a synthetic derivative to correct the deficiency.

*Myxedema* is a condition caused by hypothyroidism during later childhood and adulthood. The most prominent symptom of this disorder is edema, especially of the face and hands. The edema is caused by an increased accumulation of extracellular tissue fluid indirectly promoted by the profound alteration in general metabolism. Such metabolic derangements are also responsible for an increase in serum cholesterol, a condition which frequently gives rise to atherosclerosis in patients suffering from advanced cases of the disease.

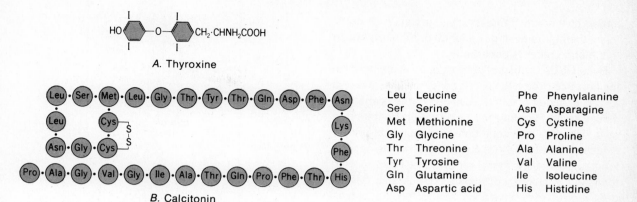

**Figure 18-4** *The molecular structures of the hormones secreted from the thyroid gland. The amino acid residues of calcitonin are identified in the key.*

In addition to an iodine deficiency and a hypofunction of the adenohypophysis, *hypothyroidism* may be caused either by infections which lead to the destruction or atrophy of the thyroid, surgical removal of the gland, the excessive use of antithyroid drugs, or developmental anomalies.

The general symptoms of pathologic hypothyroidism are the reverse of those effects attributed to the presence of too much thyroxine; namely, cardiovascular and respiratory decreases, mental depression, digestive impairments which provoke constipation, and muscular sluggishness and myasthenia. Frequently, hypothyroidism also causes extreme drowsiness and integumentary effects such as the loss of hair and a scaly skin.

A pathologic excess of thyroxine leads to a condition termed *hyperthyroidism*. In over one-fourth of such cases the eyeballs bulge. Protrusion of the eyes associated with hyperthyroidism is called *exophthalmic goiter*.

The symptoms of hyperthyroidism are exaggerations of the normal effects of thyroxine mentioned earlier. These effects include (1) a marked elevation in BMR, causing a profuse perspiration and a significant loss of weight; (2) an excessive increase in digestive functions including gastrointestinal motility, leading to diarrhea; (3) an abnormal increase in muscle tone, producing fatigue and tremors; and (4) a hyperexcitability of the nervous system, resulting in

nervousness and mental disturbances. Most of these symptoms are found in Shakespeare's characterization of Cassius, who is observed to have a "lean and hungry look" in the play *Julius Caesar*.

The cause of hyperthyroidism is obscure, but is most frequently cited as a hypersecretion of TSH. Other suspected causes include an increased blood concentration of TSH-RF or some environmental factor which influences a hereditary predisposition to the disease. Whatever the etiology of hyperthyroidism, the thyroid gland undergoes hyperplasia which increases the size between two- and threefold and elevates drastically its secretory ability.

Exophthalmic goiter causes enlargement of the eyeballs through interstitial edema. The swollen eyeballs stretch the optic nerves and prevent the lids from closing completely. As a consequence, visual damage may result from neural destruction or corneal drying. A dry cornea also increases the probability of secondary infection.

**Calcitonin** Along with parathyroid hormone and the skin hormone calciferol, calcitonin helps to regulate blood levels of $Ca^{++}$ and $HPO_4^{--}$. Calcitonin inhibits the absorption of $Ca^{++}$ from bone, which indirectly decreases the concentration of $Ca^{++}$ in blood. Calcium absorption is ordinarily prevented by a feedback mechanism when the blood concentration of $Ca^{++}$ is excessively high. An abnormally low blood concentration of this ion depresses the secretion of calcitonin, thereby removing the inhibitory brake applied to the absorption of bone $Ca^{++}$.

Because $Ca^{++}$ absorption from bone is promoted in the absence of calcitonin, this condition may give rise to brittle bones and the development of symptoms similar to $Ca^{++}$-deficiency rickets. Hyposecretion of calcitonin may also contribute to formation of urinary stones as excess $Ca^{++}$ is concentrated in the nephric tubules during the formation of urine. However, the formation of stones containing $Ca^{++}$ is also dependent upon an excessive concentration of other chemicals and occupational or idiopathic factors which retard or prevent normal excretion of such materials. The formation of urinary stones is discussed in Chap. 20.

## THE SKIN

The skin contains a hormonal precursor called 7-dehydrocholesterol. Ultraviolet radiation from sunlight activates this chemical, causing its conversion into calciferol. Although calciferol is actually a hormone, it has been called incorrectly vitamin $D_2$ or vitamin $D_3$. The structure of calciferol is illustrated in Fig. 18-5.

Calciferol   Calciferol increases the intestinal absorption of $Ca^{++}$ and $HPO_4^{--}$, thereby preventing the development of calcium- and phosphorus-deficiency rickets. Cod-liver oil is an effective antirachitic substance because it contains relatively large concentrations of calciferol.

Confinement indoors and chronic air pollution are two factors which reduce the effective dosage of ultraviolet radiation which the skin ordinarily secures from sunlight. This deficiency of ultraviolet energy causes a depression in the blood concentration of calciferol and a corresponding reduction in the intestinal absorption of dietary $Ca^{++}$ and $HPO_4^{--}$. A chronic reduction in the blood concentration of $Ca^{++}$, a condition known as *hypocalcemia*, prevents the normal deposition of this mineral in bone and initiates the development of rickets. Rickets is mentioned again under hyposecretion of parathyroid hormone and is re-

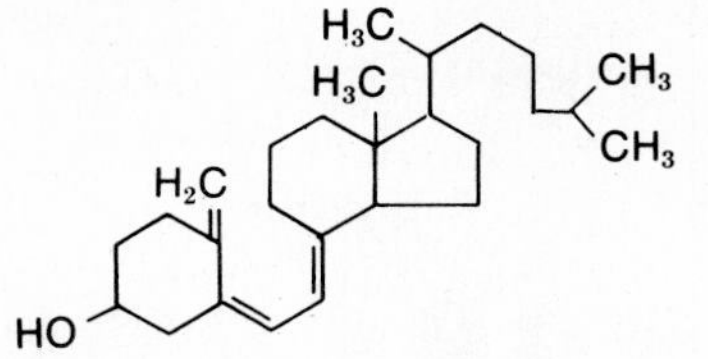

**Calciferol**

*Figure 18-5*  *The molecular structure of calciferol.*

viewed in Chap. 27, in the discussion of vitamin D deficiency.

## THE PARATHYROID GLANDS

The parathyroid glands secrete parathyroid hormone.

Parathyroid hormone   This hormone causes the bone absorption and tubular reabsorption of $Ca^{++}$, resulting in hypercalcemia. At the same time, increased concentrations of $HPO_4^{--}$ are excreted via the kidneys. A reduction in the secretion of parathyroid hormone reverses the above effects by preventing bone absorption of $Ca^{++}$ and promoting its renal excretion and the tubular reabsorption of $HPO_4^{--}$. The depression in blood $Ca^{++}$ results in hypocalcemia.

The action of parathyroid hormone is diametrically opposed to the action of calcitonin. In fact, both of these hormones, in conjunction with calciferol and vitamin D, directly or indirectly affect the concentration of blood $Ca^{++}$ and $HPO_4^{--}$. Since bone contains most of the $Ca^{++}$ in the body, the continual changes in the deposition and absorption of $Ca^{++}$ ordinarily have little effect on skeletal integrity; however, the effect on blood $Ca^{++}$ concentrations is direct and stabilizing. Therefore, the control of blood $Ca^{++}$ concentrations is parathyroid hormone's "reason for being."

A decrease in the concentration of blood $Ca^{++}$ caused by a *reduction* in the secretion of parathyroid hormone may lead to intermittent tonic spasms called *tetany*. Although muscular spasms are the result of hypocalcemia, the physiologic effects of $Ca^{++}$ deficiency are experienced first by nerve cells. This is so because the permeability of neuronal membranes to $Na^+$ increases as the extracellular concentration of $Ca^{++}$ decreases. The ease with which $Na^+$ enters nerve fibers during hypocalcemic states initiates spontaneous depolarizations and causes a hyperexcitability of the nervous system. An elevated irritability of the neurons is reflected in spasms of skeletal muscles, especially of the hands and larynx. It is not uncommon for a patient suffering from severe hypocalcemia to die of asphyxiation due to laryngospasm (see Chap. 29).

An *overabundance* of parathyroid hormone caused by a parathyroid tumor or, more rarely, lack of dietary $Ca^{++}$ results in $Ca^{++}$ absorption from bone into blood. Therefore, hypertrophy of the parathyroid glands and hypersecretion of its hormone can lead to two disorders: (1) *calcium-deficiency rickets*, causing an increased bone brittleness and probability of fracture; and (2) *hypercalcemia*, causing neural depression by decreasing the permeability of neuronal membranes to $Na^+$.

## THE THYMUS GLAND

At or shortly after birth, the thymus secretes a presumptive hormone commonly referred to as *thymic hormone*. This hormone confers immunocompetence on specialized cells in the body, thereby ensuring the recognition of self-antigens from foreign antigens and initiating the development of specific immunity.

Thymic hormone   Thymic hormone has two basic effects: (1) maturation of lymphocytes in the thymus, causing them to become capable of interacting with foreign antigens even though still uncommitted to a specific antigen; and (2) the maturation of stem cells in lymphoid tissues and organs, causing such cells to become committed lymphocytes with the capability of manufacturing antigen-specific antibodies.

A *lack* of thymic hormone during the critical period in the development of immunocompetence may give rise to a disorder called *hypogammaglobulinemia*. The condition is characterized by a depression in the number of circulating antibodies and a corresponding increase in susceptibility to infection. The disease is most often caused by a malformation of the thymus gland due to congenital defects. Hypogammaglobulinemia is discussed further in Chap. 31.

## THE PANCREAS

The islets of Langerhans secrete two hormones: insulin and glucagon. These hormones are important regulators of carbohydrate, lipid, and protein metabolism. The structures of these hormones are illustrated in Fig. 18-6*A* and *B*, respectively.

Insulin   Insulin is normally secreted in response to a rising blood-sugar level. This hormone has the following effects: (1) facilitation of glucose transport across most, but not all, cell membranes, thereby promoting hypoglycemia, or a decrease in the concentration of blood sugar, as well as increasing both the catabolism, or breakdown, of carbohydrates and the quantity of glycogen stored intracellularly; (2) a depression in the utilization of lipids for energy, since the available carbohydrate supply under these conditions is more than adequate for energy needs; and (3) facilitation of amino acid transport across cell membranes, thereby promoting intracellular protein anabolism.

Insulin secretion is reduced when the blood sugar level is subnormal. Normal *hyposecretion* brings about the following responses: (1) hyperglycemia, since the transport of blood glucose to most of the tissue cells is depressed; (2) the utilization of fats for energy, causing the release by the liver of acetoacetic acid into the blood as well as an increase in the concentration of plasma lipids, including cholesterol; and (3) a decrease in the anabolism of protein, causing an intracellular reduction in its concentration.

Pathologic hypoinsulinism results in *diabetes mellitus* and has three profound effects on the physiology of the body: (1) the hyperglycemia, or glucosemia, leads to glucosuria, or glucose in the urine, causing the excretion of large quantities of this primary energy nutrient and producing significant derangements of metabolism; (2) the utilization of lipids as an alternative energy source and the corresponding accumulation of the intermediary energy metabolite, acetoacetic acid, in blood lead to ketosis, acidosis and possibly coma, while the increase

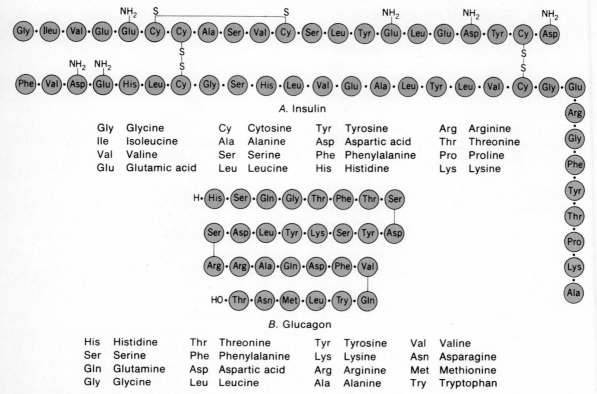

**Figure 18-6** *The molecular structures of the hormones secreted by the islets of Langerhans of the pancreas. The amino acid residues are identified in the key.*

| | | | | | | | |
|---|---|---|---|---|---|---|---|
| Gly | Glycine | Cy | Cytosine | Tyr | Tyrosine | Arg | Arginine |
| Ile | Isoleucine | Ala | Alanine | Asp | Aspartic acid | Thr | Threonine |
| Val | Valine | Ser | Serine | Phe | Phenylalanine | Pro | Proline |
| Glu | Glutamic acid | Leu | Leucine | His | Histidine | Lys | Lysine |

*A. Insulin*

| | | | | | | | |
|---|---|---|---|---|---|---|---|
| His | Histidine | Thr | Threonine | Tyr | Tyrosine | Val | Valine |
| Ser | Serine | Phe | Phenylalanine | Lys | Lysine | Asn | Asparagine |
| Gln | Glutamine | Asp | Aspartic acid | Arg | Arginine | Met | Methionine |
| Gly | Glycine | Leu | Leucine | Ala | Alanine | Try | Tryptophan |

*B. Glucagon*

in serum cholesterol is reflected in a higher-than-normal incidence of atherosclerosis; and (3) the lack of intracellular protein synthesis causes further metabolic complications.

During chronic acidosis due to an accumulation of ketone bodies in the blood, loss of consciousness occurs when the blood pH falls below 7.0. Since a diabetic patient in this situation does not respond to the usual methods of restoring consciousness, the condition is referred to as *diabetic coma*.

An *excessive secretion* of insulin results in hypoglycemia. As blood sugar levels progressively drop, the nervous system becomes initially excited and secondarily depressed. Neural hyperexcitability leads to convulsions; the depression which follows causes neural, or insulin, shock and coma. Severe cases both of hyperinsulinism and hypoinsulinism may therefore cause coma.

**Glucagon**   The secretion of glucagon from the pancreas occurs in response to hypoglycemia. Glucagon promotes the conversion of liver glycogen into blood glucose. This action, referred to as *glycogenolysis*, elevates the blood sugar level and restores the normal balance of this metabolite. The hormone also causes the conversion within the liver of amino acids into glucose, a process called *gluconeogenesis*. Through these mechanisms, glucagon prevents hypoglycemia from developing under normal conditions. Thus, glucagon is in the same functional constellation as insulin.

An *abnormal lack* of glucagon contributes to hypoglycemia and the general symptoms associated with hyperinsulinism. This type of pathophysiology is most often seen in persons engaged in prolonged physical exertion or suffering from the effects of undernutrition.

## THE STOMACH AND SMALL INTESTINE

Although a number of hormones are secreted by gland cells in the mucosa of the stomach and small intestine, only gastrin has been implicated in disease. Gastrin is produced by both the stomach and duodenum.

**Gastrin** This hormone promotes an increase in the secretion of hydrochloric acid (HCl) from the parietal cells of the gastric mucosa. Presumably, mechanical stimulation of the intestinal walls by the passage of food is responsible for stimulating the production of duodenal gastrin, while psychogenic factors initiate the release of gastric gastrin.

Pancreatic tumors or psychosomatic disorders may cause a chronic secretion of excessive amounts of gastrin. Since the presence of gastrin causes the release of HCl which activates pepsinogen, precursor of the proteolytic enzyme pepsin, the subsequent erosion of the gastrointestinal mucosa leads to peptic ulcers and gastrointestinal upsets such as diarrhea. Peptic ulcers are considered in detail in Chap. 15, and diarrhea is discussed in Chap. 27.

## THE ADRENAL GLANDS

Each of the adrenal glands consists of an inner medulla and an outer cortex. The medulla produces two hormones, epinephrine and norepinephrine, collectively referred to as catecholamines. The structures of the catecholamines are illustrated in Fig. 18-7. The cortex manufactures aldosterone, a mineralocorticoid, and a number of related hormones called glucocorticoids, the best known of which is cortisol. The structures of aldosterone and cortisol are shown in Fig. 18-8A and B, respectively.

**Figure 18-7** The molecular structures of the catecholamines secreted from the adrenal medullae.

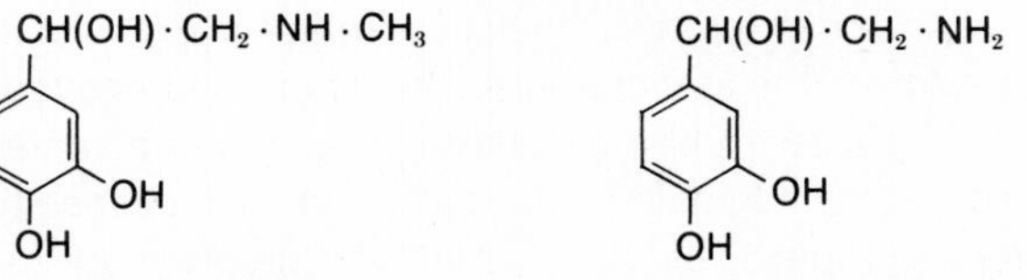

A. Epinephrine      B. Norepinephrine

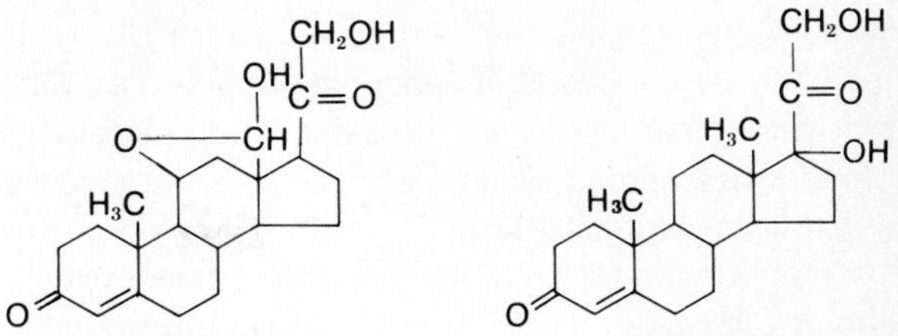

**Figure 18-8** The molecular structures of the major hormones secreted from the adrenal cortices.

**Epinephrine and norepinephrine** The medullary hormones are secreted in response to sympathetic stimulation of the medullae. The role of the catecholamines during stress is reviewed in Chap. 4. Basically, epinephrine and norepinephrine enhance the effects of the sympathetic division of the autonomic nervous system and increase overall metabolism. Sympathetic stimulation is reflected by increases in the vigor and rate of the heartbeat, mean systemic pressure, central nervous excitation, muscular efficiency, and by a decrease in digestive processes. Metabolism is promoted by an increased oxygen consumption, glycogenolysis in the liver, and an increased utilization of fat.

Because the catecholamines elevate blood pressure and increase metabolism, an abnormal and chronic excess of the medullary hormones may contribute to or cause hypertension and loss of weight. However, remember that stimulation which causes secretion of the catecholamines also directly activates the sympathetic nervous system, making it difficult to differentiate between the nonmetabolic effects caused by hormonal and neural influences. Metabolic effects are excluded from consideration here since oxygen consumption and glycogenolysis are not affected by stimulation of tissue cells innervated by sympathetic fibers.

**Aldosterone** This hormone causes the reabsorption of $Na^+$ from the renal tubules and, secondarily, an osmotic retention of water from the formative urine. Aldosterone also stimulates the secretion of ADH from the adenohypophysis. The action of ADH in water regulation is reviewed in an earlier part of this chapter. The retention of water dilutes the blood, thereby increasing its volume. The larger volume of blood is reflected by increases in venous return, cardiac output, and mean systemic pressure. The increased blood pressure and volume also elevate the amount of extracellular tissue fluid.

Coupled with $Na^+$ retention is an increased excretion of potassium ion ($K^+$). A depression in the concentration of intracellular $K^+$ is one of the mechanisms leading to muscle fatigue. Sodium retention also increases both the secretion into the urine of $H^+$ and the renal reabsorption of $Cl^-$. These ionic effects reduce the acidity of the blood.

The secretion of aldosterone is probably determined by the blood concentrations of ACTH, $Na^+$, and $K^+$ brought to the adrenal cortex. That is, high concentrations of ACTH and $K^+$ and a low concentration of $Na^+$ induce the secretion of aldosterone, and the opposite conditions inhibit its synthesis.

An *abnormal deficit* of aldosterone results in polyuria and a corresponding hypotension. The loss of body fluids reduces cardiac output and decreases mean systemic pressure. The reduction in blood pressure is drastic and, in some cases, causes circulatory shock. The effects of circulatory shock are reviewed in Chap. 31.

An *excessive amount* of aldosterone results in hypertension, edema, and oliguria (a decrease in the volume of urine output). Hypertension is an indirect effect of water retention, since an increased blood volume promotes cardiac output and a compensatory vasoconstriction. Both these effects elevate the blood pressure considerably. As the hydrostatic pressure of the blood is elevated, serum is forced from the capillaries into the extracellular tissue spaces. This condition results in edema. Oliguria, of course, is a direct consequence of water retention.

Cortisol   Cortisol is the best known of the glucocorticoids. This hormone is secreted by the adrenal cortex in response to increased blood concentrations of corticotropin releasing factor from the hypothalamus and ACTH from the anterior pituitary gland. Cortisol has the following effects: (1) suppression of the inflammatory process, possibly by preventing the disruption of intracellular lysosomes and subsequent digestion of dead and dying cells following tissue damage or by preventing the synthesis of natural products called *prostaglandins* which are essential to the inflammatory response; (2) mobilization of fatty acids and amino acids for oxidation and biosynthesis; and (3) increase in gluconeogenesis in the liver and the corresponding development of hyperglycemia. The above physiologic effects occur in response to stress.

Although an abnormal *deficit* of glucocorticoids is usually implicated as the cause of *Addison's disease*, the secretion of mineralocorticoids is also depressed. Symptoms caused primarily by a hyposecretion of cortisol include (1) a lowered resistance to stress, since raw materials such as amino acids and fatty acids are not available in adequate amounts for the synthesis of essential ingredients; (2) muscular weakness, because hypoglycemia reduces the availability of glucose for oxidation within muscle cells; and (3) an increased blood viscosity (hemoconcentration) due to the loss of body fluids through vomiting resulting from gastrointestinal upsets. Blood viscosity is also elevated due to polyuria caused by a related depression in the blood concentrations of mineralocorticoids. A lack of adrenocorticosteroids also induces an increased output of MSH and ACTH from the pars intermedia and the anterior lobe of the hypophysis, respectively. As we learned earlier, excessive amounts of MSH and ACTH are responsible for increasing the pigmentation of the skin, a prominent characteristic of adrenocortical insufficiency.

Although an abnormal *excess* of glucocorticoids causes *Cushing's disease*, the secretion of mineralocorticoids is also elevated. Symptoms caused primarily by a hypersecretion of cortisol include (1) chronic hyperglycemia due to an increase in gluconeogenesis; (2) intracellular protein catabolism and a corresponding inability to synthesize most types of intracellular proteins, resulting in a negative nitrogen balance; and (3) mobilization of fatty acids and the subsequent redistribution of fat within the body.

The effects enumerated above have profound physiologic consequences. First, hyperglycemia results in one form of sugar diabetes. The high blood-sugar levels cause a compensatory increase in the secretion of insulin from the pancreas which may ultimately damage the islet cells. Second, the loss of intracellular proteins causes muscular weakness and bone fragility. Third, alterations in fat metabolism may cause the development of a "moon" face, other signs of obesity, humpback, and atherosclerosis.

Some physiologic effects of Cushing's disease are due to a corresponding increase in the secretion of mineralocorticoids. These effects include (1) an increase in the volume of body fluids due to $Na^+$ retention, causing mild hypertension and edema; and (2) the development of mild alkalosis due primarily to a compensatory renal loss of $H^+$.

The adrenal cortex is also responsible for secreting minute amounts of androgens. Thus, an abnormally large secretion of adrenocorticosteroids usually contains some adrenal androgens. In certain female patients suffering from Cushing's disease, the androgens are responsible for various masculinizing effects such as a male distribution of body hair, development of male genitals, thickened skeleton, increased musculature, and lowered pitch of the voice. These effects are seen in female patients of all ages and in male patients prior to puberty.

## THE TESTES

The testes secrete two major androgens—testosterone and androsterone. The most important and abundant of the two is testosterone, produced by the interstitial cells of Leydig. The structures of testosterone and androsterone are illustrated in Fig. 18-9A and B, respectively.

**Testosterone** As we mentioned earlier in this chapter, the production and secretion of testosterone is promoted by an increased blood concentration of ICSH releasing factor from the hypothalamus and a correspondingly elevated concentration of ICSH, the male counterpart of LH, from the adenohypophysis. ICSH directly stimulates the interstitial cells to manufacture testosterone.

The primary sexual hormone of the male has the following effects: (1) maturation of the genitals, including the penis and scrotum, and development of accessory reproductive glands such as the seminal vesicles, bulbourethral glands, and the prostate gland; (2) development of secondary sex charac-

teristics such as a male distribution of body hair, deep pitch of the voice, increased growth of the musculature, and thickening of the bones; and (3) increase in sexual desire.

*Hypogonadism* may arise before puberty through congenital factors such as developmental anomalies, including cryptorchism, which prevent the development of normal testes (see Chap. 21). Hypogonadism can also develop after puberty due to a lack of ICSH releasing factor or ICSH because of hypothalamic or anterior pituitary damage, respectively. In some cases, the disease is induced by the removal of the testes, a procedure referred to as *castration*.

If hypogonadism appears before puberty, then the condition is called *eunuchism*. The following symptoms are characteristic of eunuchism: (1) a lack of development of the genitals and accessory reproductive glands; (2) a lack of development of the secondary sex characteristics; and (3) a lack of sexual desire.

Hypogonadism which develops after puberty depresses, but does not eliminate, sexual desire. Penile erection and ejaculation are still possible following postpuberal castration.

An abnormal *excess* of testosterone is usually due to a tumor of the testes. Prior to puberty, hypergonadism results in precocious sexual and physical development. After puberty, an excess of testosterone promotes an increase in one or more of the characteristics associated with maleness. Usually, such effects are difficult to measure due to the wide range of variability in "normal" maleness in the general population of human males. However, one useful indicator of hypergonadism is an increased excretion via the urine of degradation products of testosterone called 17-ketosteroids.

**Figure 18-9** *The molecular structures of the major hormones secreted from the testes.*

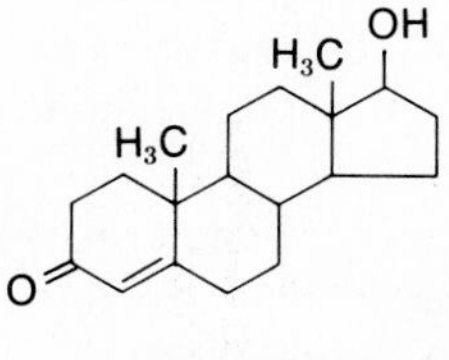

A. Testosterone

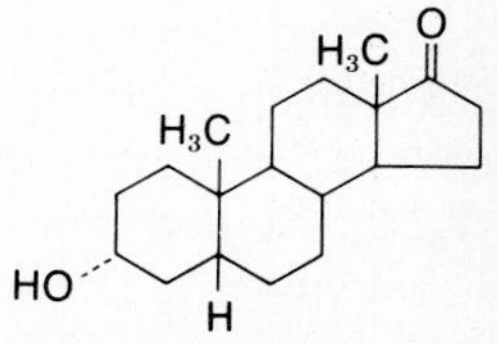

B. Androsterone

## THE OVARIES

The ovaries secrete two hormones—estrogens and progesterone. Estrogens are secreted from mature graafian follicles. The best known and most abundant estrogen is 17β-estradiol. In addition to estrogen, progesterone is manufactured by the corpus luteum following ovulation. The structures of estrogen and progesterone are shown in Fig. 18-10A and B, respectively.

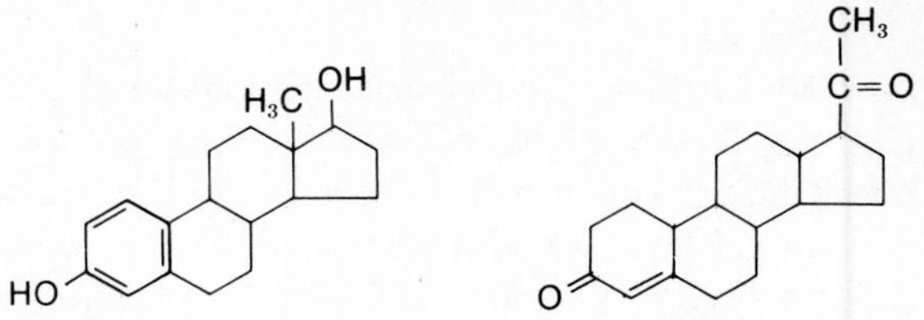

**A. Estrogen**
(17 β-estradiol)

**B. Progesterone**

**Figure 18-10** *The molecular structures of the major hormones secreted from the ovaries.*

**Estrogen** Estrogen is secreted in response to increased blood concentrations of LH releasing factor and LH. Similar in action to its male counterpart, androgen, estrogen has the following effects:

1 Maturation of the female reproductive system, including the fallopian tubes, uterus, and vagina
2 The development of secondary sex features, such as the clitoris, mammary glands, mons pubis, pubic hair, and labia
3 The deposition of increased amounts of fat in the subcutaneous connective tissues, mammary glands, mons pubis, and labia
4 Rapid bone growth following puberty and an early disappearance of the epiphyseal cartilaginous disks, causing an overall decrease in the length and girth of the bones compared to those of males matched in age
5 An increase in sexual desire

An abnormal *deficit* of estrogen may be due to the following factors: (1) developmental anomalies which result in absent or malformed ovaries; (2) lack of LH releasing factor from the hypothalamus or a deficit of LH from the adenohypophysis; or (3) castration some time after birth.

If the deficiency occurs before the onset of puberty, then typical symptoms of eunuchism are observed. These symptoms include (1) a lack of development of the female reproductive system, secondary sex characteristics, and sexual desire; and (2) tall stature due to a prolongation of bone growth.

When the hormonal deficit is experienced after puberty, there is some regression in the size and shape of the organs composing the reproductive sys-

tem and in the various secondary sex structures. Usually, a diminution in sexual desire also occurs. Furthermore, since the periodic changes in the menstrual cycle are dependent on a balance in the blood concentrations of FSH and LH, a depression in the blood concentration of estrogen causes a compensatory increase in the concentration of FSH and decreases the amount of LH, thereby delaying or preventing ovulation. Menstrual abnormalities are discussed further in Chap. 20.

An abnormal *excess* of estrogen before puberty promotes the precocious development of sex and other physical characteristics associated with femaleness. Following sexual maturation, an excess of the primary sex hormone leads to an increase in the degree of "femaleness." Since estrogen usually causes a proliferation of the uterine endometrium, hypersecretion of this hormone generally induces menstrual bleeding.

**Progesterone** As we noted previously, ovulation occurs when there is a critical balance between two of the gonadotropins in the blood, namely, FSH and LH. Under the continued influence of LH and the increasing influence of LTH, the corpus luteum begins to secrete progesterone in addition to estrogens. As LTH concentrations continue to rise while LH concentrations decrease during the 2 weeks following ovulation, estrogen and progesterone concentrations remain relatively high. For about 3 months subsequent to successful fertilization, the corpus luteum remains intact and continues to secrete large amounts of estrogen and progesterone. These functions are augmented and finally taken over by the placenta in the later stages of pregnancy. The role of placental hormones in the causation of disease states is considered below.

The secretion of progesterone is related to the development of conditions suitable for intrauterine development, should fertilization occur. More specifically, progesterone has the following functions: (1) nourishment of the early blastocyst within the fallopian tubes; (2) preparation of the endometrium of the uterus for receipt of the blastocyst; and (3) relaxation of the uterine myometrium, thereby promoting successful implantation and development. The endometrial changes result in a highly glandular and vascularized tissue lining the lumen of the uterus, a phase of endometrial development during the menstrual cycle referred to as the secretory stage. Progesterone also initiates alveolar development within the mammary glands.

A *deficit* of progesterone ordinarily results from a lack of ovulation during a menstrual cycle. An abnormal depression in the blood concentration of this hormone causes excessive menstrual bleeding. Abortion or stillbirth may occur if the concentration of progesterone drops precociously during gestation because of the development of a menstrual flow and an increase in uterine contractility.

An *excess* of progesterone is correlated with a prolongation of the menstrual cycle, renal retention of $Na^+$, and a corresponding increase in body fluids. The latter effect causes a mild increase in body weight.

## THE PLACENTA

The placenta is a compound structure formed from the chorion frondosum of the fetus and the decidua basalis of the mother. The functional organization of the placenta is described in Chap. 19.

Following its formation, the placenta is responsible for the production of several hormones including the gonadotropins, estrogens, and progesterone. In addition, the placenta ordinarily functions as an exchange membrane between the maternal and fetal bloodstreams. The roles of the placental hormones are identical to, and supplement, the actions of LH from the adenohypophysis and estrogen and progesterone from the corpus luteum of the ovary. Because the effects of estrogen and progesterone have just been discussed, we will restrict our present review to the effects of chorionic gonadotropin.

The presence of chorionic gonadotropin with a predominance of LH activity prevents the corpus luteum from involuting, thereby maintaining relatively high blood concentrations of estrogen and progesterone. A high progesterone concentration keeps the uterus in a state conducive to the continued development of the embryo (first eight weeks of prenatal development) and fetus (the remainder of prenatal development).

A *lack* of chorionic gonadotropin after the first trimester may result in spontaneous abortion, since the corpus luteum becomes atretic (degenerative). Luteal atresia initiates the termination of the secretory phase of the menstrual cycle, provided that the concentration of chorionic progesterone is insufficient to maintain pregnancy, and leads to a sloughing off of the uterine endometrium. The sloughing off, or menstrual flow, removes the implanted fetus.

An abnormal *excess* of chorionic gonadotropin may cause toxemia by indirectly promoting an increased secretory activity by the uterine endometrium. This response is brought about by increasing the amount of progesterone secreted from the corpus luteum. Alternatively, high blood concentrations of chorionic progesterone can exert the same effect by acting directly on the gland cells of the uterine endometrium.

## COLLATERAL READING

Gillie, R. Bruce, "Endemic Goiter," *Scientific American,* June 1971. Because iodine deficiency is the result of malnutrition or undernutrition in geographic areas where iodine-deficient soils prevail, the disease, according to Gillie, is now associated with poor people and underdeveloped countries.

# Part Four
# reproduction

# CHAPTER 19
# reproduction and development

Despite its importance, there is probably more misunderstanding and misinformation about the reproductive system than about any other organ system of the human body. The inaccuracies stem from inadequacies in sex education during childhood and adolescence, social prohibitions of both indiscriminate sexual activities and sexual scientific experimentation with human beings, and the inherent difficulties of extrapolating to human beings the results of experiments on animal sexuality.

The testes and ovaries produce spermatozoa and developing ova, respectively. Through sexual intercourse, spermatozoa can be transferred from the male to the female reproductive tract. Conception begins when a spermatozoon comes in contact with and successfully fertilizes a developing ovum. The hereditary characteristics of the formative individual, including genetic sex determination, are established at the time of fertilization. The early embryo must successfully embed itself within the uterine wall (*implantation*) and establish an indirect system of communication with the bloodstream of the mother (*placentation*). The genetic potential contained in the chromosomes at the time of fertilization begins to unfold during cellular and organ differentiation of the embryo and fetus, respectively. Birth, or parturition, culminates prenatal development. Significant changes in structure and function are required to cope with the change from aquatic to terrestrial habitat at the time of birth. *Organogenesis,* the development of organs, continues after birth, of course, as growth and maturation proceed.

This chapter will explore the structures and functions of the male and female reproductive systems. Specific emphasis is given to gametogenesis (spermatogenesis and oogenesis), sexual intercourse, fertilization, implantation, placentation, differentiation, organogenesis, birth, and contraception.

## THE REPRODUCTIVE SYSTEMS

### THE MALE REPRODUCTIVE SYSTEM

The reproductive system of the human male consists of (1) a scrotum containing two glands called *testes* in which spermatozoa are formed and from which the male reproductive hormone is secreted; (2) a transport system for the conduction of spermatozoa from the testes to the outside, including paired epididymides, vasa deferentia, ejaculatory ducts, and an unpaired urethra; (3) accessory reproductive glands including paired seminal vesicles and bulbourethral glands and an unpaired prostate gland, all of which produce secretions that maintain the viability of the spermatozoa before and during insemination; and (4) an organ of insemination, the penis. These structures are shown from lateral aspect in Fig. 19-1*A*.

**The scrotum** The testes are located outside and below the pelvic cavity in a sac called the *scrotum*. The scrotum is partitioned internally into right and left compartments by a septum. One testis resides in each compartment, or sac.

**Figure 19-1** *The male reproductive system.* A. *The primary and secondary reproductive structures from lateral aspect.* B. *Inferior aspect of the cylinders of erectile tissue which make up the penis.* C. *Cross section through the penis, showing the relationships among the three cylinders of erectile tissues.* (From Melloni, Stone, and Hurd.)

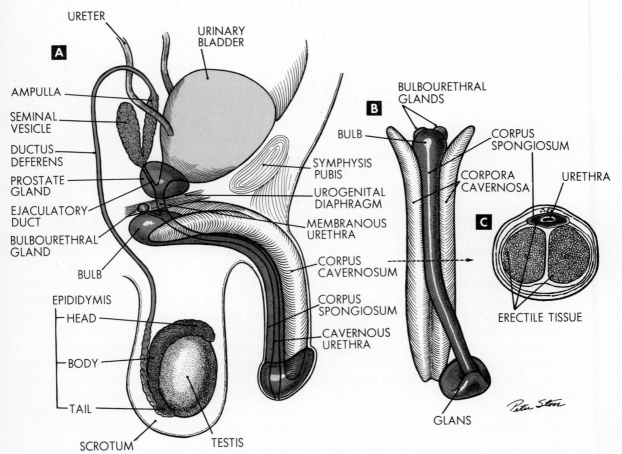

Spermatozoa cannot form at temperatures characteristic of the pelvic cavity. The temperature within the scrotum is about 4°C cooler than the temperature in the abdomen. Scrotal temperature is regulated to some degree by the action of the *dartos muscle*. When the scrotum is exposed to cold temperatures, the dartos contracts, elevating the testes and bringing them closer to the warm pelvic cavity. The dartos relaxes under the opposite temperature conditions.

Prior to or shortly after birth, the testes descend from the abdominal cavity through the inguinal canals and into the scrotal sacs. Failure of the testes to descend from their hot environment in the abdomen gives rise to a type of functional sterility called *cryptorchism*. This condition is discussed in Chap. 21.

**The testes**    Located in the scrotal sacs, the *testes* produce spermatozoa and secrete the male reproductive hormone. The structure of a testis and its associated duct system is illustrated in Fig. 19-2A.

Each testis is an oval body bounded by a fibrous capsule called the *tunica albuginea* (white coat, literally). Partitions called *septa* project inward from the tunica albuginea, dividing the testis into numerous *lobules*. *Seminiferous tubules* surrounded by intersti-

**Figure 19-2**    *The fine structure of the testes.* A. *An exploded view of the testis and epididymis from lateral aspect.* B. *Cross section through a seminiferous tubule.* C. *The structure of a human spermatozoon.* (From Melloni, Stone, and Hurd.)

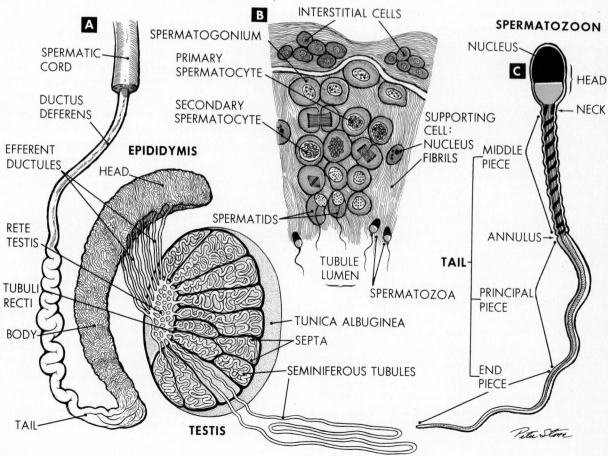

tial tissue are found in each lobule. Each highly coiled seminiferous tubule is between 30 and 60 cm in length.

*Spermatogenesis,* or the formation of spermatozoa, occurs within the seminiferous tubules. Primordial germ cells called *spermatogonia* are located in the outer borders of the walls of these tubules. The spermatogonia give rise to spermatozoa following their enlargement and two maturation divisions. Spermatogenesis is explained later in this chapter.

The male reproductive hormones are referred to as *androgens.* These hormones are secreted by the interstitial cells of Leydig found in the interstitial tissues which surround the seminiferous tubules within the lobules. The most abundant and important of the androgens is called *testosterone.* The functions of testosterone will be considered later.

Only about one-hundredth the size of mature ova, spermatozoa are the cellular products of the seminiferous tubules. The structure of a spermatozoon is illustrated in Fig. 19-2C. A mature spermatozoon consists of a head, neck, and tail. The head possesses a large nucleus containing chromosomes bearing the potential hereditary contribution of the father, in addition to a small amount of cytoplasm and an apical projection called an *acrosome.* When a spermatozoon contacts a developing ovum after insemination, an enzyme called *hyaluronidase* is secreted from the acrosome. Hyaluronidase erodes the outer covering of the developing ovum, thereby permitting entry of the spermatozoon into its cytoplasm. Called *fertilization,* this process will be described later.

The neck of a spermatozoon contains two centrioles. The centrioles will participate in the first cleavage (mitotic) division of the *zygote,* the cell formed by the fusion of the spermatozoon and ovum. This division initiates embryonic development of the new individual.

The tail of a spermatozoon consists of three parts: (1) a proximal middle piece consisting of a central core of axial filaments surrounded by a cluster of mitochondria called a *mitochondrial spiral*; (2) a middle principal piece in which the mitochondrial spiral is replaced by a fibrous sheath; and (3) a distal end piece in which the fibrous sheath disappears. Running the length of the tail, the axial filaments are the contractile elements responsible for the swimming movements of the spermatozoon. The mitochondria of the spiral provide a large amount of energy to the spermatozoon to sustain its swimming movements.

Following their formation, spermatozoa in the lumina of the seminiferous tubules are transported into straight tubules called *tubuli recti,* which drain the bases of the seminiferous tubules. The tubuli recti give rise to a plexus of tubules known as the *rete testis* which, in turn, coalesces to form a small number of tubules referred to as *efferent ductules.* The efferent ductules connect the testis with the epididymis.

Spermatozoa are transported passively from the seminiferous tubules to the epididymis through this system of tubules. Transport is accomplished by the rhythmic beating of cilia which line the lumina of these tubules.

**The epididymides** The spermatozoa undergo further physical maturation in the epididymides. Spermatozoa may spend from 1½ days to several months in this portion of the reproductive tract. Lying on the superolateral border of the testis, each *epididymis* is a tightly packed and highly coiled tubule, the extended length of which is approximately 4 m. The epididymis is divided into a head, body, and tail. The tail of the epididymis is continuous with the ductus, or vas, deferens.

**The vasa deferential** Each *vas deferens* (ductus deferens, or sperm duct) connects between the epididymis and the ejaculatory duct. The expanded distal end of the ductus deferens (see Fig. 19-1A) is called an *ampulla.* The seminal vesicle, one of the accessory reproductive glands, extends as a blind pouch from the confluence of the ampulla and ejaculatory duct. The seminal vesicle will be described later.

Spermatozoa are transported from the epididymis to the ejaculatory duct through the vas deferens. The transport is passive, accomplished by peristaltic contractions of the smooth muscular wall of the epididymis. The rhythmic contractions are initiated near the epididymis during sexual intercourse, thereby propelling spermatozoa into the ejaculatory duct. This physiologic act is largely responsible for *emission,* or the appearance of spermatozoa in the urethra prior to ejaculation.

The vas deferens runs in parallel with spermatic arteries, veins, nerves, and lymphatics. All these structures are wrapped together with fibrous connective tissue. Called the *spermatic cord* (see Fig. 19-2*A*), the entire unit is invested with peritoneum as it enters the pelvic cavity through the inguinal canal.

**The ejaculatory ducts** Each ejaculatory duct, as mentioned previously, connects the ampulla of the vas deferens and the seminal vesicle with the prostatic, or proximal, portion of the urethra. Peristaltic contractions of the smooth musculature of the ejaculatory duct aid in the emission of spermatozoa prior to ejaculation. The ejaculatory duct also transports the secretions of the seminal vesicle to the prostatic urethra.

**The urethra** The urethra begins on the inferior aspect of the urinary bladder and ultimately courses through the corpus spongiosum urethrae, one of the cylinders of erectile tissue which make up the substance of the penis. About 8 in. long, the male urethra is divided into three regions continuous with each other: (1) the proximal portion, called the *prostatic urethra,* is surrounded by the prostate gland and receives the products transported by the ejaculatory ducts as well as urine stored in the urinary bladder; (2) the middle portion, called the *membranous urethra,* receives the products transported by the prostatic duct; and (3) the distal portion, called the *cavernous urethra,* opens to the outside and receives the ducts of the bulbourethral glands.

The male urethra serves a dual role, transporting either urine from the bladder or semen consisting of spermatozoa and the secretions added by the accessory reproductive glands during ejaculation. Because of the arrangement of the reproductive and urinary tracts, ejaculation and urination are mutually exclusive acts. However, sometimes a small amount of semen minus spermatozoa may be voided immediately after urination is completed.

**The penis** Since the urethra courses through the *penis,* this organ has both reproductive and urinary functions. During sexual intercourse, the erect penis ejaculates seminal fluid into the female vagina.

The penis consists of three cylinders of erectile tissue wrapped together by connective tissue and covered by skin. The recurved foreskin is called the *prepuce.* The two superolateral cylinders are named the *corpora cavernosa penis.* The inferomedial cylinder is referred to as the *corpus spongiosum urethrae.* All three cylinders are highly vascular and contain a large number of blood sinuses which fill during sexual stimulation, resulting in an erection. The cylinders of the penis are shown from inferior aspect in Fig. 19-1*B.* A cross section of the penis is diagramed in Fig. 19-1*C.* The process of erection is discussed later in this chapter.

The corpus spongiosum urethrae is unique in several respects: (1) the cavernous urethra runs through its substance, as mentioned earlier; (2) its proximal end, called the bulb, is surrounded by the bulbocavernosus (bulbospongiosus) muscle which propels the semen to the outside by contracting spasmodically during ejaculation; and (3) its distal end consists of a cap-shaped expansion known as the *glans.*

**The accessory reproductive glands** The accessory reproductive glands consist of the following structures:

1   Paired *seminal vesicles.* Each seminal vesicle is a highly glandular blind pouch arising from the intersection of the ampulla of the vas deferens and the proximal end of the ejaculatory duct. The *ejaculatory duct* transports spermatozoa from the vas deferens and secretions of the seminal vesicles to the prostatic urethra prior to ejaculation.
2   An unpaired *prostate gland.* Shaped like an acorn, the prostate gland surrounds the prostatic urethra immediately inferior to the urinary bladder. The *prostatic duct* carries the secretions of the prostate gland to the membranous urethra.
3   Paired *bulbourethral* (Cowper's) *glands.* These pea-shaped glands lie below the prostate gland

and immediately above the bulb of the corpus spongiosum urethrae. The bulbourethral ducts empty into the proximal end of the cavernous urethra at the base of the penis.

The accessory reproductive glands have the following functions: (1) the products of all three glands contribute to the seminal fluid, or semen, which is ejaculated; (2) their secretions are believed to activate the swimming movements of the spermatozoa; (3) their secretions presumably provide nourishment for the spermatozoa; (4) phosphate and carbonate buffers in the semen reduce the acidity of the seminal fluid and the vaginal walls; and (5) their secretions are probably responsible for lubricating the inside wall of the urethra.

Ordinarily, *semen* consists of spermatozoa and the secretory products of the accessory reproductive glands. The sperm count in an average ejaculate ranges between 200 million and 400 million. Sperm counts below 50 million are likely to result in functional sterility. High sperm counts are required to ensure fertilization for two reasons: (1) the journey through the uterus and fallopian tubes will kill many spermatozoa; and (2) the antigen-antibody interaction between fertilizin of the ovum and antifertilizin of the spermatozoa will cause most of the spermatozoa which reach the developing ovum to clump together, thereby resulting in their incapacitation (see "Fertilization," below).

## THE FEMALE REPRODUCTIVE SYSTEM

The reproductive system of the human female consists of the following structures:

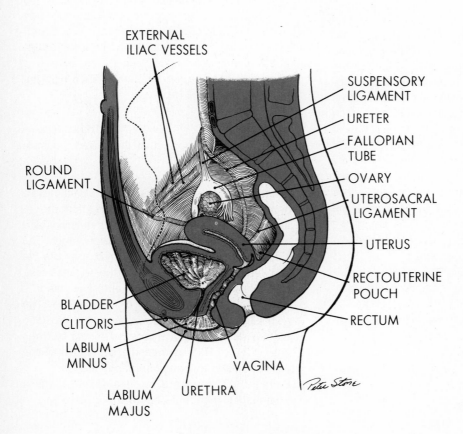

EXTERNAL
ILIAC VESSELS

SUSPENSORY
LIGAMENT

URETER

FALLOPIAN
TUBE

ROUND
LIGAMENT

OVARY

UTEROSACRAL
LIGAMENT

UTERUS

RECTOUTERINE
POUCH

BLADDER
CLITORIS

RECTUM

LABIUM
MINUS

VAGINA

LABIUM
MAJUS

URETHRA

**Figure 19-3** *The female reproductive tract. The view is of a midsagittal section through the abdominal and pelvic regions from the left side.* (From Melloni, Stone, and Hurd.)

1 Paired ovaries, from which developing ova (primary oocytes) are discharged (ovulated) and the hormones estrogen and progesterone are secreted
2 Paired fallopian tubes, in which oogenesis, fertilization, and early embryonic development occur and through which either the unfertilized developing ovum (secondary oocyte) or the early embryo after fertilization is transported to the uterus
3 The uterus, in which prenatal development occurs
4 The vagina, into which an erect penis is inserted for the transfer of spermatozoa and through which the fetus and afterbirth are expelled during birth or the menses are voided when fertilization does not occur
5 The external genitalia, or vulva, which contain

erogenous and secretory structures that assist in precoital and coital activities
6 The mammary glands, or breasts, which function in lactation and sexual attraction

The female reproductive tract and external genitalia are shown in situ from lateral aspect in Fig. 19-3. The female reproductive tract is illustrated from posterior and anterior aspects in Fig. 19-4A and B, respectively. The following descriptions are with reference to these two figures.

The ovaries   The ovaries are the counterparts of, or homologous to, the male testes. They have the following functions: (1) approximately once each month

**Figure 19-4**   *The female reproductive tract.* A. *Posterior aspect.* B. *Anterior aspect, showing details of the fallopian tubes, uterus, and vagina.* (From Melloni, Stone, and Hurd.)

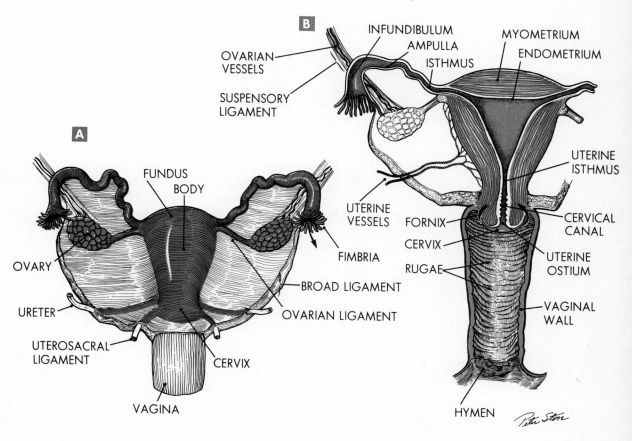

during the reproductive life of the female, the ovarian wall ruptures, releasing a developing ovum from a mature follicle into the abdominal cavity; and (2) depending upon the phase of the menstrual cycle, the follicular cells either secrete estrogens or a combination of estrogens and progesterone.

Located in the pelvic cavity, the ovaries are small oval glands measuring almost an inch in length. The ovarian wall consists of a fibrous capsule similar to that of the testicular tunica albuginea. The outside of the ovary appears bumpy because of the presence of old, degenerating follicles. Just beneath the follicular wall is the *germinal epithelium* containing *oogonia* destined to become developing ova. The cortical tissue below the germinal epithelium is called the *ovarian stroma*. Cells of the ovarian stroma will surround the developing ova to form complex structures called *follicles*. Follicular development initiates oogenesis, or the formation of ova.

The ovaries are supported by a number of ligaments. Illustrated in Fig. 19-4, these ligaments include (1) the ovarian ligaments which support their medial aspects to the body of the uterus; (2) the suspensory ligaments which support their lateral aspects; and (3) the broad ligaments which support their posterior surfaces. The ligaments are formed primarily from folds of peritoneum, but the major part of the ovary is not covered by peritoneum.

**The fallopian tubes** Called oviducts in other mammals, these tubes are continuous with the superolateral aspects of the uterus. Their proximal ends are close to, but not in contact with, the ovaries. The fallopian tubes have the following functions: (1) developing ova after ovulating are trapped in the fluid currents created by the beating cilia which line the inside walls of their multifinger-shaped borders; (2) both maturation divisions ordinarily occur in this part of the female reproductive tract; (3) fertilization also usually takes place in the upper third of one of these tubes; (4) following fertilization and prior to implantation, early embryonic development occurs here; (5) peristaltic contractions of their walls and the beating cilia which line their lumina aid in fertilization by promoting the upward migration of spermatozoa; and (6) they transport unfertilized secondary oocytes

or early embryonic stages following fertilization to the uterus.

Each fallopian tube is divided into a proximal expanded portion known as the *infundibulum,* a tapering middle portion called the *ampulla,* and a constricted distal portion called the *isthmus.* The mouth (ostium) of the infundibulum opens into the abdominal cavity; the distal end of the isthmus opens into the uterine cavity.

The tissue structure of each fallopian tube consists of an inner mucosa containing ciliated columnar cells which line the lumen, a middle layer of circular and longitudinal smooth muscle, and an outer serosa of visceral peritoneum. The mucosa of the fallopian tube is similar to the endometrium of the uterus, with which it is continuous.

**The uterus** This pear-shaped vesicular organ is located in the pelvic cavity just inferior to and between the ovaries, immediately superior to the proximal portion of the vagina, and anterior to the rectum. The uterus is ordinarily tipped forward about 90°, with reference to the position of the vagina, causing its inferior border to rest upon the superoposterior wall of the urinary bladder. The distal tip of the uterus projects into the upper part of the vagina, forming a circular recess called the *fornix* in which semen is ordinarily deposited during sexual intercourse. The position of the uterus is highly flexible, since it expands greatly into the upper part of the abdominal cavity during pregnancy and is ordinarily displaced by a distended urinary bladder or rectum.

The uterus is supported by the pelvic musculature to which the vagina is attached as well as by numerous ligaments. Illustrated in Figs. 19-3 and 19-4, these ligaments include: (1) the *broad ligaments,* which attach its lateral margin to the walls of the pelvis; (2) the *uterosacral* ligaments, which attach its cervical region to the sacrum; and (3) the *round ligaments,* which attach its lateral margins to the broad ligament before inserting on the superior aspects of the labia majora. These ligaments are either formed of peritoneum or from fibromuscular tissue.

The uterus is divided into the following regions: (1) a *proximal fundus* superior to the entrances of the fallopian tubes; (2) a *middle body* containing the major part of the uterine cavity; and (3) a *distal cervix* containing the cervical canal. The cervix projects into the upper vagina. The lumina between the fundus and body are continuous with the lumina of the fallopian tubes.

The tissue structure of the uterus is similar to, but more elaborate than, that of the fallopian tubes. The mucosa, muscular layer, and serosa of the uterus are called endometrium, myometrium, and perimetrium, respectively. The *endometrium* consists of a ciliated columnar epithelium and a richly vascularized submucous connective tissue in which are located uterine glands, spiral arterioles, and maternal venules. The *myometrium* is composed of many layers of smooth muscle fibers which run in various directions. The *perimetrium* is made up of visceral peritoneum.

The uterus has the following functions: (1) embryonic and fetal development occurs within its endometrium; (2) a portion of the endometrium called the decidua basalis becomes the maternal contribution to the placenta; (3) contractions of the myometrium expel the fetus to the outside during the expulsion stage of labor; and (4) unfertilized secondary oocytes are resorbed by its endometrium.

**The vagina** The proximal end of the vagina surrounds the uterine cervix. Its distal end, the *external vaginal orifice,* opens to the outside in the vestibular region of the vulva. The external vaginal orifice is partially covered by a membrane called the *hymen.* With intercourse this membrane becomes abraded and all but disappears.

When standing erect, the vagina is tilted at about a 45° angle, with its proximal end facing posteriorly and its distal end anteriorly. The vagina is in the midsagittal line within the pelvic cavity, sandwiched between the posterior walls of the urinary bladder and urethra and the anterior wall of the rectum. The vagina is funnel-shaped, gradually expanding from its distal to its proximal end.

The tissues of the vagina are quite distensible, and the diameter of its canal can be increased sufficiently to receive an erect penis. The tissue lining the vaginal canal consists of stratified squamous epithelium and contains many transverse *rugae.* The middle layer of smooth musculature is powerful and can facilitate

ejaculation by contracting against the erect penis during sexual intercourse. The outer covering of the vagina is composed of fibroelastic connective tissue. The base of the vagina is anchored to the pelvic musculature. The flora and fauna in the vaginal canal during reproductive life and following menopause are described in Chap. 2.

**The external genitalia** Located between the legs, the external genitalia of the human female are referred to as the *vulva* (see Fig. 19-5).

**Figure 19-5** *The female reproductive system. The external genitalia. Called the vulva, this region between the legs is viewed from inferior aspect.* (From Melloni, Stone, and Hurd.)

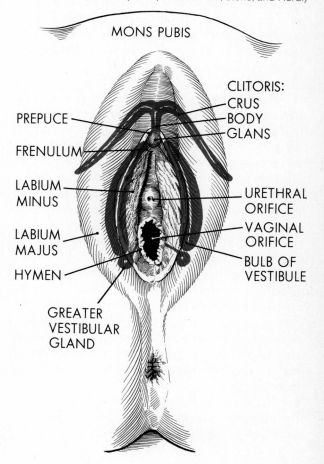

The vulva is bounded anteriorly by the mons pubis, laterally by the labia majora, and posteriorly by the perineum leading to the anus. Consisting of adipose and fibrous connective tissue, the *mons pubis* is a mound of sensitive hair-covered skin adjacent to the pubic symphysis. The *labia majora* are the homologues of the male testes and encircle smaller folds of skin called the *labia minora*. The labia minora consist of distensible tissues which surround the *vestibule* and give rise anteriorly to the *prepuce*. The glans of the clitoris lies immediately behind the prepuce. The vaginal and urethral orifices and the openings of the ducts from the greater and lesser vestibular glands are located in the vestibular region.

The *clitoris* is the female homologue of the penis. However, there are only two cylinders of erectile tissue in the clitoris, compared to three cylinders in the penis. The internal erectile tissues, called the *bulb of the vestibule,* are homologous to the corpus spongiosum urethrae. The clitoris consists of three regions—the crus, body, and glans. Only the glans is visible externally.

The vaginal orifice, partially obscured by the membranous hymen, lies between the posterior aspects of the labia minora. The external urethral orifice lies a short distance anterior to the vagina, in the central portion of the vestibule. The anus, through which the rectum opens to the outside, lies outside the vestibular region in the perineum posterior to the vagina.

The following are among the accessory reproductive glands of the female: (1) the *greater vestibular glands* (Bartholin's glands), homologues of the bulbourethral glands of the male, open on the lateral aspects of the vagina and secrete a vaginal lubricant during sexual stimulation; and (2) the *lesser vestibular glands* (Skene's glands), the ducts of which open in the vestibular region between the vaginal and urethral orifices, are probably nonfunctional. However, they might produce a sexual odor that serves as an erotic stimulant. Both glands are especially vulnerable to microbial pathogens and are frequently infected in cases of gonorrhea.

**The mammary glands (breasts)**  The breasts secrete milk for the nourishment of the infant, and are thus considered accessory reproductive glands. The mature mammary gland is illustrated in Fig. 19-6.

The breast is a compound alveolar sweat gland lying between the second and sixth ribs on the anterior aspect of the pectoralis major muscle. The skin of the breast is thinner than that in adjacent areas and contains a *nipple* surrounded by an *areola*. The nipple is just below and lateral to the center of the breast and in line with the fourth intercostal space. Both the nipple and areola are pigmented. The pigmentation deepens in a lactating (milk-producing) breast. The ducts of the mammary gland open to the outside on the tip of the nipple. The areola receives ducts from both sweat and sebaceous glands.

There are about 15 to 25 *lactiferous ducts* which drain the secretory alveoli within each mammary gland. The proximal ends of the lactiferous ducts are highly branched and drain numerous *lobules* of glandular tissue. The lobules are separated from each other by connective tissue partitions called *septa*. Partitions which attach the lobules to the subcutaneous connective tissue are called *suspensory ligaments*. All the lobules ultimately drained by one lactiferous duct are collectively referred to as a *lobe.* Thus, there are as many lobes as there are lactiferous ducts. The spaces below the skin and above the skeletal musculature not occupied by glandular material are filled with *adipose connective tissue.* Because of the septa, the inside of the mammary gland resembles a honeycomb of fatty deposits.

The hormonal control of mammary gland development and lactation is discussed in Chap. 18. Mastitis, an inflammation of the mammary glands, is reviewed in Chap. 20.

## HORMONAL CONTROL OF REPRODUCTION

Glands involved in the secretion of hormones regulating reproduction include the hypothalamus, the adenohypophysis, and the gonads (testes or ovaries). First, we will review the complex interactions between these glands in the male, then in the female.

The sex chromosomes of a genetically determined

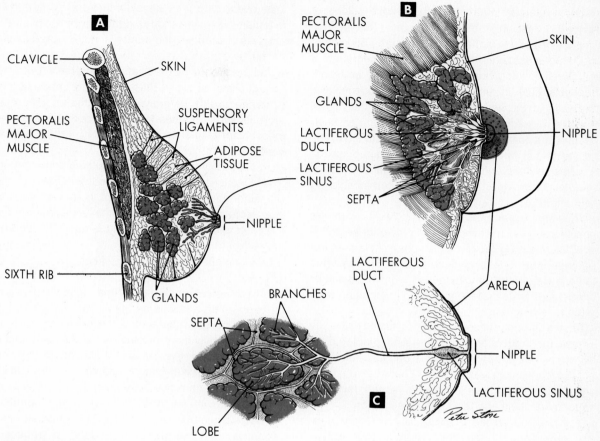

**Figure 19-6** *The mammary gland. A. The breast from lateral aspect. B. The left breast from anterior aspect. The skin and subcutaneous connective tissue have been removed from one side to show the glandular organization. C. An enlargement of a portion of the breast from lateral aspect. (From Melloni, Stone, and Hurd.)*

male are XY. In response to the presence of chorionic gonadotropins secreted from the placenta of the mother, the fetal gonads of a genetically determined male secrete a small amount of the male reproductive hormone testosterone. The presence of testosterone causes the gonads to differentiate into testes.

Following birth, and until puberty is reached, the testes secrete an almost negligible amount of testosterone. The reason for the practically total absence of testosterone is believed to be the inhibition of the hypothalamus, upon which the secretion of testosterone depends. Presumably, the prepuberal hypo-

thalamus is hypersensitive to the trace amounts of testosterone secreted by the immature testes. As puberty approaches, hypothalamic sensitivity decreases so that the trace amounts of testosterone present in the blood gradually become incapable of inhibiting the secretion of the hypothalamic hormones involved in reproduction.

From puberty on, the male hypothalamus secretes two functional reproductive hormones—follicle-stimulating hormone releasing factor (FSH releasing factor) and interstitial cell–stimulating hormone releasing factor (ICSH releasing factor). The latter is the male counterpart of luteinizing hormone releasing factor (LH releasing factor) of the female (see below). The first of these hypothalamic hormones, FSH releasing factor, stimulates the synthesis and secretion of follicle-stimulating hormone (FSH) by the adenohypophysis. FSH, in turn, initiates spermatogenesis in the walls of the seminiferous tubules of the testes. However, spermatogenesis does not proceed beyond the formation of immature spermatids. The conversion of spermatids into mature spermatozoa requires the presence of testosterone and the prior stimulation provided by interstitial cell-stimulating hormone (ICSH).

The adenohypophysis produces ICSH. Its synthesis and secretion are dependent upon the prior secretion of ICSH releasing factor from the hypothalamus. Apparently, FSH releasing factor and ICSH releasing factor are secreted simultaneously in the male so that spermatogenesis ordinarily goes to completion once puberty is reached.

As the blood concentrations of testosterone increase, the production of ICSH releasing factor in the hypothalamus is depressed. This negative feedback control mechanism (see Chap. 17) between the testes and the hypothalamus maintains normal levels of spermatogenesis and testosterone secretion.

Testosterone has the following effects: (1) the physical maturation of the male reproductive system, including the primary reproductive organs and the accessory reproductive glands; (2) the development of the secondary sex characteristics of the male including an enlarged musculature, a prolonged bone growth and an increased bone mass, an increased laryngeal development and a lowered voice, an increased distribution of body hair, and an increased basal metabolic rate and erythrocyte count; and (3) the psychological development of sexual desire. These widespread and profound effects are presumably the result of promoting specific types of protein anabolism.

The molecular formula of testosterone is shown in Fig. 18-9. The effects of a pathologic hyposecretion of this hormone are discussed in the same chapter.

The sex chromosomes of a genetically determined female are XX. In the absence of the secretion of tes-

testosterone during fetal development, the gonads differentiate into ovaries following the secretion of small amounts of estrogen, the primary female reproductive hormone.

The prepuberal ovaries secrete extremely low concentrations of estrogen. Presumably, until puberty, the presence of this hormone inhibits the secretion of reproductive hormones from the hypothalamus by a mechanism similar to that described earlier for the male.

From puberty on, the female hypothalamus secretes three functional reproductive hormones—FSH releasing factor and LH releasing factor, both of which are also secreted by the male hypothalamus, in addition to luteotropic hormone inhibiting factor (LTH inhibiting factor).

Estrogens have the following effects:

1 The physical maturation of the female reproductive system and external genitalia
2 The initial development of the mammary glands
3 The establishment of other secondary sex characteristics of the female, including a smaller average physical stature and development than the male because of an accelerated completion of bone growth following puberty and a parallel decrease in bone and muscle mass; a broadened pelvis and narrowed shoulders; a reduced laryngeal development and correspondingly elevated voice; a smooth skin and a relatively reduced distribution of body hair; the deposition of an increased amount of subcutaneous fat; and a slightly increased basal metabolic rate
4 The psychological development of sexual desire

The molecular formula of estrogen is shown in Fig. 18-10. Disorders caused by an abnormal deficit of estrogen are described in the same chapter.

## THE MENSTRUAL CYCLE

The *menstrual cycle* is a hormonally controlled reproductive cycle repeated approximately each month during the reproductive life of a human female, except when pregnant. During each natural menstrual cycle (1) a primary oocyte is ovulated from a graafian follicle within the ovary; (2) the endometrium of the uterus develops sufficiently to receive an implanting embryo called a blastocyst, if fertilization occurs; and (3) development of the mammary glands progresses in preparation for a possible future lactational function.

The menstrual cycle depends on complex interactions between hypothalamic, adenohypophyseal, and ovarian hormones. If pregnancy results, gonadotropins and ovarian hormones secreted by the placenta also play a prominent role. The target structures of the adenohypophyseal and ovarian hormones are primarily the ovarian follicles and uterine endometrium, respectively.

Basically, there are two stages during the menstrual cycle—*preovulatory* and *postovulatory*. The preovulatory stage starts on day 1 of the menstrual cycle and ends on the day of ovulation, and the postovulatory stage begins from this point and ends on the first day of menstruation. The preovulatory follicular and endometrial events are referred to as the *follicular* and *menstrual-proliferative* phases, respectively. The postovulatory follicular and endometrial events are called the *luteal* and *secretory* phases, respectively. Figure 19-7 traces the effects of the adenohypophyseal

**Figure 19-7** *The human menstrual cycle. The diagram traces the influence of the gonadotropic and ovarian hormones on follicular and endometrial development. The ovum becomes mature only after fertilization takes place.* (From Melloni, Stone, and Hurd.)

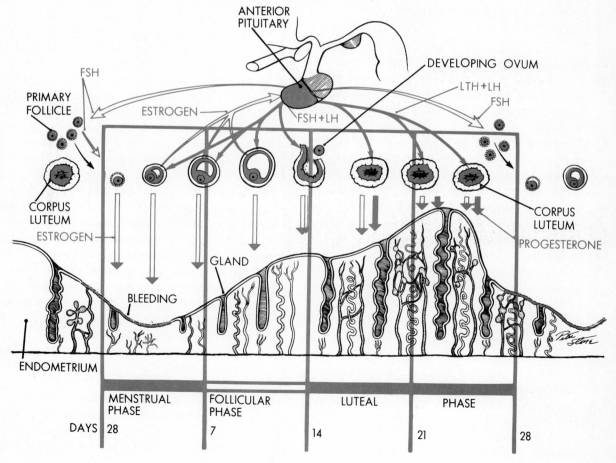

gonadotropins on the follicles within the ovaries and the effects of the ovarian hormones on the endometrium of the uterus.

## The preovulatory stage

*The follicular phase*   The development of numerous follicles within the ovaries is initiated when a hypothalamic hormone called follicle-stimulating hormone releasing factor (FSH releasing factor) promotes the synthesis and secretion of follicle-stimulating hormone (FSH) from the adenohypophysis (anterior lobe of the pituitary gland). Thus, although the immediate chemical signal for follicular induction is FSH, the hypothalamus is responsible for establishing the existence of the signal.

FSH induces the development of follicles in the following manner:

1   Oogonia enlarge to become primary oocytes.
2   The primary oocytes become surrounded by a single layer of cells from the ovarian stroma, at which point the entire unit is called a primary follicle.
3   The primary follicle continues to develop by the addition of multiple layers of stromal cells.
4   The stromal cells of each follicle form layers of differentiated tissue and secrete a fluid called liquor folliculi into an enlarging cavity called an antrum.
5   The expanding antrum displaces the primary oocyte to one side of the follicle.
6   A specialized tissue called the stratum granulosum immediately surrounding the antrum begins to secrete estrogens, the most abundant of which is 17$\beta$-estradiol.

Estrogens, of course, collectively represent the primary female reproductive hormone. The rise in the blood concentrations of estrogens as the ovarian follicles develop stimulates the hypothalamus to decrease its secretion of FSH releasing factor (negative feedback) and to cause the secretion of luteinizing hormone releasing factor (LH releasing factor), an example of positive feedback. The adenohypoph-

ysis responds to a decrease in FSH releasing factor by reducing its secretory output of FSH while the presence of LH releasing factor promotes the synthesis and secretion of luteinizing hormone (LH). During this time, a hypothalamic hormone called luteotropic hormone inhibiting factor (LTH inhibiting factor) inhibits the secretion of luteotropic hormone (LTH) from the adenohypophysis.

LH exerts the following effects on follicular development:

1   Usually only one of the numerous developing follicles is induced to mature physically and is called a *graafian follicle,* while the other developing follicles involute and are ultimately resorbed by the ovarian stroma.
2   The graafian follicle destined to ovulate migrates to a point immediately beneath the ovarian wall from a location deeper in the stroma.
3   An increase in the volume of liquor folliculi and the size of the graafian follicle (over 1/2 in. total diameter) in the days preceding ovulation cause the liquor to press against the follicular wall and the follicle to press against the ovarian wall. When the concentrations of FSH and LH are balanced critically, the primary oocyte is ovulated following local ruptures in the follicular and ovarian walls.

The mechanics of ovulation are still a matter of conjecture. Two factors are probably involved: (1) an increase in intrafollicular pressure due to the progressive accumulation of liquor folliculi, and (2) local ischemia and subsequent necrosis in the wall of the ovary next to the bulging graafian follicle. In any event, the ovarian and follicular walls rupture, resulting in an exudation into the abdominal cavity of the liquor folliculi. The primary oocyte is carried in its wake. Local hemorrhaging of the ovarian wall at the site of rupture may result in a small accumulation of blood within the antrum of the ruptured follicle. The blood will ultimately disappear.

The primary oocyte is usually trapped in a fluid current created by the rhythmic beating of cilia which line the inside of the finger-shaped border of the fallopian tube. At this time, the primary oocyte enters into and completes the first meiotic, or maturation, division. Thus, the developing ovum that passes down the fallopian tube is a secondary oocyte and not yet a mature ovum (see "Oogenesis," below).

*The menstrual-proliferative phase* The endometrial events correlated with the preovulatory ovarian events described above begin on the first day of menstruation; i.e., the onset of menstruation marks the first day of the new menstrual cycle.

The menstrual phase, or period, occurs during the sloughing off of the labile portion of the uterine endometrium. Usually menstruation lasts for the first 3 to 5 days of the menstrual cycle. During this time, about 40 ml of blood, a similar volume of serous fluid, and tissue debris of the desquamated uterine endometrium are menstruated, or passed through the vagina and voided.

Menstruation is precipitated directly by a depression in the secretion of the ovarian hormones estrogen and progesterone. Somehow, these chemical signals cause the spiral arterioles in the labile portion of the uterine endometrium to contract spasmodically. The diminution in the endometrial blood supply leads to endometrial ischemia, necrosis, and desquamation. Some of the engorged blood vessels rupture, so that local hemorrhages form in the spaces developing in the desquamating tissue. Contractions of the uterine myometrium promote the menstrual process. By the end of menstruation, the epithelial lining of the endometrium is being repaired in preparation for endometrial proliferation.

The proliferative phase of the endometrial cycle extends from the end of the menstrual phase to the time of ovulation. Proliferation of the endometrium is under the control of estrogens. During this time, the endometrium doubles in size, reaching a thickness of about 6 mm. The increase in endometrial thickness occurs through both *hypertrophy* (an increase in the size of the individual cells) and *hyperplasia* (an increase in the number of total cells). As endometrial proliferation proceeds, the uterine glands increase in size and begin to secrete uterine fluid. The vascularity of the endometrium increases as the spiral arterioles and maternal venules enlarge. Thus, at the time of ovula-

tion, the uterine endometrium has thickened substantially and become glandular and vascular.

The postovulatory stage

*The luteal phase* Following ovulation, the graafian follicle is transformed into a *corpus luteum,* or yellow body. This transformation is regulated by LH. Presumably, LH also promotes the initial secretion of progesterone by the corpus luteum. However, the maintenance of progesterone secretion is under the direct control of adenohypophyseal LTH. The corpus luteum also secretes estrogens because of the continued presence of adenohypophyseal FSH.

The chemical interplay between the ovary and hypothalamus which prompts the adenohypophysis to synthesize and secrete LTH is not completely understood, but the mechanism may operate in the following manner:

1   The secretion of both estrogens and progesterone from the corpus luteum decreases the adenohypophyseal secretions of FSH and LH by depressing the secretion from the hypothalamus of FSH releasing factor and LH releasing factor.
2   The decreased secretion of FSH releasing factor and LH releasing factor also depresses the secretion of LTH inhibiting factor, thereby removing the inhibitory brake which previously prevented the adenohypophysis from secreting LTH.

If fertilization of a secondary oocyte by a spermatozoon does not occur, menstruation will ultimately terminate the menstrual cycle. Called the *corpus luteum of menstruation,* the ovarian follicle begins to involute following the peak of its development about 12 days after ovulation. The secretions of estrogens and progesterone at the peak of luteal development are responsible for inhibiting the secretion of FSH and LH through hypothalamic inhibition of FSH releasing factor and LH releasing factor. As the corpus luteum fills with connective tissue, ultimately to be resorbed by the ovarian stroma, its output of estrogens and progesterone diminishes. Near the end of the menstrual cycle, in response to the depression in the blood concentrations of both estrogens and progesterone, the hypothalamus again secretes LTH inhibiting factor and FSH releasing factor. LTH inhibiting factor inhibits the further synthesis and secretion of LTH by the adenohypophysis. FSH releasing factor induces follicular development by promoting the synthesis and secretion of adenohypophyseal FSH.

If pregnancy results during the postovulatory stage of a menstrual cycle, the corpus luteum does not degenerate until the seventh month of gestation. Called the *corpus luteum of pregnancy,* the ovarian follicle continues to secrete high concentrations of both estrogens and progesterone. Hormonal competence is maintained because of the continued secretion of LTH from the adenohypophysis and the related lack of secretion of LTH inhibiting factor from the hypothalamus. After the placenta is established between the mother and embryo, this compound structure will begin to secrete gonadotropins as well as progesterone and will gradually preempt the functions of the corpus luteum during the latter part of pregnancy.

*The secretory phase*   In response to the secretion of the ovarian hormones, especially progesterone subsequent to ovulation, the uterine endometrium undergoes the following changes:

1   Further enlargement through hypertrophy
2   Further increase in the development of the uterine glands and a greatly increased secretion of glandular fluid
3   Increased vascularity through the further development of the endometrial blood vessels in general and engorgement of the maternal venules in particular
4   Deposition of lipids and glycogen in the endometrial stroma

In addition, progesterone inhibits contractions of the uterine myometrium, thereby tending to prevent abortion or premature birth.

The changes in the uterine endometrium described above prepare it to receive and nurture an embryo.

The glandular secretions and stored nutrients will provide sustenance to an implanting embryo until placentation is completed.

## SEXUAL INTERCOURSE

Sexual intercourse, coitus, or copulation is initiated when an erect penis is inserted into a vagina. The act of penile insertion is referred to as intromission. The continued and rhythmic back-and-forth movements of the penis in the vagina usually lead to orgasm in one or both sexual partners, during which the male ejaculates semen. *Orgasm* consists of the total sensorimotor experience that occurs as sexual climax is reached.

Human reproductive behavior is unique in several respects when compared to that of other animals. Firstly, mating between human partners often takes place face-to-face because of the positions assumed by the erect penis and the vagina. Secondly, human sexual receptivity seems to be present at all times during the reproductive lives of adults and is not limited to a particular season. Finally, orgasm seems to be a uniquely human experience, because other mammals during sexual climax do not exhibit intense physiologic and emotional responses.

### ERECTION

Penile erection ordinarily takes place in response to either tactile stimulation of the erogenous zones on and around the male genitalia or as the result of psychogenic sexual stimulation. Erection is controlled by the autonomic nervous system acting on blood vessels at the base of the penis. Erection of the clitoris occurs basically in the same way.

Erection is initiated when parasympathetic fibers cause the dilatation of arterioles and the closure of valves in veins at the base of the penis. Arterial dilation increases the blood supply to the penis. Valvular closure prevents blood from leaving the penis. The influx of arterial blood and the lack of venous drainage fill the vast number of blood sinuses in the erectile tissues of the penis, causing erection. Recall that the erectile tissues comprise the corpora cavernosa penis and the corpus spongiosum urethrae.

A variable time following ejaculation, sympathetic autonomic stimulation overwhelms the parasympathetic stimulation, resulting in arterial constriction and venous dilatation at the base of the penis. These actions drain the penile blood sinuses and, consequently, the penis gradually collapses to its unstimulated size.

## EJACULATION

Ejaculation propels the semen containing spermatozoa out of the penis and into the upper portion of the vagina. Ejaculation is divided into two phases—emission and ejaculation.

*Emission* precedes ejaculation and consists of the movement of semen into the urethra. The physical presence of semen in the penis during sexual intercourse can be perceived by the male. Further sexual stimulation of the male genitalia ordinarily leads to ejaculation.

*Ejaculation* is a complex phenomenon. Peristaltic contractions of the vasa deferentia are primarily responsible for transporting spermatozoa to the urethra. Their transport is aided by contractions of the ejaculatory ducts of the seminal vesicles. The major force which propels the semen through the cavernous urethra is provided by the bulbocavernosus muscle which surrounds the bulb at the proximal end of the corpus spongiosum urethrae.

## ORGASM

The elevated sensory stimulation and motor responses described as orgasm are experienced during sexual climax. Ordinarily the pleasurable sensations marked by climax are accompanied by increases in the rates of the heartbeat, pulse, and respiration as well as in blood pressure. Human orgasm is obviously a highly individual experience and may be depressed or otherwise altered by tiredness, sickness, psychological stress, and other factors.

## GAMETOGENESIS

Gametogenesis, or the formation of gametes, occurs in the gonads, or primary reproductive organs. It includes both spermatogenesis and oogenesis. *Spermatogenesis*, or the formation of spermatozoa, takes place within the seminiferous tubules of the testes. *Oogenesis,* or the formation of mature ova, is initiated within the ovaries and completed in the fallopian tubes. Each of the mature types of sex cells, whether spermatozoa or ova, contains 23 chromosomes, half the number characteristic of the human species. The maturation division by which this chromosomal reduction takes place is called *meiosis*. The process is basically similar in both sexes.

## SPERMATOGENESIS

The primordial germ cells from which mature spermatozoa develop are located in the peripheral portions of the walls of the seminiferous tubules. Spermatogenesis in a portion of a seminiferous tubule is illustrated in Fig. 19-2. The cellular and chromosomal changes which occur during spermatogenesis are diagramed in Fig. 19-8. The following description is with reference to these two figures.

The primordial germ cells are called *spermatogonia*. They proliferate by mitosis so that there is a continual supply within the seminiferous tubules from the onset of reproductive maturity (puberty) through most of the remainder of adult life. The spermatogonia destined to engage in meiosis undergo hypertrophy, or physical enlargement, and are called *primary spermatocytes*. Each of these cells undergoes the first meiotic division, giving rise to two secondary spermatocytes containing 23 pairs of chromosomes apiece, i.e., the species number for human beings, 46. Because of the unique way in which the chromosomes line up along the metaphase plate during the first meiotic division, the two secondary spermatocytes which form are not genetic carbon copies of the primary spermatocyte. In other words, the first meiotic division is different from a mitotic division.

Almost immediately after the first meiotic division,

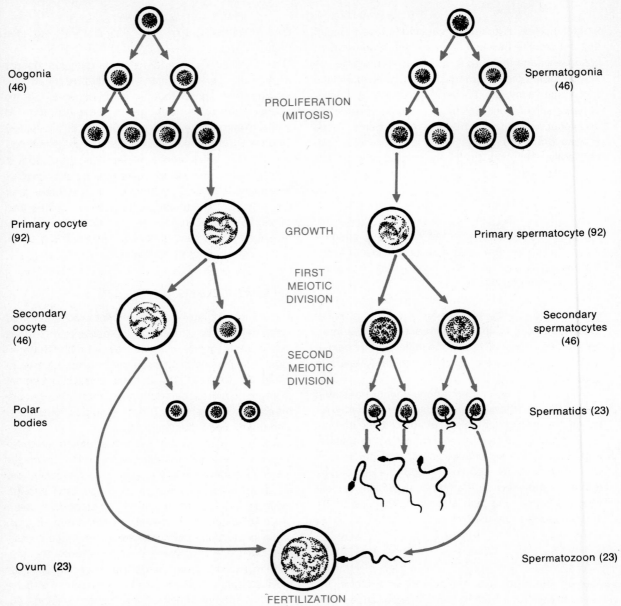

Oogonia (46)

PROLIFERATION (MITOSIS)

Spermatogonia (46)

Primary oocyte (92)

GROWTH

Primary spermatocyte (92)

FIRST MEIOTIC DIVISION

Secondary oocyte (46)

Secondary spermatocytes (46)

SECOND MEIOTIC DIVISION

Polar bodies

Spermatids (23)

Ovum (23)

Spermatozoon (23)

FERTILIZATION

**Figure 19-8** *Gametogenesis. Oogenesis and spermatogenesis are shown on the left and right, respectively.* (After Langley, Telford, and Christensen.)

the two secondary spermatocytes undergo a second meiotic division, giving rise to four physically immature spermatozoa called *spermatids*. During the second meiotic division of each secondary spermatocyte, the homologous chromosomes line up along the metaphase plate but do not replicate. Then, during anaphase, 23 chromosomes migrate to one pole, 23 to the other. Since human body (somatic) cells have 23 types of chromosomes, each in paired condition, each spermatid at the end of the second meiotic division contains half the number, but the same types, of chromosomes characteristic of the human species.

Following their formation, the immature spermatids are nourished by large supporting cells called *Sertoli cells,* to which they attach. Sertoli cells are scattered throughout the walls of the seminiferous tubules. When physically mature, the spermatids are called spermatozoa and are potentially capable of fertilizing an ovum. Notice that this transformation does not require a cellular division.

## OOGENESIS

The primordial germ cells from which mature ova develop are present in the germinal epithelium just beneath the walls of the ovaries from the time of birth. Unlike the primordial germ cells of the male, the female primordial germ cells do not ordinarily divide by mitosis. The human female is endowed with about 400,000 primordial germ cells (200,000 per ovary). Many are used in the formation of follicles during each menstrual cycle. When all are used up, the reproductive abilities of a female are ended, and she enters the *menopause* period. Menopause is preceded by a period marked by a general decline in reproductive potential, referred to as the *climacteric.* These events are correlated with the disappearance of the ovarian hormones.

Oogenesis is illustrated in Fig. 19-9, below. The cellular and chromosomal changes which occur during oogenesis are diagramed in Fig. 19-8. The following description is with reference to these two figures.

The primordial germ cells are called *oogonia.* The increased concentrations of adenohypophyseal FSH during the follicular phase of the menstrual cycle induce a number of follicles to develop within each ovary. Recall that a follicle consists of a developing ovum surrounded by one or more layers of cells derived from the ovarian stroma. As the follicle develops, the oogonium increases in size and is called a primary oocyte.

Immediately following ovulation, the primary oocyte undergoes the first meiotic division, resulting in the formation of two secondary oocytes. Each of these cells contains a full complement of chromosomes. Genetic mixing ensures that the two secondary oocytes are not carbon copies of the primary oocyte.

One of the two secondary oocytes is small, the other large because the metaphase plate formed during the first meiotic division is closer to one end of the cell than to the other. Called the first polar body, the small secondary oocyte ultimately degenerates. Essentially, the first polar body "contributes" its cytoplasm to the viable secondary oocyte, since the mature ovum will require a large amount of yolk with which to nourish the early embryo. The large secondary oocyte continues its journey down the fallopian tube accompanied by the first polar body and a star-shaped mass of surrounding cells. However, the secondary oocyte ordinarily does not undergo a second meiotic division unless fertilization takes place.

In the absence of fertilization, the secondary oocyte is resorbed by the uterine endometrium. If fertilization is successful, the secondary oocyte immediately engages in the second meiotic division. As a result of the reduction division, two cells are formed—a large ootid which in reality is a mature ovum and a small ootid called the second polar body. The differential in size of the ootids is due to a lateral displacement of the metaphase plate during the second meiotic division of the secondary oocyte. Each of the ootids contains 23 different types of chromosomes rather than 23 paired chromosomes.

The second polar body ultimately degenerates. Occasionally, the first polar body also undergoes the second meiotic division. When this happens, the resulting cells become additional second polar bodies which are also doomed to premature death. Thus, either one or three second polar bodies form as a consequence of the meiotic divisions of one primary oocyte.

A comparison of oogenesis and spermatogenesis reveals that (1) both processes are meiotic, reducing the number of chromosomes in the nuclei of the sex cells by one-half without affecting the types of chromosomes represented; (2) both processes theoretically form four sex cells from one primordial germ cell, although only one mature ovum appears from the meiotic divisions of one primary oocyte; and (3) spermatogenesis occurs in the walls of the seminiferous tubules within the testes, but oogenesis, although initiated within the follicles of the ovaries, is completed after fertilization within the fallopian tube.

## GESTATION

Gestation marks the period of pregnancy and is initiated by conception, or fertilization, and terminated by parturition, or the birth process. Human gestation lasts for approximately 274 to 280 days, counting from the first day of the last menses, or for about 267 days, counting from the day of the last ovulation.

## FERTILIZATION

Following insemination, spermatozoa are capable of fertilizing a developing ovum for about 2 days, although they exhibit active motility for at least double this length of time. Spermatozoa move at an average rate of 4 mm/min. Peristaltic contractions of the fallopian tubes accelerate this rate.

How spermatozoa reach a developing ovum in the upper third of the fallopian tube is still a matter of conjecture. The following are among the theories which have some experimental support: (1) muscular contractions of the walls of the fallopian tubes and, to a lesser extent, the uterus during and after orgasm carry groups of spermatozoa up the fallopian tubes in a manner analogous to being carried up a flight of stairs; (2) the spermatozoa move against the beat of the cilia which project from the epithelial cells that line the lumina of the uterus and fallopian tubes, thereby ensuring movement in the correct direction; and (3) the motility of the spermatozoa entering the uterus is enhanced by a hormone produced either by the developing ovum or by cells which surround it. Whatever the correct explanation, spermatozoa are certainly successful in contacting their target, as borne out by the present world population of almost 4 billion persons.

The viability of the secondary oocyte is believed to be about 1 day. The developing ovum is covered by a clear membrane called the zona pellucida and enveloped in a star-shaped cluster of cells collectively referred to as the corona radiata. The corona radiata is implicated in the nutrition of the developing ovum.

What happens next is not clear, but apparently most of the spermatozoa which make contact with the corona radiata clump together, or agglutinate, in a manner analogous to an antigen-antibody interaction. Clumping results from an interaction between a glycoprotein called fertilizin secreted by the developing ovum and another glycoprotein called antifertilizin present on the surface of the spermatozoa. Two alternative functions have been ascribed to the agglutination reaction: (1) clumping causes spermatozoa to stick to the surface of the corona radiata, thereby increasing the probability of fertilization; and (2) when fertilizin is used up upon the completion of the antigen-antibody interaction, a nearby but unneutralized spermatozoon can initiate the process of fertilization.

The process of fertilization is illustrated in Fig. 19-9. The following description is with reference to this figure.

The acrosome of a spermatozoon makes contact with the corona radiata. An enzyme called hyaluronidase is secreted from the acrosome and erodes a path through the corona radiata and zona pellucida. When the head of the spermatozoon reaches the outer boundary of the secondary oocyte, the entire spermatozoon is literally sucked into its cytoplasm. Fertilization has been accomplished.

Fertilization results in profound changes within the secondary oocyte. Its nucleus immediately enters into the second meiotic division, giving rise to a mature ovum and a second polar body. Both the first and second polar bodies appear as small bumps immediately outside the ovum, within the zona pellucida. The metabolic rate of the ovum increases, its cortex is "activated," and intracellular raw materials are mobilized in preparation for entry into mitotic divisions which will give rise to a new individual.

## EARLY EMBRYONIC DEVELOPMENT

The mature ovum with the sperm cell nucleus in its cytoplasm is called a zygote. Any supernumerary spermatozoa which may have entered ultimately degenerate. The sperm cell nucleus is now referred to as the male pronucleus, and the egg cell nucleus is called the female pronucleus. Since each nucleus contains 23 chromosomes, the species number of chromosomes (46) is reestablished in the zygote.

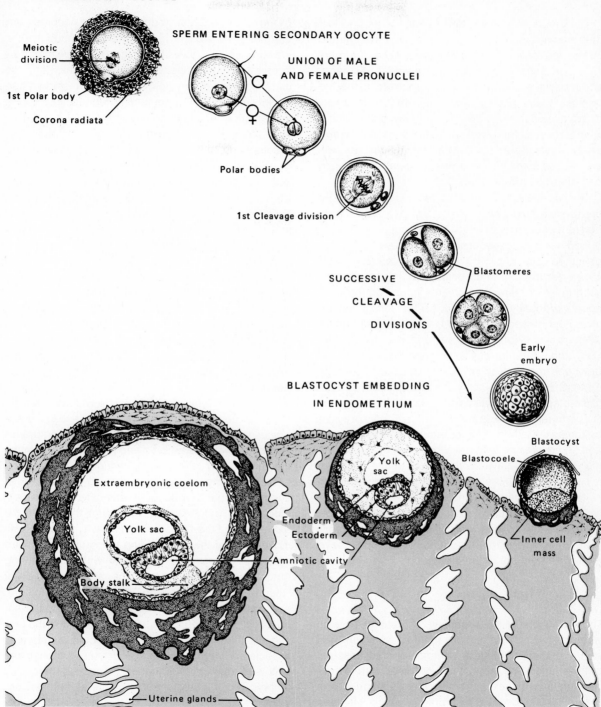

**SECONDARY OOCYTE**

Meiotic division

1st Polar body

Corona radiata

**SPERM ENTERING SECONDARY OOCYTE**

**UNION OF MALE AND FEMALE PRONUCLEI**

Polar bodies

1st Cleavage division

**SUCCESSIVE CLEAVAGE DIVISIONS**

Blastomeres

Early embryo

**BLASTOCYST EMBEDDING IN ENDOMETRIUM**

Blastocyst

Blastocoele

Inner cell mass

Yolk sac

Endoderm
Ectoderm

Extraembryonic coelom

Yolk sac

Body stalk

Amniotic cavity

Uterine glands

*Figure 19-9* *Embryogenesis*. (After Langley, Telford, and Christensen.)

The pronuclei migrate toward each other and lose their nuclear membranes. Concurrently, the chromosomes replicate. Thus, during metaphase of the first *cleavage division* of the new individual, 46 paired chromosomes line up in linear order along the metaphase plate. During anaphase, 46 chromosomes migrate to each pole, giving rise to two cells, each of which contains the same kinds and the same number of chromosomes as the zygote. Until mature sex cells are formed some 13 years after birth, all subsequent cell divisions are also mitotic.

The two-cell cleavage stage which results from the first cleavage division consists of two cells called *blastomeres* accompanied by the polar bodies acquired during oogenesis and surrounded by the zona pellucida (see Fig. 19-9, above). A series of rapid cleavage divisions then occurs, during which (1) a solid mass of many blastomeres forms; (2) the polar bodies gradually degenerate; (3) the zona pellucida ultimately disappears; and (4) the formative embryo remains approximately the same size as the zygote.

By about the fifth day after ovulation, a blastocyst develops from the solid mass of blastomeres (see Fig. 19-9). Growth in size is initiated with the appearance of the blastocyst. A blastocyst consists of two regions: (1) an outer cell mass, or *trophoblast,* part of which becomes the embryonic contribution to the placenta; and (2) an *inner cell mass* which becomes the body of the embryo. The cavity which forms in the blastocyst is called the blastocoele and will ultimately be obliterated, as we shall see, and replaced by a new cavity. The presence of a blastocoele pushes the inner cell mass toward one side of the embryo. Approximately 7 to 9 days following ovulation, the blastocyst is ready to implant in the uterine endometrium.

## IMPLANTATION

The trophoblast of the blastocyst differentiates into an outer *syntrophoblast* composed of a syncytium and an inner *cytotrophoblast* composed of a discrete cellular layer. The trophoblast together with an underlying layer of mesoderm becomes the *chorion,* one of four extraembryonic membranes which will form during embryonic development. An enzyme similar or identical to hyaluronidase and presumably secreted by the syntrophoblast erodes the uterine endometrium, permitting the blastocyst to implant or embed itself in the endometrial tissues (see Fig. 19-9). Raw materials from the eroded tissues and secretions from the uterine glands supply nourishment to the embryo during this period. The endometrium repairs itself during implantation, growing around the apical portion of, and completely covering, the blastocyst. Implantation is completed sometime between the second and third week after ovulation.

During implantation, the tissues of the embryo begin to differentiate through mass migrations referred to as *morphogenetic movements.* The inner cell mass splits into two layers, the *ectoderm* and *endoderm.* The ectodermal layer is closest to the basal portion of the blastocyst. Following their migration, each of these tissue layers gives rise to a vesicle, or sac. At the same time, cells destined to form *mesoderm* split off from the ectoderm of the cytotrophoblast and fill the former blastocoele. Cavities which form in this mesodermal tissue coalesce to give rise to one large cavity called the extraembryonic coelom, which will ultimately disappear.

The ectodermally derived vesicle with its outer lining of extraembryonic mesoderm gives rise to the *amnion,* second of the four extraembryonic membranes associated with the embryo and fetus. The fluid in the amniotic cavity is called *amniotic fluid.* This fluid bathes the embryo and fetus, serving as a hydraulic shock absorber to reduce, or attenuate, the effects of mechanical forces transmitted through the body of the mother. The fluid-filled amniotic cavity also permits movements of the fetus during the second and third trimesters of gestation. The amnion will rupture during labor, causing the loss of the amniotic fluid, or "uterine waters" (see "Birth," below).

The endodermally derived vesicle with its outer lining of extraembryonic mesoderm gives rise to the *yolk sac,* the third of the four extraembryonic membranes associated with the embryo and fetus. Although the yolk sac may participate in nourishing the early embryo, it will ultimately degenerate as prenatal development progresses and the placenta is established.

When an inner cell mass is present, the embryonic

body is disk-shaped and contains only two tissue layers—ectoderm and endoderm. A morphogenetic movement within the embryonic disk forms a middle layer of body mesoderm from a tissue layer which splits off the body ectoderm. The outer ectoderm, middle mesoderm, and inner endoderm are called *primary germ layers*. All the adult tissues, organs, and structures are derived from these three embryonic tissue layers.

At a later time, after the endodermally lined gut is formed, a protrusion will develop from the hind gut and project next to the body stalk. Formed of an inner layer of endoderm and an outer layer of mesoderm, the compound structure is called the *allantois*. The allantois is the fourth and last of the extraembryonic membranes. Although concerned with urine collection and storage in animals which develop without a placenta (such as birds), the human allantois is essentially nonfunctional. The urachus, a ligament on the apex of the adult urinary bladder (see Fig. 19-12), marks the former position of the allantois.

The amnion of the embryo is attached to the basal portion of the chorion of the blastocyst. This region of mesodermal attachment is called the *body stalk*. Blood vessels will differentiate in its mesoderm. With further growth and differentiation, the body stalk develops into the *umbilical cord,* a compound structure which will contain two umbilical arteries and one umbilical vein. The umbilical cord will link the developing body with the placenta.

When fully implanted, the chorion of the blastocyst contains an inner lining of extraembryonic mesoderm and an outer layer of extraembryonic ectoderm. The basal portion of the chorion thickens, forming projections called *villi* which will participate in placentation.

## PLACENTATION

When the blastocyst is fully implanted (1) growth in its size is rapid, forcing the endometrium around the site of implantation to project more and more into the uterine cavity; (2) the amniotic sac of the embryo grows rapidly, almost obliterating the extraembryonic coelom; (3) the chorion of the blastocyst thickens greatly, forming villi which grow more rapidly on the basal portion than elsewhere on the chorion; and (4)

the endometrium is referred to as the *decidua*. Some of these relationships are shown in Fig. 19-9, above.

The differential growth of the villi gives rise to two distinct chorionic regions: (1) the *chorion frondosum* bearing a disk of highly-branched villi on the basal portion of the blastocyst, and (2) the *chorion laeve*, or "naked" chorion, with short villi which encircle the rest of the blastocyst.

The chorion frondosum is the embryonic contribution to the placenta. The basal portion of the endometrium next to the chorion frondosum is called the *decidua basalis*. The decidua basalis is the maternal contribution to the placenta. Continuous with the decidua basalis, the rest of the endometrium which surrounds the embryo is referred to as the *decidua capsularis*. The remainder of the endometrium surrounding the uterine cavity is called the *decidua parietalis*.

Thus, the placenta is a compound structure, consisting of the chorion frondosum of the embryo and the decidua basalis of the mother. An enlargement of a portion of the placenta is shown in Fig. 19-10.

The villi of the chorion frondosum grow extensively and project to the compact layer of the decidua basalis. The tissue layers which compose the villi in the chorion frondosum, from outside inward, are the syntrophoblast, the cytotrophoblast, and the mesodermal walls of the umbilical blood vessels. Materials which diffuse between the bloodstreams of the embryo or fetus and the mother must pass through these three tissue layers, which are therefore referred to as the *placental barrier*.

The endometrial tissues in the decidua basalis form blood pools in pockets which develop from eroded tissues below the chorion frondosum (subchorial spaces) and between the villi (intervillous spaces). The *spiral arterioles* supply maternal blood to the subchorial and intervillous spaces; the blood pools are drained by *maternal venules*.

Notice that there is no direct vascular connection between the mother and the embryo or fetus. Materials which diffuse from the pools of blood in the decidua basalis through the placental barrier into the embryonic or fetal bloodstream include nutrients, $O_2$, antibodies, hormones, and viruses. Materials which diffuse from the embryonic or fetal bloodstream in the

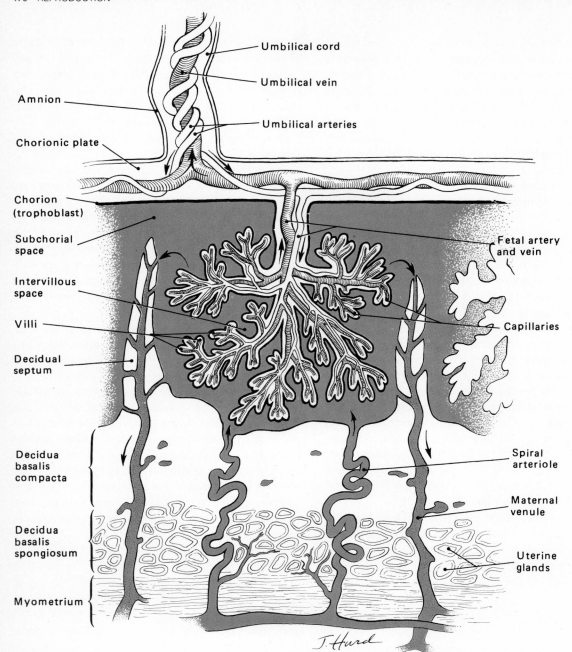

**Figure 19-10** *The placenta. This compound structure consists of the chorion frondosum of the fetus and the decidua basalis of the mother.* (From Langley, Telford, and Christensen.)

opposite direction include waste products of metabolism such as $CO_2$ and urine.

As the embryo enlarges within the uterine endometrium, the decidua capsularis approaches the decidua parietalis, gradually obliterating the uterine cavity. A mucous plug blocks the cervical canal by the eighth week of development. The decidua capsularis fuses with the decidua parietalis sometime between the tenth and seventeenth week of development. The fused deciduae are collectively referred to as the *decidua vera*.

Until the end of the second month of development, the prenatal individual is called an *embryo*. During this time, due to morphogenetic movements of various types, the major adult organs are formed and the embryo begins to resemble a human being. Thus, from the beginning of the third month until delivery, the developing individual is called a *fetus*. The development of the organ systems, or organogenesis, is accomplished during the fetal stage.

## TWINNING

Basically, there are two types of twins—*fraternal* and *identical*. Identical twins are further subdivided into normal and Siamese.

Fraternal twins develop simultaneously in utero, but they do not look exactly alike, although they may resemble each other familially. Fraternal twins result from multiple ovulation and multiple fertilization. Therefore the hereditary potentials of the zygotes are different. Each of the blastocysts which develops from cleavage divisions of the zygotes implants separately in the uterine endometrium. A separate placenta is es-

tablished between each chorion frondosum and decidua basalis. The arrangement of the placentas of fraternal twins is diagramed in Fig. 19-11A.

It is possible for each fraternal twin to be delivered on a different day. To account for this fact, one theory suggests that (1) placental progesterone diffuses into the myometrium, rather than circulates through the bloodstream, and (2) the precipitous decrease in progesterone from each placenta may occur at different times due to differences in the times of implantation.

Identical twins develop simultaneously in utero and look identical to each other because they have the same hereditary potential. Identical twins most frequently develop when blastomeres separate from each other, usually during the two- or four-cell cleavage stage. Although each of the separated blastomeres develops into a separate embryo, the embryos are enveloped in one chorion and will share a single placenta. Occasionally, identical twins develop when the inner cell mass separates along its length during the formation of the blastocyst. The arrangement of identical twins during intrauterine development is shown in Fig. 19-11B.

Siamese twins develop when the inner cell mass separates incompletely during blastocyst formation.

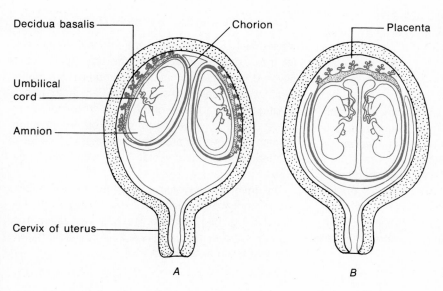

Decidua basalis

Chorion

Placenta

Umbilical cord

Amnion

Cervix of uterus

A

B

**Figure 19-11** *A comparison of the placental arrangement of fraternal and identical twins. A. Fraternal twinning. B. Identical twinning.* (From DeCoursey.)

## BIRTH

The process or birth, or *parturition,* terminates gestation. The onset of labor and subsequent termination of pregnancy are the result of the following factors:

1 The degeneration of the corpus luteum of pregnancy and precipitous decrease in the secretion of placental progesterone remove the inhibition of the myometrium to contractions.

2 Oxytocin, produced in the hypothalamus and secreted from the pars nervosa of the hypophysis in increasing concentrations near the end of gestation, promotes uterine contractility.

3 The increase in size of the fetus stretches the uterus to its limit, provoking stronger and stronger contractions as uterine irritability increases.

The interaction of these factors promotes rhythmic contractions of the uterus, or womb. Initially weak in strength, lasting about 1/2 to 1 min and separated from each other by about 10 or more min, the contractions become more forceful with time. Finally, strong contractions lasting about 1/2 to 1 min follow each other at 2-min or shorter intervals. The mucous plug which blocks the cervical canal, along with a small amount of blood, is usually voided at this time. The onset of labor has begun.

Labor is divided into three stages—dilatation, expulsion, and placental.

*Dilatation* is usually the longest of the three stages. This stage of labor begins when the cervix of the uterus begins to dilate and ends when the cervix is fully dilated. The average dilatation at the completion of this stage is 10 cm.

Following cervical dilatation, the head of the fetus is capable of passing through the uterine cervix into the vagina, or birth canal, and from there to the outside of the body. This stage of labor is called *expulsion.* Expulsion is initiated by strong waves of peristaltic contraction which pass from the fundic to the cervical portion of the uterus. These powerful contractions usually rupture the amniotic sac which surrounds the fetus, causing the loss of amniotic fluid. However, rupture of the amnion frequently precedes this stage.

Expulsion of the fetus is assisted by the mother's pushing with her thoracic and abdominal muscles and by obstetrical manipulations of the fetal head. Sometimes an incision, called an *episiotomy*, is made in the perineal region to enlarge the opening during delivery and to prevent the indiscriminate tearing of perineal tissues. After birth, the incision is stitched. The pliability of the pubic symphysis also aids the process of expulsion.

The third, or *placental*, stage of labor occurs after delivery. The umbilical cord between the newborn baby and placenta is clamped off and severed. The baby's airways are drained of mucus, and the infant is stimulated to take its first breath. The transition from prenatal to postnatal development has been accomplished.

As necessary neonatal care is being administered to the newborn baby, some 10 min after expulsion, the spiral arterioles of the mother constrict, cutting off the blood supply in the region of the placenta. The myometrium again undergoes peristaltic contractions, causing expulsion of the placenta in the same manner as expulsion of the fetus. Hemorrhaging is minimal because the uterine blood vessels are already constricted. Expulsion of the *afterbirth*, as the placenta is commonly called, terminates the third and last stage of labor.

## CHANGES IN CIRCULATORY PATTERNS

Immediately following birth, the neonate breathes air for the first time. In addition, the umbilical cord is severed, thereby eliminating the umbilical circulation which was formerly the most significant factor in sustaining the metabolism of the fetus. These changes, while dramatic, have been prepared for developmentally so that no profound disturbances occur to the heart and vascular system during the transformation.

The immediate systemic changes following birth are primarily cardiovascular. These changes are illustrated in Fig. 19-12*A* and *B*. The fetal circulation prior to birth is shown in Fig. 19-12*A*; the circulatory pattern of the neonate is depicted in *B*.

Structures in and around the fetal heart which disappear after birth are as follows:

1  The *foramen ovale.*  This opening is found in the septum between right and left atria. Thus, blood entering the right atrium from the superior and inferior venae cavae may directly enter both the right ventricle and the left atrium. As the blood supply of the fetus continues to develop, the foramen ovale ensures that the right intraatrial pressure does not increase excessively, especially since little blood is being returned to the left atrium through the pulmonary veins. A flap called the *valvula foraminis ovalis* closes when the fetal atria go into systole, thereby ensuring that the blood within each atrium moves into its respective ventricle. Once the pulmonary circulation is established after birth, the pressures in both atria equalize, resulting in the functional abandonment of the foramen ovale. Functional disuse of the foramen ovale is followed by a complete fusion of the valvula to the interatrial septum sometime during the first year after birth. The remnant left by the foramen after its closure is called the *fossa ovalis.* Sometimes on autopsy, adults who have had no prior history of heart disease are found to have an incompletely fused fossa ovalis and, therefore, an actual communication between right and left atria. Because of the pressure balance between the two atria, however, the opening never functioned postnatally in transporting blood directly from the right to the left atrium.

2  The *ductus arteriosus.* This prenatal vessel links the pulmonary artery with the arch of the aorta. Thus most of the blood entering the pulmonary artery from the right ventricle bypasses the lungs to enter the main distributing artery of the body. This shunt is necessary during prenatal development for several overlapping reasons: (*a*) the lungs, although nonfunctional, are growing and differentiating during organogenesis so that their blood supply must be tailored to their metabolic needs and physiologic capacities; (*b*) because the beat of the heart increases in strength and its stroke volume increases as hemopoiesis adds to the quantity of blood in the cardiovascular system, the ventricular load must be continually readjusted as development proceeds; and (*c*) to ensure normal physical maturation of the myocardium, a process which requires continual prenatal cardiac exercise of increasing vigor, the lungs must be protected against potential damage due to increased vascular pressures. Following the first breath after birth, the ductus arteriosus constricts, causing its functional abandonment. Presumably constriction is due to changes in hemodynamics and a high concentration of $CO_2$ accumulated metabolically before the onset of external respiration. Functional disuse of the ductus arteriosus is followed by the gradual occlusion of its lumen through an overgrowth of the intima. Ordinarily, structural obliteration is completed by the second month after birth. The remnant of the ductus arteriosus is called the *ligamentum arteriosus.*

Structures associated with the umbilical circulation which disappear after birth are as follows:

1  The *ductus venosus.*  This main channel through the liver causes most of the blood in the umbilical vein to effectively bypass the formative blood sinuses of the liver, joining instead with blood in the inferior vena cava at or near the entrance of the hepatic vein. This shunt for placental blood returning to the heart is necessary because of the physical and functional immaturity of the liver. Interruption of the umbilical circulation following delivery causes the degeneration of the ductus venosus and establishes the hepatic portal vein as the main venous pathway of blood to the sinusoids of the liver. Following its degeneration, the ductus venosus is called the *ligamentum venosum.*

2  The *umbilical vein* and *arteries.*  Severance of the umbilical cord also causes degenerative changes in these vessels which formerly led to and from the placenta. Structural obliteration of their lumina usually occurs between 3 and 5 weeks postpartum. The subsequent formation of fibrous cords from these obliterated vessels takes much longer. The umbilical vein becomes the *ligamentum teres,* which courses between the falciform ligament of

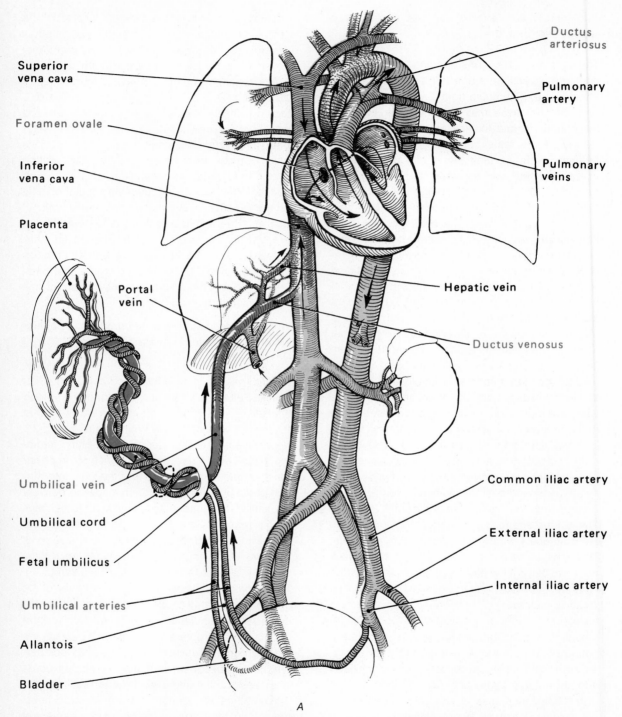

**Figure 19-12** *A comparison of human fetal and adult circulation patterns. A. Fetal circulation. B. Adult circulation.* (From Langley, Telford, and Christensen.)

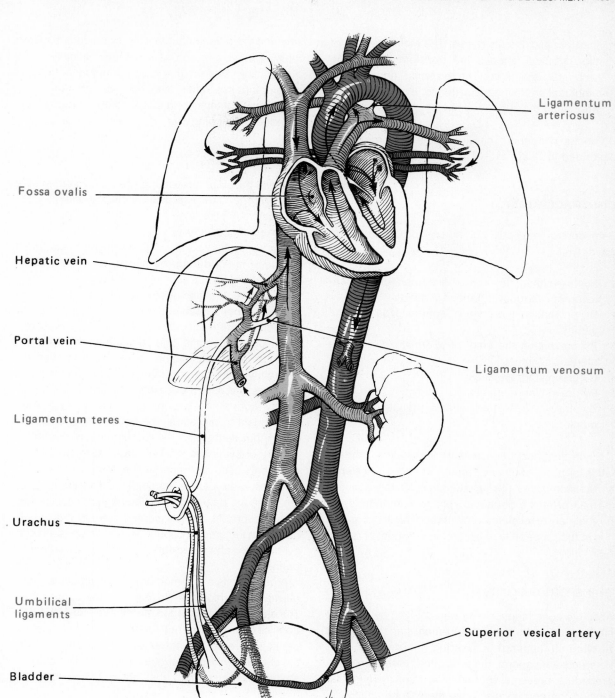

Ligamentum arteriosus

Fossa ovalis

Hepatic vein

Portal vein

Ligamentum teres

Urachus

Umbilical ligaments

Bladder

Ligamentum venosum

Superior vesical artery

*B*

the liver and the umbilicus. The main portions of the umbilical arteries are transformed into the *umbilical ligaments*. The proximal ends of the umbilical arteries remain functional after birth and are called the superior vesical arteries.

Various congenital cardiovascular disorders are described in Chap. 21.

## CONTRACEPTION

Birth control methods act in one of the following ways:

1 By preventing the normal transport of spermatozoa or ova following their formation
2 By preventing the entry of spermatozoa into the vagina
3 By preventing the entry of spermatozoa into the uterus
4 By preventing ovulation
5 By preventing implantation
6 By preventing the completion of prenatal development

Since contraception prevents fertilization, implantation, or birth, its practice results in temporary sterility. The sterility may be functional or organic, occur in the male or female, be naturally or artificially induced, or be voluntary or involuntary, but its end result is a reduction in the probability of becoming pregnant or giving birth.

### TRANSPORT BLOCKS

Absolute contraception may be achieved by surgical sterilization. In the male, sterilization is accomplished by tying off (ligating) or severing (incising) the vasa deferentia, or sperm ducts. Called *vasectomy,* the procedure prevents spermatozoa from traveling far beyond the epididymis. Thus, there is no chance of spermatozoa ever reaching the urethra. However, penile erection and the ejaculation of semen, albeit minus spermatozoa, are not affected.

Sterilization of females is achieved by tying off or severing the fallopian tubes. The procedure is called *tubal ligation*. Incision of the tubes is called *salpingectomy*. As a result of salpingectomy or ligation, developing ova after ovulation cannot reach the uterus for implantation, and spermatozoa cannot fertilize the developing ova.

### VAGINAL BLOCKS

Methods based on preventing spermatozoa from ever reaching the vagina or ovum include complete sexual abstention, the rhythm method, coitus interruptus, and the condom.

The *rhythm method* is based on having sexual intercourse at times during the menstrual cycle when a viable ovum is not and will not be present in the reproductive tract and, therefore, cannot be fertilized. Assuming that a developing ovum is released at the midpoint of the menstrual cycle, that the viability of a postovulatory ovum is 24 hr, that the viability of spermatozoa after ejaculation is 48 hr, and that the variability in the time of ovulation from cycle to cycle is 2 days, the period of sexual abstention is approximately from 3 days prior to ovulation to 2 days subsequent to ovulation. The assumptions on which the rhythm method are based, especially predicting the day of ovulation, make it the least-reliable of the commonly used contraceptive methods.

*Coitus interruptus* is accomplished by the male by removing the penis from the vagina just before ejaculation occurs. As a contraceptive method this is fallible, since the precise control of ejaculation is difficult, even when attempted by a sexually experienced partner.

A *condom* is a thin membrane of synthetic or natural material which fits closely over the entire penis. The ejaculate is trapped in the dilated end of the condom next to the external urethral orifice. Since the penis is covered, it is not as sensitive as usual to tactile stimulation. However, the method itself is efficient so long as the condom does not tear.

### UTERINE BLOCKS

Methods based on preventing spermatozoa already in the vagina from reaching the uterus include the vaginal douche, spermicidal vaginal preparations, and the diaphragm.

A *vaginal douche* consists of rinsing the vagina with a large volume of tap water to flush out the spermatozoa following sexual intercourse. However, since spermatozoa enter the uterine cervix rapidly after ejaculation, douches are relatively ineffective and often fail.

*Spermicidal vaginal preparations* are substances placed in position around the upper part of the vagina, next to the uterine cervix. Their presence forms a physical and chemical barrier, preventing spermatozoa in the vagina from entering the uterus. Frequently, the use of a spermicidal vaginal preparation is coupled with another technique such as the diaphragm.

The *diaphragm* is a rubber membrane with a stiff outer border which is used to cap the uterine cervix, thereby physically barring the entry into the uterus of spermatozoa. Ordinarily, a spermicidal vaginal preparation is spread around the rim of the diaphragm to increase its effectiveness as a prophylactic.

## OVULATION BLOCKS

Used by at least 20 million females around the world, most birth control pills function by preventing ovulation. Oral contraceptives contain synthetic derivatives of both ovarian hormones, i.e., estrogen and progesterone, in varying proportions. Their introduction into the bloodstream via the digestive system inhibits the adenohypophysis from secreting the critical concentrations of LH and FSH necessary for ovulation. Thus, menstrual cycles are anovulatory while the "pill" is being used. At the present time, oral contraceptives offer the lowest risk of unwanted pregnancy, but their side effects in some cases may militate against their use.

## IMPLANTATION BLOCKS

Called *"the morning-after"* pill, one type of oral contraceptive consists of a large concentration of a synthetic derivative of estrogen. A large concentration of estrogen prevents the normal secretion of progesterone, so that the uterus is not prepared to receive an implanting blastocyst. The method is effective, provided the pill is taken within 4 days (preferably 2 days) after sexual intercourse. However, the pill may be potentially carcinogenic, or cancer-causing.

*Intrauterine devices*, or IUDs, are believed to prevent fertilization or implantation, although their mechanism of action is still poorly understood. Alternative explanations of how IUDs work include preventing spermatozoa from passing through the uterus or, in the case of IUDs made of specific alloys, promoting the destruction of spermatozoa entering the uterus through a combination of antigen-antibody interaction and phagocytosis.

Whatever the correct explanation, the use of IUDs is rapidly gaining in worldwide popularity. IUDs are inexpensive and almost as efficient as oral contraceptives which prevent ovulation. However, some potential side effects prevent IUDs from being used by all women who wish to practice birth control by artificial means.

## GESTATION BLOCKS

The widespread practice of prophylactic abortion makes it an important method of birth control, even though conception has already occurred. In the United States, attempts to liberalize or eliminate various state abortion laws have received support from the U.S. Supreme Court and resistance from numerous state governments. The present controversy in the United States is not an issue in some other parts of the world where prophylactic abortion is a legal method of birth control.

**COLLATERAL READING**

Csapo, Arpad, "Progesterone," *Scientific American*, April 1958. Reviews the historical development of our knowledge of progesterone.

Demarest, Robert J., and John J. Sciarra, *Conception, Birth, and Contraception: A Visual Presentation.* New York: McGraw-Hill Book Company, 1969. With the extensive use of large and well-executed drawings, the authors tastefully illuminate what happens during reproduction, development, birth, and in contraception.

Djerassi, Carl, "Birth Control after 1984," *Science,* 169:941–951, 1970. Reviews the problems to be surmounted in developing and implementing new birth control agents.

Edwards, R. C., and Ruth E. Fowler, "Human Embryos in the Laboratory," *Scientific American,* December 1970. The authors removed oocytes from human ovaries, developed them to maturity in special culture media, fertilized them in vitro with human spermatozoa, and followed their development to about the 64-cell stage.

Fischberg, Michael, and Antonie W. Blackler, "How Cells Specialize," *Scientific American,* September 1961. Based on experiments with frog eggs, the authors conclude that the future embryo is fully predetermined in the unfertilized ovum because of regional differences in cytoplasmic content.

Gray, George W., "The Organizer," *Scientific American,* November 1957. The concept of an organizer which determines what organs are to be formed and when formation will occur is traced in animal experiments. The author concludes that the pattern of development is a consequence of reciprocal interactions between organizers and genes.

Harrison, Richard J., and William Montagna, *Man.* New York: Appleton-Century-Crofts, Inc., 1969. Two chapters in this paperback are excellent supplements to the material discussed in this chapter: Chap. 10, His Sexual Behavior, and Chap. 11, Man's Reproductive Patterns.

Mittwoch, Ursula, "Sex Differences in Cells," *Scientific American,* July 1963. A review of investigations which contributed to our current concepts regarding genetic and hormonal influences on sex determination during prenatal development.

Patton, Stuart, "Milk," *Scientific American,* July 1969. A valuable summary of how the mammary gland cell synthesizes, packages, and secretes its finished product.

Tyler, Albert, "Fertilization and Antibodies," *Scientific American,* June 1954. Tyler discusses "antigens" called "antifertilizin" produced by spermatozoa and "antibodies" called "fertilizin" produced by the developing ovum.

Westoff, Charles F., and Larry Bumpass, "The Revolution in Birth Control Practices of U.S. Roman Catholics," *Science,* 179:41–44, 1973. Based on data taken during the 1970 National Fertility Study, the authors conclude that two-thirds of all U.S. Catholic women in 1970 were using birth control methods disapproved by their Church in 1967.

Westoff, Leslie Aldridge, and Charles F. Westoff, *From Now to Zero: Fertility, Contraception and Abortion in America.* Boston: Little, Brown and Company, 1968. Summarizes information gathered during the 1965 National Fertility Study.

# CHAPTER 20
# Genitourinary Diseases

Genitourinary diseases upset the homeostatic balance of the reproductive and excretory systems. In the human male, portions of these two organ systems overlap both anatomically and physiologically. Therefore, an abnormality in one system may be reflected in a malfunctioning of the other system.

Disorders of the female reproductive tract affect the following functional processes: (1) the menstrual cycle, by causing an absence or prolongation of ovulation or menstruation as well as an increase in the amount of blood in, and the volume of, the menses; (2) prenatal development, by preventing the normal completion of either oogenesis, fertilization, implantation, or organogenesis and growth; and (3) lactation, by interfering with the volume or quality of the milk produced.

Abnormalities of the male reproductive tract adversely influence the following functions: (1) prenatal development, by preventing the normal completion of

spermatogenesis or fertilization; or (2) urination and accessory reproductive glandular action, through hypertrophy of the prostate gland.

Excretory disorders disrupt the following mechanisms: (1) glomerular filtration, tubular reabsorption, and secretion because of irritations, inflammations, obstructions, membrane permeability changes, or hydrostatic and osmotic pressure fluctuations which significantly increase or decrease the volume and specific content of the formative urine; and (2) urination by significantly increasing, depressing, or completely preventing its occurrence or through its involuntary or abnormal appearance. The ability to urinate can be affected without influencing kidney function, although kidney problems usually affect both processes.

This chapter explores the causes and symptoms of the better-known reproductive and excretory disorders. Refer to Chap. 21 for a detailed analysis of congenital defects and to Chap. 18 for a review of hormonal imbalances which result in abnormalities of reproduction and excretion.

## REPRODUCTIVE SYSTEM

### DISTURBANCES OF MENSTRUATION

These disorders include dysmenorrhea, menorrhagia, oligomenorrhea, and amenorrhea. These are common disturbances which occur frequently for numerous reasons. Sometimes they are indicative of more serious disease states.

**Dysmenorrhea** Difficult or painful menstruation is referred to as *dysmenorrhea*. There are two forms of the disease distinguished by when the symptoms first appear: (1) *primary,* in which the symptoms are exhibited from the onset of the first period following puberty; and (2) *secondary,* in which the symptoms show up some time after normal menstrual periods have become well established.

Many pathologic conditions which involve the pelvic portion of the abdomen can cause dysmenorrhea. Some of the specific causative agents of the disorder are ovarian cysts, tumors and infections, inflammation of the mucosae within the fallopian tubes and uterus, inflammation of the ovarian and uterine ligaments, abnormal positioning of the uterus, severe spasms of the uterine myometrium, and psychogenic factors. Pelvic inflammation, especially of the uterus, is the predominant sign associated with most of these causal factors.

**Menorrhagia** Excessive bleeding during menstruation is termed *menorrhagia*. The condition may be reflected in an abnormally large volume of menstrual blood, a prolongation in the number of days during which menstrual flow occurs, or in both factors. Menorrhagia can lead to hemorrhagic anemia.

The causes of menorrhagia are diverse. Among the more common causes of the disorder are imbalances in pituitary and ovarian hormones, inflammation of the uterus, abnormal uterine positions such as retroflex-

ion and retroversion, and certain vascular and blood abnormalities. The symptoms may also be provoked by several diseases which have widespread effects on the body or by abnormal uterine growths or tumors which cause structural changes within the endometrium.

**Oligomenorrhea** When the menses are abnormally infrequent or scanty, the condition is called *oligomenorrhea*. If one or more menstrual periods are skipped entirely, then the disorder is called *amenorrhea.*

The most common causes of amenorrhea are pregnancy and menopause, both of which terminate the menstrual cycle. Both oligomenorrhea and amenorrhea are frequently induced by psychogenic factors or a lowered resistance. Therefore, the response to psychologic and physiologic stress may be in the form of a temporary depression in, or absence of, normal menstrual flow.

### DISORDERS OF THE FEMALE REPRODUCTIVE TRACT

These diseases may involve the ovaries, fallopian tubes, uterus, and vagina, or some combination of these organs. Two of the most frequent disorders of the female reproductive tract are salpingitis and leukorrhea.

**Salpingitis** Inflammation of the fallopian tubes is referred to as *salpingitis*. The inflammation is usually caused through infection by bacteria such as gonococci, streptococci, staphylococci, tubercle bacilli, and colon bacilli. Salpingitis is not infrequently an indirect or delayed symptom of either gonorrhea, tuberculosis, or other infectious bacterial diseases.

Infections within the fallopian tubes are sometimes difficult to cure and become chronic. The formation of scar tissue during recovery from acute salpingitis may partially or completely occlude the lumina of the fallopian tubes. This condition can lead to sterility by preventing ova from reaching the uterus or spermatozoa from reaching the developing ovum. Sterility is discussed later in this chapter.

Leukorrhea  A discharge of whitish or yellowish mucus from glands of the uterine cervix or vagina is called *leukorrhea*. In moderation and in the absence of unusual signs, such discharges are completely normal when they occur just before or after menstruation. These secretions become symptomatic of disease when voluminous, greenish, flecked with blood, odorous, excessively viscous or watery, or highly acidic or alkaline, compared to the physical characteristics of a normal discharge. In addition, the discharge may contain an unusually large or different flora, compared to that ordinarily present. In other words, abnormal discharges are quantitatively different from normal secretions.

Leukorrhea usually is indicative of a pathologic condition or infection of the uterine cervix or vagina. One of the common microbes responsible for this disorder is the protozoan *Trichomonas vaginalis* (see Fig. 20-1). Leukorrhea is ordinarily accompanied by an inflammation of the reproductive tract and pelvic portion of the abdomen. The infection sometimes spreads from the vagina to the urethra, causing painful urination.

## DISORDERS OF PREGNANCY

These problems are usually caused by implantation of the blastocyst in abnormal intrauterine or extra-uterine locations, such as the lower segment of the uterus, the fallopian tubes, the abdominal cavity, or some combination of these loci. Ectopic pregnancy and placenta previa are two of the more common disorders caused by the initiation of embryonic development in peculiar locations within the mother.

Ectopic pregnancy  When the blastocyst implants or continues its development outside the uterine cavity, the pregnancy is termed *ectopic*. The most common location for implantation under these circumstances is the mucosa of a fallopian tube, although ectopic pregnancies have also been recorded in the abdominal cavity, partially in the abdominal cavity and fallopian tube, or partially within the fallopian tube and uterus.

If the pregnancy is tubal and development pro-

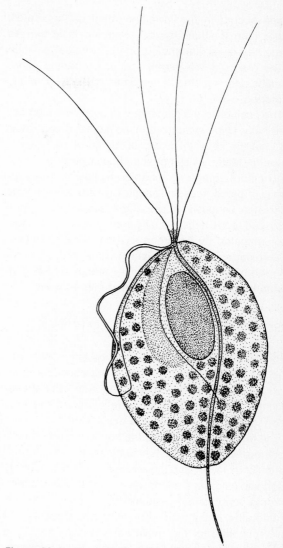

**Figure 20-1**  *The protozoan flagellate* Trichomonas vaginalis, *a causative agent of leukorrhea.*

ceeds for a sufficient period of time, the wall of the fallopian tube may rupture and cause internal hemorrhaging and severe pelvic pains. Prompt surgical intervention is usually required to arrest hemorrhaging from ruptured blood vessels inside the body.

Placenta previa  When implantation of the blastocyst occurs in the lower segment of the uterus, rather than the fundus, the condition is known as *placenta previa*. Depending upon the actual position of the placenta, the internal cervical opening either remains unobstructed or becomes partially or totally occluded.

The position of the placenta can be determined obstetrically. Overt symptoms in the mother do not show up until the seventh or eighth month of pregnancy, however. Maternal symptoms consist of mild, recurrent hemorrhaging accompanied by anemia. The rupturing of blood vessels associated with the placenta may also seriously depress the supply of oxygen provided to the fetus. Usually such pregnancies must be terminated by cesarean section because of premature separation of the placenta. In a *cesarean section*, the baby is removed from the uterus following incision through the abdominal and uterine walls.

### DISORDERS OF THE MAMMARY GLANDS

Lactation    Lactational disturbances affect milk production and secretion. Hormonal imbalances and infections are primarily responsible for lactational dysfunctions, although nutritional deficiencies and psychogenic factors may also be implicated.

The following are among the hormones which control the development of the mammary glands, the initiation of lactation, and the actual secretion of milk: (1) chorionic estrogens acting synergistically with somatotropic hormone are responsible for the growth and proliferation of the mammary ducts; (2) progesterone is responsible for the growth and maturation of the mammary lobules and secretory alveoli; (3) hypophyseal and chorionic LTH acting synergistically with estrogens and progesterone are responsible for the production by the alveoli of colostrum before birth and milk after birth; and (4) acting synergistically with antidiuretic hormone, oxytocin is responsible for constriction of the smooth musculature of the alveoli, causing milk to be expressed from the alveoli into the mammary ducts. Theoretically, at least, serious imbalances of any of these hormones at specific times relative to gestation may depress or completely prevent lactation. However, most disorders of lactation are

caused by microbial infection rather than hormonal imbalances.

Mastitis    An inflammation of the mammary glands is termed *mastitis*. Usually, the infection is contracted during lactation through bacterial contamination of various superficial lesions on the nipples and areolae formed while breast-feeding the infant.

The effects of mastitis on the mammary gland are shown in Fig. 20-2. The superficial layers of a portion of the breast have been removed to show the alveoli and ducts. Notice that the alveoli are extremely swollen, reflecting the inflammatory condition caused by microbial infection.

### DISORDERS OF THE MALE REPRODUCTIVE TRACT

Aside from sterility and certain excretory disorders which involve a portion of the reproductive tract and which will be described later in this chapter, one of the more common male disorders affects the prostate gland and the prostatic urethra.

**Figure 20-2**    *The effects of mastitis on the mammary gland.*

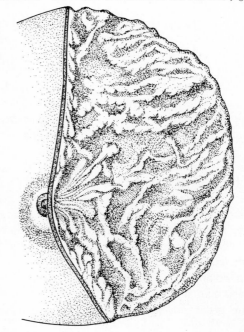

**Prostatism**  An obstruction of the prostatic urethra by any condition which results in an enlargement of the prostate gland is called *prostatism*. If the condition is inflammatory, then the disorder is more specifically referred to as *prostatitis*. Frequently, but not always, prostatitis is a complication of gonorrhea. In addition to microbial infection, prostatism can be caused by noninflammatory conditions which result in prostatic enlargement such as a carcinoma or by urinary calculi (stones) deposited within the prostate gland.

The symptoms of prostatism are determined in part by the causative agent of the disease. Glandular enlargement constricts the urethra, making urination difficult, painful, and unpredictable. Pain is also experienced in the perineal region. When bacteria are responsible for an inflammation of the prostate gland, additional symptoms may include chills and fever, nausea and vomiting, constipation, and the appearance of involuntary glandular exudations at the urethral orifice. Prostatitis may secondarily involve the urinary bladder.

## STERILITY

Any condition which prevents offspring by blocking the successful completion of gametogenesis, fertilization or conception, implantation, or prenatal development causes *sterility*, or *infertility*. The condition may be temporary or permanent, functional or structural, voluntarily or involuntarily induced, and can result from various abnormalities which affect reproductive abilities in either sex. The following are among the reasons responsible for sterility:

1  Congenital defects which result in either prenatal mortality (spontaneous abortion or stillbirth) or postnatal male sterility (cryptorchism)
2  Surgical procedures such as castration (removal of the ovaries or testes), vasectomy (excision of a portion of the vasa deferentia), salpingectomy (excision of a portion of the fallopian tubes), and hysterectomy (removal of the uterus)
3  Normal blood concentrations of the primary reproductive hormones in both sexes prior to puberty,

which prevent the maturation of the gametes, and the depletion of ova within the ovaries following menopause, which causes sterility after this time
4  Pathologic imbalances of specific reproductive hormones during reproductive life, usually by preventing the testes or ovaries from producing functional sex cells
5  The use of contraceptive devices and birth control methods to prevent ovulation, fertilization, implantation, or prenatal development
6  Various reproductive, systemic, and neurologic disorders, or drugs, radiations, and other exogenous influences which affect reproduction adversely
7  Psychogenic factors which result in either functional impotence in the male, usually by preventing penile erection or consummation of the sexual act, or frigidity in the female, usually by reducing sexual desire and impairing the physical relationship between sexual partners during coitus

Various forms of organic damage which result in sterility are illustrated in Fig. 20-3. Ovarian cysts and adhesions (Fig. 20-3A) prevent normal oogenesis and ovulation. Constriction of the lumina of the fallopian tubes, an aftermath of salpingitis caused by bacteria, interferes with the transport of ova following ovulation, thereby preventing both fertilization and implantation. Destruction of spermatozoa within a seminiferous tubule of the testis and damage to the stratum granulosum within a graafian follicle of the ovary are shown in Fig. 20-3B and C following moderate whole-body irradiation with gamma rays.

## EXCRETORY SYSTEM

Excretory, or urinary, disorders affect the formation of urine, the urination reflex, or both processes. Because the kidneys are conservatory as well as excretory, their impairment contributes to imbalances in the concentrations of electrolytes and fluids within the body. Such imbalances may be reflected in one or more of the following ways: (1) by *uremia*, an increase

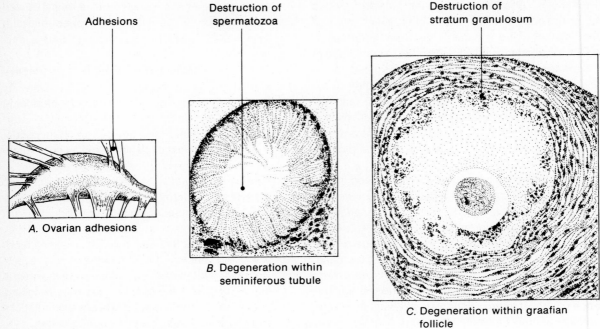

Adhesions

Destruction of spermatozoa

Destruction of stratum granulosum

A. Ovarian adhesions

B. Degeneration within seminiferous tubule

C. Degeneration within graafian follicle

**Figure 20-3** *Various forms of organic damage which result in sterility.*

within the blood of nitrogenous constituents ordinarily found in the urine; (2) by *renal (metabolic) acidosis* due to an excessive secretion of base or renal retention of phosphoric and sulfuric acids; (3) by *metabolic alkalosis* due to an excessive renal secretion of potassium or hydrogen ions; and (4) through *polyuria* and *oliguria*, leading to dehydration and edema, respectively.

Excluding urinary conditions provoked by other diseases, the origin of excretory disorders can usually be traced to one of the following factors: (1) inflammation of some portion of the urinary tract caused by microbial infection; (2) noninflammatory degeneration of kidney tissues; and (3) obstruction of some portion of the urinary tract.

## COMMON URINARY DISTURBANCES

Certain disorders affect the ease of urination or its voluntary occurrence. Among these conditions are enuresis and dysuria.

**Enuresis** Urination at inopportune times and in inappropriate ways in childhood after the third year is referred to as *enuresis*. The same condition in adult life is called *incontinence,* although sometimes both terms are used interchangeably. Enuresis may occur during the day (diurnal) or night (nocturnal), but is usually associated with bed-wetting at night. Enuresis may be induced voluntarily as an attention-getting device or may occur involuntarily usually due to subconscious desire, tactile stimuli, or pathologic conditions. Specific factors which contribute to enuresis include (1) physiologic abnormalities such as urinary infections, tiredness and lowered resistance, or illness; (2) psychogenic factors such as insecurity, fear, frustration, and other unsettling emotional states; and (3) the physical promotion of the urination reflex through episodes of laughing, crying, coughing, genital stimulation, or drinking excessive quantities of liquid.

**Dysuria** Urination which is difficult, painful, or

sometimes burning is called *dysuria*. Under these circumstances, urination is usually also frequent.

Dysuria is ordinarily symptomatic of one of the following conditions: (1) infections of the urinary tract such as cystitis and urethritis; (2) infections of the reproductive tract such as an inflammation of the uterus called *metritis;* (3) disorders of accessory reproductive structures, such as prostatism; and (4) the voiding of concentrated acid urine.

## EXCRETORY DISEASES

The more serious diseases of the urinary tract affect the kidney (nephritis and nephrosis), renal pelvis (pyelitis), ureter (nephroptosis), urinary bladder (cystitis), and urethra (urethritis). The formation of stones (lithiasis) can affect any portion of the excretory tract. Diseases of the excretory system can result in inflammation (nephritis, pyelitis, cystitis, and urethritis), noninflammatory degeneration (nephrosis), anuria (nephritis), or obstruction (lithiasis and nephroptosis).

**Nephritis and anuria** Acute or chronic inflammation of the kidney is called *nephritis*. The infection may affect the entire kidney (diffuse nephritis) or just the glomeruli (glomerulonephritis), the interstitial tissues along with the glomeruli (interstitial nephritis), the proximal and distal convoluted tubules and the loop of Henle (tubular nephritis), or the kidney and pelvis together (pyelonephritis).

Causative agents of nephritis include bacteria and viruses; toxic chemicals such as alcohol, arsenicals, mercurials, and lead; and vascular occlusions by bacterial emboli. Among the bacterial pathogens causing nephritis are *Streptococcus viridans* and other species, *Corynebacterium diphtheriae*, *Mycobacterium tuberculosis,* and *Treponema pallidum,* causative agents of scarlet fever, endocarditis, diphtheria, tuberculosis, and syphilis, respectively. Viruses are implicated in the causation of glomerulonephritis by eliciting the production of antibodies which either attack the altered tissues of the glomerulus and nephron directly or cause glomerular occlusions through precipitation reactions (see Chap. 4) within the blood.

The symptoms of nephritis reflect the type of agent responsible for the disorder. The following are among the most common symptoms:

1  Fever and lumbar pains due to the inflammatory condition
2  Either oliguria or polyuria, causing a decrease or increase in urine formation, respectively, by the affected kidney
3  Hematuria (bloody urine) due to the hemorrhaging of small arterioles into the nephrons
4  Albuminuria (albumin in the urine) because of an increase in the permeability of the glomeruli to protein and destruction of the nephrons
5  The presence in the urine of renal casts, pathologic tissues in the shape of a portion of a nephron
6  Edema due to fluid and salt retention
7  Uremia because of widespread nephric damage which permits the retention of urea, uric acid, creatinine, and other nitrogenous materials in the blood
8  Renal hypertension from the action of the renin-angiotensin mechanism initiated by an ischemic kidney (see Chaps. 22 and 31)

Because of renal overloading and extensive and widespread ischemia and necrosis of the glomeruli and nephrons, the kidneys may shut down completely. This condition is called *anuria*, or *renal failure*. Anuria leads to an exacerbation of the above-listed symptoms. The loss of plasma proteins prior to anuria decreases the plasma colloid osmotic pressure and increases the pressure in the interstitial tissue, causing *hypoproteinemic edema* through the diffusion of blood serum to the extracellular tissue spaces. Anuria also contributes to an edematous condition through the retention of salts and fluids by the body. Sometimes, fluid accumulates in the body tissues and cavities, causing dropsy and ascites. The development of renal hypertension is reflected in an elevated pulse rate. Increases in both blood volume and arterial pressure are responsible for a compensatory increase in the rate of glomerular filtration, but to no avail. The retention of nitrogenous waste products and acids in the blood leads to *urinary acidosis* and a corresponding depression in metabolism, possibly even a comatose state. If left untreated, the condition is almost always fatal.

Medical treatment involves some combination of the following steps:

1 Elimination of the agent responsible for the condition
2 Amelioration of the symptoms
3 Hygienic measures to prevent the spread of any urinary infection to other parts of the body
4 Bed rest to promote healing
5 Dietary control to restrict the intake of fluids, salt, and protein and to promote the ingestion of milk and fruits
6 Possible dialysis of the blood with the use of an artificial kidney to remove unfiltered wastes

The causes of, and vicious circle resulting from, renal failure are summarized in Fig. 20-4.

## OTHER INFLAMMATORY DISEASES

Aside from the kidney, other portions of the urinary tract can become inflamed. The resulting disorders are named for the affected structures and include (1) *pyelitis,* an inflammation of the renal pelvis and calyces; (2) *cystitis,* an inflammation of the mucosa of the urinary bladder and urethra; and (3) *urethritis,* an inflammation of one or more portions of the urethra.

The causative agents of such disorders are similar to those for nephritis. The renal pelvis frequently becomes inflamed following kidney infection. Conversely, infections of the urinary bladder commonly

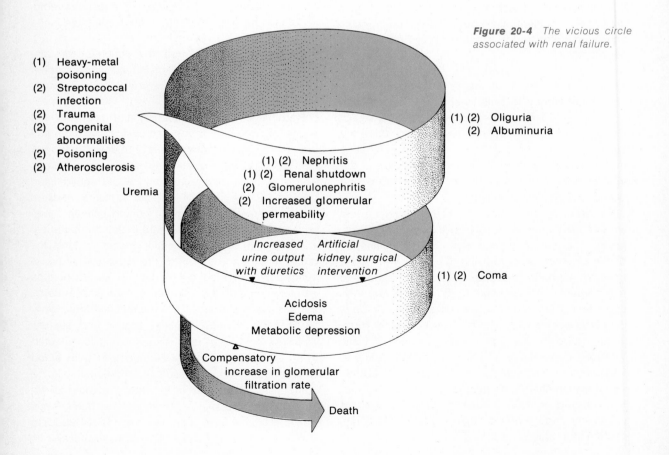

**Figure 20-4** *The vicious circle associated with renal failure.*

(1) Heavy-metal poisoning
(2) Streptococcal infection
(2) Trauma
(2) Congenital abnormalities
(2) Poisoning
(2) Atherosclerosis

Uremia

(1) (2) Nephritis
(1) (2) Renal shutdown
(2) Glomerulonephritis
(2) Increased glomerular permeability

(1) (2) Oliguria
(2) Albuminuria

*Increased urine output with diuretics*   *Artificial kidney, surgical intervention*

(1) (2) Coma

Acidosis
Edema
Metabolic depression

Compensatory increase in glomerular filtration rate

Death

spread from the urethra. Therefore, infections can spread in either direction along the urinary tract.

The symptoms of urinary tract inflammations include (1) pain, burning sensations, or discomfort in the lumbar region, urinary bladder, or urethra; (2) oliguria and difficult or painful urination; (3) hematuria, bacteriuria (the presence of bacteria in the urine), and pyuria (the presence of pus in the urine); and (4) other symptoms characteristic of the cause of the disorder such as fever caused by toxins secreted by pyretic bacteria or the excruciating pain of renal colic caused by the movement of urinary calculi from the renal pelvis to the ureter (see below).

## NONINFLAMMATORY DEGENERATION

*Nephrosis* is a noninflammatory disorder causing degeneration of the tissues of the kidneys. There are three distinct forms of the disease. In the first form, a substance called amyloid slowly infiltrates the renal arterioles and nephrons. Made of a complex mixture of proteins, amyloid gives the affected tissues a translucent appearance and causes their extensive degeneration. The second type of nephrosis has a rapid onset and may be induced by toxemia, drug intoxications, and certain infectious diseases not eliciting the inflammatory response. The third kind of nephrosis gives rise to the so-called "nephrotic syndrome" characterized by (1) albuminuria due to an increased glomerular permeability to blood proteins; (2) hypoproteinemic edema due primarily to the low plasma colloid osmotic pressure and a corresponding shift of serum from the blood to the interstitial spaces; and (3) elevated blood concentrations of serum cholesterol (hypercholesterolemia) and the related presence of fat in the urine (lipuria). Hypercholesterolemia and lipuria serve to identify the nephrotic syndrome as lipoid nephrosis.

## OBSTRUCTIVE DISORDERS

Obstructions of the urinary tract are caused by nephroptosis and lithiasis.

*Nephroptosis* is the accidental inferior displacement of a kidney within its peritoneal covering next to the posterior abdominal wall. The displacement of the kidney from its normal position causes a kinking in the ureter which blocks the flow of urine from the renal pelvis to the urinary bladder. Although the causes of nephroptosis are not well understood, abnormal intrabdominal pressures caused by either an enlarged uterus during pregnancy or unusual movements of other abdominal organs are possible predisposing factors.

The symptoms of nephroptosis are variable and may include (1) pains similar to those of renal colic, consisting of recurrent pains which radiate from the lumbar to the inguinal region; (2) oliguria and albuminuria; and (3) frequent low-volume urination.

Probably the most common obstructive disease of the excretory system is the presence of stones. Urinary stones, or calculi, can form in the kidneys, ureters, urinary bladder, and urethra, and can take place at any age, although the composition of the calculus varies with age.

The incidence of urinary stones in the general population is much higher than clinical records indicate. On an average and worldwide, one person in a thousand is treated for urinary calculi during his or her lifetime. However, routine fluoroscopic and autopsy examinations reveal that about 10 times that many people develop urinary stones which go untreated. In addition, the incidence of urinary stones differs widely between countries or geographic regions, cities or neighboring communities, or even from one isolated section of a community to another. Except for the development of some unusual stones and some isolated records which implicate dietary factors, no concrete relationship has yet been identified between the presence of stones and the nature of the diet, sex, or race of afflicted individuals. Diet is still suspected of playing an important role, however, and we shall have more to say about nutritional factors later.

Urinary calculi are complexes of minerals and organic constituents. The minerals consist of urinary salts surrounded by an organic matrix of uric acid or other substances. A stone forms when a nucleus, or "seed," is precipitated against the tissues lining the urinary tract from a supersaturated solution of urine.

The stone enlarges as an organic matrix is deposited around the nucleus.

Two basic categories of urinary calculi have been identified: (1) primary calculi formed in acid urine in the absence of inflammation; and (2) secondary calculi formed in alkaline urine as the result of inflammation, usually by streptococci.

The chemical composition of a stone is related to the age of the subject, the length of time during which the stone has been developing, and the portion of the urinary tract in which the stone is growing. The most common constituents of urinary calculi are the calcium oxalates *whewellite* ($CaC_2O_4 \cdot H_2O$) and *weddellite* ($CaC_2O_4 \cdot 2H_2O$). Calcium oxalates form the nuclei of many types of stone since oxalates are insoluble in both acid and alkaline urine.

Most urinary stones form in the kidneys (renal calculi), although bladder stones (vesical calculi) are also fairly common. In children, renal and vesical calculi are formed primarily of urates and oxalates. In adults, kidney stones are composed primarily of calcium oxalates and magnesium phosphates, but bladder stones usually consist largely of uric acid. No stones are pure; most calculi contain conglomerations of different substances.

Since oxalic acid, calcium, magnesium, and phosphorus are among the most common constituents of urinary calculi, it is prudent, insofar as possible, to avoid foods which are unusually high in these ingredients, especially in adult life. A list of forbidden foods for the patient with stones is determined by the nature of the stone and might include (1) tea and cocoa, which are high in oxalic acid; (2) most dairy products, including milk and milk substitutes, which contain a large amount of calcium; (3) spinach, since it contains an unusually large concentration of magnesium and is also high in oxalic acid and calcium; (4) boiled ham, which is high in phosphorus; and (5) white bread, which is high in phosphorus and calcium.

Implicit in dietary regulation is the premise that the probability of stone formation is somehow related to the acid and mineral content of ingested foods. The "calculogenicity" of specific nutrients is then reflected by an inability of the kidneys and the rest of the urinary tract to adequately manage the production and excretion of urine when an excessive burden of calculogenic waste products is present.

The probability of stone formation is also related to occupational status. Occupations in which there is a significantly higher-than-average incidence of urinary stones include (1) airline pilots, presumably because the occupational hazards of chronic dehydration caused by long hours in flight without an adequate fluid intake permit the urine to become highly concentrated; (2) white-collar workers, presumably because the occupational hazards of long hours of sitting at a desk permit epitaxy, or the adhesion of the surfaces of urinary crystalline substances to each other and to the inside walls of the urinary tract, precipitating nuclei from which stones grow; and (3) occupations which promote the imbibition of large volumes of alcohol, coffee, or both ingredients, presumably because these nutrients decrease the desire for water intake at the same time that they predispose toward a supersaturated acid urine. Combine a sedentary occupation with chronic dehydration and a greater-than-average consumption of coffee and alcohol in one subject and the profile presented is of the perfect candidate for the formation of urinary stones.

Some combination of the following signs and symptoms are indicative of the presence of urinary stones: (1) renal colic, characterized by excruciating pain in the lumbar region which, once experienced, is not easily forgotten; (2) renal inflammation, resulting in hematuria and pyuria due to streptococcal infection; (3) a concentrated, acid urine which contains sediments resembling gravel; and (4) frequent, scant urination or anuria followed by uremia.

Treatment of urinary stones involves elimination of the causative agent of the disease, amelioration of the symptoms, and prophylaxis against the recurrence of the same condition. Elimination of the pathogenic agent may require surgery to remove large or impacted stones, the use of smooth muscle relaxants in an attempt to void the stone naturally, or antibiotic therapy to reduce a urinary infection which may have initiated stone formation. Symptomatic treatment may necessitate the use of analgesics, antispasmodic drugs, special positioning, and other therapeutic techniques to reduce pain and promote physical comfort and tissue healing. Prophylactic methods which may be instituted include: (1) a significant increase in water ingestion to irrigate the kidneys, thereby loosening formed stones from their attachments and preventing nuclei from developing; (2) dietary control of food intake to reduce or eliminate those nutrients which are strongly suspected of contributing to the formation of stones; and (3) regular exercise to prevent epitaxy.

## COLLATERAL READING

Dixon, Frank J., "Glomerulonephritis and Immunopathology," in R. A. Good and D. W. Fisher (eds.), *Immunobiology: Current Knowledge of Basic Concepts in Immunology and Their Clinical Applications,* Stamford, Conn.: Sinauer Associates, Inc., 1971. This article, written at a somewhat advanced level but beautifully illustrated, discusses the immunogenic etiology of glomerulonephritis.

Hume, David M., "Organ Transplants and Immunity," in R. A. Good and D. W. Fisher (eds.), *Immunobiology: Current Knowledge of Basic Concepts in Immunology and Their Clinical Applications,* Stamford, Conn.: Sinauer Associates, Inc., 1971. The author discusses the successes and problems related primarily to kidney transplantation.

Lonsdale, Kathleen, "Human Stones," *Science,* 159:1199–1207, 1968. Details of the composition, rates of growth, geographic distribution, and possible causes of urinary stones are described in an engaging manner.

# CHAPTER 21 CONGENITAL DISEASES

Congenital disorders are abnormalities, malformations, and other forms of disease actually or potentially *present at birth*. This qualification is necessary since a number of congenital diseases may not appear for many years following birth. A time delay in the appearance of a disorder due to a genetic defect is determined by three factors: (1) the degree of physical maturation of affected organ systems; (2) exogenous and endogenous influences on postnatal development, i.e., the type of environment in which growth and maturation proceed and the cumulative effect of past experience; and (3) the presence of other genes which partially modify or repress the expression of the potentially harmful gene produced by a mutation.

Although they indeed appear at birth, birth injuries and prematurity are not congenital. Nevertheless, birth disorders are discussed in this chapter because of their association with the birth process.

The following general causes of congenital disease are recognized: (1) hereditary defects involving genes or chromosomes inherited from one or both parents and causing, or predisposing to, disease in either prenatal or postnatal development; (2) developmental anomalies caused by nonhereditary factors such as physiologic dysfunctions of the mother, or microbial infections, teratogenic drugs, and ionizing radiations in utero; and (3) drug addiction by the mother during pregnancy. Some disorders may occasionally be caused by any one of several of the above-listed factors; e.g., identical developmental anomalies are independently caused by either hereditary defects or environmental influences.

Many hereditary "defects" simply predispose to specific diseases contracted through nongenetic sources. In such cases, the predisposition to susceptibility is heritable but the actual disease is not.

Hereditary defects are caused in one of two ways:

(1) by mutations of genes or (2) through aberrations in the shape and number of chromosomes. Mutations which are passed from one generation to another must be present in the sex cells, i.e., in spermatozoa or ova. Although genes within somatic, or body, cells also undergo mutation, such changes are not passed from one generation to another. Aberrations, on the other hand, arise during meiosis in response to irregularities in chromosome pairing. Such dysfunctions may or may not have a heritable basis. In other words, chromosomal aberrations are either the visible expression of abnormal genes on affected chromosomes or occur in response to abnormal environmental factors.

Hereditary defects interfere ultimately with the molecular regulation of development. The interference may result in an inability to synthesize one or more specific proteins such as enzymes or may wreak havoc on the control of an entire metabolic pathway. In the former case, the products of the biochemical reaction mediated by the absent enzyme would not form, thereby blocking some aspect of metabolism; in the latter case, metabolism is more seriously impaired. However, the visible, or *phenotypic*, expression of genetic interference is usually in terms of structural malformations, functional disabilities, and behavioral aberrations.

Developmental anomalies are due to the effects of abnormal influences on morphogenetic processes. More specifically, malformations are caused by the following factors: (1) genetic defects which express themselves phenotypically during prenatal development; (2) abnormal environmental influences within the uterus such as a low oxygen tension or an insufficiency of essential nutrients in maternal blood; and (3) infectious microbes such as viruses, bacteria, and protozoa which either pass the placental barrier or spread to the child in utero from their initial loci.

Drugs constitute another category of exogenous, or external, pathogenic agent. Pharmacologically active agents cause developmental anomalies when they interfere with normal tissue formation. When addictive drugs are abused by a pregnant mother during a susceptible period of gestation, the child may be born with an addiction to the same drug.

This chapter reviews the causes and symptoms of various genetic defects, developmental anomalies, congenital infections, and birth disorders.

## HERITABLE DEFECTS

### GENE AND CHROMOSOME ABNORMALITIES

Hereditary diseases are caused by gene mutations and chromosomal aberrations. A *gene* is a portion of a chromosome which carries the potential for one hereditary characteristic. When a gene *mutates*, its molecular structure changes. Genes mutate spontaneously in response to the low levels of cosmic radiation to which everyone is exposed and to other ionizing radiations, certain chemicals, and other mutagenic agents. Since a gene consists of an unspecified length of DNA, one gene may change chemically in a number of ways. Each such change represents a mutation.

Not every mutation alters the visible expression of the normal trait controlled by the gene. One major reason is that some portions of the DNA molecule which compose a gene are genetically inactive. Furthermore, phenotypic alterations in a genetically determined trait may not be observed when the changes are not readily detectable. In fact, most viable mutations are of this variety and remain unrecognized throughout life. Indeed, several mutations on different genes are generally required before their combined effects are recognizable.

Chromosomal aberrations occur during gametogenesis, i.e., during oogenesis and spermatogenesis. An important aberration implicated in some of the more serious hereditary diseases consists of an excess or deficit of either a whole chromosome or a portion of a chromosome. Aberrations are not gene mutations because the genes have not changed chemically. The causes of chromosome aberrations are poorly understood but include gene defects and abnormal environmental influences, both of which can affect meiosis adversely. Because of meiotic dysfunctions, sex cells may wind up with an addition or deletion of genetic material.

The results of gene mutation and chromosome aberration are variable. The phenotypic expression of a gene or chromosome alteration consists of one of the following responses: (1) no demonstrable effect because a genetically inert part of DNA is changed or because the mutation enhances adaptation and is therefore impossible to distinguish from environmental variables promoting the attainment of steady states; (2) lethality or inviability exemplified by abortion or stillbirth because the phenotypic expression of the hereditary defect irreversibly disrupts homeostasis; (3) hereditary disease, the effects of which appear either at birth, during early postnatal development, or in adulthood; and (4) hereditary predispositions toward disease states, expressions of which require the impress of abnormal environmental conditions.

There are five categories of hereditary defect of concern to us:

1 Dominant defects caused by the presence of one dominant gene usually associated with chromosomes other than sex chromosomes, or *autosomes*

2 Recessive defects located on autosomes (autosomal recessives) or sex chromosomes (sex-linked recessives)

3 Defects in which the intensity of the disorder is related to the presence of one or two defective alleles (intermediate genes)

4 Defects from the combined actions of two or more genes on nonhomologous, or different, chromosomes (multiple genes)

5 A chromosomal excess (trisomy) or deficiency (monosomy) of an autosome or sex chromosome

The first four categories are due to gene mutation; the last class arises from alterations in the normal chromosomes during meiosis.

## MECHANISMS CAUSING DEFECTS

The one-celled structure formed by the fusion of a spermatozoon with an ovum is called a zygote. In human beings, a spermatozoon contacts a secondary oocyte, which then undergoes the second meiotic division to form a mature ovum. With the presence of both male and female pronuclei, the zygote is ready to enter the first mitotic division.

Both male and female pronuclei contain 23 different kinds of chromosomes. Their merger during the first mitotic division of the one-celled zygote reestablishes the chromosome number characteristic of the species, 46 in the case of Homo sapiens. In other words, each of the cells, which form by mitotic division of the zygote and all subsequent somatic cells, contains 23 pairs of chromosomes instead of 23 individual chromosomes. Of these paired chromosomes, 22 are called autosomes and the remaining pair, containing the genetic potential for sex determination, is referred to as the sex chromosomes. Since the 23 types of chromosomes are paired, each member of the pair is homologous to the other. Both chromosomal members, i.e., the homologues of a homologous pair, are similar in size, shape, and other morphologic characteristics.

The 23 pairs of human chromosomes in somatic cells are collectively termed the diploid, or 2n, number. On the other hand, a spermatozoon or ovum contains only one complete set of 23 different kinds of chromosomes. The 23 different chromosomes in human sex cells are collectively referred to as the haploid, or n, number.

The distinctiveness in size, shape, and number of the chromosomes is used in human genetic analyses. The diagnostic technique is known as karyotyping. The karyotype of a normal human male is illustrated in Fig. 21-1. The chromosomes are prepared from a body cell arrested in metaphase of mitosis. Therefore, each of the 23 paired chromosomes is already replicated. Remember that during metaphase, there are actually 92 individual chromosomes rather than 46, i.e., 46 original chromosomes plus 46 replicated chromosomes. Following cytologic preparation, the chromosomes are sorted into structural groups based on size, shape, and other morphologic characteristics, permitting a serial ordering of each chromosome pair.

Genes are lined up in linear order along the length of each chromosome. The complete *genotype*, or

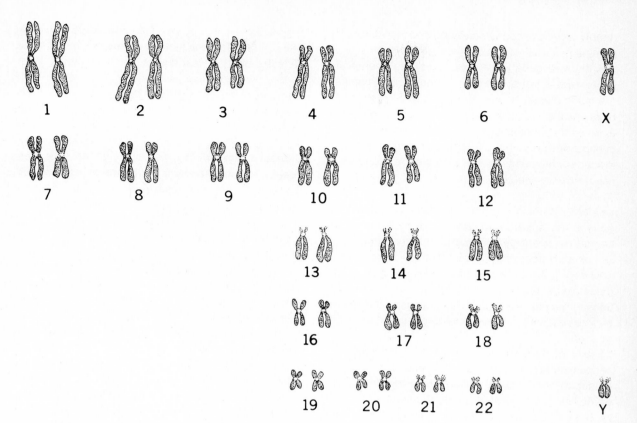

**Figure 21-1** *The karyotype of a normal human male. The chromosomes are specially prepared from the nucleus of a body cell of the subject. Each of the 23 different types of chromosomes is paired. In addition, the homologous chromosomes of each pair have replicated prior to their migration to opposite poles of the parent cell, an event which will take place during anaphase of mitosis.*

genetic constitution, of an individual is determined by the nature of the genes contained in the two sets of chromosomes with which each somatic cell is endowed. The phenotype, as we emphasized previously, is the visible expression of the genetic potential contained within the genotype.

A pair of genes controlling the potential for one hereditary characteristic is called an *allelic pair*. Each of the genes, or *alleles,* is located on a homologous chromosome. When one allele of the pair controls the phenotypic expression of a hereditary characteristic by suppressing its allelic counterpart, the controlling gene is *dominant* and the other allele is *recessive*. For

example, in an allelic pair *Aa,* where *A* suppresses *a* and thereby controls the phenotypic expression of the genotype *Aa, A* is dominant to *a.*

When both members of an allelic pair are either dominant (*AA*) or recessive (*aa*), they are called *homozygous dominant* or *homozygous recessive,* respectively. A dissimilar allelic pair of genes (*Aa*) is *heterozygous.* Since *A* is dominant to *a,* even in the heterozygous state, the phenotypic expression of *Aa* is still exclusively determined by *A.* In the homozygous recessive state (*aa*), the phenotypic expression of *a* is, of course, not repressed. In other words, the trait which shows up when the genotype is in the doubly recessive condition will be opposite that of the dominant characteristic.

A mutated gene is defective when its phenotypic expression is not adaptive in a normal environment. Although mutations are neither "good" nor "bad," most mutations possessing survival value or fitness in a given environment have already been retained evolutionarily by the process of natural selection. Practically all mutations can therefore be expected to have adverse effects.

**Defective dominant genes** A defective dominant gene is not suppressed, similar to the situation for normal dominant genes, and therefore expresses itself phenotypically during the lifetime of an individual possessing the defective gene. Usually a dominant defective genotype is heterozygous (*Aa*), since mutations in general are rare and mutations of the alleles of both parents from recessive to dominant state are rarer still.

Many dominant defects are lethal; others cause serious diseases. Generally, the effects of harmful dominant mutations are reduced by modifying genes which have accumulated during the course of evolution. Dominant defects which are fairly recent in an evolutionary sense are most likely severe. When the frequency of a defective trait in the general population is low, the defective allele most likely appeared early evolutionarily. In the latter case, the defective gene is said to have a *low penetrance*.

The inheritance of a dominant defect based on one pair of alleles is diagramed in Fig. 21-2. Homozygous dominant and heterozygous individuals exhibit the abnormal trait, but either newness of the mutation or low penetrance helps keep its incidence down.

**Defective autosomal recessive genes** The phenotypic expression of an autosomal recessive defective gene is determined solely by the genotype. For example, defective allele *a* is suppressed when paired with its allelic counterpart *A*. An individual of genotype *Aa*, where *a* is defective and *A* is dominant, is called a *genetic carrier* but does not usually exhibit symptoms of the disease caused by the recessive defective allele. However, an individual who is homozygous recessive (*aa*) for a genetic defect in gene *a* will exhibit disease symptoms. Therefore, a recessive genetic defect only expresses itself phenotypically in a doubly recessive condition, i.e., when both members of the recessive pair of alleles are defective. Inheritance of an autosomal recessive defect is also shown in Fig. 21-2.

**Defective sex-linked recessive genes** The frequency of the phenotypic appearance of a recessive defect on an X chromosome is determined by the presence or absence of a Y chromosome partner. Sex chromosomes are symbolically identified as X and Y. An individual possessing two X chromosomes is a genetically determined female. If the sex chromosome combination is XY, then the person has the genetic potential for maleness. However, the visible expression of a genetically determined sex is modified by the presence of other genes on the autosomes, hormonal imbalances during prenatal development, and nutritional status. In other words, after the X and Y chromosomes have replicated and formed a tetrad, or packet of four chromosomes, during metaphase of the first meiotic division, each of the dyads, i.e., the two attached X chromosomes or the two attached Y chromosomes, has an equal probability of migrating to one pole or the other within the primary spermatocyte or oocyte. The lack of an influence of one dyad on the other in a tetrad during this phase of meiosis is called *random assortment*.

When a spermatozoon fertilizes an ovum, four theoretical combinations of sex chromosomes are possible based on chance alone. These random combinations and the genetically determined sex of possible children resulting therefrom are indicated in Fig. 21-3. Two of the chromosome combinations (XX) give rise to genetically determined females; the other two combinations (XY) give rise to genetically determined males. However, intervening exogenous factors such as the acidity of either the seminal fluid or vagina frequently invalidate the 1:1 sex ratio predicted solely by chance combinations of sex cells.

An aberrant recessive gene on an X chromosome ordinarily expresses itself phenotypically when in the presence of a Y chromosome. In other words, in the

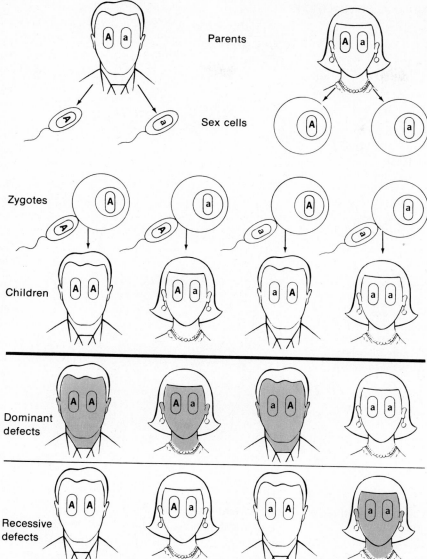

**Figure 21-2** *Inheritance of defects, the phenotypic expressions of which depend on one allelic pair of genes, e.g., Aa. The abnormal genotype may be dominant (AA or Aa) or recessive (aa).*

Parents

Sex cells

Zygotes

Children

Dominant defects

Recessive defects

male, the appearance of an abnormal, sex-linked trait is contingent on the presence of only one defective recessive allele $\underline{X}Y$, where $\underline{X}$ indicates the defective allele on the sex chromosome. When the sex chromosome combination is XX, on the other hand, the abnormal phenotype appears only when the defective allele is homozygous recessive, i.e., $\underline{XX}$. A sex-linked recessive defect in heterozygous state $\underline{X}X$ is usually masked by the normal dominant allele. Therefore, in the female, the phenotypic expression of an abnormal, sex-linked recessive genotype is dependent on the presence of both defective alleles $\underline{XX}$. When the genotype contains only one recessive allele, the female is normal in phenotype but a carrier of the

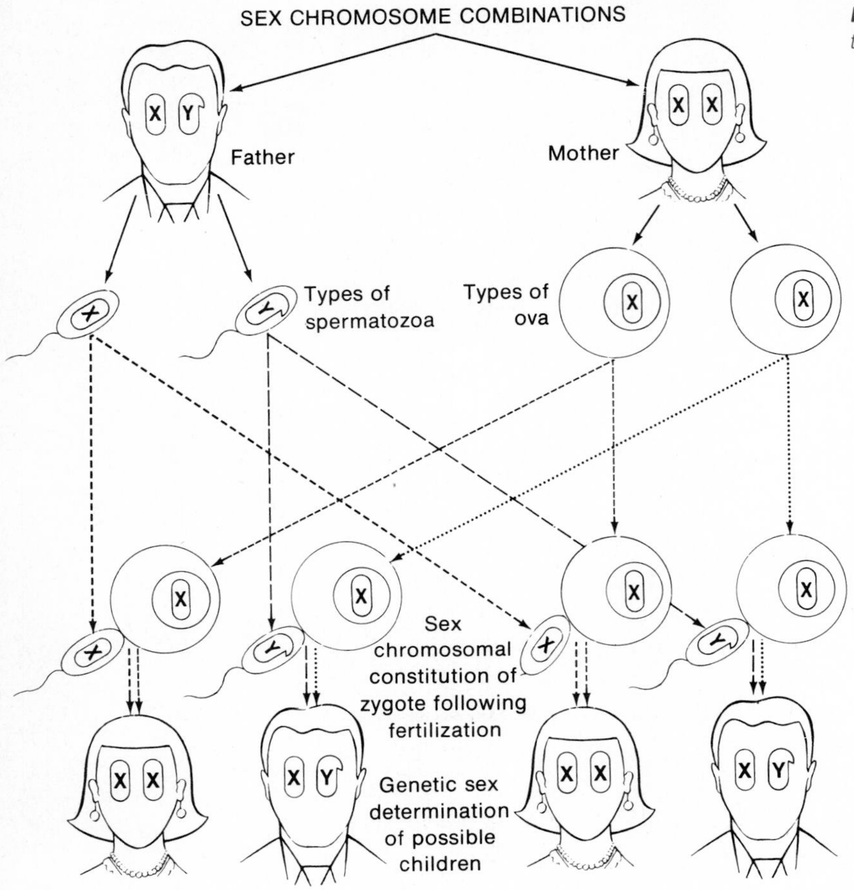

SEX CHROMOSOME COMBINATIONS

**Figure 21-3** *The possible pheno-types of children from one family.*

Father

Mother

Types of spermatozoa

Types of ova

Sex chromosomal constitution of zygote following fertilization

Genetic sex determination of possible children

defective gene. The inheritance of a sex-linked recessive defect is diagrammed in Fig. 21-4.

In this example, the father is completely normal. The mother, on the other hand, carries the harmful allele, i.e., is heterozygous (XX) with respect to the recessive defective gene. Since the father produces X chromosome and Y chromosome spermatozoa in equal frequency, half the male sex cells possess a single X chromosome and the remainder contain a solitary Y chromosome. Similarly, since the mother forms two kinds of ova in equal numbers, half the ova possess a normal X chromosome and half possess an X chromosome bearing the defective allele. Each of the spermatozoa has an equal proba-

bility of fertilizing either ovum. Thus, four sex chromosome combinations are possible in the zygote after fertilization: XX, XY, XY, and XX. In other words, the recessive allele is distributed among the children in the following ways: (1) half the daughters will be carriers of the defect (XX) and (2) half the sons will exhibit the defect (XY) phenotypically. The other children will be normal (XX and XY), with respect to the sex-linked recessive trait.

Should the son with the defective phenotype XY marry a woman carrying the same harmful allele XX, half their sons and daughters will exhibit the defect (i.e., XX and XY) phenotypically, and the rest of their children will be normal although carriers of the defec-

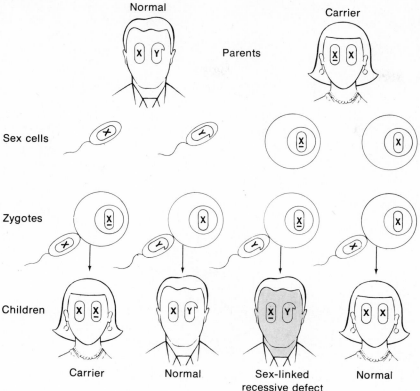

**Figure 21-4** *Inheritance of a sex-linked recessive defect. Located on an X chromosome, the defective gene is symbolized by X̲.*

tive gene if female (X̲X). On the other hand, if the afflicted son from the initial mating (X̲Y) marries a woman not carrying the defective allele (XX), then all their daughters will be carriers of the defective gene (X̲X) but all their sons will be normal (XY). These results should be verified by performing the genetic crosses on paper, using Fig. 21-4 as a model for each step of the operation.

**Defective intermediate genes**   Intermediate gene inheritance also depends on one allelic pair. The defective gene is dominant to its normal allelic alternative. In heterozygous condition, however, both alleles exhibit *partial dominance*; i.e., neither allele is dominant to the other. Because of partial dominance, three phenotypes are possible—homozygous dominant abnormals, heterozygous intermediates, and ho-

mozygous recessive normals. Under normal conditions, heterozygous individuals carry the defective trait but do not exhibit the disease. When the environment is abnormal, such carriers may show a mild form of the same disease exhibited by homozygous dominants.

The inheritance of defects which depend on intermediate genes is displayed in Fig. 21-5. In the illustration, the parents are heterozygous, possessing both defective gene *A* and its allelic counterpart, normal gene *B*. Through random combinations of sex cells during fertilization, each of their children has one chance in four of being either normal or abnormal and one chance in two of carrying the trait.

Test your knowledge of multiple gene inheritance by working the following problems: What are the probabilities of the children being normal, abnormal,

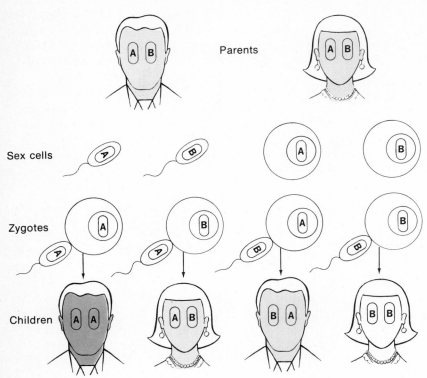

Parents

Sex cells

Zygotes

Children

**Figure 21-5** *Inheritance of defects which depend on intermediate genes. $\underline{A}$ and B are allelic alternatives. $\underline{A}$ is a defective gene and can express itself phenotypically only in homozygous condition ($\underline{AA}$). B is the normal allele of the pair; when in homozygous state (BB), such individuals are completely normal. Individuals with both alleles ($\underline{AB}$) carry the defective trait. The intensity of the abnormality is reflected by the amount of shading of the heads.*

or carriers of the trait when (1) one parent is homozygous normal BB and the other parent is homozygous abnormal $\underline{AA}$; (2) one parent is a heterozygous carrier of the trait $\underline{AB}$ and the other parent is homozygous normal BB; and (3) one parent is a heterozygous carrier of the trait $\underline{AB}$ and the other parent is homozygous abnormal $\underline{AA}$?

**Defective multiple genes** The inheritance of defects dependent on multiple genes is a marked departure from the genetic control exerted by one allelic pair, the genetic mechanism which has been our concern up to this point. The expression of multiple gene defects requires the interaction of more than one gene pair on different, or *nonhomologous,* chromosomes. The intensity of the phenotypic expression of multiple genes is determined by the following factors: (1) the number of allelic pairs which participate in the determination of the phenotype; (2) the

form of inheritance exhibited by each of the participating allelic pairs, i.e., complete dominance, intermediate genes, etc.; and (3) environmental factors which influence the expression of the genotype.

Multiple gene defects are inherited according to the model presented in Fig. 21-6. In the illustration, the defective phenotype is determined by the interaction of two allelic pairs of genes on separate sets of chromosomes (N versus $\underline{A}$ alleles on homologous chromosome pair 1 and N' versus $\underline{A}'$ alleles on homologous chromosome pair 2). Consider a mating between parents who are heterozygous for both allelic pairs ($N\underline{A}N'\underline{A}'$). Since the allelic pairs $N\underline{A}$ and $N'\underline{A}'$ assort at random during meiosis, four genetic combinations of spermatozoa or ova are possible: NN', $N\underline{A}'$, $\underline{A}N'$, and $\underline{A}\underline{A}'$. Assume that N and N' are normal, $\underline{A}$ and $\underline{A}'$ are abnormal, and N and N' are dominant to $\underline{A}$ and $\underline{A}'$. If fertilization is random, 16 combinations of zygotes are possible, one of which

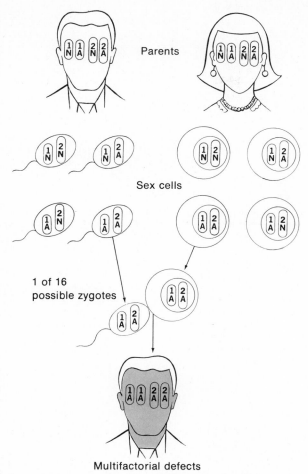

**Figure 21-6** *Inheritance of a defect determined by multiple genes. In this example, the phenotypic expression of the defect involves two pairs of alleles on nonhomologous chromosomes. Genes 1N and 2N are normal while genes 1A and 2A bear the genetic potential for an abnormality. The abnormal trait appears when the allelic pairs on chromosomes 1 and 2 are homozygous for 1A and 2A.*

Parents

Sex cells

1 of 16 possible zygotes

Multifactorial defects

contains the homozygous abnormal combination (*AAA'A'*) for both allelic pairs. An individual of this genotype will exhibit the multifactorial defect.

**Nondisjunction**    Nondisjunction occurs when a paired set of chromosomes fails to separate as anaphase begins during cellular division. This phe-

nomenon occurs occasionally during both mitosis and meiosis. However, we are concerned only with its appearance during spermatogenesis or oogenesis.

When nondisjunction occurs during the first meiotic division of a primary spermatocyte or oocyte, one of the two resulting cells contains an excess of one dyad but the other formative cell completely lacks representation of the chromosome engaged in nondisjunction. Each of these secondary spermatocytes or oocytes then engages in a second meiotic division, forming definitive spermatozoa or ova. Theoretically, two of the four possible sex cells formed from one primordial germ cell under these circumstances have an excess of one chromosome ($n + 1$), compared to the normal haploid number ($n$), and the other two gametes lack this chromosome entirely ($n - 1$). In actuality, four cells are formed during spermatogenesis but only one cell results from oogenesis. Recall that three of the four potential ova are lost as first and second polar bodies.

When these abnormal sex cells participate in fertilization with normal gametes, the zygotes which result will have a chromosomal constitution of either $2n - 1$ or $2n + 1$. Each of these abnormal chromosomal combinations has an equal probability of initiating development. Individuals with a chromosomal constitution of $2n - 1$ are called *monosomics*; those with a chromosomal constitution of $2n + 1$ are termed *trisomics*. Because excesses or deficiencies of either sex chromosomes or autosomes may occur, four types of chromosomal imbalance are possible: sex chromosome monosomy, sex chromosome trisomy, autosomal monosomy, and autosomal trisomy.

The causes of sex chromosome monosomy and trisomy are diagrammed in Fig. 21-7. In this example, nondisjunction occurs during the first meiotic division although, as we mentioned previously, chromosome separation may fail during the second meiotic division instead of the first. The range of possible ova is given in Fig. 21-7, even though we recognize that only one of these ova actually appears from one primordial germ cell. Furthermore, we note that each type of normal spermatozoon with respect to the sex chromosomes, i.e., either X or Y, fertilizes each type of ovum, i.e., either O or XX, where O indicates the absence of

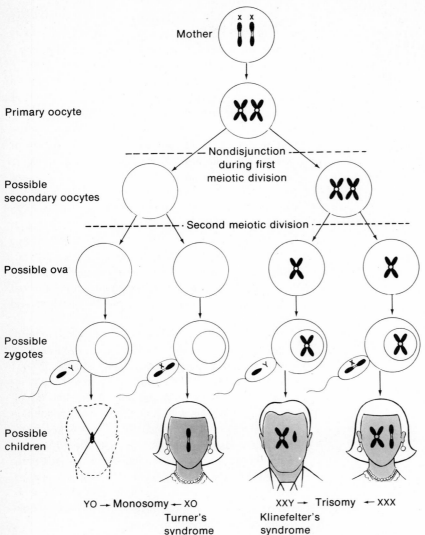

Mother

Primary oocyte

Nondisjunction during first meiotic division

Possible secondary oocytes

Second meiotic division

Possible ova

Possible zygotes

Possible children

YO → Monosomy ← XO
Turner's syndrome

XXY → Trisomy ← XXX
Klinefelter's syndrome

**Figure 21-7** *The causes of sex chromosome deficiencies and excesses caused by nondisjunction during gametogenesis.*

an X chromosome. Under these circumstances, four zygotic combinations of sex chromosomes are possible: (1) YO, which is apparently so serious a deficiency that it results invariably in prenatal mortality; (2) XO, which results in Turner's syndrome; (3) XXY, which gives rise to Klinefelter's syndrome; and (4) XXX, which results in another serious syndrome. Obviously, YO and XO are forms of sex chromosome

monosomy; XXY and XXX are types of sex chromosome trisomy.

The karyotype of a patient suffering from Turner's syndrome is shown in Fig. 21-8. The karyotype of a patient who has Kleinfelter's syndrome is illustrated in Fig. 21-9. The symptoms of both syndromes are reviewed later in this chapter.

One way in which autosomal trisomics are formed

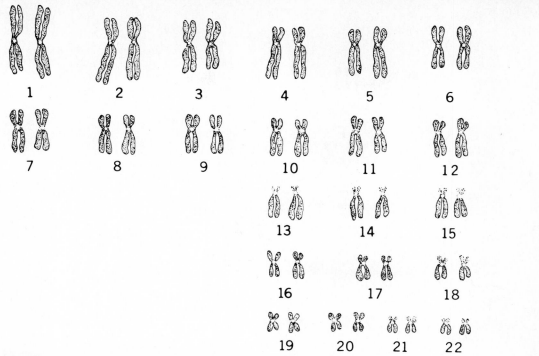

**Figure 21-8** *The karyotype of a male patient suffering from sex chromosome deficiency XO, Turner's syndrome.*

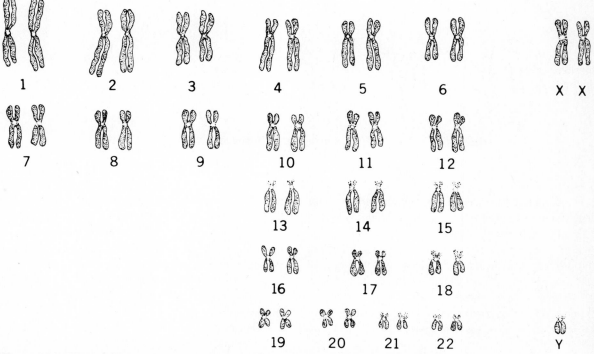

**Figure 21-9** *The karyotype of a male patient suffering from sex chromosome excess XXY, Klinefelter's syndrome.*

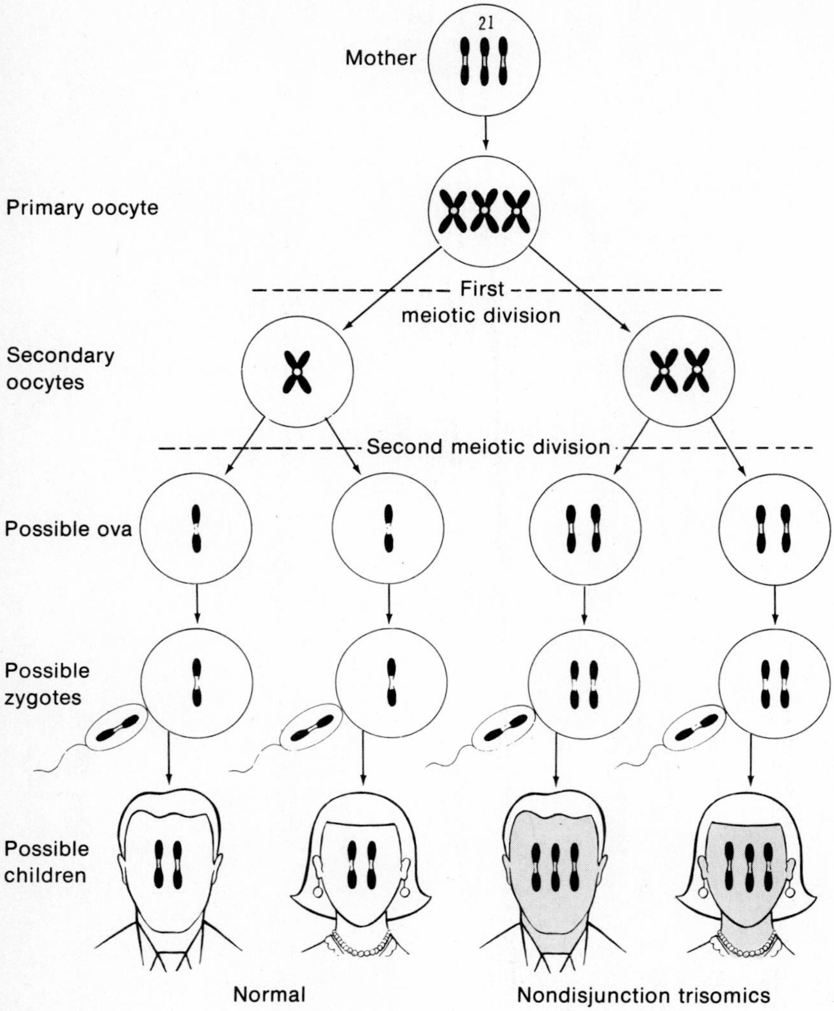

**Figure 21-10** *The causes of mongolism based on nondisjunction.*

is shown in Fig. 21-10. Here, the mother is trisomic for chromosome 21. The karyotype of a patient with this form of mongolism is illustrated in Fig. 21-11. If non-disjunction occurs during the first meiotic division, the probability is equal that an ovum which is normal for chromosome 21 ($n$) or an ovum which contains an excess of one chromosome 21 ($n + 1$) will be formed. Assuming that spermatozoa normal for chromosome 21 fertilize each of these types of ova, two kinds of progeny with regard to chromosome 21 will appear in equal frequencies: 21,21 ($2n$) and 21,21,21

($2n + 1$). Individuals with a chromosomal constitution of 21,21 are, of course, normal; those possessing a chromosome constitution of 21,21,21 suffer from mongolism, sometimes referred to as trisomy 21. The symptoms of mongolism and its alternative pathway of inheritance through the translocation of chromosome 21 onto chromosome 15 are described later in this chapter.

Referring back to Fig. 21-10, substitute a normal diploid mother with respect to chromosome 21 for the trisomic mother shown. Endow the new mother with a

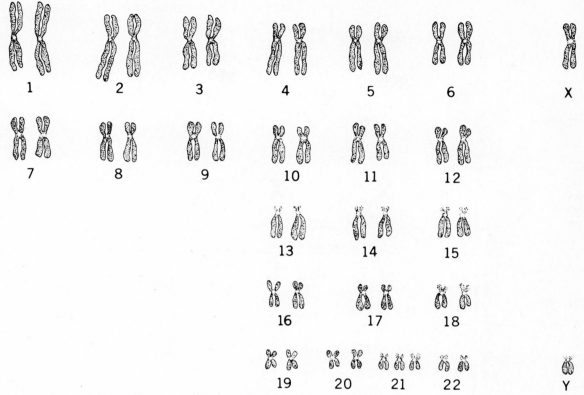

**Figure 21-11** *The karyotype of a patient suffering from trisomy 21, a form of mongolism.*

hereditary predisposition toward nondisjunction. Imagine further that the occurrence of nondisjunction increases in likelihood as the mother grows older. A similar situation is believed to exist in real life and may also give rise to trisomy 21. Nondisjunction is then due to chromosomal effects caused by the inherited predisposition and the influence of aging.

Translocation    Translocation occurs when either a portion of a chromosome or an entire chromosome attaches to a nonhomologous chromosome during cellular division. For our purposes here, we are concerned with the translocation of an entire chromosome during meiosis. Representing one type of aberration undergone by chromosomes, translocation is obviously an abnormal occurrence.

The cause of translocation trisomics is traced in Fig. 21-12. The karyotype of a patient with this form of mongolism is illustrated in Fig. 21-13. In this example, the mother is a carrier of a 15/21 translocation; i.e., chromosome 21 (or 22) is attached to chromosome 15. By tradition, either of the above chromosomes is referred to as chromosome 21 when translocated to chromosome 15. The mother is, of course, diploid for both chromosome sets and therefore phenotypically normal. Of the secondary oocytes formed as the result of translocation during the first meiotic division, one possibility contains a replicated 15/21 translocation plus a replicated chromosome 21, and another possibility contains only the replicated 15/21 translocation. Following the second meiotic division, therefore, two types of ova form from these two kinds of secondary oocytes; namely, ova with the 15/21 translocation plus chromosome 21 ($n + 1$) and ova with the 15/21 translocation ($n$). When fertilization of these ova is accomplished with normal spermatozoa, two types of

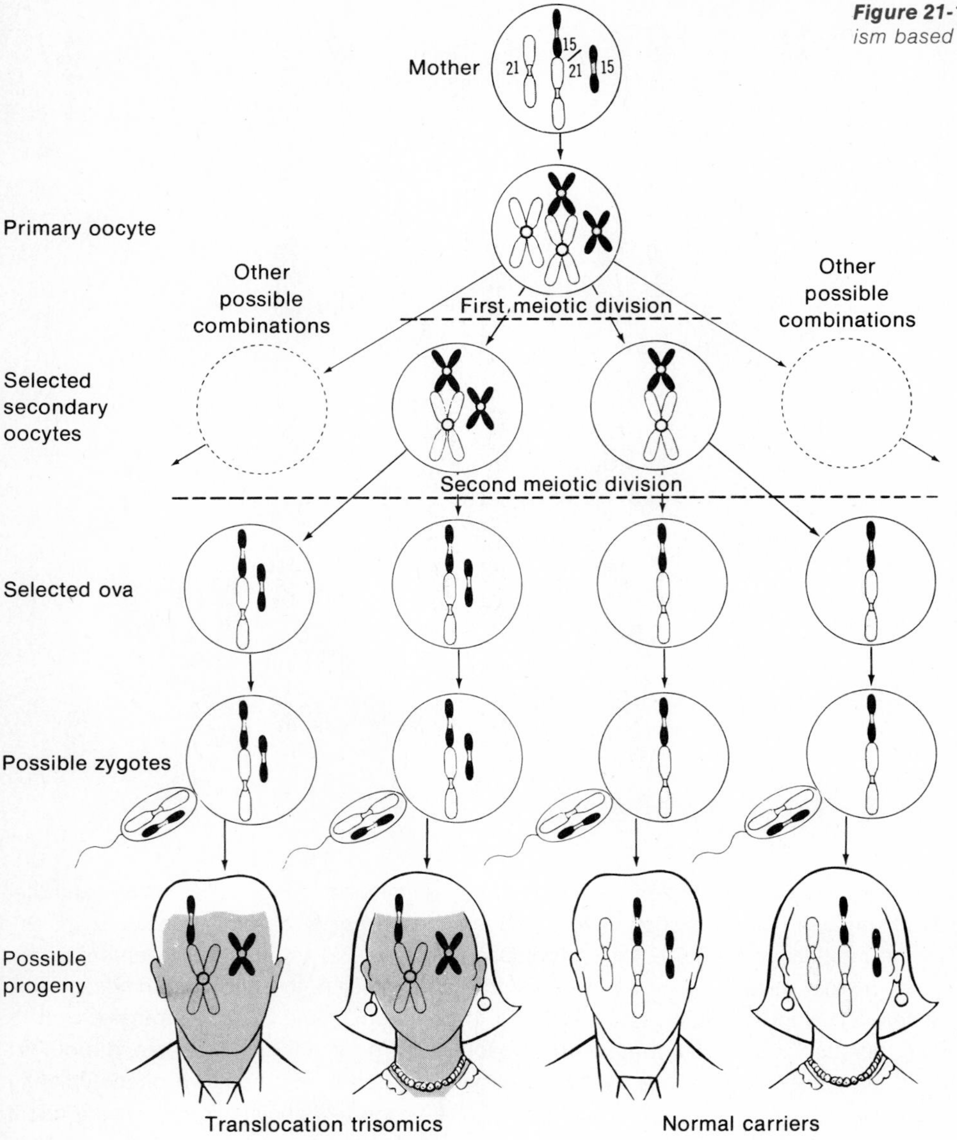

**Figure 21-12** *The cause of mongolism based on translocation.*

zygotes result: trisomics possessing a 15/21 translocation in addition to another chromosome 15 and two chromosome 21s; and carriers with a 15/21 translocation in addition to one chromosome 15 and one chromosome 21. Translocation trisomics actually possess 47 chromosomes, and translocation carriers contain 45 apparent chromosomes.

In the example described above, translocation occurs during the first meiotic division of oogenesis, although this phenomenon probably takes place just as frequently in spermatogenesis. Translocation may also take place in the second meiotic divison, rather than the first.

As an exercise, attempt to diagram the types of ova

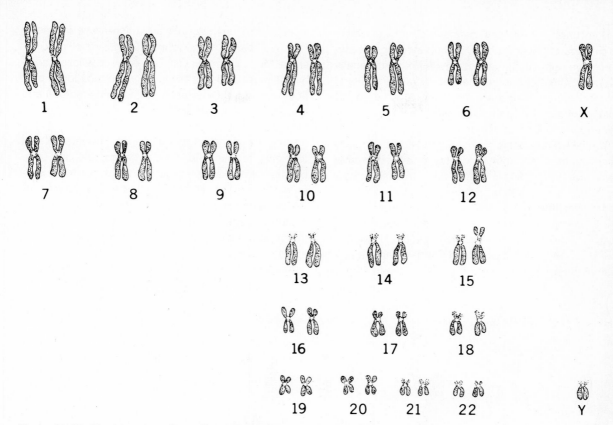

**Figure 21-13** *The karyotype of a patient suffering from a 15/21 translocation.*

formed when attachment of chromosome 21 to chromosome 15 does not take place until the second meiotic division. Assuming fertilization of these ova with normal spermatozoa, identify the possible zygotic combinations for chromosomes 15 and 21 as well as their corresponding phenotypes. Use Fig. 21-12 as a model for each of the above operations.

## HEREDITARY DISEASES

The phenotypic expression of a hereditary disease may be structural, functional, or a combination of both. Such disorders can take the form of developmental anomalies, sensory dysfunctions, mental impairments or deficiencies, hemopathies, skeletomuscular diseases, or glandular abnormalities.

## DEVELOPMENTAL ANOMALIES

Developmental anomalies are malformations which occur during embryonic or fetal development. The anomalies described here are due to heritable factors although some may also be caused by abnormal environmental influences.

Familial harelip and cleft palate  Most facial defects involve the lips, palate, or both structures. A harelip, producing one or two clefts in the upper lip, is caused by a failure of the maxillary and nasomedial processes to fuse with each other. Cleft palate, on the other hand, is due to a defect in the fusion of the palatine processes. This malformation consists of a unilateral cleft or bilateral clefts in both the hard and soft palates.

Both harelip and cleft palate are defects which occur around the end of the second month of embryonic development. Since these malformations usually overlap in time and influence closely related parts of the oral region, they are often found together in the same subject at birth. Both may be caused by genetic or nongenetic factors.

The hereditary basis for developmental anomalies involving the oral region is trisomy for either chromosome 13, 14, or 15. In other words, the hereditary mechanism is traceable to an excess of one chromosome over the normal diploid, or 2n, number in the chromosome grouping, 13–15. The karyotype of a human male trisomic for chromosome 15 is illustrated in Fig. 21-14. The subject has 47 rather than the normal 46 chromosomes characteristic of human somat-

ic cells. Presumably, this condition arose in one or the other parent through nondisjunction of chromosome 15 during meiosis. The subject whose karyotype is illustrated was born with a combination of cleft lip and cleft palate, in addition to other structural and functional abnormalities.

Anomalies of the fingers and toes    Anomalies of the limbs frequently involve the hands, feet, or a combination of both appendages. Probably the most well-known hereditary affliction of this sort is *polydactylism* (polydactyly). In this condition, an extra digit is usually appended to either the thumb or little finger or the great or little toe. The involvement may be unilateral, bilateral, or can encompass both hands and feet.

An example of bilateral polydactylism of the feet is

**Figure 21-14**   *The karyotype of a male patient trisomic for chromosome 15.*

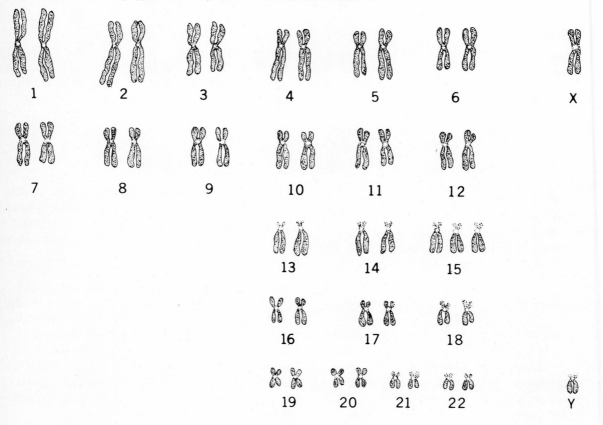

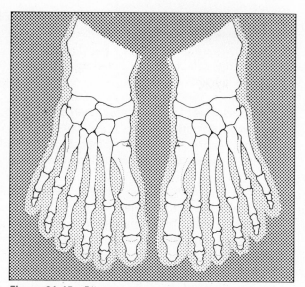

**Figure 21-15** *Bilateral polydactylism of the feet.*

shown in Fig. 21-15. The extra digit is attached to the little toe. Notice that the phalanges are abnormally club-shaped at the joints.

The inheritance of polydactylism is due to an autosomal dominant defect. The phenotypic expression of the dominant allele is modified to a considerable extent by other genes and environmental influences. An autosomal dominant defect whose phenotypic expression is highly variable is difficult to eliminate through eugenics programs, because the effects of a dominant defective gene are mild.

Other forms of dactylisms include the following:

1 Arachnodactylism, in which the fingers and toes are abnormally long and thin because of an excessive growth of the phalanges
2 Brachyphalangia, in which the fingers and toes are abnormally short because the second phalanx of each digit, although present, is poorly developed
3 Brachydactylism, in which the fingers and toes are abnormally short because the second phalanx of each digit is absent, having fused with another phalanx

4 Symphalangism, in which the fingers or toes are stiff because some of the joints between the phalanges have fused
5 Camptodactylism, in which the fingers are permanently flexed due to an abnormal shortening of the flexor tendons
6 Syndactylism, in which one or more fingers or toes are webbed

With the exception of syndactylism, all the above-named disorders arise from autosomal dominant defective genes. Syndactylism seems to be related to a genetic defect on the Y chromosome in some cases and to an autosomal recessive gene defect in other cases.

## SENSORY DISORDERS

Heritable sensory defects include color blindness and tone deafness.

Color blindness   There are several forms of color blindness. The inability to see in color is called *total color blindness*. This defect is relatively rare and is due to a sex-linked recessive gene. In addition, there are two common kinds of partial color blindness: (1) *deuteranopia*, in which there is difficulty in distinguishing between red, yellow, and green, although sensitivity to these colors is not reduced; and (2) *protanopia*, in which the ability to distinguish red is adversely affected and, therefore, sensitivity to red is reduced. In the latter case, red may be confused with yellow or green. Presumably, some portion of the red pigments or cones is lost in protanopia. Protanopia is the more common form of partial color blindness and is usually referred to as *red-green color blindness.*

The inheritance of red-green color blindness is similar to that of any sex-linked recessive defect. However, the genetics of color blindness is more complex than the simplistic model in Fig. 21-4 would lead us to believe. The differences are quantitative, rather than qualitative, and do not alter the predictability of the model as presented.

Red-green color blindness is actually caused by the interaction of two genes, rather than one, on the X chromosome. In addition, each of the two genes may contain any one of three alleles involved in the phenotypic expression of color blindness. The normal allele at each locus is dominant to the other two possible alleles. Furthermore, the most recessive alleles at the two loci combine to cause the more serious form of partial color blindness, deuteranopia. The allele at each locus which is intermediate in effect causes protanopia when in combination. These intermediate alleles are therefore dominant to the most recessive alleles but are themselves recessive to the most dominant normal alleles. Finally, when an allele at one locus is the most dominant while the allele at the other locus is either intermediate or the most recessive, their phenotypic expression may be in terms of a deficiency in color aptitude or some other functional disability with respect to color.

To facilitate your understanding of the mechanics of this form of sex-linked recessive inheritance, attempt to work the following problem, using Fig. 21-4 as a model. Consider a marriage between a normal man

$$\begin{array}{c} N \\ N' \end{array} \Big|\Big| \quad \text{and a carrier woman for protanopia} \quad \begin{array}{c} N \\ N' \end{array} \Big|\Big| \begin{array}{c} \underline{P} \\ \underline{P'} \end{array}$$

where $N$ and $N'$ = normal and $\underline{P}$ and $\underline{P'}$ = protanopia. $N$ is dominant to $\underline{P}$, and both these alleles are on the X chromosome. What are the genotypes of the spermatozoa and ova produced by the parents? What are the genotypes and phenotypes of the possible children resulting from this marriage? Did you find any difference in the final results between this form of inheritance and that presented in Fig. 21-4?

Tone deafness  Tone deafness refers to an inability to sing in tune. The genetics of this disability have not been well worked out, although multiple genes are most likely involved.

Pitch discrimination does not seem to be impaired in tone deafness. Rather, there is an apparent breakdown in the ability to perceive the harmonic relationship between individual sounds. Therefore, the defect is probably in a proprioceptive feedback

mechanism between the laryngeal and facial muscles of speech and the auditory cortex of the brain.

Other sensory defects  Other heritable sensory abnormalities include:

1 Blindness due to glaucoma inherited as an autosomal dominant defect
2 Deafness caused by two different recessive genes on nonhomologous chromosomes (multiple gene defect)
3 Myopia (nearsightedness) due to either an elongated eyeball inherited as an autosomal recessive defect or an excessive curvature of the cornea inherited as an autosomal dominant defect
4 Hyperopia (farsightedness) due to a shortened eyeball inherited as an autosomal dominant defect
5 Astigmatism due to an unequal curvature of the cornea inherited as an autosomal dominant defect
6 Nystagmus inherited as either a sex-linked recessive or autosomal dominant defect
7 Night blindness caused by a sex-linked recessive defect

Although the above-listed diseases have a heritable basis in individual cases, nutritional deficiencies, infections, psychogenic factors, and other types of abnormal environmental and experiential influences can also precipitate these disorders.

The symptoms of the more common sensory disabilities are considered in Chap. 12.

## MENTAL IMPAIRMENTS

Among the important mental impairments with a heritable basis are mongolism, amaurotic familial idiocy, Turner's syndrome, Klinefelter's syndrome, Huntington's chorea, alkaptonuria, and phenylketonuria.

Mongolism (Down's syndrome, or trisomy 21)  Mongolism results in mental deficiency and diverse structural abnormalities. Among the functional impairments are mental retardation reflected in an intelligence quotient (IQ) comparable to that of an idiot or imbecile and motor dysfunctions reflected in a shuffling gait, lack of muscle tone, and an inability to clearly articulate sounds phonetically. The latter difficulty gives rise to harsh guttural speech which may be incoherent at times.

Structural abnormalities commonly found among subjects suffering from mongolism include the following:

1  Puffed upper eyelids, the corners of which curve upward
2  A round flattened face, in addition to a superiorly and posteriorly flattened head
3  A depression in the bridge of the nose
4  A large fissured tongue which frequently protrudes from the mouth
5  Stubby and square-shaped hands and feet with an abnormally wide separation between the toes
6  Small physical stature and stunted development of brain cells, reflecting a retardation of growth during childhood
7  Various congenital anomalies of the skeletomuscular systems and visceral structures

As noted earlier, people with mongolism are trisomic for either chromosome 21 or 22. The heritable basis for trisomy 21 is either nondisjunction of autosome 21 or translocation of this autosome onto autosome 15 during meiosis. These mechanisms are reviewed in Figs. 21-10 and 21-12.

**Amaurotic familial idiocy**  This disorder leads to a degeneration of ganglionic neurons at the base of the brain, usually within the first 3 years of life. As a consequence, there is severe mental deterioration, blindness, and paralysis. The prognosis is poor, and the disease usually ends in early death. One of the important diagnostic signs is the appearance of a bright red spot in front of the macular portion of the retina following its degeneration. The disease is related to dysfunctions in the metabolism of lipids. This form of idiocy is inherited as an autosomal recessive defect and occurs most frequently in people of Jewish origin.

**Turner's syndrome and Klinefelter's syndrome**  A deficiency or excess of a sex chromosome gives rise to two discrete syndromes: (1) sex chromosome monosomy (XO), called *Turner's syndrome*; and (2) sex chromosome trisomy (XXY), called *Klinefelter's syndrome*. Other documented human trisomic combinations of sex chromosomes include XXX and XYY. All such individuals are abnormal, and most usually suffer from mental deficiency and aberrant behavior.

The inheritance of sex chromosome deficiencies and excesses is caused by nondisjunction of an X or Y chromosome during either the first or second meiotic division.

Subjects afflicted with Turner's syndrome (XO) are "partially" female in that the ovaries, if present, are abnormal and the secondary female characteristics remain undeveloped or malformed. Usually, there is a stunting of growth in the skeletal and nervous systems.

Subjects suffering from Klinefelter's syndrome (XXY) are "partially" male. Although male genitals are present, the testes do not produce spermatozoa. Testicular sterility is usually matched by the development of feminine breasts and other sexual and physical abnormalities.

**Huntington's chorea**  In this disorder affected persons undergo a progressive mental deterioration coupled with spasmodic contractions of muscles of the face, arms, and legs. The disease primarily affects the basal ganglia and cortex of the brain and is invariably fatal.

Huntington's chorea is inherited as an autosomal dominant defective gene possessing complete penetrance, so that all afflicted individuals actually display clinical symptoms of the disease. However, the age at which the disease is phenotypically expressed varies considerably. The mean age of onset is about 40.

**Alkaptonuria**  This congenital disease is caused by an inability to metabolize the amino acid tyrosine. One of the intermediate breakdown products of tyrosine is homogentisic acid. Normally oxidized completely to $CO_2$, this acid accumulates in the blood of alkaptonuric patients and is excreted in large quantities in their urine. Upon exposure to air, the urine containing homogentisic acid turns black. This color change is diagnostic for the disease. People afflicted with alkaptonuria are also highly susceptible to arthritis.

Alkaptonuria is inherited as an autosomal recessive defect which prevents the synthesis of a specific enzyme, a so-called inborn error of metabolism. The inheritance of this defect indicates a high degree of consanguinity, i.e., marriage among blood relatives in the affected portion of the population.

Phenylketonuria   This disorder results from an inability to oxidize the amino acid phenylalanine to tyrosine, reflected by the appearance of a metabolic intermediate, phenylpyruvic acid, in the urine. Phenylketonuric sufferers frequently have some form of mental deficiency due to brain damage.

Phenylketonuria is inherited as an autosomal recessive. The phenotypic expression of this genetic defect is caused by the loss of a specific enzyme. As in alkaptonuria, an inborn error of metabolism is the source of the disorder.

HEMOPATHIES

Among the more well-known blood diseases with a heritable basis are sickle-cell anemia, thalassemia, and hemophilia.

Sickle-cell anemia   Sickle-cell anemia is due to a defective intermediate gene, Hb S (hemoglobin S). Its allelic counterpart is the normal gene Hb A. Because neither allele is dominant in heterozygous state, three phenotypes are possible: (1) Hb A Hb A (normal); (2) Hb A Hb S (sickle-cell trait); and (3) Hb S Hb S (sickle-cell anemia). The appearance of erythrocytes under each of the above conditions is illustrated in Fig. 21-16. The inheritance of this intermediate defective gene obeys the pattern diagramed earlier in Fig. 21-5. Several other defective genes, each inherited independently, may also cause forms of sickle-cell anemia.

Ordinarily, erythrocytes are tire-shaped spheres. When the oxygen tension is low, however, an abnormal erythrocyte becomes C-shaped, superficially

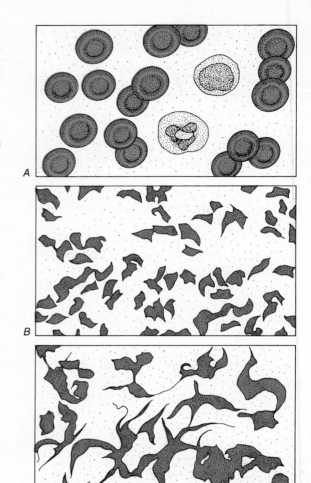

**Figure 21-16**   *The appearance of erythrocytes with normal and abnormal hemoglobin. A. Normal erythrocytes. B. Erythrocytes of subject with sickle-cell trait during hypoxia. C. Erythrocytes of subject with sickle-cell anemia.*

resembling a sickle. Sickling in individuals suffering from this form of anemia (Hb S Hb S) is caused by the complete replacement of normal hemoglobin with an abnormal form of the molecule called hemoglobin S. The oxygen-carrying capacity of hemoglobin S molecules is unimpaired, as long as the oxygen tension remains high. When the partial pressure of oxygen ($P_{O_2}$) in the blood is similar to that found in the tissue capillaries, the hemoglobin S molecules line up side-by-side or in bundles, forming stiff rods which result in sickling.

Sickling in subjects who carry the trait (Hb A Hb S) usually does not occur under normal conditions. Since only 40 percent of the hemoglobin in such subjects is abnormal, sickling is not induced until the $P_{O_2}$ of the blood is considerably lower than that of the tissues. Under these circumstances, however, sickling takes place in a manner similar to that for subjects homozygous for the abnormal hemoglobin.

Sickle-cell anemia is a chronic disease with a poor prognosis. The following are among the major symptoms of the disorder: (1) anemia caused by hemolysis of the sickle cells; (2) jaundice from the deposition of protein-bound bilirubin in the skin and other tissues of the body following the release of large amounts of hemoglobin from lysed erythrocytes (see Chap. 31); (3) tissue ischemia and the development of local necrotic areas due to vascular obstruction by clumped sickle cells; and (4) paroxysms of pain due to oxygen starvation of the body tissues (tissue hypoxia).

The incidence of both sickle-cell anemia and sickle-cell trait is higher among black people than in the general population, although by no means restricted to blacks. The disease is indigenous to Africa, where it confers resistance to malaria. Presumably, the oxygen concentration in sickle cells is not conducive to the growth and development of the malarial parasite *Plasmodium.* For this reason, the harmful mutation Hb S which results in the formation of abnormal hemoglobin has been retained in the gene pool, i.e., the total amount of heritable material of the African population. The fact that the mutant defective gene still persists in Africans after many generations in countries which lack endemic malaria indicates that an accommodation has been achieved in nature regarding the incidence of the defective gene Hb S and the normal gene Hb A. The impetus for this genetic détente arises from the natural elimination of many of the African people with the genotype Hb S Hb S, permitting those people with a selective advantage (Hb S Hb A) in a malarial environment as well as those who have a normal genotype (Hb A Hb A) to increase in frequency.

**Thalassemia**    This disorder is another form of hered-itary anemia caused by the formation of abnormal hemoglobin. The malformed respiratory pigment distorts the shape of the erythrocytes, resulting in hemolysis, anemia, and jaundice. These effects are similar to those noted for sickle-cell anemia. In addition, the liver and spleen become enlarged and the white corpuscle count increases significantly. The latter condition is called *leukocytosis* (see Chap. 31).

Thalassemia is another example of intermediate gene inheritance. There are two forms of the disease: (1) *thalassemia major,* which is homozygous for the abnormal gene; and (2) *thalassemia minor,* which is heterozygous; i.e., the defective gene exists in single rather than double dose. Subjects with thalassemia major exhibit serious symptoms, but those with thalassemia minor experience a more mild case of the disease. Subjects who are homozygous for the normal allelic alternative are, of course, completely normal. The disease is confined almost entirely to people of Italian, Greek, and Syrian origin.

**Hemophilia**    In hemophilia, the coagulation time of blood is excessively prolonged. Under normal conditions, blood coagulates within approximately 5 min; in hemophiliac people, anywhere from 10 min to 24 hr are required before clotting is complete. As a consequence, hemophiliac sufferers bleed for a considerable time following a cut or injury. In addition, small hemorrhages develop spontaneously in various portions of the body. The abnormal loss of blood coupled with multiple hemorrhaging and subsequent tissue ischemia generally lead to severe anemia and pain in affected parts of the body. In this disease, as in sickle-cell anemia and thalassemia, the prognosis is poor and the life span is greatly shortened.

Several forms of hemophilia have been described. The most common types are called hemophilia A and B. Both forms prevent the conversion of prothrombin to thrombin during hemostasis. Refer to Chap. 4 for a detailed discussion of the clotting mechanism.

*Hemophilia A* is due to the lack of a specific plasma protein called *antihemophilic globulin* (AHG) or factor (AHF). AHG is produced in the liver and may be a precursor or derivative of prothrombin. In any event, platelet factor or, more accurately, the entire system of clotting based on the presence of platelet factor remains inactive. About three-fourths of the patients who suffer from hemophilia have this form of the disease.

*Hemophilia B* is caused by the lack of another plasma protein called *Christmas factor* (CF). The factor was named for the family in which this form of hemophilia was first described. Therefore, hemophilia B is also called *Christmas disease.* CF, like AHG, is required for the activation of the molecular clotting system initiated by the presence of platelet factor. About 15 percent of all hemophiliacs suffer from this form of the disease.

Both CF and AHG act in concert with the clotting substance liberated by disintegrating blood platelets rather than with thromboplastin, the clotting substance released by dying tissue cells. In other words, the intrinsic, or intravascular, clotting mechanism associated with blood platelets is inactivated by a lack of AHG or CF, but the extrinsic, or perivascular, tissue mechanism involved in the release of thromboplastin remains unaffected.

Both hemophilias A and B are inherited as sex-linked recessive defects. This means that the defective gene for each form of hemophilia resides on the X chromosome. Presumably, the different types of hemophilia are caused by different mutations at the same gene locus on the X chromosome, giving rise to multiple alleles. In accordance with the mechanism of sex-linked inheritance described previously (see Fig. 21-4), males with a defective gene ($\underline{X}$Y) and females with two of the defective genes ($\underline{XX}$) will exhibit the disease, females with one defective gene ($\underline{X}$X) are carriers, and individuals with the remaining combinations of the allelic pair (XX and XY) are completely normal.

The reasons so few women suffer from hemophilia compared to men become readily understood when we remember that (1) those patients who suffer from the disease are usually incapacitated or do not wish to have children; and (2) women must possess the defective gene in double dose ($\underline{XX}$) to show the symptoms of hemophilia. The first point is important in reducing the incidence of the abnormal phenotype in the general population. The second point is significant in a statistical sense, since a female hemophiliac ($\underline{XX}$) must either have had a hemophiliac father ($\underline{X}$Y) and a carrier mother ($\underline{X}$X), or both parents must have suffered from the disease. In either case, such marriages are extremely rare, although they do indeed occur.

## SKELETOMUSCULAR DISEASES

Several heritable diseases affect the skeleton and muscles. Among such disorders are gout and muscular dystrophy.

**Gout** This metabolic imbalance is caused by an abnormality in the metabolism of purines, giving rise to an excessive concentration of uric acid in the blood, a condition referred to as *hyperuricemia.* Although some of the excess uric acid is excreted, most is deposited as crystalline sodium urate in and around the cartilages of the joints. Joints associated with the lower extremities in general and the big toes in particular are most frequently affected, although other parts of the body may also be afflicted.

The symptoms of gout include joint inflammation and nocturnal paroxysms of pain indistinguishable from that of arthritis. Therefore, the arthritic symptoms must be correlated with hyperuricemia before a positive diagnosis of gout can be made.

Gout is inherited as a simple autosomal dominant trait. However, many people who have the defective gene, and are therefore either heterozygous or homozygous dominant, do not exhibit symptoms of the disease. This situation arises because hyperuricemia must reach a critical threshold in order for gout to develop. The threshold is in fact exceeded in only about 10 percent of the individuals who exhibit an excess of blood uric acid. Furthermore, since men typically have a higher uric acid concentration than women, more men than women with the defective gene will develop gout by surpassing the threshold for hyperuricemia.

**Muscular dystrophy** Numerous types of muscular dystrophy (MD) have been described, and several possess or strongly suggest a heritable basis. From a genetic standpoint, the most well-understood form of muscular dystrophy is the pseudohypertrophic type.

Pseudohypertrophic muscular dystrophy, like other forms of the disease, results in a wasting away of the skeletal muscles. The disease begins early in life, commonly around the third year after birth. Initially, the muscles of the thighs and forearms hypertrophy because of an infiltration of fat and fibrous connective tissue presumably because of a defect in muscular metabolism. This aberration in muscular volume is the origin of the term *pseudohypertrophy*. As the muscles enlarge in volume, they become soft and pliable. Pseudohypertrophy leads to an increasing degeneration and weakness in the muscles of the thigh, pelvic girdle, back, and shoulder girdle. Muscular atrophy and corresponding incoordination finally result in an inability to adequately and accurately perform most kinds of voluntary muscular activity, including walking and standing erect. These deficits are reflected in the second decade of life by an erratic and rubbery gait, muscular contractures, spinal curvatures such as lordosis, and related respiratory difficulties.

Pseudohypertrophic muscular dystrophy is inherited as a sex-linked recessive defect. Although girls may theoretically exhibit the disease, in actuality there are no documented records of a female who has suffered from this affliction. The reason is that males with the disease have been either unwilling or unable to reproduce or have died before reaching reproductive age. Recall from our earlier discussions that the father of a daughter with a sex-linked recessive disease must have had the disease and the mother must either have been a carrier or also possessed the disease. The lack of reproductive competence among males with the disease is therefore the limiting factor.

## GLANDULAR DYSFUNCTION

Cystic fibrosis This disorder differentially affects the exocrine glands of the body during childhood and adolescence. Among the secretory structures attacked are the pancreas, liver, and eccrine sweat glands. The subsequent alterations in the volume and content of mucus, enzymes, and other exocrine secretions lead to pancreatic insufficiency, cirrhosis, and imbalances in the blood concentrations of specific electrolytes. These primary disturbances may in turn result in respiratory and excretory dysfunctions.

## HEREDITARY PREDISPOSITIONS

Hereditary predispositions are genetically determined tendencies toward nonhereditary diseases contracted during prenatal or postnatal development. Such diseases are ordinarily caused by nutritional deficiencies, microbial pathogens, allergens, and idiopathic factors as reflected, for example, in various mental disturbances. Although no one is completely immune to these general categories of disease, various families have a higher incidence of certain nonhereditary diseases, compared to the incidence of these diseases in the general population. Although environmental factors, cultural practices, and socioeconomic status may be solely responsible for, or major factors contributing to, the development of disease states in some families, in others, family histories or human pedigrees of the occurrence of particular diseases strongly suggest a hereditary component. Indeed, through the study of selected human pedigrees, mechanisms of inheritance of numerous predispositions have either been worked out or postulated. Sometimes, hereditary predispositions are strongly suspected, but either the familial data are too sketchy or the forms of inheritance may be too complex to permit definitive analyses of their hereditary basis.

Numerous examples abound of nonhereditary diseases in which a definite hereditary predisposition exists in given families. Some of these diseases and the hereditary mechanisms behind the predispositions are listed in Table 21-1.

The inheritance of a predisposition toward disease is relatively simple, with two predominant mechanisms: (1) predispositions determined by autosomal dominant genes toward a wide range of physical and mental diseases including cataract, rickets, arthritis, schizophrenia, manic-depressive psychosis, hemorrhoids, hypertension, diabetes mellitus, and allergies;

**Table 21-1   Hereditary predispositions to disease**

| Disease | Genetic predisposition |
| --- | --- |
| Cataract | Autosomal dominant |
| Rickets | Autosomal dominant |
| Arthritis | Autosomal dominant |
| Schizophrenia | Autosomal dominant (?) |
| Manic-depressive psychosis | Autosomal dominant |
| Hemorrhoids | Autosomal dominant |
| Hypertension | Autosomal dominant |
| Diabetes mellitus | Autosomal dominant |
| Allergy | Autosomal dominant |
| Poliomyelitis | Autosomal recessive |
| Diphtheria | Autosomal recessive |
| Scarlet fever | Autosomal recessive |
| Tuberculosis | Autosomal recessive |

and (2) predispositions determined by autosomal recessive genes toward infectious diseases such as poliomyelitis, diphtheria, scarlet fever, and tuberculosis. The list is incomplete, and there are undoubtedly other hereditary mechanisms which underlie susceptibilities to nonhereditary diseases. However, the above listing does serve to emphasize the wide variety of disease states in which a hereditary predisposition has been implicated.

## DEVELOPMENTAL ANOMALIES

Recall that developmental anomalies are malformations which occur during prenatal stages. Developmental anomalies are caused by either hereditary defects or adverse environmental influences. Hereditary mechanisms which result in developmental malformations have already been considered. Environmental factors known to disrupt the normal unfolding of prenatal organogenesis include deficiencies of oxygen and nutrients, the presence of specific drugs and infectious microbes, excesses or deficits of maternal hormones, and exposure to ionizing radiation. Factors such as drugs, microbial pathogens, and radiation are *exogenous* since they are foreign to the mother. Physiologic influences such as the concentration of nutrients, oxygen, and hormones are termed

*endogenous* because these factors determine, or are determined by, the metabolism of the mother.

The effects of some influences, including drugs, microbes, and hormones, are time-related, i.e., the presence and magnitude of the damage are determined by the stage of pregnancy during which the mother is exposed to the presence of potentially pathogenic agents or the lack of essential natural products. This means that morphogenetic processes affected by pathogenic agents or natural products are more vulnerable to alteration at one specific time, compared to any other time during development. In other cases, however, certain factors such as radiation or deficiencies of oxygen and nutrients have detrimental effects during most or all stages of prenatal development.

All developmental anomalies are traceable to a failure in the normal progression of morphogenesis. Parts fail to fuse, disappear, migrate, grow or mature, or an asymmetrical structure is displaced from its normal position during development. Among the most frequent and better-known anomalies are (1) facial defects such as cleft lip and cleft palate (discussed earlier under "Hereditary Diseases"); (2) congenital heart diseases including a patent foramen ovale, a patent ductus arteriosus, and the tetralogy of Fallot; and (3) other structural defects such as cryptorchism, spina bifida, and clubfoot.

### CONGENITAL HEART DISEASES

Congenital heart diseases affect the integrity of the heart and the surrounding great blood vessels. Congenital cardiac defects usually result in arterial hypoxia and a corresponding cyanosis in the newborn infant. The cyanotic condition causes a bluish pallor from which the term "blue baby" is derived. In fact, all forms of congenital heart disease give rise to the blue baby condition.

Patent foramen ovale   The foramen ovale is a normal embryonic opening in the interatrial septum between right and left atria (see Chap. 19). Failure of the heart to seal off the foramen ovale a reasonable time after birth gives rise to an anomaly called a patent foramen ovale. Patency means that the structure is still evident. Under these circumstances, some oxygenated blood from the left atrium passes through the foramen ovale into the right atrium, thereby decreasing both the stroke volume of the heart and the amount of oxygenated blood provided to the systemic portion of the arterial circulatory tree.

Arterial hypoxia results in dyspnea and cyanosis. The bluish pallor of the cyanotic condition is, of course, a readily detectable sign. Sometimes, however, a subject with a patent foramen ovale may remain undetected through some or all of life, especially when the life style is mostly sedentary and the defect is nominal.

Patent ductus arteriosus  The ductus arteriosus is a normal embryonic blood vessel located between the pulmonary artery and the arch of the aorta (see Chap. 19). The postnatal failure of the ductus arteriosus to close completely is referred to as a patent ductus arteriosus. Following the first breath, the aortic pressure becomes greater than the pulmonary pressure and continues to increase measurably during the first few months postpartum. When a baby has a patent ductus arteriosus, some blood is forced back from the arch of the aorta into the pulmonary artery during ventricular systole of each cardiac cycle. The amount of blood regurgitated is determined by the pressure differential between the two arteries. The increased volume of blood within the pulmonary artery due to regurgitation elevates the pulmonic blood pressure. The enlarging volume of blood within the lungs is reflected by an elevated pulmonary venous return to the left atrium. The elevated left atrial volume increases the left ventricular volume. As a consequence of aortic backflow into the ductus arteriosus, the left ventricle undergoes compensatory hypertrophy in parallel with the development of arterial hypoxia and cyanosis. The increase in pulmonary arterial pressure is also reflected by both right ventricular hypertrophy and pulmonary congestion and subsequent edema.

The altered hemodynamics are potentially serious, but the condition can be corrected by surgical ligation of the ductus arteriosus.

Tetralogy of Fallot  This disorder is caused by a persistent opening in the interventricular septum between the right and left ventricles. Called the interventricular foramen, this opening ordinarily closes during fetal development. When it does not fuse, the opening of the aorta is located close to or within the interventricular foramen. This juxtaposition between the aortic opening and the interventricular septum gives rise to an apparent shift of the aorta to the right. In addition, the semilunar valve of the pulmonary artery or the vessel itself undergoes a narrowing, or stenosis, which increases the resistance of blood flow to the lungs. As a consequence, the right ventricular pressure overcomes the pressure within the left ventricle during ventricular systole, forcing blood to enter the aorta by passing through the interventricular septum. Therefore, pulmonary stenosis and the interventricular septal defect together cause a shunting of blood from the right ventricle to the base of the aorta within the left ventricle. These conditions cause extreme arterial hypoxia, cyanosis, and a compensatory hypertrophy of the right ventricle. The tetralogy of Fallot is in fact responsible for most children born with the blue baby condition.

The abnormality is corrected either by direct rectification of the cardiac malformations during open-heart surgery or by indirect relief through the Blalock-Taussig operation in which the aorta or subclavian artery is connected to the pulmonary artery. The latter procedure returns to the lungs some of the partially oxygenated blood already within the systemic portion of the arterial circulation, permitting a greater degree of oxygenation of arterial blood than would otherwise be the case.

## OTHER STRUCTURAL DEFECTS

A number of other congenital malformations are not uncommonly seen. Such anomalies include cryptorchism, spina bifida, and clubfoot.

Cryptorchism  Cryptorchism is due to the failure of the testes to descend from the abdominal cavity into the scrotal sacs. Ordinarily, descent is accomplished shortly before birth.

Spermatozoa develop normally at temperatures within the scrotum, but spermatogenesis is inhibited at temperatures which' prevail inside the abdomen. Usually, there is a 4°C difference in temperature tween the two regions, with the scrotum at a lower temperature than the abdomen.

If the testes remain undescended following birth, then the inability to form sex cells leads to sterility when puberty is reached. Cryptorchism is usually corrected by a simple surgical procedure called *cryptorchidectomy* in which the testes are mechanically relocated to the scrotum after their ligamentous connections within the abdominal cavity are loosened.

Cryptorchism is shown in Fig. 21-17. Only the left abdominal region is exposed. Notice that the peritoneal membrane attached to the inferior end of the testes extends from the abdominal cavity through the inguinal canal into the scrotal sac. Ordinarily this membrane, called the *gubernaculum testis,* does not grow at the same rate as the developing testis to which it is attached. The growth differential permits the gubernaculum to apply traction to the inferior end of each testis as they elongate in the same direction. When the differential rates of growth of each gubernaculum and testis are not maintained, the testes assume anomalous positions in the abdominal cavity. Therefore, the cause of cryptorchism is related to abnormalities in the growth rate of each gubernaculum, compared to that of the testes.

**Spina bifida**  This defect is caused by a failure of the laminae of one or more vertebrae in the lumbosacral region to fuse. Therefore, the posterior wall of a portion of the vertebral column in the abdominal region is lacking and exposes the meninges and spinal cord within. Three different forms of spina bifida are recognized: (1) simple exposure of the meninges but no herniation of either the meninges or spinal cord; (2) herniation of a portion of the meninges; and (3) herniation of a portion of the meninges and spinal cord, with or without nerve roots and spinal nerves.

Although the symptoms of spina bifida are diverse, complex, and potentially serious, the defect in some cases is small and only shows up on routine x-ray or at autopsy. The following are some of the more important symptoms of this malformation:

1  The presence of excessive hair, excessive subcutaneous fat deposits, or skin dimpling in the posterior lumbosacral region
2  The existence of a palpable tumor in the posterior lumbosacral region
3  Central nervous system abnormalities including atrophy of muscles of the lower extremities, loss of tendon reflexes, or complete loss of sensory and motor functions below the affected region of the spinal cord
4  An accumulation of cerebrospinal fluid (hydrorrhachis) in the posterior lumbosacral region

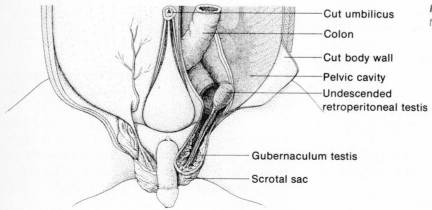

Cut umbilicus
Colon
Cut body wall
Pelvic cavity
Undescended retroperitoneal testis
Gubernaculum testis
Scrotal sac

**Figure 21-17**  *A male with cryptorchism.*

5   The presence of other commonly associated congenital defects such as clubfoot and spinal curvatures

The exposed meninges also greatly increase the threat of microbial infection and the development of meningitis.

The vertebrae are normally ossified by the third month of prenatal development. A number of centers of ossification are present in each vertebra. Of these, the lateral ossification center gives rise to the adult laminae. Spina bifida results when some of the lateral ossification centers do not ossify in accordance with the species-specific developmental plan determined by the genes. Spina bifida may have a heritable predisposition, but the disease is not believed to be caused directly by a hereditary defect.

Clubfoot (Talipes)   Talipes refers to a number of deformities of the foot which cause an abnormality in the shape of the arch or the positioning of the foot in one or more planes in space. The following are the most common forms of clubfoot:

1   Talipes varus, in which pressure during walking is differentially placed on the outside of the heel of the foot due to its inversion or adduction as well as rotation
2   Talipes valgus, in which pressure during walking is differentially placed on the medial aspect of the heel of the foot due to its eversion or abduction as well as rotation
3   Talipes calcaneus, in which pressure during walking is differentially applied to the posteromedial aspect of the heel of the foot due to its flexion and eversion
4   Talipes equinus, in which pressure during walking is differentially applied to the toes of the foot due to plantar extension
5   Talipes arcuatus (cavus), in which the plantar curvature or arch of the foot is abnormally exaggerated

In addition, one form of talipes closely parallels the symptoms of flatfoot.

Although most or all of the deformities listed above are congenital, several can also be caused by infectious diseases which involve the peripheral nerves, tendons, and connective tissues of the lower extremities. Most congenital forms of talipes are due to irregularities in the length of one or more tendons in the shank of the leg and in the amount of fascia in the heel or sole of the foot.

## CONGENITAL DEFECTS CAUSED BY EXOGENOUS FACTORS

Numerous influences arising outside the body of the pregnant mother are capable of causing developmental anomalies. Such exogenous factors are placed in the following categories: (1) microbial pathogens, such as specific viruses, bacteria, and protozoa, which are capable of causing congenital infections when contracted by mothers at specific times during pregnancy; (2) addictive drugs, such as heroin, methadone, codeine, and morphine, which cause congenital drug addiction when abused by pregnant mothers; (3) teratogenic drugs, such as thalidomide, which may cause severe malformations, especially of the limbs and head, when taken during embryonic or fetal development; and (4) ionizing radiations, such as gamma rays and x-rays, which damage the chromosomes of cells exposed to them and cause malformations, functional abnormalities, precocious aging, and a shortened life span.

### CONGENITAL INFECTIONS

Numerous infections contracted by a pregnant mother can affect the development of her embryo or fetus. The microbial pathogens responsible for congenital infections are either small enough to pass from the maternal bloodstream across the placental barrier into the umbilical vein of the child in utero or can spread to the embryo or fetus from their initial loci of infection on or in the body of the mother. Such babies may be born with both developmental anomalies and the infections contracted initially by their mothers.

In many cases, the infections are amenable to treatment with antibiotics and other types of antimicrobial agents. Damage may occur to the child in utero in the absence of antimicrobial treatment. The frequency and degree of damage are determined in part by the nature, degree of infectivity, and virulence of the microbe; the resistance and overall physiologic status of the mother; the range of time during pregnancy during which the maternal disease runs its course; and the promptness and effectiveness with which the maternal infection is treated medically.

Rubella   German measles is an acute but mild infectious disease caused by the rubella virus. The symptoms of this disease include the appearance of a slight fever, a sore throat, swollen lymph nodes, sleepiness, and reddish skin rashes or eruptions. The disease often runs its course without being recognized as other than a common cold. Its seriousness lies in the fact that when contracted by a mother during the second through the fourth month of pregnancy, the virus is capable of causing a number of developmental anomalies. The reason for the differential susceptibility of the embryo and early fetus to the rubella virus seems to reside in the ease with which the virus attacks embryonic tissue cells, compared to the mild damage caused to fully matured cells. Among the organs specifically damaged are the formative eyes, ears, brain, and heart. Commonly, rubella babies display congenital cataract as evidence of their prenatal affliction.

There are now rubella vaccines which prevent infection. To prevent pregnant women from contracting rubella, vaccinations are given to preschool children, to mothers immediately after they have given birth, and to women of childbearing age who will not conceive for at least 3 months following vaccination.

Gonorrhea   Gonorrhea is an infectious disease of the mucous membranes, especially in the urogenital regions, caused by the gonococcus *Neisseria gonorrhoeae*. If the infection persists for longer than 4 months, the disease usually becomes chronic, and subsequent damage to the reproductive organs and accessory glands may culminate in sterility.

If the mother harbors the infection (often latently),

her baby may contract congenital gonorrhea during the birth process. Usually, *N. gonorrhoeae* contaminates the cervix of the uterus and infects the conjunctiva of the eyes of a baby as its head passes through the dilatated cervix to enter the birth canal. This form of eye infection is referred to as *gonorrheal* or *gonococcal ophthalmia neonatorum*. The name is derived from the type of microbial pathogen, the structure differentially attacked, and the stage of the life cycle in which the disease occurs.

Prophylaxis against this form of congenital venereal disease is accomplished routinely by adding one drop of a 1% solution of silver nitrate or an appropriate concentration of an antibiotic such as penicillin to both eyes of every newborn baby. In the absence of such treatment, babies with undetected gonorrheal ophthalmia neonatorum run a high probability of becoming blind during infancy. In fact, at an earlier time in history, this disease was the leading cause of blindness in the early neonate.

Syphilis   Syphilis is another venereal disease; i.e., the causative agent is ordinarily contracted during sexual intercourse with a diseased person. The spirochete *Treponema pallidum* is the causative agent of syphilis. Unlike *N. gonorrhoeae*, *T. pallidum* can attack any organ or tissue of the body, including components of the circulatory and nervous systems.

The spirochetes of a syphilitic mother pass the placental barrier during pregnancy to infect the embryo or fetus. One of three fates awaits subjects infected during prenatal development: (1) miscarriage or stillbirth, which abruptly terminates pregnancy; (2) congenital syphilis of the newborn baby, with symptoms ranging from sores and rashes to chronic nasal discharges; and (3) latent congenital syphilis, in which the early neonate does not initially show any overt symptoms of the disease. In the latter case, the infection contracted congenitally crops up at some later time in life. Symptoms of latent syphilis include growth retardation, sensory deficiencies, and mental retardation.

Like gonorrhea, syphilis is amenable to penicillin therapy. Not infrequently, however, an adult subject with syphilis also suffers from gonorrhea. Since the symptoms of gonorrhea precede those of syphilis by several weeks when both microbial pathogens are contracted simultaneously, such patients may be treated for gonorrhea but not for syphilis. To avoid this error, patients with a positive diagnosis of gonorrhea either receive serologic tests for the presence of *T. pallidum* or are also treated routinely for syphilis.

**Toxoplasmosis**  This infectious disease is caused by the protozoan *Toxoplasma gondii. Toxoplasma* does not possess active methods of locomotion because it lacks cilia or flagella. Toxoplasmosis is mild in adults, presenting symptoms similar to those of a common cold. However, during pregnancy, *T. gondii* may cause serious damage to the brain and eyes of the developing child.

The following modes of transmission and portals of entry of *T. gondii* are known: (1) inspiration of airborne protozoa while cleaning cat litterboxes or similar materials containing cat feces contaminated with the protozoa; (2) entry through the skin while bitten by a contaminated arthropod vector; and (3) ingestion of rare or raw meat containing the protozoa. Apparently birds also suffer a high incidence of toxoplasmosis, forming a natural reservoir for the protozoa in nature.

Toxoplasmosis is probably a much more common human disease than is suggested by the statistics on its frequency of occurrence. Often the disease is either misdiagnosed or remains undiagnosed in adults. In such cases, the only substantive traces left in the aftermath of its clinical course in pregnant mothers are various developmental anomalies involving the eyes and brains of their babies at birth. Frequently, such abnormalities are ascribed to other causal factors because *T. gondii* is not looked for or suspected. Even when an examination for this protozoan is performed, the results may be negative. Negative results may only be indicative of a low infectivity rather than a complete absence of this microbial pathogen in the suspected human host.

## CONGENITAL DRUG ADDICTION

Among the addictive chemicals abused by human beings are the so-called hard drugs derived from opium, including morphine, codeine, and heroin. Some characteristics and effects of these and other addictive and habituative drugs are discussed in Chap. 1.

Pregnant mothers who are drug addicts frequently give birth to babies addicted to the same drug. Within the constraints imposed by their limited physical and mental maturation, addicted babies show withdrawal symptoms in the absence of the drug.

## TERATOGENIC DRUGS

Aside from being "hooked," babies suffering from congenital drug addiction exhibit an unusually high incidence of severe congenital malformations, especially of the limbs and head. Since agents which cause these grotesque deformities during prenatal development are called *teratogenic,* addictive drugs and other chemicals which act similarly are also teratogens.

**Lysergic acid diethylamide (LSD)**  One of the hallucinogens, lysergic acid diethylamide (LSD), has frequently been suggested as an agent which causes chromosome damage and subsequent malformations. The mutagenicity of LSD has been cited as the factor responsible for the apparent higher-than-normal incidence of congenital malformations among children of mothers who used the drug during their pregnancies. Recent evidence, however, indicates that LSD is a rather weak mutagen, acting only in extremely high doses. This assessment of the mutagenic effects of LSD implies that the hallucinogen is unlikely to cause chromosome damage in the range of concentrations used by human subjects. The apparent relationship between LSD and congenital malformations most likely stems from the cumulative action of contaminant drugs mixed with LSD to make it less expensive when sold in illicit drug markets of the street.

**Thalidomide**  One of the best known of the chemical teratogens is alpha-(*N*-phthalimido) glutarimide, a substance also called *thalidomide.* This chemical was widely used as a sedative in the early part of the 1960s. Thalidomide was banned in the United States when an abnormally high incidence of European babies born with rudimentary limbs was traced to the use of this drug by mothers during their first trimester. The effects of this chemical persist to this day, since "thalidomide" babies have been required to make profound adjustments to partially extricate themselves from their limbless and physically grotesque disabilities. Their extreme physical and mental suffering and the intolerable medical costs incurred during rehabilitation are part of the societal price extracted for the excessive use of chemicals by people for unsound reasons; the profit motive usually pursued by pharmaceutical industries at the expense of scientifically valid, long-term studies of drug safety and efficacy; and the relatively ineffective screening and regulation of pharmaceuticals at the federal level. All must accept part of the blame for this kind of disaster. Unfortunately, these patterns of conduct have not been altered substantially since the thalidomide episode, suggesting that other equally dangerous chemical teratogens will turn up from time to time after inflicting their damage on completely defenseless children in utero.

## IONIZING RADIATIONS

Most of our knowledge of the effects of radiation during prenatal development and on neonatal viability have come from the following sources: (1) radiation experiments on laboratory mice and other animals; (2) extensive studies of human beings exposed to gamma and neutron radiation in the Hiroshima and Nagasaki atomic explosions during World War II as well as in industrial nuclear accidents; (3) the birth records of mothers exposed to diagnostic or therapeutic x-rays during pregnancy; and (4) the fragmentary records of children born to mothers accidentally exposed to moderate or high levels of radioactivity during pregnancy. Interpretations of the data must be tempered by the realization that the findings are controversial, especially when extrapolated from laboratory mice to pregnant women and their children in utero. These constraints, however, in no way detract from the main thrust of the argument presented below.

**X- and gamma rays** ionizing radiations, in many cases at surprisingly low levels, cause increases in the frequencies of prenatal mortality and develop-

mental anomalies, compared to the average in the absence of radiation above that provided by natural background radiation. The nature of the damage is determined by the type of radiation, the total dosage, the length of exposure, portions of the body exposed, physiologic status of the subject and, most importantly for our purposes here, the exact time when exposure occurred prenatally.

Three qualitatively different effects are noted during prenatal development: (1) *prenatal mortality* through spontaneous abortion if the early embryo is exposed to radiation prior to implantation; (2) *gross structural abnormalities* when the embryo is exposed to radiation in the period of major organogenesis; and (3) *radioresistance,* except for the nervous and reproductive systems, when the fetus is exposed to radiation during the major part of its growth period following organogenesis. If the fetus exposed to radiation in utero survives to birth, then the newborn child has a significantly higher-than-average probability of dying during infancy.

Developmental anomalies recorded among children born to survivors of Hiroshima and Nagasaki include microcephaly, or diminished brain size, and a corresponding degree of mental retardation as well as bone dislocations, heart disease, and mongolism.

## DISORDERS OF BIRTH

We will broaden the definition of birth disorders to include the events which occur during the birth process and shortened periods of gestation. These disorders are not congenital.

## PREMATURITY

Prematurity results when prenatal development is incomplete at birth. The premature neonate shows physical signs of organ system immaturity and physiologic symptoms of homeostatic instability. Such babies possess abnormally small and sometimes malformed organs, and their regulatory functions are either depressed or erratic.

The above definition of prematurity is functional. In practice, prematurity is tentatively applied to infants who weigh 5½ lb (2,500 Gm) or less at birth. Other criteria, such as structural measurements, gestational time if known, physical integrity and completeness, and reflex competence and the ability to cope physically, are also considered when determining whether the infant is in fact premature or simply of low birth weight. In other words, not all babies who qualify by weight are actually premature.

Because of structural immaturity and functional instability, the premature infant is in potential danger of succumbing to numerous disorders which may develop. Special care must therefore be instituted to meet the baby's needs on a continuing basis. The amount of danger and the type of care are determined by the degree of disability.

The vicious circle associated with premature birth is presented in Fig. 21-8. The circle is initiated when birth is hastened by either a premature rupture of the amniotic sac, physical trauma in the pelvic region, multiple carriage, microbial infection of the reproductive tract, or obstetrical intervention for other reasons. The various deficits of the premature baby combine to create a vicious circle of deterioration, as reflected in the following possible disorders and their causes:

1 Indigestion and possible undernutrition due to underdevelopment of the stomach and digestive reflexes

2 Malnutrition because of an inability to absorb fatty acids and calcium from the intestine

3 Hypoproteinemic edema because of an inability of the underdeveloped liver to synthesize many proteins normally found in blood

4 Respiratory distress caused by underdeveloped lungs and a correspondingly low vital capacity and functional residual volume

5 Depression in hemostasis because the underdeveloped liver does not synthesize an adequate supply of intrinsic blood factors necessary for coagulation and clotting

6 Aplastic anemia through insufficient hemopoietic activity because of an underdevelopment of erythrocyte-forming centers, deficiencies of essential nutrients, and imbalances in metabolism

7 Decrease in resistance and a correspondingly elevated susceptibility to infection because of an inadequate storage of maternal antibodies and an inability to synthesize endogenous antibodies in effective concentrations

8 Hypothermia caused by an excessive loss of heat from the body because of an extremely high basal metabolic rate, an unusually large surface area of the body compared to its mass, thin skin, a lack of a sufficient layer of subcutaneous fat for insulation, and a general instability in the operation of the hypothalamic control mechanism for body temperature

9 Electrolyte and fluid imbalances caused by an inability of the underdeveloped kidneys to function normally

Care of the premature infant revolves around therapy for prevailing dysfunctions, prophylaxis of potential disorders, and promotion of its physical and psychological well-being. The techniques employed in prevention and treatment of the real and potential disabilities caused by prematurity require some combination of the following actions:

1 Forced feeding by means of a stomach tube, thereby avoiding both indigestion through the possible malfunction of digestive reflexes and excessive tiredness because of the necessity for frequent feeding sessions

2 Dietary control of food intake including a low fat diet to avoid malabsorption of fats, the administration of vitamin D and ultraviolet radiation to prevent the development of rickets, and calcium supplements to ensure against $Ca^{++}$-deficiency rickets and hypocalcemic tetany

3 Ventilatory assists including mild oxygenation, humidification, and gentle suctioning to improve the efficiency of external respiration and relieve respiratory distress

4 Asepsis to prevent the development of secondary infection by counteracting the depression in the specific immunologic response to foreign proteins

5 The use of an incubator to stabilize body temperature, thereby preventing the development of hypothermia

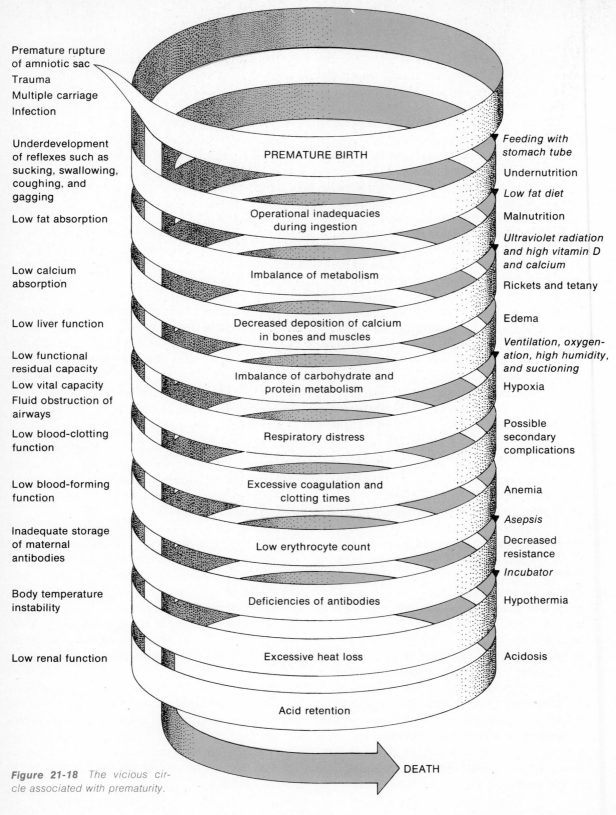

Premature rupture of amniotic sac
Trauma
Multiple carriage
Infection

Underdevelopment of reflexes such as sucking, swallowing, coughing, and gagging

Low fat absorption

Low calcium absorption

Low liver function

Low functional residual capacity
Low vital capacity
Fluid obstruction of airways

Low blood-clotting function

Low blood-forming function

Inadequate storage of maternal antibodies

Body temperature instability

Low renal function

PREMATURE BIRTH

Operational inadequacies during ingestion

Imbalance of metabolism

Decreased deposition of calcium in bones and muscles

Imbalance of carbohydrate and protein metabolism

Respiratory distress

Excessive coagulation and clotting times

Low erythrocyte count

Deficiencies of antibodies

Excessive heat loss

Acid retention

*Feeding with stomach tube*
Undernutrition
*Low fat diet*
Malnutrition
*Ultraviolet radiation and high vitamin D and calcium*
Rickets and tetany
Edema
*Ventilation, oxygenation, high humidity, and suctioning*
Hypoxia
Possible secondary complications
Anemia
*Asepsis*
Decreased resistance
*Incubator*
Hypothermia
Acidosis

DEATH

**Figure 21-18** *The vicious circle associated with prematurity.*

530

The premature baby may be placed in a rocker-type incubator to promote the normal development of the cerebellum.

## BIRTH INJURIES

Injuries sustained during birth of full-term babies are caused by abnormal pressures developed during obstetrical procedures and manipulations or by "bearing down" unduly during labor and delivery. Since babies are usually presented headfirst, such pressures differentially affect the skull, brain, and associated tissues in the cranial region. Frequently, peripheral nerve damage in the cervical and thoracic regions also occurs (see Chap. 16). Excluding the effects of nutritional deficiencies, most cases of infant mortality are the result of birth injuries.

The following types of birth injury are frequently seen: (1) tearing of the meninges or hemorrhaging of the subarachnoid blood vessels, or both; and (2) depressed skull fractures and brain damage due to cerebral lacerations or cortical hemorrhaging. When hemorrhaging occurs, the cerebrospinal fluid (CSF) in the subarachnoid space and within the brain ventricles may show evidence of blood. However, bloody CSF is also common in healthy, full-term babies and cannot by itself be taken as proof of serious birth injury.

## COLLATERAL READING

Bearn, A. G., and James L. German, III, "Chromosomes and Disease," *Scientific American,* November 1961. Various mechanisms are described by which chromosomes become aberrant.

Dishotsky, Norman I., William D. Loughman, Robert E. Mogar, and Wendell R. Lipscomb, "LSD and Genetic Damage," *Science,* 172:431–440, 1971. A comprehensive review of the literature through 1970 on the mutagenicity of LSD in human beings.

Eichhorn, Mary M., "Rubella: Will Vaccination Prevent Birth Defects?" *Science,* 173:710–711, 1971. The various objections to the rubella vaccines are placed in perspective by Eichhorn.

Friedmann, Theodore, "Prenatal Diagnosis of Genetic Disease," *Scientific American*, November 1971. Describes diagnostic techniques used in the determination of heritable disorders in utero.

Kalmus, Hans, "Inherited Sense Defects," *Scientific American,* May 1952. A broad, general discussion of color blindness, tone deafness, and several other inherited sensory abnormalities, including some interesting questions about the possible relationships between specific diseases and diet.

Singleton, W. Ralph, *Elementary Genetics.* Princeton, N.J.: D. Van Nostrand Company, Inc., 1962. Chapter 19, Biochemical Genetics in Man, discusses human diseases caused by inborn errors of metabolism.

Winchester, A. M., *Genetics: A Survey of the Principles of Heredity,* 2d ed, Boston: Houghton Mifflin Company, 1958. Chapter 26, "Survey of Human Heredity," provides short descriptions of numerous human abnormalities which are either caused directly by genetic defects or precipitated indirectly by genetic predispositions.

# Part Five
# Homeostasis

# CHAPTER 22
# CONSErVATION AND EXCReTION

The urinary system forms urine from blood serum. Urine is a liquid waste material, in contrast to feces, which are ordinarily solid, or $CO_2$, which is a gas. Although the activities of the excretory system result in the formation of urine, this is not its only reason for being. In other words, urine is like the tip of an iceberg—it is the visible portion of unseen activities of vastly greater dimensions.

The urinary system is responsible for maintaining the normal chemical composition of the extracellular fluids, including the blood, lymph, and interstitial fluids. Since the chemical composition of the extracellular fluids directly affects that of the intracellular fluids, the urinary system regulates the chemical composition of all the fluid compartments of the body. In a practical sense, this is accomplished by controlling the fluid and electrolyte balance of blood.

More specifically, the urinary system participates in the following functions:

1 The excretion or retention of water when blood becomes too dilute or too concentrated, respectively

2 The excretion of vital organic compounds when their concentrations in blood are excessive, and the retention of these compounds under normal conditions

3 The excretion of excess $H^+$ and the retention of sodium bicarbonate ($Na^+HCO_3^-$) when the pH of blood falls below its normal range, or the excretion of $Na^+HCO_3^-$ to reduce the alkaline reserves of the body when the blood pH rises above its normal range

4 The maintenance of an ionic balance between extracellular $Na^+$ and intracellular $K^+$

5 The excretion of proteinaceous N, P, and S in the form of nitrogenous compounds, phosphates, and sulfates, respectively

6 The retention of $Ca^{++}$ and the excretion of $HPO_4^{--}$ when the blood $Ca^{++}$ levels are depressed, and the opposite when they are elevated

7 The excretion of substances derived from the degradation of nucleic acids.

8 The excretion of pigments formed following the destruction of hemoglobin

9 The maintenance of renal blood pressure

Collectively, these functions help to cleanse the blood and regulate its volume, solute concentrations, and acid-base balance.

The functions of the urinary system are mediated by hormonal and neural control mechanisms as well as by control devices built into the kidneys themselves. This chapter explains the mechanisms by which the urinary system serves in its dual roles of conservation and excretion.

## ORGANIZATION OF THE URINARY SYSTEM

The urinary system consists of two kidneys, two ureters, one urinary bladder, and one urethra. The kidneys can be likened to factories in which the functions of the urinary system are accomplished. The rest of the urinary system is essentially analogous to a highly sophisticated pipeline, or plumbing network, which connects the kidneys with, and transports urine to, the outside of the body.

The urinary system is located in the abdominal and pelvic regions. The anatomic relationships among the various organs of the urinary system are illustrated in Fig. 22-1.

### KIDNEYS

The kidneys are positioned against the posterior abdominal wall. Their superior surfaces lie opposite the twelfth thoracic vertebra, and their inferior surfaces are adjacent to the third lumbar vertebra. The right kidney is somewhat inferior to the left kidney because of the larger volume of space occupied by the

liver on the right side of the body, compared to the volume which it occupies on the opposite side.

Each kidney is covered on its anterior surface by a peritoneum continuous with the inside tissue lining of the abdominal wall. Therefore, the kidneys are *retroperitoneal,* i.e., in back of, or posterior to, the peritoneum. Each kidney is embedded in a substantial amount of fatty tissue which forms a fatty capsule around it. Indeed, the kidneys are supported in place primarily by their fatty capsules and the peritoneum, in addition to the support provided by connections with the ureters and the renal arteries and veins. The adrenal glands are located on the superior and medial aspects of the kidneys (see Fig. 22-1).

Blood is transported to the kidneys by the renal arteries, which originate from the abdominal aorta. Branches of the abdominal aorta also supply the adrenal gland. Blood is drained from the kidneys by the renal veins, which are also penetrated by veins from the adrenal glands. The renal veins enter the inferior vena cava.

The gross structure of the kidney, i.e., the structure visible to the naked eye, is illustrated in Fig. 22-2. Covered by a tough fibrous capsule, the kidney consists of two regions, an outer cortex and an inner medulla. The *cortex* is richly vascularized because of its role in the filtration of blood. The first part and the coiled portions of the renal tubules, or nephrons, are located in the cortex. The *medullary region* consists of from 8 to 14 inverted triangular structures referred to as *renal pyramids,* which contain the U-shaped portions of the nephrons as well as the collecting ducts which transport fully formed urine away from the renal medulla. The spaces between the pyramids are invaded by cortical tissue referred to as *renal columns.* Blood is carried to and from the cortex by vessels which course through the columns and through the region between the cortex and medulla. The detailed circulation of blood within the kidneys will be described below.

The next functional portion of the kidney is concerned with the pooling of urine from the many individual collecting ducts. Shaped like a basin fringed with flowers on its superior surface, this portion of the kidney is called the *renal pelvis.* The "flowers" are called *calyces.* Each calyx surrounds the apex of a

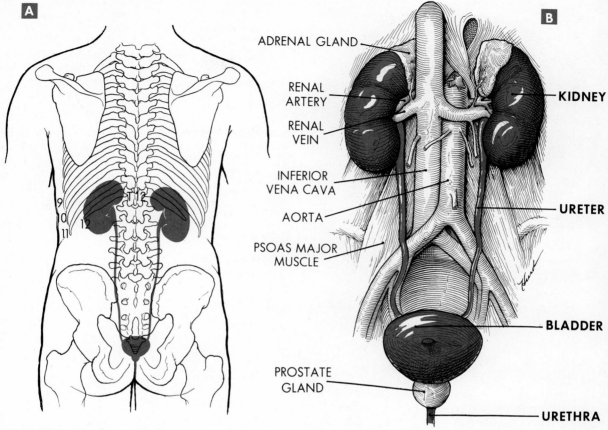

**Figure 22-1** *The gross anatomy of the urinary system. A. Posterior aspect. B. Anterior aspect.* (From Melloni, Stone, and Hurd.)

renal pyramid. Each pyramidal apex is called a *renal papilla* and opens freely into the lumen of a calyx. Since the distal ends of the collecting ducts terminate in the renal papillae, urine from the collecting ducts enters the lumina of the calyces and is in turn collected within the lumen of the pelvis. The *pelvis* exits on the medial aspect of the kidney through a renal opening called the *hilus*. The hilus also receives arteries, veins, lymphatic vessels, and nerves.

The calyces and pelvis at the base of a kidney are continuous with the *ureter,* a tube which communicates between the kidney and urinary bladder. The calyces, pelvis, and ureter are all formed embryonically in the same way and therefore have the same basic structure. The cortical and medullary por-

tions of the kidneys have a separate embryonic origin.

The *renal artery* enters the kidney through the hilus and branches in the region peripheral to the pelvis. Passing through the renal columns opposite the pyramids, these branches are called *interlobar arteries.* These arteries terminate in smaller branches, the *arcuate arteries,* which run along the bases of the pyramids at the junction between the medulla and cortex. The arcuate arteries give off still smaller branches in the cortex called *interlobular arteries.* The destination of the interlobular arteries is discussed later in this chapter, in conjunction with the blood supply associated with the nephrons.

The venous circulation runs parallel to, and is iden-

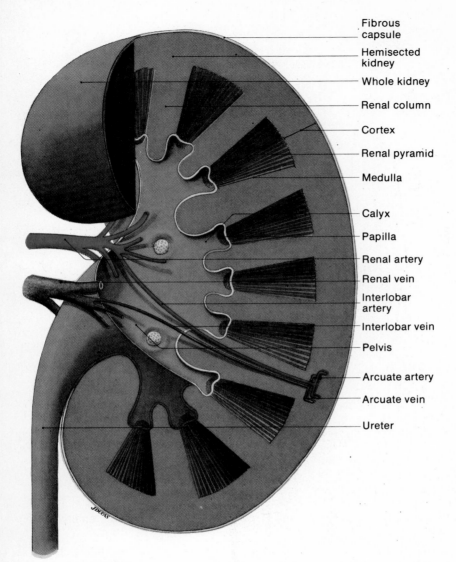

Fibrous capsule

Hemisected kidney

Whole kidney

Renal column

Cortex

Renal pyramid

Medulla

Calyx

Papilla

Renal artery

Renal vein

Interlobar artery

Interlobar vein

Pelvis

Arcuate artery

Arcuate vein

Ureter

**Figure 22-2** *Gross structure of the kidney. The left kidney is illustrated. Most of the kidney is hemisected along the midsagittal plane, but a part of the superior region is left intact. Part of the wall of the calyces and pelvis is removed to reveal the lumen.*

tical in terminology with, the arterial circulation described above.

## URETERS

The thin, elongated extension of the pelvis from each kidney is called a *ureter*. Covered anteriorly by peritoneum, the ureters transport urine from the kidneys to the urinary bladder. Their locations are shown in Fig. 22-1. Although the transport of urine is a continual process, periodic waves of peristaltic, smooth muscle contractions of their walls aid its transport. Therefore, peristalsis tends to propel the accumulated urine from the renal pelvis to the bladder in spurts. There are perhaps 5 to 10 rhythmic contractions per min. Their frequency seems to be independent of the volume of urine being transported.

## URINARY BLADDER

The gross structure of the urinary bladder is illustrated in Fig. 22-3. The bladder lies in the pelvic cavity posterior to the pubic symphysis (see Fig. 22-1). The ureters penetrate the lower bladder wall at an oblique angle on its posterior face. The ureteral orifices become covered by muscular folds when the bladder is thrown into contractions. These flaps prevent urine from entering the ureters during urination. Urine is drained from the bladder by the urethra, which extends from its neck. The opening into the urethra is called the *internal urethral orifice.*

The triangular area defined by the ureteral orifices and the internal urethral orifice, the *trigone,* is a relatively smooth region located on the internal surface of the posterior and inferior aspects of the bladder. In contrast to the trigone, the rest of the internal wall consists of pronounced muscular ridges known as *rugae,* especially when the bladder is undistended.

The tissue structure of the bladder wall has several unique features. The internal lining tissue consists of *transitional epithelium* similar to that which lines the cavities of the ureters. This type of epithelium consists of a basal layer of cuboidal cells and several other layers of cells. The most peripheral layer contains squamous cells and is therefore flattened in shape. When the bladder is distended, the transitional epithelium is stretched considerably. Only two layers of cells can be identified in the transitional epithelium of fully distended bladders—basal cuboidal cells and apical squamous cells. The stretch characteristics of the transitional epithelium are responsible for the ability of the bladder to increase greatly in volume. As the bladder fills with urine, it changes in shape from a flaccid, highly wrinkled sac to an ellipsoidal vesicle which rises appreciably in the pelvic cavity.

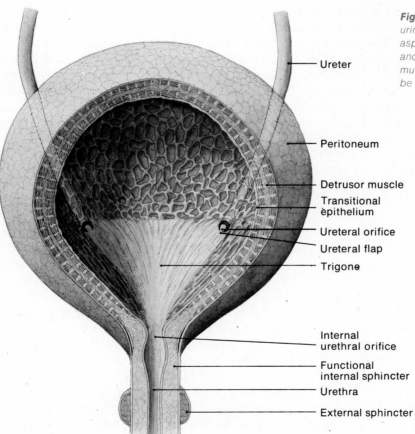

**Figure 22-3** *The gross structure of the urinary bladder, viewed from anterior aspect. The anterior wall of the bladder and urethra is removed so that the mucosa, musculature, and orifices can be observed.*

Ureter

Peritoneum

Detrusor muscle

Transitional epithelium

Ureteral orifice

Ureteral flap

Trigone

Internal urethral orifice

Functional internal sphincter

Urethra

External sphincter

A layer of connective tissue lies between the transitional epithelium and the muscles of the bladder wall. The connective tissue permits a certain degree of independence in the movements of the layers it joins. This arrangement provides a mechanism for an increase in the intravesicular volume of urine without causing a marked increase in the intravesicular fluid pressure. This characteristic plays an important role during bladder filling, a topic which we will consider further at a later time.

There are three layers of smooth muscle in the wall of the urinary bladder. The muscle of the bladder wall is called the *detrusor* and is one of the few smooth muscles in the human body which has its own name. The musculature in the trigone region, although smoother in appearance than elsewhere in the bladder, is still a portion of the detrusor.

In the neck of the bladder, the circular layer of the trigone musculature is thickened and assumes the role of a functional *internal sphincter.* Surrounding the internal urethral orifice, the inside surfaces of this thick band of circular muscle remain tightly opposed when the bladder is undistended, thereby preventing urine from entering the urethra. If the bladder contracts with enough force after filling, the hydrostatic pressure of the urine forces open the internal sphincter and permits entry of the urine into the urethra. Urination is described later in this chapter.

The outer layer of the bladder wall is composed of peritoneum.

## URETHRA

The *urethra* conducts urine from the base of the urinary bladder to the outside of the body. Its anatomic position is shown in Fig. 22-1. The opening at the distal end of the urethra is called the *external urethral orifice.* Surrounding the urethra between its proximal and distal ends is an *external sphincter* of striated muscle (see Fig. 22-3). The role of the external sphincter in urination is explained later.

There are certain anatomic differences in the urethra of the female and male. The female urethra is only about 1 in. long and is located in front of the anterior wall of the vagina, placing the external urethral orifice just anterior to the vaginal orifice. In contrast, the male urethra is about 8 in. long and is divided into three regions—the *prostatic, membranous,* and *cavernous urethrae.* The latter extends through the penis, and the external urethral orifice is located at its tip. The membranous and cavernous urethrae serve as a common passageway for both urine and spermatozoa. The divisions of the male urethra are discussed in more detail in Chap. 19.

## THE FORMATION OF URINE

Between the filtration of blood serum and the excretion of urine is a series of elegant regulatory processes that determine whether and to what extent (1) the constituents of whole blood will be filtered for processing by the kidneys; (2) the constituents of serum will be reabsorbed into the blood after filtration by the kidneys; and (3) certain constituents of serum and products of special tubular cells will be secreted by the kidneys into the urine. These three processes are called *filtration, reabsorption,* and *secretion,* respectively. The fluid which enters the renal pelvis, after processing is completed, is excreted as urine. In the following discussion, we will focus on the mechanisms by which urine forms.

### FUNCTIONAL ORGANIZATION OF THE NEPHRON

The structural unit of the kidneys is called a *nephron,* or *renal tubule.* There are about 1 million nephrons in each kidney, or about 2 million nephrons in all. The nephrons carry out the functions of the kidneys, but they cannot regulate the fluid and electrolyte balance between the fluid compartments of the body without a blood supply to filter or into which filtered constituents may be reabsorbed. Therefore, although the nephron is the structural unit of the kidneys, to be a functional unit it must be associated with blood vessels. One of these functional units is illustrated diagrammatically in Fig. 22-4.

First, let us obtain an overview of the organization of a nephron and its related blood vasculature, using Fig. 22-4 as a guide. Each nephron consists of a

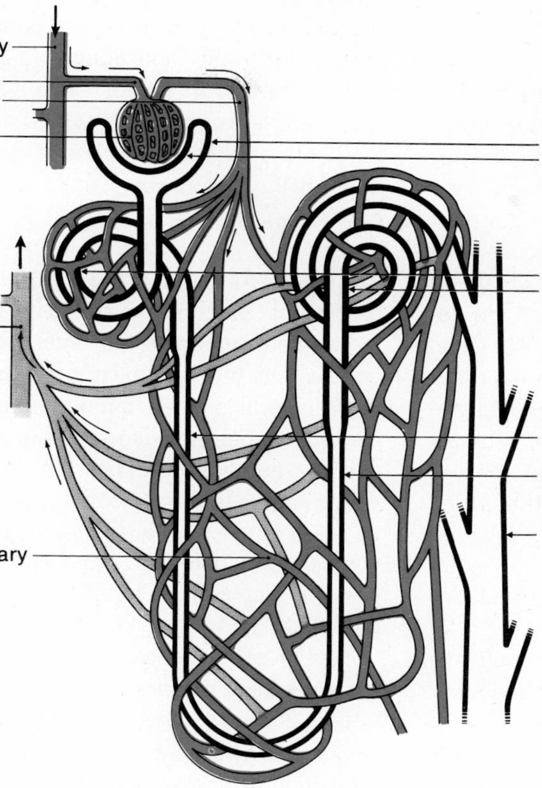

VASCULAR STRUCTURES

NEPHRIC STRUCTURES

- Interlobular artery
- Afferent arteriole
- Efferent arteriole
- Glomerulus

Parietal wall ⎫
Visceral wall ⎬ Bowman's capsule

- Proximal convoluted tubule
- Distal convoluted tubule

- Interlobular vein

Descending limb ⎫
Ascending limb ⎬ Loop of Henle

← Collecting duct

- Peritubular capillary network

**Figure 22-4** *The organization of a nephron and its related blood vessels.*

highly coiled and long tubule, the proximal and distal ends of which are in the renal cortex. The middle portion of a nephron dips either shallowly or deeply into the medullary portion of the kidney.

The proximal end of a nephron is composed of a double-walled cup called *Bowman's capsule.* The inner and outer walls are called the *visceral* and *parietal layers,* respectively. From Bowman's capsule, the tubule almost immediately becomes highly coiled and is known as the *proximal convoluted tubule.* This portion of the nephron leads into a straight, U-shaped, thin-walled tubule known as the *loop of Henle.* Its descending limb leads from the proximal convoluted tubule into the medulla, and its ascending limb leads back to the cortex, where the tubule again becomes

highly coiled. Called the *distal convoluted tubule,* this portion of the nephron leads into a *collecting duct.* Each collecting duct receives the distal ends of numerous nephrons and transports urine to the pelvic basin of the kidney.

Consisting of arterioles, capillaries, and venules, the blood supply of a nephron is highly ramifying and functionally complex. Blood is transported to each nephron through a small artery, the *afferent arteriole,* which originates from an interlobular artery, and terminates in a closely woven network of capillaries collectively referred to as a *glomerulus.* Each glomerulus resides in the well formed by the visceral layer of Bowman's capsule. A portion of the blood serum is filtered from the glomerulus into the lumen of Bowman's capsule. This *glomerular filtrate* becomes the raw material processed by the various portions of a nephron and its associated collecting duct.

Unfiltered blood within the glomerulus is transported away from Bowman's capsule by a small artery, the *efferent arteriole,* which branches into an extensive plexus of blood vessels called the *peritubular capillary network* surrounding all portions of the nephron. These vessels are contiguous with the various regions of the nephron and its associated collecting duct, forming anastomoses between the proximal and distal convoluted tubules as well as between the descending and ascending limbs of the loop of Henle. The blood in the peritubular capillary networks reabsorbs constituents from the formative urine in the nephrons and initiates the mechanisms which cause the secretion of certain constituents into the urine.

The peritubular capillary network of each nephron is drained by venules which carry the blood to an *interlobular vein.* These veins run parallel to the interlobular arteries in the cortical portions of the kidneys.

A specialized structure associated with both the nephron and its related blood supply plays an important role in the regulation of the rate of blood flow through the afferent arteriole and, therefore, the rate of formation of glomerular filtrate. Called the *juxtaglomerular apparatus,* this structure is found around the afferent arteriole next to the glomerulus as well as around a portion of the distal convoluted tubule closely opposed to the afferent arteriole. The juxtaposition of the distal convoluted tubule and the afferent arteriole is not shown in Fig. 22-4, in order to illustrate clearly the basic organization of the nephron and its associated blood vessels. How the juxtaglomerular apparatus functions will be discussed shortly.

## RENAL FUNCTION

The ways in which the nephron and its associated blood vessels interact functionally during filtration, reabsorption, and secretion are depicted in Fig. 22-5. This figure is a balance sheet which keeps track of the following changes during the formation of urine: (1) constituents of the glomerular filtrate, constituents reabsorbed from the nephrons, constituents secreted into the urine, and constituents to be excreted in the urine; (2) the flow rate of formative urine in various portions of the nephrons and their associated collecting ducts; and (3) the flow rate of blood in various portions of the blood vasculature associated with the nephrons.

Filtration    As shown in Fig. 22-5, an average of 1,200 ml of whole blood is brought to the glomeruli through the afferent arterioles. To enter Bowman's capsule, constituents of blood must pass through the walls of the glomerulus, then through a special basement membrane between the glomerulus and Bowman's capsule, and finally through the visceral wall of Bowman's capsule. This compound membrane acts as an ultrafilter, normally screening out the formed elements and virtually all proteins of the blood. Therefore, the glomerular filtrate consists of plasma minus proteins.

How much blood serum is filtered into Bowman's capsules each minute depends on the *glomerular filtration pressure* (GFP). The net GFP is 24 mmHg. Under this pressure, 125 ml of glomerular filtrate forms/min. In other words, roughly one-tenth of the total blood brought to the nephrons each minute ordinarily becomes glomerular filtrate. Since 1,200 ml of blood/min is transported to the glomeruli and 125 ml of glomerular filtrate forms in the same period of time, 1,075 ml blood/min is transported away from the glomeruli by the efferent arterioles.

The GFP depends on the balance between the glomerular and tubular pressures. The hydrostatic pressure due to the force of the blood within the afferent arteriole is 70 mmHg. However, as blood serum is filtered, the glomerular blood becomes effectively more concentrated with proteins which remain behind. The proteins impart an osmotic pressure of about 31 mmHg to the glomerulus. The *colloid osmotic pressure* of the glomerulus pulls $H_2O$ back from Bowman's capsule and therefore opposes the hydrostatic blood pressure in the afferent ar-

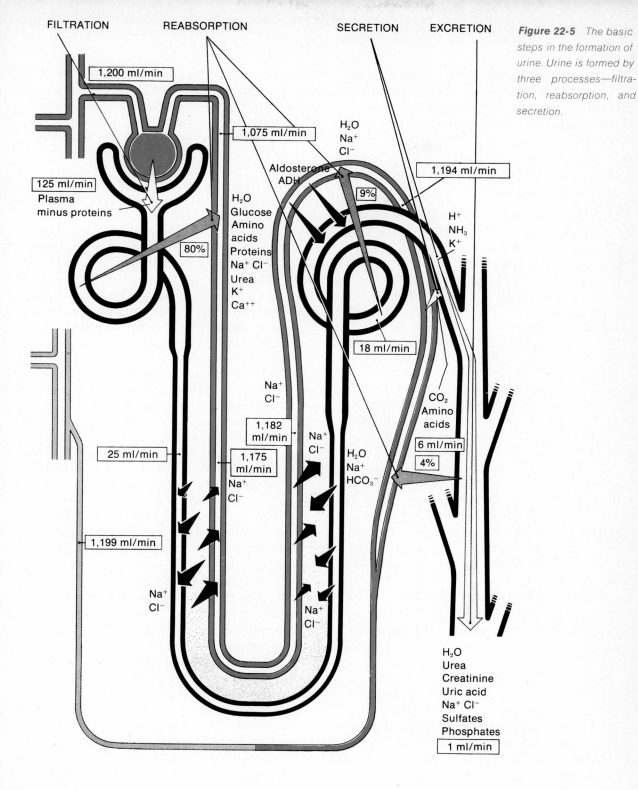

**Figure 22-5** *The basic steps in the formation of urine. Urine is formed by three processes—filtration, reabsorption, and secretion.*

FILTRATION   REABSORPTION   SECRETION   EXCRETION

1,200 ml/min

125 ml/min
Plasma
minus proteins

1,075 ml/min

$H_2O$
$Na^+$
$Cl^-$

Aldosterone
ADH

9%

1,194 ml/min

$H_2O$
Glucose
Amino
acids
Proteins
$Na^+$ $Cl^-$
Urea
$K^+$
$Ca^{++}$

80%

$H^+$
$NH_3$
$K^+$

18 ml/min

$CO_2$
Amino
acids

$Na^+$
$Cl^-$

1,182
ml/min

1,175
ml/min
$Na^+$
$Cl^-$

$Na^+$
$Cl^-$

$H_2O$
$Na^+$
$HCO_3^-$

6 ml/min

4%

25 ml/min

1,199 ml/min

$Na^+$
$Cl^-$

$Na^+$
$Cl^-$

$H_2O$
Urea
Creatinine
Uric acid
$Na^+$ $Cl^-$
Sulfates
Phosphates

1 ml/min

teriole. The hydrostatic pressure due to the presence of the tubular fluid also opposes the hydrostatic pressure of the blood. This pressure is approximately 15 mmHg. Thus, the net GFP is calculated by the following formula:

GFP = blood hydrostatic pressure − [blood osmotic pressure + tubular hydrostatic pressure]
= 70 − [31 + 15] = 24 mmHg

Any factors which alter the blood flow rate through either the afferent or efferent arterioles will affect the GFP. The normal feedback mechanism which adjusts the blood flow rate through the afferent arteriole to that of the volume or concentration of the formative urine in the distal convoluted tubule is initiated by the juxtaglomerular apparatus. In response to either an increased concentration or decreased volume of tubular fluid, specialized juxtaglomerular cells secrete the enzyme *renin,* which increases blood pressure by activating a circulating vasopressor substance. An increase in renal blood pressure elevates the blood flow rate through the afferent arterioles. Because of the operational characteristics of each nephron, a slight rise in the GFP causes a large increase in the volume of tubular fluid.

The GFP can also be modified by the action of the sympathetic division of the autonomic nervous system or by specific drugs and other chemicals which affect the diameter of blood vessels, including afferent or efferent arterioles. In general, an increase in the diameter of the afferent arterioles or a decrease in the diameter of the efferent arterioles causes an elevation in the GFP. The GFP is decreased when either of these conditions is reversed.

**Reabsorption** The glomerular filtrate becomes formative urine, or tubular fluid. Practically all the formative urine is reabsorbed from various portions of the nephron, especially from the proximal and distal convoluted tubules. The reabsorptive mechanisms include passive transport, active transport, and pinocytosis. To appreciate how these mechanisms affect the types and quantities of reabsorbed constituents, we must follow the tubular fluid as it leaves Bowman's capsule and enters the proximal convoluted tubule.

About 80 percent of the glomerular filtrate, amounting to approximately 100 ml/min, is reabsorbed from the proximal convoluted tubules. Therefore, the flow rate of blood in contiguous peritubular capillaries rises to 1,175 ml/min as the collective flow rate in this portion of the tubule is reduced to 25 ml of formative urine/min.

Among the constituents reabsorbed from the tubular fluid in the proximal convoluted tubule are $H_2O$, essential organic compounds, and electrolytes. Water is reabsorbed osmotically due to the high colloid osmotic pressure maintained in the glomerulus as a consequence of filtration. Although some of the smaller proteins with low molecular weights are filtered into the nephron, practically all the proteins remain behind in the glomerular blood. Proteins which enter the tubular fluid are reabsorbed by pinocytosis, a mechanism described in Chap. 6.

Virtually all the glucose and amino acids in the glomerular filtrate are reabsorbed by active transport mechanisms. The principles upon which these mechanisms are based are explained in Chap. 6. Only when the concentrations of these materials in the filtrate exceed the abilities of the transport enzymes in the cell membranes of the tubular cells to keep pace with their return do they appear in the urine. Under normal circumstances, this situation may occasionally arise following the ingestion of a large amount of candy or after a heavy meal rich in protein.

All the positively charged electrolytes such as $Na^+$, $K^+$, and $Ca^{++}$ are also actively reabsorbed. Negatively charged electrolytes such as $Cl^-$ are reabsorbed passively when an electrochemical gradient is established between the proximal convoluted tubule and its contiguous peritubular capillaries. Because the tubular fluid is negative to the blood, $Cl^-$ is electrostatically attracted to $Na^+$ and other positively charged ions, thereby establishing electrical neutrality between the formative urine and the blood.

About 25 ml of formative urine/min enters the loops of Henle from the proximal convoluted tubules. The primary purpose of the loop of Henle is to adjust the osmolarity of the medullary extracellular tissue fluid and, therefore, the osmolarity of the blood.

As the tubular fluid in the descending limbs enters

the deeper portions of the medulla, the salt concentration of the interstitial fluid increases markedly. In response to the concentration gradient, $Na^+$ diffuses from the interstitial spaces into the tubular fluid within the descending limb as well as into the blood of its contiguous peritubular capillaries. Interstitial $Cl^-$ also enters these fluids because of its electrostatic attraction to $Na^+$. Since the interstitial $Na^+Cl^-$ concentration normally increases from the superficial to the deeper portions of the renal medulla, an increasing concentration of $Na^+Cl^-$ builds up in the basal portions of the loop of Henle and its contiguous capillaries at the same time that the interstitial fluid becomes increasingly dilute.

These relationships are reversed in the ascending limb and its contiguous capillaries. $Na^+$ diffuses out of the tubular fluid and the blood and returns to the extracellular tissue fluid in this portion of the nephron. $Cl^-$ follows electrostatically in its wake. To adjust and maintain finely the high salt concentration of the medullary interstitial spaces, $Na^+$ is also actively transported by the tubular cells of the ascending limb from the tubular fluid to the extracellular tissue fluid. Since the walls of the ascending limb are impermeable to $H_2O$, $H_2O$ remains in the tubular fluid and therefore does not follow the osmotic gradient established by the high salt concentrations in the interstitial spaces. As a consequence of salt losses from, and $H_2O$ retention in, the tubular fluid of the ascending limb, the formative urine which enters the distal convoluted tubule in the cortical portion of the nephron is hypotonic to (less concentrated than) the cortical extracellular tissue fluid.

About 18 ml of formative urine/min enters the distal convoluted tubules. Therefore, the flow rate of blood in contiguous capillaries is 1,182 ml/min. At this point, the tubular fluid is dilute, and whether the formative urine will become concentrated or will remain dilute depends on several important reabsorptive mechanisms which operate primarily in the distal convoluted tubules. Reabsorption is facilitated by an extremely close contact which the contiguous capillaries make with the walls of the nephron in this region.

Much of the renal regulation of electrolyte and fluid balance of the blood is accomplished while the formative urine is in the distal convoluted tubules. When the osmolarity of blood must be decreased to maintain homeostasis, $H_2O$ is reabsorbed from the distal convoluted tubules. Water permeability of the tubular cells in this region increases under the influence of antidiuretic hormone (ADH). In the absence of ADH, the simple diffusion of $H_2O$ from the tubular fluid to the blood is relatively slow. The mechanism which triggers the secretion of ADH is discussed later in this chapter.

The osmotic reabsorption of $H_2O$ is increased markedly by the action of still another hormone, aldosterone. Aldosterone is therefore primarily responsible for maintaining the fluid balance of the blood. This hormone causes the active reabsorption of $Na^+$ from the distal convoluted tubules. $Cl^-$ is passively reabsorbed by electrostatic attraction to $Na^+$. Because of the subsequent increase in the osmolarity of the blood in capillaries contiguous to the distal convoluted tubules, $H_2O$ is rapidly reabsorbed. The effects of aldosterone are explained in detail later in this chapter.

When aldosterone is present and ADH is absent, only $Na^+$ is reabsorbed actively from the tubular fluid within the distal convoluted tubules. The tubular cells remain relatively impermeable to $H_2O$ under these circumstances.

About 9 percent of the formative urine, corresponding to approximately 12 ml/min, is reabsorbed from the distal convoluted tubules. Therefore, the flow rate of formative urine entering the collecting ducts is reduced to 6 ml/min as the flow rate of blood in contiguous capillaries increases to 1,194 ml/min.

Finally, about 4 percent of the formative urine, corresponding to 5 ml/min, is reabsorbed from the collecting ducts. Therefore, the flow rate of urine entering the renal calyces is reduced to 1 ml/min, and the flow rate of blood drained by venules which lead from the peritubular capillary networks to the interlobular veins rises to 1,199 ml/min. In other words, the glomerular filtrate is reduced in volume by 125-fold by the time it becomes fully formed urine at the same time that the blood volume transported to and from the 2 million nephrons fluctuates negligibly.

Among the constituents reabsorbed from formative urine in the collecting ducts are $H_2O$, $Na^+$, and $HCO_3^-$. Water reabsorption depends on osmotic gradients established by the increasing concentrations of salt in the medullary interstitial fluids as the calyces are approached. Each collecting duct courses through a pyramid on its way to a calyx. As the salt concentrations become hypertonic to the formative urine in the deeper portions of the medulla, $H_2O$ is pulled osmotically from the formative urine into the interstitial fluids. Water losses from the distal convoluted tubule and the collecting duct concentrate the formative urine. When the blood is dilute, ADH is not secreted and the medullary interstitial fluids are dilute. Under these conditions, $H_2O$ remains in the urine within the collecting ducts.

The reabsorption of other materials from the collecting ducts is inextricably intertwined with the secretion of certain constituents from the distal convoluted tubules into the formative urine. Therefore, let us look briefly at the process of secretion.

Secretion　Among the constituents secreted into the formative urine by the tubular cells of the distal convoluted tubules are $H^+$, $NH_3$, and $K^+$. $H^+$ is secreted actively in response to a high plasma concentration of $CO_2$ (hypercapnia). For every $H^+$ secreted, a $Na^+$ is actively reabsorbed as a consequence of a reciprocal exchange inside the tubular cells. $HCO_3^-$ is passively reabsorbed from the tubular cells by virtue of its electrostatic attraction to $Na^+$. When the acidity of the blood is chronically high, ammonia ($NH_3$) is secreted passively into the formative urine in increasing concentrations by the tubular cells. $NH_3$ is synthesized constantly within the tubular cells following the removal of $NH_2$ from specific amino acids. Secretion of $H^+$ and the urinary role of $NH_3$ are considered in more detail later in this chapter.

$K^+$ is secreted into the formative urine by the tubular cells of the distal convoluted tubules in response to the active reabsorption of $Na^+$.

The secreted constituents play an important role in determining the acidity or alkalinity of urine and blood as well as in maintaining the electrolyte balance of blood, as we shall see later.

## THE HORMONAL CONTROL OF FLUID AND ELECTROLYTE BALANCE .

Several hormones simultaneously adjust the volume and content of both the formative urine and the blood. These hormones include antidiuretic hormone (ADH) from the hypophysis (pituitary gland), aldosterone from the adrenal glands, and parathyroid hormone from the parathyroid glands.

### OSMOTIC HOMEOSTASIS

The blood contains many types of solute particles, both electrolytes and nonelectrolytes. To prevent either hemolysis (swelling) or crenation (shrinking) of erythrocytes, the particle concentration of the plasma must be isotonic with that of the blood corpuscles; i.e., the plasma and formed elements must have the same number of particles per unit volume.

The osmotic concentration of a nonelectrolyte is the same as its molarity, but that of an electrolyte is the product of its molarity and the number of particles into which each of its molecules dissociates or ionizes. Osmotic concentrations are expressed in osmoles or milliosmoles (1 mOsm = 1/1,000 Osm).

The osmotic pressure of a solution such as blood is measured in atmospheres (atm). The osmotic pressure of a 1 $M$ solution of any nonelectrolyte or of 1 mEq of any electrolyte is 22.4 atm, or over 22 times the atmospheric pressure at sea level and 0°C. Therefore, the osmotic pressure of a 1 mOsm solution is $1/1,000 \times 22.4$ atm, or 0.0224 atm. Since the average osmotic pressure of the total solutes in the blood is about 6.7 atm, the osmotic concentration of the blood is 6.7 atm ÷ 0.0224 atm, or approximately 300 mOsm. This osmotic concentration is maintained within narrow limits by regulating the ratio of the plasma concentration to the blood volume.

### THE MAINTENANCE OF BLOOD OSMOLARITY

Assume that blood returning to the heart from the kidneys is overly concentrated with solutes. This is a perfectly normal occurrence but must be corrected to maintain homeostasis. This concentrated blood, of course, is distributed to all parts of the body, including the hypothalamic region of the brain. In response to hypertonic blood, osmoreceptors in the anterior hypothalamus lose $H_2O$ osmotically. As a result, the osmoreceptors shrink, causing an increase in the frequency with which they generate nerve impulses. These nerve impulses are conducted by nerve fibers to the hypothalamus where they stimulate neurosecretory cells to release ADH into the posterior lobe of the pituitary gland. From here, ADH is transported through the blood to all parts of the body, including the kidneys, which are its target organs. ADH increases the water permeability of the cells of the distal convoluted tubules. In response to an increased permeability, $H_2O$ moves osmotically from the urine in the distal convoluted tubules to the blood within contiguous peritubular capillaries.

There are two consequences of an increased secretion of ADH: (1) an increase in the volume of $H_2O$ reabsorbed from the formative urine, and (2) a corresponding decrease in the volume of formative urine. Therefore, the blood becomes more dilute as the urine becomes more concentrated.

The effects of ADH on water excretion and reabsorption are summarized in Fig. 22-6A. The opposite effects of those described above occur in response to a decrease in the synthesis of ADH. An abnormal hyposecretion of ADH results in diabetes insipidus, a disease described in Chap. 18.

## THE MAINTENANCE OF BLOOD VOLUME AND NA+ AND K+ BALANCE

Aldosterone is responsible for the active reabsorption of $Na^+$ from the kidneys. When ADH is present, $H_2O$ is reabsorbed rapidly. When ADH is absent, $H_2O$ is excreted, even though $Na^+$ reabsorption continues.

Aldosterone is synthesized by, and secreted from, the cortices of the adrenal glands. The following factors increase the amount of aldosterone in the blood:

1  The secretion of aldosterone is stimulated by a decrease in the normal concentrations of $Na^+$ or an increase in the normal concentrations of $K^+$ in the blood supplied to the adrenal glands. To understand why these ions are influential, we must look more closely at the primary effect of aldosterone, which is to reabsorb $Na^+$ from the kidneys. The reabsorption of $Na^+$ is due to the aldosterone-initiated increase in its active transport from the lumina of the distal convoluted tubules to the extracellular tissue spaces. From here, $Na^+$ moves by diffusion into the surrounding blood capillaries. When $Na^+$ is reabsorbed, $K^+$ is secreted in a ratio of 1:1 by the tubular cells of the distal convoluted tubules.

2  The secretion of aldosterone is stimulated indirectly by a decrease in the normal blood pressures in the renal arteries. The secretory process is initiated when the juxtaglomerular cells secrete the enzyme renin. Although this response is normal, ischemic kidneys secrete a large concentration of renin as part of a homeostatic mechanism which restores renal blood pressures. Renin initiates a complex series of chemical reactions in the blood which activates angiotensin. This chemical elevates the blood pressure by decreasing the diameter of arterioles. Simultaneously, angiotensin stimulates the synthesis of aldosterone. The renin-angiotensin mechanism is described in Chap. 31.

3  The secretion of aldosterone is stimulated indirectly by an increase in the amount of stress experienced by a subject. Stress initiates the pituitary-adrenal axis, a mechanism discussed in Chap. 4. Basically, in response to neural excitation of a stressful nature, the hypothalamus produces a hormone called corticotropin releasing factor, which is carried through the blood to the anterior lobe of the pituitary gland (adenohypophysis). There it stimulates the production and secretion of adrenocorticotropic hormone (ACTH). ACTH is carried through the blood to all parts of the body, including the cortices of the adrenal glands, its target organs. In the adrenal cortices, ACTH promotes the synthesis of adrenal cortical hormones, including the mineralocorticoid aldosterone. Although the effect is slight, it is measurable. The results of abnormal hyper- and hyposecretions of ACTH are discussed in Chap. 18.

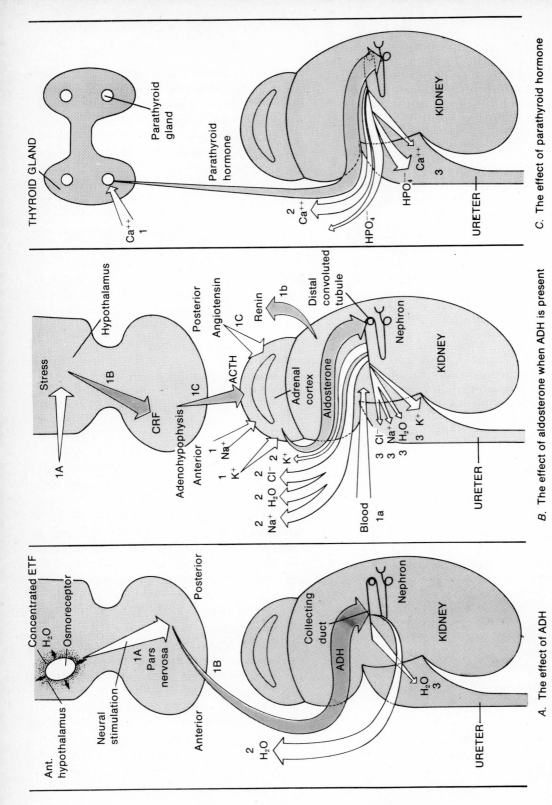

**Figure 22-6** *The hormonal mediation of fluid and electrolyte balance by the kidneys.*

*A.* The effect of ADH

*B.* The effect of aldosterone when ADH is present

*C.* The effect of parathyroid hormone

The effects of an increase in the secretion of aldosterone are summarized in Fig. 22-6B and are as follows:

1 Aldosterone causes an increase in the reabsorption of $Na^+$ and a decrease in the reabsorption of $K^+$ from the distal convoluted tubules. Thus, a balance is maintained between the ionic concentrations of $Na^+$ and $K^+$ in the blood. Active transport of $Na^+$ across the tubular cells of the distal convoluted tubules is promoted through the activation of $Na^+$ pumps. Located in the membranes of the tubular cells in this region, these molecular pumps increase the rate of $Na^+$ transport far beyond that which can be accomplished by diffusion alone.

2 Aldosterone is indirectly responsible for reabsorbing $H_2O$ from the formative urine, provided that ADH has paved the way by increasing the $H_2O$ permeability of the tubular walls.

3 Aldosterone causes an increase in the reabsorption of $Cl^-$ and a corresponding decrease in its excretion. As $Na^+$ reabsorption increases, an electrochemical gradient develops between these two fluids. To reestablish electrical neutrality, $Cl^-$ is electrostatically attracted to $Na^+$.

Retention of $H_2O$ is important in conditions which tend to dehydrate the body. Since an increase in blood volume increases cardiac output and mean systemic pressure, another secondary effect of aldosterone is to increase the blood pressure. The importance of this compensatory mechanism in certain cardiovascular diseases is discussed in Chap. 31.

## RENAL REGULATION OF BLOOD $Ca^{++}$ AND $HPO_4^{--}$

Parathyroid hormone affects the renal absorption and secretion of $Ca^{++}$ and $HPO_4^{--}$. This hormone is secreted by the parathyroid glands in response to a decrease in blood $Ca^{++}$, a condition called *hypocalcemia*. Although the primary effect of parathyroid hormone is to promote the absorption by the blood of bone $Ca^{++}$ (see Chap. 23), it also causes the reabsorption of $Ca^{++}$ from the proximal convoluted tubules and increases the urinary excretion of $HPO_4^{--}$. The action of parathyroid hormone on the kidneys has two consequences: (1) the renal reabsorption of $Ca^{++}$ aids in increasing the concentration of blood $Ca^{++}$ and prevents the depletion of bone $Ca^{++}$, an event which otherwise would occur; and (2) the urinary excretion of $HPO_4^{--}$ increases the amount of $Ca^{++}$ absorbed from bone, thereby maintaining a constant ratio between the products of these ions in the blood.

The urinary effects of parathyroid hormone are summarized in Fig. 22-6C. The opposite effects occur when the synthesis of parathyroid hormone is depressed. Pathologic conditions which result from abnormal hypo- or hypersecretions of either parathyroid hormone or other hormones which regulate $Ca^{++}$ and $HPO_4^{--}$ metabolism are discussed in Chap. 18.

## FLUID BALANCE AND THIRST

Because of $H_2O$ losses via the urine, sensible and insensible perspiration, respiration, and defecation, renal control of $H_2O$ balance is not able to maintain $H_2O$ homeostasis over the long term. In other words, without an exogenous supply of $H_2O$, the human body would eventually dehydrate. Water is periodically added to the body by drinking. The desire to drink $H_2O$ is stimulated by a mechanism similar to that which promotes the secretion of ADH.

Although not all authorities are in agreement on the details, apparently the desire to drink $H_2O$ is triggered by *hemoconcentration*. As the osmolarity of the blood increases, $H_2O$ is drawn osmotically from the tissue cells into the blood. A reduction in blood volume, for whatever the reason, will also cause cellular dehydration. Intracellular $H_2O$ restores the blood volume. Among the cells which become partially dehydrated during hemoconcentration are hypothalamic cells close to the osmoreceptors. These hypothalamic cells constitute a thirst center in the brain. The desire to drink is initiated when the thirst center is stimulated through partial dehydration.

Water imbibition increases the volume and dilutes the concentration of blood. Therefore, $H_2O$ moves from the blood back to the tissue cells, including cells of the thirst center in the hypothalamus. Although it is attractive to believe that the desire to drink is reduced when the $H_2O$ balance between the fluid compartments of the body is reestablished, we know from experiments that this is not the case. Before $H_2O$ can be absorbed from the digestive tract into the blood, the desire to drink is already reduced. The control mechanism which shuts off the desire to drink immediately after drinking is not understood. Obviously neural in nature, this shut-off mechanism is not peculiar to man, however, since a similar phenomenon has been reported in other animals too.

## THE EXCRETION OF H+

An elaborate system of molecules and organs maintains the chemical concentrations of $H^+$ ($[H^+]$), or pH, of the blood within narrow limits. The regulatory molecules are buffers such as carbonates ($H_2CO_3$ and $Na^+HCO_3^-$) and the protein hemoglobin (Hb) in the erythrocytes, phosphates ($Na^+H_2PO_4^-$ and $Na_2^+HPO_4^{--}$) in the urine, and proteinates (HPr and $Pr^-$) inside cells. The participating organs include the lungs, which excrete $CO_2$, and the kidneys, which excrete $H^+$. Any significant change in the pH of blood from 7.35 disrupts the homeostatic balance established between these components and initiates reactions which will return the entire system to normal $[H^+]$.

First we will sketch the mechanisms by which blood pH is adjusted, and then we will look more closely at the details. It will pay to review the concept of pH and the mechanism of buffering set forth in Chap. 10. The effects of an abnormally large increase in pH (alkalosis) or decrease in pH (acidosis) are discussed in Chap. 31. Here we are concerned only with routine adjustments in $[H^+]$.

## AN OVERVIEW

An increase in blood $[H^+]$ is induced by the ingestion of acidic food, as a by-product of intermediary metabolism, or by a prolonged elevation in metabolic rate. Initially, the excess $H^+$ is buffered by the $H_2CO_3/Na^+HCO_3^-$–Hb system in the blood. This system ties up excess blood $H^+$, preventing significant changes in pH, but does not get rid of excess $H^+$. Metabolically produced $CO_2$ is excreted by the lungs. Excretion of excess $H^+$ is accomplished entirely by the kidneys. In other words, the blood-buffering of excess $H^+$ is a short-term answer to increased acidity, but the goal over the longer term is to eliminate excess $H^+$ by voiding it in urine.

The basic inputs to, and outputs from, the kidneys during the excretion of $H^+$ are summarized in Fig. 22-7. In response to an elevation in the $[H^+]$ of the blood, an increasing $[H^+]$ appears in the formative urine. Urinary $H^+$ is bound to phosphates ($HPO_4^{--}$) as $H_2PO_4^-$ and to ammonia ($NH_3$) as ammonium ions ($NH_4^+$). $NH_4^+$ is electrostatically attracted to $Cl^-$ or to some other negatively charged ion and excreted. $H_2PO_4^-$ is electrostatically attracted to $Na^+$ or to some other positively charged ion and excreted. For each $H^+$ secreted, a $Na^+$ is reabsorbed. $HCO_3^-$ is also reabsorbed by electrostatic attraction. The reabsorption of $Na^+HCO_3^-$ replenishes the alkaline reserves of the body. The decrease in the $[H^+]$ of blood reduces its acidity.

## SOURCES OF H+

Ordinarily, most of the excess $H^+$ in blood is derived from a temporary increase in the blood concentration of $CO_2$, a waste product of metabolism. The concentration of $CO_2$ is expressed by its partial pressure, or $P_{CO_2}$. How an increase in the $P_{CO_2}$ gives rise to an excess of blood $H^+$ is explained by the following equation:

$$CO_2 + H_2O \rightarrow H_2CO_3 \rightarrow H^+ + HCO_3^- \qquad (22\text{-}1)$$

This equation states that $CO_2$ interacts chemically with $H_2O$, giving rise to carbonic acid ($H_2CO_3$). $H_2CO_3$ is a weak acid and dissociates to some extent into $H^+$ and bicarbonate ions ($HCO_3^-$). Since there is essentially an unlimited supply of unbound $H_2O$ available for chemical interactions, an increase in the $P_{CO_2}$ of blood increases the amount of $H_2CO_3$ by driving the reaction from left to right. Therefore, the concentration of the dissociation products, $H^+$ and $HCO_3^-$, also increases.

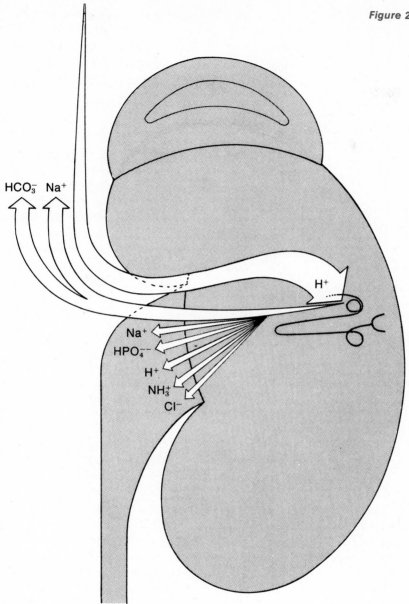

**Figure 22-7** *The excretion of H+ by the kidney.*

The ingestion of acidic, fatty, or proteinaceous foods also gives rise to an increase in blood acidity. Certain fruit juices contribute acids directly. Other foods add $H^+$ to the body during their metabolic breakdown within tissue cells. The pH of the blood tends to be lowered by the formation of fatty acids during the degradation of fats and by the formation of acids containing phosphorus or sulfur, e.g., $H_3PO_4$ or $H_2SO_4$, during the degradation of proteins containing atoms of these elements.

Finally, the formation of lactic acid during vigorous physical exercise also leads to an increase in the acidity of blood.

## BLOOD BUFFERING OF FREE H+

Recall from Chap. 10 that a buffering system consists of a weak acid and its salt. The primary buffering system of blood consists of *carbonates*. The carbonates include the weak acid $H_2CO_3$ and the salt sodium bicarbonate ($Na^+HCO_3^-$). The salt is written in ionic form because $Na^+$ and $HCO_3^-$ are held apart in blood by the solvent action of $H_2O$.

When a strong acid HA is added to blood, the following reaction takes place:

$$Na^+ + HCO_3^- + HA \rightarrow H_2CO_3 + Na^+ + A^- \qquad (22\text{-}2)$$

Notice that protons ($H^+$) from HA are accepted by $HCO_3^-$, forming the weak acid $H_2CO_3$ and leaving $A^-$, any negatively charged ion such as $Cl^-$, behind in the aqueous medium of blood. The ionization reaction summarized by Eq. (22-2) is identical in principle to that for the action of buffers described by Eq. (10-6). In other words, a strong acid is converted into a weak acid.

As we indicated in Eq. (22-1), an increase in the $P_{CO_2}$ of blood also increases the concentration of $H_2CO_3$. Therefore the addition of $H^+$ or $CO_2$ to blood leads to the same result—the formation of the weak acid $H_2CO_3$.

Although the trade of a strong acid for a weak acid prevents a gross change in blood pH, it still permits a slight increase in the acidity of blood. Although a slight decrease in pH is less disruptive homeostatically than a gross decrease in pH, the excess $H^+$ resulting from the dissociation of $H_2CO_3$ must still be buffered during the transport of blood from the tissues to the lungs.

Hemoglobin (Hb) also has an important function in the buffering of $H^+$. Located within erythrocytes, this respiratory pigment has a primary role in the transport of $O_2$ to the tissues of the body and a minor role in the transport of $CO_2$ to the lungs. As $O_2$ is unloaded to the tissue cells from erythrocytes in blood capillaries,

$CO_2$ is transported in the opposite direction. Within the erythrocytes, $CO_2$ interacts with $H_2O$, forming $H_2CO_3$. Free Hb readily accepts protons donated by $H_2CO_3$ as it dissociates, giving rise to reduced hemoglobin, or HHb. The erythrocytic buffering of $H^+$ is summarized as follows:

$$\begin{array}{l} H_2CO_3 \rightarrow HCO_3^- + H^+ \\ HbO_2^- \rightarrow O_2 + Hb \end{array} \Big\rangle \longrightarrow HHb \qquad (22\text{-}3)$$

As HHb is formed, $O_2$ diffuses into the tissue cells while $HCO_3^-$ diffuses into the plasma.

In the lungs, the reactions summarized in Eq. (22-3) are essentially reversed. When $O_2$ diffuses from the alveoli into erythrocytes within pulmonary capillaries, Hb becomes oxygenated. $H^+$ is set free as $O_2$ displaces it. $H^+$ is accepted by $HCO_3^-$, forming $H_2CO_3$. $H_2CO_3$, in turn, breaks down into $H_2O$ and $CO_2$. These products, for the most part, are transported in the plasma to all parts of the body, including the kidneys. This sequence of reactions is the reverse of that shown in Eq. (22-1).

## THE SECRETION OF H+

As the $[H^+]$ of blood increases, the equilibrium point of Eq. (22-1) shifts to the left, giving rise to an increase in the concentration of $CO_2$ and $H_2O$. Under these conditions, the $P_{CO_2}$ of blood brought to the glomeruli through the afferent arterioles is higher than normal. It is as $CO_2$ that the nephrons initially deal with an excess of blood $H^+$.

Upon entering the tubular cells by diffusion, $CO_2$ interacts chemically with $H_2O$, forming $H_2CO_3$. The reaction is accelerated by the enzyme *carbonic anhydrase*. In turn, $H_2CO_3$ dissociates somewhat into $H^+$ and $HCO_3^-$. $H^+$ is secreted by the tubular cells into the formative urine within the nephrons while $HCO_3^-$ is reabsorbed into the blood. The reciprocal exchange occurs in the cells of the distal convoluted tubules. During reabsorption, $Na^+$ combines electrostatically with $HCO_3^-$, forming $Na^+HCO_3^-$. The reabsorption of $Na^+HCO_3^-$ maintains the alkaline reserves of the blood.

So long as the $P_{CO_2}$ remains high, little or no filtered

$HCO_3^-$ appears in the urine because an increase in urinary $H^+$ drives Eq. (22-1) far to the left, forming $CO_2$ and $H_2O$ in the tubular cells. Water diffuses into the urine, and $CO_2$ either engages in the same sequence of reactions described above or diffuses into the blood to be excreted via the lungs.

If the blood is more alkaline than normal, however, the secretion of $H^+$ is diminished as filtered $Na^+HCO_3^-$ begins to appear in the urine. In other words, the presence or absence of $H^+$ in the formative urine determines whether $Na^+HCO_3^-$ will be reabsorbed or excreted, respectively.

Two urinary buffers, *ammonia* ($NH_3$) and *phosphates,* accept $H^+$, thereby setting the stage for its excretion. Formed during the removal of nitrogen from particular amino acids, $NH_3$ is secreted by tubular cells into the formative urine. There, $NH_3$ accepts $H^+$ and is transformed into ammonium ions ($NH_4^+$). $NH_4^+$ is excreted after combining electrostatically with $Cl^-$ or some other negatively charged ion. Called *ammonium chloride*, $NH_4^+Cl^-$ is a salt and does not contribute to the acidity of urine.

The excretion of $H^+$ by combination with $NH_3$ is an especially effective mechanism when blood pH is chronically depressed. Thus, the tubular cells respond to a long-term increase in blood $H^+$ by secreting greater-than-normal quantities of $NH_3$ into the formative urine. The levels of $NH_3$ secretion remain elevated as long as $H^+$ continues to be secreted into the urine.

The primary urinary buffering system consists of phosphates. The phosphates include $Na^+H_2PO_4^-$, a weak acid, and $Na_2^+HPO_4^{--}$, a salt. This buffering system is filtered into the formative urine from the glomeruli as $2Na^+$ and $HPO_4^{--}$. In the presence of $H^+$, $HPO_4^{--}$ is transformed into $H_2PO_4^-$. In other words, each $HPO_4^{--}$ acts as a base and accepts a proton. $H_2PO_4^-$ is therefore a potential acid. $H_2PO_4^-$ combines electrostatically with $Na^+$ (or with some other positively charged ion), becoming the weak acid $Na^+H_2PO_4^-$. It is in this form that $H^+$ is normally excreted in the urine. Because of its proton-donating capabilities, the presence of $Na^+H_2PO_4^-$ in the urine increases its acidity.

On the other hand, in the absence of urinary $H^+$, $HPO_4^{--}$ combines electrostatically with $2Na^+$, forming the salt $Na_2^+HPO_4^{--}$. This phosphate normally appears in urine when the pH of the blood is on the alkaline side of 7.35.

As a result of the complex interplay between the secretion of $H^+$ and the action of urinary buffers, on the one hand, and the balance between the reabsorption and excretion of $HCO_3^-$, on the other hand, the pH of urine may range anywhere from an extremely acid value of approximately 4.5 to a moderately alkaline value of about 8. These renal regulatory mechanisms are, in turn, affected by the following factors:

1 The rate of metabolism (which affects the $P_{CO_2}$ of blood)
2 The nature of the diet (which may be acidic or alkaline)
3 Hemoconcentration through dehydration due to profuse perspiration, vomiting, diarrhea, or diuresis
4 The ingestion of buffers such as bicarbonate of soda to neutralize gastric acidity
5 Vomiting (which reduces the concentration of HCl in the lumen of the stomach and causes a corresponding decrease in the acidity of the blood)
6 Diarrhea (which increases the acidity of the blood through the loss of carbonates from the body)

## URINATION

The act of urinating, also called *micturiting,* is a complex reflex mediated primarily by the spinal cord and the autonomic nervous system. The act is facilitated or inhibited by the brain, and certain thoracic and pelvic muscles may also participate in emptying the bladder. Nevertheless, urination is solely a spinal reflex under the following conditions: (1) before voluntary control of urination is established through toilet training; (2) after damage to specific regions of the brain and spinal cord or to particular peripheral nerves; (3) following extreme fatigue or fright; or (4) when the bladder is so full that urine cannot be held back voluntarily. These and other disturbances of normal urination are explored in Chap. 20. Here, we are interested in normal urination after toilet training has been well established.

## NEURAL CONTROL OF BLADDER FILLING

The bladder is capable of storing a volume of urine well in excess of 1 liter. Its ability to stretch is due in part to the transitional epithelium of its mucosa. However, the urge to urinate increases sharply when the intravesicular (inside the vesicle) volume is only 200 to 300 ml. Up to this point, the urination reflex is automatically inhibited.

The urinary bladder is filled in spurts corresponding to the periodic waves of peristaltic contraction which pass from the proximal to the distal ends of the ureters. The degree of bladder distension is monitored by stretch receptors in its wall, but the stretch characteristics (compliance) of the bladder are such that there is very little increase in the intravesicular pressure as the first 50 ml urine is collected. Therefore, the stretch receptors are initially inactive.

Between volumes of 50 to 200 ml of intravesicular urine, there is a gradual increase in intravesicular pressure. In response to distension of the bladder wall, more and more stretch receptors are stimulated. These receptors convert, or transduce, the mechanical stimulus of stretch into nerve impulses. An increase in the intensity of stimulation increases the number of nerve impulses generated each second. These nerve impulses are conducted along visceral sensory (afferent) nerve pathways to thoracic segments 10–12 and lumbar segments 1–2 of the spinal cord.

After reaching the cord, neural information on bladder status is conducted to lower brain centers by the lateral spinothalamic tract. This spinal tract is illustrated in Fig. 14-10C.

Reflex *inhibition* of urination is carried out by stimulating the *sympathetic* outflow from the thoracic and lumbar cord. Reflex *facilitation* of urination is carried out by stimulating the *parasympathetic* outflow from the sacral region of the spinal cord. Au-

tonomic control of the bladder is mediated by the hypothalamus. The nerve impulses which result in autonomic sympathetic or parasympathetic stimulation are conducted by the reticulospinal tract from the brainstem.

Urination at low intravesicular urine volumes is normally reflexly inhibited. The sympathetic autonomic nerve fibers exit from spinal cord segments T10 to T12 and L1 to L2, completing a reflex arc with the urinary bladder. Postganglionic nerve fibers of the sympathetic outflow innervate the detrusor muscle of the bladder. So long as the frequency of nerve impulse transmission from the sympathetic pathways to the bladder exceeds that of the parasympathetic pathways, contractions of the detrusor muscle are inhibited. Urine is retained within the bladder because the basal portion of the trigone musculature, which functions as an internal sphincter, remains closed.

Consisting of striated muscle, the external sphincter surrounds the urethra distal to the internal sphincter. This voluntary muscle is innervated by somatic motor fibers which exit from the sacral cord. As the intravesicular pressure increases, the somatic motor pathway to the bladder can be voluntarily inhibited by nerve impulses which travel down the lateral corticospinal tract from the cerebral cortex.

## NEURAL CONTROL OF BLADDER EMPTYING

Notice that before urination can occur (1) the bladder must be sufficiently distended; (2) the detrusor muscle must contract forcefully enough to open the internal sphincter, thereby permitting urine to enter the urethra; and (3) the external sphincter must relax in order for urine to flow through the rest of the urethra. The neural correlates of these events are diagramed in Fig. 22-8 and described below.

The intravesicular pressure increases sharply as the intravesicular urine volume rises above 150 ml. At these volumes, the stretch receptors in the walls of the bladder generate a greatly increased frequency of nerve impulses. The nerve impulses are conducted along the visceral afferent nerve pathways to the lower cord and from there to the brain by the lateral

which affect both metabolism and perpetuation. Not only are bones the primary reservoir of the important electrolyte calcium ($Ca^{++}$) in the body, but they are also constantly responding to the physical stresses and strains of normal living by subtle changes of shape which distribute applied loads more equitably. The storage functions of bones and the ability to mold themselves are inextricably intertwined with the deposition and absorption of $Ca^{++}$.

## HORMONAL INFLUENCES

Over 99 percent of the body stores of $Ca^{++}$ are located in bone. More than half the bone $Ca^{++}$ is freely exchangeable with blood $Ca^{++}$ through physical solution and chemical reactions; yet the blood concentration of $Ca^{++}$ is held constant at 10 mg/100 ml.

Since $Ca^{++}$ is vital to muscular contraction, nerve impulse conduction, and blood clotting, the maintenance of a constant blood $Ca^{++}$ concentration is a homeostatic process of paramount importance. Therefore, it is not too surprising to find that three hormones (parathyroid hormone, calcitonin, and calciferol) and one vitamin (vitamin D) are involved in its regulation.

The roles of calciferol and vitamin D are considered in Chap. 26. The effects of a deficit of vitamin D or of hyposecretions of the above hormones are reviewed in Chaps. 27 and 18, respectively. The renal effects of parathyroid hormone are explained in Chap. 22. Here, we shall examine the reciprocal actions of parathyroid hormone and calcitonin in the normal maintenance of blood $Ca^{++}$ concentrations.

*Parathyroid hormone* is secreted by the parathyroid glands. Two of these pea-shaped glands are embedded in the tissues on the posterior aspect of each lobe of the thyroid gland. When the blood concentrations of $Ca^{++}$ decrease below 10 mg/100 ml or, more correctly, when the product of the $Ca^{++}$ and $HPO_4^{--}$ concentrations in blood drop below a critical level, the parathyroid glands secrete parathyroid hormone into the bloodstream. Parathyroid hormone is transported in the blood throughout the body. Its target cells are osteocytes.

In response to parathyroid hormone, osteocytes passively accumulate an increasing concentration of $Ca^{++}$ from the extracellular tissue spaces. The effect of parathyroid hormone is therefore to increase the permeability of osteocytic membranes to $Ca^{++}$. The accumulation of intracellular $Ca^{++}$ somehow serves as the signal for the osteocytes to begin the absorption (resorption) of bone $Ca^{++}$. In other words, osteocytes are stimulated to dissolve bone as a consequence of an increase in intracellular $Ca^{++}$. The bone $Ca^{++}$ absorbed by this action comes from the hydroxyapatite and not from the freely exchangeable $Ca^{++}$ surrounding the hydroxyapatite crystals. Thus, parathyroid hormone is actually responsible for the destruction of bone, rather than an increase in the transfer of nonstructural $Ca^{++}$ from bone to blood.

Due to the influence of parathyroid hormone on osteocytes, the blood concentrations of $Ca^{++}$ increase. The excessive accumulation of blood $Ca^{++}$ somehow signals specialized cells within the thyroid gland to increase the secretion of calcitonin (thyrocalcitonin). Probably, $Ca^{++}$ promotes an increase in cyclic AMP within these specialized cells by a mechanism ordinarily employed by hormones. Calcitonin is transported in the blood throughout the body, but its effect, like that of parathyroid hormone, is differentially exerted on osteocytes.

Calcitonin activates a membrane transport system which actively expels $Ca^{++}$ from osteocytes. The reduction in intracellular $Ca^{++}$ removes the signal which originally promoted the absorption of bone $Ca^{++}$ by the osteocytes. Decrease in the absorption of bone $Ca^{++}$ results in a corresponding reduction in the concentrations of blood $Ca^{++}$.

The reciprocal interactions of parathyroid hormone and calcitonin stabilize the blood concentrations of $Ca^{++}$. The effects of these two hormones on bone are diagramed in Fig. 23-3.

## BONE MOLDING

Unlike the steel girders which form the framework of a building, the bony framework of the body is constantly being shaped, modeled, or molded in response to applied forces, all the while carrying on its

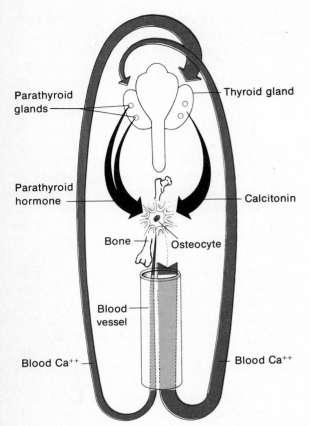

Parathyroid glands

Thyroid gland

Parathyroid hormone

Calcitonin

Bone

Osteocyte

Blood vessel

Blood Ca++

Blood Ca++

**Figure 23-3** *The reciprocal regulation of bone and blood Ca++ concentrations by parathyroid hormone and calcitonin.*

functions of hemopoiesis, Ca++ storage, support, protection, and leverage. Bones that receive heavy, prolonged usage respond dynamically to their applied loads, being shaped and reshaped constantly to provide an optimally stable platform for the weights being supported. A student may develop prominent ischial tuberosities from a habitual sitting posture, an equestrian may develop bowed legs, and an author may wind up with a slightly deformed writing finger.

If we can imagine a building endowed with the abilities of living bone, we might find a portion of the structure containing, for example, a heavily used conference room to be spontaneously enlarging over the course of many years. This fantastic ability of bone is obviously adaptive, enabling the skeleton of each person to respond physically in the most appropriate way to applied stresses.

Although molding depends upon the differential absorption and deposition of structural Ca++, the process is distinct from either ossification or the maintenance of a constant concentration of blood Ca++. Bone molding occurs throughout the life of an individual, whereas ossification is complete by approximately the twenty-fifth year. In addition, bone molding takes place irrespective of the physiologic concentrations of parathyroid hormone or calcitonin. In other words, the molding of bone must depend on an exclusive process, even though all these mechanisms involve the differential deposition and absorption of bone Ca++.

The events believed to occur during the molding of a bone are diagramed in Fig. 23-4. An unstressed bone is shown in Fig. 23-4A. A highly enlarged collagen fiber on each side of the bone symbolizes its organic matrix. Hydroxyapatite crystals are interspersed in the collagenous matrix, and numerous electrolytes and other raw materials are adsorbed to the inorganic mineral crystals from the serum which permeates all portions of the osteons. In other words, there are three distinct, albeit mixed, zones in bony tissue—crystals of *hydroxyapatite*, threads of the fibrous protein *collagen*, and a bathing medium of serum. The highly ordered organization of the hydroxyapatite crystals and collagen fibers is somehow implicated in the generation of electric currents when bone is physically stressed.

In Fig. 23-4B, the bone is stressed laterally by a force applied from the left side. Because both ends of the bone are bent toward the right, the left side of the bone is placed under tension or stretched while its right side is compressed. The side under tension assumes a convex shape and the opposite side becomes concave. This type of stress results in the concave side becoming negatively charged, compared to the opposite side.

How an electric current is generated in bone is still a matter of conjecture. Either the forces generated by adjacent collagen fibers as they slide over each other or the stresses induced when the collagen fibers rub against the hydroxyapatite crystals are capable of producing a measurable electric current. Whatever

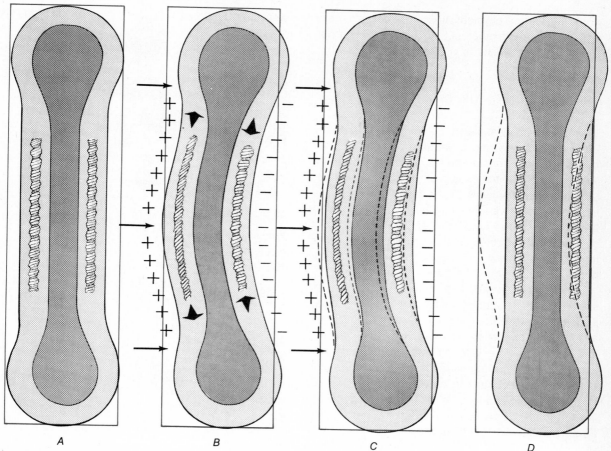

**Figure 23-4** *The molding of bone in response to stress. A. Bone prior to receiving stress. Strips on each side of the bone represent collagen fibers, the organic matrix. B. Bone is stressed by a force applied from the left side. Collagen fibers are under tension on the left side and are compressed on the right, resulting in stresses between collagen fibers and inorganic hydroxyapatite crystals. The stresses cause a buildup of negative charge on the concave side and positive charge on the convex side. C. The separation of charge is correlated with the deposition of new bone in the negative region and the absorption of bone from the positive region. D. Bone after molding. Frames show the progressive displacement of the bone.*

the explanation, *bone deposition* is initiated in the electronegative, concave region and *bone absorption* occurs in the electropositive, convex region.

Deposition and absorption continue as long as the bone remains under stress. Bone molding in response to stress is advanced in *C* and complete in *D* of Fig. 23-4. Notice that repeated stress caused by a force applied from one side has resulted in a lateral dis-

placement of the bone toward the opposite side.

This mechanism is believed to be responsible for the molding of bone in vivo, and this principle is routinely used in the treatment of malocclusion (see Chap. 27). Because of the multitudinous stresses imposed on normal living bones from all directions almost constantly, however, the shapes of bones remain essentially unchanged during adult life. In other words, the various physical stresses on bones more or less cancel each other out so that only unusual, sustained forces result in visible effects.

## ORGANIZATION OF THE SKELETAL SYSTEM

The adult human skeleton contains approximately 206 named bones and a variable number of other bones such as *Wormian bones*, associated with the sutures of the cranial bones, and *sesamoid bones*, associated with the joints of the hand and foot. The reason for some variability in the number of named bones is most often caused by an occasional lack of complete fusion of the frontal bones of the cranium or the presence of an extra coccygeal vertebra in the tail bone.

The skeleton is divided into two regions—the axial skeleton and the appendicular skeleton. The *axial skeleton* is located along the long axis of the body and includes the bones of the skull, vertebral column, and rib cage. The *appendicular skeleton* is associated with the appendages and includes the bones of the pectoral and pelvic girdles, arms and thighs, forearms and legs, as well as hands and feet.

The skeleton is viewed from anterior aspect in Fig. 23-5. The axial skeleton is colored.

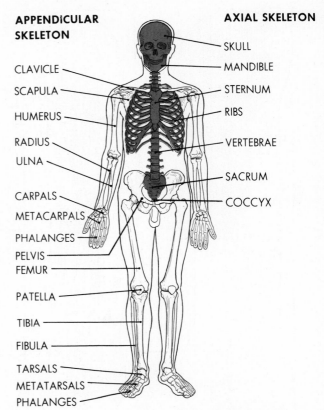

**Figure 23-5** *Anterior view of the skeleton, showing axial and appendicular divisions.* (From Melloni, Stone, and Hurd.)

### THE AXIAL SKELETON

**Bones of the skull** There are 29 named bones in the skull: 8 of these bones enclose the brain, forming a *cranium;* 14 skull bones are *facial;* and 6 specialized bones called *ear ossicles* are embedded in the substance of the temporal bones. The ear ossicles mechanically transmit air pressure disturbances to the inner ear. Called the malleus, incus, and stapes (hammer, anvil, and stirrup), they are illustrated in situ in Fig. 11-9. Another specialized, U-shaped bone called the *hyoid* is the only bone in the body which does not articulate with another bone. The hyoid bone is illustrated in Fig. 28-1.

*Cranium* Excluding the ear ossicles and hyoid bone, there are four unpaired bones (*frontal, ethmoid,*

*sphenoid,* and *occipital*) and two paired bones (*parietal* and *temporal*) in the cranium. The frontal, parietal, temporal, and occipital bones contribute primarily to the anterior, superolateral, inferolateral, and posterior portions, respectively, of the brain case. These bones are illustrated from lateral aspect in Fig. 23-6A. The ethmoid and sphenoid bones contribute primarily to the base of the cranium. The ethmoid also forms the superior and middle nasal conchae. The ethmoid and sphenoid bones are seen from superior aspect in Fig. 23-7 and in sagittal section in Fig. 23-8. Note the key location occupied by the massive sphenoid bone and the contributions to the base of the cranium made by the frontal, temporal, and occipital bones.

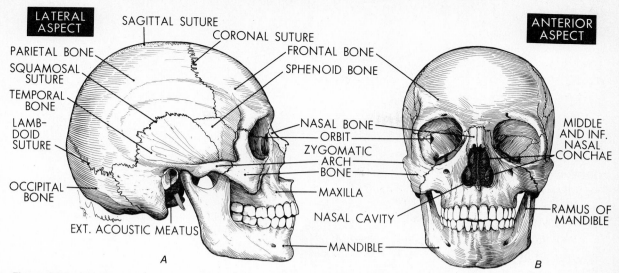

**Figure 23-6** *Structure of the skull. A. Lateral aspect. B. Anterior aspect.* (From Melloni, Stone, and Hurd.)

**Figure 23-7** *The superior aspect of the base of the skull.* (From Melloni, Stone, and Hurd.)

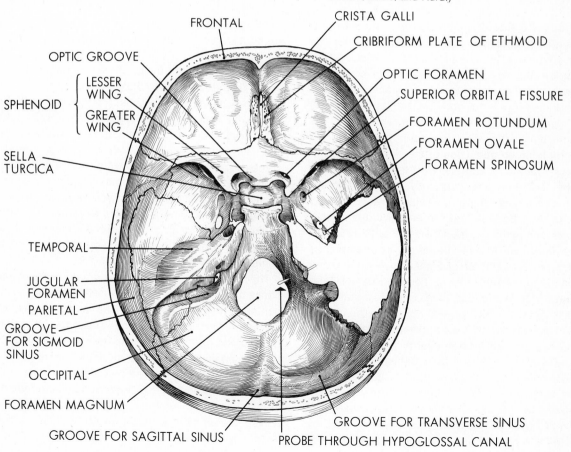

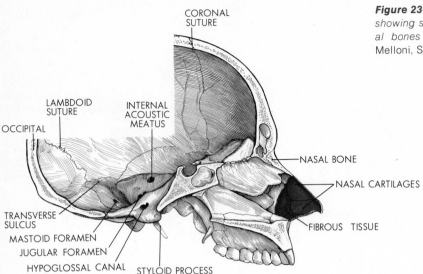

**Figure 23-8** *Midsagittal section of the skull, showing some details of the facial and cranial bones at the base of the skull.* (From Melloni, Stone, and Hurd.)

CORONAL SUTURE

LAMBDOID SUTURE

INTERNAL ACOUSTIC MEATUS

OCCIPITAL

NASAL BONE

NASAL CARTILAGES

TRANSVERSE SULCUS

MASTOID FORAMEN

JUGULAR FORAMEN

HYPOGLOSSAL CANAL

STYLOID PROCESS

FIBROUS TISSUE

The serrated articulations between adjacent cranial bones are called *sutures*. There are four major sutures of the cranium: (1) the *sagittal suture*, located along the midsagittal line between the parietal bones; (2) the *coronal suture*, located in the coronal plane between the frontal bone and the parietal bones; (3) the *squamosal suture*, located between the parietal bone and the squamous portion of the temporal bone on each side of the skull; and (4) the *lambdoid suture*, located between the parietal bones and the occipital bone. Frequently, the lambdoid suture line recurves on itself, enclosing an unnamed piece of skull bone referred to as a Wormian bone. The cranial sutures are identified in Fig. 23-6A.

The *dural sinuses* are spaces in the membranes which surround the brain and are closely opposed to the inside wall of the cranium. These venous drainage channels conduct blood and cerebrospinal fluid from the head to veins on the way to the heart. The friction generated by blood within these spaces causes the gradual erosion over the years of the inside walls of contiguous cranial bones. This continual erosion leads to the formation of bone grooves which mark the position of the underlying dural sinuses. The grooves of several dural sinuses are shown in Fig. 23-7. The grooves are those formed by the sagittal sinus, the transverse sinus, and the sig-

moid, or S-shaped, sinus. The dural blood sinuses are considered further in Chap. 14.

*Face* Excluding the hyoid, there are two unpaired bones (*vomer* and *mandible*) and six paired bones (*lacrimal, nasal, inferior nasal conchae, palatine, zygomatic,* and *maxillae*) in the facial region. The facial bones contribute to the formation of four cavities—two *orbits*, the *nasal cavity*, and the *oral cavity*.

The eyeballs are contained within orbits. Each orbit is bounded by the zygomatic, maxillary, lacrimal, and palatine bones of the face as well as by the frontal, ethmoid, and sphenoid bones of the cranium. Located on the medial aspect of the orbit, the lacrimal bone contains a groove for the nasolacrimal duct. These bones are shown in Fig. 23-6A and *B*.

The nasal cavity is bounded externally by the nasal bones and maxillae. The nasal bones form the upper portion of the bridge of the nose, and the maxillae constitute the upper jawbone. The nasal cavity is divided into right and left *nasal fossae* by a centrally located *nasal septum*. The superior portion of the nasal septum is formed by the perpendicular plate of the ethmoid bone and its inferior portion is formed by the vomer. Inside each nasal fossa are curved portions of the ethmoid bone known as the *superior* and *middle conchae* and a paired facial bone called the *inferior nasal concha*. The turbulence created when inspired air strikes these coiled bones brings the air temperature in line with body temperature. These bones are illustrated in Fig. 23-6A and B.

The anterior portion of the nasal cavity is bounded by five cartilages, several of which are shown in the midsagittal section of the skull in Fig. 23-8.

The roof of the nasal cavity is formed primarily by the frontal bone, the cribriform plate of the ethmoid bone, and the sphenoid bone. Olfactory nerve fibers from the nasal epithelium above the superior nasal conchae enter the brain case through the openings in the cribriform plate (see Fig. 11-13). The bones of the roof of the nasal cavity are shown from superior aspect in Fig. 23-7.

The floor of each nasal fossa is composed of the horizontal plate of the palatine bone and the palatine process of the maxilla. The union of right and left horizontal plates forms the posterior portion of the hard palate, and the union of right and left palatine processes forms its anterior portion. These bones are shown in midsagittal section in Fig. 23-8.

The oral cavity is bounded externally by the maxillae (upper jawbones) and the mandible (lower jawbone). With the exception of the mandible, all the facial bones articulate with the maxillae. The maxillae and mandible anchor the upper and lower teeth, respectively, in sockets called *alveoli*. The mandible is the largest and strongest of the facial bones, qualities necessitated by its functions in biting and chewing. The jawbones are illustrated from lateral and anterior aspects in Fig. 23-6A and B. The roof of the oral cavity is composed of the *hard palate*, the bony contributions of which have already been considered. The *soft palate* is attached to the horizontal plates of the hard palate.

The zygomatic, or cheek, bone forms the prominence of each cheek. The temporal process of the zygomatic bone articulates with the zygomatic process of the temporal bone, forming the *zygomatic arch*. Thus, the zygomatic arch is a compound structure formed by processes from two adjacent bones, one from the face and the other from the cranium. The zygomatic arch is shown from lateral and anterior aspects in Fig. 23-6A and B.

Named for the bones in which they reside, the *paranasal air sinuses* are cavities located in the frontal, ethmoid, and sphenoid bones of the cranium as well as in the maxillae of the face. The sinuses replace the cancellous tissues in these bones. The paranasal sinuses are lined by a mucosa continuous with the nasal mucosa and similar to that found within the mastoid sinuses of the temporal bones. The paranasal sinuses are considered further in Chap. 28.

**Vertebral column**  The *vertebral column*, backbone, or spine is ordinarily composed of 33 vertebrae divided into five regions, as follows: (1) *cervical* region (7 cervical vertebrae); (2) *thoracic* region (12 thoracic vertebrae); (3) *lumbar* region (5 lumbar vertebrae); (4) *sacral* region (5 sacral vertebrae fused into a *sacrum*); and (5) *coccygeal* region (4 coccygeal vertebrae fused into a *coccyx*, or tailbone). The vertebral column supports and assists in moving the head and trunk and encloses the spinal cord. The right side of the vertebral column is illustrated in Fig. 23-9.

The first and second cervical vertebrae are called the *atlas* and *axis*, respectively. The atlas supports the head. Flexion and extension of the head are permitted by the joint formed by the articulations of the facets of the atlas with the *occipital condyles*, knuckle-shaped processes to either side of the foramen magnum of the occipital bone. The *foramen magnum* is the opening for the exit of the spinal cord from the cranium. Right and left rotation of the head is permitted by the joints formed by the articulations of the facets between the atlas and axis as well as by the articulation of the *odontoid process*, or *dens*, with the odontoid facet of the atlas. The odontoid process is an anteriorly projecting, toothlike process of the axis.

As one can gather from Fig. 23-9, the lumbar vertebrae are the thickest and strongest vertebrae in the body. These vertebrae absorb most of the compressional forces sustained when the body lifts heavy loads. Note the disks of fibrocartilage between adjacent vertebrae in the cervical, thoracic, and lumbar regions. Called *intervertebral disks*, these pads absorb mechanical shock and provide resiliency to the movements of the vertebral column. Their resiliency is aided by the *nucleus pulposus*, the fluidlike central portion of each intervertebral disk. Finally, observe

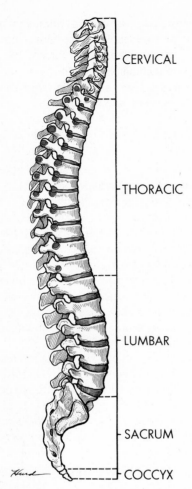

**Figure 23-9** *The structure of the vertebral column. The right side of the vertebral column is illustrated, and its anterior surface is on the right.* (From Melloni, Stone, and Hurd.)

located *arch* which encloses the spinal cord. The sides of the arch consist of *pedicles*, one of which attaches to each side of the body. The *laminae* arise from the pedicles, forming the roof of the arch. The paired *transverse processes* arise from the union of the pedicles and laminae. The single *spinous process* arises from the posterior union of the laminae. Muscles originate and insert at various points on the spinous and transverse processes. *Facets* for articulation with adjacent vertebrae, or with the ribs in the case of the thoracic vertebrae, are located at characteristic points on each vertebra. The superior aspects of the fourth cervical vertebra (C4), the sixth thoracic vertebra (T6), and the third lumbar vertebra (L3) are compared in Fig. 23-10.

There are four curvatures along the length of the vertebral column of an adult human skeleton. When the vertebral column is viewed from anterior aspect, the *cervical* and *lumbar curvatures* are convex and the *thoracic* and *sacral curvatures* are concave. The curvatures of the vertebral column are shown from lateral aspect in Fig. 23-9. Age changes in the contours of the spine are discussed later in this chapter.

**Rib cage**   The rib cage is composed of the bones in the thoracic region. Bounded posteriorly by 12 thoracic vertebrae, laterally by 12 pairs of ribs, and anteriorly by the sternum, or breast bone, the rib cage contains 37 bones. The bony thorax is illustrated from the left anterolateral aspect in Fig. 23-11. The rib cage encloses and protects the bronchi and lungs, the heart and great blood vessels, a portion of the esophagus, stomach, and liver, as well as lymphatic vessels and nerves in the thoracic region.

There are 12 pairs of *ribs* in both sexes. The vertebral borders of each pair of ribs articulate with the facets of two adjacent vertebrae. The sternal border of each vertebra is inferior to the vertebral border. Based on the nature, or absence, of sternal articulation, there are three types of ribs: (1) the first 7 pairs of ribs, or *true ribs*, articulate with the sternum by means of separate *costal cartilages* constructed of hyaline connective tissue; (2) rib pairs 8 to 10, called *false ribs*, articulate with the sternum indirectly by means of costal cartilages which attach to the costal cartilages

the openings formed by the articulations of the vertebrae. Called *intervertebral foramina*, these paired openings permit entry into and exit out of the spinal cord for the posterior and anterior spinal nerve roots, respectively. The relationships of the spinal roots to the vertebral column are shown in Fig. 14-7.

All the vertebrae are constructed similarly, although there are regional and individual structural specializations. Basically, each *vertebra* consists of a thickened, anteriorly located *body* and a posteriorly

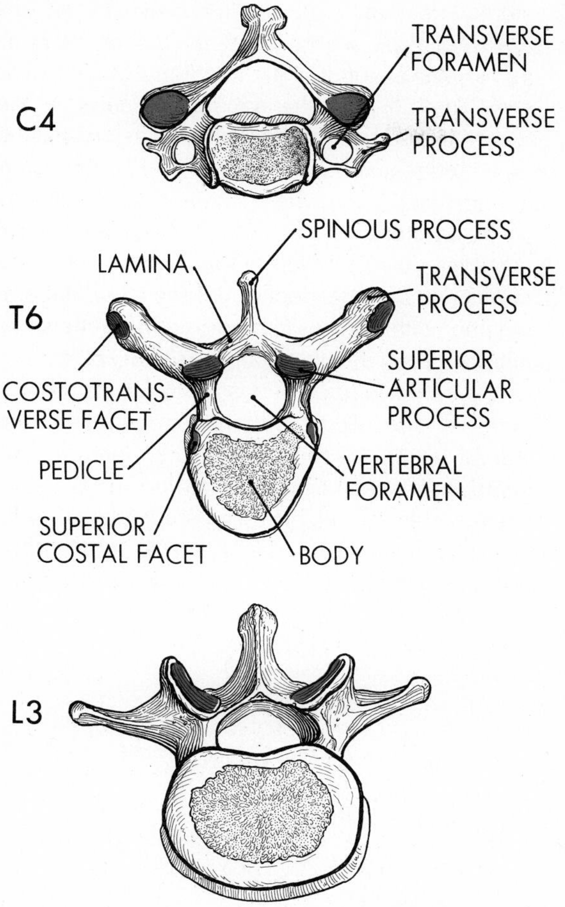

C4

TRANSVERSE
FOREMEN

TRANSVERSE
PROCESS

T6

LAMINA

SPINOUS PROCESS

TRANSVERSE
PROCESS

COSTOTRANS-
VERSE FACET

SUPERIOR
ARTICULAR
PROCESS

PEDICLE

SUPERIOR
COSTAL FACET

VERTEBRAL
FOREMEN

BODY

L3

**Figure 23-10** *The structure of representative vertebrae. The superior aspects of the fourth cervical (C4), sixth thoracic (T6), and third lumbar (L3) vertebrae are illustrated.* (From Melloni, Stone, and Hurd.)

clavicles and first pair of ribs articulate with the manubrium. The next nine pairs of ribs articulate either directly or indirectly with the body of the sternum. None of the ribs articulates with the xiphoid process. The fused bones of the sternum are illustrated in Fig. 23-11.

## THE APPENDICULAR SKELETON

The appendicular skeleton includes the bones of the appendages, i.e., of the arms and legs, as well as the bones which link the appendages to the axial skeleton. The appendicular skeleton of the shoulder, arm, forearm, and hand constitutes the bones of the *upper extremity*. The appendicular skeleton of the hip, thigh, leg, and foot makes up the bones of the *lower extremity*.

**Figure 23-11** *The structure of the rib cage. The bony thorax is viewed from the left anterolateral aspect.* (From Melloni, Stone, and Hurd.)

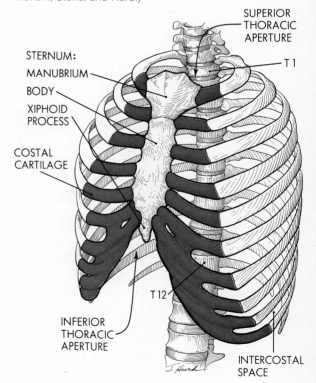

SUPERIOR
THORACIC
APERTURE

T 1

STERNUM:

MANUBRIUM

BODY

XIPHOID
PROCESS

COSTAL
CARTILAGE

T 12

INFERIOR
THORACIC
APERTURE

INTERCOSTAL
SPACE

of the preceding ribs; and (3) rib pairs 11 and 12, called *floating ribs*, lack costal cartilages and do not attach to the sternum. Coupled to the actions of the external and internal intercostal muscles, the ribs participate in respiration. The ribs, together with the sternum (described below), are also major sites of hemopoietic activity. The ribs are shown in situ in Fig. 23-11.

The *sternum* is composed of an upper *manubrium*, a middle *body*, and a lower *xiphoid process*. The

**The upper extremities** There are 32 named bones in each of the upper extremities in the human being. These bones are grouped into the following regions:

1 *Shoulder.* Each *shoulder girdle* consists of a *scapula* (shoulder blade) and *clavicle* (collar bone). The shoulder girdle links the rest of the bones of the upper extremity to the axial skeleton. The anterior aspect of the right shoulder girdle is shown in Fig. 23-12, and the joints of the shoulder girdle are identified. The clavicle articulates with the sternum at the *sternoclavicular joint*, the acromion of the scapula at the *acromioclavicular joint*, and the head of the humerus at the *shoulder joint*. The most flexible joint of the shoulder girdle and, indeed, of the body is that of the shoulder. The shoulder joint is a ball-and-socket joint formed by the articulation of the head of the humerus, an enlargement supported by a con-

stricted neck at its proximal end, with the glenoid cavity, or fossa, of the scapula, a depression on the superior aspect of its lateral margin.

2 *Arm.* The bone of the arm (upper arm) is the *humerus.* The anterior and posterior aspects of the right humerus are shown in Fig. 23-13A. The lateral and medial condyles, knuckle-like projections on the posterior aspect of the distal end of the humerus, articulate with the proximal ends of the *radius* and *ulna,* respectively. These articulations, together with another formed in part by the proximal portion of the ulna, are referred to collectively as the *elbow joint.*

3 *Forearm.* The *radius* and *ulna* are the *lateral* and *medial bones,* respectively, in each forearm. The anterior surfaces of the right radius and ulna are illustrated in Fig. 23-13C. The *olecranon process* of the ulna, a marked projection at its proximal end, forms the tip of the elbow. This process artic-

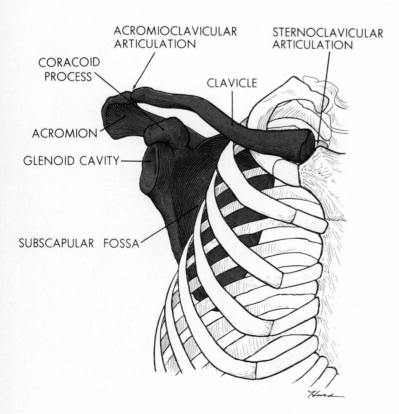

ACROMIOCLAVICULAR ARTICULATION

STERNOCLAVICULAR ARTICULATION

CORACOID PROCESS

CLAVICLE

ACROMION

GLENOID CAVITY

SUBSCAPULAR FOSSA

**Figure 23-12** *The right half of the shoulder girdle viewed from anterior aspect.* (From Melloni, Stone, and Hurd.)

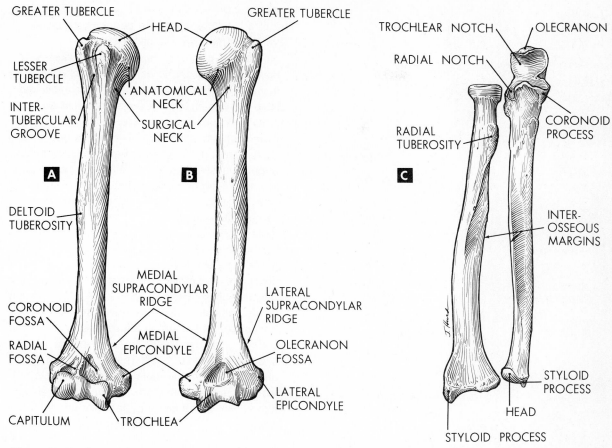

GREATER TUBERCLE

HEAD

GREATER TUBERCLE

LESSER TUBERCLE

ANATOMICAL NECK

INTER-TUBERCULAR GROOVE

SURGICAL NECK

**A**

**B**

DELTOID TUBEROSITY

MEDIAL SUPRACONDYLAR RIDGE

LATERAL SUPRACONDYLAR RIDGE

CORONOID FOSSA

MEDIAL EPICONDYLE

OLECRANON FOSSA

RADIAL FOSSA

LATERAL EPICONDYLE

CAPITULUM

TROCHLEA

TROCHLEAR NOTCH

OLECRANON

RADIAL NOTCH

RADIAL TUBEROSITY

CORONOID PROCESS

**C**

INTER-OSSEOUS MARGINS

STYLOID PROCESS

HEAD

STYLOID PROCESS

**Figure 23-13** *Bones of the right arm. A. Humerus from anterior aspect. B. Humerus from posterior aspect. C. Radius and ulna from anterior aspect. (From Melloni, Stone, and Hurd.)*

ulates with the olecranon fossa, a depression on the posterior aspect of the distal end of the humerus. This articulation is the conspicuous part of the elbow. The proximal *radioulnar joint* is formed by an articulation of the head of the radius with the radial notch of the ulna, a U-shaped depression anterior and inferior to the olecranon process. The shafts of the radius and ulna also articulate with each other by means of an *interosseous membrane.*

4 *Hand.* The bones of each hand are composed of 8 *carpals,* or wrist bones, 5 *metacarpals,* and 14 *phalanges,* or finger bones. Therefore, the hand is

constructed of 27 bones. The palmar surface of the right hand is shown in Fig. 23-14. The *wrist joint* is formed by articulations between the distal end of the radius and the scaphoid, lunate, and triquetrum bones of the wrist. Notice that the carpal bones are arranged in two transverse groupings, each of which contains four bones. The metacarpal bones radiate like the spokes of a wheel from the wrist, forming the framework of the palm of the hand. Notice that the fourth and fifth metacarpals articulate with the hamate. Observe the *sesamoid bones* which surround the distal portions of several of the metacarpals. Associated with the *me-*

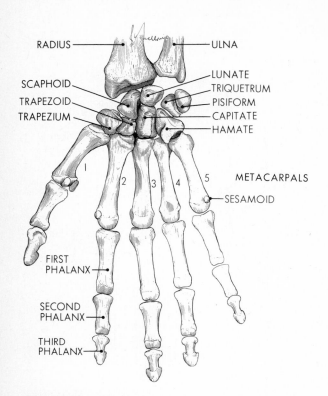

**Figure 23-14** *Bones of the palmar surface of the right hand. Note the sesamoid bones surrounding the distal portions of several metacarpal bones.* (From Melloni, Stone, and Hurd.)

tacarpophalangeal joints, these small rounded bones are contained within either tendons or fascia. *Fascia* is a fibrous connective tissue between the skin and skeletal muscles. Variable in number, sesamoid bones are believed to increase the mechanical advantage of the simple machines of which the articulating bones are members. Note that each of the fingers contains three phalanges, except for the thumb which contains only two.

**The lower extremities** There are 31 named bones in each of the lower extremities in human beings. These bones are grouped into the following regions:

1 *Hip.* Each hipbone is called an *os coxae,* or pelvic girdle. Each *pelvic girdle* is composed of three fused bones—an *ilium* which projects superiorly, an *ischium* which projects posteriorly,

and a *pubis* which projects anteriorly. The pelvic girdle links the rest of the bones of the lower extremity to the axial skeleton. The lateral aspect of the right os coxae is illustrated in Fig. 23-15. The fusion points of the three bones which compose the pelvic girdle are identified with thick lines. Note the cup-shaped depression called the *acetabulum* located around the fusion point of the ilium, ischium, and pubis. The pelvic girdles are fused anteriorly by articular cartilage between the pubic bones. This joint is called the *pubic symphysis.* Posteriorly, the pelvic girdles articulate with the sacrum at the *sacroiliac joints.* The sacrum also articulates with the fifth lumbar vertebra at the *lumbosacral joint* and with the coccyx at the *sacrococcygeal joint.* The fused pelvic girdles, together with the sacrum and coccyx, constitute the *pelvis.* The anterior aspect of a female pelvis is shown in Fig. 23-16. Since the pelvis consists of 2 hipbones (2 sets of 3 fused bones) plus a sacrum of 5 fused vertebrae and a coccyx of 4 fused vertebrae, the pelvis is constructed of 4 (15) bones. The joints of the pelvis are identified in the figure. The most flexible pelvic joint is that of the hip. The *hip joint* is a ball-and-socket joint formed by the articulation of the head of the femur, an enlargement supported by

**Figure 23-15** *The right hipbone, or os coxae. Consisting of three fused bones, the os coxae represents one-half of the pelvic girdle. The fused bones are the ilium, ischium, and pubis. Their fused articulation points are marked by solid lines.* (From Melloni, Stone, and Hurd.)

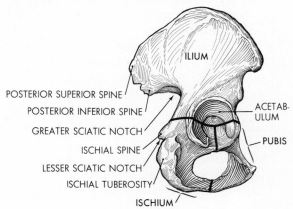

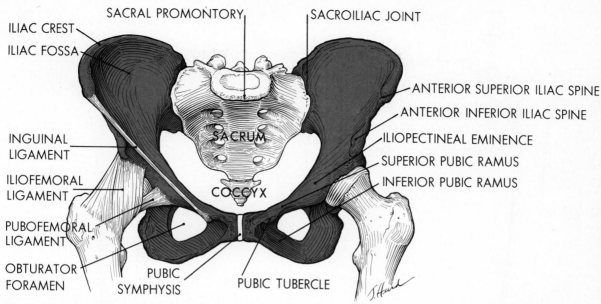

ILIAC CREST
ILIAC FOSSA
SACRAL PROMONTORY
SACROILIAC JOINT
INGUINAL LIGAMENT
ILIOFEMORAL LIGAMENT
PUBOFEMORAL LIGAMENT
OBTURATOR FORAMEN
PUBIC SYMPHYSIS
PUBIC TUBERCLE
SACRUM
COCCYX
ANTERIOR SUPERIOR ILIAC SPINE
ANTERIOR INFERIOR ILIAC SPINE
ILIOPECTINEAL EMINENCE
SUPERIOR PUBIC RAMUS
INFERIOR PUBIC RAMUS

**Figure 23-16** *Anterior aspect of a female pelvis.* (From Melloni, Stone, and Hurd.)

a constricted neck at its proximal end, with the acetabulum of the pelvic girdle.

2 Thigh. The anterior and posterior aspects of the right femur, or thighbone, are illustrated in Fig. 23-17A. The *femur* is the longest and strongest bone in the body. Because of the slight convex curvature of the femur and the breadth of the pelvis, the distal ends of the right and left femurs are closer together than their proximal ends. The lateral and medial condyles, knuckle-like projections on the posterior aspect of the distal end of the femur, articulate with corresponding condyles on the proximal end of the tibia. The patellar surface of the femur (see Fig. 23-17A) articulates with the *patella*, or "kneecap." Flat and triangular in shape, the patella is the largest sesamoid bone in the body. This bone is enclosed on the anterior aspect of the knee joint by the tendon of the quadriceps femoris muscles. Similar in function to all sesamoid bones, the patella increases the mechanical advantage of the joint around which it is located.

3 *Leg.* The *tibia* and *fibula,* or shin bones, are the *lat-*

*eral* and *medial bones,* respectively, in each leg or shank. The tibia is, for the most part, thicker than the fibula. The anterior surfaces of the right tibia and fibula are shown in Fig. 23-17B. These bones articulate with each other both proximally and distally at the *superior* and *inferior tibiofibular joints,* respectively, as well as along their shafts by means of an *interosseous membrane.* The *knee joint* is formed in part by articulations between the distal end of the femur, the proximal end of the tibia, and the posterior aspect of the patella. The knee joint is the largest and most complicated joint of the body.

4 *Foot.* The bones of each foot are composed of 7 *tarsals,* or ankle bones, 5 *metatarsals,* and 14 *phalanges,* or toe bones. Therefore, the foot is constructed of 26 bones, one less than the number of bones in the hand. The superior, or dorsal, aspect of the right foot is illustrated in Fig. 23-18. The *ankle joint* is formed by articulations between the talus and the distal ends of the tibia and fibula. Inferior to the talus is the *calcaneus,* or heel bone. Notice that the fourth and fifth metatarsals articu-

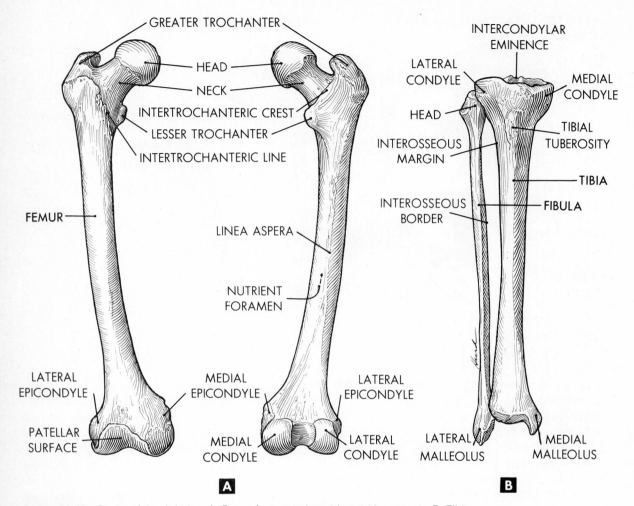

**Figure 23-17** *Bones of the right leg.* A. *Femur from anterior and posterior aspects.* B. *Tibia and fibula from anterior aspect.* (From Melloni, Stone, and Hurd.)

late with the cuboid. Observe the sesamoid bones in the region of the metatarsophalangeal joint of the great toe. As in the case of the fingers, each of the toes contains three phalanges, with the exception of the great toe, which contains only two. Notice the thickness of the first metatarsal and the phalanges of the great toe, compared to the other metatarsals and their associated phalanges. The directions of the arches of the foot are shown in Fig. 23-19. There are three arches: (a) a *lateral longitudinal arch* composed of the calcaneus,

talus, cuboid, and the fourth and fifth metatarsals; (b) a *medial longitudinal arch* shaped by the calcaneus, talus, navicular, cuneiforms, and the first, second, and third metatarsals; and (c) a *transverse arch* molded by the tarsals and, to some extent, the metatarsals. These arches are supported by tendons, ligaments, and muscles of the leg and foot. When a person stands, the load is evenly distributed between the ball and heel of the foot anterior and posterior to the transverse arch, respectively. Each of the longitudinal arches sup-

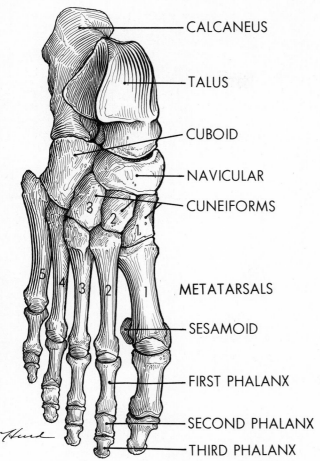

CALCANEUS

TALUS

CUBOID

NAVICULAR

CUNEIFORMS

METATARSALS

SESAMOID

FIRST PHALANX

SECOND PHALANX

THIRD PHALANX

**Figure 23-18**  *Bones of the superior, or dorsal, aspect of the right foot. Note the sesamoid bones surrounding the distal portion of the first metatarsal bone. (From Melloni, Stone, and Hurd.)*

## SKELETAL CHANGES AND VARIATIONS

From the time bones are first laid down during embryonic development to the time of death, individual bones and bone groupings undergo significant changes in shape. Figure 23-20 compares the major skeletal changes which occur.

The timing of both the onset of ossification and the completion of growth in length depends upon hereditary factors and hormonal influences. Indicative of the role of hormones is the fact that bone growth of a girl occurs more rapidly and ends at an earlier age than bone growth of a boy. Abnormalities of growth due to hormonal imbalances are considered in Chap. 18.

The following points are derived from a comparison of the fetal and adult skeletons in Fig. 23-20: (1) the axial skeleton decreases in size, relative to the appendicular skeleton; (2) the facial bones enlarge tremendously, relative to the bones of the cranium; (3)

**Figure 23-19**  *Arches of the foot. The lateral and medial longitudinal arches are represented by the lines c—b—d and c—d, respectively. The transverse arch is represented by the line c—f. The line a—b represents the axis around which the foot moves at the ankle joint. (From Melloni, Stone, and Hurd.)*

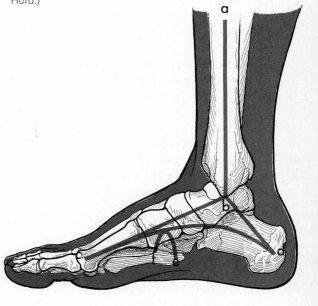

ports about half the weight borne by the ball of the foot. When a person walks, the heel is the first part of the foot to make contact with the ground. Support of the weight of the body is then passed to the ball of the foot as motion continues. Finally, the weight of the body is pushed forward when the distal phalanx of the great toe presses against the ground. The thicknesses of the first metatarsal and phalanges of the great toe are therefore related to the ability to walk.

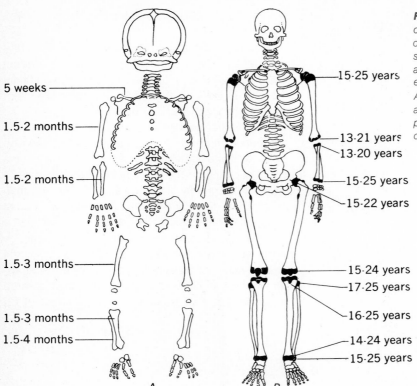

5 weeks

1.5-2 months

1.5-2 months

1.5-3 months

1.5-3 months

1.5-4 months

A

15-25 years

13-21 years
13-20 years

15-25 years

15-22 years

15-24 years
17-25 years

16-25 years

14-24 years
15-25 years

B

**Figure 23-20** *A comparison of the onset and termination of development of a fetal and adult skeleton. A. Fetal skeleton. The figures refer to the average developmental times when the epiphyses first begin to appear. B. Adult skeleton. The figures refer to the average years after birth when the epiphyseal cartilaginous disks completely disappear. (From DeCoursey.)*

growth of the cranial bones enlarges the volume of the braincase significantly; (4) the rounded rib cage becomes elliptical in shape; and (5) the pelvis enlarges, causing the distal portions of the femurs to turn inward.

At birth, the sutures between the flat bones of the cranium are not completely fused, exposing areas of fibrous membrane. These depressions can be palpated in an infant and are called *fontanels*. Fontanels mark the points on the cranium where intramembranous ossification is still incomplete. Fontanels permit a degree of overlap between the cranial bones during birth, facilitating labor and delivery. Disorders caused by premature ossification of the suture lines are discussed in Chap. 15. Brain damage caused by abnormal pressures on the skull during delivery is considered in Chap. 21.

Figure 23-21 identifies the six major fontanels in the cranium at birth. Notice that all the fontanels are associated with the borders of the parietal bones. The fontanels are named for their locations, as follows: (1)

the unpaired *anterior fontanel*, the largest of the cranial depressions at birth, is located between the frontal and parietal bones; (2) the unpaired *posterior*, or occipital, *fontanel* is located between the parietal and occipital bones; (3) the paired *sphenoidal fontanels* are located at the junctions of the frontal, sphenoidal, parietal, and temporal bones; and (4) the paired *mastoid fontanels* are located at the junctions of the parietal, occipital, and mastoid processes of the temporal bones. The anterior and mastoid fontanels ordinarily close between $1\frac{1}{2}$ and 2 years of age. The sphenoid and occipital fontanels usually close between the second and third months after birth.

At birth, the curvature of the spine is entirely concave when the vertebral column is viewed from anterior aspect. The concave thoracic and sacral curvatures are *primary*, since they are present at birth. The cervical and lumbar curvatures, on the other hand, are *secondary* because they are acquired sometime after birth. Both secondary curvatures are convex. The cervical curvature appears around 3 months after birth

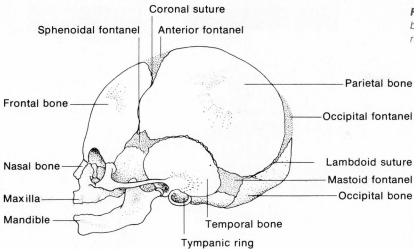

**Figure 23-21** *The skull of a newborn baby showing the fontanels from the right side.* (From DeCoursey.)

Labels on figure:
Coronal suture
Sphenoidal fontanel
Anterior fontanel
Parietal bone
Frontal bone
Occipital fontanel
Nasal bone
Lambdoid suture
Mastoid fontanel
Maxilla
Occipital bone
Mandible
Temporal bone
Tympanic ring

when infants first attempt to support their heads. The lumbar curvature appears at approximately 12 months of age when babies learn to stand and walk.

Not only do the shapes of bones change with time but the shapes of numerous individual bones and bone groupings differ when a comparison is made of male and female skeletons matched in age. In general, the bones of a male skeleton are more massive and contain more prominent projections and borders than the bones of a female skeleton. The pelvis of a female skeleton is broader and more squat, or compressed vertically, than the pelvis of a male skeleton, forcing the femurs to assume a pronounced medial direction from their proximal to their distal ends. The process of birth is aided by a large spherical pelvic inlet, broad sacrum, and shallow (obtuse) subpubic arch in the female. In an average male skeleton, there is a small oval pelvic inlet, straight sacrum, and sharp (acute) subpubic arch.

## ARTICULATIONS BETWEEN BONES

With the exception of the hyoid bone, each bone in the body articulates, or makes contact with, one or more adjacent bones. These articulation points be-

tween bones are called *joints*. Joints permit the normal operation of simple machines consisting of bones and muscles. The structure of a joint determines the nature of the movements which it permits.

Based on structural characteristics, there are three types of joint: (1) *fibrous*, or *synarthroses*; (2) *cartilaginous*, or *amphiarthroses*; and (3) *synovial*, or *diarthroses*. The structures of these joints are illustrated in Fig. 23-22.

### SYNARTHROSES

*Synarthrodial joints* permit very little, if any, movement since the bones of these joints are attached by fibroelastic connective tissue. Basically, there are two types of synarthrosis—sutures and syndesmoses. These joints are shown in Fig. 23-22A and B.

**Sutures**  The opposed, serrated edges of adjacent flat bones, e.g., the articulation joints between the bones of the cranium, are referred to as *sutures*. Notice that the entire thickness from periosteum to endosteum of each of the bones is joined together by fibrous tissue.

**Syndesmoses**  Adjacent borders of the radius and ulna and the distal ends of the tibia and fibula are

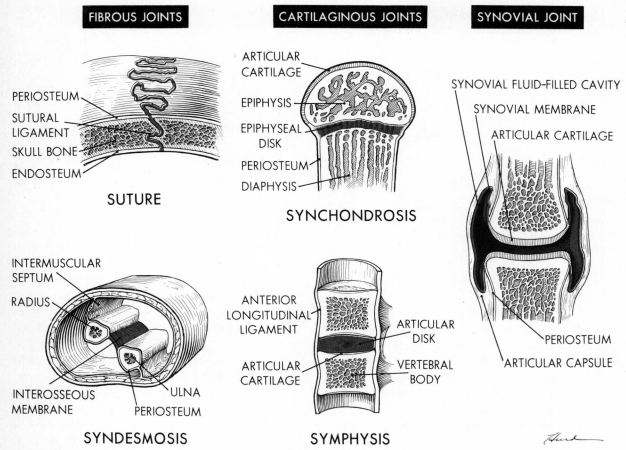

**FIBROUS JOINTS**

PERIOSTEUM
SUTURAL LIGAMENT
SKULL BONE
ENDOSTEUM

SUTURE

INTERMUSCULAR SEPTUM
RADIUS
INTEROSSEOUS MEMBRANE
ULNA
PERIOSTEUM

SYNDESMOSIS

**CARTILAGINOUS JOINTS**

ARTICULAR CARTILAGE
EPIPHYSIS
EPIPHYSEAL DISK
PERIOSTEUM
DIAPHYSIS

SYNCHONDROSIS

ANTERIOR LONGITUDINAL LIGAMENT
ARTICULAR DISK
ARTICULAR CARTILAGE
VERTEBRAL BODY

SYMPHYSIS

**SYNOVIAL JOINT**

SYNOVIAL FLUID-FILLED CAVITY
SYNOVIAL MEMBRANE
ARTICULAR CARTILAGE
PERIOSTEUM
ARTICULAR CAPSULE

**Figure 23-22** *The basic types of structural joint between bones. There are three groupings of joint—fibrous, cartilaginous, and synovial. Fibrous joints include sutures and syndesmoses. Cartilaginous joints include synchondroses and symphyses. Synovial joints are freely movable, compared to fibrous joints which are practically immovable or to cartilaginous joints which possess only limited motion. (From, Melloni, Stone, and Hurd.)*

joined together by interosseous membranes of fibroelastic connective tissue. These fibrous joints are referred to as *syndesmoses*. The interosseous membrane serves more to anchor adjacent bones in place than to permit movement.

## AMPHIARTHROSES

*Amphiarthrodial joints* permit a limited degree of movement, since the articulating portions of bones at these joints possess hyaline cartilage. There are two types of amphiarthrosis—synchondroses and symphyses. These joints are shown in Fig. 23-22C and *D*.

**Synchondroses** A *synchondrosis* is established when bone is linked to cartilage. Portions of the same bone may be joined by cartilage temporarily, as in the case of the union between the epiphysis and diaphysis by means of an epiphyseal cartilaginous disk during the growth stages of a long bone, or a bone may articulate with cartilage permanently, as in the case of the ribs and costal cartilages.

Symphyses   When two adjacent bones articulate by means of a fibrocartilaginous pad, the joint is called a *symphysis*. Examples include the articulations between either the bodies of adjacent vertebrae or the pubic bones.

## DIARTHROSES

*Diarthrodial joints* are freely movable and constitute the most common type of articulation between bones. The articulating portions of the bones of these joints are covered with articular hyaline cartilage to reduce friction during movements. A *synovial membrane* of fibroelastic connective tissue separates the articular hyaline cartilages, forming a synovial cavity between the adjacent bones. The *synovial cavity* is filled with a *synovial fluid*, which provides lubrication during movements of the joint as well as nutrients to the synovial membrane and articular cartilages. Because of the synovial cavity and fluid, diarthroses are also called *synovial joints.* Surrounding the synovial membrane and the periosteum of the articulating bones is an *articular capsule*. Reinforcing the articular capsule and the joint which it encloses are filamentous bands of tough, fibrous connective tissue called *ligaments,* which join the ends of the adjacent bones together. The general structure of a diarthrodial joint is depicted in Fig. 23-22*E.*

Functional types   As we mentioned earlier, the type of movement executed by any joint is determined by its structure. The structural determinants include (1) the nature of the connective tissue between the articulating bones; (2) accessory structures in and around the region of the joint; and (3) the specific shapes of the articulating bone surfaces. The first determinant largely establishes whether, and to what extent, a joint is capable of movement. The second determinant, characteristic of amphiarthrodial and diarthrodial joints, assists in executing the movements of a joint, provides mechanisms to absorb mechanical shock,

and supplies nourishment to the specialized connective tissues in the region of the joint. The third determinant fixes the functional capabilities of a joint.

The shape of a diarthrodial joint is determined by location and function. Joint shapes are characterized by their resemblance to physical objects or by the nature of the movement executed. Basically, there are five functional classes of diarthrodial joint—ball-and-socket, hinge, saddle, plane or gliding, and pivot. Most of these joints are shown in section in Fig. 23-23.

*Ball-and-socket joints* provide the greatest range of motion of any synovial joint in the body. In this type of joint, a cuplike depression in one or more bones articulates with the rounded head of another bone. Examples of ball-and-socket joints include the shoulder joint (Fig. 23-23*A*) and hip joint (Fig. 23-23*B*).

Notice the bursa immediately inferior to the acromion in the shoulder joint. *Bursae* are sacs lined by a synovial membrane and filled with synovial fluid; nevertheless, they *do not represent true joint cavities.* In general, bursae facilitate the movements of synovial joints by decreasing the friction between muscles or their tendinous extensions and bones or their ligamentous connections.

*Hinge joints* provide for movement in one plane, just as the hinges on a door permit it to swing open and closed. Examples of hinge joints include the elbow joint (Fig. 23-23*C*) and the knee joint (Fig. 23-23*D*).

The knee joint is the most structurally complicated joint in the body. The structural details of this joint are illustrated in Fig. 25-1, in conjunction with a discussion of a common affliction known as "game knee." Notice the numerous ligaments, bursae, and menisci, or cartilages, associated with the knee joint. The menisci facilitate movement and shape specific portions of the joint. The numerous ligaments and tendons which surround the knee joint stabilize the positions of the femur, tibia, and fibula.

A *saddle joint* enables movement in two planes, each of which is at right angles to the other. This wide range of motion is a consequence of an articulation between the convex end of one bone and the concave

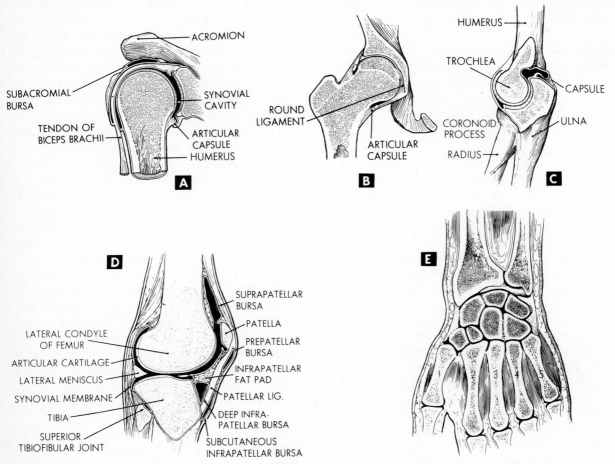

**Figure 23-23** *Examples of various functional types of synovial joint. A. Anterior section through the right shoulder joint. B. Anterior section through the right hip joint. A and B are examples of ball-and-socket joints. C. Anterior section through the right elbow joint. D. Anterior section through the right knee joint. C and D are examples of hinge joints. E. Horizontal section through the right wrist and portion of the hand viewed from superior or dorsal aspect. The metacarpophalangeal joint of the thumb is an example of a saddle joint. The intercarpal joints are examples of plane joints.* (From Melloni, Stone, and Hurd.)

end of another bone. The metacarpophalangeal joint of the thumb (Fig. 23-23*E*) is an example of a saddle joint.

A *plane,* or *gliding, joint* permits only limited motion as the surfaces of two bones held together by ligaments slide past each other. Examples of plane joints include intercarpal joints (Fig. 23-23*E*) and intertarsal joints.

A *pivot joint* permits rotational movements. Rotation is accomplished when the rounded side of one bone articulates with a lateral depression in another bone, as in the case of the proximal radioulnar joint (see Fig. 23-23*C*) or when a projection of one bone rotates in a depression of another bone, as in the case of several articulations formed by the first and second cervical vertebrae.

## SIMPLE MACHINES AND MECHANICAL ADVANTAGE

The bones at articulation points and the muscles which control their movements form simple machines. *Simple machines* change the direction of an applied force.

One type of simple machine exemplified by muscle-bone systems is called a *lever*. Determined by the ratio of the force arm to the weight or resistance arm, the mechanical advantage of a lever system es-

tablishes whether a functional grouping of muscles is used for speed or strength. When the weight or resistance arm is longer than the force arm, the lever is used for speed and the mechanical advantage is low, and vice versa. When both arms are approximately the same length, the lever system does not confer a mechanical advantage but simply changes the direction of an applied force. Various types of skeletomuscular lever systems are illustrated in Fig. 23-24A and B.

A *first-class lever* (Fig. 23-23A) is characterized by

**Figure 23-24** *Types of simple machines exemplified by skeletomuscular organization.*

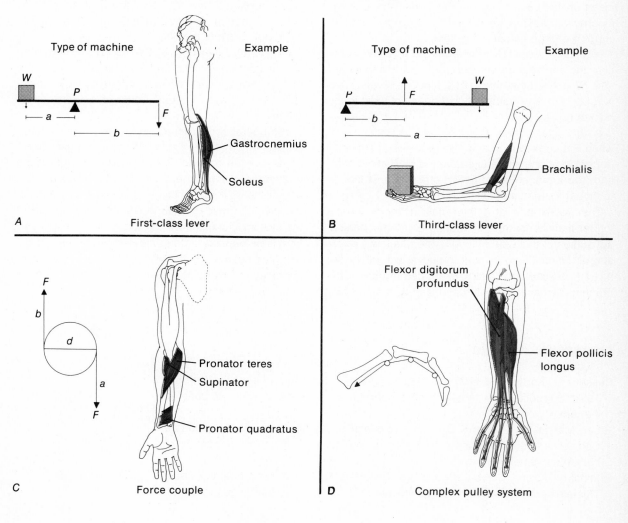

an arm supported at or near its center by a fulcrum or pivot (*P*). At one end of the arm is a resistance or weight (*W*). The length of the arm between *P* and *W* is called the *weight arm* (*A*). A force (*F*) is applied at the other end of the arm. The length of the arm between *P* and *F* is called the *force arm* (*b*). The arrangement of a first-class lever is similar to that of a seesaw.

The function of a first-class lever depends on how close *W* and *F* are to each other. When *W* and *F* are close together, the lever system is used for support or posture. For example, when standing on the toes, the weight of the body is on the ball of the foot, the pivot is the ankle joint, and the applied force of the calf muscles is transmitted mainly through tendons which insert on the calcaneus, or heel bone. Although the mechanical advantage is low, this lever system can support a large amount of weight.

A *second-class lever* is established when *W* is at or near the center of an arm supported by *F* at one end and *P* at the other end. This type of lever is exemplified by a wheelbarrow but does not exist in the human body.

A *third-class lever* (Fig. 23-24*B*) is established when *F* is applied at or near the center of an arm supporting *W* at one end and supported by *P* at the other end. The arrangement of a third-class lever is similar to that of a reel-type fishing rod.

The weight arm of a third-class lever is always longer than its force arm. Therefore, the mechanical advantage of this type of lever system is low. For example, when lifting or supporting a weight in the hand, *P* is the elbow joint, *F* is applied in part by the brachialis muscle to the forearm, and *W* is supported in the hand. Third-class levers are utilized for speed and will move a small object through a large distance.

A *force couple* (Fig. 23-24*C*) is established when equal but opposite forces are applied tangentially at different points to a pivotal axis. Force couples are responsible for the execution of rotational movements. The arrangement of a force couple is similar to the situation which would exist if a person rowing a boat were to push one oar backward and simultaneously pull the other oar forward with approximately equal force.

A force couple is found wherever rotational movements of any kind occur at the axis of a joint. A biological example of a force couple is provided when the forces applied by the pronator and supinator muscles of the forearm cause the head of the radius to rotate in the radial notch of the ulna.

A *complex pulley* (Fig. 23-24*D*) is established when the direction of *F* is changed by means of a series of pulleys. The arrangement of a complex pulley is similar to that of a compound boom of a construction crane.

The operation of a complex pulley is exemplified biologically by finger flexion. When the finger flexors of the forearm contract, the tendons pull against the phalanges of the fingers, causing their flexion. Not only is the direction of the applied force of the flexor muscles changed but the mechanical advantage of the system increases as the tendons are displaced from their joint axes during movement. In other words, complex pulleys are capable of increasing the mechanical advantage as well as changing the direction of the applied force.

## COLLATERAL READING

Bassett, C. Andrew, "Electrical Effects in Bone," *Scientific American,* October 1965. Suggests how small electric currents are generated when a bone is reshaped.

Broer, Marion R., *Efficiency of Human Movement,* Philadelphia: W. B. Saunders Company, 1960. Chapters 3 to 7 discuss the basic mechanical principles of human movement.

Brunnstrom, Signe, *Clinical Kinesiology,* Philadelphia: F. A. Davis Company, 1962. This elementary textbook applies mechanical principles to the human body.

Napier, John, "The Evolution of the Hand," *Scientific American,* December 1962. The discovery of the fossilized remains of hand bones and tools of a hominoid believed to be more than a million years old indicates that the capabilities of the modern hand were evolved at an early time in man's evolution.

Napier, John, "The Antiquity of Human Walking," *Scientific American,* April 1967. The discovery of the fossilized remains of a distal phalanx of the right great toe of a hominoid believed to be more than a million years old indicates that man's ability to stride is relatively primitive.

Rasmussen, Howard, "The Parathyroid Hormone," *Scientific American,* April 1961. Written before the discovery of calcitonin, the article identifies the actions of parathyroid hormone and vitamin D as primary in controlling both the total amount, and the ratio, of $Ca^{++}$ and $HPO_4^{--}$ in bone and blood.

Rasmussen, Howard, and Maurice P. Pechet, "Calcitonin," *Scientific American,* October 1970. Summarizes the probable molecular mechanisms by which calcitonin and parathyroid hormone reciprocally influence the activity of osteocytes.

# CHAPTER 24
# THE MUSCULAR SYSTEM

The muscular system consists of organs composed of three types of tissue: skeletal, cardiac, and smooth. All muscle tissues possess two functional qualities to a marked extent—*contractility*, or the ability of the tissue cells to shorten in length, and *elasticity*, or the ability of the tissue cells to stretch, or elongate.

There are over 400 skeletal muscles in the body. As their name implies, *skeletal muscles* are associated with bones. The ends of skeletal muscles attach to bones, other muscles, or specialized types of connective tissue such as fascia or aponeuroses. The overlapping functions of skeletal muscles include

1   The provision of support, the maintenance of posture, and the execution of movements by providing power to the simple machines of which the muscles are members
2   Facilitation of interpersonal communication through the production of sounds (phonation) and by enabling facial expressions and body gestures to occur
3   Control of pulmonary ventilation by increasing or decreasing the diameter of the rib cage and the volume of the thoracic cavity
4   Facilitation of defecation, urination, and birth by muscular compression of either the thoracic or abdominal viscera
5   Control of the volitional aspect of urination by contraction or relaxation of the external urethral sphincter
6   Control of the movements of the eyeballs by the extrinsic eyeball muscles
7   Control of the amplitude of bone conduction of sound pressures by the tensor tympani muscle of the malleus and the stapedius muscle of the stapes
8   Regulation of body temperature by the production and liberation of heat through shivering

9   Regulation of blood volume through the local storage or release of intramuscular blood
10  The return of venous blood to the heart by muscular compression of the veins in the extremities

*Cardiac muscle* is found in the wall of the heart. Called *myocardium,* the heart muscle pumps blood from the atria to the ventricles and from the ventricles to the lungs and the rest of the body.

*Smooth muscle* is found (1) in the walls of all tubular and vesicular visceral organs; (2) in the dermis of the skin associated with hair follicles; and (3) in the eye and other specialized structures. In location or function, therefore, smooth muscle is either dermal, visceral, or sensory. The functions of smooth muscles include

1   Movement by peristalsis of food, urine, spermatozoa, and blood through the digestive tract, ureters, ejaculatory ducts, and blood vessels, respectively
2   Evacuation of materials such as bile and urine during contraction of the smooth muscle walls of the gallbladder and urinary bladder, respectively
3   Maintenance of, and adjustments in, the diameters of the respiratory air passages
4   The mechanical digestion of food by means of segmentation movements executed by the walls of the small intestine
5   Maintenance of, and adjustments in, blood pressure through constriction or dilation of blood vessels
6   Regulation of body temperature by controlling heat losses through the constriction or dilation of peripheral blood vessels and by creating a rough skin surface through contraction of the muscles associated with the body hairs on the arms and legs when the skin is cold
7   The focusing of visual images (accommodation) and control of the intensity of light entering the eyeballs (the pupillary reflex) through the actions of the ciliary and iris muscles, respectively.

In this chapter, we will examine the structure of the different types of muscle tissue, the nature of the contractile process, physiologic aspects of skeletal muscle contraction, and the organization and general functions of the major skeletal muscles. Muscular disorders are considered in the next chapter.

## TYPES OF MUSCLE

### SKELETAL MUSCLE

Skeletal muscles are organs composed of striated muscle tissue, together with other types of connective tissues, blood vessels, lymphatic vessels, and a nerve supply. Striated muscle tissue, in turn, consists of specialized cylindrical cells, some of which may be as long as four inches. Called *skeletal muscle fibers*, these cells compose the primary tissue of skeletal muscles. Each muscle fiber is invested by a coat of connective tissue called *endomysium* (the prefix *my-* or *myo-* means "muscle").

Skeletal muscles contain radially arranged, internal partitions of connective tissue which divide the organ into compartments called *fasciculi*. Each fasciculus contains many muscle fibers packed closely together in parallel. The connective tissue partitions which form the fasciculi are termed *perimysium*. Skeletal muscles are enveloped by a fascia referred to as *epimysium*.

The nerve fibers which innervate skeletal muscles originate from either cranial or spinal nerves. Specialized regions in skeletal muscles and their associated tendons contain sensory structures such as muscle spindles and Golgi tendon organs. These sensory structures provide information to the brain on the exact positions of the muscles in space and their rate of displacement with time. This sensory ability, which depends upon the degree of stretch of a muscle, tendon, or ligament is called *proprioception* and is discussed further in Chap. 11.

Skeletal *muscle fibers* are the structural units of skeletal muscle. When appropriately stimulated, muscle fibers contract by shortening in length and increasing in thickness. The structure of a skeletal muscle fiber is illustrated in Fig. 24-1. A knowledge of

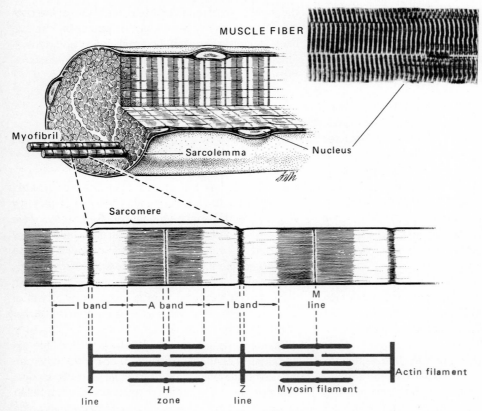

MUSCLE FIBER

Myofibril

Sarcolemma

Nucleus

Sarcomere

I band — A band — I band

M line

Z line

H zone

Z line

Actin filament

Myosin filament

**Figure 24-1** *The structure of a skeletal muscle cell. A perspective view through one muscle cell is shown in the upper left. A photomicrograph of a muscle fiber is illustrated in the upper right. The muscular banding and protein organization of the sarcomeres are diagramed in the lower half of the illustration.* (From Langley, Telford, and Christensen.)

its organization is essential for an understanding of the physical events which take place during muscular contraction. A side view of a portion of a skeletal muscle fiber observed with the use of a compound light microscope is shown in the upper right-hand corner of Fig. 24-1. Notice the regularly repeated striations which run perpendicular to the long axis of the muscle fiber. We shall have more to say about these striations shortly.

A similar portion of a skeletal muscle fiber is diagramed in three dimensions in the upper left-hand corner of the same figure. The outer boundary of a skeletal muscle fiber is composed of a *sarcolemma.* Performing the functions of a cell membrane, the

sarcolemma is equivalent to the neurilemma of a nerve fiber (see Chap. 13). Each muscle fiber is *multinucleated.* The nuclei reside on the periphery of the muscle cell, immediately beneath the sarcolemma. The presence of peripheral nuclei causes bulges in the sarcolemma. The inner substance of a muscle fiber consists of closely packed, ultramicroscopic, rod-shaped structures oriented along its long axis. Called *myofibrils,* these filamentous rods contain the contractile machinery of the muscle fiber.

Between the myofibrils is *sarcoplasm,* the muscle fiber equivalent of cytoplasm. Running through the sarcoplasm is a complex system of tubules (not shown) known as the *sarcoplasmic reticulum.* Equivalent to the endoplasmic reticulum of other types of cells (see Chap. 5), this specialized organelle is responsible for the following functions: (1) the transmission of action potentials (excitation) from the sarcolemma to the individual myofibrils, causing the release of calcium ($Ca^{++}$), the activator of muscular contraction; and (2) the intracellular distribution of enzymes and other metabolic substances which lead to the syntheses of ATP, creatine phosphate, and lactic acid. These materials are either necessary for, or produced by, muscular contraction. The roles played by these chemicals will be explained later.

As shown in Fig. 24-1, the organization of the myofibrils is responsible for the regularly repeated pattern of vertical striations seen in skeletal muscle fibers. Called a *sarcomere*, each of these repeating units of a myofibril runs from one thick, dark Z line to the next.

The molecular organization of two adjacent sarcomeres is conceptualized in the lower portion of Fig. 24-1. Two types of filamentous proteins are found in each sarcomere: thin filaments composed of a protein called *actin* and thick filaments composed of a protein called *myosin*. The actin filaments are attached to the Z lines and, in an uncontracted state of a myofibril, do not make contact with each other in the center of a sarcomere. The myosin filaments, on the other hand, are located in the central region of the sarcomere and, in an uncontracted state of a myofibril, do not make contact with the Z lines. Molecular *cross bridges* which apparently pivot on the myosin filaments make contact with adjacent actin filaments at specific binding sites. These cross bridges play pivotal roles during muscular contraction, as we shall see.

The characteristic positions occupied by the contractile filaments in a sarcomere give rise to a distinctive vertical banding when light is transmitted through the width of a sarcomere. Characterized by the presence of actin filaments and the absence of myosin filaments, the band to either side of a Z line is called an *I band*. An I band therefore encompasses portions of two adjacent sarcomeres. The region between two adjacent I bands is characterized by a partial overlap of actin and myosin filaments and is referred to as an *A band*. The central portion of an A band is called an *H*

*zone*. The H zone is characterized by the presence of myosin filaments and the absence of actin filaments. These bands and lines will become more meaningful as we discuss the physical events which transpire when sarcomeres undergo contraction.

## CARDIAC MUSCLE

Cardiac muscle (myocardium) is a specialized muscle tissue which makes up the thickest portion of the wall of the heart. Its specializations permit rapid excitation, great strength of contraction, and rhythmicity. Observed with the use of a compound light microscope, cardiac muscle tissue is illustrated in the upper portion of Fig. 24-2.

With three exceptions, cardiac muscle fibers are qualitatively similar to skeletal muscle fibers in structure. The exceptions are as follows:

1   Cardiac muscle tissue consists of cells which are highly branched and interconnected. The anastomoses between individual cells make the beginning and end of one cell in the complex difficult to identify. Tissues in which cells are indistinguishable are called *syncytial*. The myocardium is a syncytium because, literally, the cells are fused together. The arrangement is adaptive, however, in that the excitation of one region of the myocardium results in a rhythmic wave of contraction along the rest of the branching cells.
2   The ends of adjacent cardiac muscle cells are fused to each other by specialized organelles similar to desmosomes (see Chap. 5). These fusion points between the cardiac cells are marked by dark vertical bands called *intercalated disks*. Such specializations presumably contribute to an increase in the rate of conduction of the wave of excitation along the myocardium.
3   Innervated by autonomic nerve fibers, myocardium is ordinarily under involuntary control.

There are two other unique features of myocardium which we should note here. Firstly, groups of cardiac muscle fibers are oriented in several different directions in myocardium so that a wringing action results

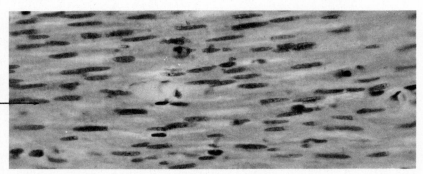

Cardiac fiber

Intercalated disk

Muscle nuclei

**Figure 24-2** *The structure of cardiac and smooth-muscle tissue. Upper photomicrograph shows cardiac muscle; lower photomicrograph shows smooth muscle. (From Langley, Telford, and Christensen.)*

from their contraction. When the ventricles contract, this wringing action is responsible for the expulsion of every drop of blood within their chambers, just as twisting a wet washcloth wrings it dry. The result is an increase in the pumping efficiency of the heart. Secondly, groups of modified cardiac muscle cells in the walls of the myocardium serve as a specialized *conduction system* to transmit the excitation from the right atrium to and around all portions of the ventricular myocardium. The arrangement of the cardiac muscle fibers and the functional organization of the conduction system of the heart are considered in detail in Chap. 30.

## SMOOTH MUSCLE

Smooth muscle cells lack the vertical striations characteristic of skeletal and cardiac muscle cells.

Whether single fibers, tissues, or multiple layers, smooth muscles are innervated by autonomic nerve fibers. Therefore, the responses of smooth muscle are ordinarily involuntary. Observed with the use of a compound light microscope, smooth muscle tissue is shown in the lower portion of Fig. 24-2.

Each smooth muscle cell is spindle-shaped. The relatively thick, central portion of each cell houses an elongated nucleus. The absence of vertical striations is explained by a lack of a specific orientation by the actin and myosin filaments within the myofibrils.

In smooth muscle tissue, the tapered end of one cell lies next to the thick, central portion of an adjacent cell, forming a staggered arrangement similar to that exhibited by bricks in a brick wall. Multiple attachment points exist between adjacent cells, presumably to facilitate the conduction of a wave of muscle excitation along the tissue. Intercellular elastic connective tissue fibers bind all the contiguous smooth muscle cells together. The surrounding elastic fiber matrix tends to cause all the cells to contract rhythmically when one portion of a smooth muscle tissue is appropriately stimulated.

## MUSCULAR CONTRACTION

The mechanism of skeletal muscular contraction has been a subject of study since man first became curious about his anatomy and muscular abilities. Today, the broad outlines of how a muscle contracts are fairly well established. Although the story is complex, its elegance may be comprehended by relating the events of muscular contraction in three parts—excitation, molecular correlates, and mechanical correlates. As the story unfolds, bear in mind that these events take place in less than a tenth of a second.

### EXCITATION

Skeletal muscles are innervated by somatic motor nerve fibers. Some muscles, such as those of the hand with which fine, precision movements are executed, have an abundant nerve supply. The axonal branches of one motor fiber from the spinal cord, for example, may innervate no more than five skeletal muscle fibers in muscles which control movements of the fingers. Other muscles, such as prime movers, are not as amply innervated. Perhaps as many as 50 muscle fibers in the superficial muscles of the back may be innervated by branches from one motor fiber.

When a motor fiber conducts a nerve impulse of at least threshold strength for contraction of a muscle, all the muscle fibers innervated by the nerve fiber are thrown into contraction simultaneously. Because the identically innervated muscle fibers act as a unit, the nerve fiber and the muscle fibers which its branches innervate are referred to collectively as a *motor unit.*

Muscles maintain their tone or resistance to stretch by the alternating actions of surprisingly few active motor units, thereby decreasing the probability of early fatigue during sustained contraction. Within limits, the strength of a muscular contraction depends upon the number of active motor units and, therefore, the number of muscle fibers which are contracting.

The intersection of an axonal branch of a somatic motor nerve fiber and the sarcolemma of a muscle fiber is called a *myoneural,* or *neuromuscular, junction.* There is an extremely small space between the axonal tip and the sarcolemma. For muscular contraction to occur, several molecular events must take place at the myoneural junction.

Let us assume that a motor fiber is conducting a nerve impulse. When the nerve impulse reaches the tip of the axon, it somehow causes the release of a neurohumor called *acetylcholine* (ACh) from vesicles in which it is stored. Acting as a neurotransmitter substance, ACh crosses the cleft between axon and muscle fiber, exciting a specialized portion of the sarcolemma called the *motor end plate.* Almost immediately, ACh is hydrolyzed into *acetate* (A) and *choline* (Ch) by the enzyme *cholinesterase* (ChE). Hydrolysis of ACh causes its inactivation, thereby preventing excessive stimulation of the motor end plate after the passage of a nerve impulse. ACh is gradually resynthesized from A and Ch and stored in axonal vesicles so that a continual supply of ACh is available, if needed. Normally, neural fatigue caused by the depletion of ACh is rare.

The excitation of the motor end plate by ACh is accomplished by changing the permeability of the sarcolemma in a manner similar to the permeability change in a neurilemma (see Chap. 13). The permeability of the motor end plate to $Na^+$ is increased, causing an influx of $Na^+$ from the extracellular spaces and a secondary leakage of intracellular $K^+$ into the extracellular spaces. These ionic fluxes change the electrical potential of the membrane and result in its *depolarization.* The electrical potential established during depolarization of the end plate is called the *end plate potential.* The end plate potential spreads rapidly in both directions along the sarcolemma. Depolarization of the sarcolemma results in an *action potential* of the muscle fiber. The basic events which occur at the myoneural junction are summarized on the left in Fig. 24-3.

The action potential of a muscle fiber is conducted

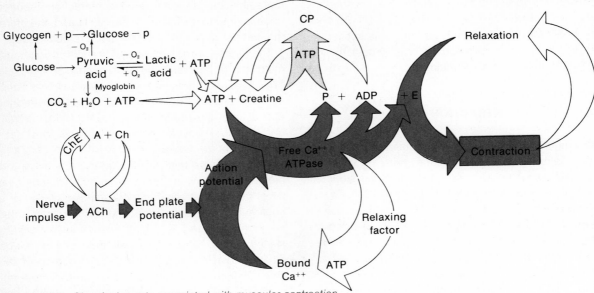

**Figure 24-3** *Chemical events associated with muscular contraction.*

through the sarcoplasmic reticulum to the individual myofibrils. The end result of the action potential is believed to be the release of bound $Ca^{++}$ from specialized portions of the sarcoplasmic reticulum. $Ca^{++}$ is the activator of muscular contraction. Almost immediately after the release of bound $Ca^{++}$, an as yet uncharacterized *relaxing factor* removes the free $Ca^{++}$ by binding it once again to the sarcoplasmic reticulum. Removal of free $Ca^{++}$ prevents sustained contractions in the absence of repetitive action potentials. These sequential relationships are shown in the middle of Fig. 24-3.

## CHEMICAL CORRELATES

To contract, muscles require a supply of energy. This energy is initially provided by *adenosine triphosphate* (ATP) molecules which are stored in specialized portions of the myosin filaments. Illustrated in Fig. 24-4, ATP is the energy currency of cells, liberating its chemical energy when split enzymatically. Free $Ca^{++}$ activates an enzyme called *adenosine triphosphatase* (ATPase), which splits ATP, forming *adenosine diphosphate* (ADP) and *phosphate* (P),

and liberating energy (E). A small portion of this energy, perhaps no more than 35 percent of the amount liberated, causes actin and myosin filaments to slide over each other by a process called *interdigitation*. Interdigitation is the molecular counterpart of muscular contraction. The energy not used for muscular contraction is lost as heat.

When a muscle is called upon to contract after a period of rest, all its energy is derived from intramuscular phosphagens, i.e., molecules containing

**Figure 24-4** *The structure of adenosine triphosphate (ATP). ATP consists of a base (adenine on the left), a sugar (ribose in the middle), and a side chain consisting of three P atoms which alternate with O atoms (on the right).*

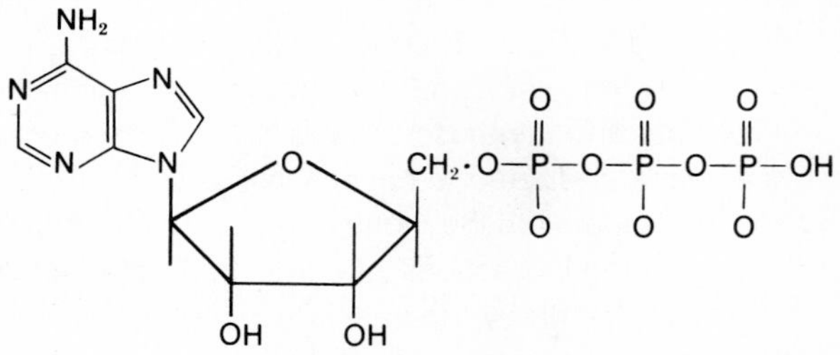

phosphate. The phosphagen drawn upon immediately, as we have already noted, is the ATP stored in specialized regions on the myosin filaments. If muscular work is to continue, the degraded ATP must be replaced.

ATP is regenerated by splitting another phosphagen called *creatine phosphate* (CP), or phosphocreatine, stored in the sarcoplasm. When CP is split in the presence of ADP, the products which form are ATP and creatine (C). The energy necessary for the synthesis of ATP is obtained from the splitting of CP. However, CP must also be regenerated. The raw materials required for its synthesis are C formed during the splitting of CP, and P formed during the splitting of ATP. These chemicals are already present in the sarcoplasm. Thus, CP is synthesized from C and P, in the presence of ATP.

The ATP necessary for this synthetic reaction is obtained from the oxidation of intermediary metabolites in mitochondria located within the sarcoplasm. In other words, the synthesis of the ATP, upon which the regeneration of CP depends, in turn relies upon aerobic respiration.

During vigorous physical exercise, the rate of depletion of the phosphagens exceeds the rate of their replacement through the action of oxidatively derived ATP. Under these circumstances, the muscle incurs an *oxygen debt*. The sequential relationships described above are shown in the upper portion of Fig. 24-3.

Intramuscular respiration is aided by a specialized respiratory pigment called *myoglobin*. Muscles rich in myoglobin, such as the arm and leg muscles, are red and are ordinarily called upon for brief spurts of rapid activity. Muscles sparse in myoglobin, such as the superficial muscles of the chest and back, are white and must remain contracted for long periods of time. Similar but not identical to hemoglobin, myoglobin has a significantly greater affinity for $O_2$.

As the concentration of $O_2$ in the intramuscular blood capillaries becomes depressed during physical exertion, sarcoplasmic myoglobin is still able to take up $O_2$ from hemoglobin and make it available to sarcoplasmic mitochondria. This extra supply of $O_2$ enables intramuscular oxidation to continue for a longer period than would otherwise be possible. The onset of fatigue is delayed by the continued synthesis of ATP.

When the energy derived from the phosphagens and from aerobic respiration decreases below a critical threshold during extended physical exercise, the muscle begins to derive energy through anaerobic respiration. During this stage of muscular metabolism, the muscle is of course in oxygen debt.

The energy for muscular contraction is ultimately derived from the degradation of glucose by a process called *glycolysis* (see Chap. 6). Glycogen is a polysaccharide, or multiple sugar, made up of many glucose residues. In the presence of phosphate and an enzyme called *phosphorylase,* a terminal glucose residue is split off the glycogen molecule and a phosphate group is added to the free glucose residue, forming a molecule called *glucose phosphate.*

In a series of enzymatically mediated reactions, glucose phosphate is degraded into two pyruvic acid molecules. In the presence of $O_2$, each pyruvic acid forms the end products $H_2O$ and $CO_2$. Aerobic respiration liberates large amounts of energy trapped in the chemical bonds of ATP. In the absence of $O_2$, glucose phosphate is degraded into the end product lactic acid. Anaerobic respiration liberates only one-tenth to one-twelfth as much energy as aerobic respiration. These sequential relationships are shown in Fig. 24-3.

The ATP formed during anaerobic respiration, although much less than that synthesized during oxidative metabolism, is still sufficient to permit contractions by a muscle in $O_2$ debt. However, as lactic acid accumulates, the intramuscular pH decreases. The acidosis which results from the presence of lactic acid and other biochemical changes are responsible for the onset of muscle fatigue, presumably by preventing the normal actions of intramuscular enzymes.

As lactic acid accumulates intramuscularly, some of it leaves the muscle fibers and enters intramuscular blood capillaries. This lactic acid is ultimately transported through the blood to the liver, where it is used in the synthesis of glycogen. However, the majority of the lactic acid produced when a muscle incurs an

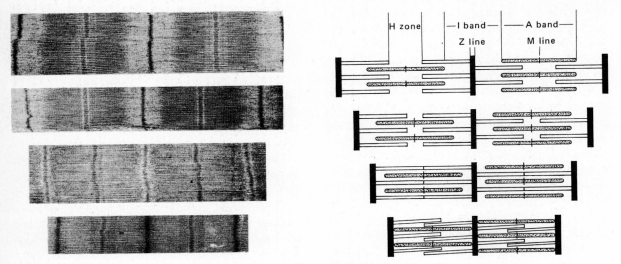

**Figure 24-5** *Physical changes in striated muscle fibers during contraction. Photomicrographs of these changes are on the left. The physical changes are diagramed on the right. Two adjacent sarcomeres are illustrated.* (From Langley, Telford, and Christensen.)

$O_2$ debt remains in the muscle fibers. Following the completion of muscular exercise, the rapid breathing characterized by panting begins to restore the normal levels of intramuscular $O_2$. The infusion of $O_2$ during the repayment of an $O_2$ debt converts the lactic acid into pyruvic acid, which is then metabolized aerobically. However, the removal of accumulated lactic acid and the restoration of aerobic respiration is relatively slow. Thus, the subjective experience of muscle fatigue continues for some time after the completion of hard physical work.

Actually, muscle fatigue is a complex phenomenon which may have a multiplicity of causes. In a physiologic sense, muscle fatigue can arise from the following factors: (1) acidosis due to an accumulation of lactic acid during anaerobic respiration, as mentioned above; (2) depletion of ACh; (3) depletion of extracellular $Na^+$ or intracellular $K^+$; or (4) lack of the mineral $Ca^{++}$. Muscle fatigue is discussed further in Chap. 25. The effects of $Ca^{++}$ deficiency on muscles are discussed in Chaps. 25 and 27.

## MECHANICAL CORRELATES

The theory that a muscle shortens when actin and myosin filaments slide over each other is supported by changes in the appearance of the H zones of the sarcomeres and the I bands between adjacent sarcomeres in a contracting myofibril. These changes are illustrated on the left in Fig. 24-5 with sequential photomicrographs of striated muscle during contraction. Diagrams of the corresponding changes in the relative positions of the actin and myosin filaments in adjacent sarcomeres are shown on the right in the same figure. Notice that the H zone decreases in thickness and is finally obliterated as the ends of the actin filaments in a single sarcomere approach, contact, and slide over each other. Also observe the disappearance of the I band spanning adjacent sarcomeres as the myosin filaments from each sarcomere approach and contact the Z line. Finally, note that the degree of contraction is limited by the lengths of the myosin filaments in each sarcomere; i.e., a sarcomere cannot shorten below the length of its individual myosin filaments.

A partial contraction occurs when the ends of the actin filaments contact each other. The muscle is fully contracted when the actin filaments of the individual sarcomeres partially overlap.

The ability of the actin and myosin filaments to slide over each other has not yet been satisfactorily explained. However, a specialized, globular portion of each myosin filament called *heavy meromyosin* is thought to swivel about its axis during interdigitation while attached to its specific binding site on an actin filament. These heavy meromyosin cross bridges are believed to use the energy released when ATP is split in order to tilt. Presumably, tilting supplies the force causing interdigitation, in a manner analogous to the force supplied by an individual ratchet of an automobile jack as it is used to raise one end of a car.

## THE PHYSIOLOGY OF MUSCULAR CONTRACTION

### SKELETAL MUSCLE

**The recoiling-spring analogy** The contractions of skeletal muscles can be compared to the behavior of a spring fastened at one end to a fixed point and at the other end to a load. The analogy is diagrammed in Fig. 24-6. Notice that the spring consists of series elastic elements capable of stretching, such as sarcoplasm, and that the spring is attached to a bar of contractile elements corresponding to filaments of actin and myosin. In *A*, the spring is allowed to stretch in response to the load before the elastic elements pull in the opposite direction; i.e., before contraction occurs. In muscles, this type of contraction is called *isotonic*. In *B*, the load is fixed, or immovable, so that the spring is not stretched before the elastic elements exert tension against the load. In muscles, this type of contraction is called *isometric*. Any muscle may contract isotonically under one set of conditions and isometrically under another set of conditions.

Assume that an isotonic muscle is stretched by a load and is then stimulated to contract, as in the stretched spring analogy of Fig. 24-6*A*. Before the muscle undergoes a visible contraction which displaces the load, energy must be expended to overcome the viscous resistance of the stretched sarcoplasm. In other words, in response to the energy liberated by ATP bound to myosin, the actin and

**Figure 24-6** *A mechanical model of a muscle undergoing (A) isotonic contraction and (B) isometric contraction.*

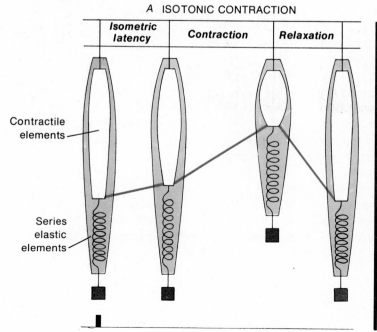
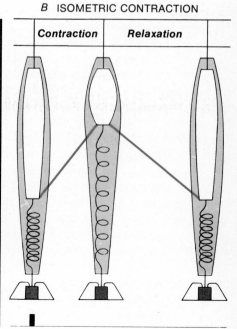

myosin filaments begin to interdigitate. In our analogy, the series elastic elements are partially stretched and the contractile elements are partially compressed. Since heat is liberated although visible contraction does not yet occur, this phase of isotonic contraction is called *isometric latency.*

Next, the muscle shortens visibly and the load is displaced from its original position. Called *contraction,* this phase of isotonic contraction is characterized by the completion of interdigitation. In our analogy, the contractile elements are completely compressed, although there is no further change in the length of the series elastic elements.

Finally, after the stimulus is removed, the muscle lengthens to its original resting position as interdigitation of the actin and myosin filaments is reversed. This phase of isotonic contraction is referred to as *relaxation.* In our analogy, the original relative dimensions of the series elastic and contractile elements are reestablished.

The entire cycle of a muscular contraction lasts about 0.1 sec (100 msec). The phase of relaxation takes about as long as isometric latency and contraction combined. Interestingly, maximum heat is liberated before peak contraction is reached, but heat evolution in lesser quantities continues for a short time after the phase of relaxation is ended.

Now assume that an isometric muscle is attached to an immovable load and then stimulated to contract, as in the unstretched spring analogy of Fig. 24-6B. Because the muscle is not stretched by the load, the actin and myosin filaments do not have to overcome the viscous resistance of stretched sarcoplasm. Therefore, the muscle is thrown into contraction without undergoing a period of isometric latency. Notice, however, that the series elastic elements are stretched considerably at the peak of contraction, compared to the degree of their stretch at the peak of isotonic contraction. The events during isometric relaxation are identical to those during isotonic relaxation. Relaxation of an isometric muscle takes a longer time than the period of contraction.

Because isometric muscles are not allowed to stretch against their loads, their efficiency as measured by the work they can perform is only about half

that of the same muscles under isotonic conditions. The reason is that the strength of a muscle contraction is proportional to the degree of muscle stretch exerted by an applied load. In other words, a muscle which is allowed to stretch against a load before contracting operates more efficiently, compared to the same muscle when prevented from stretching against the same load before contracting.

**Responses to stimuli.** Skeletal muscle behaves in characteristic ways to stimuli which vary in strength and duration or in the time interval between their individual application. First, let us turn our attention to the responses of a skeletal muscle fiber to single stimuli of graded or increasing intensity. These responses are illustrated in Fig. 24-7. Bear in mind that a muscle fiber is a cell and not an organ.

The stimulus on the left is below the response threshold of the muscle fiber. Called *subliminal,* the stimulus does not cause a contraction of the fiber.

The middle of the three stimuli is referred to as a *liminal stimulus.* The magnitude of a liminal stimulus is exactly at the response threshold of the muscle fiber. A threshold stimulus elicits a contraction of the fiber.

The stimulus on the right is above the threshold, or *supraliminal.* The supraliminal stimulus elicits a contraction of the same order of magnitude as that of the liminal stimulus, even though the former is stronger than the latter.

The ability of a striated muscle fiber to contract maximally to any single stimulus of at least threshold strength is called the *all-or-none law.* In other words, either a muscle fiber will contract maximally or it will not contract at all.

Next, let us examine the behavior of a skeletal muscle in response to stimuli of varying strengths or separated by different interstimulus intervals. These responses are illustrated in Fig. 24-8. Note that we are now dealing with a whole organ (the muscle) and not with a cell (the fiber).

Assume that a series of closely spaced subliminal stimuli is applied to the muscle, as in Fig. 24-8A. Each individual stimulus by itself is insufficient to cause a muscular contraction. However, if applied

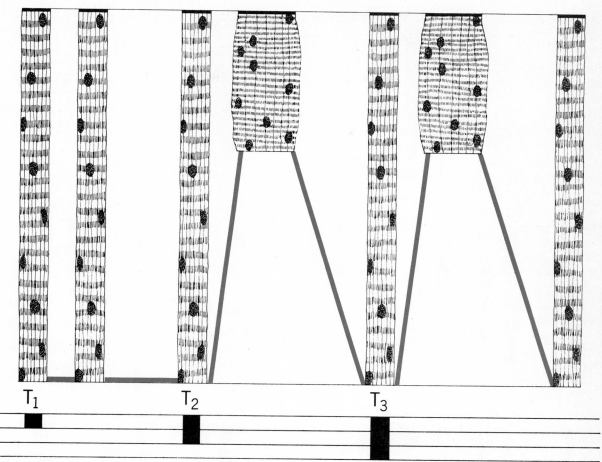

**Figure 24-7**  *The responses of a skeletal muscle fiber to single stimuli graded in strength. The fiber is suspended from its upper end; its lower end hangs free. Symbolized by vertical lines below the fiber, the stimuli are provided electrically. The lengths of the vertical lines are proportional to the intensity of the stimulus.* $T_1$. *Subliminal stimulus.* $T_2$. *Threshold stimulus.* $T_3$. *Supraliminal stimulus.*

rapidly, the individual effects of each are additive and the threshold stimulus strength for contraction is finally reached. This phenomenon is called *summation of subliminal stimuli* and elicits a single muscular contraction. An isolated contraction is called a *twitch*. Twitches may occur in particular muscles during spasms called tics. Tics are discussed in Chap. 25. Otherwise, twitches are not characteristic of muscular contraction in vivo.

Twitches can also *summate* in response to repeti-

tive supraliminal stimuli, as shown in Fig. 24-8*B*. All the stimuli are of the same strength and, in our example, are applied in groups of three. Notice that the time interval between the stimuli in each triplet of stimuli decreases, i.e., the stimuli within each grouping are closer together in the progression from left to right.

As you can observe, the magnitudes of contraction increase from left to right as the time spent in the relaxation phase of muscular contraction decreases.

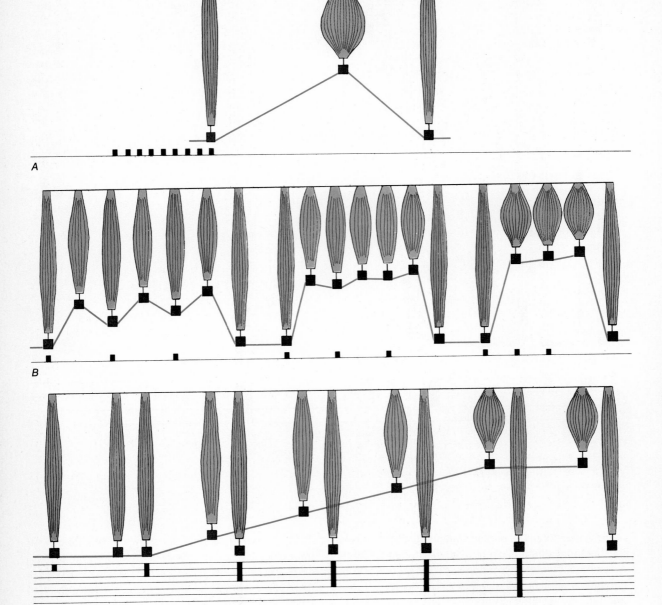

**Figure 24-8** *The responses of a skeletal muscle to electrical stimuli. The muscle is suspended from its upper end; a weight is attached to its free lower end. The application of electrical stimuli are marked by vertical lines below the muscle. A. Summation of subliminal stimuli. B. Summation of twitches. C. The responses of a skeletal muscle to single stimuli graded in strength.*

The alternation of contraction and relaxation exemplified by the responses during the first two triplets is called *clonic muscular contraction,* or *clonus.* The sustained contraction elicited by the third triplet of stimuli is referred to as a *tetanic contraction,* or *tetanus.* Muscles in tetanic contraction do not oscillate between contraction and relaxation as do muscles in clonus. Clonic and tetanic contractions are either isometric or isotonic, depending upon whether the muscle only undergoes changes in tension or actually displaces a load.

To understand why the tension developed by isometric muscles or the degree of shortening of isotonic muscles increases in response to rapidly applied supraliminal stimuli, we must go back to our recoiling-spring analogy and the molecular architecture of the myofibrils. When successive stimuli are spaced relatively far apart, part of each muscular response is occupied in stretching, or overcoming the viscous resistance of, the elastic elements. In other words, part of the interdigitation between the actin and myosin filaments is essentially wasted during separate contractions in that it does not perform useful work.

When the time interval between successive stimuli is decreased, on the other hand, each of the stimuli acts on an incompletely relaxed muscle. Since the actin and myosin filaments are partially interdigitated when the next action potential arrives at the sarcoplasmic reticulum, the expenditure of energy to overcome the inertia of elastic elements is not necessary. Since the muscle is already in an active state, the magnitude of its tension or degree of shortening can increase. This is the reason behind the summation of twitches.

When single stimuli of graded intensity are applied to an isotonic muscle, the muscle responds with graded contractions. This phenomenon is diagrammed in Fig. 24-8C. The degree of shortening of a muscle contracting isotonically or the increase in tension of an isometric muscle increases as the strength of a stimulus rises from the threshold level. In a given muscle, there is always a specific stimulus strength which causes a maximum response. Stronger stimuli do not evoke a greater shortening or tension.

When the responses to graded stimuli of a skeletal muscle (Fig. 24-8C) and a skeletal muscle fiber (Fig. 24-7) are compared, one major difference stands out—a *muscle fiber* responds all-or-none and does not exhibit a graded response, but a *skeletal muscle* shows graded responses. The explanation for this physiologic difference between a muscle cell and a muscle revolves around the response threshold for contraction. At any given time, a single muscle fiber has only one response threshold value. However, this threshold value is not the same for all muscle fibers. In other words, a group of muscle fibers possesses a spectrum of response thresholds. Some fibers respond to stimuli of relatively low intensity, others respond only to stimuli of relatively high intensity, and still others respond to stimuli of intermediate strengths. Since a skeletal muscle is made up of many muscle fibers, the increasing magnitude of contraction as the strength of a stimulus increases is due to the excitation of an increasing number of muscle fibers. Indeed, the degree of muscle shortening is proportional, or directly related to, the number of its fibers thrown into contraction. This relationship is known as *recruitment.*

During a maximal contraction, many, but by no means all, the muscle fibers are recruited. Some muscle fibers remain inactive, no matter how powerful the muscular contraction. This condition delays the onset of fatigue, since the resting fibers take over for those which become fatigued during sustained contraction.

Multiple contractions by a skeletal muscle in re-

**Figure 24-9** *The fatigue curve of a skeletal muscle in response to repetitive stimulation.*

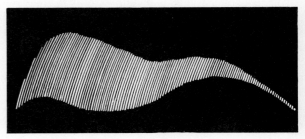

sponse to repetitive stimulation can lead to fatigue. A typical fatigue curve is shown in Fig. 24-9. The initial rise in the bottom portion of the curve is caused by a progressive increase in the degree of incomplete relaxation of the muscle during rapid and repetitive stimulation. The subsequent depression in the magnitude of contraction as stimulation continues is a consequence of progressive muscular fatigue.

## CARDIAC MUSCLE

In contrast to skeletal muscle, cardiac muscle responds to any stimulus of at least threshold strength with a maximal contraction. The all-or-none response of myocardium is due to the *syncytial* nature of the musculature. Once initiated, a wave of excitation spreads over the entire cardiac musculature. All the fibers which make up its substance contract in turn.

The magnitude of a cardiac contraction in a healthy heart is determined by the average degree of stretch imposed on the cardiac muscle fibers by the volume of blood within the heart chambers. The greater the rate of the venous return to the heart, the larger the volume of blood within the heart chambers and the greater the strength of cardiac contraction (*Starling's law*). In other words, all the blood which enters the heart during relaxation (*diastole*) of the atria leaves the heart during contraction (*systole*) of the ventricles.

During systole, the myocardium cannot increase its response magnitude to another stimulus, no matter how strong it is. This functional portion of a cardiac cycle is known as the *absolute refractory period*.

During diastole, the myocardium will respond with a slight increase in its degree of contraction to a stimulus stronger than that which caused systole. This functional portion of a cardiac cycle is known as the *relative refractory period*. The capability of a further response during the relative refractory period is a consequence of the relaxation of some of the cardiac fibers.

The myocardium has a built-in, or endogenous, rhythmicity due to the presence of specialized cardiac muscle tissue in the wall of the right atrium and in the septum between right and left ventricles (inter-ventricular septum). Because of the endogenous regulation of the cardiac cycle, the rate of the heartbeat may be altered by, but does not depend upon, the autonomic nerve supply of the heart. For example, a durable turtle heart can be completely removed from a turtle and kept beating in isolation for several days by placing it in a dish filled with physiologic saline solution. For best results, the heart should also be perfused continually with physiologic saline by forcing the solution through the heart chambers. *Physiologic saline* is a solution which is not only isotonic with a particular type of cell or tissue, in this case the myocardium, but contains all the necessary ions in proper concentrations to maintain physiologic functions, in this case the heartbeat.

## SMOOTH MUSCLE

Although the strength of a smooth muscle is proportional to the degree of its stretch, just as in the case of both skeletal and cardiac muscle, the response of smooth muscle is relatively sluggish. Indeed, when smooth muscle fibers are stretched very slowly, there is often little or no increase in their tension. The capacity of smooth muscle fibers to elongate when under tension explains this unique characteristic.

Nevertheless, all smooth muscles have the capability of responding rapidly as well as slowly. Smooth muscle movements range in quality between the extremely slow contractions exemplified by the motility of the gut, the rhythmic contractions exhibited during peristalsis of the digestive tract or ureters, the rapid contractions characterized by vasoconstriction of blood vessels in response to fright, and the spasmodic contractions of the walls of bronchial tubes provoked during certain types of allergic reactions.

## SKELETAL MUSCLES OF THE BODY

Before we analyze the movements performed by the skeletal muscles of the body, we must become familiar with the various types of muscular movements which are possible and the ways in which muscles are named.

## TYPES OF MUSCULAR MOVEMENTS

Only a limited number of movements are performed by the simple machines of which skeletal muscles are members. To facilitate your understanding, perform each movement as you read the following descriptions. Examples of the movements are illustrated in Fig. 24-10.

1  *Flexion* and *extension* are performed at hinge, pivot, and saddle joints when an angle between adjacent parts of the body is decreased or increased, respectively. Examples include flexion and extension of the head, trunk, arms, hands, fingers, legs, feet, and toes.
2  *Adduction* and *abduction* are performed at saddle and ball-and-socket joints when a part of the body is moved toward or away from the midsagittal line, respectively. Examples include medial and lateral movements of the arms, fingers, legs, and toes.
3  *Pronation* and *supination* are performed when the palm of the hand is rotated downward or upward, respectively. These movements take place at the elbow joint by the pivotal articulation between the head of the radius and the radial notch of the ulna. Other rotational movements include the transverse rotation of the head, which takes place at pivot joints established between the atlas and axis, or transverse rotation of the shank of the leg, which occurs at the knee joint by the pivotal articulation between the distal end of the femur and the proximal end of the tibia.
4  *Inversion* and *eversion* are performed when the plantar surface of the foot turns inward or outward, respectively. These gliding movements take place at plane joints established by articulations between the tarsal bones.
5  *Circumduction* is performed at ball-and-socket and saddle joints when the distal end of a body part describes an arc in space. Examples include circumduction of the arms, legs, and thumbs.

## HOW MUSCLES ARE NAMED

The names of muscles are in Latin. Frequently, the Latin names have been simplified by converting them into their English equivalents. For example, the *transversus abdominis,* a flattened muscle on the inner aspect of the lateral and anterior abdominal wall on each side of the body, may be called the *transverse abdominis* or even the *transverse abdominal.*

Muscles are assigned particular names for the following reasons:

1  Their unique *shapes*. Examples include the deltoideus and rhomboideus muscles, which are triangular and rhomboidal in shape, respectively.
2  The number of *divisions*, or heads, of which they are composed. Examples include the biceps brachii and biceps femoris, each with two heads, and the triceps brachii, with three heads.
3  The *direction* which their fibers take. Examples include the external oblique, the fibers of which are directed inferomedially; the internal oblique, the fibers of which are directed superolaterally; and the transversus abdominis, the fibers of which run transversely.
4  Their *attachment points* on various bones. The attachment points may be fixed and immovable *origins*, movable *insertions*, or a combination of both. Examples of muscles named in this manner include the sternocleidomastoid, which originates from the sternum and clavicle and inserts on the mastoid process of the temporal bone, or the supraspinatus, infraspinatus, and subscapularis, the origins of which are the supraspinous, infraspinous, and subscapular fossae of the scapula, respectively.
5  Their *location*. Examples include the tibialis anterior and tibialis posterior on the anterior and posterior aspects of the tibia, or the external and internal intercostals between adjacent ribs. The location also usually identifies an attachment point; e.g., the tibialis muscles originate in part on the tibia.
6  Their *action*, or *function*, during contraction. Most muscles are named in this way. The actions of muscles are based largely on the types of muscular movements described earlier. Thus, examples of muscles named for their actions include various

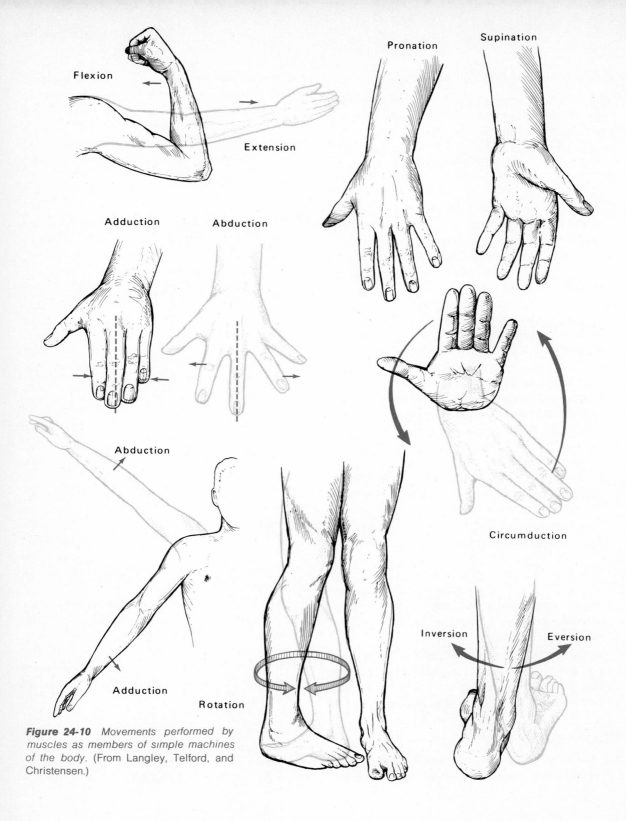

Flexion

Extension

Pronation

Supination

Adduction

Abduction

Abduction

Adduction

Rotation

Circumduction

Inversion

Eversion

**Figure 24-10** *Movements performed by muscles as members of simple machines of the body.* (From Langley, Telford, and Christensen.)

flexors, extensors, adductors, abductors, pronators, supinators, and rotators.

7 Some *combination* of the above-listed reasons. Examples include the serratus anterior, a muscle named for its shape and location; the palmaris longus, a muscle named for its location (or insertion) and shape; the extensor carpi radialis brevis, a muscle named for its action, location, and shape; and the quadratus femoris, a muscle named for its shape and insertion (or location).

## FUNCTIONAL GROUPINGS

Skeletal muscles are divided into several functional groupings, as follows:

1 *Prime movers.* Called *agonists,* prime movers initiate gross movements of either body parts or the entire body. Such muscles are usually superficial in location.

2 *Antagonists.* These oppose the action of prime movers and are usually located on the side of the body opposite their agonists. Examples of antagonistic muscular action include extensors and adductors which oppose flexors and abductors, respectively. The former are antagonists of the latter. As an example of opposed locations, the supinator on the posterior aspect of the forearm is opposed by the pronator teres and pronator quadratus on the anterior aspect of the forearm.

3 *Synergists.* These are muscles which assist, or aid, the actions of agonists by stabilizing a body part being moved or contributing to its movement. Synergists contract at the same time as agonists.

4 *Fixation muscles.* Fixation, or *postural,* muscles provide support in other regions of the body when gross movements are executed and also maintain posture and support body parts or regions in the absence of gross movements. In general, fixation muscles are located deep within the body, close to the center of gravity. Such muscles have the capacity to remain contracted for long periods of time without fatiguing.

## REGIONAL GROUPINGS

In parallel with the regions of the skeleton, skeletal muscles are divided into two regional groupings—axial and appendicular.

*Axial muscles* These striated muscles are associated with the skull, neck, vertebral column, thoracic wall, and abdominal wall. In the following discussions, however, we will restrict our comments to muscles of the face, anterior cervical region, and anterolateral abdominal wall.

*Muscles of the face* The muscles of facial expression are illustrated in Fig. 24-11. The following observations pertain to this figure:

1 There are 16 major muscles of facial expression, all except one of which are paired.

2 Most of the facial muscles originate from bones and insert on the skin or other muscles.

3 The *frontalis* muscles originate from the *epicranial aponeurosis,* a broad, flat sheet of fibrous connective tissue covering the superior aspect of the cranium.

4 The *orbicularis oculi* and *orbicularis oris* surround the orbit and lips, respectively.

5 The *frontalis* and *corrugator* muscles elevate the eyebrows and draw them together, respectively.

6 The *orbicularis oculi pars palpebralis* closes each eyelid.

7 The *frontalis* and *orbicularis oculi pars orbicularis* wrinkle and depress, respectively, the skin of the forehead.

8 The *levator labii superioris* and *levator labii superioris alaeque nasi* dilate the nostrils.

9 The *nasalis* and *procerus* act on the facial skin on and surrounding the nose, respectively.

10 The *levator anguli oris, levator labii superioris, levator labii superioris alaeque nasi, mentalis, risorius,* and *zygomaticus* elevate the lips or mouth or move them backward.

11 The *mentalis* and *orbicularis oris* cause lip protrusion and closure, respectively.

12 The *platysma,* a superficial muscle originating

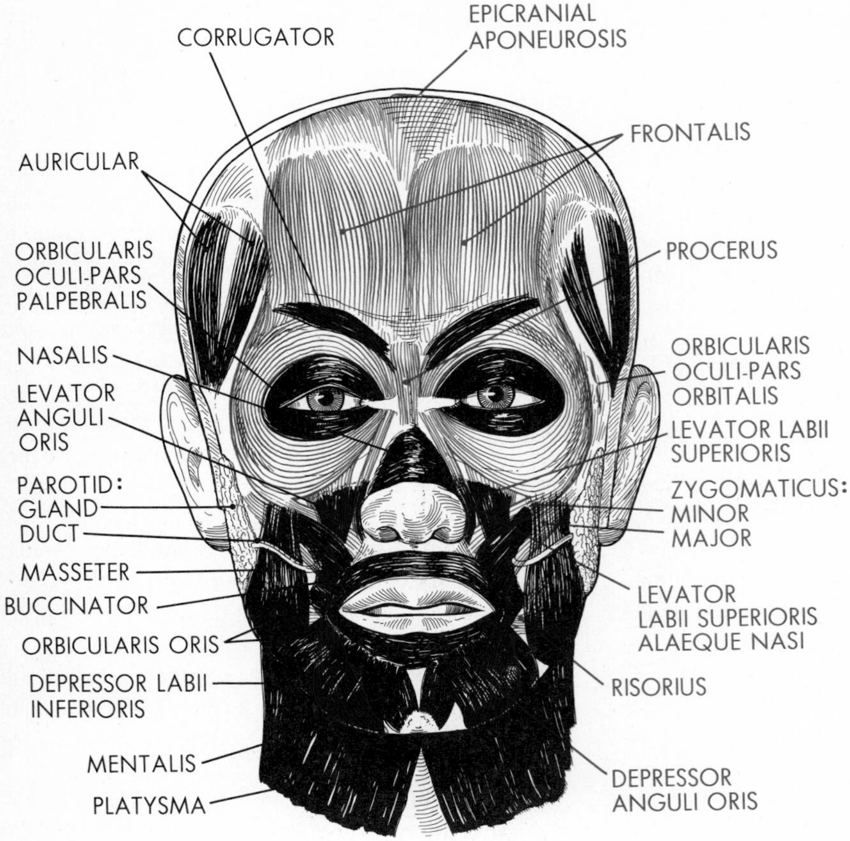

**Figure 24-11** *Muscles of the face.* (From Melloni, Stone, and Hurd.)

on the fascia of the anterior wall of the chest, tenses the skin below the chin and depresses the lower lip.

**Muscles of the anterior neck** Muscles in the anterior cervical region are shown in Fig. 24-12. The following observations pertain to this figure:

1  Since most of the anterior cervical muscles are associated with and move the hyoid bone supporting the tongue, they function in swallowing.
2  Called *suprahyoid muscles* because they are superior to the hyoid bone, the digastric, mylohyoid, and stylohyoid muscles elevate the hyoid bone or move it forward or backward.
3  The *digastric muscle* has two bellies separated by a tendinous connection which passes through a pulleylike arrangement on the hyoid bone.

4  Called *infrahyoid muscles* because they are inferior to the hyoid bone, the sternothyroid, thyrohyoid, sternohyoid, and omohyoid depress the hyoid bone.
5  The *omohyoid* has two bellies separated by a tendinous connection which passes through a pulleylike arrangement on the inner aspect of the sternocleidomastoid.
6  Rotating the head to the opposite side when contracting singly or flexing the head when contracting together, the *sternocleidomastoid* has two heads, one originating from the sternum and the other originating from the clavicle.
7  The *cricothyroid,* a laryngeal muscle, and the *sternothyroid,* an infrahyoid muscle, depress the thyroid cartilage of the larynx, while the *thyrohyoid,* another infrahyoid muscle, elevates the thyroid cartilage.

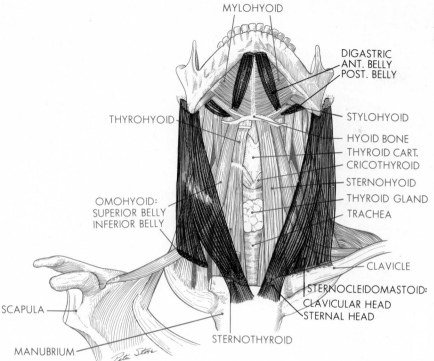

MYLOHYOID

DIGASTRIC
ANT. BELLY
POST. BELLY

THYROHYOID

STYLOHYOID

HYOID BONE
THYROID CART.
CRICOTHYROID
STERNOHYOID
THYROID GLAND
TRACHEA

OMOHYOID:
SUPERIOR BELLY
INFERIOR BELLY

CLAVICLE

STERNOCLEIDOMASTOID:
CLAVICULAR HEAD
STERNAL HEAD

SCAPULA

MANUBRIUM

STERNOTHYROID

*Figure 24-12* *Muscles of the neck.* (From Melloni, Stone, and Hurd.)

*Muscles of the anterolateral abdominal wall* The muscles of the anterolateral abdominal wall are shown in Fig. 24-13. The following observations pertain to the lower portion of this figure:

1   The *external oblique, internal oblique, transversus abdominis,* and *rectus abdominis* contribute to the major portion of the anterolateral abdominal wall.
2   The *external oblique, internal oblique,* and *transversus abdominis* are broad sheetlike muscles.
3   With the exception of the rectus abdominis which is inserted on the rectus sheath, the muscles insert at least in part on the *linea alba,* a midsagittal line formed of white fibrous connective tissue in the anterior abdominal region.
4   The anterolateral abdominal muscles compress the abdomen during contraction while the external and internal oblique muscles also rotate the vertebral column to the contracted side or flex the spine when the muscles contract in pairs.

5   The fibers of the external oblique are directed inferomedially; the fibers of the internal oblique run approximately at right angles to those of the external oblique; the fibers of the transversus abdominis course transversely around the anterolateral abdominal wall; and the fibers of the rectus abdominis are directed vertically in the anterior abdominal wall.

*Muscles of the posterior abdominal wall* The muscles of the posterior abdominal wall are illustrated in Fig. 24-19, below. The following observations pertain to the upper portion of this figure:

1   Two paired muscles, the *quadratus lumborum* and *psoas major,* contribute to the posterior abdominal wall.
2   The pelvis is pulled to the side of the contracted quadratus lumborum; the trunk is flexed when both of these muscles contract together.

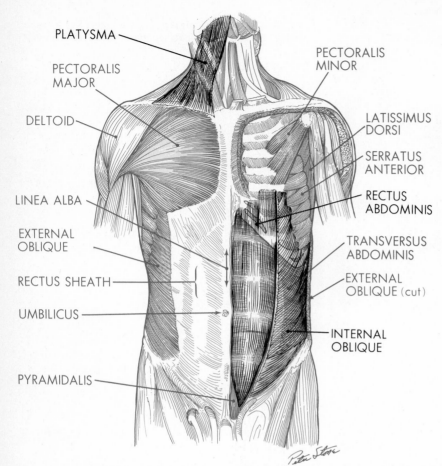

PLATYSMA

PECTORALIS
MAJOR

DELTOID

LINEA ALBA

EXTERNAL
OBLIQUE

RECTUS SHEATH

UMBILICUS

PYRAMIDALIS

PECTORALIS
MINOR

LATISSIMUS
DORSI

SERRATUS
ANTERIOR

RECTUS
ABDOMINIS

TRANSVERSUS
ABDOMINIS

EXTERNAL
OBLIQUE (cut)

INTERNAL
OBLIQUE

**Figure 24-13** *Anterior view of the thoracic and abdominal muscles. The pectoralis major has been cut on the left side to reveal the deeper thoracic muscles. The rectus sheath and external oblique have also been removed from the left side to show the deeper abdominal muscles. (From Melloni, Stone, and Hurd.)*

*3* The quadratus lumborum also depresses the lower ribs while the psoas major flexes the thigh (see below).

Appendicular muscles   Appendicular muscles are associated with movements of the upper and lower extremities. The muscles which belong to the chest and back are associated with movements of the scapula and shoulder. The muscles which belong to the shoulder, arm, and forearm are associated with movements at the shoulder joint, elbow joint, and the joints of the forearm and hand, respectively. The muscles which belong to the hip, thigh, and leg are associated with movements at the hip joint, knee joint, and the joints of the leg and foot, respectively.

**Muscles of the pectoral region**   Muscles of the

chest are shown in Fig. 24-13. The following observations pertain to the upper portion of this figure:

*1* The pectoral muscles include the pectoralis major, pectoralis minor, and serratus anterior.
*2* The *pectoralis major,* the large superficial muscle on each side of the chest, flexes, adducts, and rotates the arm medially.
*3* The *pectoralis minor* depresses the scapula while the *serratus anterior* rotates the scapula when the arm is being raised.

**Superficial muscles of the back**   The back muscles are illustrated in Fig. 24-14. The following observations pertain to the upper portion of this figure:

*1* The back muscles include the trapezius, latissimus

dorsi, levator scapulae, rhomboid major, and rhomboid minor.

2 Forming a figure which resembles a trapezoid, the *trapezius muscles* elevate the shoulders and extend the head. One contracted trapezius bends the head toward the shoulder on the side of the contracted muscle.

3 With the exception of the latissimus dorsi, these muscles act directly on the scapula.

4 The *levator scapulae* and the *rhomboid major* and *rhomboid minor* elevate the scapula.

5 The *trapezius* and *pectoralis minor* (see above) depress the scapula.

6 The *rhomboid major* and *rhomboid minor* move the scapula medially, and the *serratus anterior* moves the scapula laterally.

7 The *pectoralis minor* and the *serratus anterior* move the scapula forward.

8 The *trapezius, rhomboid major, rhomboid minor,* and *serratus anterior* rotate the scapula.

**Muscles of the shoulder** With the exception of the teres minor, the muscles of the shoulder are shown in Fig. 24-15*B* and *D*. The teres minor is illustrated in Fig. 24-15*C*. The following observations pertain to these illustrations:

1 The shoulder muscles include the deltoid, supra-

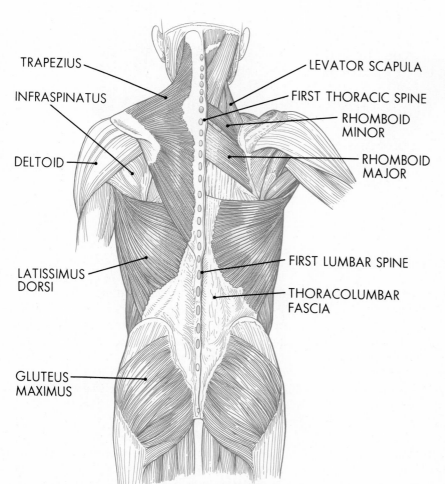

TRAPEZIUS

INFRASPINATUS

DELTOID

LATISSIMUS DORSI

GLUTEUS MAXIMUS

LEVATOR SCAPULA

FIRST THORACIC SPINE

RHOMBOID MINOR

RHOMBOID MAJOR

FIRST LUMBAR SPINE

THORACOLUMBAR FASCIA

**Figure 24-14** *Muscles of the back. The trapezius has been removed from the right side to show some of the muscles associated with the scapula. (From Melloni, Stone, and Hurd.)*

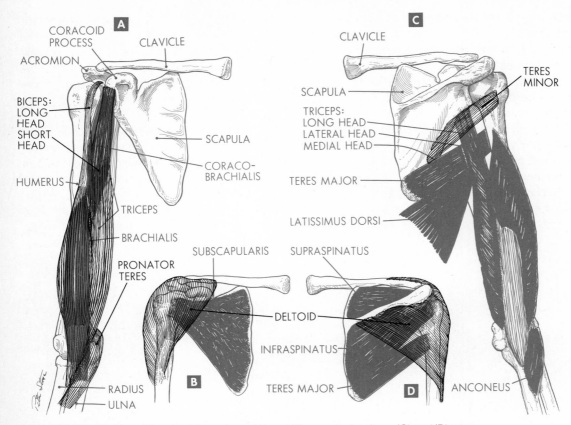

**Figure 24-15** *Muscles of the shoulder and arm.* (A) *and* (B) *are anterior views;* (C) *and* (D) *are posterior views.* (From Melloni, Stone, and Hurd.)

spinatus, infraspinatus, teres major, teres minor, and subscapularis.

2 The shoulder muscles are involved in movements of the shoulder because of their insertions on the humerus.

3 The shoulder muscles originate at least in part from the scapula.

4 The *deltoid* and *supraspinatus,* in addition to the biceps brachii (see below), abduct the arm; and the *teres major,* in addition to the biceps brachii, coracobrachialis, and triceps brachii (see below) and the pectoralis major (see above), adducts the arm.

5 The *deltoid,* in addition to the biceps brachii and coracobrachialis (see below) and the pectoralis major (see above), flexes the arm, and the *teres major* and the triceps brachii (see below) extend the arm.

6 The *deltoid, infraspinatus,* and *teres minor* rotate the arm laterally, and the *teres major* and *subscapularis,* in addition to the coracobrachialis (see below) and the pectoralis major (see above), rotate the arm medially.

**Muscles of the arm** The arm muscles are illustrated in Fig. 24-15A and C. The following observations pertain to these illustrations:

1 The arm muscles include the biceps brachii, coracobrachialis, and brachialis on the anterior aspect of the humerus, and the triceps brachii on the posterior aspect of the humerus.

2 The short head of the *biceps brachii* originates from the coracoid process of the scapula, and its long head originates above the superior border of the glenoid cavity of the scapula.

3 The arm muscles insert on the bones of the forearm, except for the coracobrachialis which inserts on the humerus.

4 The long head of the *triceps brachii* originates from the lateral margin of the scapula, and its lateral and medial heads originate from the proximal and distal portions of the posterior humerus.

5 The *biceps brachii* is the agonist, and the *brachialis*, in addition to the brachioradialis (see below), the synergist in executing forearm flexion, while the *triceps brachii* is the antagonist of the biceps brachii and causes forearm extension.

6 The biceps brachii, coracobrachialis, and triceps brachii assist in movements of the arm (see above).

**Muscles of the forearm** The forearm muscles are shown in Figs. 24-16 and 24-17. The following observations pertain to these figures:

1 The *anterior forearm muscles* include the pronators quadratus and teres, the flexors digitorum

**Figure 24-16** *Muscles of the anterior forearm.* A. *Deep layer of muscles.* B. *Middle layer of muscles.* C. *Superficial layer of muscles.* (From Melloni, Stone, and Hurd.)

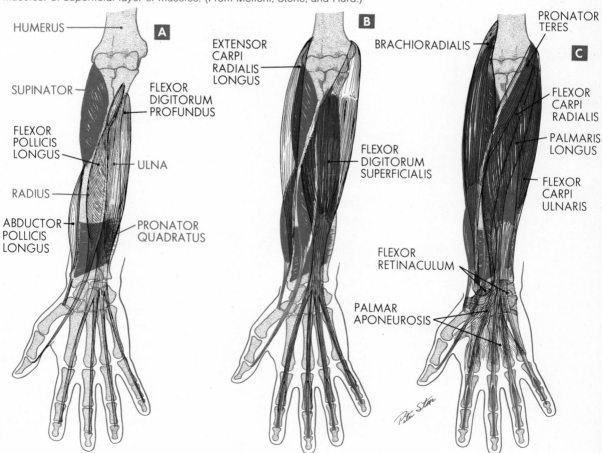

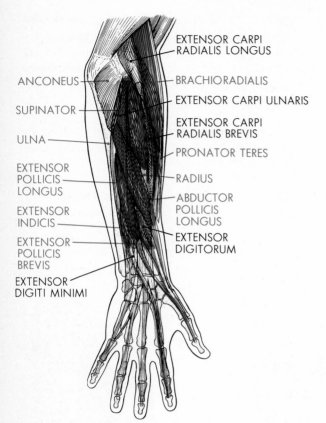

EXTENSOR CARPI
RADIALIS LONGUS

ANCONEUS

BRACHIORADIALIS

SUPINATOR

EXTENSOR CARPI ULNARIS

EXTENSOR CARPI
RADIALIS BREVIS

ULNA

PRONATOR TERES

EXTENSOR
POLLICIS
LONGUS

RADIUS

ABDUCTOR
POLLICIS
LONGUS

EXTENSOR
INDICIS

EXTENSOR
DIGITORUM

EXTENSOR
POLLICIS
BREVIS

EXTENSOR
DIGITI MINIMI

**Figure 24-17** *Muscles of the posterior forearm from anterolateral aspect. (From Melloni, Stone, and Hurd.)*

profundus, pollicis longus, digitorum superficialis, carpi radialis, and carpi ulnaris, as well as the palmaris longus.

2 The *posterior forearm muscles* include the anconeus, supinator, brachioradialis, the extensors indicis, pollicis longus, pollicis brevis, carpi radialis longus, carpi radialis brevis, digitorum communis, carpi ulnaris, and digiti minimi, as well as the abductor pollicis longus.

3 The number and complexity of the forearm musculature is reflected in the complexity and precision of the movements performed by the hand.

4 Responsible for forearm rotation, the *supinator* on the posterior aspect of the forearm (Fig. 24-16A) is opposed by the *pronators teres* (Fig. 24-16C) and *quadratus* (Fig. 24-16A) on the proximal and distal portions of the anterior forearm, respectively, and by the *biceps brachii* on the anterior arm (see above).

5 Most of the other forearm muscles are concerned with flexion, extension, abduction, and adduction of the hand, including the wrists and fingers.

6 The forearm flexors include the *pronator teres* and the *flexor carpi radialis* (Fig. 24-16C), as well as the *brachioradialis* and the *extensor carpi radialis longus* (Fig. 24-17) while the *anconeus* (Fig. 24-17) extends the forearm.

7 The wrist flexors include the *flexor carpi radialis* and *palmaris longus* (Fig. 24-16C), and the *extensor digitorum* (Fig. 24-17) extends the wrist.

8 The hand flexors include the *flexors carpi ulnaris* (Fig. 24-16C), *digitorum superficialis* (Fig. 24-16B), *digitorum profundus,* and *pollicis longus* (Fig. 24-16A), and the hand extensors include the *extensors carpi radialis longus, carpi radialis brevis,* and *carpi ulnaris* (Fig. 24-17).

9 The finger flexors include the *flexors digitorum superficialis* (Fig. 24-16B), *digitorum profundus* (Fig. 24-16A), and muscles of the hand (not shown); the finger extensors include the *extensors digitorum, digiti minimi,* and *indicis* (Fig. 24-17).

10 The thumb is flexed by the *flexor pollicis longus* (Fig. 24-16A) and by muscles of the hand (not shown) and extended by the *extensors pollicis brevis* and *pollicis longus* (Fig. 24-17).

11 The hand is abducted by the *extensor carpi radialis longus* (Fig. 24-17) and adducted by the *flexor carpi ulnaris* (Fig. 24-16C) and the *extensors carpi radialis longus* and *carpi ulnaris* (Fig. 24-17).

12 The thumb is abducted by the *abductor pollicis longus* and the *extensor pollicis brevis* (Fig. 24-17), in addition to muscles of the hand (not shown), and adducted by muscles of the hand (not shown).

13 The flexors are located on the anterior aspect of the forearm, originate from either the distal end of the humerus or the proximal end of the ulna, and

insert through long tendinous extensions either on the distal ends of the anterior radius and ulna or on the palmar surfaces of bones of the hand.

14 The extensors are located on the posterior aspect of the forearm, originate from either the distal end of the humerus or proximal ends of the radius and ulna, and insert through long tendinous extensions on the superior or dorsal surfaces of bones of the hand.

15 The specialized set of muscles associated with the thumb, together with the saddle joint at its

metacarpophalangeal articulation, endows the thumb with great freedom of movement.

**Muscles of the gluteal region** Except for the tensor fasciae latae, the muscles of, and associated with, the buttocks are illustrated in Fig. 24-18. The tensor fasciae latae is illustrated in Fig. 24-19C, below. The following observations pertain to these figures:

1 The muscles of the gluteal region include the *gluteus maximus, gluteus medius, gluteus minimus, piriformis, gemellus superior, gemellus inferior,*

**Figure 24-18** *Muscles of the gluteal region and posterior thigh.* A. *Deep layer of muscles.* B. *Middle layer of muscles.* C. *Superficial layer of muscles.* (From Melloni, Stone, and Hurd.)

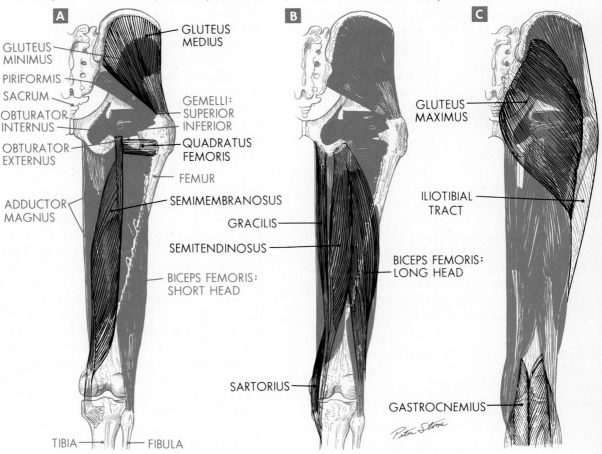

*obturator internus, quadratus femoris,* and *tensor fasciae latae.*

2 These muscles are involved in movements of the thigh, especially rotational movements.

3 The characteristic shape of the buttocks is due to the contours of the gluteus maximus.

4 The gluteus maximus and tensor fasciae latae both insert on the *iliotibial tract,* a thick band of fascia which attaches to the lateral aspect of the proximal tibia (Fig. 24-18C).

5 The thigh is rotated laterally by the gluteus maximus (Fig. 24-18C), gluteus medius, piriformis, obturator internus, quadratus femoris, gemellus superior, and gemellus inferior (Fig. 24-18A), and by the sartorius, adductor magnus, and obturator externus (see below), and rotated medially by the gluteus minimus and obturator internus (Fig. 24-18A) and the tensor fasciae latae (Fig. 24-19C), as well as by the psoas major, iliacus, pectineus, gracilis, adductor longus, and adductor magnus (see below).

6 The thigh is flexed by the tensor fasciae latae (Fig. 24-19C), in addition to the sartorius, rectus femoris, psoas major, iliacus, pectineus, adductor longus, and adductor magnus (see below), and extended by the gluteus maximus (Fig. 24-18C) and piriformis (Fig. 24-18A) and the vastus lateralis, vastus medialis, vastus intermedius, adductor magnus, biceps femoris, semitendinosus, and semimembranosus (see below).

7 The thigh is abducted by the gluteus medius, gluteus minimus, and piriformis (Fig. 24-18A), and the tensor fasciae latae (Fig. 24-19C), and adducted by the psoas major, iliacus, pectineus, gracilis, adductors longus, brevis, and magnus, in addition to the obturator externus, biceps femoris, semitendinosus, and semimembranosus (see below).

**Muscles of the thigh** The anterior and medial thigh muscles are shown in Fig. 24-19. The posterior thigh muscles are illustrated in Fig. 24-18. The following observations pertain to these figures:

1 Muscles of the *anterior thigh* include the *pectineus, vastus lateralis, vastus medialis, vastus intermedius, rectus femoris,* and *sartorius.*

2 Muscles of the *medial thigh* include the *obturator externus,* the *adductors magnus, brevis,* and *longus,* and the *gracilis.*

3 Muscles of the *posterior thigh* include the *semimembranosus, semitendinosus,* and *biceps femoris.*

4 Since they insert on the proximal end of the anterior femur, a portion of the *iliacus* and *psoas major* contributes to the anterior thigh muscles.

5 The iliacus and the psoas major are often referred to collectively as the *iliopsoas.*

6 The vastus lateralis, vastus medialis, vastus intermedius, and rectus femoris on the anterior thigh are the four heads of the *quadriceps femoris.*

7 The quadriceps femoris muscles insert on the proximal and anterior tibia by means of a common tendon called the *patellar tendon.*

8 The semimembranosus, semitendinosus, and biceps femoris on the posterior thigh are commonly called the *hamstring muscles.*

9 The long head of the biceps femoris originates from the tuberosity of the ischium, and its short head originates from the middle third of the posterior femur.

10 The hamstring muscles are largely responsible for flexion of the leg, and the quadriceps femoris muscles are largely responsible for its extension (see below).

11 Except for the vastus muscles, the thigh muscles are involved with either lateral rotation, medial rotation, flexion, extension, abduction or adduction of the thigh, or some combination of these movements (see "Muscles of the Gluteal Region," above).

12 Either because they pass over both the hip joint and the knee joint or because of their unique attachment points, numerous thigh muscles also participate in movements of the leg (see below).

13 The leg is flexed by the gracilis and all the hamstring muscles (Fig. 24-18A and B), in addition to the gastrocnemius, plantaris, and popliteus (see below), and extended by the quadriceps femoris muscles (see Fig. 24-19B and C).

14 The leg is rotated laterally by the biceps femoris (Fig. 24-18A and B) and is rotated medially by

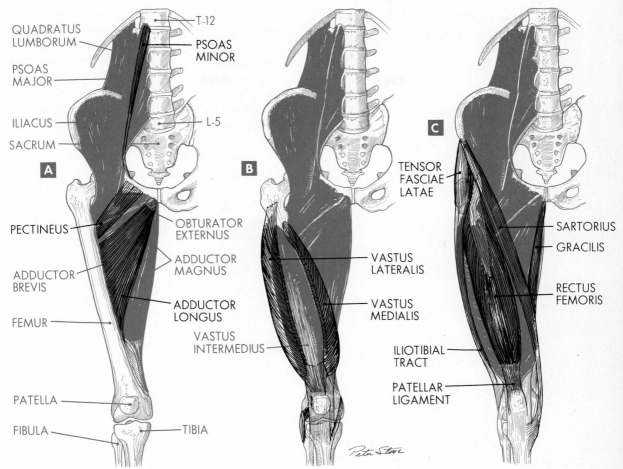

**Figure 24-19** *Muscles of the anterior hip and thigh. A. Deep layer of muscles. B. Middle layer of muscles. C. Superficial layer of muscles.* (From Melloni, Stone, and Hurd.)

the gracilis and semitendinosus (Fig. 24-18*B*), in addition to the semimembranosus (Fig. 24-18*A*) and the popliteus (see below).

**Muscles of the leg** Muscles of the anterior, lateral, and posterior leg are illustrated in Figs. 24-20 through 24-22, respectively. The following observations pertain to these figures:

1 Muscles of the *anterior leg* include the *extensors hallucis longus* and *digitorum longus*, the *peroneus tertius*, and the *tibialis anterior*.

2 Muscles of the *lateral leg* include the *peroneus brevis* and *peroneus longus*.

3 Muscles of the *posterior leg* include the *popliteus*, *tibialis posterior*, the *flexors hallucis longus* and *digitorum longus*, the *plantaris*, *soleus*, and *gastrocnemius*.

4 With the exception of the popliteus, which acts entirely on the leg, the leg muscles act primarily on the foot, including the toes.

5 The functions of the leg muscles in leg movements are explained elsewhere (see "Muscles of the Thigh," above).

6 The medial and lateral heads of the gastro-

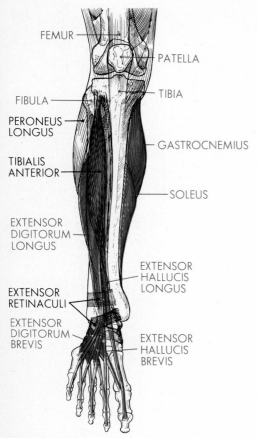

FEMUR
PATELLA
FIBULA
TIBIA
PERONEUS LONGUS
GASTROCNEMIUS
TIBIALIS ANTERIOR
SOLEUS
EXTENSOR DIGITORUM LONGUS
EXTENSOR HALLUCIS LONGUS
EXTENSOR RETINACULI
EXTENSOR DIGITORUM BREVIS
EXTENSOR HALLUCIS BREVIS

**Figure 24-20** *Muscles of the leg from anterior aspect.* (From Melloni, Stone, and Hurd.)

foot, in which the foot is actually extended, is accomplished by the peroneus longus and peroneus brevis (Fig. 24-21), the gastrocnemius, soleus, plantaris, flexors digitorum longus and hallucis longus, and the tibialis posterior (Fig. 24-22).

10  The foot is inverted by the tibialis anterior and the extensor hallucis longus (Fig. 24-20), as well as by the gastrocnemius, soleus, flexors digitorum longus and hallucis longus, and the tibialis posterior (Fig. 24-22), and everted by the extensor digitorum longus, peroneus tertius, peroneus longus, and peroneus brevis (Fig. 24-21).

11  The toes are flexed by the flexor digitorum longus (Fig. 24-22*A*), as well as by muscles of

**Figure 24-21** *Muscles of the leg from the lateral aspect.* (From Melloni, Stone, and Hurd.)

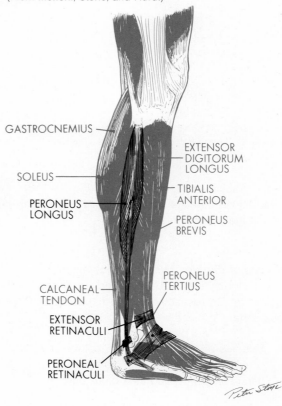

GASTROCNEMIUS
EXTENSOR DIGITORUM LONGUS
SOLEUS
TIBIALIS ANTERIOR
PERONEUS LONGUS
PERONEUS BREVIS
PERONEUS TERTIUS
CALCANEAL TENDON
EXTENSOR RETINACULI
PERONEAL RETINACULI

cnemius, or calf, muscle originate from the medial and lateral condyles of the tibia, respectively.

7  The gastrocnemius confers the curvaceous shape to the calf of the leg.

8  The soleus and gastrocnemius muscles both insert on the calcaneus by means of the *calcaneal,* or *Achilles, tendon.*

9  Dorsiflexion of the foot, in which the plantar surface is visible when viewed from anterior aspect, is accomplished by the tibialis anterior and the extensors digitorum longus and hallucis longus (Fig. 24-20), as well as by the peroneus tertius (Fig. 24-21), and plantar flexion of the

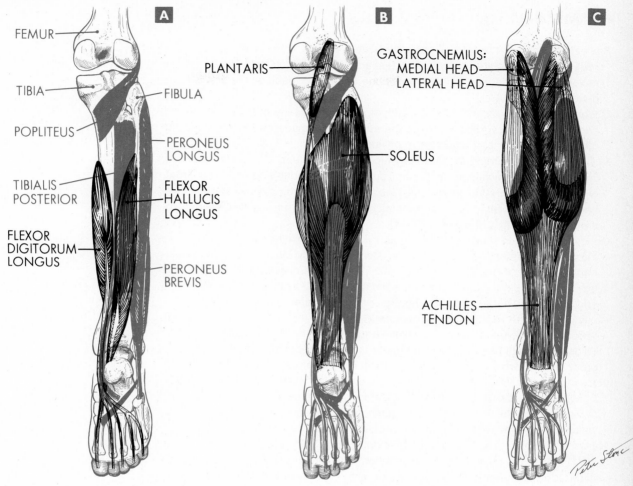

**A** **B** **C**

FEMUR

TIBIA

POPLITEUS

TIBIALIS POSTERIOR

FLEXOR DIGITORUM LONGUS

FIBULA

PERONEUS LONGUS

FLEXOR HALLUCIS LONGUS

PERONEUS BREVIS

PLANTARIS

GASTROCNEMIUS: MEDIAL HEAD LATERAL HEAD

SOLEUS

ACHILLES TENDON

**Figure 24-22** *Muscles of the leg from the posterior aspect. A. Deep layer of muscles. B. Middle layer of muscles. C. Superficial layer of muscles. (From Melloni, Stone, and Hurd.)*

the foot (not shown), and extended by the extensor digitorum longus (Fig. 24-21) and by muscles of the foot (not shown).

12 The great toe is flexed by the flexor hallucis longus (Fig. 24-22*A*) and by muscles of the foot (not shown), and extended by the extensor hallucis longus (Fig. 24-20) and muscles of the foot (not shown).

13 The flexors are located on the posterior aspect

of the leg, originate from either the distal end of the femur or the proximal end of the tibia and fibula, and insert through long tendinous extensions on the plantar surfaces of bones of the foot.

14 The extensors are located on the anterior aspect of the leg, originate from the tibia or fibula, and insert through long tendinous extensions on the superior or dorsal surfaces of bones of the foot.

## COLLATERAL READING

Basmajian, John V., "Electromyography Comes of Age," *Science,* 176:603–609, 1972. Describes electromyographic research that attempts to improve human motor performance by providing conscious neural feedback on the status of individual motor units.

Chapman, Carleton B., and Jere H. Mitchell, "The Physiology of Exercise," *Scientific American,* May 1965. Reviews the adjustments made by muscles and the other organ systems in response to physical exercise.

Di Cara, Leo V., "Learning in the Autonomic Nervous System," *Scientific American,* January 1970. Many human and animal subjects exposed to appropriate operant, or instrumental, conditioning regimes learn voluntarily to modify the rate of their heartbeats, gut motility, or other ordinarily autonomic responses. Di Cara considers the implications of these findings.

Goodwin, Guy M., D. Jan McCloskey, and Peter B. C. Matthews, "Proprioceptive Illusions Induced by Muscle Vibration: Contribution by Muscle Spindles to Perception?" *Science,* 175:1382–1384, 1972. The authors conclude that information from muscle spindles contributes to sensation as well as to automatic feedback and therefore reaches brain levels higher than the cerebellum.

Huxley, H. E., "The Mechanism of Muscular Contraction," *Science,* 164:1356–1366, 1969. Huxley points to evidence which suggests that the force causing actin and myosin filaments to slide over each other is probably generated by the tilting of cross bridges which project to the actin filaments from the myosin filaments.

Lippold, Olaf, "Physiological Tremor," *Scientific American*, March 1971. Based on the results of extensive research, Lippold believes that physiologic tremors are due to overshoots and undershoots of the automatic feedback system which controls the tone of striated muscles.

Margaria, Rodolfo, "The Sources of Muscular Energy," *Scientific American,* March 1972. Traces the sources and consumption of muscular energy and applies this knowledge to an improvement of the efficiency of individual motor performance, especially the training regimes of athletes.

Merton, P. A., "How We Control the Contraction of Our Muscles," *Scientific American,* May 1972. Concentrates on how muscle sense organs act automatically through feedback loops to ensure that muscular contractions are precise.

Porter, Keith R., and Clara Franzini-Armstrong, "The Sarcoplasmic Reticulum," *Scientific American*, March 1965. Based on electron microscopic studies, the authors conclude that the sarcoplasmic reticulum has the following functions: (1) the transmission of action potentials from the sarcolemma to the myofibrils, causing the release of $Ca^{++}$, the activator of muscular contraction; and (2) the intracellular distribution of enzymes and other metabolic substances which lead to the syntheses of ATP, creatine phosphate, and lactic acid.

Ross, Russell, and Paul Bornstein, "Elastic Fibers in the Body," *Scientific American,* June 1971. According to the authors, elastic fibers in the walls of the arteries are synthesized in arterial smooth muscle cells. However, fibroblasts synthesize elastic fibers found in ligaments.

Schmidt-Nielsen, Knut, "Locomotion: Energy Cost of Swimming, Flying and Running," *Science*, 177:222–228, 1972. The author concludes that flying consumes much less energy than running, when animals are matched in size. When man is considered to the exclusion of other animals, swimming consumes between five and ten times as much energy as running the same distance.

Van der Kloot, William G., "Muscle and Its Neural Control," *American Zoologist*, 2:55–65, 1962. Not only does an established myoneural junction affect the nature of innervated muscle, but it also apparently establishes persistent synaptic connections in the central nervous system, according to the author.

Wade, Nicholas, "Anabolic Steroids: Doctors Denounce Them, but Athletes Aren't Listening," *Science,* 176:1399–1403, 1972. In an article written for the layman, Wade critically assesses the efficacy of anabolic steroids for the improvement of weight and muscular strength.

Wilkie, D. R., *Muscle,* New York: St. Martin's Press, Inc., 1968. A succinct review of the biology of muscle.

# CHAPTER 25
# SKELETOMUSCULAR DISEASES

Skeletomuscular diseases affect bones, muscles, or components from both these systems. In addition, numerous neurologic disorders may cause muscular weakness, palsies, or paralysis and possible atrophy of muscles innervated by the affected nerves.

Skeletal abnormalities may take the following forms:

1 Dislocations and fractures, causing structural and functional disruptions of the simple machine of which the affected bone is an integral part
2 Irritation or inflammation of the periosteum, compact bone, cancellous bone, tissues which surround a joint, or some combination of these structures because of microbial infection, chronic drug intoxication, or mechanical factors
3 Bone marrow disease due to microbial infection, cancer, or congenital defects which affect hemopoiesis
4 Changes or irregularities in the mineral content of bone, causing increased pliability or brittleness through hormonal or nutritional imbalances or enzymatic erosion by infectious microbes
5 Developmental anomalies or growth defects caused by a reduction in, fusion or lack of fusion between, or loss or addition of specific bones, or because of premature or delayed ossification of joints and bones, increased internal or external pressures, hormonal and nutritional imbalances, or impaction and abnormal positioning
6 Wear and tear at joints or weaknesses in connective tissues surrounding joints due to excessive friction, compressional or rotational forces, or other forms of mechanical injuries

Muscular disorders arise from the following factors:

1  Muscular fatigue or excessive strain due to prolonged activity or overloading
2  Weakness in, or damage of, connective tissue attachments between muscles or between muscles and bones, due to repeated loading, overloading, or mechanical injuries
3  Irritation or inflammation of either muscle tissue or connective tissues associated with the muscles due to microbial infection or physical trauma
4  Mechanical tearing of muscle walls due to excessive internal pressures
5  Muscular weakness or paralysis and possible atrophy due to central or peripheral nerve disorders, hormonal or nutritional imbalances, metabolic dysfunctions, or infections and degenerative diseases
6  Muscle spasms, tremors, palsies, and fibrillation due to excessive strain, central nervous system damage, or loss of rhythmic muscular excitation

This chapter reviews the causes and effects of the more common and well-known skeletal and muscular disorders.

## SKELETAL AND JOINT DISORDERS

Skeletal diseases are called *osteopathies*. Joint disorders are referred to as *arthropathies*. Osteopathies include fractures, spinal deformities, and a miscellaneous collection of other disorders which affect the integrity of bone. Among the arthropathies are numerous joint problems, inflammatory conditions, and mechanical injuries.

### FRACTURES

Types    Fractures occur when bones break. For our purposes here, we shall recognize four types of bone fracture:

1  *Simple* fracture, in which a broken bone does not tear the skin

2  *Compound* fracture, in which a broken bone tears the skin, forming a visible wound
3  *Greenstick* fracture, in which a broken bone splinters but does not break completely because of its excessive pliability
4  *Comminuted* fracture, in which a bone breaks in more than one place

The classification of bone fractures is actually more complicated than the above list indicates. For example, the type of fracture may also be determined by the presence of bone injury to internal organs, the relationship of the broken portion of a bone to adjacent skeletal parts, and the position and direction of the fracture line.

Fractures are caused by excessive stresses which bend a portion of a bone beyond the limits of its resiliency, thereby exceeding its tensile strength. The stresses may be imposed directly by sudden and abnormal physical forces or may be transmitted indirectly by these forces to the weakest point in the adjacent skeletal framework. Although vector forces and tensile strengths of bones are limiting factors in the causation of fractures, numerous influences affect the consistency and, therefore, the strength of bone. The following factors contribute to a weakening of bone: (1) microbial infections which cause decalcification and subsequent softening of the compact tissues of long bones; (2) mineral deficiencies, as of calcium and phosphorus, which increase bone fragility directly, copper deficiencies which depress the absorption of calcium or phosphorus from bone, or fluoride deficiencies which allow the development of dental caries through microbial enzyme hydrolysis of dentin and enamel; (3) a vitamin D deficiency that depresses both the absorption of calcium ($Ca^{++}$) from the small intestine and its subsequent deposition in bone; and (4) hormonal imbalances, such as a deficiency of calciferol, which decrease the intestinal absorption of $Ca^{++}$, or a deficiency of calcitonin or an excess of parathyroid hormone, both of which promote the absorption of $Ca^{++}$ from bone.

Among the most prominent symptoms of a fracture are an impairment of posture and locomotion, severe pain or tenderness, and edema in the affected region.

Movement may be impossible, difficult, or unnatural. Injuries to the skin or visceral organs due to fractures may be exacerbated by attempting active or passive movements. When motion is attempted, grating sounds may be produced because broken portions of bone rub together.

Treatment of fractures involves relief of symptoms, prevention of further damage, promotion of healing, and restoration of function. In compound fractures, external wounds are cleansed with an antiseptic and subsequently bandaged to maintain sterile conditions. Until medical attention is secured, splints are used to immobilize limbs with broken bones. Bones that can be manipulated externally are reset by aligning their broken portions. Pain may be reduced with analgesics or narcotics. During healing, the part is immobilized in a cast or bed rest is prescribed, with or without traction. Rehabilitation strives to promote functional activities during the period of immobilization and to restore or increase passive and active motion after immobilization is no longer necessary. To protect the injured extremity, persons with fractures may resume their activities with the use of crutches for a leg cast or a sling for an arm cast.

## SPINAL DEFORMITIES

Kyphosis, lordosis, and scoliosis Spinal deformities are the result of abnormal curvatures of the vertebral column. Three types of spinal deformity are recognized: (1) an abnormal increase in the primary concave curvature of the thorax, called *kyphosis,* or humpback; (2) an excessive accentuation in the secondary convex curvature of the lumbar region, known as *lordosis,* or swayback; and (3) a lateral curvature of the vertebral column in the thoracic region, referred to as *scoliosis.* Lateral curvatures usually lean to the right.

Frequently, scoliosis is combined with either kyphosis and rotation of the spine in the opposite direction (*kyphoscoliosis*) or lordosis and similar opposite rotation (*lordoscoliosis*). Therefore, scoliosis usually involves bony changes of the thoracic vertebrae in three planes in space—increased lateral curvature, increased anterior or posterior curvatures, and a compensatory rotation of the spine in a direction opposite that of the lateral curvature. Kyphoscoliosis and, to a lesser extent, lordoscoliosis increase the stiffness of the chest wall. Therefore, spinal deformities may cause restrictive lung disease as reflected by hypoxic hypoxia, arterial hypoxia, dyspnea, and cyanosis (see Chap. 29).

The causes of spinal deformities are manifold and include

1 Congenital and growth anomalies which result in a malformed vertebral column at birth or differences in the lengths of long bones of the legs in the adult stage of development
2 Fractures of the vertebrae or hip which lead to orthopedic abnormalities during healing
3 Nutritional or hormonal imbalances which induce rickets and soften the vertebrae
4 Infectious diseases, such as tuberculosis in which the bacterium *Mycobacterium tuberculosis* invades specific vertebrae (Pott's disease), causing their compression, destruction, or in the case of syphilis caused by *Treponema pallidum,* the formation of cancerous lymphoidal or epithelial growths called gummas
5 Certain myopathies and neuropathies which involve muscles associated with the spines of vertebrae, causing their weakening, paralysis, or degeneration
6 Differential amounts of wear and tear at articulation points on each side of adjacent vertebrae, due to postural habits or conditions such as rheumatism and sciatica, causing one side to be preferred to the other

Treatment of spinal deformities is directed toward reducing and balancing abnormal spinal curvatures and compensating for their adverse physiologic effects. Orthopedic correction or reduction of spinal deformities requires complete immobilization or postural support with the use of appropriate braces. Insofar as possible during therapy, improvement of ventilation is attempted at least once a day with proper breathing exercises or manual manipulation of the rib cage.

## OTHER OSTEOPATHIES

A number of other bone diseases cause structural abnormalities and related functional disabilities. These disorders include osteitis, osteomyelitis, osteoporosis, osteomalacia, and microcephaly.

**Osteitis, osteomyelitis, osteoporosis, and osteomalacia**  An inflammation of bone is termed *osteitis*. Osteitis may lead to one of the following types of bone abnormalities:

1  Spinal deformities due to syphilis which secondarily involves the vertebral column
2  Bone softening and fractures in either childhood or adulthood because of absorption of bone $Ca^{++}$, enlargement of the haversian canals, and a corresponding increase in bone porosity (*osteoporosis*) because of mineral, vitamin, or hormonal imbalances (*rickets*)
3  Bone softening in adulthood (*osteomalacia*) for reasons similar to those for osteoporosis
4  Bone softening and the development of bone cysts or tumors due to a chronic hypersecretion of parathyroid hormone, causing absorption of bone $Ca^{++}$ and urinary secretion of $HPO_4^{--}$ and inducing imbalances of these two elements in bone and blood (*von Recklinghausen's disease*)
5  Growth deformities such as a thickening of long bones and an enlargement of flat bones due to idiopathic factors which cause hypertrophy through irregular periods of $Ca^{++}$ deposition in, and absorption from, bone (*Paget's disease*)
6  Bone marrow disease through ossification of cancellous tissue
7  Inflammation of the bone marrow exclusively or of the bone marrow, compact bone, and periosteum caused by pus-producing bacteria (*osteomyelitis*)

Note that osteoporosis, osteomalacia, and von Recklinghausen's disease all lead to a softening of compact bone tissue.

**Microcephaly**  The abnormality termed *microcephaly* is due to a premature ossification of one or more sutures between adjacent cranial bones and results in a cranium that is abnormally small. The reduced brain volume leads to severe mental retardation such as idiocy. However, whether premature ossification is a cause of small brain volume or an effect of stunted brain growth depends on the causative agent.

Microcephaly may be caused by growth anomalies. Brain development may be arrested prematurely by severe dietary deficiencies, especially of protein, during early childhood. The disease, marasmus, is one example of such a dietary deficiency and is discussed in Chap. 27. Growth retardation of the brain is then followed by precocious ossification of the cranial sutures. Congenital malformations of the skull due to premature closure of the sagittal, coronal, or other cranial sutures can also result from hereditary diseases and developmental anomalies, in which case early ossification of the suture joints precedes the stunting of brain growth. The congenital form of the disease is termed craniostenosis (craniostosis).

## COMMON DISTURBANCES OF JOINTS

A number of mechanical injuries give rise to joint abnormalities which usually are more like chronic nuisances or acute but transient disorders rather than serious diseases. Among such injuries are sprains, game knees, and bunions. These conditions primarily affect the tissues which surround a joint.

**Sprain**  A sprain is a sudden physical overloading at a joint, usually caused by twisting which irritates, inflames, tears, or otherwise damages nearby ligaments. Sometimes sprains also involve a specific muscle which is thrown into a spasm or a particular bone which is fractured by the force of the physical trauma. Joints at the ankles, wrists, fingers, and lower back are among the most frequently sprained structures.

The violence of a sprain results in an immediate but temporary functional disability. The region at the involved joint swells considerably, becomes red and hot, and as most of us have experienced, is extremely painful and tender. Internal hemorrhaging of peripheral blood vessels at the site of trauma results in a discoloration of the skin that lasts for several days and changes from purple to yellow as time progresses. The edema stretches the skin, causing an intense itching during the early stages of tissue regeneration.

Game knee   "Game knee" is a broad, nontechnical term which describes a number of mechanical injuries to, and derangements at, the knee joint. The physical trauma associated with bumping, twisting, or excessively stretching the knee may bruise, displace, slip, crush, or tear the cruciate ligaments, the collateral ligaments, the menisci (semilunar cartilages), or tendons associated with muscles of the thigh. Most of these structures are illustrated in Fig. 25-1. As in the case of sprain, muscles and bones may also be involved, especially fracture of the tibial spine.

The symptoms of game knee are similar to those of sprain, except that they appear in the region of the knee joint. Usually, these symptoms include functional impairment of standing, walking, or running and one or more of the inflammatory signs, especially knee pains.

Sprains and game knee are ordinarily treated as if they were fractures. Immobilization or the use of splints are standard procedures. As an aftermath of game knee, the knee joint may be permanently weakened. In such cases, the subjects are susceptible to dislocations of the knee, sometimes called "trick knee." The abnormality is especially prevalent in athletes. The weakened joint is treated by wearing a firm support such as an elastic bandage. Trick knee may lead to bursitis through chronic inflammation of the patellar bursa.

Bunion   A bunion is a swelling and thickening of the

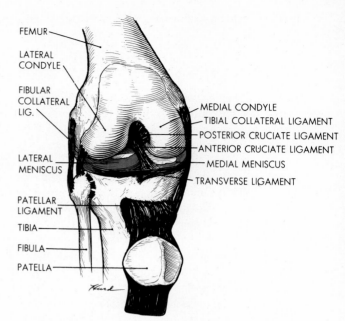

**Figure 25-1**   *The structure of the knee joint from anterior aspect.* (From Melloni, Stone, and Hurd.)

bursa at the metatarsophalangeal joint of the great toe. Usually caused by chronic rubbing or pressure due to poorly fitting shoes, the formation of a bunion is accompanied by the cardinal signs of inflammation and functional impairment such as limping.

A bunion increases the size of the metatarsophalangeal joint, forcing the great toe to move closer to the other toes so that they become bunched together. This lateral displacement of the great toe is responsible for the technical name for a bunion, *hallux valgus*. *Hallux* means "great toe" and *valgus* refers to its lateral displacement.

## INFLAMMATION AT JOINTS

Although it is difficult to make a clear-cut separation between joint injuries which are a common nuisance and those which are relatively infrequent but more serious, the disorders discussed below more often resemble the latter than the former. Included among these inflammatory disorders are bursitis and arthritic diseases such as rheumatoid arthritis and osteoarthritis.

Bursitis   Bursitis is an inflammation of one or more bursae in the connective tissues associated with diarthrodial, or synovial, joints. As you will recall from Chap. 23, such bursae are sacs or cavities which contain synovial fluid and serve to lessen the frictional resistance at joints where bony prominences, tendons, and ligaments are present. Inflammation of a bursa through microbial infection, physical trauma, or other factors may lead to its hardening due to the deposition of $Ca^{++}$. One type of bursitis, bunions, has already been described.

The treatment of bursitis depends on the location of the inflamed bursa and the length and severity of the disability. Bunions and calcified bursae may have to be removed surgically. Immobilization may be prescribed when functional incapacitation is great. Analgesics or anesthetics are sometimes required to reduce the severity of pain. Since bursitis is an inflammatory disease amenable to treatment with cortisol, this anti-inflammatory agent may be used with caution in an attempt to facilitate healing.

Rheumatism and arthritis   Bursitis and many other diseases which primarily affect the joints are lumped together under the broad heading of rheumatism. Characterized by joint pains and muscular soreness or stiffness, *rheumatism* also encompasses a complex of syndromes known as *arthritis*. All arthritic conditions have two features in common—joint inflammation and subsequent pathologic changes in the affected region. Even gout, a metabolic disorder discussed in Chap. 21, is considered a form of arthritis. Arthritis and bursitis are therefore different expressions of rheumatism which involves joints both structurally and functionally.

Two types of arthritis are of especial interest for us here. One is rheumatoid arthritis, because of its frequent and widespread occurrence; the other is osteoarthritis, because of its involvement of both joints and bones. Sometimes rheumatoid arthritis also causes pathologic changes in bones associated with the inflamed joints, so that the distinction between this form of arthritis and osteoarthritis becomes blurred.

*Rheumatoid arthritis* is an arthropathy of idiopathic origin which causes a degeneration of the collagen of connective tissues associated with the joints. This chronic but intermittent disease is characterized by multiple joint inflammation, pains, tenderness or stiffness, swelling, and bone atrophy or degeneration, especially at articulation points. In advanced stages of rheumatoid arthritis, these effects may give rise to bone deformities which cause ankylosis. This condition is described later in this chapter.

Treatment of rheumatoid arthritis is directed toward symptomatic relief, prevention of more serious disabilities, elimination of the causative agent, and promotion of healing, including physical and psychological support. Reduction of symptoms is provided by the use of analgesics such as aspirin to reduce pain and the administration of ACTH, cortisol, or the application of heat treatments, all of which reduce inflammatory symptoms. Prevention of more serious disabilities is accomplished with the use of splints in the affected region and by bed rest during the acute stage of the disorder when pain, swelling, and functional disability are greatest. Elimination of any microbial pathogen responsible for the disease is attempted with the use of antimicrobial drugs and vaccines. Other measures which promote healing include (1) diet and vitamin therapy to promote an optimal utilization of raw materials in synthetic processes taking place during tissue regeneration; (2) maintenance tasks such as daily massage and frequent positioning to promote an optimum circulation and prevent bed sores, respectively; (3) regularity through dietary control; and (4) psychological support and direction to engender positive attitudes and reduce the frequency of egocentric, or self-centered, thoughts.

Like rheumatoid arthritis, *osteoarthritis* affects bones, actually causing the formation of abnormal growths or projections called *spurs*. This form of arthritis differentially affects the articular cartilages of hinge joints active in movement and lifting. The inflammatory condition is accompanied by functional disabilities in the affected region. The pathologic changes found in advanced stages of this disorder may lead to a degeneration of affected skeletal parts.

Osteoarthritis may be induced by a severe case of osteitis which spreads to the joints. In general, arthritis usually appears as a sequel to other disorders such as (1) infectious diseases, including syphilis, gonorrhea, tuberculosis, and rheumatic fever; (2) certain degenerative and metabolic abnormalities; (3) various neuromuscular disturbances; and (4) physical and emotional trauma. Often, however, the development of arthritis is not directly traceable to any of the above-listed factors.

Many arthritic conditions may give rise to *ankylosis,* a pathologic fixation, or immobility, at the affected joints. Ankylosis is due to surface irregularities caused by degenerative changes of either the affected bone or the articular cartilage which covers its end. Ankylosis may also result from the aging process due to excessive wear at the articulating surfaces of affected joints. Finally, ankylosis may result from degenerative changes in ligaments or other tissues which surround joints.

## INJURIES TO JOINTS

Subluxation and dislocation Mechanical injuries in the region of a joint may cause direct physical derangement of its structural components or increase the probability of future displacements. Such traumatic displacements may be partial, or incomplete (*subluxation*), or may involve the displacement of an entire bone from its normal position at a joint (*complete dislocation*). Joint injuries which cause subluxation or dislocation may also involve other structures such as tendons and ligaments in or around the affected joint.

The classification of dislocations in part parallels that of bone fractures and is based on the degree of completeness, the presence or absence of an external wound at the joint, and the identity of the joint and other structures damaged. In addition, the dislocation may be named for the causative agent, the ability of the body to repair the damage, the frequency of injury at the same location, or the investigator who first described a unique form of dislocation. As in so many classification schemes, the system is based on tradition and expediency, rather than on functional utility.

In addition to physical trauma, other factors which result in dislocation include congenital defects, birth injuries, and collagen diseases which weaken the connective tissues supporting the affected joint. Muscular weakness or paralysis increases the susceptibility of a joint to mechanical injury and, therefore, to dislocation.

Functional impairment, extreme pain, and other signs of inflammation are corollaries of dislocation. Treatment emphasizes symptomatic relief and therapeutic immobilization with splints or other devices of the parts at the injured joint. Where an external wound exposes a displaced joint (compound dislocation), the wound must be cleansed with an antiseptic to prevent secondary infection, just as in the case of compound fractures.

Slipped disk In contrast to dislocation, a slipped disk is actually a herniation of an intervertebral disk. The fibrocartilaginous disk may shatter or weaken due to abnormal pressure or the aging process, causing a displacement of the gelatinous material in its center. Called the *nucleus pulposus,* this semisolid substance may be extruded enough to place mechanical pressure on the spinal cord or on one or more roots of spinal nerves in the region of damage.

Nine out of ten cases of slipped disk occur in the lower lumbar region where most of the weight is supported during heavy lifing. Much less frequently, the lower cervical region may be involved due to mechanical injuries caused by excessive compression or abnormal twisting.

Herniation of an intervertebral disk is conceptualized in fig. 25-2. In this illustration, the abnormally displaced nucleus pulposus is pressing against the root of the last lumbar nerve on the left side, causing severe sciatica (see Chap. 16).

Herniation of a lumbar intervertebral disk causes some combination of the following abnormalities: (1)

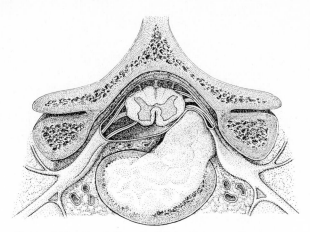

**Figure 25-2** *Herniation of a lumbar intervertebral disk*

impairment of motion at the knee and weakness of the muscles of the feet and toes; (2) depression of the knee jerk; and (3) decrease in tactile and pain sensations or complete anesthesia of portions of the leg and foot coupled with severe, burning pains which radiate along the inner aspect of the thigh (sciatica). In addition, there is a tendency to lean away from the side of the body from which sciatic pains are emanating, inducing a compensatory scoliosis.

Similarly, rupture of a lower cervical intervertebral disk can cause one or more of the following difficulties: (1) impairment of cervical motion and weakness and subsequent degeneration of the muscles of the arm; (2) decrease in the biceps and triceps reflexes ordinarily elicited by sudden mechanical stretching; and (3) paresthesias and pain in portions of the upper extremity supplied by the injured nerve roots.

Treatment of a slipped disk is directed toward relief of symptoms and avoidance of future relapses. Symptomatic relief may be obtained by (1) reducing or eliminating the usual loads applied to the damaged portion of the vertebral column with traction or bed rest; (2) obtaining extra physical support, proper vertebral alignment, and immobilization of specific vertebrae with the use of collars, belts, braces, and bed boards; and (3) reducing pain and inflammation with

analgesics and heat treatments. Prophylaxis is approached through moderation in the weight of loads which are lifted or carried and by instruction in the proper methods of supporting weight while moving heavy objects from one place to another.

## MUSCULAR DISORDERS

Many myopathies give rise to common disturbances through excessive muscle usage and corresponding fatigue, sudden overloading and corresponding abnormalities of muscular contraction, and through neuromuscular effects which result in muscular weakness or fixation. Other myopathies occur in response to inflammation or excessive internal pressures. Neuromuscular disorders lead to muscular paralysis and possible degeneration.

### COMMON DISTURBANCES

**Fatigue** Muscle fatigue is a form of physical tiredness accompanied by muscular aches and pains and a loss of muscle tone. The condition is due to a prolonged use or vigorous activity of skeletal muscle. Muscle fatigue is normally induced by alterations in the chemistry of muscles or the nerves which innervate them.

Muscle fatigue is much more common than the neuronal form. In addition to a muscular depletion of extracellular $Na^+$ or intracellular $K^+$, fatigue based on muscle metabolism may occur because of a depletion or deficit of calcium ($Ca^{++}$), a reduction in the amount of high-energy phosphate stored as creatine phosphate, and an accumulation of intermediary metabolites, including lactic acid, during anaerobic respiration.

The subjective experience of muscle fatigue may also result from emotional factors or from various abnormalities which cause metabolic imbalances, toxic effects, or $O_2$ starvation. More specifically, pathogenic agents responsible for fatigue include (1) deficiencies of essential nutrients, which give rise to malnutrition or undernutrition (see Chap. 27); (2) oxygen deficiencies or underutilization due to respiratory abnormalities that cause hypoxic hypoxia, circulatory disorders that induce stagnant and anemic hypoxia (see Chap. 29), or metabolic disorders that create histotoxic hypoxia; (3) microbial pathogens that secrete or liberate toxins, or introduced chemicals that cause intoxication; (4) psychogenic factors that bring about emotional weariness or promote physiologic exhaustion; and (5) hormonal imbalances, such as hypothyroidism, through a decrease in the basal metabolic rate, hypoinsulinism (diabetes mellitus), by causing hyperglycemia and glucosuria, or a hyposecretion of cortisol (Addison's disease), by depriving muscle cells of glucose through the induction of hypoglycemia.

**Flatfoot**    Flatfoot usually results from chronic fatigue of the muscles and ligaments that support the inner longitudinal and anterior transverse metatarsal arches of the foot. Technically referred to as *pes planus,* fallen arches cause the head of the talus to move inferiorly and medially (see Fig. 25-3). This abnormal displacement and the corresponding flattening of the foot elicit a characteristic plodding gait.

Although flatfoot is ordinarily caused by chronic fatigue, the condition may also be precipitated suddenly by spasmodic contractions of the peroneal muscle. This action results in inversion of the foot.

**Spasm**    Spasms are sudden, involuntary contractions of either striated or smooth muscle. Skeletal muscle spasms are caused by straining or overloading the muscles. Smooth muscle spasms can be induced by allergens and irritants.

Spasms are of two types—tonic and clonic. *Tonic spasms* are extended or prolonged involuntary contractions. The muscle remains contracted and does not relax during a tonic spasm. *Clonic spasms,* on the other hand, consist of alternating contractions and relaxations; i.e., the cycle of contraction and relaxation is repeated many times.

Strong and painful tonic spasms are referred to as *cramps.* Many forms of cramps are named for the specific occupations which elicit them. In such cases,

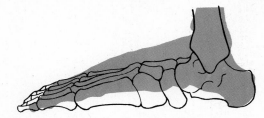

**Figure 25-3**    *The appearance of the arches in flatfoot.*

only when particular tasks are performed do the active muscles tighten up and harden, causing functional impairments and muscular pains called *myalgia.* The most well-known type of cramp is associated with extended periods of writing with a pen or pencil, a condition called writer's cramp.

Tonic and clonic spasms can be expected to cause most or all of the following symptoms: (1) decreased range of motion, stiffness, or complete functional immobility of parts housing the affected muscles; (2) myalgic pains which emanate from the afflicted region; and (3) hypertonicity or repetitive twitching of the involved muscle. Spasmodic contractions of smooth muscle cause various types of visceral discomforts or disturbances such as wheezing, vomiting, and colic.

Treatment of spasm is directed toward the relief of its symptoms and a reduction in the degree of spasm. Since myalgia is believed to be caused by an interruption of intramuscular blood circulation during a spasmodic contraction, massage can promote circulation and reduce ischemia and associated pain. Symptomatic relief may also be obtained with analgesics and heat treatments. Prevention of the recurrence of a spasm requires the elimination of muscle strain by temporarily abstaining from, or reducing the extent of, activities liable to induce spasmodic contractions.

**Tics**    Tics are tonic or clonic spasms of the face, although spasmodic contractions of the head, neck, and shoulders are generally included in the category. Caused by fatigue or psychogenic factors associated

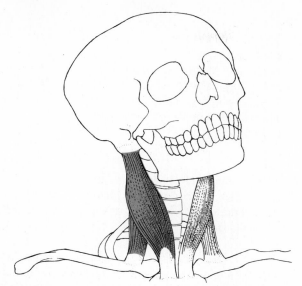

**Figure 25-4**  *Torticollis caused by a spasm of the sterno-cleidomastoid muscle.*

with tension, eyelid tics are probably among the most common clonic spasms in the facial region. Tics of the clavicular portion of the sternocleidomastoid muscle or of the deltoid muscle of the shoulder and arm are also fairly common. Tic douloureux, or trigeminal neuraglia, is considered in Chap. 16.

**Torticollis**  Stiff neck and wry neck are common names for the intermittent, spasmodic contractions of the neck muscles, especially the sternocleidomastoid muscle. Spasms in the neck region displace the head to one side, causing muscular rigidity and myalgia. The chin usually deviates to the side opposite that of the affected muscle. The condition is illustrated in Fig. 25-4.

Such spasms may be induced by straining or abnormally flexing the sternocleidomastoid muscle on one side, as in bending over a laboratory microscope for an extended period. Since the spinal accessory nerve (XIth cranial nerve) innervates the sternocleidomastoid muscle, various lesions of this nerve also give rise to torticollis. In addition, torticollis may

be a consequence of rheumatism, abnormal swellings or growths, or degenerative conditions in the cervical region.

**Muscle strain**  Muscle strain resembles spasm in symptomatology, although involuntary, spasmodic contractions of the affected muscle may not occur. The experience is more like that of bruising due to physical trauma. Muscle strain is ordinarily caused by excessive stretching or by abnormally violent contractions during vigorous exercise.

*Shin splints* is a term descriptive of a condition resulting from strain to the flexor digitorum longus muscle located on the posterior and medial aspects of the shank of the leg. This muscle is illustrated in Fig. 25-5. The abnormality is characterized by pain, tenderness, soreness, and stiffness in the region of the shank associated with the tibia. Shin splints is an occupational hazard of athletes in training.

## OTHER MYOPATHIES

Several disorders involving muscles result from inflammation or mechanical injury. Among these abnormalities are myositis and hernia.

**Myositis**  An inflammation of skeletal muscle is called *myositis*. Muscle cells may be affected exclu-

**Figure 25-5**  *The flexor digitorum longus muscle from posterior aspect. Strain of this muscle gives rise to "shin splints."*

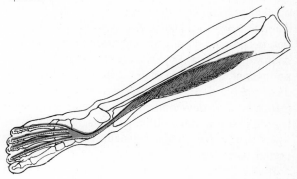

sively or together with surrounding connective tissues. As a result of myositis, the muscle tissue may harden or the affected connective tissue may increase in size through hyperplasia. Muscle inflammation causes myalgia and edema, in addition to muscular dysfunctions. Among the causative agents of myositis are microbial infections, including encystment by the trichina worm (see Chap. 3), physical trauma, and direct irritation due to alcoholism.

**Hernia**   When a muscle wall ruptures, forcing an internal structure or organ into or through the opening, the condition is referred to as a *hernia*. In most cases of herniation, the displaced viscera are contained in a saclike covering of peritoneum, the *hernial sac*. During herniation, the hernial sac is forced through a muscle wall at the point of weakest resistance. In most cases, the weak point is the inguinal canal. The hernial sac containing a portion of the intestine protrudes at, within, or through the inguinal canal in this form of herniation. Other less common forms of herniation include protrusion of the hernial sac containing a portion of the intestine through the umbilicus, or navel, or through the diaphragm into the pleural cavities, protrusion of the stomach through the esophageal hiatus of the diaphragm into the mediastinal cavity, or protrusion of a portion of the diaphragm through the same hiatus into the stomach. Sometimes, hernias occur when organs such as the urinary bladder, uterus, vagina, or a portion of the large intestine protrude into or through natural tubular structures and orifices in the pelvic region.

The most common cause of herniation is the development of sudden, excessively large internal pressures or compressional forces against a visceral wall due to straining. Straining may result from coughing, sneezing, or lifting. Physical trauma, pregnancy, and abdominal tumors may also be responsible for muscular rupturing. Additional factors which cause herniation include developmental anomalies, degenerative diseases, and the aging process, i.e., conditions which lead to a weakness in muscles surrounding the abdominal cavity.

The treatment of hernia requires some combination of the following procedures: (1) restoring the original anatomic relationships by physical manipulation of the hernial sac, if possible; (2) physical support and prevention of further damage with the use of trusses; and (3) surgical intervention for serious hernias, a technique which corrects the positioning of abnormally twisted or constricted herniated viscera.

## NEUROMUSCULAR DISORDERS

A number of myopathies are caused by various neural disturbances. Such disorders include spastic and flaccid paralysis, wristdrop and foot drop, and myasthenia gravis.

**Paralysis**   Paralysis usually denotes the inability to contract skeletal muscle voluntarily. However, reflex contractions of smooth muscle may also be adversely affected. The classification of paralysis is based on structural and functional criteria, although causative agents, investigators' names, temporal patterns, or other factors may take precedence. The following are among the structural and functional criteria:

1   Level of neural involvement. The damage can be in one or more of the following structures:
   a In motor area 4 of the cerebrum
   b In the corticospinal, or pyramidal, tracts from the cerebral cortex to lower motor neurons in the anterior horns of various levels of the spinal cord, pathways used for the mediation of voluntary movements
   c In upper motor neurons of the extrapyramidal tracts from the bulb of the brain to lower motor neurons in the anterior horns of various levels of the spinal cord, pathways used for the mediation of mostly reflex movements
   d In lower motor neurons from the anterior horns of various levels of the spinal cord to one or more muscles innervated by them
   e At myoneural junctions between the axonal ends of the lower motor neurons and the innervated muscles
2   Extent of motor involvement. The amount of func-

tional disability is expressed by the suffix -*plegia,* which means "paralysis," and the appropriate prefix for the disablement of some combination of the upper and lower extremities. This classification includes:

a Monoplegia, or paralysis of one upper or one lower extremity

b Diplegia, or paralysis of both upper or both lower extremities

c Paraplegia, or paralysis of both lower extremities only

d Hemiplegia, or paralysis of the upper and lower extremity on one side of the body

e Quadriplegia, or paralysis of both upper and both lower extremities

3 Degree of motor involvement. The magnitude of motor disability is expressed in terms of either partial paralysis (paresis) or complete paralysis of the affected muscles. As pointed out in Chap. 16, peripheral nerve disorders may cause sensory "paralysis," or anesthesia, as well as motor paralysis.

The level or region of neural involvement is primarily responsible for the dichotomy of spastic versus flaccid paralysis.

*Spastic paralysis* is an involuntary tightening, or hypertonicity, of a functional grouping of skeletal muscles. Such spastic muscles are tense and resist passive movement. Spasticity is caused by lesions of the extrapyramidal tracts, the fibers of which all belong to upper motor neurons, or by blockage of the synaptic enzyme cholinesterase at myoneural junctions. Extrapyramidal lesions cause spasticity by exaggerating deep-tendon reflex activity. Simultaneously, superficial reflex activity is diminished. However, reflex contractions may still be elicited by stimulating the intact efferent limb of the reflex arc either electrically or mechanically.

*Flaccid paralysis* consists of a marked hypotonicity of one or more skeletal muscles. Such flaccid muscles are soft and blubbery and may degenerate or become atrophic with time. Flaccidity is induced by (1) lesions of motor area 4 of the cerebral cortex,

removing the facilitating effects of central nervous system control; (2) damage to the pyramidal tracts for the same reason; (3) injury to the lower motor neurons which innervate the involved muscles after exiting the spinal cord; and (4) chemical inactivation of the neurotransmitter acetylcholine at myoneural junctions. Of these factors, lower motor neuron damage is the most frequent cause of flaccidity. Since damage in the latter case is to the efferent limb of the reflex arc, there is a depression in, or complete absence of, deep-tendon reflexes in the affected region.

The causative agents of paralysis include physical trauma, microbial infections, degenerative diseases, intoxications with heavy metals or alcohol, poisons introduced by parasitic arthropods during feeding (see Chap. 3), decompression sickness (see Chap. 29), complete exhaustion, severe ischemia caused by stagnant hypoxia (see Chap. 29), or psychogenic factors.

The treatment of paralysis is directed toward prevention of further injury, reduction of symptoms, increase in functional capability, and psychological support during retraining. Prevention of further injury may require bed rest, positioning to avoid bedsores, and splinting to reduce possible contractures. Reduction of symptoms is accomplished by massage and heat treatments to improve circulation and reduce inflammation. Other appropriate treatments include exercises and activities to increase the range of motion by manipulating specific functional groups of muscles. Increase in functional capability is achieved through (1) appropriate passive and active exercises; (2) neuromuscular facilitation by increasing the functional capabilities of noninvolved muscles surrounding the paralyzed region; (3) activities of daily living which improve motor abilities and coordination in routine maintenance tasks; (4) avocational and recreational or other forms of therapy to achieve similar objectives; and (5) the use of wheelchairs, crutches, and other devices which assist in reducing the difficulty of performing specific tasks. The above techniques are administered by physical and occupational therapists. Psychological support is contributed by the entire health team, including therapists, nurses, psychiatrists, social workers, the family doctor, and auxiliary personnel, as well as the patient's family.

Wristdrop and foot drop   Wristdrop and foot drop are usually symptomatic of peripheral nerve disorders. *Wristdrop* is due to lesions of the radial nerve, a branch of the brachial plexus. The hand is abnormally flexed in a pronated position at the wrist and cannot be extended voluntarily due to a paralysis of the extensor muscles (see Fig. 25-6). Under these circumstances, only with extraordinarily vigorous contractions of the flexors of the digits may the wrist be extended.

*Foot drop* is the counterpart of wristdrop for the lower extremity. Foot drop is caused by lesions of the common peroneal nerve, a branch of the sacral plexus. The foot hangs downward in a flexed and adducted position. Vigorous flexion of the toes may cause extension of the foot, analogous to the effect of flexing the fingers during wristdrop. The condition is brought about through a paralysis of the extensor and abductor muscles of the foot.

Myasthenia gravis   This neuromuscular disorder is characterized by serious muscular weakness and fatigue. Paralysis may occur, but atrophy is not a part of the syndrome. Although the disease may affect skeletal muscles throughout the body, the muscles innervated by motor fibers of the IIId through VIIth and IXth through XIIth cranial nerves and of spinal nerves belonging to the upper cervical plexus are most likely to be involved. These muscles include the extrinsic eyeball muscles and muscles of facial expression, chewing and swallowing, speech, and respiration, including the diaphragm. Death frequently follows a respiratory crisis precipitated by diaphragmatic paralysis. Ordinarily, however, muscle fatigue is reversible following a period of rest.

As a corollary of these symptoms, there is usually a depression in deep-tendon reflexes. The disease is usually exhibited after the second decade of life and more frequently by women than by men.

Apparently, the disorder is precipitated by some abnormality at myoneural junctions probably induced by metabolic dysfunctions. The following are among

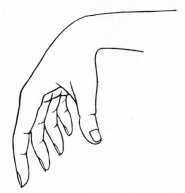

**Figure 25-6**   *The appearance of the hand in wristdrop.*

the explanations which have been proposed to explain the origin of this disease: (1) a marked and sustained depression in the secretion of the neurotransmitter acetylcholine at synapses; (2) an excessive and sustained degradation of acetylcholine by the synaptic enzyme cholinesterase; and (3) an autoimmune response in which anti–skeletal muscle antibodies manufactured by the patient attack the patient's own muscle proteins. Most of the recent evidence indicates that the latter explanation is probably closest to the truth. The autoimmune hypothesis is supported by the demonstration of anti–skeletal muscle antibodies in the blood serum of numerous patients suffering from myasthenia gravis. Additional support is provided by the observation that the thymus gland enlarges during this disease. (See Chap. 4 for a discussion of the role of the thymus gland in immunity.)

Treatment of myasthenia gravis is mostly symptomatic. Drugs such as neostigmine or its analogues are used to inactivate cholinesterase, thereby prolonging and intensifying the action of acetylcholine. This treatment is effective in relieving the symptoms of fatigue in most cases of the disease. Respiratory crises are treated by increasing the dosages and frequencies of administration of the anticholinesterase drug as well as by the use, when required, of various respiratory assists. Where conventional treatment proves ineffective, removal of the thymus gland (thymectomy) or x-ray therapy to differentially destroy this gland may be tried,

## COLLATERAL READING

Brunnstrom, Signe, *Clinical Kinesiology,* Philadelphia: F. A. Davis Company, 1962. Although emphasizing normal movements, the textbook considers various aspects of the pathologic motor behavior of skeletal muscles, especially of paralysis due to peripheral nerve injury.

# PART SIX
# METABOLISM

# CHAPTER 26
# DIGESTION AND
# NUTRITION

Although we are not justified in asserting "you are what you eat," the nature of the foods you consume certainly contributes to your state of health. To support metabolism and perpetuation, the human body requires minimum concentrations of specific raw materials. Many of these nutrients are secured in foodstuffs.

When the consumption of calories derived from foodstuffs exceeds their utilization, weight is gained. Weight is lost under the opposite conditions. Energy homeostasis requires a balance between caloric input and output.

The nutrients utilized by the body include proteins, fats or lipids, carbohydrates, water, minerals, and vitamins. Employed in biosynthetic and degradative reactions, these nutrients are required because either they cannot be synthesized within the body or their body stores must be continually replenished.

The processing of nutrients begins with *ingestion* (taking food in). The tastes of ingested foods and drinks are sampled by taste buds found mostly on the tongue. The nutrients are then processed mechanically and chemically to render them in forms which can be readily absorbed into blood and lymphatic vessels from the lumen of the small intestine. Called *digestion,* this step transforms insoluble foodstuffs into a soluble form through progressive decreases in particle size during their journey through the digestive tract. Following digestion, the soluble foodstuffs can be absorbed directly into the body without further processing.

The assimilation of the soluble nutrients requires their intracellular utilization throughout the body. Many soluble nutrients are prepared within the liver for utilization by other tissue cells. The liver also regulates the blood concentrations of numerous nutrients. Nutrients in excess are either stored within the liver and fat depots or excreted via the urine. The indigestible residues remaining in the digestive tract following digestion are eliminated as feces.

The more specific functions of the digestive system are as follows:

1 The ingestion, digestion, and absorption of nutrients to provide a continuous supply of raw materials for both biosynthetic and degradative reactions in all the cells of the body
2 The acceptance or rejection of potential foods or drinks based on gustatory, or taste, information
3 Body defense through the removal of foreign particles by fixed phagocytes in the small intestine and liver
4 The elimination through defecation of indigestible food residues remaining after digestion is completed
5 The further processing of various soluble foodstuffs by the liver prior to their entry into the intracellular pathways of metabolism

In this chapter, we will examine the structures of the alimentary canal and the digestive glands as well as their specific contributions to ingestion, digestion, absorption, postabsorptive processing, and defecation.

## THE DIGESTIVE SYSTEM

The digestive system consists of the alimentary canal, or digestive tract, and associated digestive structures. The *alimentary canal* is a grouping of organs extending from the mouth to the anus, and includes the oral cavity, pharynx, esophagus, stomach, small intestine, and large intestine. The teeth, tongue, salivary glands, liver, gallbladder, and pancreas are accessory digestive structures. The organs of the digestive system are illustrated from anterior aspect in Fig. 26-1.

## MOUTH

The mouth is the first portion of the alimentary canal and encloses a chamber called the *oral cavity*. The oral cavity (see Fig. 26-1) is bounded anteriorly by the lips, laterally by the cheeks, inferiorly by the tongue,

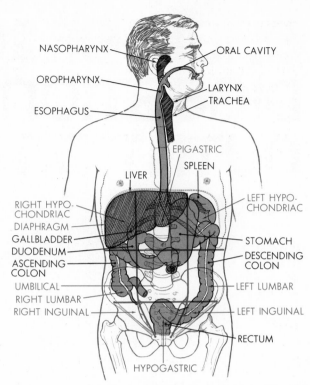

**Figure 26-1**  *The digestive system viewed from anterior aspect. The anatomic regions of the abdomen are superimposed over the viscera of the abdominal cavity. (From Melloni, Stone, and Hurd.)*

superiorly by the hard and soft palates, and posteriorly by arches which mark the entrance into the pharynx.

Foodstuffs are ingested by the mouth. Located in or opening into the oral cavity are the teeth, salivary glands, and tongue. These organs participate in the digestion of foodstuffs.

**Teeth**  The teeth are used primarily in biting and chewing (*mastication*). The teeth participate in mechanical digestion by tearing food apart. Because mastication thoroughly mixes saliva with food in the oral cavity, the teeth indirectly aid in chemical digestion as well as in swallowing, or *deglutition*.

**Structure**  Each tooth is located in an *alveolar socket* in either the maxilla or mandible. A tooth in its alveolar socket within the mandible is illustrated in Fig. 26-2. A tooth is divided into three regions—crown, neck, and root. The *crown* is the apical portion of the tooth projecting beyond the gum. The *neck* is the constricted region between the crown and root. The *root* is the basal region embedded in the bony socket.

From outside inward, there are three layers to each tooth—enamel, dentin, and pulp. Possessing only 3 percent organic matter, *enamel* is the hardest and most resistant substance in the body. Enamel caps the outside of the crown and neck. The outer portion of the root is lined by *cementum* composed of a material harder than compact bone. The *dentin,* or ivory, is the middle layer of a tooth. Although almost one-third organic matter in composition, dentin is a calcified material which, like enamel and cementum, is harder than bone. The *odontoblast layer* on the inner aspect of the dentin contains dentin-producing cells. Located in the central *pulp cavity,* the *dental pulp* is a loose connective tissue containing blood capillaries, lymphatics, and nerves. The pulp cavity extends into each root as a *root canal.* The *apical foramen* is the entrance into the root canal at the base of each root.

The gum and periodontal membrane anchor the tooth in the alveolar socket. The *gum,* or *gingiva,* lines the outside of the maxilla and mandible and consists of fibrous connective tissue covered by a mucous membrane. The *periodontal membrane,* or *alveolar periosteum,* consists of fibrous connective tissue which unites the roots of each tooth to the jawbone. In Fig 26-2 notice the canal within the jawbone through which blood vessels, lymphatics, and nerves are carried to and from the root canals and the surrounding periodontal membrane.

*Sets*   Two sets of teeth develop in human beings: (1) a temporary set of 20 teeth called deciduous, or milk, teeth; and (2) a permanent set of 32 teeth.

1   *Deciduous teeth.* Although these teeth begin to develop by the second month of prenatal development, the first deciduous teeth, ordinarily incisors,

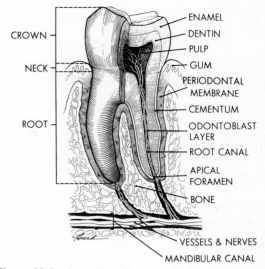

**Figure 26-2**   *A section of a tooth in its alveolar socket.* (From Melloni, Stone, and Hurd.)

do not begin to erupt until about six months of age. The full set of deciduous teeth is usually attained by the end of the following two years. There are 10 deciduous teeth in each jaw. Each quadrant, or half jaw, contains the following sequence of deciduous teeth, starting from the midsagittal line of the face: 2 incisors, 1 canine, and 2 molars.

2   *Permanent teeth.* Ordinarily erupting during the sixth year of life, the first permanent teeth are usually the first molars. The last of the permanent teeth to be cut are the third molars. Because they may not appear until age twenty-five, the third molars are commonly called "wisdom" teeth. There are 16 permanent teeth in each jaw. Each quadrant contains the following sequence of permanent teeth: 2 incisors, 1 canine, 2 premolars, or bicuspids, and 3 molars, or tricuspids. The incisors, canines, premolars, and molars are adapted for biting, tearing, grinding, and crushing, respectively. The permanent teeth and their positions both in the maxilla and in the mandible are illustrated in Fig. 26-3.

**Salivary glands**   Three pairs of salivary glands—the

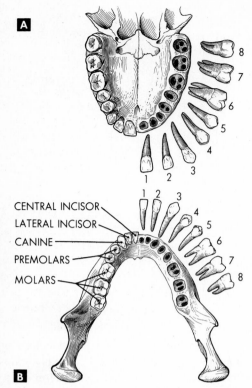

CENTRAL INCISOR
LATERAL INCISOR
CANINE
PREMOLARS
MOLARS

*Figure 26-3* *The locations and shapes of the permanent teeth. A. Teeth associated with the maxilla. B. Teeth associated with the mandible. (From Melloni, Stone, and Hurd.)*

the external ear and overlapping the posterior margin of the masseter muscle, the parotid gland produces a serum containing salivary amylase.

*Stensen's duct* (parotid duct) drains the parotid gland. This duct passes anteriorly along the lateral aspect of the masseter muscle, pierces the buccinator muscle, and opens in the oral cavity immediately above the upper second molar.

**Submandibular gland**   This gland lies on the inferior aspect of the inner border of the mandible. Because it contains both serous and mucous cells, the submandibular gland is a mixed gland.

*Wharton's duct* (submandibular duct) drains the submandibular gland. This duct passes anteriorly beneath the tongue, opening on the lateral aspect of the frenulum at the anterior edge of the sublingual fold.

**Sublingual gland**   The sublingual gland is the smallest of the salivary glands. Located between the lateral aspect of the mandible and the tongue, the sublingual gland produces a mixed secretion of both serum and mucus. The sublingual gland is drained by about 20 small ducts which open along the margin of the sublingual fold.

**Control of salivary secretion**   The parotid gland is

parotid, submandibular, and sublingual glands—are associated with the facial and upper cervical regions. These glands secrete approximately 1,300 ml of saliva a day through ducts which open into the oral cavity. Consisting primarily of water mixed with mucin and salts, *saliva* serves as a lubricant during chewing and swallowing. Saliva also contains an enzyme called *salivary amylase,* or *ptyalin,* which initiates the chemical digestion of starch. Finally, saliva acts as a buffer, tending to prevent large-scale changes in pH which may damage the oral mucosa. The salivary glands on the right side of the face are shown in Fig. 26-4.

**Parotid gland**   The parotid gland is the largest of the salivary glands. Lying on the inferoanterior margin of

*Figure 26-4* *The right salivary glands and ducts viewed from lateral aspect in a parasagittal section of the face. (From Melloni, Stone, and Hurd.)*

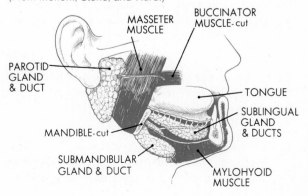

supplied by sensory and motor fibers from the glossopharyngeal nerve (IXth cranial nerve). The submandibular and sublingual glands are supplied by sensory and motor fibers from the facial nerve (VIIth cranial nerve). These cranial nerves also innervate the tongue. The nuclei in the brainstem from which the motor fibers arise are in turn controlled by collateral fibers from the food appetite center in the hypothalamus. The cranial nerves are reviewed in Chap. 14.

*Salivation,* or the secretion of saliva, can be induced by the thought or sight of food, mechanical stimulation within the oral cavity, or the taste of foods. Conversely, salivation can be decreased or inhibited by distasteful foods, colds which adversely affect the sense of taste, or other factors which dull the senses.

**Role in chemical digestion**   Chemical digestion begins in the mouth during chewing. Salivary amylase partially hydrolyzes starch, producing *maltose,* a double sugar, or disaccharide. The reaction is summarized as follows:

$$\text{Starch} \xrightarrow{\text{salivary amylase}} \text{maltose} \qquad (26\text{-}1)$$

Secreted primarily by the parotid glands, salivary amylase acts best in the slightly acidic environment of saliva.

Maltose is formed during the enzymatic hydrolysis of starch only if salivary amylase is allowed to act for a long enough time. Ordinarily, however, food is swallowed too quickly for salivary amylase to have a significant effect on starch digestion. Therefore, salivary amylase continues to act on the starch in the bolus of food as it travels peristaltically through the esophagus and for some time following its entry into the stomach. Finally, the inhospitable acid pH within the gastric lumen inactivates the starch-splitting enzyme.

Starch digestion is reinitiated by another enzyme as the food mass enters the small intestine. The enzymatic degradation of maltose will be followed later during the discussion of chemical digestion in the small intestine.

**Tongue**   The tongue is a compound muscular organ located in the oral cavity and oropharynx (see Fig. 26-1). The striated muscles which compose the tongue are divided into right and left halves by a septum of fibrous connective tissue and covered by a mucous membrane. Papillae containing taste buds mediating the salty, sweet, sour, and bitter modalities of taste are on the superior aspect of the tongue (see Fig. 11-12) and on the roof of the oral cavity. The ducts of the submandibular and sublingual salivary glands open on the sublingual folds on the side of the tongue or on the frenulum attached to the base of this organ (see Fig. 26-4). The lingual tonsil, an aggregation of lymph nodules, is located at the root of the tongue.

The tongue participates in the following functions: (1) *mastication,* or the thorough chewing and mixing of food in the oral cavity; (2) *gustation*, or taste, through stimulation of the taste buds; (3) *deglutition*, or swallowing, by forming a bolus, or semisolid mass of food, which is then pushed into the oropharynx; and (4) *phonation*, or speech, by shaping sounds. Gustatory stimulation also contributes to the production of gastric secretions. The smell, sight, or thought of food can also initiate gastric activity.

## PHARYNX

The oropharynx is the posterior portion of the oral cavity. This chamber is illustrated in Fig. 26-1, and is described in detail in conjunction with the respiratory system in Chap. 28.

**Deglutition**   The *first stage* of deglutition is initiated when a bolus of food is pushed voluntarily by the tongue into the oropharynx. During the passage of food through this region, touch receptors are stimulated in the mucous membrane of the pharyngeal wall. Excitation of these receptors is necessary for the involuntary reflex which follows.

Directing food into the esophagus, the *second stage* of deglutition is involuntary and requires the following reflex actions:

1   Elevation of the tongue against the hard palate, preventing food from returning to the oral cavity
2   Contraction of the pharyngeal musculature, forcing the food posteriorly

3 Relaxation of the pharyngoesophageal sphincter guarding the entrance between the pharynx and esophagus

4 Elevation of the soft palate, preventing food from entering the nasopharynx

5 Closure of the nasopharyngeal orifices of the eustachian tubes

6 Elevation of the larynx, causing the epiglottis to cover the glottis and thereby blocking the entrance into the larynx

7 Constriction of the true vocal folds, further decreasing the possibility of food entering the larynx

The *third stage* of deglutition is completed when the bolus enters the stomach following its peristaltic transport through the esophagus and passage through a relaxed cardiac sphincter guarding the gastric entrance. Peristalsis will be explained shortly.

## TISSUE ORGANIZATION OF THE DIGESTIVE TRACT

All portions of the digestive tube from the esophagus through the large intestine have the same basic tissue organization. From the lumen of the digestive tract to the outside surface of the wall, these layers, or coats, consist of a mucous membrane, or *mucosa*, a *submucosa*, a *muscle layer*, and a serous layer, or *serosa*. These layers are shown in a generalized cross section of the digestive tube in Fig. 26-5.

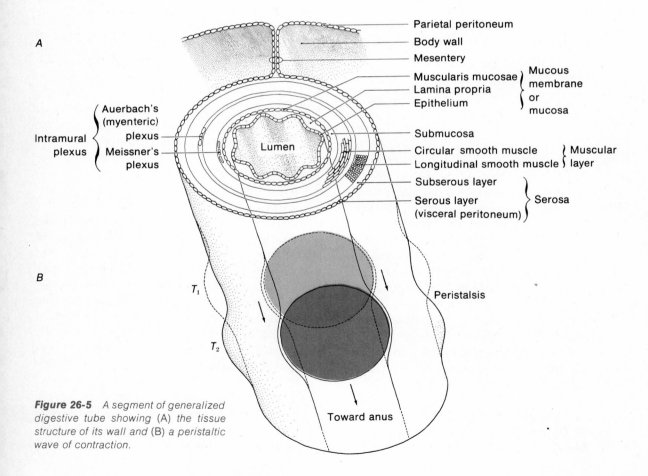

**Figure 26-5** *A segment of generalized digestive tube showing* (A) *the tissue structure of its wall and* (B) *a peristaltic wave of contraction.*

**Mucosa** The mucosa lines the lumen of the alimentary canal. This coat contains an epithelial lining which immediately surrounds the lumen; a lamina propria of connective tissue (see Figs. 26-7 and 26-8); and a muscularis mucosae composed of a thin layer of smooth muscle fibers. The epithelium is secretory, providing chemicals which participate primarily in digestion. The lamina propria contains solitary and aggregated lymph nodules which function in body defense.

**Submucosa** The submucosa consists of fibroelastic connective tissue. The flexibility of this layer permits adjustments in the size of the digestive tube. Within the submucous layer are extensive plexuses of blood vessels, lymphatic vessels, and nerve fibers.

The submucous nerve plexus is called *Meissner's plexus* and consists of both sympathetic and parasympathetic nerve fibers. Sympathetic stimulation inhibits secretory activity. Parasympathetic stimulation has the opposite effect.

**Muscle layer** The smooth muscle coat consists of an inner layer of circular muscle fibers and an outer layer of longitudinally arranged muscle fibers. Since the inner layer of smooth muscle encircles the digestive tube, its contraction narrows and elongates the tube, thereby decreasing its diameter. Conversely, contraction of the outer longitudinal musculature causes the digestive tube to thicken and shorten, because these muscle fibers are arranged along its long axis. The nature of the motility exhibited by the gut is due to the antagonistic actions of these two bands of muscle. The circular muscle layer is modified in various portions of the digestive tract into *sphincters* which guard the openings between adjacent regions.

A nerve plexus is located in the connective tissue between the circular and longitudinal muscle layers. Called *Auerbach's,* or the *myenteric, plexus,* this grouping of nerve fibers is formed from both divisions of the autonomic nervous system. In this respect, Auerbach's plexus is similar to Meissner's plexus. Indeed, both plexuses together are called the *intramural plexus.* However, Auerbach's plexus is involved with the excitation and inhibition of the smooth muscle layer. Sympathetic stimulation inhibits contraction of the muscle layer. Excitation of parasympathetic fibers promotes gut motility by causing contractions of the muscle layer. The role of Auerbach's plexus in peristalsis will be explained shortly.

**Serosa** The outside of most portions of the digestive tube is covered by a thin membrane of squamous cells called *visceral peritoneum.* The visceral peritoneum is attached to the muscular coat by a thin layer of fibrous connective tissue referred to as the *subserous layer,* which is part of the *serosa.*

A double membrane of visceral peritoneum attaches the digestive tube to the posterior body wall. The double membrane of peritoneum is called a *mesentery* and is given specialized names in various portions of the digestive tract. The peritoneum also lines the inside of the body wall and is termed the *parietal peritoneum.* The space between the parietal and visceral peritoneum is called the peritoneal cavity, or *coelom.* The coelom is identical in origin to the pericardial cavity which surrounds the heart (see Chap. 30) and the pleural cavities which surround the lungs (see Chap. 28).

## ESOPHAGUS

The esophagus is a musculomembranous tube about 9 in. long and 1 in. in diameter connecting the oropharynx with the stomach. Descending in the mediastinum anterior to the vertebral column and posterior to the trachea, the esophagus courses through the esophageal hiatus of the transverse diaphragm at the level of the tenth thoracic vertebra. The esophagus opens into the cardiac end of the stomach less than an inch below the diaphragm. The esophagus is illustrated from anterior aspect in Fig. 26-1.

The tissue organization of the esophageal wall differs little from the basic coats described above. The lining of the mucous membrane consists of stratified squamous epithelium that has the ability to distend greatly. Small glands scattered throughout the mucous membrane secrete mucus to provide a lubricant for the bolus as it passes through. The proximal portion of the esophageal wall contains striated muscle fibers which aid in propelling the bolus through the esophagus. These striated fibers gradually disappear and are replaced by a smooth musculature in the lower half of the esophagus. Lacking a visceral peritoneum, the serosa in this region of the digestive tract is composed of a fibrous tissue. Circular smooth muscle at each end of the esophagus is modified into a sphincter. A *pharyngoesophageal sphincter* guards the entrance to the esophagus from the pharynx, as was pointed out earlier. A *gastroesophageal sphincter* is located in the wall at the gastric end of the esophagus.

The esophagus transports ingested materials by peristalsis from the oropharynx to the cardiac portion of the stomach. Since the peristaltic movement of food through the esophagus is the last stage of deglutition, the esophagus functions in this process.

**Peristalsis**  Peristalsis is the basic movement exhibited by the wall of the esophagus and the rest of the digestive tract distal to the esophagus. *Peristalsis* is a wave of smooth muscle contraction which moves for some distance along the length of the digestive tract. As a consequence of peristalsis, the contents within the lumen of the motile portion of the gut are pushed closer to the anus.

In the esophagus, peristalsis ordinarily passes without stop from the pharyngeal to the gastric extremity in about 6 sec. A peristaltic esophageal wave moves at an average velocity of 1.5 in./sec. The peristaltic speed varies from less than 1 in./sec to about 4 in./sec, depending upon the region of the digestive tract exhibiting motility.

Other types of peristaltic movement are seen in the stomach, small intestine, and large intestine. The nature and functions of these movements will be discussed in appropriate sections later in this chapter.

Peristalsis is induced by distension of the wall in a portion of the digestive tract due to the presence of food in the lumen. Although the smooth musculature of the gut wall will contract rhythmically when stimulated mechanically in the absence of a nerve supply,

the rate of peristaltic contraction is markedly increased when the parasympathetic innervation of the intramural plexus remains intact. In other words, the intrinsic contractility of the smooth muscle layer is facilitated by parasympathetic stimulation, when the musculature is stretched.

Peristalsis ordinarily transports materials toward the anus. This directional transport requires contraction of the smooth muscle layer about one inch above the location of the food and relaxation of the smooth muscle layer in the region immediately below the food mass. Presumably aided by the action of the intramural plexus, the region of relaxation promotes the movement of the food mass when the musculature contracts on the opposite side.

This stage of peristalsis is conceptualized at time $T_1$ in Fig. 26-5. As the mass descends in the lumen, the bolus distends the next portion of the wall, causing corresponding shifts in the regions of contraction and relaxation. Therefore, at $T_2$, the peristaltic wave is inferior to its position at $T_1$. In this manner, the peristaltic wave moves progressively along the wall in a portion of the digestive tract.

## STOMACH

In its undistended form, the stomach is a J-shaped organ of the digestive tract located between the esophagus and small intestine. The stomach lies in the left hypochondriac and epigastric regions of the abdomen. As it expands with food, however, it may extend into the umbilical and right hypochondriac regions. The position of the stomach is shown in Fig. 26-1.

A bolus entering the stomach from the esophagus must pass through the *cardiac orifice*. This orifice is surrounded by a band of circular smooth muscle called the *cardiac sphincter*. The cardiac sphincter tends to prevent food from being regurgitated from the stomach into the esophagus.

Longitudinal ridges called *rugae* line the lumen of an undistended stomach. As food enlarges the volume of the stomach, the rugae tend to disappear. The concave *lesser curvature* on the right side of the stomach and the convex *greater curvature* on the opposite

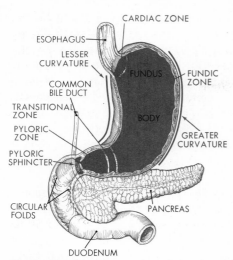

*Figure 26-6* A cutaway view of the stomach showing its gross morphology and relationships to the esophagus, small intestine, and pancreas. (From Melloni, Stone, and Hurd.)

side also diminish in magnitude as the stomach fills with food, so that a full stomach appears as a straight sac.

Food in the gastric lumen is digested both physically and chemically. As a result of these actions, each semisolid bolus is transformed into a semiliquid mass of pastelike consistency called *chyme*. Chyme is periodically forced by peristaltic movements through the *pyloric orifice* at the distal end of the stomach. The pyloric orifice is surrounded by a band of circular smooth muscle called the *pyloric sphincter.* This sphincter contracts reflexly when sufficient chyme collects in the duodenum.

Based on the nature of the glandularity of the mucosa and the muscularity of the muscular coat, the stomach is divided functionally into several regions:

1 Located on the right superior portion of the stomach, the *cardiac zone* possesses coiled glands containing mucous neck cells which secrete mucus.
2 Located in the fundus and body of the stomach, the *fundic zone* possesses simple branched gastric glands containing chief cells which secrete

pepsinogen and the "intrinsic factor," facilitating absorption of vitamin $B_{12}$ (cyanocobalamin, see Chap. 27), parietal (oxyntic) cells which secrete HCl, and mucous cells which secrete mucus.
3 Located between the body and pylorus of the stomach, the *transitional zone* possesses a mixture of gastric glands and pyloric glands.
4 Located at the inferior end of the stomach, the *pyloric zone,* or *antrum,* possesses glands containing mucous neck cells which secrete mucus.

The zones, orifices, and curvatures of the stomach are shown from anterior aspect in Fig. 26-6.

A section through the gastric wall is diagramed in Fig. 26-7. The tissue organization is basically similar to that of the generalized digestive tube presented earlier, although there are some distinctive modifications. Notice the mucous coat and the gastric pits, or glands, of the mucosa from which the gastric juice enters the lumen. The lamina propria of the mucosa is well supplied with isolated lymph nodules for the filtration of bacteria and other foreign materials which may be absorbed from the gastric lumen. In addition to circular and longitudinal muscle layers, a portion of the stomach also contains an innermost layer of oblique smooth musculature.

*Figure 26-7* The basic organization of the gastric wall. The lumen of the stomach is uppermost, next to the mucosa. (From Melloni, Stone, and Hurd.)

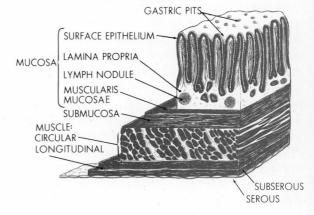

**Movements**   Two movements are characteristic of the stomach: (1) *mixing movements,* which promote digestion by mechanically churning together the gastric juice and the bolus; and (2) *peristaltic movements,* which propel small quantities of chyme from the stomach into the duodenum.

**Mixing movements**   These movements promote the transformation of the semisolid bolus within the stomach into semiliquid chyme. Originating in an excitable portion of the cardiac region of the stomach, called the *"pacemaker,"* and passing for variable distances along the length of the stomach, mixing movements cause the gastric juice containing pepsin and hydrochloric acid to blend with all portions of the bolus. As a consequence, the efficiency of gastric chemical digestion is increased.

Mixing movements occur sporadically. Their frequency and intensity are increased by an accumulation of food in the stomach. Contraction is strongest in the pyloric antral region because of the presence of thick muscular walls. As the food content within the stomach increases, some of the mixing movements become unusually intense. These contractions are called peristaltic movements.

**Peristaltic movements**   Each of the peristaltic waves forces a small amount of chyme from the stomach into the duodenum by increasing the intragastric pressure beyond the intraduodenal pressure. Peristaltic movements are required to empty the stomach.

Peristaltic movements of the stomach are reduced or blocked by either neural inhibition of the gastric smooth muscle layer by an accumulation of chyme in the small intestine or chemical inhibition by the hormone *enterogastrone* secreted by the crypts of Lieberkuhn in the small intestinal mucosa. The neural mechanism is referred to as the *enterogastric reflex.* Simultaneously, the pyloric sphincter constricts, preventing the regurgitation of duodenal chyme. The chemical mechanism is promoted by the presence of lipids in the lumen of the duodenum. The secretion of enterogastrone permits lipid digestion to continue for a longer time than would otherwise be possible.

**Chemical digestion**   The hydrolysis of proteins and emulsified fats is initiated in the stomach. Protein hydrolysis is begun by the enzyme *pepsin* and hydrochloric acid (HCl). Both these chemicals are secreted by gastric glands. The hydrolysis of emulsified fats is accomplished by *gastric lipase,* another enzyme secreted by the gastric glands. Mucus secreted by the pyloric glands protects the gastric mucosa from the potentially destructive actions of pepsin and HCl.

**Protein hydrolysis**   Chief cells in the gastric glands of the gastric mucosa synthesize an inactive precursor of pepsin called *pepsinogen.* When food is not present in the stomach, the inactivity of this proteolytic enzyme prevents the gastric mucosa from being digested by its own secretory product.

When appropriately stimulated, the chief cells secrete pepsinogen into the gastric lumen. The activation of pepsinogen requires the secretion of HCl by the parietal cells of the gastric glands. Ordinarily, the same stimuli responsible for the liberation of pepsinogen also cause the simultaneous release of HCl. Therefore, when pepsinogen is secreted, it is automatically converted into its active form, pepsin. This activation reaction is summarized as follows:

$$\text{Pepsinogen} \xrightarrow{\text{HCl}} \text{pepsin} \qquad (26\text{-}2)$$

The presence of pepsin in the gastric lumen also contributes to the activation of pepsinogen:

$$\text{Pepsinogen} \xrightarrow{\text{pepsin}} \text{pepsin} \qquad (26\text{-}3)$$

Thus, both HCl and pepsin are activators of pepsinogen.

Pepsin breaks down proteins into polypeptides, molecules consisting of more than two but less than 50 amino acid residues. The general reaction is summarized as follows:

$$\text{Proteins} \xrightarrow{\text{pepsin}} \text{polypeptides} \qquad (26\text{-}4)$$

The polypeptides are degraded by other proteolytic enzymes in the small intestine.

Since HCl is a strong acid, it also contributes to the chemical degradation of proteins. In addition, the acidity of the gastric lumen created by the presence of HCl is required for the normal functioning of pepsin. The pH within the lumen of the stomach during digestion is about 2, i.e., highly acidic. Pepsin becomes inactivated when the pH increases beyond 4 and the acidity correspondingly decreases. Such conditions prevail when the stomach is empty or after the chyme mixed with pepsin is forced from the acid stomach into the slightly alkaline duodenum.

The secretion of pepsinogen and HCl is stimulated by the following factors: (1) the thought, sight, smell, or taste of food causes parasympathetic excitation of the gastric glands by vagal nerve fibers; (2) mechanical stimulation of the pyloric region of the stomach through distension results in hormonal stimulation of the gastric glands due to the secretion of *gastric gastrin;* and (3) mechanical stimulation of the duodenal region of the small intestine through distension also results in hormonal stimulation of the gastric glands due to the secretion of *intestinal gastrin.*

The secretion of pepsinogen and HCl is inhibited by enterogastrone. In addition to reducing the secretory output of the gastric glands, enterogastrone inhibits gastric motility.

*Lipid hydrolysis*  An enzyme called *gastric lipase* is also secreted by specialized cells of the gastric glands. This enzyme is capable of acting only on lipoidal emulsions such as milk. However, the contribution of gastric lipase to lipid digestion is minimal. The digestion of most lipids must wait until chyme enters the small intestine.

*Chemical protection*  The gastric mucosa, like all protoplasm, is an aqueous solution of proteins and therefore vulnerable to the proteolytic action of pepsin and the oxidative action of HCl. However, the columnar epithelium is protected from damage by a layer of mucus. The mucus is secreted by mucous cells of the pyloric glands located primarily in the antral region of the gastric mucosa.

Pepsinogen and HCl are ordinarily secreted only when needed, i.e., when food is ingested. Under these circumstances, the mucous layer is not exposed to the destructive action of these chemicals for a long enough time to be eroded. Indeed, even if localized erosion does occur, the regenerative capacity of the mucosa is sufficient to routinely repair the damage before more serious consequences develop. The causes of pathologic conditions which result in ulcerations of the gastric or duodenal mucosa are described in Chap. 27.

*Gastric absorption*  Very few commonly ingested materials can breach the mucosal barrier formed by the tightly woven gastric columnar epithelial cells and its overlying layer of secreted mucus. Nevertheless, water is freely diffusible and will move osmotically from the lumen of the stomach across the mucosal barrier into the bloodstream. Ethyl alcohol ($CH_3CH_2OH$) can also breach the mucosal barrier and therefore is absorbed directly from the stomach. The active ingredient in aspirin, acetylsalicylic acid, is still another chemical absorbed through the gastric mucosa.

The acidity of the gastric lumen and the alcoholic content of the stomach are important factors in the absorption of aspirin. Ordinarily aspirin is soluble in water. Were it not for the acidity within the stomach, acetylsalicylic acid would not be able to move across the lipoid portion of the plasma membranes. The acid environment makes aspirin lipid-soluble. Interestingly, the presence of ethyl alcohol in the stomach also promotes the absorption of salicylates. The ingestion of aspirin and alcohol together is dangerous, however, because of the potentially damaging effects of aspirin on the gastric mucosa.

## SMALL INTESTINE

The small intestine is a long, coiled, tubular portion of the alimentary canal between the stomach and large intestine. About 23 ft in length, the small intestine decreases in diameter from the pyloric orifice at its proximal end to the ileocecal valve at its distal end. The relationship of the small intestine to the rest of the alimentary canal is shown in Fig. 26-1.

The small intestine is divided into three continuous regions: a proximal duodenum; a middle jejunum; and a distal ileum. These regions are demarcated by slight variations in the tissue organization of their walls, but these differences are not obvious externally.

Duodenum . The duodenum extends for a distance of about 10 in. from the inferior end of the stomach. The pancreas is located in part between the duodenum and the stomach. Approximately 3 in. below the pyloric sphincter, the duodenum receives a common duct draining pancreatic juice from the pancreatic duct and bile from the common bile duct. The relationship of the duodenum to these ducts and the pancreas is shown in Fig. 26-9.

Jejunum   The jejunum constitutes the next 6½ ft of the small intestine. This region extends from the duodenum to the ileum.

Ileum   The transition between the jejunum and the ileum is gradual. Measuring almost 16 ft in length, the ileum is the longest portion of the small intestine. The ileocecal valve is located at the junction between the ileum and the large intestine.

Tissue organization   The layers in the wall of the small intestine follow the basic plan of the generalized digestive tract described earlier. Specializations of the small intestinal wall are as follows:

1   Mucous membrane
a  *Brunner's glands* (duodenal glands). Found only in the duodenum, these highly branched and coiled submucosal glands secrete mucus only. Presumably, the mucous covering prevents erosion of the duodenal epithelium by acid chyme entering from the stomach.
b  *Crypts of Lieberkühn* (intestinal glands). These simple tubular glands are formed by depressions of the columnar epithelium into the muscularis mucosae. The crypts are distributed over the surface of the small and large intestines. The crypts in the small intestine secrete an intestinal juice called *succus entericus*. The succus entericus contains various digestive enzymes in aqueous solution. The crypts of Lie-

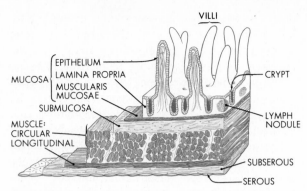

**Figure 26-8**   *The basic organization of the duodenal wall of the small intestine. The lumen of the small intestine is uppermost, next to the mucosa. (From Melloni, Stone, and Hurd.)*

berkühn are illustrated in a portion of the small intestinal wall in Fig. 26-8.
c  *Villi.* Four to five million tiny fingerlike processes called villi project into the lumen from all portions of the small intestinal mucosa. Some villi are shown in Fig. 26-8. The villi absorb soluble nutrients from the intestinal lumen. The columnar epithelium which lines the villi and the rest of the inside wall contain minute folds called *microvilli* in the plasma membrane next to the lumen. The microvilli increase the absorptive area of the intestinal mucosa. The inside of each villus contains projections from the muscularis mucosae. Centrally located in each villus is a *lacteal,* a lymphatic vessel which functions in fat absorption. Surrounding each lacteal is a blood capillary plexus which is responsible for the absorption of all other soluble nutrients.
d  *Circular folds.* The mucosa and submucosa are thrown into transverse circular folds which project for varying distances into the lumen. The villi project into the lumen from all points on the surface of the circular folds. Running partially or completely around the circumference of the small intestine, the circular folds do not disappear when the wall is distended by chyme. The

circular folds are prominent in the duodenum and jejunum. The folds taper from the middle of the jejunum to the middle of the ileum and are absent completely in the lower portion of the ileum. The circular folds increase the surface area for the absorption of soluble nutrients. The surface area provided for absorption by the circular folds, villi, and microvilli is estimated in the vicinity of 10 m². This area is the size of a room almost 10 ft on each side. Since food takes an average of 7 hr to pass through the small intestine, the efficiency of digestion and absorption is great. In other words, few undigested or soluble nutrients ordinarily enter the large intestine.

    *e Lymph nodules.* Solitary and aggregated lymph nodules are scattered throughout the lamina propria in the small intestine. Although these nodules are most concentrated in the ileum, they are also found in the jejunum and duodenum. The aggregated lymph nodules are commonly called *Peyer's patches.* Both types of nodule are associated with lymphatic plexuses and function in the filtration of bacteria and the neutralization of other antigens.

2 Submucosa. This region consists of richly vascularized connective tissue.

3 Muscular coat. The inner circular smooth muscle layer is thick, the outer longitudinal layer thin. This relationship is shown in Fig. 26-8.

4 Serosa. Visceral peritoneum covers the distal portion of the duodenum and all the jejunum and ileum. A double layer of peritoneum called the mesentery attaches the jejunum and ileum to the posterior body wall. The fan-shaped mesentery is richly vascularized and contains many lymph nodes scattered about its surface.

**Movements**  The small intestine exhibits three different types of movement: (1) segmentation movements, which promote digestion by mechanically dividing chyme into smaller units; (2) peristaltic movements, which conduct chyme slowly toward the large intestine; and (3) mucous membrane movements, which increase the efficiency of absorption by promoting the motility of the intestinal folds and villi.

*Segmentation movements*  Segmentation movements are various types of rhythmic contractions involving localized regions of the small intestinal wall. Compared to the jejunum or ileum, these movements are more regular and have a higher frequency in the duodenum.

    The chyme in the lumen of the small intestine is divided into smaller and smaller masses through the actions of these intermittent constrictions of the wall. As the mass of chyme is reduced, the surface area increases. Thus, the efficiency of enzymatic digestion of foodstuffs in the small intestine is increased by this mechanism. The effect of particle size on the rate of chemical reactions is discussed further in Chap. 10.

*Peristaltic movements*  Peristaltic movements differ from segmentation movements in that they travel a short distance along the length of the small intestine. However, these intermittent waves of contraction move relatively slowly and die out quickly.

    Peristaltic waves in the small intestine are promoted by distension of the stomach induced by the entry of a bolus. This response is referred to as the *gastroenteric reflex.*

*Mucous membrane movements*  When excited by sympathetic stimulation, Meissner's plexus in the small-intestinal wall is responsible for an increase in both the degree of mucosal folding and the motility of the villi. The enlarged mucosal folds increase the surface area for the absorption of soluble nutrients. The waving villi eventually bring their absorptive walls in contact with all portions of the chyme as well as promote circulation of the lymph within the lacteals. These actions increase the efficiency of absorption.

**Chemical digestion**  By the time chyme enters the small intestine, the foodstuffs are in the following forms: (1) proteins have been degraded into polypeptides; (2) emulsified lipids have been broken down into fatty acids and monoglycerides but neutral fats, which represent the majority of ingested lipids, are essentially unchanged; and (3) some starch has been converted into maltose. Digestion of the foodstuffs is completed in the small intestine.

Earlier, we mentioned that digestion is the process by which insoluble materials are made soluble so that they can be readily absorbed into the body. Solubility requires a reduction in the size of the particles of food to that which can be transported across the plasma membranes of the epithelial cells which line the small-intestinal mucosa. Mechanical digestion by itself cannot decrease the size of the food particles sufficiently to permit their passage across the differentially permeable membranes of cells. Solubility of the foodstuffs is attained in the intestinal lumen primarily through the actions of enzymes secreted by the pancreas and small intestine.

Chemical digestion in the intestinal lumen is initiated, sustained, and shut off by a complex interplay between chemical and neural influences. The chemical regulators of enzymatic secretions are primarily hormonal. The secretion of the hormones is finely tuned to the presence of specific nutrients in particular portions of the digestive tract. In the sections below, we will follow these reciprocal interactions during the intestinal digestion of proteins, lipids, carbohydrates, nucleic acids, and their derivatives.

*Proteins*   Most proteins in chyme entering the small intestine are either still completely undigested or partially digested polypeptides due to the action of pepsin. The chemical digestion of proteins and their derivatives is mediated by a series of enzymes produced in both the pancreas and small intestine.

The pancreas, an elongated gland frequently sold in supermarkets as "sweetbread," is nestled for the most part between the stomach and duodenum. The pancreas consists of three continuous portions—head, body, and tail. The head is attached to the duodenum; the tail runs to the spleen. The relationship of the pancreas to the duodenum is shown in Fig. 26-9.

The pancreas is both an exocrine and an endocrine gland. Its exocrine secretion is called *pancreatic*

*juice* and consists of numerous digestive enzymes mixed in aqueous solution. The pancreatic juice is produced by the main mass of the pancreas consisting of lobules of pancreatic *acinar cells* drained by ductules. The lobules form larger groupings of glandular tissue called *lobes*. Because the gland contains many outpouchings termed acini, the exocrine portion of the pancreas is referred to as a *compound alveolar gland*. Each lobe is drained by a branch of one of the two pancreatic ducts which penetrate the duodenum. The relationship of the pancreatic ducts to the small intestine will be explained shortly.

Scattered between the lobes throughout the pancreas are isolated patches of cells called *islets of Langerhans*. The islets of Langerhans are the endocrine portions of the pancreas, secreting the hormones insulin and glucagon. These hormones play important roles in the regulation of carbohydrate metabolism (see Chap. 17).

The pancreatic enzymes which participate in the digestion of proteins include *trypsin, chymotrypsin,* and *carboxypeptidase*. These enzymes are normal constituents of pancreatic juice. Because these en-

**Figure 26-9**  *The system of ducts draining products from the liver, gallbladder, and pancreas into the duodenum.* (From Melloni, Stone, and Hurd.)

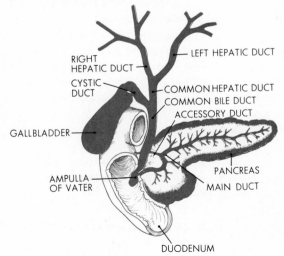

zymes are proteolytic, they are secreted in inactive form only when needed. To follow the synthesis, secretion, and subsequent activation of these enzymes, we must look at the factors which stimulate the pancreas to produce pancreatic juice.

The volume and bicarbonate ($HCO_3^-$) content of the pancreatic juice, but not its enzymatic concentration, are increased when the pancreatic cells are stimulated by the intestinal hormone *secretin*. Under the influence of acid chyme, *prosecretin* is converted into secretin within the gland cells of the crypts of Lieberkühn. Although the major target organ of secretin is the pancreas, this hormone also affects the liver (see below).

The $HCO_3^-$ in pancreatic juice reduces the acidity of chyme in the small intestine by tying up excess hydrogen ions ($H^+$). This neutralization reaction is summarized as follows:

$$H^+ + HCO_3^- \rightarrow H_2CO_3 \rightarrow H_2O + CO_2 \qquad (26\text{-}5)$$

The action of $HCO_3^-$ not only eliminates the potentially harmful effect of gastric HCl on the relatively unprotected duodenal mucosa but, in addition, promotes the slightly alkaline environment required for the optimal action of the enzymes which normally function within the intestinal lumen.

The enzymatic content of the pancreas, including the concentration of its proteolytic enzymes, is increased when the pancreatic cells are stimulated by the intestinal hormone *pancreozymin*. When polypeptides are present in intestinal chyme, pancreozymin is secreted by the epithelial cells of the crypts of Lieberkühn. Vagal stimulation of the pancreas when a person eats also increases the synthesis of zymogen granules in the pancreatic acinar cells. The intracellular mechanism by which zymogen granules are manufactured is described in Chap. 5.

When secreted into the pancreatic acini, the proteolytic enzymes are in their inactive forms—*trypsinogen, chymotrypsinogen,* and *carboxypropeptidase*. These inactive enzymes and the rest of the pancreatic juice are transported primarily through the main pancreatic duct (*duct of Wirsung*) and to a minor extent through the accessory pancreatic duct

(*duct of Santorini*) to the duodenum. After uniting with the *common bile duct,* the main pancreatic duct opens into the lumen of the small intestine through the *ampulla of Vater.* When functional, the accessory pancreatic duct penetrates the duodenal wall a short distance above this point. The relationships of the pancreatic ducts to the ampulla of Vater and the biliary system are shown in Fig. 26-9.

The activation of the pancreatic proteolytic enzyme precursors occurs in the lumen of the small intestine. The presence of intestinal chyme stimulates the secretion of the enzyme *enterokinase* from the crypts of Lieberkühn. Present when the pancreatic juice enters the duodenum, this enzyme converts trypsinogen into its active form. The reaction is summarized as follows:

$$\text{Trypsinogen} \xrightarrow{\text{enterokinase}} \text{trypsin} \qquad (26\text{-}6)$$

Once trypsin is present, it also activates trypsinogen entering the duodenum in the pancreatic juice. This reaction is summarized as follows:

$$\text{Trypsinogen} \xrightarrow{\text{trypsin}} \text{trypsin} \qquad (26\text{-}7)$$

Thus, enterokinase activates the initial increments of trypsinogen while trypsin itself sustains proteolytic activity in the lumen of the small intestine by continuing to activate further increments of trypsinogen.

Indeed, trypsin also serves as an activator of chymotrypsinogen. This reaction is stated below:

$$\text{Chymotrypsinogen} \xrightarrow{\text{trypsin}} \text{chymotrypsin} \qquad (26\text{-}8)$$

Therefore, chymotrypsinogen cannot be activated without the presence of trypsin in the small intestine.

The activator of carboxypropeptidase is not known with certainty. However, carboxypeptidase is present following the entry of pancreatic juice into the intestinal lumen. Presumably a chemical is responsible for the activation of this proteolytic enzyme.

Proteolytic enzymes are also secreted by the gland cells of the intestinal mucosa in response to mechanical stimulation. These enzymes include *aminopeptidase* and *dipeptidase*.

Proteins are hydrolyzed into amino acids by the sequential actions of the proteolytic enzymes from the pancreas and small intestine. The reactions mediated by the proteolytic enzymes are enumerated below:

$$\text{Proteins and large polypeptides} \xrightarrow[\text{trypsin}]{\text{chymotrypsin}} \text{small polypeptides} \qquad (26\text{-}9)$$

$$\text{Small polypeptides} \xrightarrow[\text{aminopeptidase}]{\text{carboxypeptidase}} \text{dipeptides} \qquad (26\text{-}10)$$

$$\text{Dipeptides} \xrightarrow{\text{dipeptidase}} \text{amino acids} \qquad (26\text{-}11)$$

The absorption of the soluble end products of protein digestion will be discussed later in this chapter.

*Lipids*  Most lipids in chyme entering the small intestine are in the form of triglycerides, or neutral fats. These lipids are contained in fat globules of various sizes. The fat globules must be reduced in size physically and then digested chemically before absorption can occur. The physical decrease in the size of the fat globules is promoted by the action of bile in the small intestinal lumen. The chemical digestion of lipids is mediated by the enzyme pancreatic lipase.

*Bile* is neither an enzyme nor a hormone. It is a secretion produced by the liver, stored in the gallbladder, and released upon appropriate stimulation into the lumen of the duodenum. Bile acts as a wetting agent on fat globules, thereby decreasing their surface tension. This action causes large fat globules to break up physically into smaller fat globules, a process called *emulsification*. Emulsification occurs when chyme is fragmented by segmentation movements of the small intestinal wall. Bile is therefore referred to as an emulsifying agent.

Bile is secreted continually by the hepatic cells of the liver into collecting ducts called *bile canaliculi*. The bile canaliculi, in turn, are drained by many hepatic ducts which coalesce into one main *hepatic duct* which leaves the liver. During periods of digestive inactivity, the secreted bile is stored in the gallbladder through retrograde filling via the *cystic duct*. The biliary system of ducts from the liver to the small intestine is illustrated in Fig. 26-9.

Bile just secreted by the liver contains $H_2O$, electrolytes, cholesterol, bilirubin, fatty acids, and bile salts derived from cholesterol. The gallbladder stores approximately 50 ml of bile. During storage within the gallbladder, bile is concentrated at least fivefold by the absorption of most of the $H_2O$ and electrolytes through the wall of the gallbladder. The liver produces about 1 Gm bile salt each day. As we shall see, the bile salts are important in both the digestion and absorption of lipoidal derivatives within the lumen of the small intestine.

The presence of chyme in the duodenum has two important biliary effects. Firstly, acid chyme promotes an increase in the secretion of bile by inducing the secretion of secretin. The hormone secretin affects the liver by promoting a moderate increase in the synthesis and secretion of bile. Vagal stimulation of the hepatic cells during eating is also believed to play a minor role in the secretory excitation of the hepatic cells.

The other biliary effect of chyme in the small intestine is triggered by the presence of lipids. Lipoidal constituents in chyme stimulate the intestinal crypts to secrete a hormone called *cholecystokinin.* Carried throughout the body in the bloodstream, cholecystokinin differentially affects the walls of the gallbladder and cystic ducts, causing them to contract spasmodically. As a consequence, the stored bile is forced from the gallbladder into the lumen of the duodenum by way of the cystic and common bile ducts (see Fig. 26-9).

During the passage of bile into the small intestine, the *sphincter of Oddi* surrounding the ampulla of Vater must be relaxed. Relaxation of the sphincter occurs when the gallbladder wall is thrown into contraction. This response is presumably regulated reflexly.

Once the neutral fats in the small intestinal lumen are emulsified, they can be digested chemically by the enzyme lipase. Most of the lipase is secreted in the pancreatic juice. A small fraction of the total lipase is produced as a normal constituent of the succus entericus.

The secretion of pancreatic lipase requires stimulation of the pancreas by the intestinal hormone pancreozymin. As we have already learned, pancreozymin increases the enzymatic content of the pancreatic juice by increasing the synthesis of zymogen granules.

Distension of the small intestine due to mechanical stimulation causes the crypts of Lieberkühn to secrete succus entericus containing intestinal lipase. This enzyme acts in a manner similar to that of pancreatic lipase.

Pancreatic and intestinal lipase hydrolyze emulsified fats, forming monoglycerides, fatty acids, and glycerol. This digestive reaction is summarized as follows:

$$\text{Emulsified fats} \xrightarrow[\text{intestinal lipase}]{\text{pancreatic lipase}} \text{monoglycerides,}$$
$$\text{fatty acids, and glycerol} \qquad (26\text{-}12)$$

Following the adsorption of bile salts to the monoglycerides and fatty acids, the products of lipid digestion can be absorbed by the epithelial cells of the intestinal mucosa. Absorption of the lipoidal derivatives is then possible because the bile salts are ionic and therefore water-soluble. Thus, bile salts confer water solubility to the products of lipid digestion. The role of bile salts in absorption is discussed further below.

Most of the bile salts are reabsorbed from the small intestine. These bile salts are reconcentrated within the gallbladder. Thus, there is a continual recycling of bile salts between the gallbladder and the small intestine. The small amounts of bile salts lost in the feces are readily restored by new bile secreted by the hepatic cells.

**Carbohydrates** Carbohydrates in chyme entering the small intestine are either completely undigested or in the form of maltose due to the action of salivary amylase. Mediated by a series of enzymes produced in both the pancreas and small intestine, the chemical digestion of carbohydrates and their derivatives is completed in the small intestine.

The pancreas secretes a carbohydrase called *pancreatic amylase* in the pancreatic juice. Factors which promote the production and secretion of pancreatic juice and the pathways by which this secretion enters the lumen of the duodenum have al-

ready been reviewed during the discussion of protein digestion.

Carbohydrate-digesting enzymes are also secreted by gland cells of the intestinal mucosa in response to mechanical stimulation. Among these enzymes are maltase, lactase, and sucrase.

Carbohydrates entering the duodenum are hydrolyzed into disaccharides. The double sugars, in turn, are degraded enzymatically into monosaccharides. The general reactions mediated by the carbohydrate-digesting enzymes are given below:

$$\text{Carbohydrates} \xrightarrow{\text{pancreatic amylase}} \text{disaccharides} \qquad (26\text{-}13)$$

$$\text{Maltose} \xrightarrow{\text{maltase}} \text{glucose} + \text{glucose} \qquad (26\text{-}14)$$

$$\text{Lactose} \xrightarrow{\text{lactase}} \text{glucose} + \text{galactose} \qquad (26\text{-}15)$$

$$\text{Sucrose} \xrightarrow{\text{sucrase}} \text{glucose} + \text{fructose} \qquad (26\text{-}16)$$

Maltose, lactose, and sucrose are all disaccharides. Each is degraded into two monosaccharides. Notice that only three simple sugars are obtained during the degradation of these disaccharides—glucose, galactose, and fructose.

**Nucleic acids** The degradation of nucleoproteins in the small intestine gives rise to polypeptides and nucleic acids. The polypeptides are hydrolyzed by proteolytic enzymes, as explained earlier. The breakdown of the nucleic acids is mediated by several enzymes secreted by the pancreas and small intestine.

The pancreas secretes an enzyme called *nuclease* in the pancreatic juice. Nucleic acid–digesting enzymes are also secreted by gland cells of the intestinal mucosa in response to mechanical stimulation. These enzymes include *nucleotidase* and *nucleosidase.*

Nucleic acids consist of many nucleotides linked together. Each nucleotide contains a purine or pyrimidine base, a pentose sugar called *ribose* or *deoxyribose*, and a phosphate group. The structure of nucleic acids is discussed further in Chap. 6.

Nucleic acids in the duodenum are hydrolyzed into *nucleotides.* The nucleotides, in turn, are degraded into smaller molecules called *nucleosides.* The general reactions mediated by the nucleic acid–splitting enzymes are summarized below:

Nucleic acids $\xrightarrow{\text{pancreatic nuclease}}$ nucleotides          (26-17)

Nucleotides $\xrightarrow{\text{nucleotidase}}$ nucleosides
   + phosphoric acids          (26-18)

Nucleosides $\xrightarrow{\text{nucleosidase}}$ purines or pyrimidines
   + pentose sugars          (26-19)

**Absorption**   Practically all the ingested nutrients, including most of the $H_2O$, are absorbed from the lumen of the small intestine. These nutrients include (1) amino acids derived from protein digestion; (2) fatty acids, glycerol, and monoglycerides derived from lipid digestion; (3) simple sugars derived from carbohydrate digestion; and (4) nonenergy nutrients including nucleic acids, vitamins, minerals, and $H_2O$.

*Amino acids*   Amino acids are the soluble end products of protein digestion. These acids are actively transported across the plasma membranes of the columnar epithelial cells of the villi and ultimately enter blood capillaries in the plexuses which surround the lacteals. Some peptides, polypeptides, and proteins are also occasionally transported across the mucosa by pinocytosis.

*Fatty acids*   Fatty acids are the soluble end products of lipid digestion. Because these acids and monoglycerides (one fatty acid residue attached to a glycerol backbone—see Chap. 6) are lipid-soluble, they are made water-soluble by bile salts adsorbed to them and are then transported by diffusion into the columnar epithelial cells which line the villi. By a synthetic process described in Chap. 6, these raw materials are used in the manufacture of triglycerides. The triglycerides then undergo further processing, giving rise to minute fat globules known as *chylomicrons*. The chylomicrons enter the lacteals from the interstitial fluid beneath the columnar epithelial cells of the villi. The motility of the villi during digestion squeezes the lacteals, promoting a constant circulation of lymph within these vessels.

Most of the fatty acids absorbed through the villi enter the lacteals in the form of chylomicrons. However, some of the fatty acid molecules enter the bloodstream directly through the capillary plexuses surrounding the villi without undergoing conversion into chylomicrons. Some free fatty acids also enter the lymphatic circulation directly. Therefore, while most of the absorbed fatty acids are transported in the lymphatic circulation to the venous blood returning to the heart, some of the fatty acids are carried in the blood to the liver for immediate processing.

*Simple sugars*   Monosaccharides are the soluble end products of carbohydrate digestion. Monosaccharides, like amino acids, are actively transported from the lumen of the small intestine into the villi. These simple sugars enter the bloodstream through the walls of the capillary plexuses surrounding the lacteals.

*Nonenergy nutrients*   The fat-soluble vitamins A, D, E, and K are transported through the lymphatic circulation after entering the lacteals of the villi. The vitamins of the B complex series and vitamin C are water-soluble and enter the bloodstream following their absorption into the body. Although at least one vitamin is partially hydrolyzed prior to its absorption, most vitamins are absorbed without being digested enzymatically. However, many vitamins, especially those which are water-soluble, require some form of processing during or shortly after absorption. The nature of this processing is discussed later in this chapter.

The absorption of minerals and other electrolytes is primarily through active rather than passive transport. The $Na^+$ pump mechanism (see Chap. 6) is responsible for the active transport of this ion. Vitamin D and calciferol (see Chap. 23) participate in the active transport of $Ca^{++}$. Iron ($Fe^{++}$), $Mg^{++}$, and other charged particles are also actively transported through the wall of the small intestine. Chloride ($Cl^-$) is the only ion which probably moves entirely passively in response to an electrical gradient established by the active transport of $Na^+$.

Water is absorbed osmotically from the lumen of the small intestine because the transport of soluble nutrients and electrolytes into the blood increases its particle concentration. As the blood becomes hypertonic to the chyme in the lumen of the small intestine, water diffuses from the lumen into the blood. This action restores osmotic homeostasis. Osmosis is considered in detail in Chap. 6.

## LARGE INTESTINE

The large intestine is shorter and thicker than the small intestine. Its relationship to the small intestine is shown in Fig. 26-1. This portion of the digestive tract measures about 5 ft in length.

The large intestine is divided into the cecum, including the vermiform appendix; the colon, including ascending, transverse, descending, and sigmoid portions; and the rectum, including the anal canal and anus. Figure 26-10 illustrates the large intestine from anterior aspect.

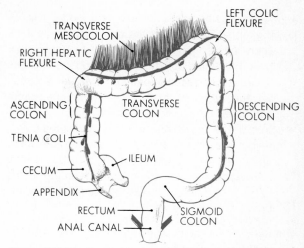

**Figure 26-10** *The large intestine from anterior aspect. (From Melloni, Stone, and Hurd.)*

Cecum   The cecum, the first portion of the large intestine, is a blind pouch about 2 in. long. The ileum opens at right angles into the large intestine at the junction between the cecum and ascending colon. An *ileocecal valve* guards the orifice, preventing the regurgitation of cecal chyme into the ileum.

The *vermiform appendix,* a 3-in.-long blind pouch, projects from the end of the cecum. This nonfunctional remnant of the digestive tract is richly supplied with lymphoidal tissue, but its blood supply is limited. Intestinal flora frequently becomes trapped in the lumen of the appendix and, in spite of the actions of the lymph nodes, causes an inflammation referred to as *appendicitis.* Appendicitis is discussed in Chap. 27.

Colon   The *ascending colon* courses upward on the right side of the abdomen from the cecum to the inferior surface of the liver. At the right colic, or *hepatic, flexure,* the colon bends 90° to the left, giving rise to the *transverse colon.* This portion of the colon is supported to the posterior abdominal wall by a double fold of peritoneum called the *transverse mesocolon.* The transverse colon runs from right to left across the

abdominal cavity. At the left colic, or *splenic, flexure,* the colon again bends, this time in an inferior direction, and gives rise to the *descending colon.* At the brim of the pelvis, the colon turns medially, forming an S-shaped segment termed the *sigmoid colon.* At the pelvic midline, the sigmoid colon merges with the rectum.

The tissue structure of the colic wall, while basically similar to that of the generalized digestive tube, is unique in several respects:

1   Mucosa. The mucosal folds and villi found in the small intestine are not present in the colon. The crypts of Lieberkühn secrete mucus primarily. The lamina propria contains isolated lymph nodules, but Peyer's patches are absent.
2   Submucosa. The submucosa of fibroelastic connective tissue is similar to that found in other portions of the digestive tract.
3   Muscular coat. The inner circular muscular layer composes most of the wall. The outer longitudinal muscular layer is compacted into three discrete longitudinal bands called *teniae coli.* Each muscular band is called a tenia because of its super-

ficial resemblance to a tapeworm (see Chap. 3). The teniae extend from the cecum to the rectum, where they fan out to form a complete longitudinal layer. The teniae coli divide the colon into a series of transverse sacculations, or small sacs, called haustra. Adjacent haustra are partitioned from each other by internal *semilunar folds.*

4  *Serosa.* The serosa only partially covers the colon, giving rise to the transverse mesocolon, mentioned earlier, and to outpouchings called *epiploic appendages* on either side of the teniae coli. The epiploic appendages contain fat deposits.

**Rectum**   The rectum continues from the sigmoid colon at the level of the third sacral vertebra to the *anal canal* at the level of the coccyx, a distance of about 5 in. As mentioned previously, the teniae coli form a complete longitudinal muscular layer in the rectal wall. The disappearance of the teniae coli results in the corresponding absence of epiploic appendages in this region of the alimentary canal.

The rectum merges with the anal canal, the distal portion of the large intestine. Between 2 and 3 in. long, the anal canal does not contain a serous coat or a longitudinal musculature. The circular smooth muscle layer is modified into an *internal sphincter* at the distal end of the anal canal. The internal sphincter surrounds the *anus,* the opening of the large intestine to the outside of the body. Surrounding the internal sphincter is a layer of striated muscle, the *external sphincter,* which permits voluntary control of the defecation reflex (see below).

**Movements**   The large intestine exhibits two characteristic movements—haustral churning and mass peristalsis. *Haustral churning* promotes water reabsorption. *Mass peristalsis* leads to evacuation.

*Haustral churning*   The chyme in the large intestine is churned when inch-wide bands of circular smooth muscle contract in the walls of the ascending and transverse colons. Movement of the chyme toward the anus is aided by contractions of the longitudinal smooth musculature which composes the teniae coli. The haustral movements are periodic, appearing in different regions at various times. Similar in principle to the segmentation movements of the small intestine, the haustral movements churn the chyme, exposing all portions of it to the epithelial lining of the colic mucous membrane. This action increases the efficiency of water reabsorption by the colon. As a consequence of water reabsorption, fluid chyme entering the cecum is transformed into semisolid feces by the time the material reaches the distal end of the transverse colon.

*Mass peristalsis*   Mass peristalsis consists of an almost simultaneous contraction of a large portion of either the transverse or descending colon. Mass movements propel the feces into the rectum, thereby initiating the defecation reflex.

Mass movements are promoted reflexly by the appearance of food in the stomach or the duodenum of the small intestine. The response is called the *gastrocolic reflex* when induced by gastric stimulation, or the *duodenocolic reflex* when initiated by distension of the small intestine.

**The defecation reflex**   About two or three times a day, mass peristalsis of the colon moves a small amount of feces into the rectum. The presence of feces in the rectum applies pressure against the inside rectal walls, stimulating stretch receptors. The excitation of these stretch receptors initiates the *defecation reflex.*

The afferent nerve impulses from the stretch receptors are integrated at the level of the sacral spinal cord. However, defecation may be facilitated or inhibited by neural influences which originate in the cerebral cortex. The feces can be voluntarily retained by tonic contraction of the external sphincter.

Defecation is voluntarily facilitated by reducing the volume of the abdominal cavity. This is accomplished by contraction of the anterior abdominal muscles and the transverse diaphragm. The diaphragm descends in response to a marked increase in the intrapulmonic pressure (see Chap. 28) due to forced expiration while the epiglottis remains closed. As a consequence of these actions, the abdomen is compressed. Defecation is promoted because of an increase in intrarectal pressure caused by these efforts.

*Feces* The feces, or stools, are the waste materials eliminated from the alimentary canal via the anus. The feces are composed of the following constituents:

1 Water secreted by the intestinal mucosa and making up a majority of the fecal volume

2 Indigestible and undigested food residues

3 The lubricant mucin secreted by glands of the large intestinal mucosa

4 Chemical products of bacterial putrefaction, such as skatole and indole

5 A cellular component made up of intestinal flora, sloughed off epithelial cells of the intestinal mucosa, and discarded phagocytes

6 Excess ions, such as $Mg^{++}$, $Fe^{++}$, and $HPO_4^{--}$, secreted by the large intestinal mucosa

7 Constituents of bile, including derivatives of bilirubin and cholesterol

Because of normal water reabsorption in the ascending and transverse colons, the feces are ordinarily semisolid by the time they reach the descending colon. The characteristic brown coloring of the feces is due to the presence of bilirubin derivatives and $Fe^{++}$. The nitrogenous odor of the feces is a consequence of the bacterial putrefaction of proteins and peptides which have escaped digestion in the small intestine. On an average, less than 1/2 lb of feces is defecated each day.

## POSTABSORPTIVE PROCESSING

The soluble nutrients are absorbed primarily from the small intestine but also to a limited extent from the stomach and large intestine. These nutrients are either (1) stored in the liver, fat depots, or intracellularly; (2) excreted primarily in the urine; (3) utilized as raw materials, activators, or coenzymes in various synthetic or degradative reactions; or (4) used as metabolic fuels from which energy is extracted. The body closely regulates the concentrations of various nutrients in the blood to meet the metabolic demands of the body, permit efficient utilization of the nutrients, and maintain osmotic equilibrium between the fluid compartments.

The liver plays an important role in the postabsorptive processing of nutrients. Most of the soluble nutrients absorbed by the digestive tract are transported to the liver by a specialized venous circulation known as the *hepatic portal system*. Based on their blood concentrations, the nutrients may be stored, released unchanged into the venous blood returning to the heart, or converted into other nutrients required by the body. The liver also neutralizes antigens and potentially toxic drugs.

## TRANSPORTATION

The nutrients routed directly to the liver include a vast majority of the amino acids, perhaps 10 percent of the fatty acids, all the monosaccharides, $H_2O$, electrolytes, including minerals, and the water-soluble vitamins.

Among the nutrients transported directly to the heart are neutral fats, some free fatty acids, and variable amounts of amino acids and other nutrients which enter the interstitial fluid surrounding the lacteals of the villi.

Hepatic portal system This venous system drains blood primarily from the abdominal portions of the digestive tract and spleen, diverting the blood to the liver instead of carrying it directly back to the heart. Portal systems are invariably venous and transport blood to the organs for which they are named—hence, *hepatic* for "liver."

The hepatic portal system is shown from anterior aspect in Fig. 26-11. Note that the portal vein is formed by a confluence of the following vessels: (1) the superior mesenteric vein, which drains the jejunum, ileum, ascending colon, and the proximal portion of the transverse colon; (2) the inferior mesenteric vein, which drains the distal portion of the transverse colon, the descending and sigmoid colons, and the rectum; and (3) the splenic vein, which drains the spleen. Veins from the duodenum and pancreas enter the superior mesenteric vein. The

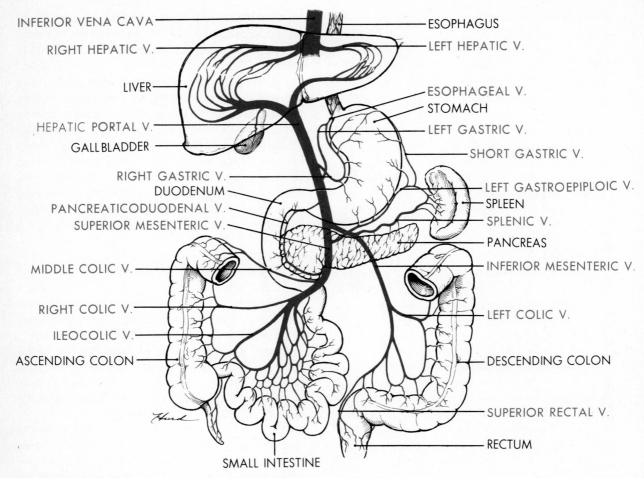

INFERIOR VENA CAVA

RIGHT HEPATIC V.

LIVER

HEPATIC PORTAL V.

GALLBLADDER

RIGHT GASTRIC V.

DUODENUM

PANCREATICODUODENAL V.

SUPERIOR MESENTERIC V.

MIDDLE COLIC V.

RIGHT COLIC V.

ILEOCOLIC V.

ASCENDING COLON

ESOPHAGUS

LEFT HEPATIC V.

ESOPHAGEAL V.

STOMACH

LEFT GASTRIC V.

SHORT GASTRIC V.

LEFT GASTROEPIPLOIC V.

SPLEEN

SPLENIC V.

PANCREAS

INFERIOR MESENTERIC V.

LEFT COLIC V.

DESCENDING COLON

SUPERIOR RECTAL V.

RECTUM

SMALL INTESTINE

**Figure 26-11** *The hepatic portal system.* (From Melloni, Stone, and Hurd.)

stomach and distal portion of the esophagus are also drained by veins which empty into the hepatic portal vein.

The hepatic portal vein branches extensively within the substance of the liver. These branches ultimately drain into *hepatic sinusoids*. Nutrients and other materials in the sinusoidal blood are processed in the hepatic lobules. Following processing, the blood is collected by venules which lead into several hepatic veins. Blood from the *hepatic veins* empties into the *inferior vena cava* for transport to the heart.

Lymphatic circulation    The lacteals drain interstitial

fluid containing chylomicrons, free fatty acids, and some other nutrients from the villi. The lacteals are continuous with the lymphatic circulation of the inferior portion of the body. The lymphatic vessels draining the lacteals lead into an expanded lymphatic storage vessel in the midtrunk region called the *cisterna chyli*. Passive compression of the cisterna chyli pumps the stored nutrient-rich lymph through the *thoracic duct* and into the venous circulation at the union between the *left subclavian* and *internal jugular veins*. The blood in these vessels is returned to the right atrium of the heart. The lymphatic circulation described above is illustrated in Fig. 30-12.

## LIVER

The liver is the largest gland in the body, usually weighing between 2.5 and 3.5 lb. The liver is located in the right hypochondriac and epigastric regions and sometimes extends into the left hypochondriac region (see Fig. 26-1). The superior convex surface of the liver lies immediately beneath the diaphragm and closely follows its contours.

The liver is illustrated from anterior aspect in Fig. 26-12. This remarkably complex gland is divided into four lobes: (1) the right lobe, the largest of the hepatic lobes; (2) the left lobe; (3) the square-shaped quadrate lobe on its inferior surface; and (4) the tail-shaped caudate lobe on its posterior surface. The right and left lobes are separated from each other by the *falciform ligament.* Formed from peritoneal folds, the falciform ligament links the diaphragm and the liver with the anterior abdominal wall.

Most of the surface of the liver is enveloped by a fibrous tissue known as *Glisson's capsule.* Overlying Glisson's capsule is a serous layer of visceral peritoneum.

Five ligaments support the liver. One of these ligaments, the falciform ligament, has already been mentioned. This and three other ligaments are derived from folds of peritoneum. The fifth ligament, the ligamentum teres (see Fig. 26-12), is a remnant of the former umbilical vein of fetal development (see Chap. 19). The ligamentum teres courses from the umbilicus to the liver.

The liver is supplied by two vascular circulations—the hepatic portal and hepatic circulations. The hepatic portal system has already been de-

scribed. About two-thirds of the total blood transported to the liver is carried through this venous shunt. The remainder of the blood is brought to the liver by the hepatic artery. The *hepatic artery* is a branch of the *celiac artery* from the aorta. Branches of the hepatic artery supply the hepatic tissues with nutrients and $O_2$. Blood from the hepatic capillary plexuses is returned to the heart through the inferior vena cava after being drained from the liver by the hepatic veins (see Fig. 26-11). There is also an extensive lymphatic circulation within the liver.

The gallbladder (see Fig. 26-9) is located in a fossa, or depression, in the lower margin of the right hepatic lobe. Attached to the liver by an investment of connective tissue, this pear-shaped organ possesses a relatively thin wall consisting of a mucous membrane, fibromuscular coat, and a partial covering of visceral peritoneum.

**Nutritive functions**   The liver processes nutrients either stored there or transported to it through the hepatic portal system or the hepatic artery. Some of the more specific nutritive functions of the liver are as follows: (1) *glycogenolysis,* or the conversion of stored glycogen into blood glucose when blood glucose levels are depressed (hypoglycemia); (2) *glycogenesis,* or the conversion of blood glucose into liver glycogen when blood glucose levels are excessive (hyperglycemia); (3) *gluconeogenesis,* or the conversion of amino acids into blood glucose, to provide an extra source of metabolic fuel; (4) *deamination,* or the removal of amino groups ($NH_2$) from amino acids, so that the amino acid residues can be utilized for energy; and (5) *beta oxidation,* or the removal of successive two-carbon fragments from the carboxyl end (COOH) of fatty acids, to provide a source of metabolic fuel during hypoglycemia.

**Other functions**   The liver has numerous other functions. Many of these functions are related either directly or indirectly to nutrition and metabolism. Such functions include storage, synthesis, body defense, and excretion.

***Storage***   The liver serves as a storage depot for the following materials: (1) glycogen, stored in hepatic

**Figure 26-12**   *The anterior aspect of the liver.* (From Melloni, Stone, and Hurd.)

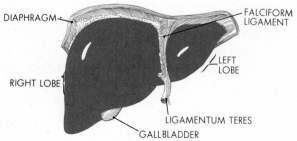

DIAPHRAGM

FALCIFORM LIGAMENT

LEFT LOBE

RIGHT LOBE

LIGAMENTUM TERES

GALLBLADDER

cells during hyperglycemia; (2) neutral fat, stored in the adipose connective tissue associated with the liver; (3) many fat-soluble vitamins and some water-soluble vitamins (see "Vitamins," below); and (4) certain minerals, such as $Cu^{++}$ and $Fe^{++}$ in the form of ferritin. In addition, 10 percent of the total blood volume and, under special circumstances, up to 25 percent of the blood in the body can be stored in the hepatic vasculature.

*Synthesis* The liver synthesizes a wide variety of substances for various purposes. Some of the more specific synthetic functions in which the liver engages are as follows:

1 *Transamination,* or the formation of a new amino acid following deamination of an available amino acid and the transfer of its amino group to another molecule
2 Glycogenesis, a process mentioned earlier
3 The synthesis of fat from excess deaminated amino acids in the presence of the raw material glycerol
4 The manufacture of bile by the hepatic cells (see "Lipid Digestion," above)
5 The synthesis of vitamin A from carotene
6 The production of urea from ammonia in the presence of $CO_2$ (see Chap. 22)
7 The synthesis of blood factors important in hemostasis, such as prothrombin, fibrinogen, and heparin (see Chap. 4).
8 The synthesis of immunogenic proteins called *globulins* (see Chap. 4)
9 The manufacture of practically all the serum albumins (see Chap. 4)

*Body defense* The liver is involved in both nonspecific and specific immunologic functions. In addition to its roles in hemostasis and the immune response, the liver promotes body defense through the phagocytic engulfment and neutralization of bacteria, protein toxins, and cellular debris by hepatic reticuloendothelial cells (see Chap. 4) and through the detoxification of hemoglobin (see Chap. 31) and various potentially toxic drugs.

*Excretion* The liver participates in the processing of various materials for excretion from the body. Some of the major excretory functions of the liver are the formation of urea, uric acid, and other nitrogenous waste products of metabolism for subsequent excretion via the urine, and the formation of bilirubin during the breakdown of hemoglobin molecules for subsequent excretion in the urine and feces.

## NUTRITION

Nutrition refers to all those processes which ensure a continuous and adequate supply of raw materials with which persons maintain normal metabolism. Nutrition is accomplished through the ingestion, digestion, absorption, and assimilation of at least the minimum daily concentrations of the wide range of nutrients required for cellular homeostasis. It follows that malnutrition results from an inability of the body to ingest, digest, absorb, or assimilate the minimum daily requirements for one or more nutrients. Malnutrition is discussed in Chap. 27. Here, we shall emphasize normal nutrition.

### NUTRIENTS

Nutrients are substances required for normal metabolism and, in addition to water, include (1) proteins, fats, and carbohydrates, or their soluble end products of digestion, for use in energy metabolism and biosynthesis; (2) essential elements, or minerals, for use as structural components or activators of vitamins, hormones, enzymes, or other biologically active molecules; and (3) vitamins for use as coenzymic, structural groups necessary for the functioning of numerous enzymes.

Excluding essential minerals and vitamins, 3 fatty acids and 10 amino acids are also classified as essential in that they must be obtained mostly or entirely from exogenous, or outside, sources. These essential acids are soluble end products of digestion. They are used in a wide variety of anabolic, or synthetic, processes. Although the body cannot synthesize essential nutrients from available raw materials or can only manufacture them in inadequate amounts, a number of vitamins can be manufactured if certain precursors are available or may be produced by the synthetic activities of normal intestinal flora.

Calories and nutrition The energy content of foodstuffs is measured in kilocalories (kcal). As you will recall from Chap. 8, 1 kcal is the quantity of heat energy which must be added to 1 kg $H_2O$ to raise its temperature from 14.5 to 15.5°C. The caloric value of a foodstuff is obtained by measuring the heat energy liberated during the complete combustion of 1 Gm of the nutrient in a special furnace. Because there are many different types of proteins, lipids, and carbohydrates, each yielding somewhat different caloric values when mass is held constant, the caloric content expressed for each type of foodstuff represents an average value. The average caloric values for the three classes of energy nutrient are as follows:

| Foodstuff | Caloric value |
| --- | --- |
| Carbohydrates | 4.1 kcal/Gm |
| Fats | 9.5 kcal/Gm |
| Proteins | 5.7 kcal/Gm |

Considerably more energy is expended in the oxidation of amino acids derived from proteins, compared to the energy consumption required during the oxidation of the soluble end products derived from the digestion of fats or carbohydrates. This phenomenon is called the *specific dynamic action* (SDA) of food. In a practical sense, this means that the caloric values of proteins and carbohydrates are about the same when oxidized in the body. However, notice that the caloric value of fats is about double those of the other foodstuffs. Because of its large energy yield per unit of mass, fat is an ideal energy nutrient for storage purposes.

The average number of kilocalories required by the body per day is determined by the number of kilocalories consumed during metabolism in the same time period. Obviously, more energy is utilized by a physically active person than by someone who leads a sedentary existence. Although authoritative estimates of the minimum daily caloric requirements differ, 3000 kcal is a fairly good yardstick under average conditions. About 1500 kcal/day, or half the total, is derived from carbohydrate. The energy obtained from lipids accounts for approximately 1200 kcal/day, or 40 percent of the total caloric expenditure. The remainder of the energy, or 300 kcal, is secured from the oxidation of amino acids derived from the digestion of proteins.

Essential amino acids Of the 20-plus amino acids commonly found in foods eaten by people, 10 cannot be synthesized within the body or are made in amounts insufficient to meet the requirements of normal metabolism. These 10 amino acids are essential to life. Their structures are shown in Fig. 26-13. Pathologic conditions resulting from dietary deficiencies of the essential amino acids are discussed in Chap. 27.

Many of the essential amino acids are structural components of numerous biologically important proteins. Histidine, isoleucine, and lysine are used in the synthesis of hemoglobin. A dietary deficiency of one or more of these amino acids can cause anemia. Tryptophan is a structural component of niacin, one of the B-complex vitamins. A deficiency of niacin causes pellagra.

A number of amino acids are used in the synthesis of other amino acids and carbon compounds. The sulfur of methionine is used in the synthesis of the amino acid cystine, and phenylalanine is employed in the manufacture of the amino acid tyrosine. Although cystine and tyrosine can replace some of the dietary requirement for methionine and phenylalanine in man, neither amino acid is completely effective in the absence of their essential amino acid precursors.

Tyrosine is utilized in the synthesis of the skin pigment melanin and the hormones epinephrine and thyroxine. Methyl ($CH_3$) groups from methionine interact with the amino acid arginine and other intermediates to form creatine. Phosphorylated creatine (creatine phosphate) tends to maintain a supply of ATP in active muscles until an increase in glycolysis provides sufficient ATP to meet the increased metabolic demands (see Chap. 24).

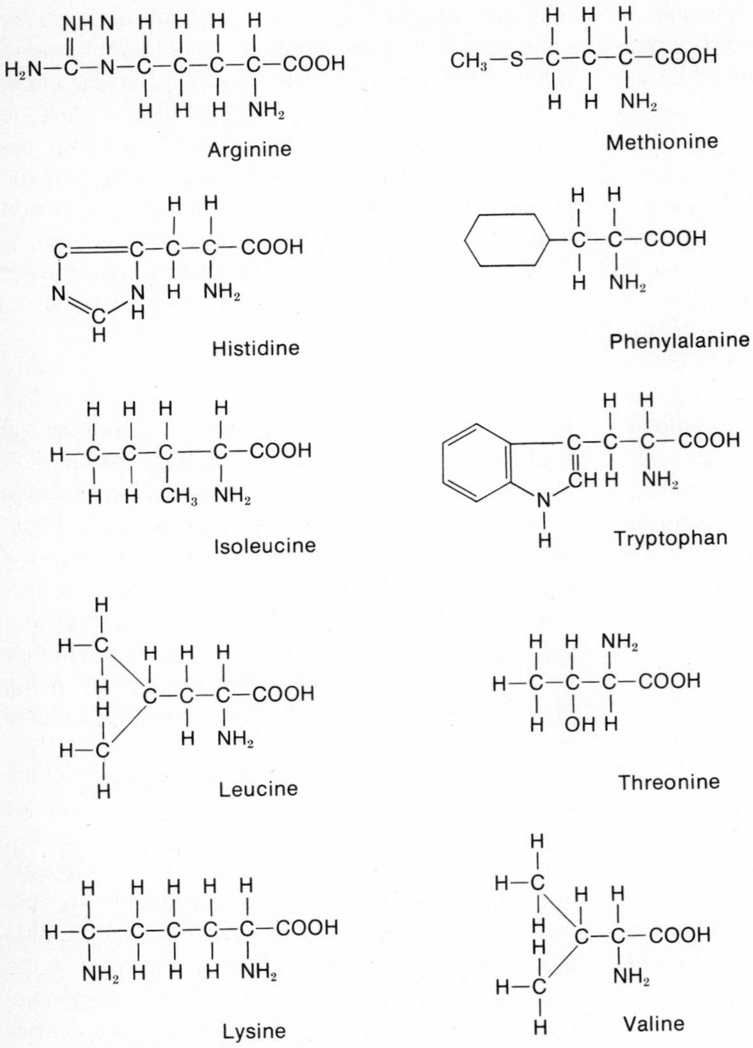

**Figure 26-13** *The structures of the essential amino acids.*

Essential fatty acids *Linoleic, linolenic,* and *arachidonic acids* are also essential. The structures of these unsaturated fatty acids are shown in Fig. 26-14. The absence of two hydrogen atoms at each point of unsaturation gives rise to a double bond (C=C). The dietary absence of these acids results in increases in metabolic rate and water consumption and a retardation of growth. The disorders associated with deficiencies of essential fatty acids are discussed further in Chap. 27.

Linoleic acid is the precursor of arachidonic acid, the essential fatty acid of demonstrated functional significance. However, the body can probably use any of the three unsaturated fatty acids interchangeably.

Vitamins play a vital role in the metabolism of essential fatty acids. Vitamin E preserves their structural integrity by preventing their oxidation. Biotin, part of the B-complex series, converts inactive mixtures of arachidonic acid into their active form.

**Vitamins** Vitamins are essential dietary ingredients which combine with, and are necessary for the functions of, many enzymes involved in intermediary metabolism and biosynthesis. Because vitamins aid the actions of enzymes after becoming nonprotein portions of enzyme molecules, they are called *coenzyme prosthetic groups,* or simply *coenzymes.* In the absence of their coenzymes, various enzymes remain inactive, causing metabolic derangements of varying magnitude. The effects of various vitamin deficiencies are described in Chap. 27. The structures and metabolic roles of the vitamins are given in Fig. 26-15.

Vitamins A, D, E, and K are fat-soluble. They are absorbed readily through the plasma membranes of the columnar epithelial cells which line the lumen of the small intestine. Vitamin A is hydrolyzed to an alcohol prior to absorption. The absorption of vitamin D requires the presence of bile. Following their absorption, the fat-soluble vitamins enter the lymphatic circulation through the lacteals of the villi. The liver stores excesses of the fat-soluble vitamins to a limited extent.

The B-complex vitamins and vitamin C are water-soluble. In general, the water solubility of these vitamins is an impediment to cell membrane transport. The membrane transport of vitamin $B_{12}$ (cyanocobalamin) is facilitated by a mucoprotein called *intrinsic factor* produced by the epithelial lin-

ing of the gastric mucosa. Several other water-soluble vitamins including thiamine and riboflavin are phosphorylated in the small intestinal mucosa before entering the bloodstream. All the water-soluble vitamins enter the blood capillary plexuses in the villi for transport to various parts of the body. Cyanocobalamin is bound to a specialized protein while it is carried through the blood. The liver can store small amounts of most of the B-complex vitamins. The body does not seem to store vitamin C, although moderate concentrations of this vitamin are found in the adrenal cortices.

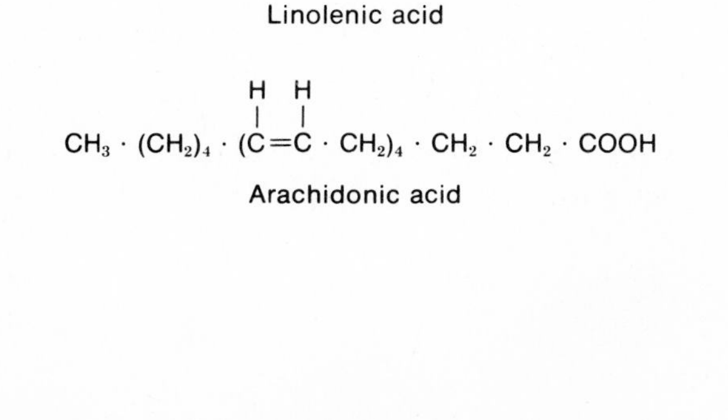

**Minerals** Nutritionally, all 25 chemical elements vital to life are called minerals. These elements are identified in Fig. 10-1. Minerals in ionic form are also referred to as *electrolytes*. The 13 elements in the lowest natural concentrations in the body are termed *trace elements*. The distribution of the vital elements in the periodic table as well as their abundance and importance in the body are reviewed in Chap. 10. Pathologic conditions which result from nutritional deficiencies of the vital elements are described in Chap. 27.

Over 99 percent of the bulk of the human body is composed of four elements—hydrogen, carbon, nitrogen, and oxygen. Therefore, atoms of these elements make up the bulk of all biologically important molecules. $Ca^{++}$ and phosphorus in the form $HPO_4^{--}$ are utilized in the normal metabolism of bone. The normal deposition of these elements in bones requires the presence of trace amounts of $Co^{++}$ as well as vitamin D and a proper balance between the hormones parathyroid hormone, calcitonin, and calciferol. An inadequate deposition of $Ca^{++}$ or $HPO_4^{--}$ in bones gives rise to rickets.

The propagation and maintenance of bioelectrical potentials associated with nerve impulse conduction and muscular contraction are dependent upon a proper balance between extracellular $Na^+$ and intracellular $K^+$ (see Chap. 13). An imbalance of these ions can lead to neuromuscular fatigue. $Na^+$ and $K^+$, along with chloride ($Cl^-$), the ionic form of chlorine, also contribute importantly to the regulation of osmot-

**Figure 26-14** *The structures of the essential fatty acids.*

$$CH_3 \cdot (CH_2)_4 \cdot \overset{H}{\underset{}{C}} = \overset{H}{\underset{}{C}} \cdot CH_2 \cdot \overset{H}{\underset{}{C}} = \overset{H}{\underset{}{C}} \cdot (CH_2)_7 \cdot COOH$$

Linoleic acid

$$CH_3 \cdot CH_2 \cdot \overset{H}{\underset{}{C}} = \overset{H}{\underset{}{C}} \cdot CH_2 \cdot \overset{H}{\underset{}{C}} = \overset{H}{\underset{}{C}} \cdot CH_2 \cdot \overset{H}{\underset{}{C}} = \overset{H}{\underset{}{C}} \cdot (CH_2)_7 \cdot COOH$$

Linolenic acid

$$CH_3 \cdot (CH_2)_4 \cdot (\overset{H}{\underset{}{C}} = \overset{H}{\underset{}{C}} \cdot CH_2)_4 \cdot CH_2 \cdot CH_2 \cdot COOH$$

Arachidonic acid

**Figure 26-15** *The structures and functional roles of some vitamins.*

| Vitamin | Structure |
|---|---|
| Lipid soluble A | 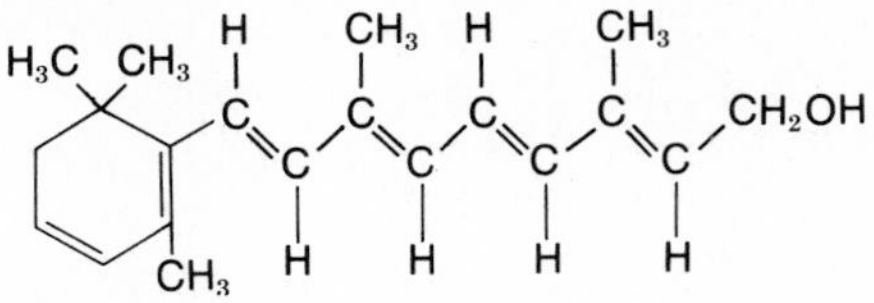 |
| D | 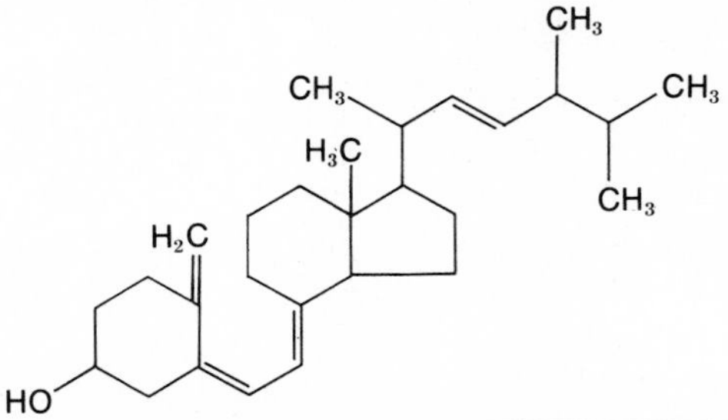 |
| E (alpha tocopherol) | 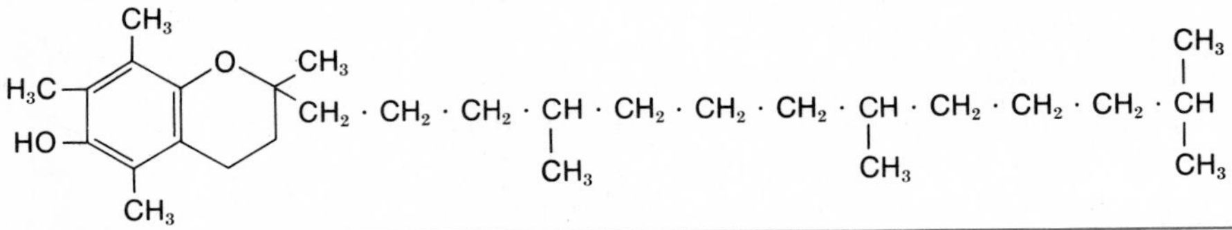 |
| K (antihemorrhagic factor) | 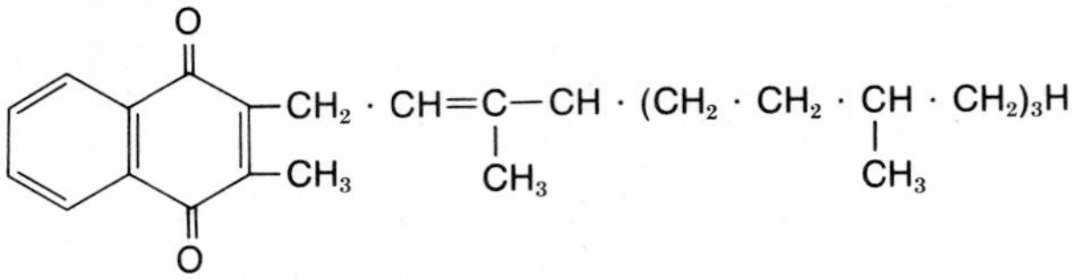 |
| Water soluble B₁ (thiamine) | 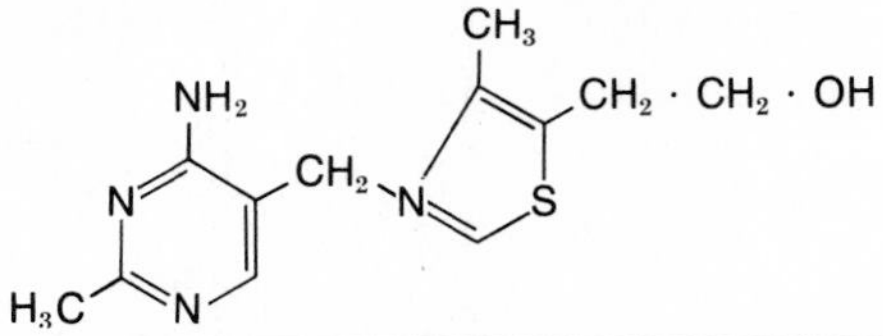 |
| B₂ (riboflavin) | 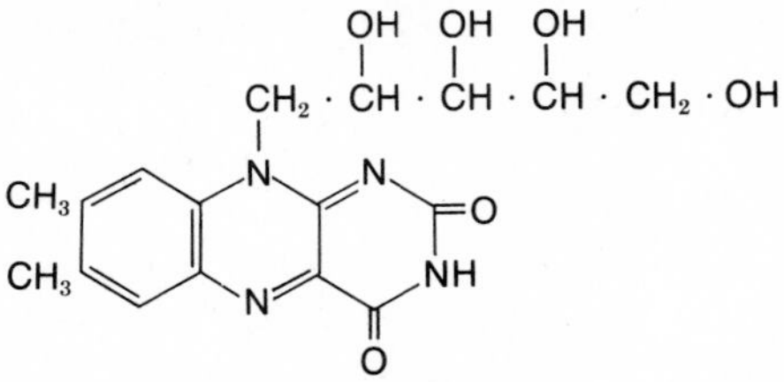 |
| Niacin (nicotinic acid) | 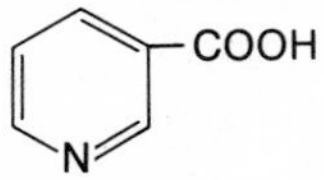 |
| Folic acid | 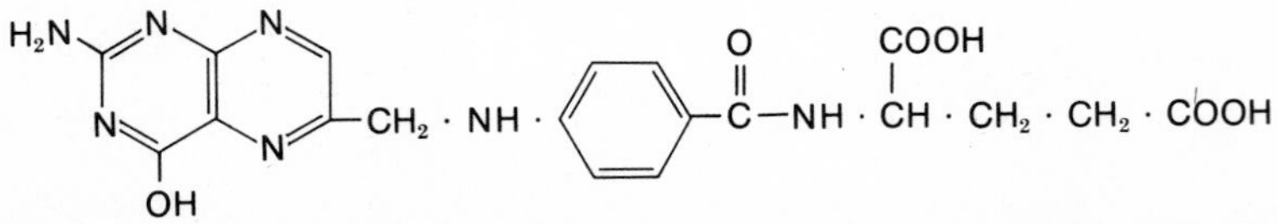 |

| Function | Mechanism of action |
|---|---|
| Maintains normal night vision and the integrity of epithelial tissues | Required for resynthesis of visual pigment rhodopsin following bleaching |
| Promotes intestinal absorption of $Ca^{++}$ and controls its deposition in bone | |
| | Required for oxidation of essential fatty acids; utilized in electron transport system within mitochondria |
| Maintains normal hemostatic ability (coagulation) | Required for synthesis of prothrombin in liver |
| Sustains normal carbohydrate metabolism | Incorporated into coenzyme, cocarboxylase, it removes carboxyl groups (COOH) from pyruvic acid |
| Promotes normal intracellular energy liberation | Incorporated into coenzymes flavin adenine dinucleotide (FAD) and flavin mononucleotide (FMN); flavoproteins serve as hydrogen carriers in citric acid cycle and electron transport system within mitochondria |
| Promotes normal intracellular energy liberation | Incorporated into coenzymes nicotinamide adenine dinucleotide (NAD) and nicotinamide adenine dinucleotide phosphate (NADP); coenzymes serve as hydrogen carriers in glycolysis, citric acid cycle, and hexose monophosphate shunt |
| Maintains normal cellular growth | Required for synthesis of specific nucleotides |

| Vitamin | Structure |
|---|---|
| Vitamin B₃ (pantothenic acid) | |
| Biotin | |
| Vitamin B₆ (pyridoxine) | |
| Vitamin B₁₂ (cyanocobalamin) | |
| Vitamin C (ascorbic acid) | |

## Vitamin B₃ (pantothenic acid)

$$HOCH_2-\underset{\underset{CH_3}{|}}{\overset{\overset{CH_3}{|}}{C}}-\underset{\underset{OH}{|}}{CH}-CO\cdot NH\cdot CH_2\cdot CH_2\cdot COOH$$

## Biotin

$$H_2C-S-CH\cdot (CH_2)_4COOH$$

## Vitamin B₆ (pyridoxine)

## Vitamin B₁₂ (cyanocobalamin)

## Vitamin C (ascorbic acid)

| Function | Mechanism of action |
|---|---|
| Sustains normal carbohydrate and lipid metabolism | Incorporated into coenzyme, acetyl CoA, it transfers two-carbon fragments called acetyl groups from cytoplasm to mitochondria, forming citric acid from oxaloacetic acid |
| Promotes synthesis of fatty acids | Required for $CO_2$ fixation of CoA intermediates, thereby increasing number of carbon atoms in their chains |
| Promotes synthesis of proteins and normal metabolism of amino acids | Incorporated into coenzyme, pyridoxyl phosphate, it transfers amino groups ($NH_2$) from one amino acid to another (transamination) |
| Sustains normal nucleic acid synthesis and is required for erythrocytic maturation | Incorporated into several cobamide coenzymes, it is required for conversion of several intermediary metabolites into other products and for synthesis of several deoxyribonucleotides |
| Maintains integrity of connective tissues, including collagen and bone | Required for conversion of proline into hydroxyproline, an amino acid found in collagen; participates in hexose monophosphate shunt; required for removal of $Fe^{++}$ from plasma ferritin and its transfer to liver ferritin |

ic pressures between the fluid compartments of the body.

Phosphorus is required in the synthesis of ATP, the energy currency of cells. Because high-energy phosphate is required in synthetic and degradative reactions in cells, a nutritional deficiency of phosphorus is reflected in a depression in metabolism or fatigue.

$Ca^{++}$ is the activator of muscular contraction by initiating the interdigitation reaction between the proteins actin and myosin (see Chap. 24). A dietary deficiency of $Ca^{++}$, called hypocalcemia, causes tetany.

Trace elements are essential for the normal functions of various vitamins. Cobalt actually constitutes about 4 percent of cyanocobalamin (vitamin $B_{12}$), which is required for the maturation of erythroblasts. Therefore, a dietary deficiency of cobalt can precipitate pernicious anemia.

Although not a structural component of a vitamin, manganese ($Mn^{++}$) activates an enzyme in which the vitamin thiamine serves as a coenzyme. The disease beriberi results from a dietary deficiency of this vitamin or its enzymatic activator.

Trace elements are also vital structural components of several hormones. Iodine is a constituent of thyroxine, a hormone secreted by the thyroid gland. A dietary inadequacy of iodine causes simple goiter, a form of hypothyroidism. $Zn^{++}$ is incorporated into the structure of the hormone insulin secreted by the pancreas. Diabetes mellitus results from a pathologic hyposecretion of insulin.

Finally, trace elements serve as activators of many enzymes. $Mn^{++}$ is necessary for the normal catalytic activities of peptidases, phosphatases, and carboxylases. A dietary deficiency of $Mn^{++}$ can impair digestion, absorption, and assimilation, functions in which $Mn^{++}$-containing enzymes participate. $Cu^{++}$ activates an enzyme needed for the incorporation of $Fe^{++}$ into the molecular structure of hemoglobin. A lack of $Cu^{++}$ can therefore contribute to iron-deficiency anemia. $Zn^{++}$ is responsible for activating the enzyme carbonic anhydrase. In erythrocytes, this enzyme catalyzes a reaction in which $H_2O$ and $CO_2$ form $H_2CO_3$ (see Chap. 30). Carbonic acid dissociates into $H^+$ and $HCO_3^-$, which are then processed in the plasma. If the activities of carbonic anhydrase are depressed, then $CO_2$ accumulates in the blood (hypercapnia).

$Mg^{++}$ and ATP or ADP form active complexes with at least four glycolytic enzymes (see Chap. 6). A lack of $Mg^{++}$ causes metabolic depression by reducing the efficiency of glycolysis.

Minerals other than trace elements are also structural components of biologically important molecules. The heme groups of the respiratory pigment hemoglobin and the iron porphyrin groups of the oxidative enzymes called *cytochromes* both contain iron. A deficit of iron causes iron-deficiency anemia. Because of its role in cellular respiration, a lack of iron can cause metabolic depression.

Sulfur is a constituent of cystine, an important amino acid in the fibrous proteins of which human hairs are composed. Therefore, a dietary inadequacy of sulfur can result in abnormalities of hair growth.

Water  The body loses approximately 2,400 ml of $H_2O$ per day. Over 1,400 ml, or considerably more than 50 percent of this loss, occurs through the urine. Sensible and insensible perspiration account for another 600 ml, or 25 percent, of the $H_2O$ loss. Most of the remainder is lost as a consequence of respiration, and some $H_2O$ is ordinarily lost in the feces. Water losses can be reduced automatically through the reabsorption of $H_2O$ from the kidneys.

Water homeostasis requires that $H_2O$ losses be balanced by $H_2O$ gains. This means that the average $H_2O$ intake must also approximate 2,400 ml/day. Water is added to the body by (1) drinking $H_2O$ directly; (2) drinking beverages and eating foods which contain $H_2O$; and (3) forming $H_2O$ intracellularly as an end product of metabolism. Probably more than a third of the total $H_2O$ intake is obtained by drinking.

Water is required by the body for the following reasons:

1  $H_2O$ is incorporated structurally into all the organs and tissues.
2  An aqueous medium is necessary for the dissolution of protoplasmic substances.

*3* $H_2O$ serves as a solvent in which biochemical reactions can take place.

*4* $H_2O$ participates in a host of biochemical reactions.

*5* $H_2O$ maintains an osmotic equilibrium between the blood, interstitial fluid, and the tissue cells.

*6* $H_2O$ participates in the maintenance of a constant

body temperature through evaporative heat losses and by resisting temperature changes because of its high specific heat.

The qualities of $H_2O$ which are essential to life are discussed more fully in Chap. 7.

## COLLATERAL READING

Bodenheimer, F. S., *Insects as Human Food: A Chapter of the Ecology of Man.* The Hague: Dr. W. Junk, Publishers, 1951. A historical and contemporary review of the dietary uses of specific insects made by peoples and tribes throughout the world.

Brown, Lester R., and Gail W. Finsterbusch, *Man and His Environment: Food.* New York: Harper & Row, Publishers, 1972. This highly informative paperback deals with the world's food problems.

Davenport, Horace W., "Why the Stomach Does Not Digest Itself," *Scientific American*, January 1972. According to Davenport, aspirin and certain other chemicals can breach the mucous barrier, causing negligible to severe hemorrhaging.

Frieden, Earl, "The Biochemistry of Copper," *Scientific American*, May 1968. Reviews the metabolic reactions in which the trace element copper serves as an enzymatic activator.

Kretchmer, Norman, "Lactose and Lactase," *Scientific American*, October 1972. The lactose in milk is digested by lactase in human beings up to four years of age. Most human adults cannot digest lactose because lactase is absent. The intolerance of adults to lactose should be considered in programs which distribute milk powder to the peoples of underdeveloped countries, according to Kretchmer.

Majumder, Sanat K., "Vegetarianism: Fad, Faith or Fact?" *American Scientist*, 60:175–179, 1972. Although there is not enough scientific evidence to assess the impact of a vegetarian diet on basic personality, the author indicates that such diets can be balanced to meet human requirements.

Neurath, Hans, "Protein-Digesting Enzymes," *Scientific American,* December 1964. Emphasizes the specificity of proteolytic enzymes for specific substrates, or reactant molecules.

# CHAPTER 27
# DIGESTIVE and nutritive Diseases

Digestive and nutritive diseases are associated with indigestion and malnutrition. *Indigestion* is the result of abnormal digestion and originates from one of the following factors: (1) psychogenic factors adversely affecting specific digestive reflexes; (2) structural or functional defects due to congenital or postnatal factors, by impairing ingestion, mechanical or chemical digestion, absorption, and defecation; (3) abnormalities in the movement of foodstuffs and digestive juices through the alimentary canal; or (4) inflammation or irritation due to microbial infection or physical trauma of any portion of the digestive system or surrounding membranes.

*Malnutrition* is caused by a lack of specific foodstuffs. The lack may be due to dietary deficiencies or an inability to adequately digest, absorb, or utilize particular nutrients. The following types of malnutrition are recognized: (1) a low-protein diet, with or without a deficit of essential amino acids; (2) a low-carbohydrate diet and, therefore, a deficit of calories;

(3) a combination of a low-protein, low-carbohydrate diet; and (4) deficiencies in either essential fatty acids, minerals, or vitamins, or some combination of these nutrients.

## DIGESTIVE DISORDERS

### TEMPORARY DIGESTIVE DISTURBANCES

The lumen, or cavity, of the digestive tract meets the outside environment through the mouth at one end and the anal orifice at the other. Whether or not the lumen is a part of the body is a moot point, but the fact remains that the walls which surround it are potentially vulnerable to the actions of the substances which pass through as well as to the physiologic and psychological states of the subject.

**Malocclusion**    When the bite is imperfect, the facets

of the maxillary and mandibular teeth do not meet properly. This condition, referred to as *malocclusion*, impairs the mechanical digestion of ingested foodstuffs.

Although the causes are diverse and include congenital defects and abnormal growth, orthodontic correction of the condition with braces is relatively straightforward. The braces apply a gentle force against the teeth to be shifted in position. An electrical effect caused by stress on bone (see Chap. 23) leads to the absorption of jawbone adjacent to the teeth on the side stretched by the braces and the deposition of new bone on the opposite side under tension. (See "Bone Molding" in Chap. 23.) The maintenance of pressure applied by the brace over an extended period of time results in the desired displacement of the teeth.

**Gases**   Gases in the alimentary canal are collectively referred to as *flatus*. Most of the gas consists of air which is swallowed and therefore has the same composition as the atmosphere. Much of the air is absorbed directly into the tissues and fluids of the body from the alimentary canal. Another fraction of the total is expelled by belching, or *eructation*. The remainder, a rather small amount, enters the small intestine. Microbial action in the intestine is responsible for the buildup of carbon dioxide, hydrogen, nitrogenous compounds, and methane, although contributions also come from gases which diffuse directly from blood. Flatus may accumulate, causing the walls of the large intestine to distend and thereby provoking varying degrees of abdominal discomfort and pain. The energy created by some mixtures of intestinal gases causes their forceful expulsion through the anal orifice, a phenomenon sometimes referred to as *crepitation*.

**Diarrhea, nausea, and vomiting**   Diarrhea and vomiting consist of the forceful expulsion from the digestive tract of its liquefied or semisolid contents. Vomiting ejects the contents of the stomach and duodenum through the mouth; diarrhea results in the washing out through the anus of the contents of the intestines. Diarrhea arises from one of two possible causes: (1) an infection termed *gastroenteritis* in some portion of the digestive tract; or (2) psychogenic factors which overexcite the parasympathetic nervous system. Vomiting is a reflex act caused by gagging, microbial infection, excessive gaseous distension, or psychogenic factors such as fear or dread. The subjective sensation of nausea usually precedes vomiting. Therefore, the same stimuli which cause vomiting also elicit nausea.

Diarrhea occurs in response to excessive secretions of water and mucus into the lumen of the large intestine. Coupled with an increase in intestinal motility, diarrhea promotes evacuation of the feces. *Constipation,* on the other hand, may occur by preventing the feces from entering the rectum because of a spastic contraction of its musculature or through a lack of smooth muscle tone in the wall of the rectum, a condition which prevents the feces from being expelled within a reasonable time. Under either of these circumstances, the desire to defecate is gradually diminished. However, water within the fecal mass continues to be reabsorbed through the intestinal walls or lost to the atmosphere by evaporation. As the feces becomes harder, they also become more difficult to expel.

**Cardiospasm, heartburn, and colic**   During the processing of food through the digestive tract, certain reflex mechanisms may not operate properly. These malfunctions interfere with the processing of its contents and create characteristic sensations. Examples of these malfunctions include (1) *cardiospasm*, in which the food seems to "stick in the throat" because the cardiac sphincter guarding the entrance to the stomach remains closed during swallowing; (2) *heartburn*, because of the accidental entry into the esophagus of acidic secretions containing proteolytic enzymes from the stomach; and (3) *colic*, because the enterogastric reflex is inhibited, permitting chyme to enter the small intestine prematurely.

**Mottling and dental caries**   The mottling of teeth is a structural abnormality due to the ingestion of water containing fluoride in excess of 2.0 parts per million (ppm). The enamel becomes discolored and exhibits pitting and chalking, although the functional capabilities of the teeth are ordinarily not impaired. Conversely, the lack of sufficient fluorides in drinking water can promote the development of cavities, or dental caries.

The mechanism by which small amounts of fluorine contribute to resistant enamel is still conjectural, but fluorine may interfere with the action of proteolytic enzymes secreted by bacteria in the mouth. According to this theory, fluorides prevent bacterial enzymes from dissolving the dentin and enamel.

## DIGESTIVE DISEASES

Diseases of the liver, gallbladder, biliary tract, and pancreas  Over half of all digestive disorders are due to inflammatory conditions usually caused by microbial infection or physical trauma. The liver, gallbladder, and pancreas are particularly susceptible in this regard because of a rich vascular supply to the liver from the small intestine and direct or close communication between these organs via the biliary tract. The jaundiced condition seen in the hepatitises (see Chap. 3) and certain other disorders involving the liver is caused by the deposition of bile pigments throughout the body. Their presence is associated with the skin only because its peripheral position permits the pigment to be seen.

Jaundice, itself, is of two types: hemolytic and obstructive. *Hemolytic jaundice* is caused by an excessive destruction of erythrocytes and the corresponding release of large amounts of hemoglobin. As usual, the reticuloendothelial system detoxifies hemoglobin, removing its heme group for use in the synthesis of the pigment bilirubin. Bilirubin becomes bound to blood proteins on its way to the liver for solubilization and excretion into the bile. However, the amount of bilirubin to be processed now greatly exceeds the ability of the liver cells to handle the pigment; consequently the blood concentrations of protein-bound bilirubin rise precipitously.

In *obstructive jaundice*, on the other hand, the liver is usually able to process bilirubin normally, removing the conjugated proteins and subsequently causing solubilization of the pigment. However, blockage of the hepatic ducts leading away from the liver prevents excretion of the formed bilirubin. Due to the hepatic congestion, many bile canaliculi rupture and release soluble bilirubin into the plasma.

The gallbladder may also become inflamed, a condition known as *cholecystitis.* Causative agents include *Escherichia coli,* streptococci, and staphylococci, as well as alcohol and gallstones. Chronic infections or other factors may lead to the formation of gallstones (*cholelithiasis*).

Gallstones form from the crystallization and subsequent enlargement of cholesterol against the inside wall of the gallbladder. Cholesterol is ordinarily soluble because of the presence of bile salts, water, and other constituents in bile. Any factor which increases the cholesterol content of the body or causes an abnormally large absorption of bile salts and water tends to initiate the precipitation of cholesterol in the gallbladder. Hence, a diet rich in fat may cause gallstones to form through a conversion of excess fat into cholesterol by the liver. Similarly, chronic cholecystitis may lead to gallstones by gradually altering the permeability characteristics of the wall of the gallbladder in a manner which causes a progressive increase in the amount of bile salts and water reabsorbed from its lumen.

The main pancreatic duct of Wirsung may also become obstructed by gallstones located in the ampulla of Vater, causing pancreatic juice to collect within. Microbial infection or physical injury can also lead to the same disorder. Whatever the cause, some inactive trypsinogen may become activated while still within the pancreatic duct. The trypsin which forms is then capable of activating more trypsinogen as well as chymotrypsinogen. These proteolytic enzymes begin to digest the tissues of the duct and the pancreas itself, causing acute or chronic *pancreatitis.* Severe pancreatic damage can result from the continued intracellular actions of trypsin and chymotrypsin.

*Cirrhosis* may also result from obstruction of the common bile duct by gallstones, although its most common causes are nutritional deficiencies and alcoholism. The hepatic cords of the liver lobules are destroyed and replaced by scar tissue, causing a damming up of blood in the hepatic portal vein. The resulting fluid congestion may affect the spleen, intestine, and abdominal cavity.

**Diseases of the intestine**   Appendicitis, diverticulitis, and hemorrhoids are some of the potentially more serious intestinal disorders. The first two are inflammatory in nature.

Inflammation of the vermiform appendix is not uncommon and results in gastric indigestion, nausea, and vomiting. Acute inflammation may cause the rupture of the walls of the appendix and the subsequent formation of an abscess in the ileocecal region.

*Appendicitis* is believed to arise from the putrefactive actions of intestinal bacteria, including streptococci, staphylococci, and spore-forming and non-spore-forming anaerobes, on the foods which lodge in the appendiceal lumen. These actions would have little effect were it not for the narrow orifice and poor blood supply characteristic of the appendix.

The causes and effects of *diverticulitis* are similar to those of appendicitis, with the exception that the pains are usually experienced on the left side and obstruction of the intestine may result from a gradual thickening of its walls. The inflammation of diverticula of the colon reduces intestinal motility, a condition which results in constipation.

The condition of *hemorrhoids*, or piles, usually has a noninflammatory cause. Hemorrhoids are varicosities, or swellings, of the hemorrhoidal veins in the mucous membranes of the rectum and anal canal. During difficult defecation, these swollen veins may rupture, causing the appearance of blood in the stools and accompanying defecatory pains.

## NUTRITIONAL IMBALANCES

### ESSENTIAL ELEMENT DEFICIENCIES

Disorders associated with deficiencies of essential elements include rickets, metabolic derangements, neuromuscular fatigue, vitamin deficiencies, anemia, hypercapnia, and various abnormalities of bone, muscle, and skin. Many of these disorders may also be caused by vitamin and hormonal deficiencies or by the actions of microbial pathogens.

**Bone abnormalities**   Practically all the calcium and phosphorus in the body is found in bones and teeth. Therefore, even though rickets is ordinarily caused by a deficiency of vitamin D or the hormone calciferol, bones may become similarly pliable or fragile through either a lack of dietary calcium or phosphorus or an inadequate reabsorption of these two minerals by the kidney tubules.

Deficiencies of several other elements also affect the skeletal system. A diet lacking in copper, for example, may cause the absorption of calcium and phosphorus from bones, thereby increasing bone fragility and the threat of fractures. The increased risks of dental caries because of a lack of an adequate concentration of fluoride in the water supply is another example already noted.

**Metabolic derangements**   Deficiencies in one or more essential elements lead to metabolic derangements. The following examples are indicative of the variety of abnormalities of this nature:

1   A lack of phosphorus reduces the ability to trap and transport energy released during metabolism, because this element is required for the generation of ATP.
2   A lack of magnesium (Mg) causes a depression in glycolysis, since $Mg^{++}$ and ATP or ADP form active complexes with at least four glycolytic enzymes.
3   A lack of manganese depresses the actions of peptidases, phosphatases, carboxylases, and other enzymes, thereby impairing digestion, absorption, and assimilation.
4   A lack of iron reduces the operation of the electron transport system, because the cytochromes contain nonprotein groups similar to the heme portion of hemoglobin.
5   A lack of iodine depresses the synthesis of thyroxine (hypothyroidism), since iodine is a vital constituent of this hormone.
6   A lack of zinc contributes to hypoinsulinism, since

the element is a structural part of the insulin molecule.

**Muscular disturbances** Muscular contraction is dependent upon a proper balance between extracellular $Na^+$ and intracellular $K^+$. Sodium ions along with $Cl^-$ also contribute importantly to the regulation of blood pH and osmotic pressures between the fluid compartments of the body, factors which may influence the abilities of the muscles to contract. Sodium deficiency occurs under conditions of intense perspiration; $K^+$ deficits develop as a consequence of extreme diarrhea or severe vomiting.

Calcium ($Ca^{++}$) and magnesium ($Mg^{++}$) also affect muscles. A severe lack of $Ca^{++}$ causes tetany and cardiac abnormalities. The paroxysms of tonic spasms of the arms and legs during tetany are related to an increase in neuromuscular excitability. Excluding acid-base imbalances, causes of tetany include various hormonal, dietary, and excretory imbalances which have the effect of reducing the body stores of $Ca^{++}$. Deficiencies of $Mg^{++}$ can also lead to tetany, in addition to cardiac disorders and convulsions.

**Vitamin deficiencies** Deficiencies in manganese and cobalt affect the utilization of thiamine and cyanocobalamin, respectively. Cobalt actually constitutes about 4 percent of an active molecule of cyanocobalamin. In contrast to the structural role assigned to cobalt, manganese most likely acts by activating an enzyme in which thiamine serves as the nonprotein component. The causes and effects of vitamin deficiencies are reviewed later in this chapter.

**Anemia** Sometimes, anemia may result from deficiencies of iron, copper, or cobalt. Iron is a constituent of the heme portion of the hemoglobin molecule to which $O_2$ attaches. Iron-deficiency anemia may also result from an acute or chronic loss of blood or as a by-products of new and increased nutritional demands, such as those experienced by pregnant women. Trace amounts of copper are required for the utilization of iron during the synthesis of hemoglobin. Copper probably activates an enzyme needed for the incorporation of iron into the molecular structure of hemoglobin.

As mentioned earlier, cobalt is a constituent of vitamin $B_{12}$. Cyanocobalamin is stored in the liver and used during hemopoiesis. Therefore, a cobalt deficiency can markedly depress the production of erythrocytes, giving rise to pernicious anemia. In another form of pernicious anemia, the body may lose the ability to absorb vitamin $B_{12}$ from the digestive tract, even though cyanocobalamin is present in adequate amounts. As we noted in Chap. 26, the chemical responsible for absorption of vitamin $B_{12}$ is called intrinsic factor. This natural product is produced by the gastric mucosa. Damage to the mucosa followed by the subsequent lack of absorption of vitamin $B_{12}$ are the usual causes of pernicious anemia, since frank vitamin $B_{12}$ deficiency is rare.

**Hypercapnia** Since almost 10 percent of the $CO_2$ in blood is carried as carbaminohemoglobin, a lack of sufficient hemoglobin or erythrocytes may also contribute to an increase in the plasma $P_{CO_2}$. Elevated concentrations of $CO_2$ bound as carbonic acid in blood (hypercapnia) may lead to acidosis by significantly lowering the pH of blood. Iron deficiencies can therefore initiate the development of hypercapnia.

The lack of zinc may also induce hypercapnia, but in a quite different manner. Zinc activates the enzyme, carbonic anhydrase, which converts carbon dioxide and water into carbonic acid within erythrocytes, i.e.,

$$CO_2 + H_2O \quad \xrightarrow[\quad Zn^{++} \quad]{\text{carbonic anhydrase}} \quad H_2CO_3 \qquad (27\text{-}1)$$

In the absence of zinc the reaction occurs slowly, just as it does ordinarily in the plasma. A depression in the rate of action of carbonic anhydrase leads to an excess of $CO_2$ in plasma.

**Abnormalities of hair** The role of essential elements in the normal growth of hair is at best conjectural. Two elements may be involved: sulfur and zinc. Sulfur is a constituent of the amino acid cystine, which is found in large concentrations in human hair and is therefore presumed to play a physiologic role in its development and growth. However, cystine may only be a by-product of biochemical reactions associated with the formation of hair and not a vital prerequisite. Zinc is also implicated in abnormalities of hair growth, but here too, the evidence is indirect and therefore only suggestive.

## ESSENTIAL FATTY ACID DEFICIENCIES

*Linoleic, linolenic*, and *arachidonic acids* are essential, since their dietary absence results in increases in metabolic rate and water consumption and a retardation of growth. Their structures are given in Fig. 26-14. The essential fatty acids are long-chain carbon compounds that possess a carboxyl (COOH) group at one end and at least one double bond between two adjacent carbon atoms in the chain (—C=C—), so that the structure of one molecule has the general formula R . . . —$\overset{|}{C}$=$\overset{|}{C}$— . . . COOH. The absence of two hydrogen atoms at each point of unsaturation confers an instability which can lead to the oxidation of the molecule, causing it to turn rancid. Vitamin E prevents rancidity, thereby preserving the structural integrity of the essential fatty acids.

Linoleic acid is the precursor of arachidonic acid, the essential fatty acid of demonstrated functional significance. However, the body can probably use any of the three unsaturated fatty acids interchangeably. Pyridoxine (vitamin $B_6$) is essential for the conversion of linoleic acid into arachidonic acid, and biotin, part of the B-complex series, converts inactive mixtures of arachidonic acid into their active form.

## ESSENTIAL AMINO ACID DEFICIENCIES

Of the 20-plus amino acids which are commonly found in foods eaten by human beings, 10 cannot be synthesized within the body or are made in amounts insufficient to meet the requirements of normal metabolism. These 10 amino acids are regarded as essential. Their structures are shown in Fig. 26-13.

**Anemia**   Deficiencies of several of the amino acids, including histidine, isoleucine, and lysine, impair the manufacture of hemoglobin and give rise to anemia. All three amino acids are important structural components of hemoglobin. Their absence leaves the formative molecule in an unfinished, inactive state.

**Hypoproteinosis**   The inability to synthesize proteins because of a lack of essential amino acids is reflected by a corresponding biochemical incompetence to constructively utilize the many other amino acid building blocks which go into their makeup. These amino acids are deaminated in the liver; i.e., the amino ($NH_2$) groups are removed from the rest of the molecule. As a result, the nonprotein nitrogen concentration of the blood rises measurably. The nitrogen is excreted in the form of urea, ammonium salts, and other nitrogenous compounds via the urine, feces, and sweat. When the total elimination of nitrogen exceeds the amount contained in the proteinaceous foodstuffs ingested, the body experiences a negative nitrogen balance. Deficiencies of valine, leucine, or tryptophan are especially liable to cause this condition. Because protoplasm is proteinaceous, a chronic negative nitrogen balance is likely to be reflected in a loss of weight and a retardation of growth.

**Biosynthetic depression**   A number of amino acids are used in the synthesis of other amino acids and carbon compounds. The sulfur of methionine is used in the synthesis of the amino acid cystine, and phenylalanine is employed in the manufacture of the amino acid tyrosine. Tyrosine is utilized in the synthesis of melanin, epinephrine, and thyroxine. Methyl ($CH_3$) groups from methionine interact with the amino acid arginine and other intermediates to form creatine. Phosphorylated creatine (creatine phosphate) is usually able to maintain the supply of ATP in active muscles until an increase in glycolysis provides sufficient ATP to take care of demands.

**Pellagra and phenylketonuria**   Deficiencies in, or an inability to metabolize, several amino acids may lead to discrete syndromes. A lack of the amino acid tryptophan, for example, contributes to the development of pellagra. Tryptophan is used in the biosynthesis of niacin, one of the B-complex vitamins, although the majority of niacin is usually supplied from exogenous sources. The lack of niacin is ultimately responsible for causing the symptoms of pellagra. Still another example is the disorder phenylketonuria. This disease is produced by a genetic defect which prevents the utilization of phenylalanine. Phenylketonuria is discussed further in Chap. 21.

## PROTEIN AND PROTEIN-CARBOHYDRATE DEFICIENCIES

Two important disease entities result from a deficit of protein or a combination of protein and carbohydrate. The diseases are called *kwashiorkor* and *marasmus*. These disorders are extremely common in the under-developed countries of the world and appear in young children who have recently been weaned.

**Kwashiorkor** This disease is brought on by a diet low in protein and high in carbohydrate. In such a diet, calories are available but the proteins required for biosynthesis are grossly insufficient to meet the needs of the body. Under these circumstances, growth ceases or is depressed. The brain appears stunted, giving rise to a doleful lethargy and a lack of interest that may in fact be early signs of mental retardation. The hypoproteinosis gives rise to a hypoproteinemia which is reflected in an enlarged liver and a general edematous condition. The deficit of protein is also responsible for a spotty deposition of pigment in hair. The synergistic relationship which exists between this type of malnutrition and secondary infection makes the presence of both in a very young patient much more serious than would be expected from the additive effects of the two, and death usually results from secondary infection.

**Marasmus** This disease is induced by a diet low in both protein and carbohydrate. The lack of sufficient calories combines with protein deficiency to make the symptoms of marasmus quite different from those of kwashiorkor. In marasmus, the body of the child seems to waste away. "Skin and bones" is an apt description. The elasticity of the skin is markedly reduced, causing it to appear thin and wrinkled. The hair is dull and dry. However, these integumentary symptoms are overshadowed by the seriousness of the gastrointestinal disturbances seen in most cases of this disease. These symptoms include severe diarrhea and vomiting and almost inevitably lead to an irreversible dehydration. Secondary infection is also a significant factor in causing death.

## VITAMIN DEFICIENCIES

Vitamins are essential dietary ingredients which act as coenzymic (nonprotein) groups of many enzymes involved in intermediary metabolism and biosynthesis. In the absence of their coenzymes, various enzymes remain inactive, causing metabolic derangements of varying magnitude. The degree of malfunction is determined by the extent of the deficiency and the affected metabolic pathway.

The identity and functions of vitamins are determined by observations and experiments on animals and human beings. Substances so designated must withstand the rigors of two critical tests somewhat reminiscent of Koch's postulates described earlier (see Chap. 2):

1 The symptoms and signs of deficiency must be experimentally induced within a reasonable time following the reduction or elimination of the dietary ingredient.
2 The symptoms and signs of deficiency must be ameliorated after a reasonable time following an addition to the diet of therapeutic amounts of the missing ingredient.

The steps themselves are practical compromises for the following reasons:

1 It is usually not possible or morally acceptable to perform deficiency experiments on human beings and, indeed, most nutritional data are obtained from tests with rats or other animals.
2 Species and individual variations in physiologic responses to deficiencies of vitamins make interpretations and extrapolations of results difficult.
3 Similar symptoms and signs are produced in individual cases by deficiencies of essential minerals or amino acids, by other causative agents, or by an interaction of these factors, complicating the task of ensuring that a vitamin deficiency is responsible for the effects.
4 The biochemical reactions in which suspected vitamins act are incompletely understood, a dilemma which in some cases leaves an element of uncertainty as to whether the ingredient in ques-

tion is a coenzyme or a molecule which acts in some other capacity.

With regard to vitamins C and E, and perhaps several others, some studies suggest that massive amounts may be required in particular circumstances, a concept which upsets traditional views of the concentration of a "physiologic" dose. Proponents of this school claim wide-ranging beneficial effects from the ingestion of extremely large concentrations of these vitamins.

Although the above compromises are necessary to pursue the practical study of *hypo-* and *hypervitaminosis*, they leave open some fundamental questions. As examples, consider the following:

1 How relevant to the health of human beings are dietary studies performed on rats or insects?
2 Why do some human beings respond to vitamin therapy in clinical trials while others suffering from the same apparent disability do not?
3 Under what conditions may we be justified in diagnosing vitamin defiencies solely on the basis of symptoms?
4 As our knowledge increases, might some accepted vitamins turn out to be other types of biologically active molecules such as hormones, necessary precursors of important metabolic pathways, or structural components of molecules other than enzymes?
5 How sure are we that the present inventory of vitamins includes all those which are found in human beings?
6 Why do the physiologic demands for certain vitamins sometimes seem to exceed the normal supply?
7 Given no contraindications, how safe are vitamins to take in "large" concentrations?
8 How many vitamins are normally synthesized by the intestinal flora of human beings?
9 Are the synthesized vitamins identical to the natural products within the body?

The above list is by no means exhaustive but does give some idea of the no-man's-land treaded upon when dealing with vitamin deficiencies.

The picture is obviously not as clear as one might wish, considering the importance of this group of nutrients to the health and well-being of human beings and animals. Nevertheless, our knowledge of the effects of vitamin deficiencies is considerable, ranging over at least 7 decades of study, and the information to be assimilated is growing at an exponential rate. Out of the conflicting claims and contradictory results, the jumble of hard data and interpretations, clinical reports and case studies, and the wealth of anecdotal evidence in this field, we can piece together the beginnings of a coherent story of what vitamins do by looking at what happens to the body in their absence.

Visual abnormalities Visual abnormalities are caused by a lack of either vitamin A or $B_2$ (riboflavin). *Vitamin A* is a precursor of rhodopsin, the visual pigment in rods. Only rods are used in night vision, as you learned in Chap. 11. A deficiency of vitamin A depresses the regeneration of rhodopsin after the pigment is partially bleached by exposure to bright light. Under these circumstances, rods are less sensitive in dim illumination than they were formerly. This condition is known as *night blindness*. In addition to night blindness, vitamin A deficiencies can lead to corneal damage and total blindness, apparently because epithelial structures, including the cornea, respond to a deficit of vitamin A by thickening and hardening.

The action of riboflavin in causing visual abnormalities is quite different from that of vitamin A. Riboflavin acts as nonprotein portions of both flavine adenine dinucleotide (FAD) and flavine mononucleotide (FMN). These enzymes serve as hydrogen carriers during the oxidation of intermediary metabolites. A depression in oxidation due to a deficiency of riboflavin shows up quickly in poorly vascularized structures such as the cornea. Riboflavin also participates in the biochemical pathway that regenerates rhodopsin from vitamin A.

Bone abnormalities Bone abnormalities can be

caused by a lack of vitamins A, C, and D. Vitamin D is probably the most well-known of the three because of its role in the prevention of rickets. Vitamin D promotes the absorption of $Ca^{++}$ from the intestine and controls its deposition in bone. However, the body has a supplementary mechanism which reinforces the action of vitamin D; namely, the hormone calciferol in the skin. The precursor of calciferol is activated by ultraviolet radiation from sunlight. Ultraviolet radiation converts the precursor into calciferol, a derivative of vitamin D. Vitamin D and calciferol act in a similar fashion.

During winters in northern temperate climates, sunlight is often at a premium because of the diminished time which people spend outdoors. Vitamin D supplements are usually necessary, especially during the growth stages of development, to make up for the lack of both ultraviolet radiation and calciferol.

New growth of bone from epiphyseal cartilaginous disks is depressed in the absence of vitamin A, apparently by inhibiting the growth of new cells. This conclusion is consistent with the finding, previously referred to, that vitamin A promotes the development of epithelial tissues.

A lack of vitamin C also depresses bone growth by hindering the normal development of bone matrix. Under these conditions, the matrix remains unossified, even though calcification occurs. The resulting fragility increases the susceptibility of bones to fractures.

**Beriberi**  Beriberi is believed to be caused by a lack of vitamin $B_1$ (thiamine), which acts as part of a molecule called *thiamine pyrophosphate* (TPP). TPP is the nonprotein group of the enzyme *carboxylase,* which oxidizes derivatives of sugars and amino acids. Thiamine deficiencies lead to a depression in oxidation that is reflected by a significant reduction in the amount of energy available for cellular metabolism. The energy reduction is felt primarily by the peripheral nerves and the cardiac and smooth muscles, structures which are particularly susceptible to poor nutrition. The peripheral nerves become irritated and their myelin sheaths may subsequently degenerate. The heart weakens, giving rise to fluid congestion and

edema. In addition, there is a loss of smooth muscle tone in the walls of peripheral veins and the digestive tract. These effects decrease the volume of blood returned to the heart and cause gastrointestinal upsets.

**Pellagra**  Although pellagra is a vitamin-deficiency disease, the disorder, as we noted above, may also be provoked by a deficiency of tryptophan. The primary cause of pellagra is a lack of niacin, and tryptophan is required for its biosynthesis. Niacin, as nicotinamide adenine dinucleotide (NAD), accepts and transports hydrogen during cellular oxidation. Its lack is revealed by a depression in both oxidation and energy release similar to that caused by thiamine. The resemblance in causation, albeit by different biochemical mechanisms, is reinforced by the presence of similar neurologic and gastrointestinal symptoms. However, full-blown pellagra is more severe than beriberi. Brain lesions, psychotic reactions, and serious inflammations of the digestive system are not uncommon. Lesions and other abnormalities, such as drying and scaling, appear in the epithelial tissues of the skin due to poor nutrition and inadequate circulation. Considered as a whole, the symptoms have been appropriately described by the four D's: diarrhea, dermatitis, dementia, and death.

**Scurvy**  Scurvy is caused by a deficiency of vitamin C (ascorbic acid). The actual function of vitamin C in the body remains unknown, although the theory that it serves in a wide variety of oxidation-reduction (redox) reactions seems to predominate. Ascorbic acid also has a regulatory role within mitochondria. A deficit in the formation of intercellular substance probably accounts for most of the effects seen in scurvy. The following symptoms are characteristic of scurvy: (1) capillaries rupture because of their fragility; (2) bone growth is depressed because the organic matrix is not laid down correctly (see "Bone Abnormalities," above); and (3) wound regeneration is impaired because collagen formation is retarded or inhibited. Chronic bleeding also invites secondary infection.

Generally speaking, then, the clinical picture is one where the structures affected contain large amounts of connective tissue. In an attenuated form, the symptoms of scurvy are superficially similar to some of the age-specific changes seen over a normal lifetime. These similarities have encouraged proponents of vitamin C therapy to refer to aging as a subclinical form of scurvy.

**Lowered resistance**   In Chap. 4, we pointed out a possible relationship between the presence of adequate concentrations of vitamin C in the body and the ability to withstand physiologic and psychological stresses. Among the organs which contain large concentrations of the vitamin are the adrenal glands. Even though the role of ascorbic acid in the adrenal glands has not been established unequivocally, vitamin C is suspected of playing some part in the formation of cortisol. At least the ascorbic acid content of the adrenal glands declines in subjects under stress. Since cortisol and its hormonal activator, ACTH from the hypophysis, have widespread beneficial effects in the body, it is attractive to ascribe a central role to ascorbic acid in these actions. There is, in fact, a growing volume of empirical evidence to support the claim. Whether the virtues of taking large doses of vitamin C each day (2 to 20 Gm) include protection against the common cold and other diseases, as claimed by its proponents, awaits rigorous experimental testing. In any event, the theoretical argument behind this idea is intriguing and worthy of careful investigation.

**Anemia**   In addition to the causes given previously, anemia may be caused by deficiencies of three vitamins: $B_6$ (pyridoxine), $B_{12}$ (cyanocobalamin), and folic acid (part of the B-complex series). Pyridoxine acts as part of a coenzyme required for transamination reactions in which the amino groups ($NH_2$) of certain amino acids are removed and used in the synthesis of other amino acids. Pyridoxine is also implicated in the metabolism of folic acid. Although an absence of vitamin $B_6$ reduces the quantity of tryptophan and niacin and may therefore contribute to the development of pellagra, *microcytic anemia* is one of the most common symptoms of pyridoxine deficiency. This type of anemia is marked by characteristically small

erythrocytes and an increase in the blood serum concentration of iron. Presumably, pyridoxine interferes with hemopoiesis and consequently prevents iron from being incorporated into hemoglobin.

As we have seen above, a lack of vitamin $B_{12}$ results in a failure of the erythroblasts to mature. The erythrocyte count is therefore depressed. The erythrocytes that do form become macrocytic, undergoing compensatory enlargement in an attempt, so to speak, to make up for the undersupply of oxygen carriers in the blood. Anemia caused by a deficiency of vitamin $B_{12}$ is referred to as *pernicious*.

We have already noted the contribution made by deficiencies of cobalt to a depression in the amount of vitamin $B_{12}$ in the body. The cobalt in cyanocobalamin forms a chemical group similar to that of the iron porphyrin of hemoglobin, but the mechanism by which the vitamin contributes to the maturation of erythroblasts is still an open question.

A deficiency of folic acid may cause *megaloblastic anemia*. In this condition, not only is the blood deficient in erythrocytes but it also contains embryonic megakaryocytes, large irregularly shaped cells from which platelets normally form by fragmentation. However, megakaryocytes are ordinarily found only within the bone marrow. Folic acid deficiencies depress the maturation of erythroblasts, apparently because this vitamin aids the action of vitamin $B_{12}$. Nevertheless, the therapeutic use of folic acid in the treatment of pernicious anemia does not relieve all its symptoms, especially those of a neurologic nature.

The functional significance of any vitamin is determined by its biochemical activities. Folic acid is involved in the synthesis of portions of DNAs and RNAs and of several amino acids, at least one of which is considered essential. Its functions in the manufacture of genes places this vitamin in the class of growth-promoting factors essential for normal development. It is probably in this role that folic acid acts during erythrocytic maturation.

**Skin abnormalities**   Skin abnormalities have been attributed to deficiencies of a number of vitamins, including biotin, riboflavin, niacin, pyridoxine, and vitamin A. Biotin acts as a cocarboxylase during portions of the citric acid cycle (see Chap. 6) and is also involved in the metabolism of fatty acids. Although the absence of biotin may contribute to the development of anemia, as mentioned earlier, skin abnormalities are among the more common symptoms of biotin deficiency. Human volunteers who have eaten diets deficient in biotin exhibited skin lesions and a grayish pallor. The functional capabilities of sebaceous glands are usually affected, a condition referred to as *seborrhea*.

A lack of riboflavin also affects sebaceous glands, causing a greasy dermatitis, scaling of the skin, and cracking at the lips and corners of the mouth called *cheleiosis*. When the deficiency is chronic, the epidermis finally atrophies.

The early integumentary symptoms of niacin deficiency resemble sunburn (erythema). A dilatation of blood vessels in the dermis causes the redness. Chronic deficiencies of niacin involve the sebaceous glands and epidermis, resulting in seborrhea, a grayish pallor, and scaling. These symptoms in part overlap those elicited by deficiencies of biotin and riboflavin.

The cheleiosis seen in riboflavin deficiency is also part of the syndrome when vitamin $B_6$ is lacking. A localized edema usually accompanies the dermatitis.

Skin abnormalities caused by a deficiency of vitamin A include lesions, hardening, and drying. The atrophic changes are especially prevalent in specialized epithelial tissues with poor mitotic abilities.

**Growth abnormalities** Growth is adversely affected by a lack of several vitamins, including folic acid, riboflavin, and vitamin A. Folic acid, as you will remember, is a growth-promoting substance by virtue of its role in the manufacture of genetic material. Riboflavin promotes growth because of its energy-releasing functions during cellular oxidation. Vitamin A aids the development of epithelial cells in ways which are not well understood. A serious lack of one or more of these three vitamins may result in a retardation of growth and subsequent keratinization and atrophy of affected tissues.

**Circulatory abnormalities** An absence of several vitamins causes abnormalities of the heart, blood vessels, and blood. The list includes vitamins E, K, and C, as well as thiamine and folic acid. The tocopherols, especially alpha tocopherol, apparently are capable of regulating the flow of oxygen to the tissues of the body, a function wich has given rise to the employment of alpha tocopherol in the treatment of a wide variety of cardiovascular diseases. Medically supervised therapy with vitamin E has achieved remarkable success in numerous patients with coronary disease, valvular disease, aneurysms, hypertension, and varicosities. These diseases are discussed in Chap. 31. In other cases, negative results have been reported. The essence of the theory espoused by proponents of vitamin E therapy is that this vitamin provides an optimum concentration of oxygen to diseased or damaged tissues, thereby relieving pain and other signs of inflammation, arresting further necrotic changes, and maintaining conditions conducive to healing and regeneration. The biochemical mechanisms by which alpha tocopherol accomplishes these feats, if indeed they are true, remain largely unknown.

Deficiencies of vitamin K adversely affect hemostasis (the ability of the blood to clot). Vitamin K is required for the formation by the liver of prothrombin and perhaps several other factors which have roles in coagulation. A diet deficient in vitamin K increases the clotting time of blood, a function which earned vitamin K its reputation as "the antihemorrhagic factor."

A lack of vitamin C increases capillary fragility, part of the syndrome of scurvy discussed above. The capillaries rupture easily, causing multiple hemorrhages throughout the body. The increased fragility of the walls of blood vessels is related to the inability to elaborate intercellular substance.

Thiamine deficiency can cause cardiac arrhythmias such as palpitations. The heart muscle weakens and undergoes compensatory hypertrophy. Venous pooling increases as the cardiac output diminishes, placing a further strain on the heart. Circulatory stagnation leads to a general congestion and edema and continues a vicious circle that may result in cardiac failure and circulatory shock. These symptoms are part of the syndrome of beriberi.

Folic acid deficiencies lead to megaloblastic and pernicious anemias, conditions previously described under "Anemia." In addition, a deficiency of this vitamin causes leukopenia, i.e., a depression in the production of leukocytes. Leukopenia is considered in Chap. 31.

**Digestive disturbances**  Digestive disturbances are caused by deficiencies of thiamine, niacin, pantothenic acid, biotin, and vitamin $B_6$.

Digestive symptoms of thiamine deficiency include the inability to swallow easily (dysphagia), loss of appetite (anorexia), colic, nausea, and vomiting. The gastritis which accompanies thiamine deficiency may reduce the secretion of HCl, a condition referred to as hypochlorhydria. These gastrointestinal symptoms are part of the syndrome of beriberi and may occur because the smooth muscles of the viscera are weakened by an inadequate energy supply, as pointed out earlier. Niacin and pyridoxine deficiencies both result in anorexia, nausea, and vomiting, symptoms which are similar to those caused by a deficiency of thiamine. In the case of niacin, the digestive disturbances are part of the syndrome of pellagra and occur in response to an intense inflammation of the digestive tract. The inflammatory response is triggered by a marked depression in both cellular respiration and energy release. Vitamin $B_6$ deficiencies, on the other hand, are related to an inability to metabolize amino acids.

Among the gastrointestinal symptoms of pantothenic acid deficiency are anorexia and constipation. The symptoms also include gastritis and hypochlorhydria, conditions similar to those produced by a deficiency of thiamine. In the case of pantothenic acid, the effects are traceable to a depression in the metabolism of carbohydrates and fats.

The nutritional disorders caused by biotin deficiency in people consist of anorexia and an exacerbation of the poor nutrition already present. The loss of appetite brought about by a lack of biotin or other vitamins and by microbial pathogens emphasizes the general vulnerability of the digestive tract to homeostatic disruptions.

**Nervous system abnormalities**  Neurologic symptoms are produced by deficiencies of thiamine, niacin, pantothenic acid, and vitamin $B_6$, i.e., all the vi-

tamins that cause digestive disturbances, with the exception of biotin.

The neurologic symptoms of thiamine deficiency include polyneuritis, loss of extension reflexes such as the knee jerk, personality disorders such as instability, intellectual deficits such as forgetfulness and disorganization, as well as possible brain lesions. These impairments of nervous system function are the result of poor cellular nutrition.

With the exception of brain lesions, a deficiency of niacin elicits neurologic symptoms similar to those obtaining in thiamine deficiency. In addition, a lack of thiamine may cause headaches and vertigo. These symptoms are part of the clinical syndrome of pellagra.

An absence of pantothenic acid in the diet affects the nervous system by causing polyneuritis, paresthesias such as numbness and tingling in the hands and feet, and vertigo. For the most part, these symptoms overlap those produced by deficiencies of thiamine and niacin.

In the case of a lack of vitamin $B_6$, sensory nerves become inflamed, a condition called neuralgia. The paresthesias resemble those caused by a deficiency of pantothenic acid.

## MALNUTRITION, UNDERNUTRITION, AND OBESITY

**Malnutrition**  Diseases caused by deficiencies of essential substances are specific examples of malnutrition; however, the effects of malnutrition can be more diffuse. The signs may be subtle rather than obvious or may overlap those seen in a number of other unrelated disorders, making diagnosis difficult. This problem is compounded when the complex causal factors of malnutrition are considered.

Because malnutrition is the result of an imbalance in the types of nutrients in the body, its cause is not restricted to dietary deficiencies; e.g., the condition may be brought about by an inability to synthesize essential nutrients from raw materials. This functional impairment originates either in the intestine where the microbial flora which engages in synthesis is lacking or within the tissue cells as a consequence of hereditary defects reflected in metabolic derangements.

Sometimes, malnutrition may be due to an inability

to absorb soluble foodstuffs or essential nutrients from the intestine. Malabsorption results from one of the following factors:

1 Necessary digestive enzymes, hormones, or activators are not present in adequate amounts or are lacking as a consequence of metabolic disease.
2 Adequate concentrations of nutrients are prevented from reaching the intestine because of difficulties in chewing, swallowing, or food transport.
3 Digestive disturbances, such as vomiting and diarrhea, prevent nutrients from remaining in the intestine long enough to permit adequate digestion and absorption.
4 Pathologic conditions affect the absorptive efficiency of the intestinal mucosa by decreasing its surface area for absorption, increasing its thickness, or altering its permeability characteristics.

Finally, the inability of tissue cells to utilize available nutrients for energy release or biosynthesis can also cause imbalances of energy nutrients. The source of this form of malnutrition is usually traced to intoxications or congenital defects.

The effects of general malnutrition are traced in Fig. 27-1, which shows a vicious circle of progressive deterioration. Among the early symptoms are loss of weight and evidence of mild vitamin deficiency. Depending upon the degree and nature of the nutritional imbalance, stored fat or body proteins may be degraded for their energy content. If the condition remains unchecked over an extended period of time, all aspects of metabolism may be depressed, including vital signs. Other symptoms, such as tiredness, apathy, and a decreased resistance to secondary infection, develop from the depression in metabolism, leading to even more serious consequences.

**Undernutrition**  Referring back to Fig. 27-1, we will now trace the effects of undernutrition in its most extreme form, complete starvation, and compare its symptoms to those of malnutrition. Although both types of nutritive disorder are often considered identical, *undernutrition* explicitly denotes a lack of enough food to adequately sustain the normal functions of the body. However, in many cases of severe malnutrition and borderline undernutrition, the theoretical distinction is blurred and one condition may merge imperceptibly into the other.

Compared to malnutrition, undernutrition is usually more serious and requires less time before symptoms appear. However, the effects of both disturbances are more or less similar. The resemblances are particularly close during the early course of these disorders.

During the first several days of complete starvation, the glycogen reserves of the body are completely consumed. If the nutritional deficit continues, stored fats are increasingly utilized for energy. Since both carbohydrates and fats are oxidized simultaneously, as the reserves of carbohydrates decrease, the oxidation of fats increases. In the absence of food, the fat depots are depleted after 1 or 2 months. The rate of utilization depends in part on the amount of adipose connective tissue in, and the size of, the subject. After the stored fats are significantly reduced, body proteins, especially of the muscles, are broken down and used for energy. This cytolysis is, of course, an extreme physiologic effort to continue existence and postpone death.

The symptoms of starvation are indicative of the type of foodstuff being differentially oxidized for energy. Initially, when carbohydrates are being depleted, the most obvious symptom is a loss of weight. At the stage of starvation when fats are the most widely used energy resource, symptoms of vitamin deficiencies appear in varying degrees of severity. As consumption of body proteins increases, gastrointestinal and neurologic symptoms predominate. Nausea and hallucinations are common. Frank expressions of mineral deficiencies show up in the latter stages of complete starvation and are compounded somewhat later by a complex of symptoms associated with an absence of other essential nutrients. Muscles become weak and atrophic because their proteins are being

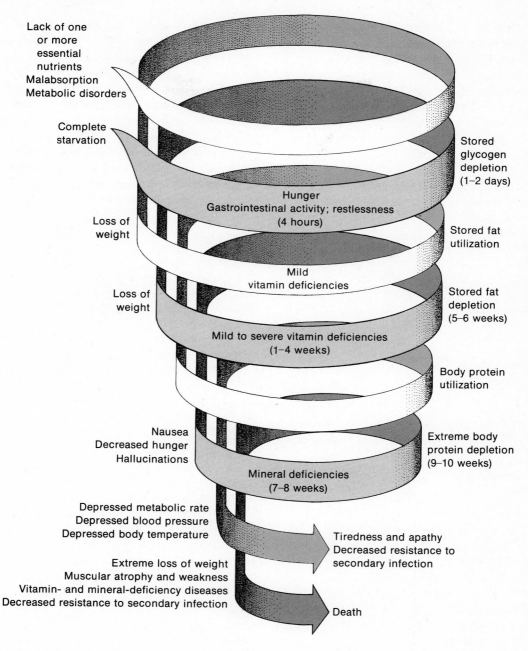

Lack of one or more essential nutrients
Malabsorption
Metabolic disorders

Complete starvation

Stored glycogen depletion (1–2 days)

Hunger
Gastrointestinal activity; restlessness
(4 hours)

Loss of weight

Stored fat utilization

Mild vitamin deficiencies

Loss of weight

Stored fat depletion (5–6 weeks)

Mild to severe vitamin deficiencies
(1–4 weeks)

Body protein utilization

Nausea
Decreased hunger
Hallucinations

Extreme body protein depletion (9–10 weeks)

Mineral deficiencies
(7–8 weeks)

Depressed metabolic rate
Depressed blood pressure
Depressed body temperature

Tiredness and apathy
Decreased resistance to secondary infection

Extreme loss of weight
Muscular atrophy and weakness
Vitamin- and mineral-deficiency diseases
Decreased resistance to secondary infection

Death

**Figure 27-1**  *The vicious circles associated with malnutrition and undernutrition.*

oxidized for energy. An extreme loss of weight in the last stages of starvation is reflected by the correspondingly emaciated appearance of the subject. The ensuing depression of metabolism dulls the senses, incapacitates the body physically, and reduces the effectiveness of internal defense mechanisms. Vulnerable to the depredations of microbial pathogens, the body experiences secondary complications which usually precipitate death.

Obesity   When nutrient intake exceeds energy output, the body increases in weight. If the weight is at least 30 percent above the average for the age, height, and sex of the subject, the condition is referred to as *obesity*. This disease, like malnutrition, has multiple causes. Although they are not well understood, obesity is related to the following factors: (1) psychogenic factors and possible hypothalamic lesions which create a compulsive desire for food; (2) hereditary defects which express themselves either as central nervous system abnormalities affecting hypothalamic control of food intake or as biochemical derangements of metabolism, causing an excessive deposition, or inadequate breakdown, of fat in the

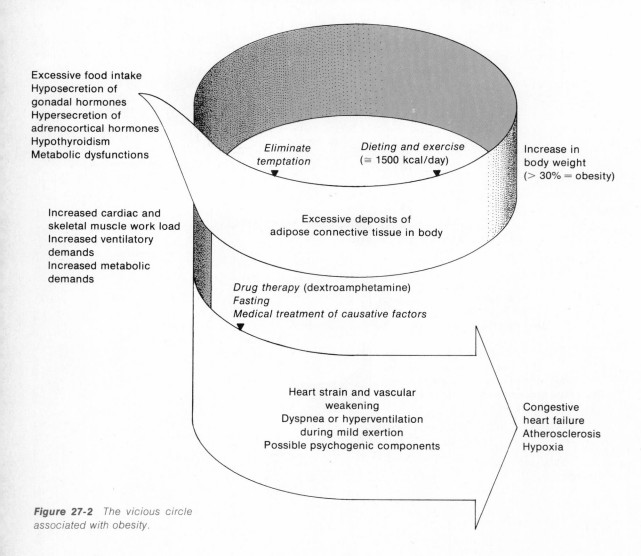

Excessive food intake
Hyposecretion of gonadal hormones
Hypersecretion of adrenocortical hormones
Hypothyroidism
Metabolic dysfunctions

*Eliminate temptation*   *Dieting and exercise ($\cong$ 1500 kcal/day)*

Increase in body weight (> 30% = obesity)

Excessive deposits of adipose connective tissue in body

Increased cardiac and skeletal muscle work load
Increased ventilatory demands
Increased metabolic demands

*Drug therapy* (dextroamphetamine)
*Fasting*
*Medical treatment of causative factors*

Heart strain and vascular weakening
Dyspnea or hyperventilation during mild exertion
Possible psychogenic components

Congestive heart failure
Atherosclerosis
Hypoxia

**Figure 27-2** *The vicious circle associated with obesity.*

adipose tissues of the body; and (3) hormonal imbalances such as a deficiency of thyroxine which depresses energy output, an excess of adrenocorticosteroids which causes a redistribution of fat in the body by altering fat metabolism, or deficiencies of androgens or estrogens and a corresponding depression in growth regulation, which cause the deposition of nutrients as fats.

The vicious circle associated with obesity is outlined in Fig. 27-2. Basically, excess weight places greater-than-normal demands on the metabolic machinery of the body. The increased workload of the heart, especially under physical exertion, may weaken both the heart and the great blood vessels associated with it. The correspondingly elevated requirements for $O_2$ may result in respiratory complications such as hyperventilation or shortness of breath (dyspnea).

The thickness of the fat depots serves as an insulator which tends to retain heat within the body (see Chap. 8). When the body temperature rises during exercise, the perspiration of an obese person becomes profuse. Perspiration, as we are well aware, serves to cool the body by dissipating excess heat.

Obese people are particularly prone to congestive heart failure, in which the left ventricle is differentially damaged, causing asphyxiation from fluid congestion within the lungs, and atherosclerosis, in which fatty deposits constrict the lumina of various arteries, drastically reducing the supply of nutrients and oxygen to tissues supplied by the affected vessels. Cardiovascular diseases are discussed in Chap. 31.

The treatment of obesity is as complex as its causes are diverse. First and foremost is a regimen of dieting and exercise which reverses the lop-sided ratio of food intake to energy output. Because of the psychological support provided by others undergoing the same treatment, ideally the dieting should be reinforced in a group setting. The recommended caloric consumption per day during dieting is usually close to 1500 kcal, although the amount will vary with age, height, sex, and the nature of the dietary regime. Important adjuncts to successful dieting require the elimination of gastronomic temptations between meals and the strict regulation of food consumption during meals. These measures will ensure an optimum intake necessary for adequate nutrition as well as a dietary minimum required for necessary energy production. If obesity is extreme, fasting may be ordered under medical supervision. To depress the desire for food, drug therapy with amphetamines may be prescribed.

In obesity and in many other diseases, symptomatic treatment is doomed to failure. This is so because the factors which originally caused the disease have not been eliminated, even if weight returns to near-normal through a combination of the above treatments. After diet therapy stops, weight again increases, following a predictable pattern familiar to most of us. Therefore, where physiologic abnormalities or psychogenic factors are the basic causes of the disease, treatment should concentrate on their identification and elimination.

## COLLATERAL READING

Loomis, W. F., "Rickets," *Scientific American,* December 1970. Loomis dispels the myth that rickets is due solely to a dietary deficiency of vitamin D and establishes this disease as the first one traced directly to air pollution.

Mitchell, Helen S., Henderika J. Rynbergen, Linnea Anderson, and Marjorie V. Dabble, *Cooper's Nutrition in Health and Disease,* 15th ed., Philadelphia: J.B. Lippincott Company, 1968. Part 2 of this book provides useful information on the nutritive requirements and restrictive diets of patients suffering from the major diseases which afflict man.

Scherp, Henry W., "Dental Caries: Prospects for Prevention," *Science,* 173: 1199–1205, 1971. Considerable progress in reducing the incidence of dental caries will most likely be accomplished in the following ways: (1) by protecting teeth with fluorides and sealants; (2) by modifying the diet through the reduction or elimination of sucrose and its replacement with other sugars and starchy foods; and (3) by combating cariogenic bacteria through brushing the teeth, decomposing the dextran of dental plaque by incorporating a dextranase preparation in the diet and drinking water, or preventing cariogenic bacterial growth with intraoral applications of suitable antimicrobial agents.

Spitznagel, Edward L., Jr., "Lognormal Model for Ascorbic Acid Requirements in Man," *BioScience,* 21(19):981–984, 1971. The author points out some of the difficulties in interpreting relevant nutritional experiments and provides a brief but comprehensive bibliography supporting the contention that daily vitamin C requirements may be in the order of grams rather than milligrams.

Starzl, Thomas E., "*The Current Status of Liver Transplantation,*" in R. A. Good and D. W. Fisher (eds.), *Immunobiology: Current Knowledge of Basic Concepts in Immunology and Their Clinical Applications*, Stamford, Conn.: Sinauer Associates, Inc., 1971. According to Starzl, the findings of graft acceptance after initial rejection may pave the way for the development of a technique to prepare for acceptance of the graft in advance of transplantation.

Young, Vernon R., and Nevin S. Scrimshaw, "The Physiology of Starvation," *Scientific American,* October 1971. In an interesting review of studies using human volunteers and obese patients on a starvation diet, the authors identify the physiologic alterations made by the body in response to a lack of energy nutrients.

# CHAPTER 28
# respiration

Respiration provides a mechanism for the exchange of gases between the body and the external environment. Human life quickly ceases without the ability to respire because cellular metabolism is severely impaired.

The more specific functions of the respiratory system are as follows: (1) the transport of oxygen ($O_2$) from the outside air to the blood; (2) the transport of carbon dioxide ($CO_2$) from the blood to the outside air; (3) participation in the production of sounds (phonation); (4) the removal of foreign particles which enter the respiratory tract through the nose or mouth; and (5) participation in the detection of odorous substances, a sensory process called olfaction.

Once in the blood, $O_2$ is transported to the tissue cells, where it is used in an intracellular chemical pathway which liberates most of the chemical energy locked in metabolic fuels. Carbon dioxide, a waste product of cellular metabolism, is transported through the blood to the lungs, where it is expelled to the outside.

Thus, there are two basic types of respiration based on where the process occurs in the body—external and internal. *External respiration* refers to all those mechanisms which culminate in gaseous exchanges between the lungs and blood. *Internal respiration* applies to the diffusion of gases between the blood and tissue cells. The utilization of $O_2$ within tissue cells is called *cellular respiration* (see Chap. 6).

This chapter explores the structure and functions of the respiratory system and the mechanisms by which the respiratory gases are transported through blood for exchange in the lungs and tissue cells.

## ORGANIZATION OF THE RESPIRATORY SYSTEM

The respiratory system is located in the facial, cervical, and thoracic regions. The respiratory system includes the lungs and a series of airways which link the lungs with the outside of the body by way of the nose and mouth.

## UPPER RESPIRATORY STRUCTURES

Upper respiratory structures compose all portions of the respiratory system except the lungs and the airways within the lungs. The upper respiratory system is comprised of the nose and paranasal sinuses; the mouth and pharynx; the larynx, or voice box; the trachea, or windpipe; and the primary bronchi.

Nose and paranasal sinuses   Ordinarily, inspired air enters and expired air leaves the respiratory tract through the nose. The nose and its relationships to other portions of the upper respiratory tract are diagramed in Fig. 28-1.

As the inspired air passes through the nostrils, or *external nares,* any large particles are usually trapped by small hairs which line the vestibular region of the nasal cavity. Next, the air enters the main portion of the nasal cavity, rushing past the nasal mucosa and

its pseudostratified ciliated columnar epithelial lining capped by a layer of mucus. Secreted by goblet cells in the epithelial lining, the mucus traps many of the finer particles carried passively in the air current. The beating cilia continually propel the overlying mucus toward the throat, where it is swallowed. Because mucus is moist, it also adds water vapor to the inspired air. Called *humidification,* this process will be continued as the air passes through the airways to its alveolar destination in the substance of the lungs.

Following partial cleansing and humidification, the air strikes against the superior, middle, and inferior conchae (see Fig. 28-1), coiled bones covered by nasal mucosa on each side of the nasal septum. These bones disrupt the smooth flow of the inspired air, setting up turbulence called eddy currents. The eddy currents reduce the velocity of the airstream, thereby promoting both humidification and a change in the temperature of the air to that of the body. If the air temperature is initially cooler than that of the body, then the air is warmed as it is temporarily trapped in the whorls of the conchae. The temperature of the air is reduced under the opposite conditions.

Finally, some of the inspired air strikes the nasal epithelium above the superior nasal conchae. The presence of odorous materials is detected by specialized olfactory hair cells in the epithelium (see Fig. 11-13). These chemoreceptors convert the chemical stimulation into electric signals which are conducted to the brain by olfactory neurons. The nature and intensity of the odorous stimulation are interpreted in the olfactory cortex. The olfactory sense is explained in detail in Chap. 11.

Thus, by the time inspired air reaches the posterior portion of the nasal cavity, it has been partially cleansed, moistened, warmed or cooled as required, and sampled for the presence of odorous materials.

Paranasal sinuses in the frontal, maxillary, ethmoid, and sphenoid bones (see Fig. 28-1) communicate through small ducts with the nasal cavity. Because their mucous membranes are continuous with the lining of the nasal cavity, an inflammation of the nasal mucosa is likely to involve the paranasal sinuses as well. The paranasal sinuses and the bony structure of the nose are considered further in Chap. 23.

**Figure 28-1** *Sagittal section through the face and neck, showing the respiratory cavities and airways.* (From Melloni, Stone, and Hurd.)

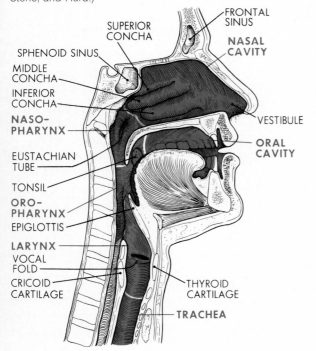

Mouth and pharynx    Air can also be inspired and expired through the mouth, since the oral and nasal cavities communicate with each other in the region of the pharynx. The pharyngeal region is demarcated in Fig. 28-1.

The *pharynx,* or throat, is a chamber which communicates with both the esophagus and larynx. Therefore, for air to reach the larynx, it must pass through the pharynx. Based on anatomic position, the pharynx is divided into two regions—oropharynx and nasopharynx.

The *nasopharynx* lies in the posterior portion of the nasal cavity, superior to the soft palate. The posterior portion of the nasopharynx contains lymphatic tissue called the *pharyngeal tonsil,* or *adenoids.* Because of their strategic position in the respiratory tract, the adenoids may become inflamed by foreign materials screened from the inspired air. The inferior orifice of the eustachian tube opens on the lateral wall on each side of the nasopharynx. The nasopharynx is lined by the same type of epithelium which covers the nasal mucosa.

Beginning in the posterior portion of the oral cavity below the soft palate, the *oropharynx* merges with the nasopharynx at a level corresponding to the base of the tongue. Clearly seen when the mouth is opened widely, the *palatine tonsils* or, simply, tonsils reside between arches in the lateral walls in the first portion of the oropharynx. These lymph glands, like their counterparts in the nasopharynx, are vulnerable to inflammation, since they filter out foreign particles carried in food or air entering through the mouth.

As mentioned earlier, the inferior portion of the pharynx leads into the esophagus posteriorly and the larynx anteriorly.  The epiglottis closes during swallowing, thereby preventing solids and liquids from entering the larynx. The mechanics of swallowing are discussed in Chap. 26.

Airways    The airways of the upper respiratory tract include the larynx, trachea, and primary bronchi. Figure 28-1 shows the structure of the larynx from lateral view. The gross anatomy of the larynx, trachea, and bronchi are illustrated from anterior aspect in Fig. 28-2.

*Larynx*    The inferior aspect of the pharynx leads directly into the larynx. The larynx is formed of nine cartilages supported by ligaments or membranes and contains both an intrinsic and extrinsic musculature. The hyoid bone, which supports the tongue, is associated with the superior portion of the larynx. The base of the larynx leads directly into the trachea.

Three of the nine laryngeal cartilages are especially prominent: (1) the *thyroid cartilage,* the largest of the laryngeal cartilages and more massive in males than in females, forms a prominence in the anterior region of the neck commonly called an Adam's apple; (2) the *cricoid cartilage* forms a ring-shaped support which encircles the lower portion of the larynx; and (3) the *epiglottis,* one end of which is attached to the thyroid cartilage, lies suspended over the glottal opening into the larynx.

The inside lining of the larynx consists of an epithelium similar to that which covers the mucosa of the nasal cavity and nasopharynx. This epithelium is thrown into folds on the inside lateral walls of the larynx, immediately below the glottis. These structures are called *vocal folds,* or false vocal cords. Between the vocal folds and in line with the glottis are fibroelastic bands known as *true vocal cords,* or *folds.* The true vocal cords participate in phonation.

The *intrinsic laryngeal musculature* is associated with cartilages which regulate the diameter of the glottis and move the vocal folds. Accordingly, the intrinsic musculature is responsible for opening and closing the glottis and for controlling the tension of the true vocal cords. Both mechanisms are continually adjusted during phonation.

Sounds are produced when air is expired through the larynx and resonates against bones in the pharyngeal, oral, and nasal cavities. The pitch, or *frequency,* of the sounds produced is determined by the width of the glottal opening and the tension exerted on the vocal cords. The narrower the glottal slit and the greater the tension on the vocal cords, the higher the frequency of the sounds generated. The amplitude, or *intensity*, of the sounds created is directly proportional to the volume of air passed through the larynx per unit of time. The *quality,* or timbre, of the

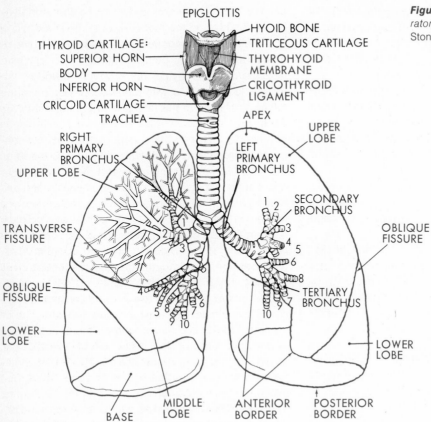

EPIGLOTTIS
HYOID BONE
THYROID CARTILAGE:
TRITICEOUS CARTILAGE
SUPERIOR HORN
THYROHYOID MEMBRANE
BODY
INFERIOR HORN
CRICOTHYROID LIGAMENT
CRICOID CARTILAGE
TRACHEA
APEX
UPPER LOBE
RIGHT PRIMARY BRONCHUS
LEFT PRIMARY BRONCHUS
UPPER LOBE
SECONDARY BRONCHUS
TRANSVERSE FISSURE
OBLIQUE FISSURE
OBLIQUE FISSURE
TERTIARY BRONCHUS
LOWER LOBE
LOWER LOBE
BASE
MIDDLE LOBE
ANTERIOR BORDER
POSTERIOR BORDER

**Figure 28-2** *Anterior view of the respiratory tract and lungs.* (From Melloni, Stone, and Hurd.)

sounds elicited is determined by the nature of the resonating chambers in the cervical and facial regions.

*Trachea* The trachea is a cylindrical tube about 5 in. long and almost 1 in. in diameter which connects the larynx with the primary bronchi that enter the lungs (see Fig. 28-2). The superior end of the trachea is located at the level of the sixth cervical vertebra; its base is in line with the fifth thoracic vertebra.

The walls of the trachea are composed of fibroelastic connective tissue. The walls are prevented from collapsing by a series of 16 to 20 C-shaped bands of hyaline cartilage which straddle the outside of the trachea. The cartilaginous supports end before reaching the posterior wall, thereby accommodating the anterior portion of the esophageal wall. Due to this region of contact, the unsupported posterior wall of

the trachea is depressed slightly. The tracheal mucosa next to the lumen is lined by a pseudostratified ciliated columnar epithelium identical to that which lines all upper portions of the respiratory tract.

The respiratory epithelium functions as an air cleaner. The goblet cells in the respiratory epithelium secrete mucus. The mucus traps foreign particles carried in the inspired air. Together with its contaminants, the mucus is transported by the beating cilia of the epithelial cells at a speed of about 1 in./min to the throat, where it is swallowed. The transportation provided by the constantly moving platform of mucus confers the name "respiratory escalator" to this mechanism.

*Primary bronchi* The right and left primary bronchi originate at the inferior end of the trachea and enter the lungs (see Fig. 28-2). Since the right primary bronchus is shorter, wider, and more vertical than its counterpart on the left side, any microbial pathogens that manage to enter the airways have a greater probability of reaching the right lung than the left lung.

The basic structure and tissue organization of the primary bronchi are the same as those of the trachea. However, the cartilaginous supports of the primary bronchi are O-shaped, or complete. The diameters of the primary bronchi also taper from their origins at the end of the trachea to their terminations within the lungs.

## LOWER RESPIRATORY STRUCTURES

The lower respiratory structures include the lungs and the airways within them. These airways consist of the following regions: (1) secondary bronchi; (2) segmental bronchi; (3) terminal bronchioles; (4) respiratory bronchioles; (5) alveolar ducts; and (6) alveolar sacs containing alveoli.

Lungs    The lungs are the organs in which gas exchange occurs between the outside air and the blood. The lungs are illustrated from anterior aspect in Fig. 28-2.

*Location*    The lungs are located within the rib cage, on either side of the heart. The superior apexes of the lungs are located at or slightly above the level of the clavicles. The bases of the lungs lie at a level corresponding to the origin of the eleventh pair of ribs. The bases of the lungs curve upward from posterior to anterior. The anterior margins of the bases are in line with the sternal borders of the eighth pair of ribs.

*Gross anatomy*    The lungs are cone-shaped, or pyramidal. The internal airways and rich vascularization are responsible for the sponginess and reddish color of these organs. Called the *apex,* the superior aspect of each lung is tapered. The inferior aspect of each lung is flattened and referred to as the *base.* At all times, the bases are in contact with the superior surface of the transverse diaphragm and the lateral surfaces lie next to the inside walls of the rib cage.

The *right lung* is larger than the left lung and con-

tains *three lobes—upper, middle,* and *lower.* The upper and middle lobes are separated externally by a *transverse fissure* and the middle and lower lobes are divided by an *oblique fissure.*

The *left lung* is composed of only *two lobes—upper* and *lower.* An oblique fissure separates the upper lobe from the lower lobe. The position of the ventricles of the heart is defined by the pulmonary region below the anterior border.

The outside surface of each lung is covered by a membrane called *visceral pleura* and the inside surface of the rib cage is covered by *parietal pleura.* The positions of the parietal and visceral pleurae are shown in thoracic cross section in Fig. 28-3. These membranes are identical to the visceral and parietal peritoneum (see Chap. 26). A space about 0.02 in. thick separates the two pleural membranes. Called the *pleural cavity,* this space contains serous fluid. The thin layer of fluid creates a surface tension between the parietal and visceral pleurae which prevents them from separating from each other during inspiration and expiration. The fluid is also a lubricant, preventing abrasions of the lungs as their surfaces slide against the inside walls of the rib cage during each respiratory cycle.

The parietal and visceral pleurae form vertical partitions on the medial aspect of each lung between the thoracic vertebrae and the sternum. The thoracic space defined by these partitions is called the *mediastinum.* The relationship of the mediastinum to the lungs is diagramed in Fig. 28-3. The relationships of the heart and other thoracic structures to the mediastinum are described in Chap. 30. The mediastinal partitions are absent only at the point where the primary bronchi enter the lungs. These medial regions are termed the *roots* of the lungs. A vertical slit, or *hilus,* at each lung root provides a pulmonary entrance or exit for the primary bronchi as well as blood vessels, lymphatics, and nerves.

*The bronchial tree*    The bronchial tree is a highly ramifying group of airways which begins with the primary bronchus that enters a lung and terminates blindly in air sacs called *alveoli* deep within the substance of the lung. The trunk of each bronchial tree is represented by a primary bronchus. The leaves are

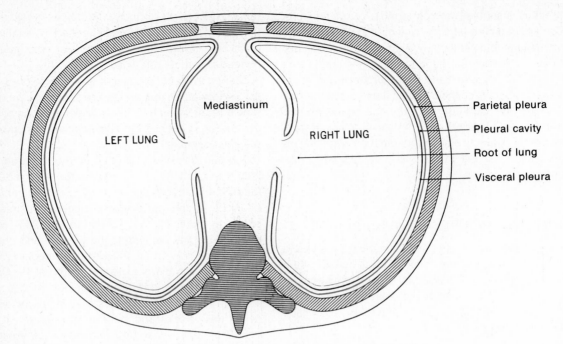

**Figure 28-3** *Cross section through the midthorax viewed from superior aspect, showing the relationship of the lungs to the mediastinum.*

represented by the alveoli. Because the airways between the primary bronchus and alveoli progressively decrease in diameter, they are analogous to the branches and twigs of a tree.

Within the lung, the primary bronchus divides, sending a branch to each lobe. These branches are called *secondary*, or *lobar, bronchi*. There are three secondary bronchi in the right lung and two in the left lung. Each of the secondary bronchi in turn subdivides, giving off 2 to 4 branches apiece. These branches are referred to as *tertiary*, or *segmental, bronchi*. There are approximately 10 tertiary bronchi in each lung. Each tertiary bronchus supplies 1 of the 10 bronchopulmonary segments into which each lung is divided internally.

The walls of the secondary and tertiary bronchi are supported by incomplete cartilaginous plates, rather than by O-shaped or C-shaped cartilaginous supports. Beginning with the subdivisions of the tertiary bronchi, the nature of the airway wall changes. The

cartilaginous plaques disappear altogether as the walls acquire a smooth musculature interlaced with elastic and collagenous connective tissue fibers. The epithelial lining in this portion of the bronchial tree consists of pseudostratified ciliated columnar cells, although goblet cells are in reduced numbers or completely absent.

Each tertiary bronchus gives rise to 50 to 80 branches called *terminal bronchioles*. The epithelium gradually changes from columnar to cuboidal in this portion of the bronchial tree. A terminal bronchiole and its subdivisions are shown diagrammatically in Fig. 28-4. A small number of *respiratory bronchioles* branch from each terminal bronchiole. Notice the anastomosing bands of smooth muscle and interlaced elastic fibers in the walls of the terminal and respiratory bronchioles in Fig. 28-4. Although their walls are too thick for gaseous diffusion, a respiratory function is subserved by isolated alveoli along them.

Two or more *alveolar ducts* branch from each respiratory bronchiole. Each alveolar duct terminates in an expansion called an *atrium,* which gives rise to an *alveolar sac* containing many alveoli separated from each other by internal partitions. The walls of the alveoli are composed of thin, squamous epithelium supported by a basement membrane. No other components are found in the walls which compose the alveolar sacs. It is through the squamous epithelium of

the alveolar walls that gas exchange occurs between the outside air and the blood.

There are about a third of a billion alveoli in both lungs. If these alveoli could be removed, sliced in half, and laid flat on a floor, their membranes would cover a surface area of 70 to 80 m² (about the size of a room 27 ft on each side). The rapidity with which respiratory gases are exchanged is related directly to the

**Figure 28-4**  *A diagrammatic view of a pulmonary lobule, showing a respiratory bronchiole, alveolar ducts, atria, and alveolar sacs, along with their associated blood supply.* (From Melloni, Stone, and Hurd.)

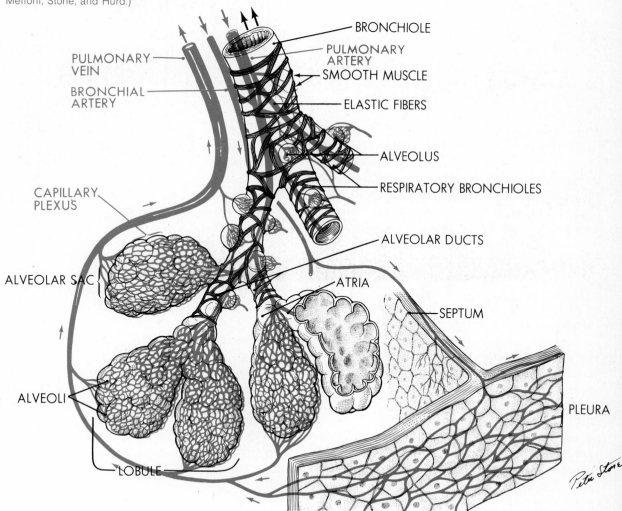

surface area of exchange membrane. Therefore, the tremendous surface area presented by the alveolar walls permits the attainment of an optimal efficiency of gaseous diffusion.

Collectively, the alveoli are referred to as the *functional pulmonic membrane.* The airways which precede these structures are called *dead space,* since gas exchange cannot take place through their walls. Pathologic conditions which decrease the amount of the functional pulmonic membrane are discussed in the next chapter.

Several other cell types are interspersed among the squamous epithelial cells of the alveoli. Probably the most important type consists of cuboidal cells which secrete a lipoprotein *surfactant.* The surfactant forms an extremely thin coating on the inside walls of the alveoli next to the lumen, lowering the surface tension of the alveoli, reducing the force required to inflate them at the end of expiration, and thereby preventing their imminent collapse.

*The blood supply* There are two circulatory pathways between the heart and the lungs: (1) the *pulmonary circuit,* through which unaerated blood from the right ventricle is transported to the lungs, aerated, and then returned to the left atrium of the heart; and (2) the *bronchial circuit,* which is the blood supply for the tissues of the lungs. Although some bronchial venous blood may return to the heart through the pulmonary veins, the pulmonary and bronchial circuits are for the most part separate.

In the pulmonary circuit, unaerated blood in the right ventricle is pumped into the pulmonary artery. The relationship of the pulmonary artery to the heart is illustrated in Fig. 30-6. As much blood is pumped into the pulmonary artery as into the ascending aorta. The pulmonary artery bifurcates into right and left pulmonary arteries and these vessels enter their respective lungs. Within each lung, the pulmonary artery branches extensively. The main arterial branches run in parallel with the airways of the bronchial tree, following the internal segmentation of the lung. The arteries taper into smaller arteries called *arterioles* and finally form networks of tightly woven capillaries, through the walls of which gaseous exchange occurs.

The capillary beds closely surround the outside wall of each alveolus. These vascular branches are shown in Fig. 28-4. An *alveolus* and its surrounding *capillary plexus* make up the functional unit of the lung. It is estimated that each alveolus is surrounded by a plexus of several thousand capillaries and that, collectively, the pulmonary capillaries present a gas exchange surface area at least double that of the 300 million alveoli, i.e., approximately 160 m². Normally, about 15 percent of all the blood in the body is held in the lungs. It ordinarily takes slightly less than 1 sec for blood to pass through a pulmonary capillary. The amounts of gases exchanged during this time are more than adequate to sustain life under most circumstances. The reasons revolve around structural specializations that increase the efficiency of gaseous exchange. First, the compound gas exchange membrane consisting of alveolar squamous cells, alveolar basement membrane, interstitial fluid, capillary basement membrane, and capillary endothelial cells is collectively considerably less than 1 $\mu$m in thickness, enabling the rapid diffusion of respiratory gases. Second, since the pulmonary capillaries are actually slightly smaller in diameter than the erythrocytes which pass through them, these corpuscles are forced during their single-file passage to flex as they rub against the inside walls of the capillaries, thereby permitting $O_2$ uptake directly from the endothelial cells rather than from the plasma. The pulmonary capillaries change imperceptibly into venules which drain into pulmonary veins. Each lobe of the right lung and left lung gives rise to a pulmonary vein. However, the pulmonary veins from the right upper and middle lobes ordinarily merge near their points of origin. Therefore, paired pulmonary veins exit both right and left lungs. The four pulmonary veins carry aerated blood to the left atrium. The relationship of the pulmonary veins to the heart is illustrated in Fig. 30-6.

In the bronchial circuit, the bronchial blood vessels supply blood to, and drain blood from, the pleura and the walls of the bronchial trees up to the respiratory bronchioles. The bronchial arteries branch off the aorta. The lobar branches of the bronchial arteries for the most part accompany the pulmonary arterial branches. Arteriolar branches penetrate the visceral pleura, forming capillary plexuses within its substance. Most of the blood is drained into bronchial venules, although some of this capillary blood finds its way into pulmonary venules, thereby slightly diluting the degree of $O_2$ saturation of the pulmonary venous blood. The bronchial venules empty into bronchial veins which exit the lungs. These relationships are shown in Fig. 28-4. The bronchial veins are drained into the superior vena cava by way of the azygous and hemiazygous veins located on the inside surface of the thoracic wall.

## THE RESPIRATORY CYCLE

Initiated during birth and ceasing permanently at death, respiration is a rhythmic process sustained by a never-ending series of respiratory cycles. A respiratory cycle consists of the events which occur during one complete inspiration and expiration. During *inspiration,* outside air is sucked into the alveoli of the lungs. The alveolar air is forced out of the lungs during *expiration.*

The average frequency of respiration of an adult is 14 respiratory cycles per minute. Almost 2 sec of each respiratory cycle are spent in inspiration; slightly less than 3 sec are spent in expiration.

The respiratory rate and the length of time spent in inspiration, expiration, or between respiratory cycles are affected by many physical, physiologic, and psychological factors. These factors will be considered later in this chapter. An increase in the ventilatory rate is marked by a decrease in the time spent in both inspiration and expiration, while ventilatory depression is characterized primarily by an increase in the time interval between respiratory cycles.

The object of respiration is twofold: (1) to replenish the blood supplies of $O_2$ and (2) to eliminate the accumulations of $CO_2$ in the blood. The efficiency of gas exchange in the lungs and the nature of the diet are reflected by the differences in the inspiratory and expiratory concentrations of these two respiratory gases. The approximate percentage concentrations of $O_2$ and $CO_2$ in both inspired and expired air are given in Table 28-1. The net $O_2$ input and $CO_2$ output

are listed in the right-hand column of the table. Although almost 79 percent of the inspired air is nitrogen ($N_2$), this gas has no known respiratory function. Therefore, the percentage concentration of $N_2$ in the expired air differs little from that in the inspired air.

**Table 28-1** Inspiratory and expiratory differences in $O_2$ and $CO_2$ concentrations

| Respiratory gas | Inspired air, % | Expired air, % | Difference,% |
|---|---|---|---|
| $O_2$ | 20.9 | 15.8 | 5.1 |
| $CO_2$ | 0.04 | 3.8 | 3.76 |

The differences between the inspiratory and expiratory concentrations of $O_2$ and $CO_2$ are instructive. Notice that only about 25 percent of the inspired $O_2$ is ordinarily extracted during one respiratory cycle, whereas the concentration of $CO_2$ in the expired air increases about 80 times beyond that of the inspired air. Furthermore, observe that approximately five parts of $O_2$ are consumed (5.1 percent) for every four parts of $CO_2$ evolved (3.76 percent). The ratio of the $CO_2$ output to the $O_2$ uptake is discussed later in this chapter.

## MECHANICS

Air rushes into the lungs when their volumes increase and is forced out of the lungs when their volumes decrease. The degree of inflation and deflation of the lungs is determined by the following factors: (1) the action of the respiratory muscles; (2) the elasticity of the lungs and the thoracic wall; and (3) the resistance to airflow offered by the airways.

Inspiration  Immediately prior to inspiration, the air pressure in the alveoli, or the *intrapulmonic pressure,* is the same as the pressure of the atmosphere. At 0°C and sea level, the *atmospheric pressure* is equal to 760 mmHg (see Chap. 7). Therefore, the intrapulmonic pressure is also 760 mmHg. The pressure in the pleural cavities immediately surrounding the lungs is about 756 mmHg, or about 4 mmHg less than that of the atmosphere. Because 760 mmHg is used as a standard of reference, the *intrapleural pressure* at this time is expressed as −4 mmHg. The negative pressure in the pleural cavities prevents the alveoli from collapsing despite their large surface tensions. The action of the surfactant described earlier plays a major role in reducing the surface tension created by a decrease in the average diameter of the alveoli.

Air will flow in response to its pressure gradient. In other words, air will move from a region in which it is at a higher pressure to a region in which it is at lower pressure. For outside air to flow into the lungs, the intrapulmonic pressure must be reduced below the pressure of the atmosphere. This function is accomplished by the contraction of the inspiratory muscles which enlarge the volume of the thoracic cavity. Since the lungs adhere to the inside walls of the rib cage, an increase in thoracic volume causes the lungs to stretch. Therefore, the intrapulmonic volume also increases. Because of the inverse relationship between volume and pressure (see Chap. 8), the intrapulmonic pressure decreases to about −3 mmHg (757 mmHg) by the end of normal inspiration. The differential between the atmospheric pressure and the intrapulmonic pressure literally sucks air into the lungs.

The primary inspiratory muscles include the transverse diaphragm and the external intercostals. During forced inspiration, these muscles are augmented by the sternocleidomastoids and pectoralis minors. The respiratory muscles are shown in Chap. 24.

The transverse diaphragm is illustrated in Fig. 28-5. When this compound muscle contracts, it descends, or lowers, in the trunk region, thereby increasing the superoinferior dimensions of the thoracic cavity. Because of the tremendous surface area of the diaphragm, a small change in its position is reflected by a large change in the volume of the thoracic cavity. It is estimated that the diaphragm is responsible for approximately 80 percent of the total increase in volume undergone by the thoracic cavity during normal inspiration. Therefore, the diaphragm is the most important muscular regulator of respiratory volume. Breathing regulated by the diaphragm is called *diaphragmatic respiration*.

The external intercostals are located between the ribs. When these muscles contract, they elevate and

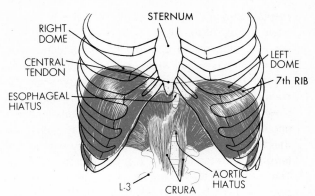

**Figure 28-5** *The transverse diaphragm.* (From Melloni, Stone, and Hurd.)

evert the ribs as well as tilt the sternum, or breastbone, outward, thereby increasing the anteroposterior dimensions of the thoracic cavity. The volume increase due to the action of the external intercostals is estimated at 20 percent of the total increase during inspiration. Breathing regulated by the intercostal muscles is referred to as *costal respiration*.

As the intrapulmonic pressure decreases during inspiration, so does the intrapleural pressure. Because the lungs stretch, the pleural cavities which surround them increase in volume. The volume increase is responsible for the decrease in intrapleural pressure. This pressure normally changes from −4 mmHg prior to the onset of inspiration to about −7 mmHg at the end of inspiration, i.e., from 756 mmHg to 753 mmHg. The increasingly negative intrapleural pressure during inspiration tends to offset the tendency of the lungs to collapse due to the tension exerted by the stretched elastic fibers in the pulmonary tissues.

**Expiration**   At the end of inspiration, the differential in air pressure between the atmosphere and the alveoli is neutralized. Therefore, the atmospheric and intrapulmonic pressures are again equal. To force air out of the lungs, the intrapulmonic pressure must be increased beyond that of the atmosphere. This is accomplished primarily by relaxing the inspiratory muscles. Because the thoracic wall and rib cage have been stretched during inspiration, their rebound decreases the volume of the thoracic cavity. As the dimensions of the thoracic cavity diminish, the lungs elastically recoil from their stretched positions. Since these events are triggered by relaxation of the inspiratory muscles, *expiration* does not require an expenditure of energy and is therefore a passive phenomenon.

However, even quiet expiration is assisted to some slight extent by the action of the expiratory muscles. Illustrated in Chap. 24, these muscles include the internal intercostal and anterior abdominal muscles. The synergistic effects of these muscles are especially prevalent in forced expiration. The action of the internal intercostal muscles is antagonistic to that of the external intercostal muscles, while compression of the abdominal muscles raises the position of the diaphragm beyond that of its normal relaxed state. As a consequence of the normal respiratory activities of these muscles, the intrapulmonic pressure usually rises to +3 mmHg (763 mmHg) by the end of expiration. When the expiratory muscles relax between respiratory cycles, the differential in pressure between the atmosphere and the alveoli is eliminated.

An increase in the intrapulmonic pressure during expiration also increases the intrapleural pressure because deflation of the lungs decreases the volume of the pleural cavities. Therefore, the intrapleural pressure rises from −7 mmHg at the end of inspiration to about −3 mmHg at the end of expiration. Following relaxation of the expiratory muscles, the value decreases to approximately −4 mmHg. This value, you will recall, is the intrapleural pressure immediately preceding the onset of inspiration.

The increase and subsequent decrease in the volume of the thoracic cavity during one respiratory cycle is summarized in Fig. 28-6. Notice that the normal intrapulmonic pressure decreases to −3 mmHg during inspiration and increases to +3 mmHg during expiration. The intrapleural pressure decreases to −7 mmHg and then increases to −4 mmHg during the same time interval.

## NEURAL CONTROL

Three brain centers are involved in the control of respiration—the respiratory center, the apneustic center, and the pneumotaxic center. The respiratory center is influenced either directly or indirectly by central and peripheral chemoreceptors, stretch receptors in the lungs, and stimulation from higher brain centers.

The rhythmicity of the respiratory cycle is controlled predominantly by the *respiratory center* located in the reticular formation of the medulla oblongata. The respiratory center is a loose aggregation of many nerve cell bodies. Some of the neurons are active only during inspiration; others fire only during expiration. The two functional types of neurons seem to be distributed randomly throughout the respiratory center.

Excitation of the inspiratory neurons results in two consequences: (1) the conduction of excitatory nerve impulses to the inspiratory muscles, causing their contraction and a corresponding increase in the volume of the thoracic cavity; and (2) the conduction of inhibitory nerve impulses to the expiratory neurons, preventing them from initiating expiration.

The inspiratory neurons cease firing at the peak of inspiration. This action causes the external intercostal muscles and the transverse diaphragm to relax and simultaneously removes the inhibitory brake on the excitability of the expiratory neurons. The ramifica-

**Figure 28-6**  *The mechanics of respiration. A diagrammatic section of the thorax is shown from lateral aspect.* (From Melloni, Stone, and Hurd.)

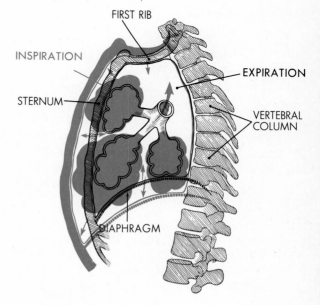

tions of these effects are that (1) the lungs elastically recoil; (2) excitatory nerve impulses are conducted by expiratory neurons to the expiratory muscles, causing their contraction and further decreasing the volume of the thoracic cavity; and (3) inhibitory nerve impulses are conducted to the inspiratory neurons, preventing them from initiating inspiration.

The expiratory neurons become inactive at the height of expiration. This event marks the onset of the inspiratory phase of the following respiratory cycle.

The two other respiratory centers in the reticular formation of the brainstem are located in the pons. Called the *apneustic* and *pneumotaxic centers,* they superimpose their influences on the basic rhythmicity of the respiratory center.

The apneustic and pneumotaxic centers receive collateral nerve fibers from the cerebral cortex and hypothalamus of the diencephalon. These centers also receive afferent vagal nerve fibers from stretch receptors in the lungs and peripheral chemoreceptors in the carotid and aortic bodies. Stimulation of the apneustic center promotes inspiration. Excitation of the pneumotaxic center modifies the rate of respiration. The normal rhythmicity of the respiratory cycle is also dependent upon the integrity of the pneumotaxic center.

## FACTORS WHICH MODIFY THE RESPIRATORY CYCLE

Although the rhythmicity of respiration is established primarily by neural interactions within the respiratory center of the medulla oblongata, numerous exogenous factors are capable of modifying the frequency and depth of respiration. Among these modifying influences are the concentrations of $CO_2$, $H^+$, and $O_2$ in blood, the degree of stretch of the lungs, the nature of the electric signals from the motor cortex and the hypothalamus, and the presence and degree of skeletal muscular activity.

Changes in blood $P_{CO_2}$ The partial pressure of $CO_2$ in the blood ($P_{CO_2}$) is by far the most important natural chemical stimulus to affect the respiratory rate. When the $P_{CO_2}$ of blood increases excessively, the rate of

respiration undergoes a compensatory increase which blows off the excess $CO_2$. The respiratory rate decreases under the opposite conditions.

The concentration of blood $CO_2$ acts both directly and indirectly on the pneumotaxic, apneustic, and respiratory centers in the brainstem. Blood $CO_2$ may stimulate the respiratory center directly. There is also evidence to indicate that $CO_2$, after readily crossing the blood-brain barrier and entering the cerebrospinal fluid, stimulates chemoreceptors located on the lateral aspects of the medulla oblongata. Peripheral chemoreceptors sensitive to $CO_2$ are located in the *carotid body* at the junction between the internal and external carotid arteries and the *aortic bodies* in the arch of the aorta. These chemoreceptors are close to the pressoreceptors which regulate blood pressure (see Chap. 30). Excitation of the peripheral chemoreceptors results in the conduction of nerve impulses through sensory fibers of the vagus nerve (Xth cranial nerve) to the respiratory center.

Changes in blood $[H^+]$ The presence of an excessive concentration of $CO_2$ in blood decreases the blood pH by increasing its hydrogen ion concentration ($[H^+]$). An elevation of the $[H^+]$ causes an increase in the ventilatory rate, even when the $P_{CO_2}$ of blood is held constant. $H^+$ forms through a chemical reaction in which $CO_2$ combines with $H_2O$ mostly within erythrocytes, forming carbonic acid ($H_2CO_3$). A small portion of the $H_2CO_3$ dissociates, giving rise to $H^+$ and bicarbonate ion ($HCO_3^-$). This erythrocytic reaction is discussed in detail in Chap. 30. The meaning of pH and significance of dissociation reactions are considered in Chap. 10.

An increase in blood $[H^+]$ causes an elevation in respiratory rate, and vice versa. Under normal conditions, however, the effects are much less dramatic than those caused by a change in the $P_{CO_2}$ of blood. In any event, it should be clear that an increase in the $P_{CO_2}$ of blood will also increase its $[H^+]$. Therefore, an elevation in the former will stimulate chemoreceptors sensitive to both $CO_2$ and $H^+$.

Changes in blood $P_{O_2}$ Ironically, the $P_{O_2}$ of blood is the least important natural chemical stimulus which influences respiration, although its effects are measurable. A depression in the $P_{O_2}$ of blood elevates the respiratory rate, thereby increasing the supply of alveolar $O_2$. The rate of ventilation is decreased under the opposite conditions.

Practically all the respiratory effects of $O_2$ are mediated by the peripheral chemoreceptors in the carotid and aortic bodies. Little if any influence is exerted centrally. The reason that the role of $O_2$ in the modification of the respiratory cycle is minor has to do with the chronically high degree of $O_2$ saturation of hemoglobin, even under conditions of temporary $O_2$ deprivation. The amount of $O_2$ still bound to hemoglobin under various conditions is considered in Chap. 30.

Degree of lung stretch Stretch receptors associated with the lungs are capable of altering the respiratory rate by inhibiting either the inspiratory or expiratory neurons in the respiratory center. These receptors are located in the visceral pleura and in the walls of the bronchial tree. The increasing excitation of the stretch receptors as the lungs reach optimal inflation results in the inhibition of the inspiratory neurons. These neurons are linked to the stretch receptors by afferent fibers of the vagus nerve. Deflation of the lungs reduces excitation of the stretch receptors, thereby removing their inhibitory influence on the inspiratory neurons. This rhythmic respiratory control mechanism is called the *Hering-Breuer reflex*.

Input from higher brain centers Stimulation from the limbic cortex and the hypothalamic region of the diencephalon can also modify the rate of respiration. Projection fibers from these higher brain regions innervate the pneumotaxic and apneustic centers in the brainstem. The limbic cortex regulates emotions such as fright, anger, and excitement. Fear, such as is created by the possibility of imminent disaster, for example, inhibits the apneustic center, causing cessation of respiration at the peak of inspiration. Anger, on the other hand, causes shallow, rapid breathing through the influence of the limbic cortex on the pneumotaxic center. The limbic cortex is discussed further in Chap. 14.

The hypothalamus also influences the respiratory

center through its effects on the pneumotaxic center. One of the numerous functions of the hypothalamus is to sense and regulate the internal temperature of the body. Respiration is reflexly modified when body temperature changes abruptly. That one's "breath is taken away" when stepping into an ice-cold shower is a dramatic illustration of this fact.

The higher brain centers are involved in many other aspects of the reflex and volitional control of respiration. Respiration temporarily stops while swallowing, speaking, singing, whistling, blowing, coughing, sneezing, and vomiting. Breathing can also be arrested voluntarily at any stage of the respiratory cycle. The rate of ventilation is also affected by other sensory stimuli, such as pain or even tactile stimulation.

Initiation of physical exercise The respiratory rate increases markedly within a few seconds after initiating vigorous muscular activity. The effect of physical activity on the rate of respiration occurs in advance of appreciable changes in the $P_{CO_2}$, $P_{O_2}$, or $[H^+]$ of blood. Therefore, we must look at neural, rather than chemical, factors when attempting to explain this rapid increase in the rate of ventilation.

Two neural influences on the respiratory centers in the brainstem are presumably involved: (1) excitation of the pneumotaxic center by the motor cortex via collateral fibers, while initiating skeletal muscular activity; and (2) stimulation of the muscular, tendinous, and ligamentous proprioceptors in the regions of the body which are moving is believed to indirectly excite the pneumotaxic center.

## RESPIRATORY VOLUMES AND CAPACITIES

It takes work to breathe. Indeed, about 7 percent of the total energy in the body is normally expended by the respiratory muscles during inspiration. Respiratory energy is utilized for the following purposes: (1) to stretch the chest wall and increase the diameter of the thoracic cavity; (2) to overcome the viscous resistance of the lungs to stretch; and (3) to overcome the resistance offered by the airways to the flow of air. The ease with which this work is accomplished is known as *pulmonary compliance*. The work of breathing, or the pulmonary compliance, can be determined by measuring the volumes of air passed into and out of the lungs under specified conditions.

Both lungs of an average adult are capable of holding a maximum of 6 liters (6,000 ml) of air. This maximum volume is called the *total lung capacity*. Ordinarily, however, the volume of air in the lungs at the end of inspiration is only half the total lung capacity. The total lung capacity is divided into several volumes and capacities based on inspiratory and expiratory abilities.

The average volume of air passed into or out of the lungs during a respiratory cycle is called the *tidal volume* and is equal to about 500 ml of air. Almost one-third of the tidal volume does not reach the alveoli during respiration because some air at the end of inspiration remains in the airways which precede the terminal bronchioles. Gas exchange cannot take place through the walls of these airways.

The volume of air which can be forcibly inspired following a normal inspiration is termed the *inspiratory reserve volume*. The inspiratory reserve volume approximates 3,000 ml of air. When the inspiratory reserve volume is coupled with the tidal volume, the total volume (3,500 ml) is referred to as the *inspiratory capacity*. The inspiratory capacity represents the maximum amount of air which can be inspired during a respiratory cycle.

The amount of air which can be forcibly expired after a normal expiration is called the *expiratory reserve volume*. The expiratory reserve volume is about 1,000 ml of air, or only one-third of the inspiratory reserve volume.

Following forced expiration, there is still approximately 1,500 ml of air within the respiratory system. This quantity of air is known as the *residual volume.* When the residual volume is added to the *expiratory reserve volume,* the total volume (2,500 ml) represents the amount of air in the lungs after normal expiration. This quantity of air is called the *functional residual capacity.*

The *vital capacity* is the quantity of air which can be expired forcibly following a forced inspiration. Therefore, the vital capacity includes the tidal volume plus inspiratory and expiratory reserve volumes. These combined volumes measure in the vicinity of 4,500 ml of air. Since the vital capacity depends upon the maximum elasticity and compressibility of the lungs, rib cage, and thoracic wall, it is an excellent measure of pulmonary compliance.

The respiratory volumes and capacities cited above are interrelated in Fig. 28-7.

The quantitative values of the vital capacity and other measures of pulmonary compliance are dependent upon the size of the person, the strength of the respiratory muscles, and the elasticity of the lungs. Since adult males are usually taller in stature and physically stronger than females of comparable age, their vital capacities are also correspondingly larger. The vital capacity decreases with age, due to age-specific changes in the nature of the collagenous connective tissue in the lungs and elsewhere in the body (see Chap. 4). The respiratory volumes and capacities are influenced by a variety of physical, chemical, and psychological factors which affect the diameter of the airways, the total amount of the functional pulmonic membrane, and the distensibility of the lungs, rib cage, and thoracic wall.

## ALTERATIONS IN RESPIRATORY RHYTHM

The normal respiratory rhythm is called *eupnea*. Eupnea can undergo the following modifications: (1) an increase in rate (*hyperpnea*); (2) an increase in both rate and depth (*hyperventilation*); (3) gasping, or labored breathing (*dyspnea*); and (4) a temporary cessation of breathing (*apnea*). Hyperpnea and hyperventilation are often used to denote an increase in the rate as well as the depth of respiration, so that the distinction is frequently blurred.

A pneumographic recording of hyperventilation is shown in Fig. 28-8A. The recording shows a relative increase in both rate and depth of respiration. Vigorous physical exercise is initiated at the arrow pointing upward. During the brief burst of activity shown, the rate is increased approximately $1\frac{1}{2}$ times normal and the depth almost doubles the resting value.

If hyperventilation persists, then increasing amounts of $CO_2$ are expelled from the blood to the atmosphere, producing a condition called *acapnia,* or

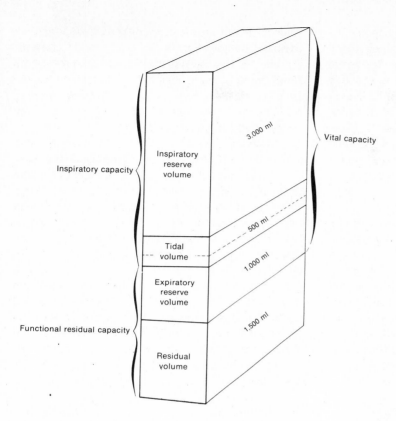

**Figure** *28-7* *Pulmonary volumes and capacities.*

Inspiratory
reserve
volume

3,000 ml

Vital capacity

Inspiratory capacity

500 ml

Tidal
volume

1,000 ml

Expiratory
reserve
volume

1,500 ml

Functional residual capacity

Residual
volume

**Figure 28-8**   *Graphic records of modifications in the normal respiratory rhythm.* A. *Hyper-*
*ventilation.* B. *Apnea.* C. *Dyspnea.* .

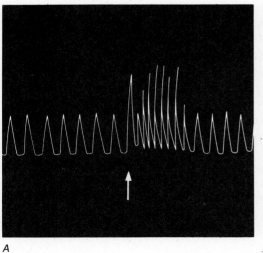

*A*

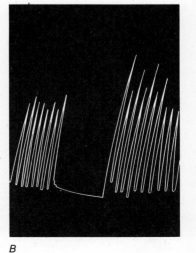

*B*

*C*

more correctly, *hypocapnia*. The significantly reduced concentrations of plasma $CO_2$ during hypocapnia are too low to trigger the inspiratory neurons of the medulla oblongata, and respiration temporarily ceases. A pneumographic recording of apnea due to hypocapnia is shown in Fig. 28-8B. Respiration resumes when concentrations of plasma $CO_2$ accumulating from normal metabolism again become sufficient to stimulate the inspiratory neurons. Notice from the recording that respiration increases in depth while the $O_2$ debt is being repaid immediately following apnea.

A graphic record of dyspnea is displayed in Fig. 28-8C. Notice that the rate and depth of respiration are irregular during labored breathing.

Hyperpnea and hyperventilation can be induced by physical exertion, which increases the tissue demands for $O_2$, psychogenic factors such as excitement and fright, or stimuli such as heat, cold, and pain because these stimuli mobilize the pituitary-adrenal axis which mediates responses to stress (see Chap. 4). By contrast, apnea is caused by hypocapnia, as described above, or by the cessation of excitation of the inspiratory neurons when the breath is withheld voluntarily. Intermittent, brief periods of apnea may also occur normally during deep sleep, especially in very young or old people.

## GAS TRANSPORT AND EXCHANGE

### OXYGEN

There are two modes of $O_2$ transport through blood: (1) about 3 percent of the total $O_2$ in blood is dissolved physically in plasma; and (2) the remainder is in chemical combination with hemoglobin (Hb) in the erythrocytes. Since 97 percent of all the $O_2$ in blood is bound to Hb molecules, in the following discussion we can justifiably disregard the small amount of $O_2$ dissolved physically in solution.

How much $O_2$ is a unit volume of blood capable of transporting? Each gram of hemoglobin is capable of binding 1.34 ml of $O_2$, and there are approximately 15 Gm of Hb/100 ml of whole blood. Therefore, a maximum of about 20 ml of $O_2$ can be transported in chemical combination in 100 ml of blood (1.34 ml $O_2$/Gm $\times$ 15 Gm Hb/100 ml of blood). Expressed as 20 vol percent, this concentration represents 100 percent, or complete, saturation of Hb with $O_2$. When Hb is combined with $O_2$, it is called *oxyhemoglobin* and abbreviated $HbO_2^-$.

The pressure exerted by $O_2$ in completely aerated arterial blood is about 100 mmHg. Because this pressure is created only by the concentration of $O_2$ and does not include the pressure contributions of other blood gases, it is called a partial pressure. In other words, blood containing 20 vol percent $O_2$ has a partial pressure of $O_2$ ($P_{O_2}$) of 100 mmHg. Factors which determine gas pressures are reviewed in detail in Chap. 7.

Oxygen is unloaded from $HbO_2^-$ in the blood capillaries which surround the tissue cells of the body. The walls of arteries and veins are too thick for an appreciable amount of $O_2$ diffusion to occur through them. Oxygen therefore diffuses through the thin walls of the blood capillaries to the tissue cells in response to its concentration gradient. Since the pressure of a gas is proportional to its concentration, provided that gas temperature and volume are held constant, the concentration gradient gives rise to a pressure gradient. Stated in another way, the $P_{O_2}$ in the blood capillaries is greater than that in the tissue cells, resulting in the differential movement of $O_2$ from the region in which it exerts a higher pressure (blood) to the region in which it exerts a lower pressure (tissues).

When a person is resting, blood entering the venules from the capillaries exerts a $P_{O_2}$ of 40 mmHg, even though its Hb is still 75 percent saturated with $O_2$. In other words, a reduction in the $P_{O_2}$ of blood by more than half reduces its total volume of $O_2$ by only one-fourth. The reasons are related to the nature of the chemical combination between Hb and $O_2$ and the characteristics of erythrocytic metabolism (see Chap. 30). The important point to remember is that there are still 15 ml $O_2$/100 ml, or 15 vol percent $O_2$, left in venous blood.

The difference in the $O_2$ content of saturated arterial blood (20 vol percent) and venous blood (15 vol percent), or 5 vol percent, represents the $O_2$ consumption by the tissues during one tour of blood through the body. During rest, this $O_2$ consumption approximates 5 ml $O_2$/100 ml whole blood.

With vigorous physical exercise, however, the $O_2$ consumption of the tissues increases markedly. The heightened metabolic demands for $O_2$ can reduce the $P_{O_2}$ of venous blood to as low as 15 mmHg. Nevertheless, blood with a $P_{O_2}$ of 15 mmHg is still approximately 25 percent saturated with this gas and therefore contains 5 vol percent $O_2$ (5 ml $O_2$/100 ml whole blood).

Changes in the $O_2$ content of circulating blood are plotted graphically in Fig. 28-9. Called the $O_2$ *dissociation curve of Hb,* the graph is read from upper right to lower left. The upper and lower horizontal lines on

the left reflect the $P_{O_2}$ of venous blood during rest and exercise, respectively.

## CARBON DIOXIDE

Carbon dioxide is carried in the blood by three different methods: (1) about 10 percent of the total $CO_2$ in blood is physically dissolved in plasma; (2) approximately 15 percent of the total $CO_2$ is combined with Hb as carbhemoglobin (carbaminohemoglobin, or $HbCO_2$); and (3) the remainder, and the most important mode of $CO_2$ transport, is in the form of the bicarbonate ion ($HCO_3^-$). Appreciable quantities of $CO_2$ are transported by each of these three mechanisms, and none can be ignored when interpreting changes in the $CO_2$ content of circulating blood.

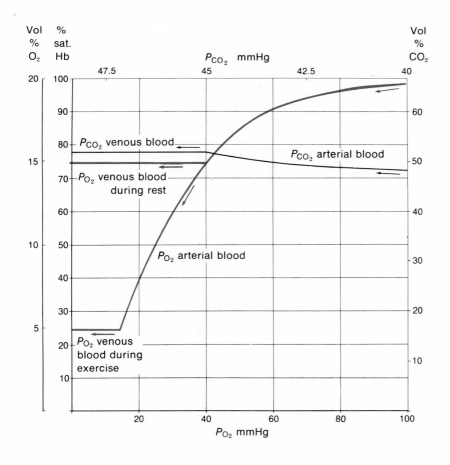

**Figure 28-9** *Changes in the concentrations of $O_2$ and $CO_2$ in circulating blood. The $O_2$ dissociation curve of Hb is in color. Its scales are on the left and bottom of the graph. The $CO_2$ dissociation curve for whole blood is in black. Its scales are on the top and right of the graph.*

Before examining the nature of these changes, we must appreciate how $HCO_3^-$ forms from $CO_2$. Carbon dioxide interacts chemically with $H_2O$, giving rise to $H_2CO_3$. A small fraction of the carbonic acid dissociates into $H^+$ and $HCO_3^-$. The entire reaction is summarized as follows:

$$H_2O + CO_2 \rightarrow H_2CO_3 \rightarrow H^+ + HCO_3^- \qquad (28\text{-}1)$$

Dissociation reactions are discussed in Chap. 10.

Equation (28-1) occurs in both the plasma and the erythrocytic portions of blood. However, the reaction occurs thousands of times more rapidly in erythrocytes, compared to plasma, because of the presence of an erythrocytic enzyme called *carbonic anhydrase*. Since the erythrocytes generate most of the $HCO_3^-$ in response to an excess of blood $CO_2$, $HCO_3^-$ diffuses into the plasma. As more and more $HCO_3^-$ enters the plasma, the electronegativity of the plasma increases correspondingly. To preserve electrical neutrality between the plasma and the erythrocytes, plasma chloride ($Cl^-$) diffuses into the erythrocytes.

The pressure exerted by $CO_2$ in arterial blood is approximately 40 mmHg. This $P_{CO_2}$ corresponds to 48 vol percent (48 ml $CO_2$/100 ml whole blood).

When the arterial blood enters the blood capillaries, it takes up the $CO_2$ generated metabolically. Carbon dioxide diffuses from the tissue cells to the blood capillaries because the $P_{CO_2}$ of the former is higher than that of the latter.

As a consequence of the capillary uptake of $CO_2$, blood entering the venules possesses a $P_{CO_2}$ of 45 mmHg under normal conditions. The rise in the $P_{CO_2}$ from 40 to 45 mmHg is reflected by an increase in the $CO_2$ content of blood from 48 to 52 vol percent.

Therefore, each 100 ml of venous blood returning to the right atrium contains 52 ml of $CO_2$.

The difference in the $CO_2$ content of venous blood (52 vol percent) and arterial blood (48 vol percent), or 4 vol percent, represents the uptake of $CO_2$ by the blood during one complete circuit through the body. This $CO_2$ uptake normally approximates 4 ml/100 ml whole blood.

Changes in the normal $CO_2$ content of circulating blood are plotted graphically in Fig. 28-9. Called the *$CO_2$ dissociation curve* of whole blood, the graph is read from lower right to upper left.

## RESPIRATORY QUOTIENT

A comparison of the dissociation curves for $CO_2$ and $O_2$ reveals that 4 vol percent $CO_2$ are taken up by the blood for every 5 vol percent $O_2$ unloaded by Hb in the tissue capillaries. The ratio of the $CO_2$ produced to the $O_2$ consumed in the same period of time is called the *respiratory quotient.* This ratio is written as follows:

$$\text{Respiratory quotient (RQ)} = \frac{CO_2 \text{ produced}}{O_2 \text{ consumed}} \qquad (28\text{-}2)$$

The RQ ordinarily ranges between 0.7 and 1.0 and depends upon the nature of the metabolic fuels being oxidized. When the diet is rich in carbohydrates, the RQ is very close to 1.0 because one part of $CO_2$ is evolved for each part of $O_2$ utilized. Conversely, the RQ approaches 0.7 when lipids are used as the primary energy source because the amount of $CO_2$ liberated is less than the concentration of $O_2$ consumed. On a mixed diet consisting of protein, lipid, and carbohydrate, the RQ is about 0.83.

## COLLATERAL READING

Clements, John A., "Surface Tension in the Lungs," *Scientific American,* December 1962. A surfactant containing lecithin promotes an even distribution of pressure between alveoli of different diameters and prevents fluids from accumulating in them, according to Clements.

Comroe, Julieu H., Jr., "The Lung," *Scientific American,* February 1966. An excellent review of the basic structure and functions of the respiratory system.

Lippold, Olaf, *Human Respiration: A Programmed Course,* San Francisco: W. H. Freeman and Company, 1968. Reviews the basic anatomy and physiology of the respiratory system.

Mead, Jere, and Harry Martin, "Principles of Respiratory Mechanics," in H. J. Hislop and J. O. Sanger (eds.), *Chest Disorders of Children, Proceedings of a Symposium,* American Physical Therapy Association, 1968. A detailed explanation of basic respiratory mechanics, emphasizing the elastic and flow-resistive properties of the respiratory system and associated structures.

Nye, Robert E., Jr., "The Control and Distribution of Ventilation," in H. J. Hislop and J. O. Sanger (eds.), *Chest Disorders of Children, Proceedings of a Symposium,* American Physical Therapy Association, 1968. The author compares the respiratory system to an ideal control system whose tasks are to maintain an $O_2$ supply to the tissues, remove $CO_2$ from the tissues, and adjust other physiologic activities to achieve these goals with optimum efficiency.

# CHAPTER 29
# respiratory diseases

Respiratory diseases impair the ability to secure, transport, or utilize $O_2$, or to excrete $CO_2$. Respiratory disorders may therefore affect either external or internal respiration.

More specifically, respiratory disorders are induced by one or more of the following factors:

1 Significant alterations in gas tensions in the inspiratory air, due to an excess or deficit of $O_2$, $CO_2$, or $N_2$ in the atmosphere

2 An inability to adequately ventilate the lungs because of airway obstruction, chest wall stiffening, or muscle paralysis

3 A marked decrease in the efficiency of gaseous exchange between the alveoli and pulmonary blood capillaries because of a thickening in, or reduction of, the functional pulmonic membrane

4 An inability to transport adequate amounts of oxygen in blood because of circulatory stagnation or reduction in the effective concentration of erythrocytes or hemoglobin

5 An inability of tissue cells to obtain a sufficient concentration of $O_2$ to maintain aerobic oxidation because energy consumption exceeds the storage and regenerative capacities for ATP

6 An inability of tissue cells to utilize available $O_2$ in cellular respiration because of the presence of metabolic poisons or the absence of essential nutrients

**RESPIRATORY DISORDERS**

COMMON RESPIRATORY DISTURBANCES

Sneezing and coughing Sneezing and coughing are reflex acts induced by irritants which enter the nose and mouth. Irritants which trigger the *sneeze reflex* are inhaled during inspiration. Such particles stimulate sensory nerve endings in the nasal mucosa. The motor impulses are conducted by a branch of the Vth cranial nerve, the trigeminal nerve.

A *sneeze* is initiated by a variable degree of inspiration following irritation. Then the cavity of the throat constricts as the uvula of the soft palate descends, effectively walling off the mouth and providing a relatively unobstructed passageway to the nose. Next, a blast of air is forced from the lungs through the nose. The force of the airstream may eject mucus on which the irritants are trapped. Expiration is controlled by a number of muscles, including the internal intercostal muscles, diaphragm, and abdominal muscles, as you will recall from Chap. 28. Their combined actions serve to decrease the volume of the thoracic cavity, thereby increasing the intrapulmonic pressure and causing the lungs to recoil elastically.

*Coughing* is usually induced by an irritation to the mucous membranes of any portion of the respiratory tract. The irritants may be inspired or ingested. The *cough reflex* is elicited because the respiratory escalator proves incapable of removing the irritants by the ordinary method of trapping them in mucus and transporting them on cilia to the throat. Coughing may also be elicited by stimulation of other parts of the body innervated by sensory fibers of the Xth cranial nerve, the vagus nerve, which mediates the cough reflex, or psychogenic factors such as suggestibility and voluntary desire.

Coughing is a complex reflex initiated by a brief inspiration. The epiglottis then closes, and the vocal cords tighten. Next, a forced expiration ensues, in the middle of which the epiglottis suddenly opens while the vocal cords relax. The air held under pressure in the passageways of the bronchial tree is finally forced out through the mouth and nose. The offending particles may be ejected via the mouth or swallowed.

Nosebleed is technically known as *epistaxis*. The hemorrhaging is due to the rupture of subcutaneous blood vessels in the nose. Epistaxis can be initiated by physical trauma, induced by violent sneezing, mechanical irritation, and low atmospheric pressures at high altitudes. The latter factor has the effect of increasing the hydrostatic pressure of blood and weakening vascular walls.

**Hiccups**  Hiccups are initiated in the following ways: (1) stimulation of sensory endings of the Xth cranial nerve; (2) the presence of irritants within the gastrointestinal tract; or (3) spasms of laughing which induce irritation. Hiccups occur reflexly when the phrenic nerve from the cervical plexus conducts motor impulses to the diaphragm, causing it to contract spasmodically. The ensuing inspiratory effort is abruptly terminated when the glottis closes in response to motor stimulation from the laryngeal nerve. The characteristic sounds during hiccuping are produced when the inspiratory air strikes the suddenly closed glottis. A forced expiration follows production of the sound.

**Sighing and yawning**  The mechanisms of sighing and yawning are basically similar. *Sighing* is initiated by a strong inspiration, followed by a slower, more shallow, and usually audible expiration. What sighing accomplishes is not clear, although one currently popular theory suggests that sighing prevents the alveoli from collapsing, since their surface tension is decreased during the act. Ordinarily, surfactant prevents alveolar collapse by maintaining a relatively uniform surface tension. During shallow respiration, however, when the diameters of the alveoli are maximally diminished, the surface tension can increase beyond the limits of the surfactant to depress it. Collapse of the alveoli may still be averted by the rush of inspiratory air created during the first stage of sighing.

*Yawning* usually consists of a series of deepening inspirations, each of which is interrupted by a partial expiration. The final expiration in the sequence is long and deep and is followed by a brief period of apnea. Sounds of a voluntary nature may accompany the expirations. Tears may also be produced during particularly deep yawns.

Yawning is attributed to weariness, since yawning stimulates the circulation and somewhat offsets fatigue, and to psychogenic factors, because the sight of someone else yawning often triggers the response.

The stimulatory effect of yawning may have some limited usefulness in preventing localized hypoxic conditions from developing by increasing the supply of $O_2$ to the tissues.

**Laryngospasm and aphonia**   The major effects of laryngospasm and aphonia are restricted to the larynx. *Laryngospasm* stems from a spasmodic contraction of the vocal folds, bringing them tightly together. The disorder more frequently occurs in children than adults. The malfunction is caused by an irritation to, or inflammation of, the larynx.

*Aphonia* stems from laryngitis, which causes loss of the voice. Although aphonia may be brought about by more serious factors such as laryngeal cancer or certain personality disturbances, this form of laryngeal inflammation is usually due to upper respiratory infections. Fatigue of the laryngeal musculature through excessive talking is also a common cause of the disturbance.

## OXYGEN DEFICIT

**Types of hypoxia**   Dyspnea may be an expression of $O_2$ starvation. The condition is referred to as *anoxia* when there is a complete absence of $O_2$ and *hypoxia* when there is an insufficiency in the amount of this gas either provided to, or utilized by, the tissues of the body. Because the former condition is rare and inevitably fatal, we will use the term *hypoxia* to describe several disorders that result in a depression in the amount of $O_2$ supplied to the body.

An $O_2$ deficit caused by a low $O_2$ tension in arterial blood is called *hypoxic hypoxia*. The deficiency in the arterial supply of $O_2$ may be brought about by a depression in the atmospheric concentration of $O_2$, e.g., at high altitudes, or by a decrease in the diameter of the airways, reduction in the amount of the functional pulmonic membrane, or decrease in the extensibility of the lungs, thorax, or both, because of respiratory disease.

A lack of $O_2$ resulting from an impairment in circulation is termed *stagnant hypoxia*. This condition is caused by (1) heart diseases which effectively reduce cardiac output; (2) vascular diseases, such as arteriosclerosis, which increase the peripheral resistance, or atherosclerosis, thrombosis, and aneurysm, all of which decrease the amount of blood flow to specific regions; and (3) developmental anomalies, such as a patent ductus arteriosus, a patent foramen ovale, and the tetralogy of Fallot, all of which jeopardize the pumping ability of the heart.

An $O_2$ deficiency caused by a diminution in the $O_2$-carrying capacity of blood is called *anemic hypoxia*. This condition is a consequence of the following factors:

1   Hemorrhages which result in an extreme loss of blood
2   Skeletal diseases which affect the bone marrow and diminish erythropoiesis, i.e., the manufacture of erythrocytes
3   Deficiencies in essential nutrients which affect either the synthesis of hemoglobin or hemopoiesis
4   Hereditary defects such as sickle cell anemia and thalassemia, both of which result in the formation of abnormal hemoglobins
5   Leukemia, by snuffing out of existence erythrocytes already present in blood
6   Carbon monoxide (CO) poisoning which inhibits $O_2$ from combining with hemoglobin
7   Hemotoxins which cause hemolysis (erythrocytic destruction)

*Histotoxic hypoxia* results from a lack of $O_2$ utilization because of derangements in cellular oxidation. This condition is caused by metabolic poisons, such as cyanide, which prevents the utilization of $O_2$ during mitochondrial oxidations; by hereditary defects, which result in a lack of specific oxidative enzymes; and by deficiencies in essential nutrients, such as the vitamins required for aerobic oxidation.

With the exception of histotoxic hypoxia, any of the hypoxic conditions discussed above may lead to dyspnea and cyanosis. The bluish pallor is due to the color of reduced hemoglobin in excessive concentrations in peripheral blood. The effects of histotoxic hypoxia originate from an inability of tissue cells to utilize $O_2$ already provided in normal amounts by blood. Whatever the cause, the lack of sufficient $O_2$ not only deprives the body of this necessary resource but also depresses the ventilatory ability of the lungs and leads directly to hypercapnia and acidosis.

The development of pulmonary hypertension is also possible, due to homeostatic mechanisms which cause a compensatory increase in cardiac output and a corresponding increase in the rate of pulmonary blood flow. An elevation in pulmonary arterial blood pressure creates a strain on the left atrium because of the augmented venous return. Pulmonary edema may result when serum is driven from the pulmonary blood capillaries into extracellular tissue spaces surrounding the alveoli.

The vicious circle of deterioration in respiratory and circulatory functions may prevent adequate oxygenation of the neurons of the brain and, if serious enough, can result in a state of unconsciousness from which the patient cannot be aroused by ordinary methods. This form of unconsciousness is referred to as a *coma*. Death may terminate the vicious circle when the effects of further degenerative changes become incompatible with life. The homeostatic disruptions caused by hypoxia are summarized in Fig. 29-1.

The body responds to a diminution in the amount of blood $O_2$ by compensatory increases in both the rate

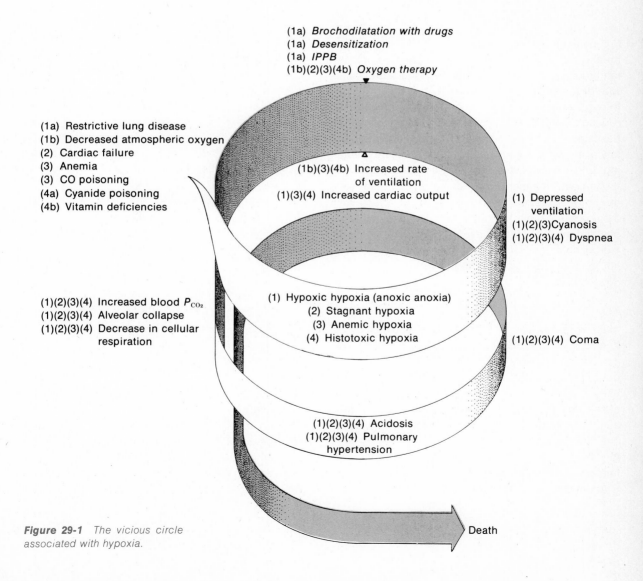

(1a) *Brochodilatation with drugs*
(1a) *Desensitization*
(1a) *IPPB*
(1b)(2)(3)(4b) *Oxygen therapy*

(1a) Restrictive lung disease
(1b) Decreased atmospheric oxygen
(2) Cardiac failure
(3) Anemia
(3) CO poisoning
(4a) Cyanide poisoning
(4b) Vitamin deficiencies

(1b)(3)(4b) Increased rate of ventilation
(1)(3)(4) Increased cardiac output

(1) Depressed ventilation
(1)(2)(3) Cyanosis
(1)(2)(3)(4) Dyspnea

(1)(2)(3)(4) Increased blood $P_{CO_2}$
(1)(2)(3)(4) Alveolar collapse
(1)(2)(3)(4) Decrease in cellular respiration

(1) Hypoxic hypoxia (anoxic anoxia)
(2) Stagnant hypoxia
(3) Anemic hypoxia
(4) Histotoxic hypoxia

(1)(2)(3)(4) Coma

(1)(2)(3)(4) Acidosis
(1)(2)(3)(4) Pulmonary hypertension

Death

**Figure 29-1** *The vicious circle associated with hypoxia.*

of ventilation and the cardiac output. Hyperpnea increases the volume of air brought to the lungs and thus brings more $O_2$ in contact with the functional pulmonic membrane. The increases in stroke volume and arterial blood pressure ensure that the tissue cells will continue to receive a relatively constant supply of $O_2$, even in the face of a considerable $O_2$ deficit.

Treatment of hypoxia   However, the natural mechanisms of the body by themselves may be inadequate to compensate for a lack of $O_2$. External intervention is then required. Therapy is based on amelioration of the symptoms or elimination of the causative agents.

Reduction in symptoms may be brought about by using some combination of the following methods: (1) inhalation therapy or artificial respiration to increase the amount of air and, therefore, $O_2$ reaching the blood and tissues; (2) dilation of the bronchioles with drugs or the elimination of accumulated mucus through enzymatic digestion or postural drainage, to reduce or eliminate sources of obstruction to the free passage of air in and out of the lungs; and (3) respiratory exercises or intermittent positive pressure breathing (IPPB) to improve or assist the ventilatory ability of the lungs. Treatment of the causative agent may require the destruction of microbial pathogens with the use of antibiotics and other drugs, the elimination of allergens through their avoidance, or a reduction in the concentration of reagins, or incomplete antibodies, built up from allergic reactions, by undergoing desensitization.

Acute respiratory failure   The progressive deterioration in bodily functions induced by hypoxia may be due to a chronic respiratory disease. Although patients suffering from respiratory disorders have some degree of functional limitation, most are able to live fairly normal lives. However, in a minority of such cases and in patients suffering from several other acute or chronic diseases, breathing suddenly stops. The clinical syndrome is referred to as *acute respiratory failure*. The vicious circle associated with this immediate threat to life is diagrammed in Fig. 29-2.

Acute respiratory failure is caused by some combination of the following factors: (1) pulmonary capillary collapse; (2) alveolar obstruction; and (3) ventilatory depression. The collapse of blood capillaries surrounding the alveoli may be induced by shock associated with thoracic surgery, severe hemorrhages, or burns. The effect of capillary collapse is to increase the amount of functional dead space within the lungs, because the collapsed capillaries cannot participate in gaseous exchange with the alveoli which they surround. Since the usable volume of inspired air is decreased, the respiratory work load of subjects suffering from collapse of pulmonary capillaries is significantly increased in an attempt to offset the development of arterial hypoxia.

Reduction in the diameter of alveoli or their airways may be due to emphysema, pneumonia, or severe obesity. Such restrictions usually cause a compensatory enlargement, or dilatation, of the affected portion of the bronchial tree. The increased volume of air space further compounds the difficulties already experienced from the diminished supply of air by reducing the alveolar $P_{O_2}$. Since the capillary blood supply is capable of withdrawing more $O_2$ than is provided by the alveoli, this condition also leads to arterial hypoxia.

Depression in the ability to ventilate the lungs can be caused by overdoses of drugs which affect the central nervous system, chest injuries which stiffen the chest wall, neuromuscular diseases which paralyze the respiratory muscles, or surgical shock which accompanies thoracic operations. Because the alveoli are underinflated, their surface tensions increase and ultimately cause alveolar collapse.

The type of damage wrought by acute respiratory failure is determined by its cause and the length of time the patient is anoxic. For example, since $CO_2$ can diffuse about 20 times as readily as $O_2$, hypercapnia and acidosis are potential dangers only in the event of capillary collapse. Carbon dioxide retention is therefore usually not a problem so long as there is at least some functional pulmonic membrane left for diffusion of gases. On the other hand, hypoxic hypoxia resulting from the presence of alveolar obstructions or ventilatory depression usually leads to cyanosis and

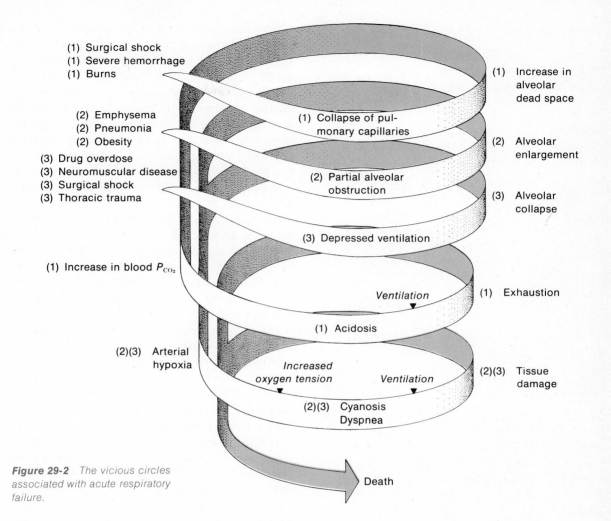

(1) Surgical shock
(1) Severe hemorrhage
(1) Burns

(2) Emphysema
(2) Pneumonia
(2) Obesity

(3) Drug overdose
(3) Neuromuscular disease
(3) Surgical shock
(3) Thoracic trauma

(1) Collapse of pulmonary capillaries

(2) Partial alveolar obstruction

(3) Depressed ventilation

(1) Increase in alveolar dead space

(2) Alveolar enlargement

(3) Alveolar collapse

(1) Increase in blood $P_{CO_2}$

(2)(3) Arterial hypoxia

*Ventilation*

(1) Acidosis

*Increased oxygen tension*   *Ventilation*

(2)(3) Cyanosis Dyspnea

(1) Exhaustion

(2)(3) Tissue damage

Death

**Figure 29-2** *The vicious circles associated with acute respiratory failure.*

dyspnea. Whatever the pathway of deterioration, death is precipitated by irreversible tissue damage due to hypoxia.

In spite of its seriousness, acute respiratory failure is not necessarily fatal, and special teams of medical personnel trained in emergency procedures have been organized in many hospitals to cope with this specific emergency. Their effectiveness is attested to by the countless patients who have been revived after they stopped breathing. Basically, emergency treat-

ment requires the provision of $O_2$, control of the rate and depth of respiration, or both. In patients suffering from cyanosis and dyspnea, an increase of a few percent in the $O_2$ tension of the plasma by physically dissolving the gas under pressure may mean the difference between life and death. Therapy during respiratory failure from any cause requires external control of the degree of pulmonary ventilation, to inflate alveoli which have collapsed and provide an optimum volume of air to all portions of the bronchial tree.

# RESPIRATORY DISEASES

The causes of ventilatory depression are diverse and include respiratory infections, thoracic trauma, various types of shock, and certain neuromuscular and allergic disorders.

**Bronchitis**  Bronchitis refers to an acute or chronic inflammation of the mucous membranes of the bronchi. In addition to infection by microbes, the tissues lining the bronchi can become irritated by substances such as dusts, volatile chemicals, and allergens which are carried in the inspired air, or by food particles which inadvertently lodge in the upper portion of the respiratory tract. The inflammation provokes mucous cells to secrete excessive amounts of mucus, plugging up some of the airways and causing episodes of coughing and dyspnea. Coughing is also provoked by irritants which the respiratory escalator is not able to handle, as we noted previously. Bronchial inflammation is accompanied by a discomfort or soreness in the midchest region. Fever results when pyrogenic bacteria are responsible for the disease.

**Bronchiectasis**  Chronic bronchitis may lead to bronchiectasis, a pathologic dilatation of the bronchi, although other respiratory diseases or congenital defects can also induce this pathology. Fatigue and malnutrition are implicated as predisposing factors. Also, there is a high incidence of bronchiectasis among people living in urban ghettos. However, crowding also plays an important role in the ease of dissemination of microbes which cause communicable respiratory diseases and the subsequent development of bronchiectasis.

The symptoms of bronchiectasis are generally similar to those of bronchitis. In addition, pus is produced in copious amounts. A purulent condition obtains from pathologic changes undergone by the bronchial mucosa in response to chronic and deep-seated infection.

**Bronchial asthma**  Bronchial asthma is a disease usually caused by a sudden contraction of the smooth musculature of the bronchi and bronchioles. The spasms are brought about by allergic reactions, psychogenic factors, or a combination of both. The spasms restrict the diameters of the airways. This condition is further exacerbated by an excessive production of mucus due to an irritation of the mucosa plus the compressive force normally exerted by the lungs against the bronchial tree during expiration. The ventilatory depression is aggravated by pulmonary edema caused by chronic inflammation and the subsequent development of pulmonary hypertension. Pulmonary hypertension is discussed further in Chap. 31.

The typical wheezing sounds produced during an asthmatic episode are due to the rush of air passing through the partially occluded airways. Because of the additional pressure placed on the walls of the bronchi and bronchioles during elastic recoil of the lungs, wheezing is usually more prominent in expiration than in inspiration.

The following categories of chemicals are employed in the symptomatic treatment of asthma: (1) epinephrine, ephedrine, aminophylline, and other sympathomimetic agents which act through autonomic pathways to dilate the spastic smooth muscles of the bronchial tree; and (2) adrenocortical hormones, their synthetic derivatives such as prednisone, or other anti-inflammatory agents which relieve chronic irritation, thereby reducing mucosal congestion and the edematous condition. Therapy may also be based on the general methods of treating respiratory diseases previously enumerated under the treatment of hypoxia.

**Atelectasis**  When the airways are completely occluded due to birth defects or restrictive lung disease, the terminal alveoli remain unexpanded or collapse, respectively. The appearance of the alveolar epithelium is shown in Fig. 29-3A. Alveolar collapse may also result from the following factors: (1) an inadequate secretion of surfactant and a corresponding inability to lower appreciably the alveolar surface tension; and (2) ventilatory depression caused by excessive physical pressure exerted against the walls of the bronchial tubes by lung tumors or pulmonary aneurysms. Called *atelectasis*, the collapse of alveoli is responsible for the simultaneous collapse of ad-

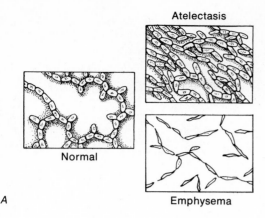

Atelectasis

Normal

Emphysema

*A*

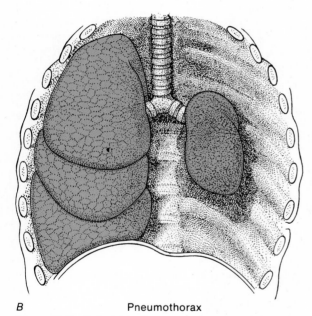

*B* Pneumothorax

**Figure 29-3** *Pathologic changes during some serious respiratory disorders. A. Alveolar epithelium in atelectasis and emphysema. B. Affected lung in pneumothorax.*

from cigarette smoking, atmospheric pollution, or respiratory diseases, the alveoli may enlarge and rupture, a disorder referred to as *emphysema*. Alveolar distension and destruction effectively reduce the amount of the functional pulmonic membrane. The nature of this tissue damage is illustrated in Fig. 29-3. The damaged tissues are replaced by fibrous connective tissue, decreasing the elasticity and extensibility of the lungs. The ventilatory depression may give rise to dyspnea and cyanosis.

In response to chronic hypoxia, the lungs undergo compensatory hypertrophy, initiating bony changes which may cause the development of the so-called barrel chest. This sign is characteristic of patients in advanced stages of chronic emphysema. Because of thoracic enlargement and a corresponding increase in the amount of alveolar dead space, the volume of air remaining within the respiratory tract at the end of expiration, i.e., the functional residual capacity, increases measurably.

**Pneumothorax** The edematous condition accompanying emphysema may cause irregularities on the surfaces of the lungs. If these tissue malformations rupture the visceral pleura immediately covering its outside surface, the affected lung collapses (see Fig. 29-3*B*) because inspired air is no longer prevented from entering the pleural cavity. This leakage neutralizes the pressure differential between the inside and the outside of the lung. Leakage of air into the pleural cavity may also result from the following factors: (1) respiratory diseases (e.g., tuberculosis) that give rise to pulmonary lesions (e.g., abscesses); (2) chest wounds that expose the pleural cavity to the pressure of the atmosphere; (3) therapeutic procedures that are sometimes intentionally instituted to collapse a tuberculous or otherwise diseased lung, to promote healing by providing rest, something not possible when adhesions exist; and (4) idiopathic factors that result in an extremely slow leakage of air into the pleural cavity.

jacent blood capillary networks, due to a decrease in the volume of the lung and the vascular response to local hypoxia. We have already noted the general effects of hypoxic hypoxia caused by alveolar collapse, namely, dyspnea and cyanosis.

**Emphysema** As a consequence of chronic irritation

**Hemoptysis** When blood is coughed up from any portion of the respiratory tract, excluding the nose and pharynx, the condition is referred to as *hemoptysis*.

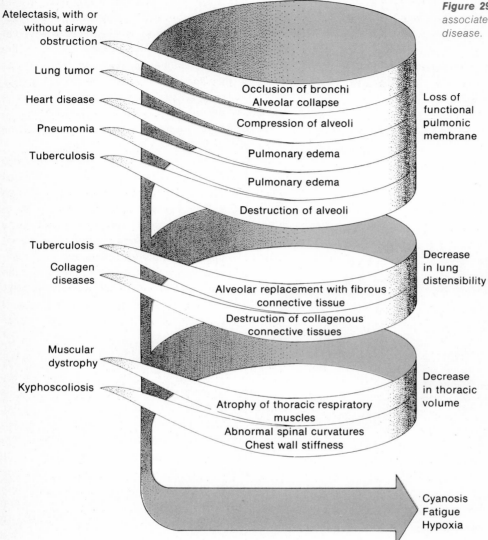

**Figure 29-4** *The vicious circle associated with restrictive lung disease.*

The presence of fresh blood in sputum is indicative of hemorrhaging usually provoked by severe episodes of coughing. Among the most common causes of hemoptysis are serious respiratory diseases of an inflammatory nature, e.g., tuberculosis and pneumonia.

**Restrictive lung disease** The diseases that are reviewed above, along with others yet to be men-tioned, prevent $O_2$ in inspired air from reaching the blood in sufficient concentrations to maintain normal metabolism. The entire complex of disorders that act in this manner is known clinically as *restrictive lung disease*. Although diverse in both origin and nature of tissue damage, restrictive lung disease follows a predictable course. The vicious circle associated with this clinical entity is shown in Fig. 29-4.

Regardless of cause, restrictive lung disease leads to a depression in ventilation in the following ways: (1) the surface area of the functional pulmonic membrane is decreased, either because the bronchial tubes are obstructed, the alveoli are compressed or destroyed, or the lungs are congested; (2) lung distensibility is reduced because of destruction of collagenous tissues or replacement of necrotic alveoli with scar tissue; and (3) thoracic distensibility is depressed because of spinal deformities which cause a stiffening of the chest wall or neuromuscular diseases which impair the action of the respiratory musculature.

The effects of ventilatory depression are identical to those discussed previously and include dyspnea, cyanosis, episodes of coughing, and chest pains. The multiple causal factors, diverse forms of both structural damage and functional loss, as well as the similarities in symptomatology make restrictive lung disease an ideal summary statement of respiratory damage caused by diseases in general.

## HIGH GAS TENSIONS

Up to this point, we have dealt with diseases in which $O_2$ deprivation is the dominant theme. The other side of the coin is represented by diseases caused by high gas tensions or the release of gases dissolved under pressure. Such disorders are induced by (1) $O_2$ toxicity due to the presence of excessive amounts of $O_2$ in the inspiratory air; (2) acidosis due primarily to an inability to excrete sufficient quantities of $CO_2$ via the lungs; (3) CO poisoning through the inhalation of abnormally large quantities of CO; (4) $N_2$ narcosis because of the inspiration of air under pressures considerably higher than that of the atmosphere; and (5) decompression sickness due to the release of gases, especially $N_2$, dissolved under pressure throughout the body.

The vicious circles initiated in the above-listed ways are summarized in Fig. 29-5.

**Oxygen toxicity** The following factors may be responsible for a high alveolar $P_{O_2}$: (1) $O_2$ therapy, where pure $O_2$ or $O_2$ in an admixture of gases is supplied under high pressure in inspiratory air; and (2) an artificial supply of $O_2$ used at high altitudes and in outer space or in the depths of the oceans. Although the use of $O_2$ or gases containing $O_2$ under the above conditions is routine, the effects of $O_2$ toxicity at high pressures present an inherent danger. The risk is due to cellular disruption, especially to nerve cells. This type of damage differentially affects the brain, causing convulsions and some degree of unconsciousness such as stupor or coma.

The second danger, applicable to the administration of any respiratory gas under high pressure and not just to $O_2$ alone, is due to the direct effects of high gas tensions on alveoli. Extreme intrapulmonic pressures destroy the integrity of the air sacs, thereby decreasing the amount of the functional pulmonic membrane and causing hypoxic hypoxia.

**Acidosis** Acidosis develops because of an increase in the total blood $P_{CO_2}$. This condition may be caused by circulatory stagnation, collapse of pulmonary capillaries, inspiration of large quantities of $CO_2$, or other factors which appreciably increase the numerator or decrease the denominator of the ratio, $H_2CO_3/Na^+HCO_3^-$. This expression summarizes the conditions which must be met to increase or decrease the acidity of blood. Respiratory acidosis is considered in detail in Chap. 31 under "Acid-Base Imbalances."

**Carbon monoxide poisoning** Most deaths from CO poisoning result from the leakage of gas fumes containing CO, due to defects in the exhaust systems of automobiles or through the accumulation of CO during the operation of automobiles, space heaters, or other fuel-powered appliances in a closed or inadequately ventilated space.

Carbon monoxide combines over 200 times as readily, and at the same point on the hemoglobin molecule, as $O_2$. At sea level and 0°C, the $P_{O_2}$ is approximately 160 mmHg (760 mmHg × 0.21). Because CO has at least 200 times as great an affinity for hemoglobin as does $O_2$, the $P_{CO}$ required to completely neutralize the combining powers of $O_2$ is only about 0.8 mmHg (160 mmHg/200). At this pressure, half the

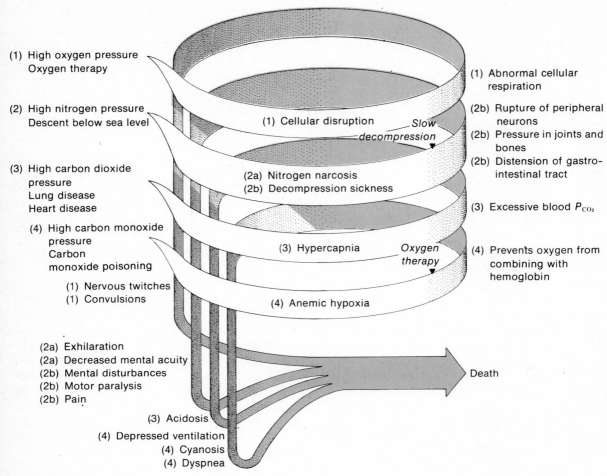

(1) High oxygen pressure
Oxygen therapy

(2) High nitrogen pressure
Descent below sea level

(3) High carbon dioxide
pressure
Lung disease
Heart disease

(4) High carbon monoxide
pressure
Carbon
monoxide poisoning

(1) Nervous twitches
(1) Convulsions

(2a) Exhilaration
(2a) Decreased mental acuity
(2b) Mental disturbances
(2b) Motor paralysis
(2b) Pain

(3) Acidosis
(4) Depressed ventilation
(4) Cyanosis
(4) Dyspnea

(1) Cellular disruption          Slow
                            decompression

(2a) Nitrogen narcosis
(2b) Decompression sickness

(3) Hypercapnia                  Oxygen
                                 therapy

(4) Anemic hypoxia

(1) Abnormal cellular
respiration

(2b) Rupture of peripheral
neurons
(2b) Pressure in joints and
bones
(2b) Distension of gastrointestinal tract

(3) Excessive blood $P_{CO_2}$

(4) Prevents oxygen from
combining with
hemoglobin

Death

**Figure 29-5** *The vicious circles associated with high gas tensions within the body.*

available hemoglobin is combined with CO and $O_2$ is bound to the rest of the respiratory pigment.

Since CO ties up hemoglobin and greatly depresses the $O_2$-carrying capacity of blood, the immediate effects of CO poisoning are similar to those of severe anemic hypoxia. A lack of $O_2$ causes a compensatory increase in the rate of respiration and frequency of the heartbeat. The respiratory rhythm becomes labored, turning into dyspnea as $O_2$ deprivation progresses. These factors cause dizziness or faintness. The increased cardiac rate is reflected in a pounding heart, a rapid pulse, and a throbbing at the temples which may precipitate a headache.

If an individual is still breathing following CO poisoning, then the prognosis for recovery is good, provided that one or more of the following procedures is employed promptly: (1) artificial respiration to ventilate the lungs; (2) pure $O_2$ therapy to drive off CO from hemoglobin, thereby permitting its reoxygenation; and (3) whole-blood transfusions to provide a supplementary source of hemoglobin, thereby increasing the $O_2$-carrying capacity of the poisoned blood. The most effective procedure is to administer pure $O_2$, since the chemical combination between hemoglobin and CO is reversible and dependent entirely on the partial pressures of CO and $O_2$ in the blood.

Nitrogen narcosis   When the pressure on the outside of the body increases appreciably, as for example during a deep-sea dive, air must be supplied under correspondingly greater pressures to prevent the chest from caving in and the lungs from collapsing. Consequently, more and more gas dissolves in the tissues and fluids of the body. Since approximately 79 percent of air is $N_2$, most of the dissolved gas consists of $N_2$. For every 33 feet in depth, compared to sea level, an additional liter of $N_2$ is dissolved in the body. No harm is done to the tissues by the dissolved $N_2$, so long as the outside pressures do not change abruptly.

Because of its lipid solubility, $N_2$ dissolves differentially in fatty tissues and layers. The myelin sheaths of nerve fibers are among the structures rich in lipids and lipoidal substances such as cholesterol, phospholipids, and fatty acids. An increase in the amount of dissolved $N_2$ in the myelin sheaths reduces the excitability of nerve fibers and causes $N_2$ narcosis.

The symptoms of $N_2$ narcosis are identical to those of central nervous system depression and include drowsiness, impaired judgment, lack of concentration, motor incoordination, muscular weakness, and unconsciousness.

Decompression sickness   If an underwater ascent is too rapid, $N_2$ gas bubbles percolate through the tissues, disrupting or destroying cells in their paths. The effect is similar to that which occurs with $CO_2$ when a bottle or can of carbonated beverage is suddenly opened. Having been dissolved under pressure, $CO_2$ remains in solution as long as the container is closed. When the container is opened, exposing the solution to an atmosphere at a lower pressure, the gas is released violently.

Although bubbles of $N_2$ damage tissue cells indiscriminately, the large concentrations of fatty materials and, therefore, dissolved $N_2$ in the myelin sheaths of nerve fibers make neurons particularly vulnerable to cytolysis. All parts of the nervous system, including the brain, spinal cord, and peripheral nerves, may be impaired, depending upon the amount of $N_2$ dissolved and the rate of its evolution.

The symptoms of decompression sickness are related to the type and amount of neural damage. Involvement of the central nervous system may result in mental disturbances and motor paralysis. Peripheral nerve damage leads to polyneuritis or inflammation, including severe pain along the affected neural pathways.

## COLLATERAL READING

Gold, Warren M., "Restrictive Lung Disease," in H. J. Hislop and J. O. Sanger (eds.), *Chest Disorders in Children: Proceedings of a Symposium*, American Physical Therapy Association, 1968. Describes the pathophysiology of pulmonary and extrapulmonary restrictions in lung volume.

Leith, David E., "Cough," in H. J. Hislop and J. O. Sanger (eds.), *Chest Disorders in Children: Proceedings of a Symposium*, American Physical Therapy Association, 1968. Reviews the mechanical events that take place during coughing and clinical aspects related to the relative effectiveness of the "cough pump" in lung clearance.

Levine, David B., "The Influence of Spinal Deformities on Chest Function," in H. J. Hislop and J. O. Sanger (eds.), *Chest Disorders in Children: Proceedings of a*

*Symposium*, American Physical Therapy Association, 1968. The pulmonary and cardiovascular disorders that may arise secondarily from spinal deformities are considered.

Winter, Peter M., and Edward Lowenstein, "Acute Respiratory Failure," *Scientific American*, November 1969. Discusses the finely honed, emergency treatments and causes of respiratory failure.

Wohl, Mary Ellen Beck, "Respiratory Problems Associated with Chest Wall Abnormalities," in H. S. Hislop and J. O. Sanger (eds.), *Chest Disorders in Children: Proceedings of a Symposium*, American Physical Therapy Association, 1968. Reviews abnormalities of the chest wall which result secondarily in abnormal pulmonary function. The article emphasizes manipulations which prevent the development of atelectasis and infection.

Wohl, Mary Ellen Beck, "Atelectasis," in H. J. Hislop and J. O. Sanger (eds.), *Chest Disorders in Children: Proceedings of a Symposium*, American Physical Therapy Association, 1968. The mechanisms which lead to the development of atelectasis and its treatment rationale are lucidly reviewed.

# CHAPTER 30
# transport and exchange

The metabolic and perpetuative activities of the 5 to 10 trillion cells in the body require a mechanism for the rapid transport of raw materials and waste products from one region to another, based on the differential needs of the organs and tissues. Rapid and regulated transport and exchange are accomplished by the circulatory system. The transport and exchange vehicle is, of course, blood. In addition, a number of other homeostatic functions take place in blood because of its transport and exchange capacities.

The following are the more specific functions of the circulatory system:

1   The transport of $O_2$ from the lungs to the tissue cells and the transport of $CO_2$ in the reverse direction (see Chap. 28)
2   The transport of energy and nonenergy nutrients from the small intestine, liver, or fat depots to the tissue cells or from the small intestine to the liver or fat depots (see Chap. 26)
3   The transport of nitrogenous and other waste products of metabolism from various regions of the body to the kidneys (see Chap. 22)
4   The transport of hormones and other information-rich substances from an endocrine tissue or gland to one or more target structures (see Chap. 17)
5   The transport of circulating antibodies, the carriage of foreign materials to components of the reticuloendothelial system, and the regulation of blood clotting as a result of vascular damage (see Chap. 4)
6   The conduction of heat from the core of the body to the surface and the regulation of radiative heat losses through peripheral vasoconstriction and vasodilatation (see Chap. 8)
7   Participation in other homeostatic functions, such as the detoxification of hemoglobin (see Chap. 31) and the maintenance of both a constant pH and a fluid and electrolyte balance (see Chap. 22)

The circulatory system consists of the cardiovascular and lymphatic divisions, or systems. Responsive to neural and chemical signals, all their component parts are designed to facilitate and maintain the transport and exchange of materials between the fluid compartments of the body. This chapter examines the contributions of the various organs of the circulatory system toward these ends.

## THE CARDIOVASCULAR SYSTEM

Consisting of the heart and blood vessels, the cardiovascular system is responsible for the transport of blood throughout the body and participates in the exchange of various materials between the blood and the tissue cells.

## BLOOD

Blood is a fluid connective tissue, consisting of cells separated from each other by an intercellular fluid matrix and enclosed within the heart and blood vessels. There are approximately 6 liters of blood in the circulatory system of an adult. Between 40 and 45 percent of the blood volume consists of cells and cellular fragments. Because most of the cells lack nuclei and fragments of cells intermingle with whole cells, the components in the "cellular" fraction of the blood are commonly referred to as *formed elements*. The remainder of the blood is composed of a fluid called *plasma*. The plasma contains various molecules and ions in closely regulated concentrations.

The formed elements of blood include erythrocytes, or red blood corpuscles; leukocytes, or white blood corpuscles; and blood platelets, or thrombocytes. Circulating *erythrocytes* lack nuclei and are the most numerous of the formed elements in blood. Erythrocytes are primarily concerned with the transport of $O_2$ and $CO_2$.

Based on the presence or absence of granules in their cytoplasm, there are two types of *leukocyte* —granulocytes and agranulocytes. The *granulocytes* are called *eosinophils, basophils,* or *neutrophils,* depending upon whether their cytoplasmic granules stain with acidic or alkaline dyes. The circulating

*agranulocytes* are referred to as *lymphocytes* and *monocytes*, depending upon the relative volume proportions of nucleus and cytoplasm.

Blood *platelets* are sloughed-off fragments of large red bone-marrow cells called *megakaryocytes*. Blood platelets function during hemostasis, or blood clotting.

The plasma portion of blood contains $H_2O$, proteins, electrolytes, informationally rich macromolecules, energy and nonenergy nutrients, and waste products of metabolism. About 90 percent of the plasma by volume is $H_2O$, which is the basic solvent of the body. The proteinaceous portion of plasma consists of antibodies, enzymes, glycoproteins, polypeptides, and other protein derivatives. Mineral electrolytes within plasma include $Na^+$, $Cl^-$, $K^+$, $Mg^{++}$, and $Ca^{++}$. Buffers such as $HCO_3^-$ and trace elements in ionic form are also found in plasma. Plasma-borne hormones secreted by endocrine tissues and glands influence the physiology of target structures in distant parts of the body. Soluble end products of digestion, such as simple sugars, fatty acids and glycerol, and amino acids, are transported through the plasma portion of blood to sites of storage or utilization. Many vitamins are also absorbed directly from the small intestine into the blood. Carbon dioxide is transported to the lungs and nitrogenous waste products, such as urea, creatine, and creatinine, are carried to the kidneys. Respiratory gases are dissolved physically in small concentrations and are also carried in chemical combination. Since all the above-listed materials are located in the fluid portion of blood, the plasma is obviously an important medium of exchange between the blood and the tissue cells.

*Whole blood* consists of the formed elements and plasma together. The physical characteristics of whole blood are normally maintained within narrowly defined ranges, thereby ensuring homeostasis. The *viscosity* of whole blood is almost four times greater than that of water. Its *specific gravity* is approximately 1.05 (see Chap. 7). The *pH* of blood varies between 7.35 and 7.45, depending upon whether the blood is arterial or venous.

The concept of pH is explained in Chap. 10, and its vascular regulation is described in Chap. 31. Other aspects in the homeostasis of blood, such as the excretion of $CO_2$ and the regulation of the content and concentration of its other fluid and particulate constituents, are discussed in Chap. 22.

Erythrocytes  Erythrocytes, or red blood corpuscles, are the most numerous cellular components in blood. There are approximately 5.5 million erythrocytes/mm³ in males and about 4.5 million erythrocytes/mm³ in females. The sex difference in number is accounted for by the fact that erythrocytes carry oxygen ($O_2$), and males require more $O_2$ than females, because of their higher average basal metabolic rate.

Each erythrocyte is a flexible biconcave disk approximately 7.5 $\mu$m in diameter, or slightly larger than that of the smallest capillaries. Since erythrocytes lack nuclei in their adult stage, to call them "red blood cells" is somewhat of a misnomer. The characteristic shape of erythrocytes is illustrated in Fig. 30-1A. The dumbbell shape which an erythrocyte presents on end provides for an optimal diffusion of gases to any point in its interior. Within each erythrocyte are several million hemoglobin molecules dispersed in a fluid stroma.

*Hemoglobin* is a complex protein molecule which differentially binds with $O_2$ in pulmonary capillaries and with $CO_2$ in tissue capillaries. The binding sites for these gases are associated with the four iron porphyrin groups in the hemoglobin molecule. The presence of iron confers a reddish tinge to erythrocytes. Because hemoglobin functions in gas transport and is colored, it is referred to as a *respiratory pigment*. When hemoglobin is loaded with $O_2$, as in arterial blood, the blood appears bright red. Following the unloading of $O_2$ from hemoglobin in tissue capillaries, the blood takes on a darker reddish hue.

Because hemoglobin is a protein and therefore charged, it acts as a blood buffer. The buffering action of hemoglobin prevents changes in blood pH during the transport of blood from the tissue cells to the lungs. The buffering role of hemoglobin is reviewed in Chap. 22.

Since the characteristics of hemoglobin determine many of the structural and functional qualities of erythrocytes, are we justified in stating that erythro-

cytes are simply bags filled with hemoglobin? On the affirmative side is the fact that the colloid osmotic pressure of plasma would be quite different if the plasma membranes surrounding the hemoglobin molecules of each erythrocyte were to simply disappear. The resulting hemoconcentration due to the countless billions of proteins added to the blood would pull water osmotically from the tissue cells, thereby increasing tremendously the blood volume and pressure. On the negative side is the fact that erythrocytes engage in a unique form of carbohydrate metabolism which apparently promotes the ability of hemoglobin to unload $O_2$ under conditions of tissue $O_2$ deprivation (ischemia). In other words, erythrocytic metabolism enhances the respiratory functions of hemoglobin. Clearly, an erythrocyte is more than a mere bag of specialized protein.

Erythrocytes are manufactured by a process called *erythropoiesis* in the red bone marrow from cells called *erythroblasts*. Vitamin $B_{12}$, folic acid, essential amino acids, iron, copper, manganese, and cobalt are all required during erythropoiesis. The erythropoietic effects of deficiencies of these essential materials are discussed in Chap. 27.

Because circulating erythrocytes lack nuclei, they age rapidly, living approximately 120 days. Death is due to *hemolysis*, a condition in which the plasma membrane ruptures. Following hemolysis, the hemoglobin spews into the plasma. It is believed that hemolysis is a final consequence of an increasing fragility of the plasma membrane during the aging process. Since literally millions of erythrocytes are dying and being replaced all the time, there are mechanisms to detoxify the hemoglobin released in the plasma and to salvage the iron for reuse in erythropoiesis. These mechanisms are explained in the next chapter.

Erythrocytes carry countless antigens. Obviously, we do not manufacture antibodies which are complementary to our own antigens. However, an antigen-antibody interaction may occur in a recipient of a transfusion of whole blood obtained from an incompatible donor. The antigen-antibody interaction consists of erythrocytic clumping, or *agglutination*, and results when a recipient possesses a circulating antibody, or

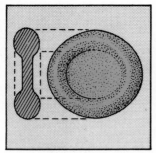 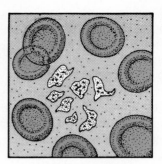

*A* Erythrocyte    *B* Blood platelets

Granulocytes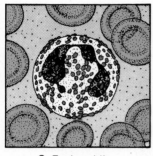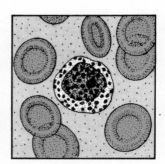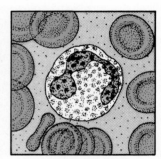

*C* Eosinophil    *D* Basophil    *E* Neutrophil

Agranulocytes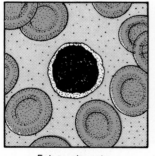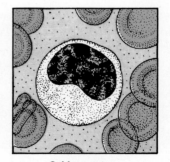

*F* Lymphocyte    *G* Monocyte

**Figure 30-1** *Types of formed elements found in blood. The characteristic shape of an erythrocyte (A); aggregations of blood platelets are shown among a grouping of erythrocytes (B); Leukocytic granulocytes consist of eosinophils (C), basophils (D), and neutrophils (E). Leukocytic agranulocytes include lymphocytes (F) and monocytes (G).*

*agglutinin*, complementary to an erythrocytic antigen, or *agglutinogen*, introduced from donor blood.

Probably the most well-known erythrocytic antigens are called A, B, and AB. The presence of these antigens permits the classification of blood into four groups, or types, based on the presence or absence of these erythrocytic antigens. Blood of type O is nonantigenic but contains circulating antibodies complementary to both type A and type B antigens. Blood of type A contains type A antigens and antibodies complementary to type B antigen. Blood of type B contains type B antigens and antibodies complementary to type A antigen. Blood of type AB contains both A and B antigens but lacks antibodies complementary to either of these antigens.

During whole blood transfusions, the blood of a recipient will agglutinate if it possesses an agglutinin complementary to an agglutinogen on the erythrocytes of the donor blood. Complementarity is indicated symbolically when the agglutinogen and agglutinin are of the same letter. Figure 30-2 gives the expected results of whole blood transfusions from donors to recipients of the various blood types. Incompatibility is shown by erythrocytic clumping of the recipient blood. Notice that type O blood is compatible when donated to a recipient of any blood type simply because type O blood is nonantigenic. For this reason, people with type O blood are called *universal donors*. Furthermore, observe that blood of type AB will not agglutinate when blood of any type is added to it. The reason, as you will recall, is that type AB blood does not contain agglutinins for blood type agglutinogens. Therefore, people with blood of type AB are referred to as *universal recipients*.

RECIPIENT'S BLOOD TYPE

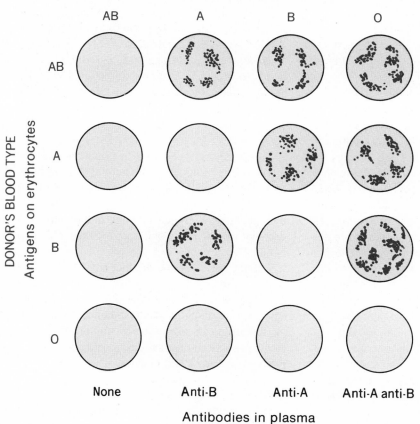

**Figure 30-2** *The range of potential agglutination reactions obtained during the transfusion of whole blood. Agglutination occurs when an erythrocytic antigen of the donor matches by letter a plasma antibody of the recipient. (From DeCoursey.)*

Another important erythrocytic antigen is the *Rh factor.* Chapter 31 discusses the inheritance and potential pathologic effects of this factor in children following the first-born of parents who are Rh incompatible.

**Leukocytes** As we have already noted, many types of leukocytes, or white blood corpuscles, are found in blood. However, the number of leukocytes is about 700 times less than that of erythrocytes, ranging between 5,000 and 9,000/mm³.

The granulocytes include eosinophils, basophils, and neutrophils. These types of leukocyte are illustrated in Fig. 30-1. Their most distinctive structural features consist of a lobated nucleus and the presence of granules in the cytoplasm. Granulocytes are about 1½ times larger in diameter than erythrocytes. Approximately 60 percent of the total number of leukocytes in the blood are neutrophils. Because their nuclei are composed of multiple interconnected lobes, neutrophils are called *polymorphonuclear.* Granulocytes are manufactured in the red bone marrow from cells called *myelocytes.* The life span of granulocytes is believed to be considerably shorter than that of erythrocytes.

Lymphocytes, monocytes, and their variants are referred to collectively as *agranulocytes.* These leukocytes are shown in Fig. 30-1. Lymphocytes are about the same size as granulocytes. Most of the intracellular volume of a lymphocyte is occupied by its nucleus. In other words, a lymphocyte contains very little cytoplasm. A monocyte, on the other hand, is the largest of the leukocytes, with a diameter about double that of an erythrocyte. The nucleus of a monocyte is somewhat U-shaped and a proportionately greater amount of cytoplasm surrounds it, compared to the nuclear-cytoplasmic ratio characteristic of a lymphocyte. Lymphocytes and monocytes together constitute approximately 30 percent of the total leukocyte concentration in the blood, with lymphocytes accounting for more than two-thirds of this number. Most of the agranulocytes in circulating blood are manufactured in the lymphoid tissues of the body, especially in the lymph nodes and spleen. Some lymphocytes are also manufactured in red bone marrow. The lifespan of agranulocytes approximates that of erythrocytes.

All leukocytes are motile. Locomotion is accomplished by means of *ameboid movement,* i.e., the formation of *pseudopodia,* or "false feet." The pseudopodia not only serve in locomotion but also enable leukocytes to engulf foreign particles that have entered the tissues or blood through a wound. Called *phagocytosis,* this ameboid ability is illustrated in Fig. 4-5. The phagocytic actions of leukocytes are part of the inflammatory response, a defense mechanism in which these cells neutralize potentially harmful materials introduced into the body during injury. The roles of the various leukocytes during the inflammatory response are described in Chap. 4.

Therefore, the primary function of leukocytes is to participate in the inflammatory response through their phagocytic actions, in the event that foreign materials gain entry into the tissues. For this reason, the roles of leukocytes are played out primarily in the tissues, rather than in the blood. In addition, many lymphocytes in widely dispersed lymphoid tissues become plasma cells committed to the production of specific antibodies. These antibodies are synthesized when complementary antigens are introduced into the body. As a result of their interaction, the physical threat posed by the presence of introduced antigens is ordinarily eliminated. Antigen-antibody interactions are described in detail in Chap. 4.

**Platelets** Platelets are usually about one-quarter the diameter of erythrocytes. The normal number of blood platelets generally ranges between 200,000 and 400,000/mm³. The granules in the cytoplasm of platelets appear similar to those in the cytoplasm of granulocytes. A group of blood platelets is illustrated in Fig. 30-1.

Platelets have two major functions: (1) they participate in blood clotting, or hemostasis, following internal hemorrhaging or external bleeding; and (2) they have an important role in intravascular clotting, following the rupture or severation of smaller blood vessels. Hemostasis and intravascular clotting are protective mechanisms reviewed in Chap. 4.

**Plasma** Plasma is that portion of the blood which remains after the formed elements are removed. Plasma constitutes 55 to 60 percent of whole blood by volume. Approximately 90 percent of the plasma is $H_2O$. This percentage concentration of $H_2O$ in blood is held fairly constant by balancing $H_2O$ gains and losses and regulating its distribution among the fluid compartments of the body.

Most of the rest of the plasma consists of proteins dispersed in solution. Excluding enzymes and certain hormones which are proteins, there are three classes of plasma protein—albumins, globulins, and fibrinogen.

Synthesized in the liver, *serum albumins* are nutritive proteins which contribute to the colloid osmotic pressure of the blood. Serum albumins constitute almost two-thirds of the total concentration of plasma proteins.

The *serum globulins* are a diverse grouping of immune substances known as *antibodies*. Figure 4-11 illustrates the various fractions into which serum globulins are divided. In the blood, the globulins engage in several types of reactions which result in antigenic neutralization.

*Fibrinogen* is the inactive precursor of fibrin, which is necessary for the development of a blood clot. Fibrinogen is synthesized in the liver. The activation of fibrinogen and its role in clot formation are explained in Chap. 4.

## HEART

The heart is a double pump which forces blood to circulate through the blood vessels. The right side of the heart is a pulmonary pump, propelling blood to the lungs. The left side of the heart is a systemic pump, driving blood through the vascular tree in the rest of the body. Each pump consists of two chambers—a superior atrium and an inferior ventricle. The superior or inferior chambers of both pumps contract and relax simultaneously.

**Morphology**  The heart receives relatively unaerated blood laden with nutrients from the veins and pumps nutrient-laden, well-aerated blood through the arteries. All the structural specializations of the heart are designed to provide an optimal pumping efficiency, based on the metabolic demands placed on it. Because the heart beats continuously throughout life, one of its structural specializations includes a unique vascular circuit, or blood supply, of its own.

*Location*  About the size of a large clenched fist, the heart is located in the center of the rib cage, between the lungs. The heart is tilted toward the left from the atria to the ventricles. The right atrium straddles the inside borders of the third and fourth ribs on the right side of the thoracic cavity. The left atrium lies mostly in back of the third rib on the left side. The apex of the ventricles is located behind the fifth rib on the left side, in line with the nipple.

*Gross structure*  The heart is a cone-shaped muscle inside of which are four chambers—*two atria* and *two ventricles*. The right and left atria and ventricles are separated from each other by muscular partitions called *septa*. The *interatrial septum* separates right and left atria. The right and left ventricles are separated by an *interventricular septum*. *Atrioventricular valves* guard the openings between the atria and the ventricles. The *tricuspid valve* is located between the right atrium and right ventricle. Its counterpart between the left atrium and left ventricle is called the *bicuspid (mitral) valve*. The chambers and valves are illustrated in Fig. 30-6, below.

Major veins and arteries enter and leave the heart, respectively. These arteries and veins are collectively referred to as the *great blood vessels*. The *superior* and *inferior venae cavae* open into the *right atrium*. The *pulmonary artery* exits the heart from the *right ventricle*. A *semilunar valve* is located at the base of the pulmonary artery. The *pulmonary veins* enter the *left atrium*. The *ascending aorta* exits the heart from the *left ventricle*. A *semilunar valve* guards the entrance of the ascending aorta. These relationships are shown in Figs. 30-4 and 30-5.

*Tissue organization of the wall*  Three tissue layers compose the walls of the atria and ventricles. From inside outward, these tissues are called *endocardium, myocardium,* and *epicardium (visceral pericardium)*.

The endocardium is a thin tissue which lines the

cavities of the heart chambers. It is continuous with the endothelium, which lines the lumina of the great blood vessels, and contains elastic and smooth muscle fibers and other supporting elements which reinforce the orifices and valves of the heart.

The myocardium, the middle muscular portion of the heart wall, performs the work of the heart. Consisting of striated muscle fibers woven into a syncytial network held together by elastic and reticular connective tissue fibers, the myocardium differs quantitatively in the walls of the various heart chambers. The atrial myocardium, for example, is thin, compared to the ventricular myocardium. In addition, the wall of the left ventricle is about three times as thick as that of the right ventricle. Finally, there are at least three different layers of ventricular myocardium, based on fiber direction, whereas the atrial myocardium does not exhibit this layering effect. These structural differences are reflections of differences in the functional activities of the various chambers. Since the atria pump blood to the ventricles, their thin myocardium is adequate. Because the left ventricle pumps blood through the entire systemic circulation, whereas the right ventricle propels blood a relatively short distance to the lungs, the left ventricular myocardium is stronger and, therefore, thicker than the right ventricular myocardium. The multiple layers of the ventricular myocardium are required for the unique wringing action which literally squeezes out almost every drop of ventricular blood during ventricular systole. The tissue structure of the myocardium is reviewed in Chap. 24.

The epicardium, or visceral pericardium, is attached to the outside surface of the heart. The epicardium consists of a single layer of cells supported by connective tissue elements.

The visceral pericardium makes up the inside layer of a double walled sac which surrounds the heart. The outer layer of the sac is called the *parietal pericardium*. A small amount of serous fluid in the *pericardial cavity* between the visceral and parietal pericardia prevents these membranes from adhering to each other. The double-walled structure is referred to as the *pericardial sac*.

*Valves*   The valves associated with the heart have al-

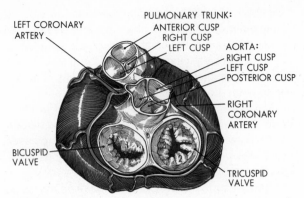

**Figure 30-3**  *The valves of the heart viewed from superior aspect. The atria have been removed, revealing the superior portions of the ventricles. The flaps of the tricuspid and bicuspid valves are numbered. The pulmonary and aortic semilunar valves are also exposed.* (From Melloni, Stone, and Hurd.)

ready been identified. The atrioventricular and semilunar valves are illustrated from superior aspect in Fig. 30-3. In this drawing, the atria have been removed, exposing the bases of the ventricles. The anterior aspect of the heart is at the top of the illustration. The cusps in each of the atrioventricular valves are numbered. The cusps in each of the semilunar valves are identified by name.

The structures attached to the cusps of the atrioventricular valves are complex, although the operation of the valves is entirely passive. Basically, the tricuspid and bicuspid valves open passively when the pressure in the atria exceeds the pressure in the ventricles. Conversely, the valves are forced closed when the intraventricular pressure rises above the intra-atrial pressure. However, because the cusps of the atrioventricular valves offer practically no physical resistance, a mechanism is required to prevent them from being pushed upward into the atrial cavities during ventricular systole.

The mechanism consists of muscular fibers called *chordae tendineae* (tendinous cords) attached to the cusps at their superior ends and continuous with either specialized finger-shaped *papillary muscles* or the ventricular myocardium at their inferior ends. Between 2 and 5 chordae tendineae are attached to each papillary muscle.

During ventricular systole, the papillary muscles also contract. They, in turn, exert tension on the muscular chordae tendineae. When the chordae are taut, the cusps of the atrioventricular valves remain closed and cannot be pushed backward into the atrial chambers.

The operation of the semilunar valves is also passive. The three cusps of each valve offer no resistance and therefore open when the intraventricular pressure exceeds the arterial pressure. The valves close under the opposite conditions. The architecture of the cusps and the pressure relationships between the ventricles and great arteries hold the cusps of each valve together during ventricular diastole.

*The conduction system*   That a heart can beat in the absence of neural and chemical stimulation is indicative of the presence of an intrinsic mechanism for rhythmic myocardial excitation. The mechanism consists of modified bundles of muscle tissue embedded in the septa and walls of the heart.

The conduction system of the heart consists of the following structures: (1) a *sinoatrial* (SA) *node* located in the posterior wall of the right atrium beneath the entrance of the superior vena cava; (2) an *atrioventricular* (AV) *node* located in the medial wall of the right atrium beneath the entrance of the inferior vena cava; and (3) *Purkinje fibers* located in the myocardium of the interventricular septum and ventricular walls. The conduction system is illustrated in Fig. 30-4.

The Purkinje fibers are extensions of the AV node. These muscle fibers course through the superior aspect of the interventricular septum as a single tract called the common atrioventricular *bundle of His*. The bundle of His divides into right and left *branch bundles* in the interventricular septum at a level where the chordae tendineae terminate. Branches from the branch bundles course through all portions of the ventricular myocardium. The Purkinje fibers terminate on the myocardial syncytium next to the endocardium. Excitation from the terminal Purkinje fibers is transmitted to the myocardium. Myocardial excitation spreads from the region next to the endocardium to the region next to the epicardium.

All portions of the conduction system are self-excit-atory. However, the SA node has a higher frequency of endogenous excitation than either the AV node or the interventricular bundle of His. Therefore, the SA node sets the pace of the heart and is appropriately termed the *pacemaker*. Called a *myogenic rhythm*, the intrinsic frequency of excitation of the SA node causes the heart to beat 70 to 80 times per minute in the adult. Each excitation of the SA node is transmitted to the atrial myocardium. Since the myocardium is syncytial, the contraction of one fiber causes a wave of excitation to spread to all portions of the atrial myocardium.

The AV node transmits the wave of excitation from the atrial to the ventricular myocardium. Because of the nature of conduction through the AV node, the wave of excitation is delayed during its passage from the atria to the ventricles. This delay allows the ventricles to fill completely before entering systole.

The wave of excitation is then conducted rapidly through the ventricular myocardium by the Purkinje fibers which originate from the interventricular bundle of His. Ventricular systole is initiated when the excitation is transmitted from the terminal Purkinje fibers to

**Figure 30-4**   *The muscular conduction system of the heart.* (From Melloni, Stone, and Hurd.)

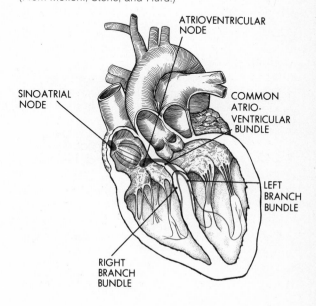

the ventricular myocardium. Ventricular systole begins at the apex of the ventricles and spreads through the compound myocardial wall from the endocardium to the epicardium. Passage of the excitation through the thickness of the myocardium is slow, compared to the velocity of excitation along the Purkinje fibers.

*The coronary circulation* Consisting of arteries, capillaries, and veins, the *coronary circulation* is the blood supply of the heart musculature. A portion of the coronary circulation is shown on the anterior surface of the heart in Fig. 30-5.

The *right* and *left coronary arteries* emerge from the ascending aorta immediately superior to the aortic semilunar valve. Several branches from each of these arteries are shown in Fig. 30-5. Notice that the right and left coronary arteries course along the circumference of the heart in the groove formed between the atria and ventricles. Embedded in fat, branches of these arteries supply the ventricular myocardium with blood. An extensive collateral circulation is established through anastomoses between the coronary arterial branches.

Since the coronary blood is tapped from the ascending aorta, the heart receives the richest supply of nutrients and $O_2$ of any organ in the body. However,

blood circulates through the coronary circulation only during ventricular diastole, because the compressional forces exerted during ventricular systole collapse these vessels.

The ventricular myocardium is drained by *cardiac veins* which run in parallel with the coronary arterial tree. The cardiac veins unite with the *coronary sinus*, a large vein running along the base of the ventricles on the posterior aspect of the heart. The coronary sinus empties directly into the right atrium.

The heart is vulnerable when the coronary blood supply is decreased through occlusions. Diseases of the coronary arteries are explained in the next chapter.

**Physiology** In prose and poetry, the heart has long been considered the seat of emotions as varied as love, compassion, and sorrow. Although there is no hard physiologic evidence to link the heart with emotions, the belief persists and is perpetuated by expressions such as "I love you from the bottom of my heart" and "heartbroken." Perhaps these and other similar expressions are an implicit acknowledgement of the central role which the heart plays in the maintenance of life.

However, that emotions can profoundly affect the heart cannot be denied. The heart rate can accelerate, run away in a galloping rhythm, or even stop temporarily through the actions of the psyche. Other influences also affect the rate and strength of the heartbeat.

*The cardiac cycle* The sequence of events which occurs during one full beat of the heart is called a *cardiac cycle*. During each cardiac cycle, first the atria transport blood to the ventricles, next the ventricles propel blood through the pulmonary artery and the ascending aorta, and finally the atria refill with blood from the venae cavae and pulmonary veins. The average cardiac cycle lasts 0.8 sec in the adult stage of life.

The events during each cardiac cycle are highly synchronized. Right and left atria contract and relax together. Atrial contraction and relaxation are termed *atrial systole* and *diastole*, respectively. About 0.3 sec

**Figure 30-5** *Anterior aspect of the heart showing the coronary circulation.* (From Melloni, Stone, and Hurd.)

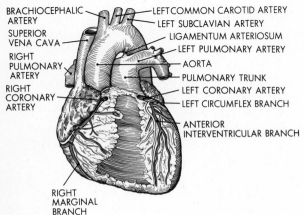

BRACHIOCEPHALIC ARTERY
SUPERIOR VENA CAVA
RIGHT PULMONARY ARTERY
RIGHT CORONARY ARTERY
LEFT COMMON CAROTID ARTERY
LEFT SUBCLAVIAN ARTERY
LIGAMENTUM ARTERIOSUM
LEFT PULMONARY ARTERY
AORTA
PULMONARY TRUNK
LEFT CORONARY ARTERY
LEFT CIRCUMFLEX BRANCH
ANTERIOR INTERVENTRICULAR BRANCH
RIGHT MARGINAL BRANCH

before the onset of atrial systole, the tricuspid and bi-cuspid valves are forced open by the hydrostatic pressure of the blood filling the atrial chambers, and blood begins to trickle into the relaxed ventricles. With the onset of atrial systole, blood is actively pumped from the atrial to the ventricular chambers. The atrioventricular valves close and venous blood again begins to fill the atria as these chambers enter diastole, 0.1 sec following the onset of atrial systole.

The transition between atrial systole and diastole is correlated with the onset of the synchronous contraction of both ventricles. Contraction of the ventricles is called *ventricular systole*. The atrioventricular valves close at the onset of ventricular systole, thereby preventing ventricular blood from backing up into the atria. The increasing intraventricular blood pressures created as ventricular systole progresses then force open the semilunar valves at the bases of the pulmonary artery and ascending aorta. The time lapse between the closure of the atrioventricular valves and the opening of the semilunar valves is less than 0.1 sec. Opening of the semilunar valves permits ventricular blood to be pumped through the great arteries while ventricular systole is in progress.

Ventricular systole lasts more than three times as long as atrial systole. Following the completion of ventricular systole, both ventricles relax in unison. Ventricular relaxation is referred to as *ventricular diastole*. At this point, both atria and ventricles are in diastole and will remain together in diastole for about 0.4 sec. Almost immediately following the onset of ventricular diastole, the semilunar valves close, preventing blood in the great arteries from flowing back into the ventricles. A fraction of a second after the closure of the semilunar valves, the atrioventricular valves open in response to the increasing hydrostatic pressure of the blood filling the atria. However, the onset of atrial systole is still about 0.3 sec away, even though the atrioventricular valves are open. Ventricular diastole is terminated approximately 0.1 sec after the initiation of the succeeding cardiac cycle.

### The circulation of blood through the heart    Even though the events of the cardiac cycle occur simultaneously on both sides of the heart, the heart consists functionally of two separate pumps. One pump is *pulmonary*, the other *systemic*. Therefore, it is instructive to trace the separate paths of circulation through the heart and its great blood vessels.

The circulatory pathway through the right side of the heart is traced in Fig. 30-6A. The right atrium and ventricle and their associated blood vessels are viewed from anterior aspect. Unaerated venous blood enters the right atrium through the superior and inferior venae cavae and the coronary sinus during atrial diastole. This blood is actively pumped through the opened tricuspid valve into the right ventricle during atrial systole. When the right ventricle enters systole, the tricuspid valve closes and the pulmonary semilunar valve is forced open. Ventricular blood is therefore propelled through the pulmonary artery toward the lungs.

The circulatory pathway through the left side of the heart is traced in Fig. 30-6B. The left atrium and ventricle and their associated blood vessels are viewed from posterior aspect. Aerated venous blood from the lungs enters the left atrium through the pulmonary veins during atrial diastole. This blood is actively pumped through the opened bicuspid valve into the left ventricle during atrial systole. When the left ventricle enters systole, the bicuspid valve closes and the aortic semilunar valve is forced open. Ventricular blood is therefore propelled through the ascending aorta to the rest of the body.

### Modification of the heartbeat    Although the cardiac rhythm is myogenic, the rate and strength of the heartbeat are modified by a variety of physical and psychological factors which result in appropriate neural or chemical stimulation. These stimuli affect the conduction system, the myocardium, or specialized receptors in the vascular walls. The heartbeat is influenced by autonomic stimulation, hormones secreted by the adrenal medullae, changes in the volume and pressure of blood entering and leaving the heart, a depression in blood $O_2$, and an excess of blood $CO_2$.

1   *Autonomic stimulation.* The heart is innervated by sympathetic and parasympathetic fibers of the

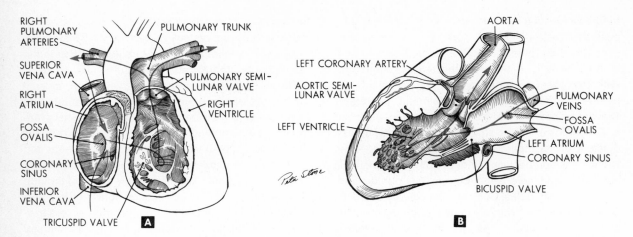

**Figure 30-6** *The circulation of blood through the heart.* A. *Right side of the heart.* B. *Left side of the heart.* (From Melloni, Stone, and Hurd.)

autonomic nervous system. The preganglionic *sympathetic* fibers of the cardiac plexus originate from spinal cord segments T1 to T5. The postganglionic fibers of the cardiac plexus innervate the right and left ventricular myocardium. The synaptic transmitter norepinephrine is released by the postganglionic fibers. This transmitter substance causes an increase in the rate and strength of the heartbeat. The *parasympathetic* innervation of the heart, on the other hand, is supplied by fibers of the *vagus nerve* (Xth cranial nerve). Most postganglionic fibers which form synaptic connections with the right vagus nerve innervate the SA node. The cardiac distribution of postganglionic fibers from the left vagus nerve is mostly to the AV node. Parasympathetic stimulation of the vagus nerve causes the release of the neurotransmitter acetylcholine (ACh) from the terminal axons of the postganglionic fibers. ACh decreases the excitability of the SA and AV nodes, thereby decreasing the heart rate or even temporarily arresting the heartbeat. Stimulation of the right vagus differentially affects the SA node, slowing the heart rate. Left vagal stimulation acts differentially on the AV node, prolonging the conduction time between the atria and ventricles. The quantitative effect depends upon the magnitude of stim-

ulation. Because sympathetic and parasympathetic excitation occur simultaneously, the cardiac effect of autonomic stimulation is determined by the relative amounts of norepinephrine and ACh stimulating the myocardium.

2   *Hormonal stimulation.* The hormones epinephrine and norepinephrine are secreted by the adrenal medullae during stress or excitement. Called catecholamines, these hormones are *sympathomimetic*, i.e., they mimic the action of the sympathetic division of the autonomic nervous system. Their sympathomimetic action is not too difficult to understand when we recall that norepinephrine is the neurotransmitter liberated by the sympathetic nerve fibers which innervate the myocardium. Epinephrine increases the frequency and vigor of the heartbeat. Norepinephrine causes cardioacceleration only.

3   *Blood pressure. Stretch receptors* are located in the walls of the right atrium, the great blood vessels, and vessels which compose the proximal anastomosis. These receptors are also called *pressoreceptors*, or *baroreceptors*, because of their response to pressure or tension. The receptors are excited when stretched by the lateral pressure exerted by an increasing volume of blood. A decreased volume of blood results in a

decreased excitability of the receptors. The stretch receptors therefore transduce mechanical energy into nerve impulses. Change in the degree of stimulation of the stretch receptors affects the excitability of the heart rate and vasomotor centers in the medulla oblongata of the brain. When the stretch receptors therefore transduce mechanical stimulated, the heart rate center is excited. In addition, an increased intraatrial pressure places the myocardium under a greater tension. Therefore, the response to an elevated venous return is an increased rate and strength of the heartbeat. This automatic response is called the *right heart reflex*. When aortic receptors are stretched in response to an increased cardiac output, the heart rate undergoes a compensatory reduction, while the blood vessels dilate. This action tends to reduce rising blood pressure, thereby maintaining blood pressure homeostasis. This response is referred to as the *aortic reflex*. Similarly, stretch of the wall of the carotid sinus at the base of the internal carotid artery due to an increase in blood pressure also depresses the heart rate and dilates the blood vessels. Modulation of blood pressure through the actions of these receptors is termed the *carotid sinus reflex*. All three circulatory reflexes are transient and cannot sustain their homeostatic influences on blood pressure for long.

4 *Nonhormonal chemical stimulation.* The heart rate is affected by the blood concentrations of $O_2$ and $CO_2$. The chemoreceptors which monitor the concentrations of these substances are located in the vascular walls of the aortic arch and carotid sinus. The chemoreceptors are stimulated by a depression in blood $O_2$ and an excess of blood $CO_2$. Nerve impulses from the chemoreceptors mainly stimulate the respiratory center which, in turn, excites the heart rate center. The subsequent increase in blood flow to the lungs promotes its ability to correct for the blood imbalances and also tends to compensate for the diminution in blood $O_2$.

**The electrocardiogram (ECG)** The electrocardiogram (ECG) is a graphic record of the electrical activity of the heart during each cardiac cycle. Minute electric currents are generated by the ionic fluxes of $Na^+$ and $K^+$ moving into and out of the myocardium during systole and diastole, respectively. These rhythmic changes in voltage in the region of the heart can be detected on the surface of the body with appropriate equipment.

The equipment used to record ECGs includes skin electrodes and an electrocardiograph. The skin electrodes are attached to the wrists, one ankle, and to specified positions across the left chest. The electrodes transduce, or convert, the electrical potentials generated by the heart into electric signals. The signals are conducted along wires connecting the electrodes with the electrocardiograph. The electrocardiograph processes and displays the variations in the electrical activity of the beating heart. Serial recordings are made from different electrode combinations to provide a better diagnostic profile of heart function.

A typical ECG is displayed in Fig. 30-7. Electrocardiograms (ECGs) are usually recorded on graph paper ruled in millimeters and moving at a speed of 2.54 cm (1 in.)/sec. Notice that the variations in electrical activity shown in Fig. 30-7 correspond to one cardiac cycle lasting 0.8 sec. The electrocardiograph is calibrated so that 10 mm (1 cm) in the vertical direction equals 1 millivolt (mV). The elapsed times are given in tenths of a second from the onset of atrial systole.

There are three basic waves in an ECG—the P wave, the QRS complex, which consists of separate Q, R, and S waves, and the T wave. The most prominent wave in the complex is the R wave.

The *P wave* marks the onset of atrial systole. During atrial systole, an increasing concentration of $Na^+$ moves from the interstitial fluid surrounding the heart into the syncytial fibers of the atrial myocardium. The P wave is a reflection of this ionic influx. Since the interstitial fluid is initially electropositive to the intracellular fluid, the progressive movement of extracellular $Na^+$ into the atrial muscle fibers decreases the charge differential across their membranes. Called *depolarization*, this phenomenon also occurs in neurons and is discussed in Chap. 13. Once a portion of the membrane is sufficiently depolarized, a progressive wave of depolarization passes from point

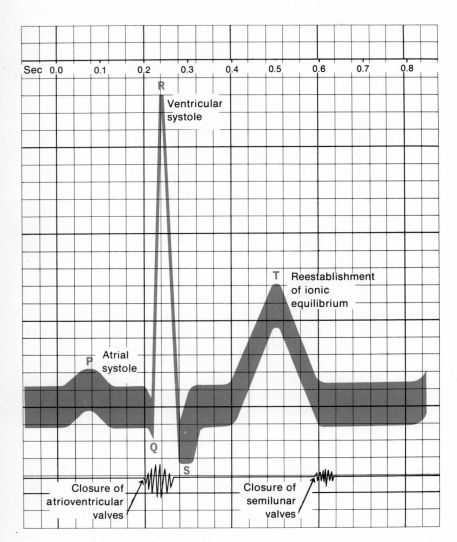

**Figure 30-7** *The electrocardiogram (ECG) and phonocardiogram.*

to point along the entire length of the muscle fiber membrane. As a consequence, an action potential spreads to the interior of the fiber. (Intramuscular action potentials are explained in detail in Chap. 24.) The chemical changes caused by the action potential result in contraction of the fiber. Because the myocardium is syncytial, the excitation of one fiber spreads throughout the entire syncytium. Due to progressive excitation, the atrial myocardium enters systole.

In response to the initial influx of $Na^+$ during depolarization, intracellular $K^+$ secondarily moves across each of the muscle fiber membranes and enters the interstitial fluid. Called *repolarization*, the efflux of $K^+$ restores the electrical balance disrupted during depolarization. The reestablishment of an electrical equilibrium marks the onset of atrial diastole. Since repolarization is caused by an ionic flux, an electric current is generated. However, because the completion of atrial systole is correlated with the initiation of ventricular systole, a repolarization wave is not detected during the waning moments of atrial systole.

The *QRS complex* marks the onset of ventricular systole. Since depolarization initiates systole, the QRS complex is a depolarization grouping of waves. The ionic events which occur during ventricular systole are exactly like those which take place during atrial systole. The complexities of the wave forms and the magnitudes of the voltages generated during ventricular systole are caused by the spatial and temporal progression in which depolarization spreads through the myocardium.

The *T wave* occurs at the end of ventricular systole, just as the ventricles enter diastole. Caused by the movement of $K^+$ from the syncytial fibers of the ventricular myocardium to the interstitial fluid during the reestablishment of electrical equilibrium, the T wave is a repolarization wave. Unlike the repolarization wave of the atrium, the T wave is not obscured by other electrical events, because the atria and ventricles remain in diastole for a short time following ventricular repolarization.

The ECG provides information on cardiac rhythmicity and the sequence of, and relationships among, the events during a cardiac cycle. The findings are indicative of the integrity of the myocardium, its coronary blood supply, and the cardiac conduction system.

*Heart sounds*  The "lub-dub" sounds produced by a beating heart are familiar to everyone. By-products of the cardiac cycle, these sounds are created when blood slaps against the just-closed atrioventricular and semilunar valves at the onset of ventricular systole and diastole, respectively.

The major heart sounds are correlated with the ECG in Fig. 30-7. The intensity or magnitude of each sound is suggested by the relative heights of the waves.

The "lub," or first heart sound, is loud and low in pitch. Produced during the QRS complex, the lub sound is caused by the backlash of blood against the undersurfaces of the just-closed atrioventricular valves when the ventricles enter systole. The ventricular walls reverberate mechanically and the sound waves spread to the thoracic wall.

The "dub," or second heart sound, is sharper and

higher in pitch than the lub sound. Produced at the end of the T wave, the dub sound is created when blood in the pulmonary artery and ascending aorta strikes the distal surfaces of the just-closed semilunar valves, as the ventricles enter diastole. Here, the arterial walls are responsible for the reverberations which initiate the spread of the sound waves to the surface of the chest.

Heart sounds are employed in the diagnosis of valvular dysfunctions. Valvular diseases are reviewed in the next chapter.

## BLOOD VESSELS

Blood vessels are porous and variably distensible tubes which carry blood throughout the body as well as participate in the exchange of energy nutrients, $H_2O$, respiratory gases, hormones, electrolytes, and certain other materials between the blood and tissue cells. Blood vessels consist of arteries, capillaries, and veins.

Blood vessels usually interconnect with several other vessels, giving rise to arterial, capillary, and venous anastomoses, or plexuses. The networks of vessels established through anastomoses form collateral circulatory beds surrounding the main vascular channels. The *collateral vessels* are capable of carrying a greater blood volume or of opening up, in the event of a primary vascular occlusion in a localized region of the body. Blood pressure fluctuations in the tissues are also reduced because of extensive interconnections between blood vessels.

Generally, one or more veins and a nerve run next to, and in parallel with, each artery. The artery, veins, and nerve are wrapped together in one unit by a connective tissue covering which provides some degree of support.

The smooth muscle walls of most arteries and veins are innervated primarily by fibers of the sympathetic division of the autonomic nervous system. Some blood vessels also receive a parasympathetic nerve supply. Vasoconstriction and dilation are regulated by neural stimulation and chemical mechanisms.

Only the innermost tissue layer of a blood vessel is nourished by the blood bathing it. This arrangement meets the metabolic needs of the single layer of squamous cells that make up the walls of blood capillaries but is inadequate for the sustenance of the tissues which compose the walls of arteries and veins. Instead, blood is supplied to, and drained from, the smooth muscle walls of larger blood vessels by a specialized portion of the vascular system known collectively as the *vasa vasorum.*

Based on the paths described by blood following its departure from, and prior to its reentry into, the heart, there are three main vascular circuits in the body: (1) the *pulmonary circuit*, which transports blood from the right ventricle to the left atrium by way of the lungs; (2) the *systemic circuit*, which transports blood from the left ventricle to the right atrium by way of the vascular tree throughout the rest of the body; and (3) the *coronary circuit*, which transports blood from the left ventricle to the right atrium by way of the heart musculature, or myocardium. Each of these circuits consists of arteries, capillaries, and veins.

**Arteries** Arteries transport blood *away* from the heart to capillaries in intimate association with the tissue cells of the body. The vessels which carry blood away from the heart include the pulmonary artery from the right ventricle and the ascending aorta from the left ventricle. The bases of these vessels are in excess of 1 in. in diameter and contain semilunar valves (see above). The pulmonary artery carries blood to the lungs. The aorta distributes blood to the rest of the body.

Passing through the pericardium, the ascending aorta leads directly into the arch of the aorta. The right and left coronary arteries are branches of the ascending aorta. The aortic arch courses to the left of and posterior to the heart, ultimately directing the systemic blood in an inferior direction. A grouping of major arteries leading to the head and upper extremities arises from the arch of the aorta close to the heart. Called the *proximal anastomosis*, these blood vessels include the *brachiocephalic artery* on the right side of the arch and the *left common carotid artery* and *left subclavian artery* on the left side of the arch. The brachiocephalic artery gives rise to the *right common carotid artery* and *right subclavian artery*. In the arms, the subclavian arteries are called *axillary arteries*.

The continuation of the arch of the aorta is called the *descending aorta*. The aorta is the main distributing artery of the body. The descending aorta is named for the regions of the body in which it is located, i.e., the aorta in the thoracic region is called the *thoracic aorta*, and that portion in the abdominal region is referred to as the *abdominal aorta*. The abdominal aorta gives off numerous distributing arteries to the abdominal viscera; namely, the celiac, superior mesenteric, adrenal, renal, gonadal (spermatic or ovarian), and inferior mesenteric branches. The aorta branches in the pelvic region, forming the left and right common iliac arteries. These vessels supply arterial blood to the lower extremities.

The major arteries mentioned above and their branches are illustrated in Fig. 30-8.

The smaller arteries which change almost imperceptibly into capillaries are called *arterioles*. Although the largest arteries are more than 1 in. in diameter, the smallest arterioles are no larger than about $14\mu m$ in diameter. Arteries are distensible because of the presence of a large amount of elastic fibers mixed in with their smooth muscle walls. Arteriolar walls, on the other hand, contain proportionately less elastic connective tissue than those of arteries and therefore resist stretch more forcefully. The structural properties of arterioles enable them to exert primary control in the regulation of blood pressure through vasoconstriction and vasodilatation. Arteriolar regulation of blood pressure is discussed later in this chapter.

A comparison of arteries with veins reveals the following differences: (1) arteries possess more rigid, thicker walls than veins; (2) the lumen of an artery is smaller than that of a corresponding vein; (3) an artery will not collapse when the blood pressure drops excessively, whereas a vein under similar conditions will collapse; and (4) an artery is more or less spherical, whereas a vein possesses an irregular circumference. These differences are due to the greater amount of smooth muscle tissue in the wall of an artery, compared to that in its corresponding vein.

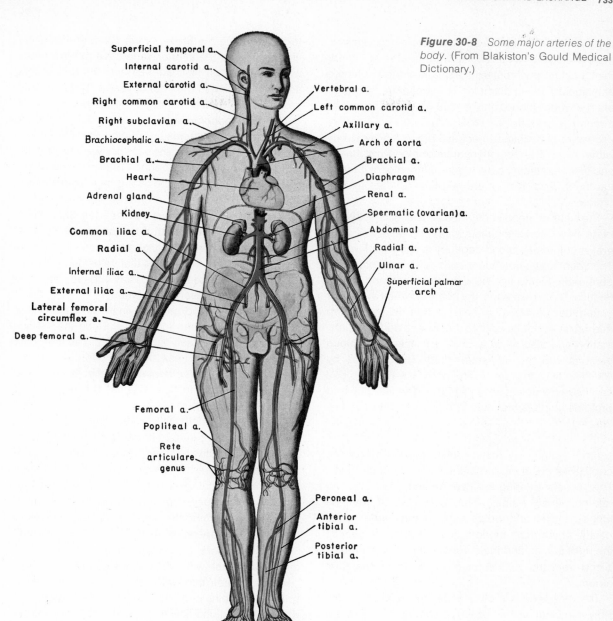

Superficial temporal a.

Internal carotid a.

External carotid a.

Right common carotid a.

Right subclavian a.

Brachiocephalic a.

Brachial a.

Heart

Adrenal gland

Kidney

Common iliac a.

Radial a.

Internal iliac a.

External iliac a.

Lateral femoral circumflex a.

Deep femoral a.

Femoral a.

Popliteal a.

Rete articulare genus

Vertebral a.

Left common carotid a.

Axillary a.

Arch of aorta

Brachial a.

Diaphragm

Renal a.

Spermatic (ovarian) a.

Abdominal aorta

Radial a.

Ulnar a.

Superficial palmar arch

Peroneal a.

Anterior tibial a.

Posterior tibial a.

**Figure 30-8**  *Some major arteries of the body.* (From Blakiston's Gould Medical Dictionary.)

**Capillaries** Capillaries connect arteries to veins or, more precisely, link arterioles with venules. No larger than a single erythrocyte, the smallest capillaries are only about 7 $\mu$m in diameter. The exchange of materials between blood and tissue cells takes place through the walls of blood capillaries. *Capillary exchange* is facilitated because blood moves slowly within them. The hemodynamics of blood capillaries and the interactions between capillary blood, interstitial fluid, and lymph are explained later in this chapter.

The walls of blood capillaries consist of one layer of large, flattened cells called *squamous cells*. The general structure of a blood capillary is shown in Fig. 4-4. The *squamous endothelial cells* form a differentially permeable membrane between the blood and the interstitial fluid. In addition, the pores between every four contiguous endothelial cells permit leukocytes to migrate from the blood into the tissues during inflammation and also allow a continual leakage of blood proteins into the interstitial fluid. The process by which leukocytes crawl through the walls of blood capillaries is described in Chap. 4. The fate of blood proteins in interstitial fluid is considered later in this chapter.

**Veins** Veins carry blood *toward* the heart from the capillaries that surround the tissue cells of the body. The vessels entering the heart include the right and left pulmonary veins, which drain blood from the lungs to the left atrium, the superior and inferior venae cavae, which drain blood from the rest of the body to the right atrium; and the coronary sinus, which drains blood from the ventricular myocardium to the right atrium.

The venous ends of capillaries transform imperceptibly into small veins known as *venules*. The smallest venules are approximately 14$\mu$m in diameter. The venules in each of the organs of the body coalesce, forming larger vessels called *veins*. As veins proceed toward the heart, their diameters progressively increase.

Blood from the lower extremities is drained by the *common iliac veins*. These veins merge in the pelvic region, forming the *inferior vena cava*. The common iliac veins and the inferior vena cava run in parallel with the common iliac arteries and the descending aorta, respectively. The inferior vena cava is the main vein draining the inferior region of the body.

The *brachiocephalic veins* receive blood from the head and upper extremities. These veins merge with the *superior vena cava*. The superior vena cava is the main vein draining the superior region of the body. The major veins mentioned above and their branches are shown in Fig. 30-9.

Veins are more numerous than arteries. Since veins are larger in diameter than their corresponding arteries, their total cross-sectional area is also greater. In a practical sense, this means that veins can hold significantly more blood than arteries. Indeed, the veins are an important reservoir of blood.

The average blood pressure in veins is more than three times lower than in arteries. Since the venous blood pressure by itself cannot maintain an adequate return of blood to the heart, the driving force of venous blood must be supplemented by other factors. The major contribution to this external driving force is called the *venous pump*.

The venous pump is a mechanical apparatus which assists the return of blood to the heart. The venous pump is activated every time a skeletal muscle is contracted, thereby pressing against an adjacent vein. The physical pressure from an active skeletal muscle partially or completely collapses one or more nearby veins, forcing the blood mechanically toward the heart. The operation of the pump is dependent upon three factors: (1) skeletal muscular activity; (2) the relative ease with which veins can be collapsed because of their thin walls; and (3) one-way valves, primarily in veins of the upper and lower extremities, which permit blood flow toward the heart and prevent backflow.

Numerous valves are found in each peripheral vein. A venous valve consists of two semilunar flaps which are attached to the inside of the vascular wall. The free ends of the flaps project into the vascular lumen, pointing toward the heart. A drop in venous blood pressure causes the blood within the vein to slap

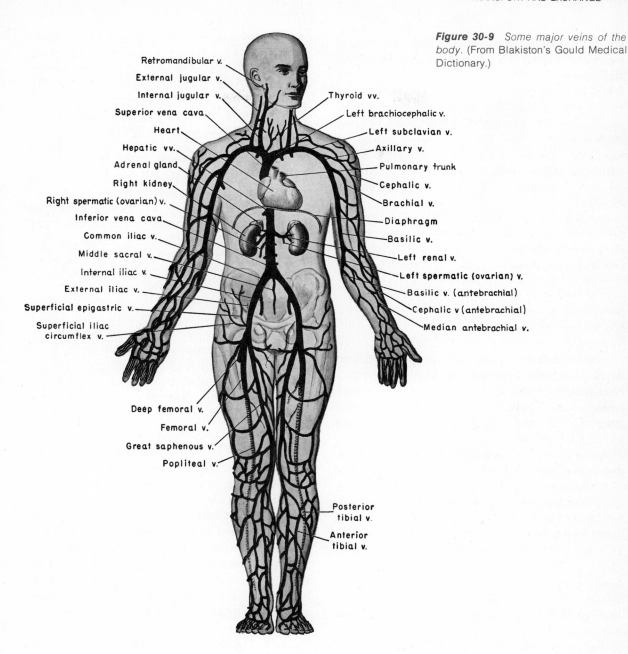

**Figure 30-9** *Some major veins of the body.* (From Blakiston's Gould Medical Dictionary.)

Retromandibular v.
External jugular v.
Internal jugular v.
Superior vena cava
Heart
Hepatic vv.
Adrenal gland
Right kidney
Right spermatic (ovarian) v.
Inferior vena cava
Common iliac v.
Middle sacral v.
Internal iliac v.
External iliac v.
Superficial epigastric v.
Superficial iliac circumflex v.

Thyroid vv.
Left brachiocephalic v.
Left subclavian v.
Axillary v.
Pulmonary trunk
Cephalic v.
Brachial v.
Diaphragm
Basilic v.
Left renal v.
Left spermatic (ovarian) v.
Basilic v. (antebrachial)
Cephalic v. (antebrachial)
Median antebrachial v.

Deep femoral v.
Femoral v.
Great saphenous v.
Popliteal v.

Posterior tibial v.
Anterior tibial v.

against the proximal surfaces of the flaps. The force of the blood closes the valves, thereby preventing appreciable backflow.

Not all veins have valves. In general (1) the larger veins do not contain valves; (2) the most peripheral veins, where the blood pressure is lowest, possess valves; and (3) veins with the greatest number of valves are located in the most physically active peripheral regions of the body, i.e., where the need is greatest for the maintenance of an adequate venous return.

Significant changes in portions of the vascular system occur during the transition from prenatal to postnatal life. Initiated by the process of birth, these changes are reviewed in Chap. 19.

## HEMODYNAMICS

The cardiovascular system is essentially a closed system of porous, distensible tubes connected to a double pump. The fluid is forced through the system under pressure generated by the pulsating action of the pump. The 5 to 7 liters of fluid make one complete circuit each minute.

Unlike a circulating hot water heating system, however, the cardiovascular system is not a grouping of rigid pipes within which is contained an invariable quantity of fluid. The blood in the cardiovascular system is subject to intrinsic and extrinsic influences which may increase or decrease its total volume and viscosity. Similarly, changes in the diameter of blood vascular walls can increase or decrease the *blood velocity*, i.e., the distance traversed in a unit of time, and the *blood flow rate*, i.e., the total volume passing a given point in a unit of time. These factors in turn affect the blood pressure of the system.

Hemodynamic studies attempt to assess the interrelationships among the diverse *physical* factors which influence the flow rate and velocity of blood. Because these factors affect blood pressure, our knowledge of hemodynamics must of necessity begin with an appreciation of the meaning of blood pressure.

## BLOOD PRESSURE

Blood pumped from the heart is under a pressure greater than that in any other portion of the circulatory system; otherwise, blood would not flow from the arteries to the veins. The kinetic energy causing blood to flow is imparted to arterial blood leaving the heart during ventricular systole. However, the pressure of arterial blood fluctuates with the phases of the cardiac cycle. Although there are wide variations due to differences in age, height, sex, physiologic status, and degree of physical activity, the arterial pressure during the peak of ventricular systole in an average adult is 120 mmHg. Called the *systolic pressure*, this value reflects the height to which a column of Hg would rise in a manometer during systole if its base were inserted through an arterial wall into the circulating blood. The column of Hg rises until its weight exactly counterbalances the pressure of the blood. At the peak of ventricular diastole, on the other hand, the arterial pressure diminishes to an average minimum of only 80 mmHg. This value is called the *diastolic pressure*. The difference between peak systolic and diastolic pressures, i.e., 120 − 80 mmHg, or 40 mmHg, is the *pulse pressure*.

Physical factors which influence blood pressure The blood pressure varies between individuals matched in age, height, sex, weight, and physical activity, as well as in one individual from time to time. The major physical determinants of blood pressure are the volume of blood pumped by the heart in a unit of time, i.e., the *cardiac output*, and the rate of dissipation of the kinetic energy imparted by the heart to the blood entering the circulation, i.e., the *peripheral resistance*. Each of these determinants will now be considered in turn.

*Cardiac output* Let us first calculate the volume of blood that enters the aorta from the left ventricle during each beat of the heart in an average adult. Assume that the resting heart beats 72 times a minute, that the average volume of circulating blood is 6 liters (6,000 ml), and that the entire volume of blood is pumped through the heart once each minute. Under these conditions, the cardiac output per beat is slightly over 80 ml (6,000 ml/min divided by 72 beats/min). The average quantity of blood ejected during systole is called the *stroke volume*. The quantity of blood pumped each minute, 6 liters in our example, is referred to as the *minute volume*.

The cardiac output is influenced by the volume of venous blood entering the heart during atrial diastole as well as by the frequency of the heartbeat. The blood flow rate through the venous circulation is controlled by the venous blood pressure or, more precisely, by the difference between the venous and right atrial pressures. The venous blood pressure is in turn related to the average diameter of the veins. As the veins increase in diameter, a larger quantity of blood can be stored temporarily within them, thereby reducing the blood pressure and flow rate to the heart. A decrease in the diameter of the veins has the opposite effect.

*Vasoconstriction*, or a decrease in the diameter of blood vessels, and *vasodilatation*, or an increase in their diameters, are regulated by neural and chemical influences. Most blood vessels are innervated only by fibers of the sympathetic division of the autonomic nervous system. An increase in the frequency of nerve impulses increases vasoconstriction, and vice versa. The autonomic innervation of blood vessels is explored further in Chap. 14. The hormone norepinephrine and the protein angiotensin, both of which mimic the action of the sympathetic division of the autonomic nervous system, also cause vasoconstriction. The vascular effects of these natural products are discussed in Chaps. 18 and 22, respectively.

Within normal physiologic limits, the flow rate of venous blood entering the right atrium passively determines the strength with which the myocardium contracts. As the myocardium is stretched by the pressure of the blood against the inside walls, it is placed under increasing tension. The strength of the myocardium is proportional to its degree of stretch. Thus, an increasing volume of blood within the heart results in more forceful cardiac contractions. The heart therefore adjusts its output to its load. This relationship is of such wide general applicability at various heart rates that it is called *Starling's law* of the heart, in

honor of the investigator who first formulated it. The relationship between the strength of striated muscle fibers and their degree of tension is considered further in Chap. 24.

*Peripheral resistance* The kinetic energy imparted to arterial blood by the beating heart is dissipated as the blood flows through the vascular tree. The factors which resist the flow of blood are collectively called *peripheral resistance*. Peripheral resistance is a function of the viscosity of blood and the diameter of the vessels through which the blood flows.

The aphorism that "blood is thicker than water" is certainly correct in a literal sense. Indeed, blood is approximately 3.8 times more viscous than water. Viscosity is reviewed in Chap. 7. The viscosity of blood is due primarily to the concentration of its formed elements and secondarily to the quantity of its plasma proteins. Therefore, factors which influence the viscosity of blood include the level of hemopoietic, or blood cell manufacturing, activity and the osmolarity (see Chap. 10), or relative concentration, of the blood. The important concept to remember is that the *flow rate of blood is inversely proportional to its viscosity.*

The percentage of formed elements in a given volume of whole blood is the *hematocrit*. It is measured by centrifuging a specified quantity of whole blood placed in a calibrated tube and observing the reading at the separation between the layer of formed elements and the layer of plasma. A normal hematocrit ranges between 40 and 45. An elevated hematocrit is reflected in an increased viscosity of blood; a depressed hematocrit is indicative of a decreased viscosity of blood. Pathologic factors which affect the hematocrit are cited in the next chapter.

The major component of peripheral resistance is regulated by adjustments in the diameters of the blood vessels. Vasoconstriction and dilatation are controlled by the tone of the smooth musculature in the walls of the blood vessels. The resiliency or tension is determined by the nature and amount of the collagenous and elastic fibers in the connective tissue of the blood vascular walls.

The physical factors which influence the rate of dissipation of the kinetic energy of blood are diagramed

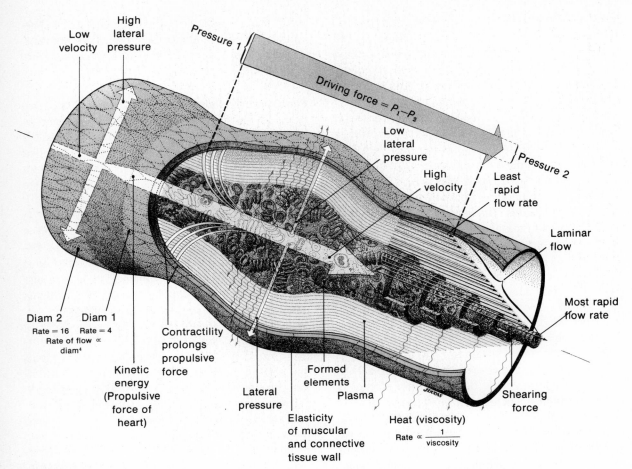

**Figure 30-10** *Some physical characteristics of blood flow. The passage of a pulse of blood causes the surrounding wall of an artery to elastically recoil, prolonging each pulsation and evening out pressure fluctuations. During laminar blood flow, concentric layers of blood exert shearing forces as they move past one another. The highest flow rates are located in the center of a vessel, where the formed elements are concentrated; the lowest flow rates are found at the periphery, where plasma is concentrated.*

in Fig. 30-10. As we have previously observed, the heart ejects a pulse of about 80 ml of blood into the ascending aorta with a pressure of 120 mmHg during ventricular systole. The kinetic energy transferred to the pulse of blood is gradually dissipated during circulation by two factors: (1) the *lateral pressure* exerted against the inside walls of the blood vessels, and (2) the *heat energy* lost through friction.

In Fig. 30-10, notice that a circulating pulse of blood actually has two pressures—a forward component and a lateral component. The forward component is determined by the residual kinetic energy in the pulse of blood and the diameter of the vessel in which the pulse is located. The lateral component is inversely proportional to the velocity of the blood. In other words, as the velocity increases, the lateral pressure decreases, and vice versa. To understand the reasons behind these relationships, we must look more closely at the effects of changes in the diameter of a blood vessel on circulating blood contained within it.

Figure 30-10 shows that the *flow rate* of blood, i.e., the volume of blood passing a given point in a unit of time, is *proportional to the fourth power of the diameter of the vessel in which it is located.* For example, if the vessel is originally 1 cm in diameter and could double its diameter, the flow rate of blood within it would be 16 times greater than initially. What the fourth power function means in a practical sense is that slight changes in the diameter of a blood vessel cause dramatic changes in its blood flow rate. When considering the effects of changes in vascular diameter on flow rate, we of course must assume that the cardiac output and blood viscosity and volume remain constant.

Blood flows in an essentially closed system under a relatively constant head of pressure. Therefore, a decrease in the diameter of a blood vessel will increase the velocity of its blood, i.e., the distance blood moves in a unit of time, so that the same average flow rate is maintained. Conversely, the velocity of blood will decrease within a vessel as its wall undergoes dilatation.

Thus, we conclude that *blood pressure is proportional to the tension exerted by the walls of a vessel and inversely proportional to the vascular radius.* This relationship is summarized as follows:

$$P\text{(pressure)} = \frac{T\text{(tension)}}{r\text{(radius)}} \quad \text{or} \quad P = \frac{T}{r} \qquad \text{(30-1)}$$

Equation (30-1) states that blood pressure is equal to the tension divided by the radius. If the driving force of the blood remains constant, a decrease in the radius of a blood vessel causes a corresponding increase in its tension and resistance to blood flow. In other words, the smallest vessels exert the greatest tension at a given pressure. This fact explains why capillaries close when the blood pressure falls below a critical value—the tension of the vascular wall overcomes the pressure exerted by the blood.

The lateral pressure of the blood decreases as a vessel undergoes vasoconstriction because of an increased tension exerted by the vascular wall. A decrease in vascular diameter therefore also reduces the amount of kinetic energy dissipated through lateral pressure. The result is an increase in the forward pressure and velocity of the blood. However, now an increasing amount of kinetic energy is dissipated through frictional resistance encountered during blood flow.

Frictional resistance is greatest next to the inside walls of a blood vessel, where moving fluid encounters a stationary solid surface, and least in the center of the vessel. Assuming that the flow of blood is smooth, a condition called *laminar flow*, the flow rate at any cross-sectional location in the vessel will be decreased in proportion to the magnitude of the frictional resistance encountered by the blood. Because of differential flow rates, the blood is separated into a series of concentric layers called *laminae*, as shown in Fig. 30-10. As the laminae slide past each other, kinetic energy is dissipated as heat. Heat is evolved because of shearing forces generated between adjacent layers. Notice that the formed elements of blood tend to accumulate in the most rapidly moving laminae in the center of the vessel, while the plasma tends to occupy the more peripheral, slowest moving laminae.

Frictional resistance to blood flow increases during vasoconstriction because of a progressive reduction in the distance between the center and the periphery. All the blood in a minute vessel encounters a uniformly high frictional resistance because very little distance separates the center of the vessel and its wall.

Blood flow is ordinarily smooth, or laminar, even though the heart pump is pulsatile. The maintenance of a constant head of pressure in arterial blood is due to the distensibility of the vascular walls, which tends to dampen pressure fluctuations caused by peak systolic and diastolic pressures. Vascular elasticity contributed by smooth muscle and elastic connective tissues enables each portion of an arterial wall to stretch in turn as a pulse of blood passes by. Vascular contractility provided by the smooth musculature in response to passive stretch pushes the pulse of blood smoothly along the length of an artery, in spite of the fact that the pulse pressure averages 40 mmHg. The action of a vascular wall in maintaining the propulsive force of the blood is conceptualized in Fig. 30-10. Notice that some of the kinetic energy dissipated by the lateral pressure of the blood is converted into a passive wavelike contraction of the vascular wall which maintains the propulsive force of the blood between beats.

The pressure of blood after leaving the heart depends on the part of the vascular tree in which the blood is located. The *mean arterial pressure,* i.e., the average pressure generated between the peak of systole and diastole, is near 100 mmHg. By the time blood is returned to the heart, its pressure has been reduced to zero. However, because very little frictional resistance is encountered by blood in arteries, the blood pressure is still about 80 mmHg when blood reaches the arterioles. The arterioles are the most important vascular regulators of blood pressure, since they offer the greatest resistance to flow of any single grouping of blood vessels. Thus, when blood reaches the arteriolar ends of blood capillaries, its pressure measures approximately 30 mmHg. In other words, the blood pressure has dropped by 50 mmHg during the transport of blood through the arterioles. As blood passes from the arteriolar to the venular ends of the blood capillaries, its pressure falls from 30 mmHg to about 10 mmHg, for a total pressure decrease of 20 mmHg. The remainder of the blood pressure is dissipated during the journey of blood through the venules and veins. Therefore, based on a mean arterial pressure of 100 mmHg, blood pressure decreases by an average of 50 percent in arterioles, 20 percent in both arteries and capillaries, and 10 percent in the venous portion of the circulatory system. These relationships are summarized in Fig. 30-11.

Earlier, we learned that the velocity of blood under a constant pressure is inversely related to the diameter, or cross-sectional area, of the vessel in which the blood is flowing. The situation is analogous to the velocity of water flowing through the nozzle of a garden hose before and after a portion of the hose is partially constricted. Assuming that the water pressure does not change, a decrease in the diameter of the hose will increase the velocity and, therefore, the pressure of the water leaving the nozzle. The faster-moving but more-concentrated stream of water compensates for

the reduced flow rate due to partial constriction. The result is that the same average flow rate is maintained, regardless of the diameter of the hose.

The velocity of a fluid is inversely proportional to the diameter of the tube in which it is traveling, provided that the fluid moves through a single tube of variable diameter. However, the actual flow rates of a fluid in a series of interconnected tubes of different diameters revolve around the total cross-sectional area in each section. More specifically, the total cross-sectional areas of the arteries, capillaries, and veins must be compared when determining the relative rates of flow through these regions. Since the capillaries and veins have total cross-sectional areas which are, respectively, 700 times and 2 times that of the arteries, blood moves most rapidly through the arteries and least rapidly through the capillaries. Differences in the average velocity of blood in the various portions of the vascular system compensate for differences in their total cross-sectional areas, thereby maintaining the same average flow rate throughout the system.

**The measurement of blood pressure** The most commonly employed method of measuring blood pressure is indirect, in that it does not necessitate the insertion of a manometer into a blood vessel. Instead, the method uses a *sphygmomanometer* and a *stethoscope.* The sphygmomanometer is either of a mercurial or aneroid type.

Basically, the clinical determination of blood pressure requires the *auscultation,* or stethoscopic detection, of characteristic sounds known as *Korotkov sounds.* These sounds are made by blood flowing through a peripheral artery under different pressures applied with a sphygmomanometer. Controlled pressures are applied through an inflatable cuff fastened around the arm in the region of the brachial artery. A compressible bulb with an adjustable escape valve links the cuff of the sphygmomanometer to its mercury column or air pressure meter.

The object in blood pressure measurements is to determine the systolic and diastolic pressures. The cuff is first inflated to a pressure which completely collapses the brachial artery. Since blood is pre-

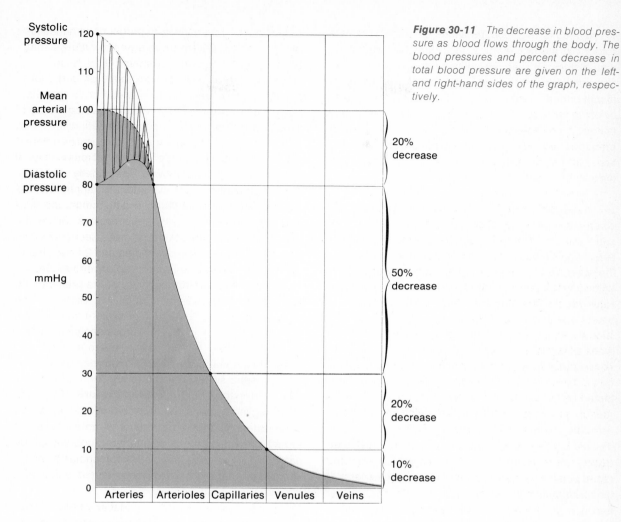

Systolic pressure

Mean arterial pressure

Diastolic pressure

mmHg

Arteries | Arterioles | Capillaries | Venules | Veins

**Figure 30-11** *The decrease in blood pressure as blood flows through the body. The blood pressures and percent decrease in total blood pressure are given on the left- and right-hand sides of the graph, respectively.*

20% decrease

50% decrease

20% decrease

10% decrease

vented from flowing through the brachial artery when the applied pressure is in excess of the peak systolic pressure, no sounds are detected when a stethoscope is placed over the portion of the brachial artery in front of the cuff. To prevent blood flow, usually the cuff is initially inflated to a pressure of 150 mmHg. Then the escape valve of the bulb is adjusted so that the cuff pressure decreases at an average rate of 3 mmHg/sec. The presence and nature of the Korotkov sounds are auscultated as the cuff pressure is decreased.

When the cuff pressure drops just below the peak systolic pressure, blood spurts through the brachial artery in pulses. The sounds produced by this type of blood flow are discrete, intermittent, and metallic in quality. The pressure at which these sounds are first auscultated corresponds to the *systolic pressure*. These sounds persist as the pressure is decreased until the cuff pressure falls just below the diastolic pressure. At this point, the discrete sounds become muffled. The pressure at which muffled sounds first appear corresponds to the *diastolic pressure*. As the cuff pressure is further reduced, the muffled sounds will ultimately die away completely. Usually, the Korotkov sounds disappear at a pressure about 10 mmHg lower than the diastolic pressure.

The results of blood pressure measurements are subject to many variables, including the immediate past history and various physical and psychological characteristics of the subject, the skills of the technician performing the measurements, and the accuracy of the sphygmomanometer. Because of these many factors, blood pressure readings are usually considered to reflect a range of systolic and diastolic pressures which may be 10 percent higher or lower than the actual values.

Arterial pulse pressure waves    Each pulse of blood is ejected into the ascending aorta under pressure from the heart. As we have already learned, blood flows smoothly through the arteries because of the wavelike contractions of the vascular walls which follow the course of each successive pulse of blood. However, the sharp rise in pressure generated by the heart during ventricular systole is conducted as *pressure waves* through the walls of the arteries about 10 times as fast as the pulses of blood within the arteries. These pressure waves are transmitted mechanically through all portions of the arterial tree and can be detected by palpating almost any peripheral artery.

In most clinical situations in which the pulse is used diagnostically, the *radial artery* is palpated and the number of pulse beats in a unit of time and their quality are recorded. The measurement is called the *radial pulse*. The frequency, strength, and evenness of the radial pulse are indicative of the rate, rhythmicity, and strength of the heart and the integrity of the circulatory system.

## THE LYMPHATIC SYSTEM

The lymphatic system is composed of (1) a system of highly anastomosed lymph vessels which run in parallel with the blood vessels of the body; (2) localized, widely distributed aggregations of lymph nodes supplied and drained by lymph vessels; (3) two major ducts which drain lymph from the lymph vessels into the venous circulation; and (4) other lymphoid tissues and organs, such as the thymus gland, spleen, and reticular connective tissue in the red bone marrow and certain other parts of the body.

Functionally, the lymphatic system is an accessory circulatory system primarily responsible for the maintenance of normal interstitial fluid pressures. In addition, the lymphatic system, because of its strategic functional relationships with both the cardiovascular system and the interstitial fluid, has assumed important roles in body defense.

The more specific functions of the lymphatic system are as follows: (1) the removal of proteins and other plasma constituents which pass from the blood capillaries to the interstitial fluid and their return to the bloodstream; (2) the neutralization of foreign particles which have entered the lymph from the interstitial fluid; and (3) the formation and transport of lymphocytes, antibodies, and other proteins to the bloodstream. The lymphatic system is a part of the reticuloendothelial system because of the presence of fixed macrophages (see Chap. 4).

## LYMPH

Lymph is the fluid within the vascular components of the lymphatic system. Lymph forms during the drainage of interstitial fluid into lymph vessels. Although reflecting the composition of interstitial fluid, the actual constituents of lymph are determined by the body region from which the interstitial fluid is drained, the nature of the diet, and the specific products added by specialized parts of the lymphatic system.

Formation and constituents    To understand how lymph forms, we must review the mechanism by which blood gives rise to interstitial, or extracellular, tissue fluid. The constituents of blood which enter the interstitial fluid are filtered through the endothelial walls at the arteriolar ends of blood capillaries. About 90 percent of the filtered blood serum is returned to the bloodstream at the venous end of blood capillaries.

The force causing filtration is due to the difference between the arteriolar blood pressure, a value of about 30 mmHg, and the hydrostatic pressure of the interstitial fluid, which is actually a negative value. For simplicity, let us assume that the pressure forcing blood serum out of the capillaries is 30 mmHg. This force is partially counterbalanced by the increasing colloid osmotic pressure of the capillary blood as serum enters the interstitial fluid but most blood proteins remain behind. The colloid osmotic pressure in a blood capillary is approximately 23 mmHg. Since this pressure tends to pull serum back into the blood, the net filtration pressure is only 7 mmHg (30 mmHg − 23 mmHg).

However, the net filtration pressure is great enough to cause a continual leakage of some of the blood proteins through the capillary walls into the interstitial fluid. The proteins pass through the pores formed at the intersection of every four adjacent endothelial cells in the walls of the capillaries. About one-fourth of the total concentration of blood proteins is lost to the interstitial fluid every 6 hr. If there were no way to return these proteins to the blood, the loss would be fatal in a relatively short period of time. In addition, as blood proteins accumulate in the interstitial fluid, the interstitial colloid osmotic pressure increases, thereby tending to pull a correspondingly increased volume of serum from the capillary blood into the interstitial space. In other words, as proteins accumulate in the interstitial fluid, so does serum.

Because of the pressure differential between capillary blood and the interstitial fluid, once proteins have leaked into the interstitial fluid, they cannot be returned directly to the bloodstream at the venous end of the blood capillaries. However, structural specializations in the walls of the lymphatic vessels do permit the entry of proteins from the interstitial tissue. Thus, as interstitial protein concentrations increase, a tide of serum pulled osmotically from the blood washes the proteins into the lymphatic vessels. Of course, any excessive amount of interstitial fluid enters the lymphatics along with the proteins.

How can the tide of interstitial fluid and protein enter the lymph vessels when the average interstitial pressure is negative? The interstitial fluid pressure must fluctuate between slightly negative and slightly positive values, depending on the local amounts of interstitial fluid and the degree of muscular compression of tissues in localized regions of the body. When the interstitial pressure is slightly positive due to a greater-than-normal volume of interstitial fluid or a compressional force applied to the tissues, the fluid will be forced into the lymph vessels. Since interstitial pressure fluctuations are normally based on random variations in the net capillary filtration pressure and the amount of mechanical pressure in each region of the body, lymph is more or less continually formed. This mechanism ensures that the quantity and constituents of interstitial fluid remain constant.

Combined with certain products of the lymphatic system such as lymphocytes, antibodies, and other proteins, the interstitial fluids and the leaked proteins make up the constituents of lymph. Since the interstitial fluid composition varies from time to time and from one region of the body to another, the composition of lymph varies similarly. Lymph is ultimately drained into the blood by specialized ducts which enter at the junctions of the left and right subclavian and internal jugular veins. Thus, like cerebrospinal fluid (see Chap. 14), lymph is continually derived from, and returned to, the blood. Indeed, the lymphatic system acts as a bridge between the interstitial fluid and the blood.

## ORGANIZATION AND DISTRIBUTION

The lymphatic system consists of a highly anastomosed, widely distributed network of vessels interlaced with regional groupings of nodes and drained by ducts continuous with venous blood vessels. The organization and distribution of the lymphatic system are shown in Fig. 30-12.

Plexuses and vessels    In certain regions, groups of lymph vessels form extensive interconnections called *lymph plexuses*, e.g., the mammary and palmar plexuses. In other regions, the lymph vessels simply run in parallel with the blood vessels.

*Lymph vessels* possess a tissue structure similar to that of capillaries but contain a series of closely spaced internal valves which point toward the heart. The valves are actually cellular flaps formed from the wall. At each flap there is an opening in the wall large enough to admit interstitial fluid containing leaked proteins and foreign particles.

When a lymph vessel is compressed mechanically, the lymph within is forced toward the heart. Backflow is prevented by the action of the one-way valves. Although muscular compression of the lymph ves-

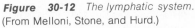

**Figure 30-12** *The lymphatic system.* (From Melloni, Stone, and Hurd.)

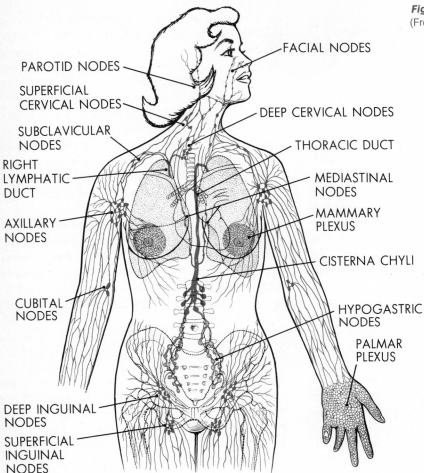

PAROTID NODES

SUPERFICIAL CERVICAL NODES

SUBCLAVICULAR NODES

RIGHT LYMPHATIC DUCT

AXILLARY NODES

CUBITAL NODES

DEEP INGUINAL NODES

SUPERFICIAL INGUINAL NODES

FACIAL NODES

DEEP CERVICAL NODES

THORACIC DUCT

MEDIASTINAL NODES

MAMMARY PLEXUS

CISTERNA CHYLI

HYPOGASTRIC NODES

PALMAR PLEXUS

sels—the so-called "lymphatic pump"—is primarily responsible for the flow of lymph, other factors which contribute to lymph flow include the pressure provided by the entry of interstitial fluid and gravitational forces.

Nodes   Lymph nodes are ordinarily aggregated in regions of the body where exposure to infection is relatively high. In Fig. 30-12 note the accumulation of lymph nodes in the facial, carotid, cervical, axillary, mediastinal, cubital, hypogastric, and inguinal regions. Solitary and aggregated lymph nodes, or Peyer's patches, are found within the inside wall of the

small intestine. The palatine, lingual, and pharyngeal tonsils are large aggregations of lymph nodes strategically located in the pharyngeal region.

Each lymph node is supplied by afferent, and drained by efferent, lymph vessels. Foreign materials transported within the lymph are neutralized within the lymph nodes by reticuloendothelial cells. At the same time, lymph nodes manufacture lymphocytes, antibodies, and certain other proteins which are added to the lymph before it is drained into the efferent lymph vessels.

Ducts   The right lymphatic and the left thoracic ducts

empty the lymph into the venous bloodstream. These two ducts are illustrated in Fig. 30-12.

The *right lymphatic duct* drains lymph from the right side of the face, neck, and thorax, as well as from the right arm. The duct opens into the bloodstream at the junction of the right subclavian and internal jugular veins.

The *left thoracic duct* drains lymph from the left side of the face, neck, and thorax, in addition to the left arm and the entire inferior region of the body. In other words, this duct drains all portions of the body not handled by the right lymphatic duct. Lymph drained from the lower extremities and abdomen accumulates in the *cisterna chyli*, a lymphatic reservoir in the mid-trunk region. Since the cisternal chyli is continuous with the thoracic duct, lymph is forced sporadically from the former to the latter. The left thoracic duct opens into the bloodstream at the junction of the left subclavian and internal jugular veins.

The lymphatic constituents of the cisterna chyli include lipids which have entered through the lacteals in the small intestinal villi. In addition, lymph within the cisterna chyli may contain a higher concentration of protein than that entering the right lymphatic duct, due to a greater leakage of protein into the hypogastric interstitial fluid following a meal rich in this nutrient. Therefore, the lymph drained by the thoracic duct is more variable than that drained by the lymphatic duct, reflecting variations in the dietary intake of specific nutrients.

## OTHER LYMPHATIC STRUCTURES

Other lymphoid tissues and organs are distributed throughout the body. The major lymphoid organs are the thymus gland and spleen. Since the role of the thymus gland is considered in detail in Chap. 4, the following discussion is restricted to the spleen.

**Spleen**  The spleen is situated on the left side of the abdominal cavity, just below the diaphragm. Its reddish appearance is due to its high degree of vascularization. Because the functions of the spleen may be taken over by other organs, it can be removed for medical reasons without endangering or shortening the life of the patient.

The lymphoid tissues of the spleen give rise to lymphocytes and antibodies and filter foreign materials from the blood. Other splenic tissues serve as a reservoir for living erythorcytes as well as in the destruction of those which are dead or dying.

## COLLATERAL READING

Adolph, E. F., "The Heart's Pacemaker," *Scientific American,* March 1967. The embryonic heart has an intrinsic rhythm which becomes responsive to neural influences as the autonomic nerves which innervate the heart become mature, shortly before and after birth.

Brewer, George J., and John W. Eaton, "Erythrocyte Metabolism: Interaction with Oxygen Transport," *Science,* 171:1205, 1971. The authors cite evidence indicating that the ability of hemoglobin to unload $O_2$ at relatively high $O_2$ tensions in the tissue capillaries is due to the presence of certain chemicals utilized in the erythrocytic metabolism of carbohydrates.

Ebert, James D., "The First Heartbeats," *Scientific American,* March 1959. In human embryos, the formation of the heart and blood vessels is initiated during the third week and almost entirely completed by the eighth week of development. This article reviews these embryonic events.

Longmore, Donald, *The Heart.* New York: McGraw-Hill Book Company, 1971. This attractively illustrated paperback provides a thorough review at an introductory level of the anatomy, physiology, and pathology of the heart and the diagnosis and treatment of cardiac disorders.

Mayerson, H. S., "The Lymphatic System," *Scientific American,* June 1963. A basic review of the anatomy and physiology of the lymphatic system.

Perutz, M. F., "The Hemoglobin Molecule," *Scientific American,* November 1964. Consisting of approximately 10,000 atoms distributed among four helical chains folding among each other, hemoglobin contracts when loaded with $O_2$ and expands under the opposite conditions. Perutz discusses the significance of these observations.

Salk, Lee, "The Role of the Heartbeat in the Relations between Mother and Infant," *Scientific American,* May 1973. Since about 80 percent of both right- and left-handed people hold a baby on the left side of the chest or shoulder, Salk believes that the heartbeat is soothing to both parent and child.

Scholander, P. R., "The Master Switch of Life," *Scientific American,* December 1963. Based on experiments with diving mammals and human volunteers, Scholander concludes that potential asphyxiation causes (1) the heart rate to slow down (bradycardia); (2) peripheral capillary collapse, due to arteriolar constriction; and (3) the maintenance of normal or slightly elevated arterial pressures, so that the heart and brain continue to receive adequate supplies of blood.

Wiggers, Carl J., "The Heart," *Scientific American,* May 1957. Reviews the structure and functions of the heart.

Wood, J. Edwin, "The Venous System," *Scientific American,* January 1968. The author cites research findings indicating that the veins constrict actively to maintain the venous blood pressure as well as to redistribute blood from the periphery to the central portions of the body.

Zweifach, Benjamin, "The Microcirculation of the Blood," *Scientific American,* January 1959. According to Zweifach, the mechanisms which cause capillary collapse revolve around a deficit of corticosteroids from the adrenal glands, a deficit of metabolic substances from nearby tissues, the presence of norepinephrine secreted by adrenergic nerve fibers, and the presence of certain chemicals secreted by nearby mast cells.

# CHAPTER 31
# cardiovascular Diseases

Disorders of the circulatory system impair the structural integrity and functional capabilities of the heart, blood vessels, blood, or some combination of these components. Basically, cardiovascular diseases give rise to one or more of the following functional impairments: (1) reduction in the transport of nutrients and $O_2$ to tissue cells, causing a depression in metabolism and stagnant hypoxia, respectively; (2) decrease in the excretion of $CO_2$ via the lungs and nitrogenous compounds by the kidneys, resulting in hypercapnia and acidosis or possible uremia, respectively; and (3) homeostatic disruptions in the balance of electrolytes and fluids between the various compartments of the body, leading to congestion and edema.

Certain parts of the body are more vulnerable to the effects of circulatory disorders than others. For instance, the brain and heart are particularly susceptible to a blood $O_2$ deficiency. These organs are also profoundly affected by local vascular disturbances.

The lungs and liver may become swollen with stagnant blood, due to cardiac or circulatory shock. Kidney functions may become impaired by marked alterations in blood pressure. Thus, the possibility of neural, respiratory, and excretory abnormalities is not unexpected in cases of cardiovascular involvement.

## CARDIAC DISORDERS

The following components of the heart may be damaged by disease: (1) the wall, consisting of the endocardium, myocardium, and pericardium; (2) the conduction system, including the SA and AV nodes, interventricular bundles of His, and Purkinje fibers; (3) the valvular system, containing the atrioventricular and semilunar valves, papillary muscles, and chordae tendineae; or (4) some combination of these components.

## THE HEART WALL

**Carditis** Inflammation of the heart wall is referred to as *carditis*. An inflammation of the endocardium is called *endocarditis*. The condition is usually caused by microbial infections. The most commonly implicated pathogens are staphylococci and diplococci. Although any part of the endocardium can become infected, damage is usually confined to the membrane lining the atrioventricular valves. Because such infections ordinarily impair valvular functions, further discussion of this disability will be postponed until diseases of the valvular system are considered later in this chapter.

When the muscle tissue of the heart is irritated, usually as a consequence of other disorders, the condition is termed *myocarditis*. The causative agents include microbes such as staphylococci, diplococci, gonococci, streptococci, and the influenza virus. A number of extreme inflammatory conditions originating from nonmicrobial factors may also cause myocardial inflammation. Damage is usually restricted to the muscle cells or the interstitial tissues but may encompass the entire myocardium.

An inflammation of the peritoneum surrounding the heart is called *pericarditis*. These tissues, as you will recall from Chap. 30, comprise the pericardial sac and include the visceral pericardium, or epicardium, and the parietal pericardium. Because both membranes are usually inflamed in cases of pericarditis, the pericardial cavity is often secondarily infected and may become partially or totally obliterated by subsequent adhesions. As in other forms of carditis, pericarditis is generally a secondary complication of other diseases, such as respiratory disorders caused by bacteria or fungi, myocardial infarcts caused by a coronary thrombosis and other types of coronary disease, or severe physical trauma to the region of the chest anterior to the heart, the *precordium*.

The effects of the various types of cardiac inflammation are variable. The immediate effect is a cardiac insufficiency quite similar to that caused by heart attacks due to other factors. Depression in cardiac output results from the effects of myocardial hypoxia and subsequent irregularities in cardiac rhythm.

Cardiac insufficiency gives rise to dyspnea and possible cyanosis, for want of an adequate oxygenation of the blood (stagnant hypoxia). Pulmonary, systemic, or general edema may also occur because of reduced blood pressures and correspondingly poor venous return. The hypoxic condition in the tissues of the heart wall results in pain, tenderness, or distress in the precordial region. The functional impairments include abnormal heart sounds, a weak and rapid beat at the apex of the ventricles, and heart palpitations. These abnormalities are mirrored by a weak and irregular radial pulse. If the heart undergoes compensatory hypertrophy following myocarditis, then the apex of the heart is displaced to the left. In many cases of pericarditis, a palpable bulge or edematous area is located in the precordial region. When pyrogenic bacteria are responsible for the inflammation, a high fever accompanies the infection.

**Angina pectoris** When a weakened heart is strained through overexertion, intense pain is referred to the precordium and may radiate to the left shoulder and arm. This syndrome is termed *angina pectoris*.

The cause of the pain experienced in angina pectoris is not completely understood. Presumably, the hypoxic tissues of the heart wall release chemical substances which accumulate in sufficient amounts to stimulate visceral afferent end-organs for pain in the affected regions. The ensuing nerve impulses are conducted to the brain. The subjective sensation of pain is then referred to somatic regions of the body containing tissues which originated embryonically in the affected segment. Referred pain is discussed in detail in Chap. 11.

Aside from severe chest pains which may radiate to the left shoulder and arm, the main subjective sensations experienced during a paroxysm of angina pectoris are intense feelings of suffocation and pressure in the region of the heart. These subjective symptoms create a fear of imminent demise which may exacerbate the dyspnea caused by stagnant hypoxia and the perspiration created as a by-product of the compensatory increase in the work load of the ventilatory muscles.

The decrease in cardiac output is homeostatically

corrected by strong sympathetic excitation and simultaneous parasympathetic inhibition almost immediately following the onset of an attack of angina pectoris. The pronounced increase in sympathetic stimulation almost doubles the vigor of the heartbeat and the volume of blood returning from the veins to the heart. These effects result in a significant elevation of arterial pressure that temporarily more than compensates for the initial circulatory depression. The variations in blood pressure during an attack of angina pectoris are reflected by a variable and frequently rapid pulse.

Coronary occlusions   The most frequent cause of damage to the heart wall is a decrease in its blood supply. Because the myocardium is supplied by the coronary arteries, the most serious threat to the heart is posed by the presence of *coronary occlusions*. Obstructions of the coronary arteries result from the following factors: (1) an atherosclerotic thrombosis, or embolus, which lodges in a coronary artery; (2) atherosclerotic deposition of lipids containing large amounts of cholesterol within and beneath the intima which lines the inside wall; and (3) arteriosclerotic changes in the vascular wall, resulting in a thickening, hardening, and decrease in its elasticity.

An *atherosclerotic thrombus* is probably the most frequent cause of coronary thrombosis. An intravascular clot, or thrombus, forms when atherosclerotic plaques build up sufficiently to project into the arterial lumen. The shear forces encountered by the formed elements of the blood, including the blood platelets as they stream past the rough surfaces of the plaques, disrupt their membranes and change the sign of their electric charge. These effects result in clumping and clotting. The formation of an intravascular clot is referred to as *thrombosis*. A small portion of the clot may loosen to become an *embolus* transported passively through the bloodstream. An embolus may finally come to rest as a thrombus in smaller branches of the coronary arterial tree.

Thus, thrombotic emboli can become intravascular clots trapped in small blood vessels such as the coronary arteries after being dislodged from distant locations. Other types of emboli include air bubbles, metastasizing cancer cells, clumps of bacteria and phagocytes—in fact, any mass of undissolved solid, liquid, or gaseous matter capable of occluding a blood vessel. The obstruction of a vessel by an embolus is termed *embolism*. Because embolism may be initiated by intravascular clots or foreign materials, it is a more general term than thrombosis. However, both coronary embolism and thrombosis lead to cardiac infarcts, or tissue damage, because of their obstructive effects on the coronary circulation.

*Atherosclerosis*, as we noted previously, refers to the deposition of fatty materials, mostly cholesterol, along the inside walls of blood vessels. The plaques thus formed grow in size through continued deposition until their rough edges project into the lumina of the affected blood vessels. Interaction of their jagged borders with the blood initiates intravascular clotting, while the barriers which they erect partially occlude the lumina which accommodate them.

With the progressive accumulation of calcium salts in such fatty deposits over a period of years, the atheromatous plaques become calcified and hardened. In addition, fibrous connective tissue begins to replace the elastic tissue within the walls of the abnormal vessels, gradually making them weaker and less resilient. The hardening and loss of elasticity cause a marked increase in the blood pressure within the affected vessels. These sclerotic changes are collectively referred to as *arteriosclerosis*. The atherosclerotic and arteriosclerotic conditions are frequently found in varying degrees in the same patient. For this reason, "hardening of the arteries," or arteriosclerosis without qualification, usually refers to both.

The role of ingested, saturated animal fats in the causation of atherosclerosis is controversial. On the one hand are studies which find a positive correlation between the amount of animal fat ingested over a period of time and the subsequent development of atherosclerosis. These observations are supported by data which indicate that obesity significantly increases the probability of developing the disease. Furthermore, other data suggest that diets low in saturated fat prevent the accumulation of cholesterol in the body.

On the other hand is a growing body of physiologic evidence which indicates little, if any, relationship between the concentration of saturated fats ingested and the actual levels of cholesterol in blood. The contradictory evidence is based on the following findings: (1) for each 100 mg of cholesterol eliminated from an average diet containing between 600 to 700 mg of cholesterol per day, the serum cholesterol levels fall only by about 3 mg, suggesting endogenous control of serum cholesterol concentrations; and (2) the body manufactures approximately 2,000 mg of cholesterol a day, an amount which is essentially unaffected by the average quantity of saturated fat ingested during the same period of time.

The conflicts in interpretation of these results revolve around the significance of many other variables which also affect serum cholesterol concentrations. Serum cholesterol concentrations are increased by (1) stress and the hypertensive condition created by stress, perhaps suggesting that cholesterol plays a role in energy production during emergency situations; (2) lack of exercise, indicating that accumulations of fatty materials are much more likely to build up in sedentary situations; (3) diabetes mellitus, possibly because of the profound alterations in fat metabolism which occur secondarily in this disease; and (4) a hereditary defect or predisposition, although the mechanisms which induce *hypercholesterolemia* remain unknown. In addition, the probability of developing atherosclerosis is increased by virtue (or liability) of being male, old, or obese. No one seriously attempts to refute the existence of these relationships. However, in the final analysis, high serum cholesterol levels may turn out to be an effect rather than a cause.

When a coronary artery is occluded, the collateral circulation associated with it becomes functional by increasing in diameter. Although this natural homeostatic mechanism is usually not sufficient to prevent cardiac ischemia, the rapidity of its establishment tends to retard more serious myocardial infarctions. In fact, during the recovery stages following a coronary attack, the collateral circulation generally increases once more. The new anastomoses usually take over the functions of the occluded blood vessels. The myocar-dial infarct is then the sole evidence of vascular disaster.

**Heart failure and circulatory shock** A *cardiac infarct* consists of dead and dying muscle fibers in the center of an ischemic region of nonfunctional myocardium. As the collateral circulation replaces a damaged coronary artery, the peripheral portion of the infarct once again becomes functional, leaving a central area of necrotic tissue. This portion of the infarct is gradually replaced by fibrous connective tissue, forming a scar in the myocardium. In the years that follow recovery from a heart attack, the weakened portion of the heart wall usually becomes stronger as the scar tissue turns thicker and more dense. Simultaneously, the uninvolved myocardium undergoes compensatory hypertrophy. The changes in the nature of the scar tissue combine with the enlargement of heart muscle to effectively reduce the relative size of the nonfunctional portion of the heart. These long-term changes minimize the amount of permanently incompetent cardiac muscle. However, the physiologic reverberations of a myocardial infarct may persist for a lifetime.

A marked depression in the ability of the heart to pump blood is termed *cardiac failure.* The vicious circle associated with this condition is given in Fig. 31-1.

Obviously, the initial effects of cardiac insufficiency are a decrease in cardiac output and a corresponding drop in mean systemic pressure. Kidney function is drastically affected by a decrease in the volume of blood coursing through the renal arteries. Anuria results when the rate of glomerular filtration is cut in half. If continued, renal depression can lead to degenerative changes in the epithelial tissues of the nephrons, a condition known as *tubular necrosis.* In time, the necrotic tissue may obstruct the lumina of the kidney tubules, causing an increase in the amount of nitrogenous waste products in the blood. This uremic condition is potentially fatal.

The body quickly responds to the acute effects of cardiac failure by invoking the following homeostatic mechanisms: (1) sympathetic stimulation of both the heart and blood vessels, to increase cardiac output

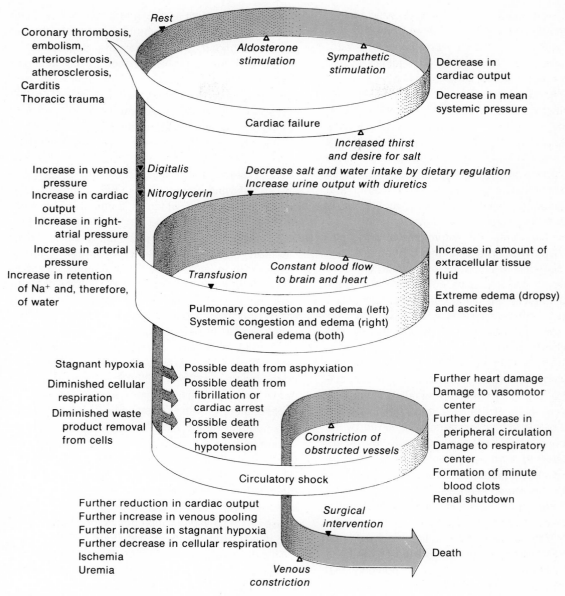

**Figure 31-1** The vicious circle associated with cardiac failure.

and systemic venous pressure, respectively; and (2) aldosterone secretion by the adrenal cortex, to promote the retention of sodium by the kidneys. The reabsorption of $Na^+$ causes $H_2O$ to be retained by osmosis, thereby increasing the volume of blood and extracellular tissue fluid. The elevated blood volume further contributes to the compensatory increase in systemic venous pressure induced by constriction of the blood vascular walls following a heart attack. As a consequence, more blood returns to the heart, causing an increase in the right atrial pressure. These homeostatic adjustments reestablish a near-normal cardiac output, even though the myocardium is weakened by the effects of ischemia.

However, the increase in the volume of blood elevates its hydrostatic pressure and contributes to a vascular congestion. When the hydrostatic pressure of blood exceeds its plasma colloid osmotic pressure, serum leaves the blood capillaries and enters the extracellular tissue spaces, causing edema.

The natural adjustments to cardiac failure described above, albeit vital to survival, have built-in risks which may ultimately negate their usefulness. The blessings provided by an increased cardiac output through strong sympathetic stimulation immediately following heart failure may ultimately be swamped by the compound damage incurred by the heart through continued pumping activities. A severely weakened heart wall, for example, may rupture by the stretch experienced in its ischemic portion during systole. A ruptured heart wall permits the entry of blood into the pericardial cavity. Death results from a further depression in cardiac output due to the following factors: (1) loss of blood from the damaged heart chamber; (2) fluid compression exerted on the heart from the outside; and (3) increase in nonfunctional myocardium and appearance of serious cardiac arrhythmia due to the tear in the wall of the heart.

Aside from its beneficial effects in augmenting cardiac output and mean systemic pressure, $Na^+$ and, therefore, $H_2O$ retention has one potentially disastrous consequence: congestion and edema through any increase in the capillary blood pressure, which permits serum to enter the interstitial tissues. This condition leads to stagnant hypoxia and tissue ischemia.

Deprived of nutrients and $O_2$ and unable to eliminate accumulating waste products of metabolism, the tissue cells undergo an irreversible depression in cellular respiration. Toxins released by the damaged tissue cells cause blood in adjacent capillary networks to clump. Thus, the ischemic condition results in masses of necrotic tissue and the development of minute blood clots throughout the body. If left unabated, this vicious circle of deterioration leads rapidly to death.

The location of the congestion and subsequent edema following a heart attack is determined by the portion of the heart which is damaged. The most frequent occurrence is left ventricular heart failure, since the normal work load of the left ventricle is considerably greater than that of the right ventricle. Because a weakened left ventricle pumps diminished quantities of blood into the aorta, the volume of blood returning to the left atrium from the lungs is correspondingly reduced. However, the unaffected right ventricle usually pumps normal quantities of blood through the pulmonary artery to the lungs for some time following left ventricular failure. As more and more blood accumulates within the lungs, the pulmonary hydrostatic pressure gradually increases. In response to the growing resistance created by pulmonary congestion, the work load of the right ventricle also increases. If this condition remains unchecked, the right ventricle finally succumbs to the overload. When this point is reached, both left and right ventricles are physiologically impaired. Frequently, however, pulmonary congestion and edema combine to cause death by asphyxiation before total heart failure can develop subsequent to left-sided failure.

The situation is reversed in cases of right ventricular heart failure, although the effects are somewhat different. A weakened right ventricle does not pump adequate volumes of blood to the lungs for oxygenation, but the output of the left ventricle remains unaffected for a while. Because blood differentially leaves the lungs and enters the systemic circulation, the congestion and edema which result from this condition are restricted to the systemic circulation. Because the total volume of blood in the lungs is only about one-seventh of that in the rest of the body, however, the additional increase in systemic blood volume following right ventricular failure is usually not large enough to cause a significant amount of congestion and edema. Nevertheless, blood begins to dam up in the systemic circulation because of the stagnation caused by right ventricular insufficiency. In response to the inertial resistance created by the mass of blood pooled in the systemic veins, the vigor of the left ventricle increases until the strain overwhelms its pumping abilities. The ensuing left-sided failure further depresses the cardiac output originally caused by right ventricular failure.

By now, it should be evident that unilateral heart failure usually leads to a pathologic involvement of the initially unaffected ventricle. The primary result of this chain of events is a profound depression in cardiac output. The same result accrues in cases of heart failure where both ventricles are damaged from the start. Cardiac failure leads to *cardiac shock* because of a corresponding depression in circulation. The failure of one or both ventricles leads to a pooling of blood in the venous portions of the circulatory system, as mentioned previously. As vascular congestion and interstitial edema worsen, the tissues become extremely bloated with fluids, a condition known as *dropsy.*

*Ascites* is noted when serous fluids spill into the abdominal cavity from hepatic sinusoids. Ascites is a condition which occurs when systemic congestion obstructs the normal flow of blood in hepatic veins from the liver to the inferior vena cava. Because it originates within the liver, ascitic fluid contains large concentrations of plasma proteins. The great number of proteinaceous particles increases the colloid osmotic pressure of ascitic fluid, causing it to osmotically absorb water from tissues which compose the walls of the intestine and other visceral organs. The osmotic uptake of water further increases the volume of fluid in the abdominal cavity, contributing to a pronounced distention of the belly. Dropsy and ascites combine to increase the stagnation of blood by effectively reducing its volume.

Marked depression in circulation leads to circulatory shock and another round of degenerative changes similar to those described above. Increase

in the pooling of venous blood further reduces cardiac output and arterial blood pressure and serves to intensify the existing heart damage, by depriving the coronary arteries of the reduced blood supply still trickling through. A lack of coronary blood widens the ischemic portion of the myocardium and may lead directly to death as a consequence of rapid, uncoordinated contractions of the cardiac musculature, a condition called *fibrillation.* During extreme hypoxia, death may result from a complete cessation of the heartbeat (*cardiac arrest*).

If death does not occur at this point, the mounting stagnant hypoxia continues to create more ischemic areas containing necrotic tissue in other parts of the body. As more and more tissue cells die, their toxins join those already present in the bloodstream and induce the formation of an increasing number of intravascular clots. Because the tissues supplied by the affected vessels no longer receive blood, the ischemic areas enlarge in size and increase in number.

If tissue destruction continues because of ischemia, brain tissues may also die. When the vasomotor center is damaged, blood vessels throughout the body lose their tone and enlarge. Vasodilatation effectively reduces the relative volume of blood in the body because the available blood occupies considerably more intravascular space than before. As a consequence, the mean systemic pressure plummets and death comes rapidly from the effects of severe hypotension. Conversely, if the respiratory center is damaged, the resulting inability to ventilate the lungs reflexly can lead to asphyxiation.

In addition to the natural homeostatic mechanisms mentioned previously— namely, sympathetic stimulation and $Na^+$ retention—the body has a virtual arsenal of other adjustments which may be invoked in an attempt to negate the progressive cardiac and circulatory depression which follows heart failure. Their appearance is determined by the degree of circulatory stagnation. These steady-state, regulatory processes include (1) an increased ingestion of $H_2O$ and salt; (2) an increased uptake by the blood of $H_2O$ from the digestive tract and tissue cells; (3) constriction of blood vessels not adequately supplied by blood; and (4) venous constriction following the pooling of blood within them. All the adjustments enumerated above promote an increase in the volume of circulating blood, thereby attenuating the potentially damaging effects of hypotension and improving the $O_2$-carrying capacity of the blood.

Another natural mechanism is called into play when the blood pressure has dropped to almost one-half its original value. This adaptive response consists of vascular adjustments which maintain a relatively constant blood pressure to both the brain and heart muscle, despite a markedly reduced cardiac output. Within wide limits, the uninterrupted flow of blood to these organs ensures that their vital functions are not deranged by random blood pressure fluctuations.

The natural compensatory mechanisms of the body are not a panacea, as we pointed out previously. They improve cardiac output and arterial pressure but cannot substantially increase the strength of the beat of a weakened heart. Sodium retention indirectly increases the volume of circulating blood, but subsequent edema can lead to potentially serious consequences. The maintenance of a constant cranial and coronary blood pressure cannot increase the amount of $O_2$ to, or reduce the amount of ischemia in, tissues supplied by stagnating blood elsewhere in the body. Obviously, medical intervention is required to provide symptomatic relief from pain caused by an ischemic myocardium and to reverse the circle of deterioration described above.

The treatment of patients suffering from cardiac insufficiency requires some combination of the following techniques and procedures:

1    The administration of *digitalis*, or a derivative of this compound, to improve the vigor of the heartbeat by increasing the permeability of the cardiac muscle cells to $Ca^{++}$, the activator of actin and myosin during interdigitation (see Chap. 24)
2    The administration of *nitroglycerin* to dilate the blood vessels, thereby providing a larger volume of blood to hypoxic tissues
3    The *transfusion* of whole blood, plasma, or plasma substitutes to increase directly the volume of blood in the body

4    *Bed rest*, to aid healing by reducing the work load of the ventricles
5    A reduction in the intake of salt and $H_2O$ through dietary control and an increase in the production of urine through the administration of *diuretics* to relieve the edematous condition
6    Surgical intervention to remove coronary occlusions, assist the pumping action of the heart and relieve its work load, or replace an irreparably damaged heart either in part or completely through cardiac transplantation

*Circulatory shock* is the direct result of coronary occlusion and subsequent cardiac shock. However, circulatory depression may originate from causes that only secondarily impair the functional abilities of the heart, such as a decrease in the volume of blood or, conversely, an increase in the volume of intravascular space, inducing hypovolemic shock and subsequent venous pooling, and the presence in the blood of toxins released by gram-negative bacteria, causing blood poisoning, or septic shock. The vicious circles associated with these forms of circulatory shock are traced in Fig. 31-2.

*Hypovolemic shock* may arise from the following factors: (1) severe hemorrhages in which over one-third of the total volume of blood in the body is lost; (2) third-degree burns covering a large portion of the body, because of the profound exudation of extracellular tissue fluid; and (3) loss of vasomotor tone, inducing hypotension. The loss of blood proteins normally present in the interstitial spaces in cases of severe cutaneous burns depresses the plasma colloid osmotic pressure and leads to hypoproteinemic edema. Loss of vasomotor tone may occur in response to brain damage, anesthesia, or antigen-antibody interactions. Hypovolemic shock from any cause reduces the mean systemic blood pressure and retards the return of blood to the heart from the veins, resulting in *venous pooling shock*.

*Septic shock* may be induced by bacterial infections that cause septicemia and from peritonitis caused by the spread of a bacterial infection originating in the female reproductive tract or from an intesti-

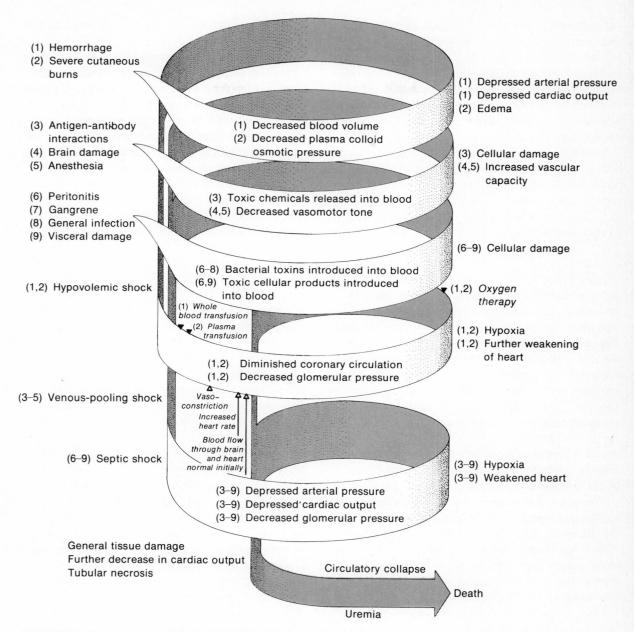

**Figure 31-2** *The vicious circles associated with circulatory shock.*

(1) Hemorrhage
(2) Severe cutaneous burns

(3) Antigen-antibody interactions
(4) Brain damage
(5) Anesthesia

(6) Peritonitis
(7) Gangrene
(8) General infection
(9) Visceral damage

(1,2) Hypovolemic shock

(3–5) Venous-pooling shock

(6–9) Septic shock

General tissue damage
Further decrease in cardiac output
Tubular necrosis

(1) Decreased blood volume
(2) Decreased plasma colloid osmotic pressure

(3) Toxic chemicals released into blood
(4,5) Decreased vasomotor tone

(6–8) Bacterial toxins introduced into blood
(6,9) Toxic cellular products introduced into blood

(1) *Whole blood transfusion*
(2) *Plasma transfusion*

(1,2) Diminished coronary circulation
(1,2) Decreased glomerular pressure

*Vaso-constriction*
*Increased heart rate*
*Blood flow through brain and heart normal initially*

(3–9) Depressed arterial pressure
(3–9) Depressed cardiac output
(3–9) Decreased glomerular pressure

(1) Depressed arterial pressure
(1) Depressed cardiac output
(2) Edema

(3) Cellular damage
(4,5) Increased vascular capacity

(6–9) Cellular damage

(1,2) *Oxygen therapy*

(1,2) Hypoxia
(1,2) Further weakening of heart

(3–9) Hypoxia
(3–9) Weakened heart

Circulatory collapse

Death

Uremia

nal perforation. The toxins released by the bacteria and necrotic tissues cause extensive tissue damage, impairing the functions of the cardiovascular system. Eventually, the progressive deterioration in the functional capacity of the circulatory system leads to a drop in the mean systemic pressure which, as we are well aware by now, is the antecedent of venous pooling shock.

Therefore, hypovolemic shock and septic shock both lead to venous pooling. The significance of this observation resides in the appreciation of the fact that venous pooling is a relatively advanced stage in all forms of circulatory shock, including that which follows cardiac shock.

## VALVULAR DISEASE

Valvular disease refers to the structural damage of one of the valves associated with the heart. The damage decreases the operational efficiency of the heart. Valvular impairments usually result from bacterial infections, especially by streptococci, although developmental anomalies may also be responsible. Where infection is the cause, the inflamed membranous tissue dies and is replaced by a tougher and more dense fibrous connective tissue.

**Stenosis and regurgitation**  The repercussions of valvular inefficiency are determined by the condition of the valvular orifice. If the opening is permanently narrowed by fibrous ingrowths, a disorder referred to as *stenosis*, then the volume of blood passing through the stenosed valve during each cardiac cycle is diminished. On the other hand, when a valve does not close completely, some blood is free to flow backward through its orifice. Called *regurgitation*, this condition also results in a decreased cardiac output.

Although the homeostatic mechanisms of the body compensate in part for the chronic decrease in arterial pressure by permanently increasing blood volume, cardiac efficiency gradually deteriorates over the long term. Where the left side of the heart is involved, i.e., in mitral or aortic stenosis and regurgitation, an elevated left atrial pressure resulting from inefficient valvular functioning retards blood flow from the lungs to the left atrium, causing pulmonary vascular congestion and interstitial edema.

The symptoms and signs of stenosis and regurgitation reflect on the inefficient pumping abilities of the heart and the congestive effects induced by the development of high left atrial pressures. At worst, pulmonary congestion leads to asphyxiation; at best, it leads to hypoxic hypoxia and anemia. The latter condition occurs because of the inefficiency of gaseous exchange through a pulmonic membrane swollen by congestive fluids. In either event, the skeletal muscles usually suffer from chronic fatigue because of a lack of sufficient $O_2$, while the reduction in circulation may adversely affect kidney function by causing tubular necrosis, as explained previously.

## CARDIAC ARRHYTHMIA

The other potential cardiac debility results from irregularities in the rhythm of the heart, or *cardiac arrhythmias*. In some cases, arrhythmias are a matter of temporary changes in rate or synchrony. In other, more serious, cases, the abnormalities consist of gross alterations in the normal progression of myocardial excitability. We will now examine the various types of cardiac arrhythmia.

**Tachycardia and bradycardia**  The most common disturbances to the normal heart rhythm involve significant changes in the frequency of its beat.

When the heart rate exceeds 100 beats/min, the condition is termed *tachycardia*. Increase in heart rate is regulated by the number of impulses conducted to the myocardium through the sympathetic cardioaccelerator nerves and by the blood concentrations of catecholamines secreted by the adrenal medullae. A rapid heartbeat is induced by muscular exertion, excitement, or fright. Such factors place a physical strain on the heart through tissue demands for an increased supply of $O_2$ and nutrients. Although tachycardia is a common transient irregularity in normal healthy subjects, it may also be a sign of some other pathology, e.g., hemorrhage or hypoxia.

*Bradycardia* occurs when the heart rate falls below 50 beats/min. This condition is brought about through excessive stimulation of the myocardium by fibers of the Xth cranial nerve, the vagus nerve. A slow heartbeat is normal in many well-trained athletes and is also prevalent during deep sleep in many people. Bradycardia is not considered pathologic but is introduced here for use as a reference point in the following discussion.

Large variations in rate, as exemplified by tachycardia and bradycardia, have little effect on cardiac output. This fact becomes evident when the cardiac output of a normal subject is plotted against the heart rate over the range of heartbeat frequencies discussed above. Such a graph is presented in Fig. 31-3.

At an average heart rate of about 72 beats/min, the cardiac output is approximately 5 liters/min. The minute volume increases by 0.7 liter, or about 13 percent, at a frequency of 100 beats/min, the minimum rate for tachycardia. At 50 beats/min, the uppermost limit for bradycardia, the minute volume decreases by an almost identical amount. Of course, above and below these extremes of rate, the volume of blood pumped by the heart drops precipitously, as suggested by the extrapolated, dotted portions of the curve.

Extrasystole    *Premature beat*, or *extrasystole*, is an-

other common form of cardiac arrhythmia. Extrasystole occurs when an irritable, or ectopic, focus in either the atria or ventricles initiates a contraction before the SA node normally triggers systole. Because the wave of excitation from an ectopic focus travels slowly through the myocardium, usually without affecting the conduction system, the subsequent beat is delayed.

A sample ECG showing these relationships is illustrated in Fig. 31-4. The ECG is of a premature atrial contraction. This record graphically portrays the diminution in the diastolic time of the beat preceding extrasystole and the prolongation of time required for the initiation of the cardiac cycle following it.

*Ectopic foci* may appear transiently in normal subjects due to cardiac irritation. The following factors may be responsible for irritating the heart: (1) stimulants such as nicotine and caffeine; (2) subclinical vascular disturbances which involve the myocardium; (3) psychogenic factors such as neurotic breakdown; and (4) fatigue from overexertion or loss of sleep. Extrasystole may also be indicative of serious cardiac pathologic conditions, since myocardial infarcts can function as ectopic foci.

Heart block    Heart block is one of the more serious cardiac arrhythmias. This condition occurs when a

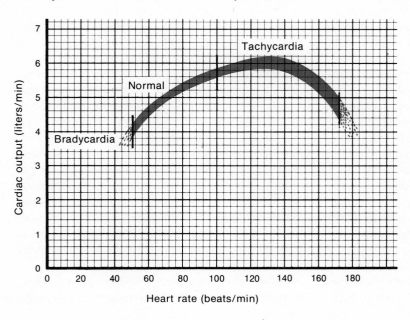

**Figure 31-3**  *The effect of variations in heart rate on cardiac output.*

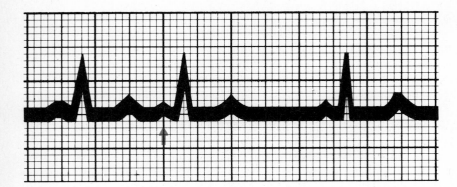

**Figure 31-4** *Extrasystole. The ECG shows a premature atrial contraction at the arrow.*

wave of excitation originating at the SA node is depressed at some point along the myocardial conduction system. This specialized system of muscular nodes and fibers is diagramed in Fig. 30-4. Usually, the blockage is between the SA and AV nodes, although depression of excitation is also common in the interventricular bundles of His. As a result of heart block, an ectopic focus develops in the ventricular myocardium. The ectopic focus sets its own frequency of ventricular contraction, completely separating ventricular and atrial beats. Although the rate of the atrial beat remains unaffected, the frequency of contraction of the ventricles drops to about one-fourth its former rate.

Heart block is usually caused by one of the following factors: (1) inhibitory stimulation of the myocardium through the vagus nerve; (2) parasympathomimetic drugs, which have a similar action; (3) microbial infections which cause myocarditis; or (4) cardiac infarcts caused by coronary occlusions. The resulting depression in cardiac output gives rise to the usual syndrome of cardiac insufficiency, including circulatory shock, stagnant hypoxia, and cardiac ischemia due to a decrease in the coronary blood supply. The slow ventricular beat is reflected in a correspondingly languid pulse rate.

Flutter and fibrillation   Two other forms of cardiac arrhythmia are also potentially serious disorders. The

first condition is called *flutter* and consists mostly of coordinated contractions of the atria or ventricles in frequencies up to about 400 beats/min. The other condition, termed *fibrillation*, results in extremely rapid and uncoordinated contractions occurring simultaneously in various portions of either the atrial or ventricular myocardium.

Flutters are initiated when a rhythmic wave of excitation initiated by the SA node is not extinguished after passing over or around the syncytial myocardium. The excitatory wave then persists for an extended period of time. Often, flutters appear following enlargement of a heart chamber, especially the atrium. Enlargement occurs because of the engorgement of one or more heart chambers with blood and is generally a consequence of rheumatic fever and other forms of endocarditis. These conditions lead to cardiac insufficiency and subsequent intravascular congestion and venous pooling.

A wave of excitation passing around the wall of a dilated heart chamber requires a longer time than usual to traverse all the myocardium. Brought about by an increase in the conductive area of excitable tissue, this delay is one of several factors which can cause flutter. In an undilated heart, flutter may originate from a delay in conduction time due to factors similar to those which cause heart block.

Because of its causes, atrial flutter is not uncommon. However, patients with this condition are generally not seriously debilitated under normal circumstances. Although the affected atrium is essentially incompetent, the hydrostatic pressure of the blood in the venae cavae and pulmonary veins is usually sufficient to overcome the resistance of the closed atrioventricular valves. Thus, blood fills the ventricles, despite a reduction in the force of atrial systole. The volume of blood entering the ventricles during a cardiac cycle is reduced by about one-third, however, causing a decrease in cardiac output which may become significant in cases where overexertion places extraordinary demands on the heart. In any event, the rhythmicity of flutter usually deteriorates with the passage of time and is replaced by the asynchronous and very rapid contractions characteristic of fibrillation.

Fibrillation is brought about by the same factors which cause heart block and flutter. The condition is induced when certain areas of the myocardium are excitable at the same time that other areas of the same tissue are refractory. Under these circumstances, some portions of an excitatory wave initiated by the SA node are neutralized when they collide with refractory regions while other parts of the same wave progress without impediment. Such partial wave fronts persist because they divide into smaller waves as they reach an island of refractory tissue. In this way, some portions of the original excitatory wave divide and subdivide repeatedly until the myocardium becomes finely laced with a patchwork of excitatory and inhibitory zones. Because the irritable and refractory areas of the myocardium become completely mixed, the myocardium is not capable of undergoing a coordinated contraction.

If fibrillation occurs in the ventricular myocardium, death ensues in a brief time, in the absence of external intervention. However, atrial fibrillation, like its precursor atrial flutter, is not incapacitating by itself but can lead to long-range complications similar to those described above.

If treated quickly, fribrillation may be eliminated. Treatment consists of a combination of closed cardiac massage to maintain blood flow, especially to the coronary circulation, and the passage of a strong electric current across the heart to bring most or all the myocardium to an identical refractory state, thereby eliminating the immediate cause of fibrillation. The shock is applied by chest electrodes attached to an instrument called a *defibrillator*. If the shock treatment is successful, the SA node takes over as the natural pacemaker, following a brief period of cardiac arrest.

The vicious circles associated with serious cardiac arrhythmias are summarized in Fig. 31-5. The diagram compares the causes and effects of heart block, flutter, and fibrillation. Although their causes vary, their repercussions differ only in severity. Indeed, as we have seen, one form of arrhythmia may give way to another as cardiac deterioration progresses.

## VASCULAR DISORDERS

### HEMORRHAGING

Rupture of a blood vessel is the most common type of vascular damage. Rupture is caused by excessive internal or external pressure against the vascular wall or weakening of the vascular wall through disease. The ensuing bleeding is referred to as a *hemorrhage*, as explained previously. The forms of hemorrhage already discussed include epistaxis, or nosebleed, and hemoptysis, or the presence of blood in sputum expectorated from the lower respiratory tract.

Blood released from a ruptured vessel either pools within the body or is lost through a natural or unnatural portal or exit. These alternatives are termed *internal* and *external hemorrhage,* respectively. If hemorrhaging is external and severe, then the syndrome associated with hypovolemic shock may set in. Some of the symptoms and signs of hypovolemia include dyspnea, a rapid pulse, and cold and clammy skin. These symptoms and others described previously are indicative of cardiac impairment.

**Cerebrovascular accident** When blood vessels rupture or become occluded within the brain or its meningeal covering, the disorder is commonly referred to as a *stroke,* or apoplexy, and is known medically as a *cerebrovascular accident* (CVA). Hemorrhaging may occur from the rupture of either

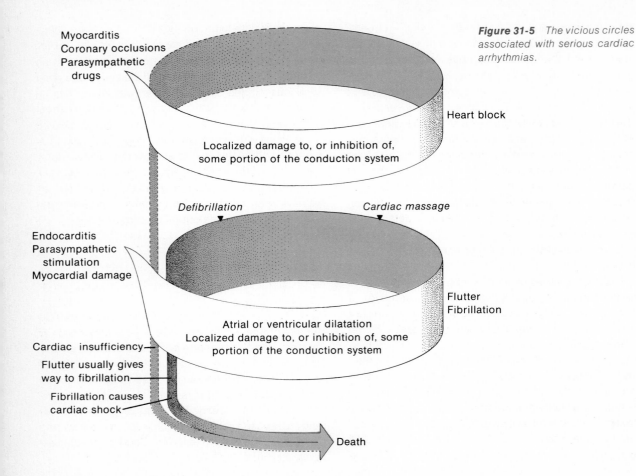

Myocarditis
Coronary occlusions
Parasympathetic drugs

Heart block

Localized damage to, or inhibition of, some portion of the conduction system

*Defibrillation*

*Cardiac massage*

Endocarditis
Parasympathetic stimulation
Myocardial damage

Flutter
Fibrillation

Atrial or ventricular dilatation
Localized damage to, or inhibition of, some portion of the conduction system

Cardiac insufficiency—

Flutter usually gives way to fibrillation—

Fibrillation causes cardiac shock—

Death

**Figure 31-5** *The vicious circles associated with serious cardiac arrhythmias.*

cerebral or subarachnoid arteries or veins. A massive hemorrhage in the right cerebral hemisphere is diagramed in Fig. 31-6.

In addition to cerebral and meningeal hemorrhages, CVAs are caused by (1) cerebrovascular occlusions, due to thrombosis, embolism, or atherosclerosis, and (2) arteriosclerosis of the internal carotid or vertebral arteries, causing a weakening of the vascular wall which subsequently ruptures. Whether through hemorrhage or occlusion, the brain tissues supplied by the degenerating or abnormal blood vessels become ischemic and soft. The necrotic tissue of the cerebral infarct is gradually replaced by scar tissue. When a CVA is caused by occlusions in cerebral arteries, the tissues surrounding the infarct

become edematous, due to the overflow of blood from the obstructed vessels. Therefore, we see the same progression of damage in the brain that was previously observed in the heart; namely, stagnant hypoxia, ischemia, infarction, as well as local edema. In addition, we note that some causes of stroke are identical to the factors which initiate heart disease —thrombosis, embolism, atherosclerosis, and arteriosclerosis. In fact, cerebral embolism, a common cause of a CVA, usually occurs as a secondary complication of heart diseases such as atrial fibrillation or coronary thrombosis.

The symptoms following a CVA include headache and some form of unconsciousness, such as stupor, as well as motor paralysis, speech impairment, and sensory depression. Usually the arm and leg on the side opposite the affected part of the cerebrum appear flaccid. Recovery is usually gradual, with the paralyzed skeletal muscles becoming secondarily spastic.

**Purpura**  Purpura is another form of hemorrhaging in which the blood may pool anywhere in the body. Especially prominent is a tendency to bleed from mucous and serous membranes. Cutaneous hemorrhages initially appear bright red and progress through purple to yellow with the passage of time. The accumulation of blood in extracellular tissue spaces leads to edema and pain, as well as to other symptoms characteristic of impairments in the affected region.

The hemorrhages result from an increase in the coagulation time of blood. Depression in hemostasis is due to fibrinolytic enzymes which degrade fibrin monomers as they form or a plasma substance which causes agglutination of blood platelets, thereby reducing their number in circulating blood.

**Aneurysm**  An aneurysm is a vascular dilatation resulting from a weakened area in the wall of a blood vessel. The dilated wall may subsequently pulsate, changing its shape from time to time, or may ulti-mately rupture and give rise to a massive hemorrhage. The size and shape of aneurysms vary widely, depending in part on their anatomic location and cause. Aneurysms of large size compress other organs, erode bony tissues through frictional forces, and exert pressure on brain tissues when located intracranially.

*Aortic aneurysm* is probably the most frequently occurring type. Although aneurysms of the thoracic aorta used to be common because of the higher incidence of syphilitic infection, aneurysms of the abdominal aorta predominate at the present time. Because an aortic aneurysm is *fusiform*, i.e., a simple expansion of a portion of the vessel, blood flow distal to the abnormality is usually not significantly impaired. Aortic aneurysms are dangerous mainly because they may rupture and cause massive internal hemorrhaging.

*Intracranial aneurysms* are next in importance to aneurysms of the aorta because of their direct and indirect effects on brain tissues. The cause of a CVA is sometimes traced to the rupture of an aneurysm of the internal carotid or basilar artery. The symptoms are similar to those of cranial and meningeal hemorrhages. In other cases, as already pointed out, damage results from physical compression by an aneurysm of surrounding cerebral tissues and nearby cranial nerves.

**Figure 31-6**  *A cerebrovascular accident shown in a coronal section through the cerebral hemispheres.*

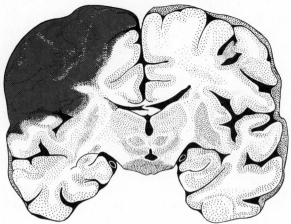

## ABNORMALITIES OF VASOMOTOR TONE

**Hypotension**  Hypotension is delineated clinically by a resting systolic pressure below 100 mmHg. The condition is normal in certain individuals possessing a significantly greater-than-average blood volume. Only in cases where the blood pressure drops below 100 mmHg from some higher value is hypotension pathologic.

Hypotension can arise from (1) vasodilatation, due to a decrease in vasomotor tone caused by brain damage, anesthesia, or anaphylaxis, and (2) metabolic diseases such as hypothyroidism. An increase in intravascular capacity, as we learned previously, leads to effects similar to those of hypovolemic shock. The reduction in mean systemic pressure causes blood to pool in the venous reservoirs of the body. *Anaphylaxis* is an allergic reaction within the circulatory system which elicits a profound vasodilation. The result is a dramatic decrease in venous return. Therefore, anaphylactic shock causes circulatory shock through venous pooling.

Hypertension    When arterial blood pressure is significantly elevated, whatever the reason, the condition is referred to as *hypertension*. To qualify, the systolic pressure must be in excess of 160 mmHg. As we learned in Chap. 30, arterial blood pressure is determined by the product of the cardiac output and peripheral resistance. Thus, any factor which increases the stroke volume of the heart or the resistance to blood flow within the blood vessels, or causes a combination of these conditions, will elevate arterial pressure and may bring about hypertension.

More specifically, hypertension has the following origins:

1   Psychogenic factors which cause constriction of arterioles and possible renal damage, through extreme sympathetic stimulation originating from the vasomotor center
2   Chronic use of sympathomimetic drugs such as epinephrine and norepinephrine, both of which induce vasoconstriction
3   Kidney diseases which impair tubular excretion of $Na^+$ and $H_2O$, since their retention contributes to an increased volume of blood and a correspondingly elevated cardiac output
4   Renal ischemia brought about through a markedly reduced renal blood flow, causing secretion of renin which activates the blood protein angiotensin to initiate arteriolar constriction
5   Vascular diseases which depress arterial blood pressure, thereby causing a compensatory increase in cardiac output

The vascular disorders which precipitate hypertension include some of those reviewed previously, especially arteriosclerosis, atherosclerosis, embolism, and aneurysm. In addition, a constriction of the aorta, termed *aortic coarctation*, may be responsible for specific cases of hypertension.

The majority of hypertensive patients do not show any obvious evidence of either vascular or renal disease. Such individuals are said to be suffering from "*essential*" hypertension. Because this specific disorder is associated with psychogenic factors, the topic is more fully discussed under "Psychosomatic Disorders" in Chap. 15.

When hypertension develops in response to renal ischemia, the condition is termed *renal hypertension*. Recall that an ischemic kidney liberates renin into the bloodstream. There, this protein engages in a complex series of biochemical reactions which ultimately results in the activation of a plasma protein called *angiotensin*. Within seconds after its formation, angiotensin causes intense arteriolar constriction, thereby provoking a profound elevation in blood pressure. Angiotensin also promotes the secretion of aldosterone, a hormone from the adrenal cortex. As we learned previously, aldosterone causes the reabsorption of $Na^+$ from the distal portions of the kidney tubules. The retention of salt increases the osmotic uptake of $H_2O$, thereby enlarging the blood volume and augmenting the increase in blood pressure homeostatically adjusted through vasoconstriction.

Increase in arterial blood pressure caused by any form of hypertension may lead to either heart failure from cardiac overload or internal hemorrhaging from excessive intravascular pressures. Subsequent depression in both cardiac output and arterial pressure, followed by congestion and edema, cardiac shock, and the ensuing vicious circle of progressive degeneration, are by now an all-too-familiar story. The complete pattern of circulatory deterioration associated with hypertension is summarized in Fig. 31-7.

VENOUS ABNORMALITIES

Several other common vascular disorders differentially affect the venous portion of the circulatory system. These diseases include phlebitis and varicosities.

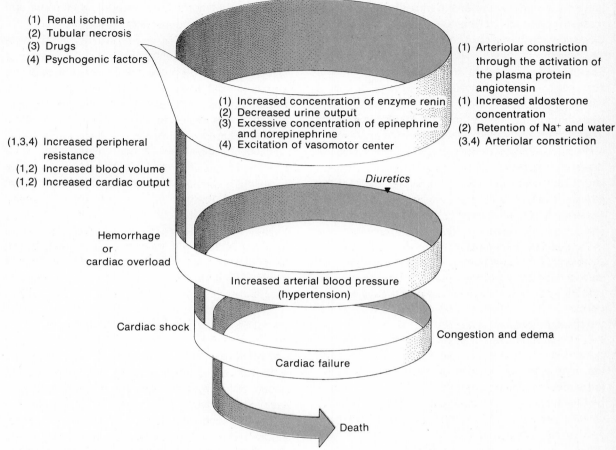

(1) Renal ischemia
(2) Tubular necrosis
(3) Drugs
(4) Psychogenic factors

(1) Increased concentration of enzyme renin
(2) Decreased urine output
(3) Excessive concentration of epinephrine and norepinephrine
(4) Excitation of vasomotor center

(1) Arteriolar constriction through the activation of the plasma protein angiotensin
(1) Increased aldosterone concentration
(2) Retention of Na⁺ and water
(3,4) Arteriolar constriction

(1,3,4) Increased peripheral resistance
(1,2) Increased blood volume
(1,2) Increased cardiac output

*Diuretics*

Hemorrhage or cardiac overload

Increased arterial blood pressure (hypertension)

Cardiac shock

Congestion and edema

Cardiac failure

Death

**Figure 31-7** *The vicious circle associated with hypertension.*

**Phlebitis** When a vein becomes inflamed, usually through physical trauma, microbial infection, or idiopathic factors, the condition is known as *phlebitis*. The inflammation usually leads to a localized acute edema, locally elevated temperature or general fever, a redness or other form of discoloration in the surrounding skin, and pain in nearby joints. The inflammation may be highly circumscribed, may migrate from one region to another, or may disappear and reappear in the same area.

The effects of phlebitis are variable. In some cases, adhesions develop within the affected vein, causing its lumen to become obliterated. In other cases, the vascular walls become hardened. Sometimes an aneurysm develops in a portion of an abnormal vein because of a weakening in its wall. Venous aneurysms generally ulcerate rather than rupture because of the lower pressure of blood within veins, compared to arteries. Since an ulcer is an open, bleeding sore, the effect of ulceration is similar to that of hemorrhaging.

**Varicosities** Varicosities, or varicose veins, are tortuous, swollen vessels usually visible along the periphery of the thigh or leg. Hemorrhoids, discussed in Chap. 27, are also a common type of varicosity. Distended veins develop when the valves within them are weakened and cannot maintain a one-way flow of blood toward the heart. The hydrostatic pressure of the column of blood overlying weakened valves swamps their collective ability to remain closed. Thus, the hydrostatic pressure of the blood forces the valves open and a portion of the blood flows backward through the valvular orifices. This action results in an extreme dilatation of the affected vein. Vascular congestion inevitably leads to edema. The increased interstitial fluid pressures in the edematous region stretch the overlying skin at the same time that the efficiency of the peripheral circulation is decreased. Therefore, the taut skin becomes extremely painful and patches of necrotic epithelium begin to develop.

Varicosities are caused by the following factors: (1) congenital defects which result in weakened venous valves; (2) occupations which require prolonged standing, causing venous valves to gradually weaken from the effects of gravity; (3) pregnancy, because of the pressure which an enlarged uterus exerts against the inferior vena cava or other large veins in the abdominal and pelvic regions; and (4) obesity, because the continual compression of the venous walls by large masses of surrounding tissues combines with a general hypertension characteristic of this disorder to weaken venous valves.

## BLOOD DISORDERS

Blood diseases, or *hemopathies*, are classified according to their causes and effects. For our purposes, we shall recognize the following types:

1 *Anemias,* or more precisely, disorders which give rise to anemic hypoxia by impairing the ability of blood to transport sufficient quantities of $O_2$ to tissue cells
2 *Hemorrhagic disorders,* which depress hemostasis through a lack of blood platelets, platelet factor, or plasma proteins required for clotting
3 *Neoplastic disease,* which causes hemolysis, ane-

mia, and hemorrhaging through a cancerous proliferation of agranular leukocytes
4 *Hemocytogenetic disorders,* which cause a dramatic *increase* in one of the formed elements in blood due to the presence of foreign materials or toxic substances
5 *Hemoblastic hemopathies,* which cause a significant *decrease* in one of the formed elements in blood due to a depression of hemopoiesis or an excessive destruction of blood corpuscles
6 *Immunologic disorders,* which lower specific resistance and increase susceptibility to infection through a depression in serologic responsiveness to antigens
7 *Acid-base imbalances,* which cause the pH of blood to fluctuate to either side of the normal range due to disturbances in, or the swamping of, the normal buffering mechanisms

## THE ANEMIAS

**Classification** Anemia is caused by any factor which decreases the number of erythrocytes or the amount of hemoglobin in blood, or has both effects. Basically, two types of anemia are recognized: primary and secondary. Each of the anemias is further subdivided according to the type of abnormality responsible for the hypoxic condition.

*Primary anemias* are caused by malfunctions in the tissues and organs responsible for normal erythropoiesis and hemolysis. These structures include the red bone marrow, liver, and spleen. Among the primary anemias are the following: (1) *aplastic anemia,* which results in abnormal or reduced numbers of erythrocytes because of damage to the red marrow due to ionizing radiation (see Chap. 9), osteopathies, or hemotoxins; (2) *pernicious anemia,* which results in a depression of erythropoiesis either because of damage to the gastric mucosa and a corresponding absence of its intrinsic factor, a natural product required for the absorption of vitamin $B_{12}$, or because of a vitamin $B_{12}$ deficiency (see Chap. 27); and (3) *iron-deficiency anemia,* which results in a reduction in both the concentration of hemoglobin and the size of erythrocytes.

Among the secondary anemias are the following: (1) *hemorrhagic anemia*, which results in a depression in both the number of erythrocytes and the concentration of hemoglobin because of a loss of blood; and (2) *hemolytic anemia*, which results in the destruction of circulating erythrocytes and the release of their hemoglobin into plasma due to the effects of toxins and microbial pathogens.

**Symptoms**    The symptoms of nonspecific anemia are identical to those of other forms of hypoxia and include muscle fatigue and weakness, apathy, dizziness, and pains originating from local necrotic areas arising from ischemic tissues. Diminution in the concentration of erythrocytes or hemoglobin, or both, also gives rise to a decrease in blood viscosity and a corresponding increase in flow rate. A greater rate of flow is reflected in an increased venous return and correspondingly increased cardiac output. The greater vigor of the heartbeat may cause a strain which induces the vicious circle of degenerative changes associated with cardiac failure. A reduction in blood viscosity also decreases the efficiency of gaseous exchange, since the time available for unloading $O_2$ by hemoglobin in blood moving rapidly through tissue capillaries is less than that for blood flowing through capillaries at normal speeds.

In addition to an increase in cardiac output, the body  responds to these homeostatic disruptions by increasing both erythropoiesis and the volume of blood. Heartbeat frequencies and blood flow rates are controlled according to the mechanisms outlined earlier. These regulatory adjustments tend to offset anemic hypoxia by increasing the number of erythrocytes and the concentration of hemoglobin in the circulating blood. Depending on the nature of the specific anemia, these steady-state mechanisms may be supplemented by an increased ingestion of proteins, folic acid, cobalt, vitamin $B_{12}$, or iron to further promote the manufacture of erythrocytes or the biosynthesis of hemoglobin. The role of each of these nutrients in anemia is discussed in Chap. 27.

In advanced cases of anemia, the compensatory mechanisms mentioned above may still be insufficient to prevent the development of serious hypoxia and subsequent heart strain. Cardiac overloading may induce heart failure, the symptoms of which have already been reviewed. The vicious circle associated with severe anemia is summarized in Fig. 31-8.

**Sickle cell anemia and thalassemia**    Sickling    of cells also gives rise to anemia. These disorders are due to hereditary defects. Both diseases are characterized by malformed erythrocytes, because they contain abnormal types of hemoglobin. Hemolysis during sickle cell anemia causes a rapid release of hemoglobin into the plasma and its subsequent conversion into bilirubin by reticuloendothelial cells. The rate of formation of bilirubin overwhelms the ability of the hepatic cells responsible for its destruction to excrete the pigment in adequate amounts into the bile. The result is an elevation in the plasma concentration of protein-bound bilirubin and the development of hemolytic jaundice.

**Erythroblastosis fetalis**    This type of hemopathy causes erythrocytic destruction during prenatal or early postnatal development. Hemolysis results from an *antigen-antibody reaction* determined mainly by heredity. The blood antigen responsible for the potential agglutination of erythrocytes is called the "rhesus," or Rh, factor and is referred to technically as *Rh agglutinogen*. Most people in the general population, i.e., 85 to 90 percent, are Rh-positive and possess the rhesus factor; the rest of the population is Rh-negative and does not possess the Rh antigen. Only Rh-negative people manufacture antibodies (*anti-Rh agglutinins*) for the rhesus factor.

Potential future problems may arise from the effects of seemingly normal childbirth terminating the first pregnancy under conditions where the mother is Rh-negative and the father is Rh-positive. If the child inherits a dominant gene for Rh factor from the father, he or she will be Rh-positive. During birth, while the placenta is being expelled as the "afterbirth," a few drops of Rh-positive fetal blood from branches of the

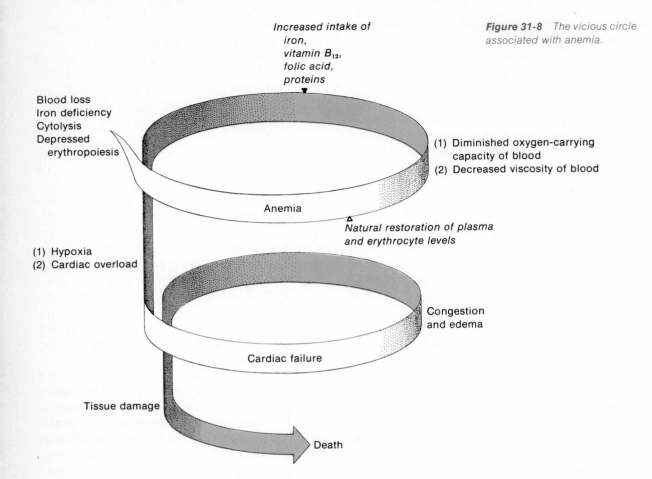

Blood loss
Iron deficiency
Cytolysis
Depressed
  erythropoiesis

Increased intake of
iron,
vitamin $B_{12}$,
folic acid,
proteins

(1) Diminished oxygen-carrying
    capacity of blood
(2) Decreased viscosity of blood

Anemia

Natural restoration of plasma
and erythrocyte levels

(1) Hypoxia
(2) Cardiac overload

Congestion
and edema

Cardiac failure

Tissue damage

Death

**Figure 31-8** *The vicious circle associated with anemia.*

umbilical blood vessels may enter the maternal bloodstream through ruptured vessels within the uterine endometrium. The introduction of Rh agglutinogens triggers the production of anti-Rh agglutinins by the mother. This kind of antigen-antibody interaction, known as a *primary immunologic response*, is covered thoroughly in Chap. 4. As a consequence of the primary response, the introduced antigens are neutralized, while the immunologic machinery of the mother responsible for the rapid synthesis of anti-Rh agglutinins is primed for possible future mobilization. The priming is due to the long-term presence of immunologically competent plasma cells in the lymphoid tissues of the body.

The above actions ordinarily do not have any effect on the first-born child because anti-Rh agglutinins are synthesized by the mother after the child is born. Occasionally, however, placental bleeding may occur early during the first pregnancy, so that a primary immunologic response is established in the mother even before birth. Under these circumstances, the first-born may indeed suffer from the effects of erythroblastosis fetalis. In most cases, however, the main problems usually come in subsequent pregnancies, when the child being carried is again Rh-positive. If this condition pertains, the Rh agglutinogens of the child in utero challenge the mother to manufacture large quantities of anti-Rh agglutinins. Because the mother's immunologic machinery is already primed, the rapid synthesis of maternal antibodies which results is referred to as a *secondary immunologic response* (see Chap. 4). These anti-Rh agglutinins pass the placental barrier and enter the fetal bloodstream, causing hemolysis and intravascular occlusions, anemic hypoxia and cyanosis, as well as hemolytic jaundice. These effects are serious enough to cause stillbirth or neonatal death.

Where the possibility of a future occurrence of erythroblastosis fetalis is great, e.g., due to obstetrical procedures known to cause transplacental bleeding, prophylactic treatment dictates the immunization of Rh-negative mothers within 2 days following the birth of first-born Rh-positive children. Immunization is accomplished by an intramuscular injection of a small amount of gamma globulin containing anti-Rh agglutinins. The immunization is successful if the introduced antibodies neutralize the antigens before the antigens establish a primary immunologic response. Alternatively, the blood of a newborn Rh-positive child can be completely replaced by Rh-negative blood, thereby eliminating the circulating anti-Rh agglutinins originally introduced from the maternal bloodstream.

Several factors combine to reduce the frequency of erythroblastosis fetalis. First, of course, is the fact that immunologic prophylaxis has been developed to combat this hemopathy. Second, only in a small percentage of all cases of Rh incompatibility does the mother actually experience transplacental bleeding prior to or during childbirth. Third, in some cases where transplacental bleeding occurs, a primary immunologic response does not actually materialize. Fourth, and finally, because many Rh-positive fathers are only partially dominant, or genetically heterozygous, the probability of producing an Rh-positive child in a situation where the father is Rh-positive and the mother is Rh-negative is considerably less than 100 percent.

ABO incompatibility The third point mentioned above requires further amplification. The reason for the frequent lack of natural immunization of Rh-negative mothers against Rh-positive children following their birth under conditions where transplacental bleeding does occur is to be found in the phenomenon of ABO incompatibility. Type O blood, you will recall, contains anti-A and anti-B agglutinins for A and B type blood, respectively. Of course, type AB blood would also be agglutinated by blood of type O.

Following transplacental bleeding where Rh incompatibility is a factor, Rh agglutinogens are introduced along with A, B, or AB agglutinogens into the maternal bloodstream. Since agglutinins for blood type antigens are already present in maternal blood of type O, intravascular agglutination and hemolysis of the introduced fetal erythrocytes are accomplished before a primary immunologic response can be triggered against the rhesus factor. In other words, the formation of anti-Rh agglutinins by the mother is prevented by the presence of other genetically determined antigens in the introduced fetal blood.

ABO incompatibility also occurs during an unmatched transfusion of whole blood. ABO incompatibility is explained in Chap. 30.

## HEMORRHAGIC DISORDERS

Hemophilia Hemophilia, also frequently called "bleeder's disease," results in a marked increase in the time required for the coagulation and clotting of blood. Because hemostasis is depressed, wound healing in a hemophiliac is a slow and difficult process accompanied by considerable pain. Frequently, blood tumors develop below the arachnoid mater. Such tumors are called *subdural hematomas* and exert pressure on tissues of the central nervous system and the peripheral nerves.

Hemophilia is due to a genetic defect discussed under "Hereditary Diseases," in Chap. 21. The defect causes a lack of one or more plasma proteins required for the formation of thrombin. In the absence of thrombin, fibrinogen cannot be converted into fibrin. Hemostasis is explained in detail in Chap. 4.

Thrombopenia When the platelet count is greatly depressed, the symptom is referred to as *thrombopenia*. This condition is caused by disorders such as microbial infection, bone marrow diseases, malnutrition, and septicemia. The effects of thrombopenia are similar to those of hemophilia, with one important exception. Thrombopenic hemorrhages are widespread in tissue capillaries, while those caused by hemophilia are usually restricted to larger, more localized blood vessels. These differences in degree and kind of hemorrhaging are reflected in blotchy skin discolorations characteristic of thrombopenia but which are ordinarily absent in hemophilia.

## NEOPLASTIC DISEASE

**Leukemia**  Leukemia is often called "cancer of the white blood cells." The extreme *leukocytosis* characteristic of this disease, up to 600,000 white blood corpuscles per mm³, may involve all types of leukocytes or only agranulocytes. This cancerous proliferation of new but malformed leukocytes is obviously an abnormal type of hyperplasia in the hemopoietic tissues. These tissues are found primarily in the red bone marrow of the vertebrae, ribs, and sternum and in lymph nodes throughout the body. When red bone marrow is involved, the number of circulating myelocytes is increased. Myelocytes, as you remember, are large cells from which leukocytes are derived. This form of hyperplasia is called *myelogenous leukemia*. If the lymphatic tissue is hyperplastic, then the disease is referred to as *lymphogenous leukemia*. In both forms of leukemia, the liver, spleen, and lymphatic tissues become enlarged through their engorgement with leukocytes. The great susceptibility of these tissues and organs is a consequence of their rich vascularization.

As the bloodstream becomes choked with white blood corpuscles, the circulating erythrocytes are pushed out of existence. Hemolysis causes the development of severe anemic hypoxia and initiates a vicious circle involving increasing amounts of tissue ischemia and necrosis. Damage to the formed elements of the blood combine with widespread cytolysis to markedly increase susceptibility to secondary infection. The other important effect in leukemia is created by the extraordinary demand for raw materials, especially amino acids, made by leukocytogenetic tissues in which cellular divisions have run amok. The demand is met at the expense of body proteins,

thus creating profound metabolic derangements, malnutrition, and muscular weakness or atrophy.

## HEMOCYTOGENETIC DISORDERS

Hemocytogenetic conditions are reflected in an absolute or relative increase in the number of formed elements in blood. In most cases, *hemocytogenesis* is a homeostatic adjustment made in response to some physiologic disequilibrium. Hemocytogenetic conditions include polycythemia, leukocytosis, and mononucleosis.

**Polycythemia**  Polycythemia, or *erythrocytosis*, refers to an excessive increase in the number of circulating erythrocytes. Usually, the concentration is between 2 and 3 million erythrocytes per mm³ above the average. The volume of blood may also, but does not always, increase. Strictly speaking, the disorder is due to hyperplasia within the erythropoietic tissues of the red bone marrow to compensate mainly for a low plasma $P_{O_2}$. However, a relative increase in the number of erythrocytes can also occur as a result of hemoconcentration due to an extensive loss of fluids from the body. Hemoconcentration may be brought about by profuse sweating or severe bouts of diarrhea, vomiting, or diuresis.

**Leukocytosis**  Leukocytosis refers to an excessive increase in the number of circulating leukocytes. The increase may pertain to all leukocytes or only to eosinophils. The number of leukocytes per mm³ may go as high as 60,000. To appreciate the difference between leukocytosis and leukemia, compare this figure with the concentration of leukocytes in leukemia cited previously.

Most cases of leukocytosis are part of an inflammatory response to a microbial infection, cardiovascular disorder, or toxemia. In chronic inflammatory conditions and parasitic infections, leukocyte hyperplasia specifically involves eosinophils.

Not all infections result in leukocytosis. To be sure, where nonspecific resistance is good and the infection is serious, leukocytosis is an early, pronounced symptom most of the time. Nevertheless, leukocytosis is not observed in cases of measles, mumps, or influenza. In addition, highly virulent infections, such as those which give rise to septic shock, run their course so rapidly that leukocyte manufacture remains essentially unaffected.

**Mononucleosis** Mononucleosis is reflected in an increased concentration of circulating agranular leukocytes such as monocytes and lymphocytes. The disease is undoubtedly infectious, although its exact cause remains unknown.

The histopathology of mononucleosis follows a course similar to that seen in leukocytosis. Swollen lymph nodes, especially noticeable in the cervical region, are correlated with their hyperplasia in response to microbial inflammation. The adenoids may become enlarged enough to interfere with normal swallowing. Other symptoms of the disease include a sore throat, fever, malaise, and extreme tiredness.

## HEMOBLASTIC DISORDERS

The opposite of hemocytogenesis occurs as the result of hemoblastic disorders which cause a depression in the number of formed elements in blood. Such disorders reduce or inhibit hemopoiesis, although the same effect may result from the destruction of blood corpuscles or platelets following their formation.

**Erythropenia** A significant reduction in the number of circulating red blood corpuscles is termed *erythropenia*. A lack of an adequate supply of erythrocytes is caused by aplastic or pernicious anemia, as explained previously. Erythropenia is due either to deficiencies of essential nutrients or the effects of ionizing radiations. Hemolysis due to septicemia may also cause erythropenia, giving rise to hemolytic anemia.

**Leukopenia** When the white blood corpuscle count drops below 5,000 per mm³, the condition is called *leukopenia*. Bone marrow diseases which depress erythrocyte production usually cause a corresponding reduction in the leukocyte concentration in circulating blood. Alternatively, a decrease in leuko-

cyte synthesis may be caused by dietary deficiencies which deny hemopoietic tissues an adequate source of essential nutrients. In some cases, the direct destruction of circulating leukocytes is caused by acute septicemia or toxemia.

**Thrombopenia** In this hemoblastic disorder, the number of blood platelets may dip as low as 50,000 per mm³, compared to a normal average value of around 300,000 per mm³. Refer to the discussion of hemorrhagic disorders for the causes and effects of thrombopenia.

## IMMUNOLOGIC DISORDERS

**Hypogammaglobulinemia** We have already mentioned two hemopathies due to specific antigen-antibody interactions; namely, erythroblastosis fetalis and ABO incompatibility. Occasionally, however, a hemopathy is reported in which there is a reduction in, or absence of, circulating antibodies or immunoglobulins. Because the most common type of antibody is gamma globulin (see Chap. 4), the disease is called *hypogammaglobulinemia*. The disease appreciably decreases specific resistance to antigens, thereby greatly increasing susceptibility to infection.

The reasons for a lack of circulating antibodies are to be found in either a lack of secretion of thymic hormone at or shortly after birth or an inability to acquire a sufficient concentration of maternal antibodies during gestation. If absence of thymic hormone is responsible for the disorder, then immunologic competence is not conferred to thymocytes in lymphatic tissues throughout the body.

## ACID-BASE IMBALANCES

Acid-base imbalances develop when the pH of blood falls below 7.3 or rises above 7.45, the normal range of variation. As you will remember from our discussion of buffers in Chap. 10, pH is a shorthand expression of the concentration of $H^+$ in aqueous solution. In the present discussion the solution, of course, is blood.

**Determinants** In a practical sense, the pH of blood depends on the ratio of the concentration of carbonic acid ($H_2CO_3$) to the concentration of plasma bicarbonates, made up mainly of sodium ions and bicarbonate ions ($Na^+HCO_3^-$):

$$\left[\frac{1}{20}\frac{H_2CO_3}{Na^+HCO_3^-}\right] \tag{31-1}$$

Normally, this ratio is 1:20, as indicated above, meaning that there is usually about one molecule of $H_2CO_3$ for every 20 molecules of $Na^+HCO_3^-$ in circulating blood.

Any factor which shifts the ratio, $H_2CO_3/Na^+HCO_3^-$, alters blood pH. Because $HCO_3^-$ must be available to remove free $H^+$ from plasma, according to the equation:

$$H^+A^- + Na^+HCO_3^- \rightarrow H_2CO_3 + Na^+A^- \tag{31-2}$$

Strong acid    Base        Weak acid    Salt

the amount of plasma $HCO_3^-$ is referred to as the *alkaline reserve.* If $H_2CO_3$ accumulates in blood, then blood ph is correspondingly lowered. From the above remarks and a reexamination of the ratio $H_2CO_3/Na^+HCO_3^-$, we note that any factor which has the effect of increasing the numerator $H_2CO_3$ also increases acidity, or reduces alkalinity. Conversely, any condition which causes either an absolute or relative increase in the denominator $Na^+HCO_3^-$ increases alkalinity, or reduces acidity.

**Acidosis and alkalosis** More specifically, blood pH may be altered in the following ways: (1) by increasing the concentration of $H_2CO_3$ without a compensatory increase in the amount of $Na^+HCO_3^-$; (2) by decreasing the concentration of $Na^+HCO_3^-$ without a compensatory decrease in the amount of $H_2CO_3$; (3) by increasing the concentration of $Na^+HCO_3^-$ without a compensatory increase in the amount of $H_2CO_3$; (4) by decreasing the concentration of $H_2CO_3$ without a compensatory decrease in the amount of $Na^+HCO_3^-$. The first two conditions enumerated above result in *acidosis,* or acidemia; the last two situations give rise to *alkalosis,* or alkalemia. These conditions are summarized as follows:

$$(1)\ \frac{H_2CO_3 \uparrow}{Na^+HCO_3^-} \qquad (2)\ \frac{H_2CO_3}{Na^+HCO_3^- \downarrow}$$

$$(3)\ \frac{H_2CO_3}{Na^+HCO_3^- \uparrow} \qquad (4)\ \frac{H_2CO_3 \downarrow}{Na^+HCO_3^-} \tag{31-3}$$

where an arrow pointing upward $\uparrow$ refers to an increase in concentration, while an arrow directed downward $\downarrow$ indicates a decrease in concentration.

Condition (1) arises from an increase in the $P_{CO_2}$ of plasma. The elevation in the $CO_2$ concentration of blood is referred to as hypercapnia and develops when the body is unable to excrete adequate amounts of $CO_2$ from the lungs. Therefore, a number of factors which depress ventilation may lead to hypercapnia and the subsequent development of acidosis. These factors include respiratory disorders, which collapse the pulmonary capillaries, and cardiovascular disorders, which reduce cardiac output and arterial pressure, thereby creating stagnant hypoxia. This form of acidosis may also result from breathing air containing a large concentration of $CO_2$. For example, respiring while the head is covered with a paper bag may precipitate a drop in the pH of blood to a seriously low value.

Condition (2) occurs when $HCO_3^-$ is given up in large quantities during diarrhea or diuresis. The loss is especially pronounced in severe cases of diarrhea since $Na^+HCO_3^-$, an abundant secretion in the intestine, is lost in copious amounts. This form of acidosis also arises following an elevation in the concentration of blood acids used in metabolism. The above situation is characteristic of advanced stages in diabetes mellitus.

In diabetes, the initial derangement in carbohydrate metabolism leads to a utilization of fats. The soluble fatty acids derived from lipids are degraded to acetoacetic acid (ketone bodies) which can be oxidized for energy. Thus the plasma concentrations of ketone bodies rise as fat metabolism supplants that of carbohydrates. Ketone bodies are then excreted in the urine, along with $Na^+$ and $K^+$, further depressing the alkaline reserves of the blood. In other words, an increase in fixed acids within the blood combines with a loss of alkaline reserves to increase blood acidity.

Condition (3) may develop in several ways: (1) following the ingestion of large quantities of $Na^+HCO_3^-$ to neutralize an excess secretion of acids, e.g., in cases of peptic ulcers; and (2) following a severe episode of vomiting in which large quantities of HCl are expelled from the body in vomitus.

Condition (4) is brought about by an increased excretion of $CO_2$ via the lungs. Hypocapnia, or a depression in the $P_{CO_2}$ of blood, usually results from hyperventilation. Due to vigorous hyperventilation, the pH of blood may rise temporarily to a value in excess of 7.6.

Conditions (1) and (4) are referred to as *respiratory acidosis* and *alkalosis*, respectively. *Metabolic acidosis* and *alkalosis* correspond to conditions (2) and (3), respectively. Thus, both types of acid-base imbalance may be caused by metabolic or respiratory derangements.

The effects of acidosis are in part determined by its cause. Respiratory acidosis is usually accompanied by dizziness caused by a homeostatic attempt to reduce the concentration of $H_2CO_3$ through hyperventilation. In both types of acidosis, an excessive increase in the $[H^+]$ in blood is reflected in a depression of neural excitability, as revealed by symptoms such as disorientation, stupor, or coma. Symptomatic treatment consists of the administration of sodium salts of organic compounds. The $Na^+$ combines with $HCO_3^-$ to increase the alkaline reserves of blood while the organic portion of the compound is metabolized.

The primary symptom of alkalosis is hyperactivity of the nervous system, especially the peripheral nerves. Spontaneous stimulation of motor fibers may throw various groupings of skeletal muscles into intermittent tonic spasms, a condition known as tetany. Other symptoms of tetany include nervousness, irritability, and paresthesias, such as numbness or tingling in the hands or feet. The excitatory symptoms of alkalosis are diametrically opposed to the neural depression seen in acidosis. The intake of large amounts of $HCO_3^-$ or the elimination of similar quantities of acids preceding metabolic alkalosis gives rise to an alkaline urine. Because the elimination of $CO_2$ from the lungs precipitates respiratory alkalosis, a prominent effect associated with this condition is *respiratory depression*. The gaseous imbalances created by hyperventilation may also cause light-headedness or even loss of consciousness.

The symptoms of alkalosis are relieved by chemicals which indirectly increase the acidity of blood. The idea is to promote production of $H_2CO_3$ at the same time that $HCO_3^-$ is reduced, thereby neutralizing the effects of alkalosis. Neutralization is accomplished by administering ammonium chloride which is transformed by the liver into urea, according to the following equation:

$$2NH_4Cl + 2Na^+HCO_3^- \rightarrow$$

Ammonium    Sodium
chloride      bicarbonate

(31-4)

$$(NH_2)_2CO + 2Na^+Cl^- + H_2CO_3 + 2H_2O$$

Urea     Sodium     Carbonic   Water
       chloride     acid

## COLLATERAL READING

Allison, Anthony C., "Sickle Cells and Evolution," *Scientific American,* August 1956. Shows how the sickle cell trait protects against malaria in African populations.

Clarke, C. A., "The Prevention of Rhesus Babies," *Scientific American,* November 1968. A lucid description of the development of immunoprophylaxis against the rhesus factor.

Congdon, C. C., "Bone Marrow Transplantation," *Science,* 171:1116, 1971. Describes the use of bone marrow transplants in cases where marrow is destroyed by physical trauma or radiation or where its functions are seriously impaired by hereditary defects.

Kolff, Willem, "An Artificial Heart Inside the Body," *Scientific American*, November 1965. Kolff believes that artificial devices offer the best long-range hope to patients with severe and irreversible cardiac damage.

# INDEX

Page numbers in *italic* indicate figures; in **boldface** indicate definitions.